Vampires

ENCOUNTERS WITH THE UNDEAD

Edited by

David J. Skal

BLACK DOG
& LEVENTHAL
PUBLISHERS
NEW YORK

Cataloging-in-Publication Data available.

ISBN:1-57912-487-9

Published by
Black Dog & Leventhal Publishers, Inc.
151 West 19th Street, New York, NY 10011

Distributed by
Workman Publishing Company
708 Broadway, New York, NY 10003

Typography and Production
Harriet Eckstein Graphic Design
Santa Barbara, CA

i g e c a b d f h j

Printed in Finland

For Bob and Fati

Contents

Part Four: Postmodern Vampires

Alan. Thou lovest me, thy soul is mine. Come to my heart, thou can'st not escape the spell my spirit has cast upon me. Why do you repulse —

Ada. Because that breast upon which you press me, seems to be the bosom of a corpse, and from the heart within I feel no throb of life!

Alan. Ah! dost thou know me, then?

Ada. Away — phantom! demon! — thy soul is dark, thy heart is cold.

Alan. Ada — thy life must pass into that heart.

Ada. Avaunt! — leave me! — my father — Edgar — oh! my voice is choked with fear — avoid thee, fiend! abhorrent spectre!

[*Retreats into room*]

Alan. She is mine.

Dion Boucicault, *The Vampire: A Phantasm* (1852)

Vampires

Painting by Mary Woronov.

INTRODUCTION

"There Are Such Things!"

WHEN THE ACTOR-MANAGER Hamilton Deane barnstormed the English provinces with his wildly successful stage adaptation of Bram Stoker's *Dracula* in the late 1920s, he added a memorable curtain speech, spoken by himself in the role of the great vampire hunter, Abraham Van Helsing:

> Just a moment, Ladies and Gentlemen! Just a word before you go! We hope the memories of Dracula won't give you bad dreams, so just a word of reassurance. When you get home tonight, and the lights have been turned out, and you are afraid to look behind the curtains, and you dread to see a face appear at the window... why just pull yourself together and remember that after all — THERE ARE SUCH THINGS!

It has been reported that Deane's audiences, especially in venues where Dracula was making one of its frequent return engagements, took to shouting out "THERE ARE SUCH THINGS!" even before Deane himself could deliver the words. Although everything about the up-to-date century told theatergoers that Dracula was sheer hokum, audience members nonetheless found a pleasurable *frisson* in the possibility that medieval legends about the thirsty dead just might have a basis in fact.

Do vampires really exist? The question itself may be naïve. It should be obvious that vampires prowl the very fabric of modern consciousness: in movies, on television, in best-selling books. The ultimate consumers themselves, vampires have become staples of consumer culture, lending their images and endorsements to every kind of product, from breakfast cereals to pornography. There are web sites (happy phrase) solely and lovingly consecrated to all things undead. And, as this anthology will demonstrate, there are even real people who practice the essential act of vampirism — the actual drinking of human blood — and others who so deeply believe in the reality of supernatural vampires that they will risk the wrath of the law with time-honored, if grisly, graveside remedies.

Drawing upon both fact and fiction, *Vampires: Encounters with the Undead* reflects the unique interdependence of reality and unreality in the vampire realm. Here you will encounter historical vampires, folkloric vampires, Hollywood vampires, psychic vampires, metaphorical vampires, paintings of vampires, city vampires, country vampires, vampire aristocrats and vampire peasantry, vampires in therapy, vampire families, gay and lesbian vampires, and just about every other permutation imaginable. I have included extensive sidebar commentary throughout — if only to make the point that vampire stories never exist in a vacuum; rather, they reflect and refract a rich and ever-shifting cultural context.

But that pesky question remains: do vampires really exist? You will have to draw your own conclusions, but I hope this book will provide ample food for thought. And skeptics may be admonished to remember one of Professor Van Helsing's shrewdest observations: "The strength of the vampire is that people will not believe in him."

Disbelieve, then, at your own peril.

DAVID J. SKAL

PART ONE

The Historical Evidence

CHEIROPTERA.
VAMPYRUS SPECTRUM.

FROM

The Book of Vampires

DUDLEY WRIGHT
(1914-26)

WHAT IS A vampire? The definition given in *Webster's International Dictionary* is "A blood-sucking ghost or reanimated body of a dead person; a soul or reanimated body of a dead person believed to come from the grave and wander about by night sucking the blood of persons asleep, causing their death."

Whitney's Century Dictionary says that a vampire is: "A kind of spectral body which, according to a superstition existing among the Slavic and other races on the Lower Danube, leaves the grave during the night and maintains a semblance of life by sucking the warm blood of living men and women while they are asleep."

"Vampires," says the learned Zopfius, "come out of their graves in the night time, rush upon people sleeping in their beds, suck out all their blood and destroy them. They attack men, women, and children, sparing neither age nor sex. Those who are under the malignity of their influence complain of suffocation and a total deficiency of spirits, after which they soon expire. Some of them being asked at the point of death what is the matter with them, their answer is that such persons lately dead rise to torment them."

Not all vampires, however, are, or were, suckers of blood. Some, according to the records, despatched their victims by inflicting upon them contagious diseases, or strangling them without drawing blood, or causing their speedy or retarded death by various other means.

Messrs. Skeat and Blagden, in *Pagan Races of the Malay Peninsula* (vol. i. p. 473), state that "a vampire, according to the view of Sakai of Perak, is not a demon — even though it is incidentally so-called — but a being of flesh and blood," and support this view by the statement that the vampire cannot pass through walls and hedges.

The word *vampire* (Dutch, *vampyr*; Polish, *wampior* or *upior*; Slownik, *upir*; Ukraine, *upeer*) is held by Skeat to be derived from the Servian *wampira*. The Russians, Morlacchians, inhabitants of Montenegro, Bohemians, Servians, Arnauts, both of Hydra and Albania, know the vampire under the name of *wukodalak*, *vurkulaka*, or *vrykolaka*, a word which means "wolf-fairy," and is thought by some to be derived from the Greek. In Crete, where Slavonic influence has not been felt, the vampire is known by the name of *katakhand*. Vampire lore is, in general, confined to stories of resuscitated corpses of male human beings, though amongst the Malays a *penangglan*, or vampire, is a living witch, who can be killed if she can be

ABOUT THE AUTHOR

Dudley Wright (1868-1949) had the distinction of writing the twentieth century's first popular chronicle of vampires. *The Book of Vampires* was first published under the title of *Vampires and Vampirism* in 1914, and revised in 1924. The far more scholarly Montague Summers, whose *The Vampire: His Kith and Kin* (1928) became a standard text on the subject, dismissed Wright's work as "insipid" and "slovenly," but Wright had a knack for writing accessibly for the general public, something that eluded Summers (whose own work will be cited here in sidebars).

Wright's deep interest in esoteric matters is well reflected in the titles of his other books, including *A Manual of Buddhism* (1912), *The Eleusinian Mysteries & Rites* (1913), *Psychical and Supernormal Phenomena: Their Observation and Experimentation*, and several works on the subject of freemasonery.

caught in the act of witchery. She is especially feared in houses where a birth has taken place, and it is the custom to hang up a bunch of thistle in order to catch her. She is said to keep vinegar at home to aid her in re-entering her own body. In the Malay Peninsula, parts of Polynesia and the neighbouring districts, the vampire is conceived as a head with entrails attached, which comes forth to suck the blood of living human beings. In Transylvania, the belief prevails that every person killed by a *nosferatu* (vampire) becomes in turn a vampire, and will continue to suck the blood of other innocent people until the evil spirit has been exorcised, either by opening the grave of the suspected person and driving a stake through the corpse, or firing a pistol shot into the coffin. In very obstinate cases it is further recommended to cut off the head, fill the mouth with garlic, and then replace the head in its proper place in the coffin; or else to extract the heart and burn it, and strew the ashes over the grave.

The *murony* of the Wallachians not only sucks blood, but also possesses the power of assuming a variety of shapes, as, for instance, those of a cat, dog, flea, or spider; in consequence of which the ordinary evidence of death caused by the attack of a vampire, viz. the mark of a bite in the back of the neck, is not considered indispensable. The Wallachians have a very great fear of sudden death, greater perhaps than any other people, for they attribute sudden death to the attack of a vampire, and believe that anyone destroyed by a vampire must become a vampire, and that no power can save him from this fate. A similar belief obtains in Northern Albania, where it is also held that a wandering spirit has the power to enter the body of any individual guilty of undetected crime, and that such obsession forms part of his punishment.

Some writers have ascribed the origin of the belief in vampires to Greek Christianity, but there are traces of the superstition and belief at a considerably earlier date than this. In the opinion of the anthropologist Tylor, "the shortest way of treating the belief is to refer it directly to the principles of savage animism. We shall see that most of its details fall into their places at once and that vampires are not mere creations of groundless fancy, but causes conceived in spiritual form to account for specific facts of wasting disease." It is more than probable that the practice of offering up living animals as sacrifices to satisfy the thirst of departed human beings, combined with the ideas of the Platonist and the terms of the learned Jew, Isaac Arbanel, who maintained that before the soul can be loosed from the fetters of the flesh it must lie some months with it in the grave, may have influenced the belief and assisted its development. Vampirism found a place in Babylonian belief and in the folklore and traditions of many countries of the Near East. The belief was quite common in Arabia, although there is no trace of it there in pre-Christian tinges. The earliest references to vampires are found in Chaldean and Assyrian tablets. Later, the pagan Romans gave their adherence to the belief that the dead bodies of certain people could be allured from their graves by sorcerers, unless the bodies

VAMPIRES, BY ANY OTHER NAME

The word *vampyre* entered the English language in 1732, according to *The Oxford English Dictionary*, perhaps derived from the Turkish word for witch, *uber*, and transformed in Slavic languages as *upior*, *upir*, *oupire*, *upyr*, and, penultimately, *vampyr* and *vampir*. Matthew Bunson, author of *The Vampire Encyclopedia*, suggests that another source may have been the Lithuanian *wempti* — "to drink." Throughout the world, vampires and vampire-like creatures are known by many other names. Here is a sampling:

ASWANG (Philippines)
ASANBOSAM (Ivory Coast, Ghana, Togo)
ALGUL (Arabia)
ALP (Germany)
ADZE (Southeastern Ghana)
ALP (Germany)

BRUXSA (Portugal)
BRAMAPARUSH (India)
BLUTSAUGER (Germany)
BIBI (Gypsy)
BEBARLANGS (Philippines)
BHEAN SIDHE (Ireland)
BAITAL (see VETALA)
BAJANG (Malaysia)
BHUTA (India)

CALLCANTZARO (Greece)
CHORDEWA (Bengal)
CHUPACABRA (Mexico/Carribean)
CHUREL / CHURAIL (India)
CIVATATEO (Mexico)

DANAG (Philippines)
DEARG-DUE (Ireland)
DHAMPIR (Gypsy)
DOPPELSAUGER (Germany)

had actually undergone decomposition, and that the only means of effectually preventing such "resurrections" was by cremating the remains. In Grecian lore there are many wonderful stories of the dead rising from their graves and feasting upon the blood of the young and beautiful. From Greece and Rome the superstition spread throughout Austria, Hungary, Lorraine, Poland, Roumania, Iceland, and even to the British Isles, reaching its height in the period from 1723 to 1736, when a vampire fever or epidemic broke out in the south-east of Europe, particularly in Hungary and Servia. The belief in vampires even spread to Africa, where the Kaffirs held that bad men alone live a second time and try to kill the living by night. According to a local superstition of the Lesbians, the unquiet ghost of the Virgin Gello used to haunt their island, and was supposed to cause the deaths of young children.

Various devices have been resorted to in different countries at the time of burial, in the belief that the dead could thus be prevented from returning to earth-life. In some instances, e.g., among the Wallachians, a long nail was driven through the skull of the corpse, and the thorny stem of a wild rose-bush laid upon the body, in order that its shroud might become entangled with it, should it attempt to rise. The Kroats and Slavonians burned the straw upon which the suspected body lay. They then locked up all the cats and dogs, for if these animals stepped over the corpse it would assuredly return as a vampire and suck the blood of the village folk. Many held that to drive a white thorn stake through the dead body rendered the vampire harmless, and the peasants of Bukowina still retain the practice of driving an ash stake through the breasts of suicides and supposed vampires — a practice common in England, as far as suicides were concerned, until 1823, when there was passed "An Act to alter and amend the law relating to the interment of the remains of any person found *felo de se*," in which it was enacted that the coroner or other officer "shall give directions for the private interment of the remains of such person *felo de se* without any stake being driven through the body of such person." It was also ordained that the burial was only to take place between nine and twelve o'clock at night.

The driving of a stake through the body does not seem to have had always the desired effect. A Strigon or Indian vampire, who was transfixed with a sharp thorn cudgel, near Labach, in 1672, pulled it out of his body and flung it back contemptuously. Bartholin, in *de Causa contemptus mortis*, tells the story of a man, named Harpye, who ordered his wife to bury him exactly at the kitchen door, in order that he might see what went on in the house. The woman executed her commission, and soon after his death he appeared to several people in the neighbourhood, killed people while they were engaged in their occupations, and played so many mischievous pranks that the inhabitants began to move away from the village. At last a man named Olaus Pa took courage and ran at the spectre with a lance, which he drove into the apparition. The spectre instantly vanished,

EMPUSA (Greece)
ERETICA (Russia)
ESTRIE (Hebrew)

GAYAL (India)
GHOUL (Arabic)

HANNYA (Japan)

IMPUNDULU (Africa)
INCUBUS (Medieval Europe)

JARACACAS (Brazil)

KOZLAK (Dalmatia)
KRESNIK / KRSNIK / KRUVNIK (Slovenia)
KRVOIJAC (Bulgaria)
KUANG-SHI (China)
KUDLAK (Slovenia)
KUKUDHI (Albania)

LAMIA (Greece)
LAMPIR (Bosnia)
LANGSUIR (Malaysia)
LEANHAUM-SHEE (Ireland)
LIDERC / LUDVERC (Hungary)
LOBISHOMEN (Brazil)
LOUP-GAROU / LOOGAROO (West Indies)
LUGAT (Albania)

MANDURAGO (Philippines)
MARA (Scandinavia)
MASAN/MASANI (India)
MATI-ANAK (Malaysia)
MORA (Czechoslovakia)
MORMO (Greece)
MOROII (Romania)
MOTETZ DAM (Hebrew)
MULLO / MULI (Gypsy)
MURONI / MURONY / MURONUL (Wallachia)

NACHZEHER (Silesia, Bavaria)
NELAPSI (Slovak)

taking the spear with it. Next morning Olaus had the grave of Harpye opened, when he found the lance in the dead body, which had not become corrupted. The corpse was then taken from the grave, burned, and the ashes thrown into the sea, and the spectre did not afterwards trouble the inhabitants.

To cross the arms of the corpse, or to place a cross or crucifix upon the grave, or to bury a suspected corpse at the junction of four cross-roads, was, in some parts, regarded as an efficacious preventive of vampirism. It will be remembered that it was at one time the practice in England to bury suicides at four cross-roads. If a vampire should make its appearance, it could be prevented from ever appearing again by forcing it to take the oath not to do so, if the words "by my winding-sheet" were incorporated in the oath.

One charm employed by the Wallachians to prevent a person becoming a vampire was to rub the body in certain parts with the lard of a pig killed on St. Ignatius's Day.

In Poland and Russia, vampires make their appearance from noon to midnight instead of between nightfall and dawn, the rule that generally prevails. They come and suck the blood of living men and animals in such abundance that sometimes it flows from them at the nose and ears, and occasionally in such profusion that the corpse swims in the blood thus oozing from it as it lies in the coffin. One may become immune from the attacks of vampires by mixing this blood with flour and making bread from the mixture, a portion of which must be eaten; otherwise the charm will not work. The Californians held that the mere breaking of the spine of the corpse was sufficient to prevent its return as a vampire. Sometimes heavy stones were piled on the grave to keep the ghost within, a practice to which Frazer traces the origin of funereal cairns and tombstones. Two resolutions of the Sorbonne, passed between 1700 and 1710, prohibited the cutting off of the heads and the maiming of the bodies of persons supposed to be vampires.

In the German folk-tale known as *Faithful John*, the statue said to the king: "If you, with your own hand, cut off the heads of both your children and sprinkle me with their blood, I shall be brought to life again." According to primitive ideas, blood is life, and to receive blood is to receive life: the soul of the dead want to live, and, consequently, loves blood. The shades in Hades are eager to drink the blood of Odysseus's sacrifice, that their life may be renewed for a time. It is of the greatest importance that the soul should get what it desires, as, if not satisfied, it might come and attack the living. It is possible that the bodily mutilations which to this day accompany funerals among some peoples have their origin in, the belief that the departed spirit is refreshed by the blood thus spilt. The Samoans called it an "offering of blood" for the dead when, the mourners beat their heads till the blood ran.

The Australian native sorcerers are said to acquire their magical

OBAYIFO (West Africa)
ODOROTEN (Russia)
OHYN (Poland)

PENANGGALAN (Malaysia)
PIJAVICA / PIJAWICA (Slovenia, Croatia)

STRIGOII / STRIGOICA (Romania)
SUCCUBUS (Medieval Europe)
SWAMX (Myanmar)

TLACIQUE (Mexico)

UBOUR (Bulgaria)
UPIOR / UPIER (Poland)
UPYR / OUPYR (Russia)
USTREL (Bulgaria)

VAMPIR / VAMPYR (Eastern Europe)
VETALA (Idia)
VOLKODLAK (Slovenia)
VOPYR (Russia)
VOURDALAK / WURDALAK (Russia)
VRYKOLAKAS / VROUKALAKAS / VRYKOLATIOS (Greece)
VUKODLAK (Serbia)

WIESZCZY (Poland)

XLOPTUNY (Russia)

ZMEU (Moldavia)

induenoe by eating human flesh, but this is done once only in a lifetime. According to Nider's *Formicarius*, part of the ceremony of initiation into wizardry and witchcraft consisted in drinking in a church, before the commencement of Mass, from a flask filled with blood taken from the corpses of murdered infants.

The methods employed for the detection of vampires have varied according to the countries in which the belief in their existence was maintained. In some places it was held that, if there were discovered in a grave two or three or more holes about the size of a man's finger, it would almost certainly follow that a body with all the marks of vampirism would be discovered within the grave. The Wallachians employed a rather elaborate method of divination. They were in the habit of choosing a boy young enough to make it certain that he was innocent of any impurity. He was then placed on an absolutely black and unmutilated horse which had never stumbled. The horse was then made to ride about the cemetery and pass over all the graves. If the horse refused to pass over any grave, even in spite of repeated blows, that grave was believed to shelter a vampire. Their records state that when such a grave was opened it was generally found to contain a corpse as fat and handsome as that of a full-blooded man quietly sleeping. The finest vermilion blood would flow from the throat when cut, and this was held to be the blood he had sucked from the veins of living people. It is said that the attacks of the vampire generally ceased on this being done.

In the town of Perlepe, between Monastru and Kiuprili, there existed the extraordinary phenomenon of a number of families who were regarded as being the offspring of *vrykolakas*, and as possessing the power of slaying the wandering spirits to which they were related. They are said to have kept their art very dark and to have practised it in secret, but their fame was so wider spread that persons in need of such deliverance were accustomed to send for them from other cities. In ordinary life and intercourse they were avoided by all the inhabitants.

Although some writers have contended that no vampire has yet been caught in the act of vampirism, and that, as no museum of natural history has secured a specimen, the whole of the stories concerning vampires may be regarded as mythical, others have held firmly to a belief in their existence and inimical power. Dr. Pierart, in *La Revue Spiritualiste* (vol. iv. P. 104), wrote: "After a crowd of facts of vampirism so often proved, shall we say that there are no more to be had, and that these never had a foundation? Nothing comes of nothing. Every belief, every custom, springs from facts and causes which give it birth. If one had never seen appear in the bosom of their families, in various countries, beings clothed in the appearance of departed ones known to them, sucking the blood of one or more persons, and if the deaths of the victims had not followed after such apparitions, the disinterment of corpses would not have taken place, and then would never have been the attestation of the otherwise incredible fact of persons buried for several years being found with the body soft and flexible, the

BLOOD MAGIC

Blood, the ultimate human symbol, has the power to assume almost endless metaphorical forms — much like the vampire itself. As the primary vital fluid, blood has been held in awe since prehistoric times, and is prominent in the imagery of most religious and folk traditions. Blood represents the life force, the emotions and sexuality — but the sight of blood paradoxically signifies death. Blood is our connection to the atavistic past, as well as our immediate bond of kinship and fealty. The ancient belief that there is no essential difference between the physical reality of blood and the less tangible qualities of spirit, courage, and consciousness — that "the blood is the life" — is notably at the root of cannibalism, blood sacrifice and vampire legends in a wide variety of cultures.

Blood, in many traditions, is believed to absorb or transmit evil; the removal of blood from the body, therefore, can provide a cathartic cleansing. Literal blood sacrifice is now rare, but the torrential presence of simulated bloodletting in popular culture may serve to exorcise diffuse cultural anxieties in a postmodern age of image and artifice. Next to the Catholic ritual of the mass, vampire legends and entertainment provide one of our richest modern repositories of blood-related themes, ceremonies and obsessions.

eyes wide open, the complexion rosy, the mouth and nose full of blood, and the blood flowing fully when the body was struck or wounded or the head cut off."

Bishop d'Avranches Huet wrote, "I will not examine whether the facts of vampirism, which are constantly being reported, are true, or the fruit of a popular error; but it is beyond doubt that they are testified to by so many able and trustworthy authors, and by so many *eye-witnesses*, that no one ought to decide the question without & good deal of caution."

Dr. Pierat gave the following explanation of their existence: "Poor, dead cataleptics, buried as if really dead in cold and dry spots where morbid causes are incapable of effecting the destruction of their bodies, the astral spirit, enveloping itself with a fluidic ethereal body, is prompted to quit the precincts of its tomb and to exercise on living bodies acts peculiar to physical life, especially that of nutrition, the result of which, by a mysterious link between soul, and body which spiritualistic science will some day explain, is forwarded to the material body lying still within its tomb, and the latter is thus helped to perpetuate its vital existence."

Apart from the spectre vampire there is, of course, the vampire bat in the world of natural history, which is said to suck blood from a sleeping person, insinuating its tongue into a vein, but without inflicting pain: Captain Steadman, during his expedition to Surinam, awoke early one morning and was alarmed to find his hammock steeped almost through and himself weltering in blood, although he was without pain. It was discovered that he had been bitten by a vampire bat. Pennant says that in some parts of America they destroyed all the cattle introduced by the missionaries.

FROM

The Phantom World

AUGUSTIN CALMET
(1746/1850)

ABOUT THE AUTHOR

One of the first scholars to systematically examine the belief in vampires, the Benedictine monk Augustin Calmet (1672-1757) was himself a contemporary of the vampire hysteria which swept central and Eastern Europe in the 1720s and 1730s. Near the end of his life, the celebrated French biblical scholar published a two-volume treatise on ghosts, vampires, and other revenants; a century later, in 1850, it appeared in English as *The Phantom World: or, the Philosophy of Spirits and Apparitions*. Calmet was essentially skeptical about the existence of vampires (he snorted at claims that "the dead have been heard to eat and chew like pigs in their graves"); nonetheless, he dutifully recorded a wide range of official vampire reports from sources in Hungary, Poland, and elsewhere, which he then subjected to analysis from natural and theological perspectives. Three of his accounts are presented here in their entirety.

The Vampires of Moravia

I HAVE BEEN told by the late Monsieur de Vassimont, counsellor of the Chamber of the Counts of Bar, that have been sent to Moravia by his late Royal Highness Leopold, first Duke of Lorraine, for the affairs of the Prince Charles his brother, bishop of Olmutz and Osnaburgh, he was informed by a public report, that it was common enough in that country to see men who had died some time before, present themselves in a party, and sit down to table with persons of their acquaintance without saying any thing; but that nodding to one of the party, he would infallibly die some days afterwards. This fact was confirmed by several persons, and amongst others by an old curé, who said he had seen more than one instance of it.

The bishops and priests of the country consulted Rome on so extraordinary a fact: but they received no answer, because, apparently, all those things were regarded there as simple visions, or popular fancies. They afterwards bethought themselves of taking up the corpses of those who came back in that way, of burning them, or of destroying them in some other manner. Thus they delivered themselves from the importunity of these spectres, which are now much less frequently seen than before. So said that good priest.

These apparitions have given rise to a little work, entitled, *Magia Posthuma*, printed at Olmutz, in 1706, composed by Charles Ferdinand de Schertz, dedicated to Prince Charles of Lorraine, Bishop of Olmutz and Osnaburgh. The author relates, that in a certain village, a woman being just dead, who had taken all her sacraments, she was buried in the usual way in the cemetery. Four days after her decease, the inhabitants of the village heard a great noise and extraordinary uproar, and saw a spectre, which appeared sometimes in the shape of a dog, sometimes in the form of a man, not to one person only, but to several, and caused them great pain, grasping their throats, and compressing their stomachs, so as to suffocate them. It bruised almost the whole body, and reduced them to extreme weakness, so that they became pale, lean and attenuated.

The spectre attacked even animals, and some cows were found debilitated and half dead. Sometimes it tied them together by their tails. These animals gave sufficient evidence by their bellowing of the pain they suffered. The horses seemed overcome with fatigue, perspired profusely, principally on the back; were heated, out of breath, covered with foam, as after a long and rough journey. These calamities lasted several months.

The author whom I have mentioned examines the affair in a lawyer-like way, and reasons much on the fact and the law. He asks, if, supposing that these disturbances, these noises and vexations, proceeded from that person who is suspected of causing them, they can burn her, as is done to other ghosts who do harm to the living. He relates several other instances of similar apparitions, and of the evils which ensued; as of a shepherd in the village of Blow, near the town of Kadam, in Bohemia, who appeared during some time, and called certain persons, who never failed to die within eight days after. The peasants of Blow took up the body of this shepherd, and fixed it in the ground with a stake which they drove through it.

The dead return: a European woodcut.

This man when in that condition derided them for what they made him suffer, and told them they were very good to give him thus a stick to defend himself from the dogs. The same night he got up again, and by his presence alarmed several persons, and strangled more amongst them than he had hitherto done. Afterwards, they delivered him into the hands of the executioner, who put him in a cart to carry him beyond the village and there burn him. The corpse howled like a madman, and moved his hands and feet as if alive. And when they again pierced him through with stakes he uttered very loud cries, and a great quantity of bright vermilion blood flowed from him. At last he was consumed, and this execution put an end to the appearance and hauntings of this spectre.

The same has been practised in other places, where similar ghosts have been seen; and when they have been taken out of the ground they have appeared red, with their limbs supple and pliable, without worms or decay; but not without great stench. The author cites diverse other writers, who attest what he says of these spectres, which still appear, he says, very often in the mountains of Silesia and Moravia. They are seen by night and day; the things which once belonged to them are seen to move themselves and change their place without being touched by any one. The only remedy for these apparitions is to cut off the heads and burn the bodies of those who come back to haunt their old abodes.

At any rate they do not proceed to do this without a form of justicial law. They call for and hear the witnesses; they examine the arguments; they look at the exhumed bodies, to see if they can find any of the usual marks which lead them to conjecture that they are the parties who molest the living, as the mobility of the limbs, the fluidity of the blood, and the flesh remaining uncorrrupted. If all these marks are found, then these bodies are given up to the executioner, who burns them. It sometimes happens that the spectres appear again for three or four days after the execution. Sometimes the interment of the bodies of suspicious persons is deferred for six or seven weeks. When they do not decay, and their limbs remain as supple and pliable as when they were alive, then they burn them. It is affirmed as certain that the clothes of these persons move without any one living touching them; and within a short time, continues our author, a spectre was seen at Olmutz, which threw stones, and gave great trouble to the inhabitants.

Dead Persons in Hungary Who Suck the Blood of the Living

About fifteen years ago, a soldier who was billeted at the house of a Haidamaque peasant, on the frontiers of Hungary, as he was one day sitting at table near his host, the master of the house saw a person he did not know come in and sit down to table also with them. The master of the house was strangely frightened at this, as were the rest of the company. The soldier knew not what to think of it, being ignorant of the matter in question. But the master of the house being dead the very next day, the soldier inquired what it meant. They told him that it was the body of the father of his host, who had been dead and buried for ten years, which had thus come to sit down next to him, and had announced and caused his death.

The soldier informed the regiment of it in the first place, and the regiment gave notice of it to the general officers, who commissioned the Count de Cabreras, captain of the regiment of Alandetti infantry, to make information concerning this circumstance. Having gone to the place, with some other officers, a surgeon and an auditor, they heard the depositions of all the people belonging to the house, who attested unanimously that the ghost was the father of the master of the house, and that all the soldier had said and reported was the exact truth, which was confirmed by all the inhabitants of the village.

In consequence of this, the corpse of the spectre was exhumed, and found to be like that of a man who has just expired, and his blood was like that of a living man. The Count de Cabreras had his head cut off, and caused him to be laid again in his tomb. He also took information concerning other similar ghosts; amongst others, of a man dead more than thirty years, who had come back three times to his house at meal-time. The first time he had sucked the blood from the neck of his own brother, the second time from one of his sons, and the third from one of the servants in the house; and all three died of it instantly and on the spot. Upon this deposition the commissary had the man taken out of his grave, and finding that, like the first, his blood was in a fluid state, like that of a living person, he ordered them to run a large nail into his temple, and then to lay him again in the grave.

He caused a third to be burnt, who had been buried more than sixteen years, and had sucked the blood and caused the death of two of his sons. The commissary having made his report to the general officers, was deputed to the court of the emperor, who commanded that some officers, both of war and justice, some physicians and surgeons, and some learned men, should be sent to examine the causes of these extraordinary events.

CALMET'S VAMPIRE PARADOXES

Augustin Calmet was the first scholar to enumerate the paradoxes inherent in the physical dynamics of vampirism. "How," he asked, "can a corpse which is covered with four or five feet of earth, which has no room even to move or to stretch a limb, which is wrapped in linen cerements, enclosed in a coffin of wood, how can it, I say, seek the upper air and return to the world walking upon the earth . . . [c]an it be maintained that these corpses pass through the earth without disturbing it, just as water and the damps which penetrate the soil or exhale therefrom without perceptibly dividing or cleaving the ground?"

Calmet considered the possibility of vampiric astral projection, but this, too, presented conundrums. "Let us suppose that these corpses do not actually stir from their tombs, that only the ghosts or spirits appear to the living, wherefor do these phantoms present themselves and what is it that energizes them? Is it actually the soul of the dead man which has not yet parted to its final destination, or is it a demon which causes them to be seen in an assumed and phantastical body? And if their bodies are spectral, how do they suck the blood of the living? We are enmeshed in a sad dilemma when we ask if these apparitions are natural or miraculous."

The person who related these particulars to us had heard them from the Count de Cabreras, at Fribourg in Brigau, in 1730.

Arnald Paul, Vampire

In a certain canton of Hungary, named in Latin Oppida Heidanum, beyond the Tibisk, *vulgo* Teiss, that is to say, between that river which waters the fortunate territory of Tokay and Transylvanian, the people known by the name of *Heyducqs* believe that certain dead persons, whom they call vampires, suck all the blood from the living, so that these become visibly attenuated, whilst the corpses, like leeches, fill themselves with blood in such abundance that it is seen to come from them by the conduits, and even oozing through the pores. This opinion has just been confirmed by several facts which cannot be doubted, from the rank of the witnesses who have certified them. We will here relate some of the most remarkable.

About five years ago, a certain Heyducq, inhabitant of Madreiga, named Arnald Paul, was crushed to death by the fall of a wagon-load of hay. Thirty days after his death four persons died suddenly, and in the same manner in which, according to the tradition of the country, those die who are molested by vampires. They then remembered that this Arnald Paul had often related that in the environs of Cassovia, and on the frontiers of Turkish Servia, he had often been tormented by a Turkish vampire; for they believe also that those who have been passive vampires during life become active ones after their death, that is to say, that those who have been sucked, suck also in their turn; but that he had founds means to cure himself by eating earth from the grave of the vampire, and smearing himself with blood; a precaution which, however, did not prevent him from becoming so after his death, since, on being exhumed forty days after his interment, they found on his corpse all the indications of an arch-vampire. His body was red. His hair, nails, and beard had all grown again, and his veins were replete with fluid blood, which flowed from all parts of his body upon the winding-sheet which encompassed him. The Hadnagi, or bailli of the village, in whose presence the exhumation took place, and who was skilled in vampirism, had, according to the custom, a very sharp stake driven into the heart of the defunct Arnald Paul, and which pierced his body through and through, which made him, as they say, utter a frightful shriek, as if he had been alive: that done, they cut off his head, and burnt his whole body. After that they performed the same on the corpses of four other persons who died of vampirism, fearing that they in their turn might cause the death of others.

All these performances however, could not prevent the recommencement of similar fatal prodigies towards the end of last year, (1732) that is to say, five years after, when several inhabitants of the same village perished miserably. In the space of three months seventeen persons of different sexes and different ages died of vampirism; some without

MONTAGUE SUMMERS, VAMPIRE HUNTER

Augustin Calmet's eighteenth-century exploration of vampires prefigured the twentieth-century research of British scholar Montague Summers (1880-1947), whose books *The Vampire: His Kith and Kin* (1928), *The Vampire in Europe* (1929) and *The Werewolf* (1933) have become indispensible (if sometimes over-credulous) texts for modern students of vampirism and related phenomena, and include exhaustive documenation.

Summers affected clerical robes and attitudes, but never seems to have been actually ordained, possibly because of his membership in the quasi-Masonic circle of Victorian homosexuals known as the Order of Chaeronea.

As for the story of Arnold (Arnald) Paul (or Paole, as he spells the name in his own account, drawing upon Calmet), Summers regards it as "One of the best-attested, most detailed and most famous histories of vampirism" on record.

Montague Summers

being ill, and others after languishing two or three days. It is reported, amongst other things, that a girl named Stanoska, daughter of the Heyducq Millo, who had been dead nine weeks, had nearly strangled in her sleep. She fell into a languid state from that moment, and at the end of three days, she died. What this girl had said of Millo's son made him known at once for a vampire: he was exhumed, and found to be such. The principal people of the place, with the doctors and the surgeons, examined how vampirism could have sprung up again after the precautions they had taken some years before.

They discovered at last, after much search, that the defunct Arnald Paul had killed not only the four persons of whom we have spoken, but also several oxen, of which the new vampires had eaten, and amongst these the son of Millo. Upon these indications they resolved to disinter all those who had died within a certain time, &c. Amongst forty, seventeen were found with all the most evident signs of vampirism; so they transfixed their hearts and cut off their heads also, and then cast their ashes into the river.

All the informations and executions we have just mentioned were made juridically, in proper form, and attested by several officers who were garrisoned in the country, by the chief surgeons of the regiments, and by the principal inhabitants of the place. The verbal process of it was sent towards the end of last January to the Imperial Council of War at Vienna, which had established a military commission to examine into the truth of all these circumstances.

Such was the declaration of the Hadnagi Barriarar and the ancient Heyducqs, and it was signed by Battuer, first lieutenant of the regiment of Alexander of Wurtemburg, Clickstenger, surgeon-in-chief of the regiment of Frustemburch, three other surgeons of the company, and Guoichitz, captain at Stallach.

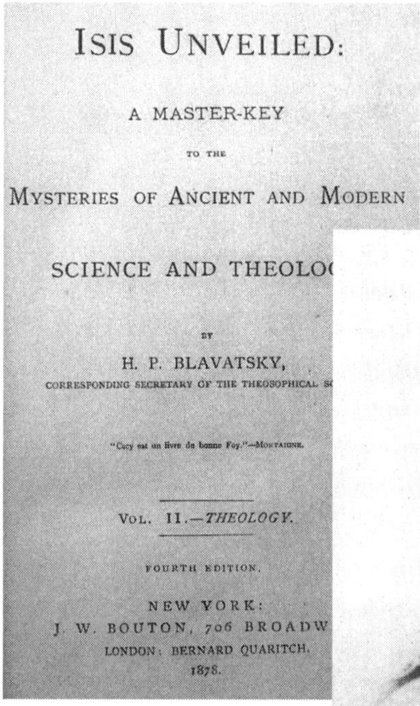

ISIS UNVEILED:

A MASTER-KEY

TO THE

MYSTERIES OF ANCIENT AND MODERN

SCIENCE AND THEOLO

BY

H. P. BLAVATSKY,

CORRESPONDING SECRETARY OF THE THEOSOPHICAL S

"Cecy est un livre de bonne Foy."—MONTAIGNE.

VOL. II.—*THEOLOGY.*

FOURTH EDITION.

NEW YORK:
J. W. BOUTON, 706 BROADW
LONDON: BERNARD QUARITCH.
1878.

Helena Petrovna Blavatsky, with the original title page of Isis Unveiled.

Tale of a Russian Vampire

HELENA PETROVNA BLAVATSKY
(1877)

ABOUT THE BEGINNING of the present century, there occurred in Russia, one of the most frightful cases of vampirism on record. The governor of the province of Tch—— was a man of about sixty years, of a malicious, tyrannical, cruel and jealous disposition. Clothed with despotic authority, he exercised it without stint, as his brutal instincts prompted. He fell in love with the pretty daughter of a subordinate official. Although the girl was betrothed to a young man whom she loved, and, the tyrant forced her father to consent to his having her marry him; and the poor victim, despite her despair, became his wife. His jealous disposition exhibited itself. He beat her, confined her to her room for weeks together, and prevented her seeing any one except in his presence. He finally fell sick and died. Finding his end approaching, he made her swear never to marry again; and with fearful oaths, threatened that, in case she did, he would return from his grave and kill her. He was buried in the cemetery across the river, and the young widow experienced no further annoyance, until, nature getting the better of her fears, she listened to the importunities of her former lover, and they were again betrothed.

On the night of the customary betrothal-feast, when all had retired, the old mansion was aroused by shrieks proceeding from her room. The doors were burst open, and the unhappy woman was found lying on her bed, in a swoon. At the same time a carriage was heard rumbling out of the courtyard. Her body was found to be black and blue in places, as from the effect of pinches, and from a slight puncture on her neck drops of blood were oozing. Upon recovering, she stated that her deceased husband had suddenly entered her room, appearing exactly as in life, with the exception of a dreadful pallor; that he had upbraided her for her inconstancy, and then beaten and pinched her most cruelly. Her story was disbelieved; but the next morning, the guard stationed at the other end of the bridge which spans the river, reported that, just before midnight, a black coach and six had driven furiously past them, toward the town, without answering their challenge.

The new governor, who disbelieved the story of the apparition, took nevertheless the precaution of doubling the guards across the bridge. The same thing happened, however, night after night; the soldiers declaring that the toll-bar at their station near the bridge would rise of itself, and the

ABOUT THE AUTHOR

Helena Petrovna "Madame" Blavatsky (1831-1891) was, and remains, a highly controversial figure in the history of philosophy and religion. Born in Russia to an aristocratic family, Blavatsky showed an early talent for languages, music, and art, but, after a failed marriage, embarked on extensive spiritual pilgrimages throughout the world. Settling in New York, she cofounded the Theosophical Society in 1875 and rapidly became a fixture of Manhattan society, then enthralled with the spiritualist craze sweeping the country. Blavatsky attempted to give spiritualism a philosophical pedigree through an eclectic synthesis of her esoteric studies and personal claims of occult power.

Although she was widely denounced as a fraud for her carnivalesque exhibitions of table rapping, ectoplasm, and other mediumistic displays, she nonetheless was a major force in the introduction of Eastern philosophy to America, especially concepts of reincarnation, karma, and the presence of life and consciousness in all matter.

The selection presented here is taken from Blavatsky's first major work, *Isis Unveiled*, first published in 1877.

spectral equipage sweep by them despite their efforts to stop it. At the same time every night, the coach would rumble into the courtyard of the house; the watchers, including the widow's family, and the servants, would be thrown into a heavy sleep; and every morning the victim would be found bruised, bleeding and swooning as before. The town was thrown into consternation. The physicians had no explanations to offer; priests came to pass the night in prayer, but as midnight approached, all would be seized with the terrible lethargy. Finally, the archbishop of the province came, and performed the ceremony of exorcism in person, but the following morning the governor's widow was found worse than ever. She was now brought to death's door.

The governor was finally driven to take the severest measures to stop the ever-increasing panic in the town. He stationed fifty Cossacks along the bridge, with orders to stop the spectre-carriage at all hazards. Promptly at the usual hour, it was heard and seen approaching from the direction of the cemetery. The officer of the guard, and a priest bearing a crucifix, planted themselves in front of the toll-bar, and together shouted: "In the name of God, and the Czar, who goes there?" Out of the coach-window was thrust a well-remembered head, and a familiar voice responded: "The Privy Councillor of State and Governor, C——!" At the same moment, the officer, the priest, and the soldiers were flung aside as by an electric shock, and the ghostly equipage passed by them, before they could recover breath.

The engraving of Blavatsky that appeared as the original frontispiece to Isis Unveiled.

The archbishop then resolved, as a last expedient, to resort to the time-honored plan of exhuming the body, and pinning it to the earth with a stake driven through its heart. This was done with great religious ceremony in the presence of the whole populace. The story is that the body was found gorged with blood, and with red cheeks and lips. At the instance that the first blow was struck upon the stake, a groan issued from the corpse, and a jet of blood spurted high into the air. The archbishop pronounced the usual exorcism, the body was reinterred, and from that time no more was heard of the vampire.

How far the facts of this case may have been exaggerated by tradition, we cannot say. But we had it years ago from an eye-witness; and at the present day there are families in Russia whose elder members will recall the dreadful tale.

FROM

Vampires of Roumania

AGNES MURGOCI
(1926)

The Girl and the Vampire

ONCE IN A village there were a girl and a youth who were deeply in love, their parents did not know, and when the relations of the youth approached the parents of the girl with a proposal of marriage they were repulsed because the youth was poor. So the young man hanged himself on a tree, and became a vampire. As such he was able to come and visit the girl. But, although the girl had loved the man, she did not much like to have to do with an evil spirit. What could she do to escape from danger and sin? She went to a wise woman, and this wise woman advised her what to do. The vampire came one evening to make love to the girl and stayed late. When he knew that it was about time to leave, he said, — "Good night," and made ready to go. The girl, following the advice of the wise old woman, fixed into the back of his coat a needle, to which was attached one end of the thread from a large ball of thread. The vampire went away, and the ball unrolled and unrolled for some time and then, all at once, it stopped. The girl understood what had happened, and followed the clue given by the thread. She traced it along the road, and found that it entered into the churchyard, and went straight to a grave. There it entered the earth, and that was the end. She came home, but the next night, as twilight came on, she hastened to the churchyard, and stood some distance from the grave to see what would happen. It was not long before she saw the vampire coming out, going to another grave, opening it, eating the heart of the dead man buried there, and then setting out towards the village to visit her. She followed him as he left the churchyard.

"Where were you this evening, and what did you see?" asked the vampire after he had greeted her. "Where was I? Nowhere, I saw nothing," said the girl. The vampire continued, — "I warn you that, if you do not tell me, your father will die." "Let him die, I know nothing, I've seen nothing, and I can say nothing." "Very well," said the vampire, and indeed in two days the girl's father was dead. He was buried with all the due rites, and it was some time before the vampire again came to the girl.

One night, however he came and made love to her as usual, but before leaving he said, — "Tell me where you were that evening, because, if you will not, your mother will die." "She may die nine times. How can I speak when I know nothing?" answered the girl.

After two days the mother died. She was duly buried. Again some

ABOUT THE AUTHOR

In 1926, folklorist Agnes Murgoci published her translations of traditional Romanian vampire stories in the British journal *Folklore*. The folktales had originally appeared in the Romanian publication *Ion Creanga*, named for one of the country's great traditional storytellers. Although stories from the oral tradition do not constitute historical evidence per se, they are nonetheless the foundation of many historical behaviors and customs, which Murgoci also fully documents. Her fascinating essay is reprinted in its entirety in *The Vampire: A Casebook*, edited by Alan Dundes (University of Wisconsin Press, 1998).

The Romanian vampire is most commonly called a *strigoi* (or the feminine *stregoica*), and frequently, though less often, a *moroii*. (The problematic word *nosferatu* is often cited as Romanian, but does not appear in Romanian dictionaries or folk stories, and is not, in fact, a Romanian word at all. For more, see the commentary accompanying "Dracula's Guest" in Part Two.)

time passed, and the vampire reappeared, and now he said, —"If you do not tell me what you saw that evening, you shall die too." "What if I do?" said she, "it will be no great loss. How can I invent a story, if I know nothing and have seen nothing?" "That is all very well, but what are you going to do now, for you are about to die?" replied the vampire.

CAUSES OF VAMPIRISM

"Roumanians believe that a man born with a caul becomes a vampire within six weeks after his death," Murgoci writes; "similarly, people who were bad and had done evil deeds in their lifetime, and more especially women who have had to do with the evil one and with spells and incantations. It is known that a man is a vampire if he does not eat garlic; this idea is also found among the South Slavs. When a child dies before it is baptized, it becomes a vampire at seven years of age, and the place where it is buried is unholy. Men who swear falsely for money become vampires six months after death. If a vampire casts its eye on a pregnant woman, and is not disenchanted, her child will be a vampire. If a pregnant woman does not eat salt, her child will be a vampire. When there are seven children of the same sex, the seventh will have a little tail and be a vampire. A dead man becomes a vampire if a cat jumps over him, if a man steps over him, or even if the shadow of a man falls over him."

In the Romanian tradition, there are both living vampires and dead vampires; the living variety tend to overlap with the traditional witch, and are not bound by the rising and setting of the sun. Needless to say, it is the dead vampire, whether a bloodsucking ghost or a reanimated corpse, that has had the greatest impact on literature and popular culture.

On the advice of the wise old woman the girl called all her relations together and told them that she was going to die soon. When she was dead they were not to take her out by the door or by the window, but to break an opening in the walls of the house. They were not to bury her in the churchyard, but in the forest, and they were not to take her by the road but to go right across the fields until they came to a little hollow among the trees of the forest and here her grave was to be. And so it happened. The girl died, the wall of the house was broken down, and she was carried out on a bier across the fields to the margin of the forest.

After some time a wonderful flower, such as has never been seen, either before or after, grew up on her grave. One day the son of the emperor passed by and saw this flower, and immediately gave orders that it should be dug up well below the roots, brought to the castle, and put by his window. The flower flourished, and was more beautiful than ever, but the son of the emperor pined. He himself did not know what was the matter, he could neither eat nor drink. What was the matter? At night the flower became again the maiden, as beautiful as before. She entered in at the window, and passed the night with the emperor's son without his knowing it. However, one night she could contain herself no longer, and kissed him, and he awoke and saw her. After that, they were married, and they lived very happily together. There was only one drawback to their happiness. The wife would never go out of the house. She was afraid of the vampire.

One day, however, her husband took her with him in a carriage to go to church, when there, at a corner, who should be there but the vampire. She jumped out of the carriage and rushed to the church. She ran, the vampire ran, and just had his hand on her as they both reached the church together. She hid behind a holy picture. The vampire stretched out his hand to seize her, when all at once the holy picture fell on his head, and he disappeared in smoke. And the wife lived with the emperor's son free from all danger and sin for the rest of her life.

A Vampire Story from Botosani

There was once a time when vampires were as common as leaves of grass, or berries in a pail, and they never kept still, but wandered round at night among the people. They walked about and joined the evening gatherings in the villages, and, when there were many young people together, the vampires could carry out their habit of inspiring fear, and sucking human blood like leeches. Once, when an evening gathering was in full swing, in came an univited guest, the vampire. But no one knew that he was a vampire. He was in the form of a handsome youth, full of fun. He said "Good day" very politely, sat down on a bank beside the girls, and began to talk, and all the girls imagined that he was a youth from another part of the village. Then the vampire began to tell stories and jokes, so that the girls did not know what to do for laughter. He played and jested and bandied words with them without ceasing. But there was one girl to whom he paid special attenton, and teased unmercifully. "Keep still, friend. Have I done anything to annoy you?" said she. But he still kept on pinching her, till she was black and blue. "What is it, friend? You go too far with your joke. Do you want to make an end of me?" said the poor girl. At the moment her distaff fell. When she stooped to pick it up, what did she see? The tail of the vampire. Then she said to the girl next to her, — "Let's go. Run away. The creature is a vampire." The other girl was laughing so much that she did not understand. So the girl who knew the dreadful secret went out alone into the yard, on the pretext that she had to take some lengths of woven linen to the attic. Frightened out of her wits, she ran away with the linen, she ran into a forest, old as the world and black as her fear.

Her companions at the gathering awaited her return. They looked and waited until they saw that she was not coming back. Where could she be? "You must fetch her wherever she is," roared the vampire, with bloodshot eyes and hair standing on end. As the girl could not be found, the vampire killed all the rest of the merrymakers. He sucked their blood, he threw their flesh and bones under the bed, cut off their lips, and put their heads in a row in the window. They looked as if they were laughing. He strung up their intestines on a nail, saying they were strings of beads, and then he fled away. He arrived at the forest where the girl had taken refuge, and found her under a beech-tree. "Why did you come here, little girl? Why did you run away from the gathering?" The girl, poor thing, was so frightened that her tongue clove to her mouth, and she could say nothing. "You are afraid, little girl. Come home with me. You will feel better there." Then, involuntarily, she asked, — "Where?" "Here in the forest. Come quicker," said the vampire.

They arrived at a hole in the depth of the forest, and she saw that this was the home of the vampire. He pressed her to enter first. "No, no. I don't want to. You go first." So the vampire went in, and began to sweep and clear up. The girl, however, stopped up the hole with lengths of linen, and fled quickly toward the east. In her flight she saw a little light a long way off. She ran toward the light, came to a house, and found it empty, except for a dead man, who was lying stretched out on a table, with a torch at his head, and his hands crossed on his breast. What was she to do? She entered the house, climbed up on to the stove, and went to sleep, worn out by suffering and fear. And she would have rested well, had not the terrible vampire pursued her. He had thrown aside the linen, and rushed after her, mad with rage. He came into the house, and the dead man rose,

PRECAUTIONS AGAINST VAMPIRES

According to Murgoci, when there is a possibility of vampirism, "precautions must be taken at burial or soon after. As suicides are potential vampires, they should be dug up at once from their graves, and put into running water. A man may know that he was born with a caul, and leave word what is to be done to save his family from disaster." The most common method is to drive a stake through the heart or the navel (although in Bulgaria, Murgoci noted, a red-hot iron is used in place of the stake). Alternately, the heart can be cut out and "burnt on charcoal, or in a fire; it may be boiled, or cut to bits with a scythe. If the heart is burnt, the ashes must be collected. Sometimes they are gotten rid of by throwing into a river, but they are usually mixed with water and given to sick people to drink."

In cases where the entire body is dismembered and burned, "it is emphasized that the burning must be complete. If the smallest piece of bone remains unburnt, the vampire can grow up again from it."

and they fought and wrestled till the cock crowed and the girl awoke. Now the light was out, the dead man was gone, and the only sound was the song of the little cricket. The girl was left alone with her guardian angel. The dead man and the vampire both vanished at cock-crow, for both were vampires. Waking up in the darkness, the girl looked round the house three times, thought she was at home and had a horrible dream, and then fell asleep again calmly and fearlessly. When she woke again, and saw all the beauties of the forest, and heard all the songs of the birds, she was amazed and thought herself in heaven. She did not stop long in wonder, but set out for her parents' house, hoping to bring them back with her.

She reached her home, and began to tell about the vampire and how he had gone, and what beautiful things she had seen in the woods of paradise. The parents looked at her, and, full of amazement and doubt, made the sign of the cross. The girl sank into the ground, deeper and deeper, for she too had become a vampire, poor thing. The vampire had bewitched her, and the beauty of the dwelling in the wood had enchanted her too much.

THE ROLE OF GARLIC

Murgoci notes the special efficacy of garlic as a vampire repellant in the Romanian tradition (Montague Summers, who cites Murgoci's research, also reports a similar use of the plant in China, Malaysia, Sumatra and the West Indies). Garlic seems to have been an all-purpose vampire prophylactic. It could be placed in the mouth of a vampire's corpse, used to seal keyholes, doors and windows, worn as a necklace, even rubbed on livestock.

Garlic, of course, has been prized for centuries for its well-known (if poorly understood) blood-purifying and immune-boosting properties; modern science points to garlic's high concentration of sulphurlike compounds which make it effective as both an antibacterial and antifungal agent. It is easy to see how prescientific societies, which often superimposed evil spirits onto disease and mortality, might extend garlic's powers as a natural remedy into the supernatural realm as well.

It is perhaps not surprising in our new age of body-fluid anxiety and reawakened vampire awareness that garlic has recently reasserted itself as a popular standby of alternative medicine. In the age of AIDS, garlic sales have soared everywhere as vague, quasi-medical claims (e.g.,"the goodness of garlic") appeal to a free-floating fear of blood contamination, encroaching death and a general sense of cultural dread.

Romania's Castle Bran is a major tourist attraction, promoting itself as "Dracula's Castle" to capitalize on the public's fascination with ancient superstitions. (Photo courtesy of Lokke Heiss)

PART TWO

Romantic & Victorian Vampires

Polidori's story was dramatized by James Robinson Planché as The Vampire: Or, the Bride of the Isles *(1820).*

The Vampyre
A Tale

JOHN POLIDORI
(1819)

IT HAPPENED THAT in the midst of the dissipations attendant upon London winter, there appeared at the various parties of the leaders of the ton a nobleman more remarkable for his singularities, than his rank. He gazed upon the mirth around him, as if he could not participate therein. Apparently, the light laughter of the fair only attracted his attention, that he might by a look quell it and throw fear into those breasts where thoughtlessness reigned. Those who felt this sensation of awe, could not explain whence it arose: some attributed it to the dead grey eye, which, fixing upon the object's face, did not seem to penetrate, and at one glance to pierce through to the inward workings of the heart; but fell upon the cheek with a leaden ray that weighed upon the skin it could not pass. His peculiarities caused him to be invited to every house; all wished to see him, and those who had been accustomed to violent excitement, and now felt the weight of ennui, were pleased at having something in their presence capable of engaging their attention. In spite of the deadly hue of his face, which never gained a wanner tint, either from the blush of modesty, or from the strong emotion of passion, though its form and outline were beautiful, many of the female hunters after notoriety attempted to win his attentions, and gain, at least, some marks of what they might term affection: Lady Mercer, who had been the mockery of every monster shewn in drawing-rooms since her marriage, threw herself in his way, and did all but put on the dress of a mountebank, to attract his notice — though in vain; — when she stood before him, though his eyes were apparently fixed upon hers, still it seemed as if they were unperceived; — even her unappalled impudence was baffled, and she left the field. But though the common adultress could not influence even the guidance of his eyes, it was not that the female sex was indifferent to him: yet such was the apparent caution with which he spoke to the virtuous wife and innocent daughter, that few knew he ever addressed himself to females. He had, however, the reputation of a winning tongue; and whether it was that it even overcame the dread of his singular character, or that they were moved by his apparent hatred of vice, he was as often among those females who form the boast of their sex from their domestic virtues, as among those who sully it by their vices.

About the same time, there came to London a young gentleman of the name of Aubrey: he was an orphan left with an only sister in the possession

ABOUT THE AUTHOR

John William Polidori (1795-1821) published very little and died very young, but his literary conception of the vampire still lives, rivaling (and in some ways overshadowing) the contributions of Bram Stoker. Polidori was educated at Edinburgh, where he wrote his dissertation on the subject of mesmerism, somnambulism, and nightmares. He earned a medical degree at the precocious age of nineteen, and was able to combine his literary and medical interests as physician and traveling companion to George Gordon, Lord Byron, during the summer of 1816.

While residing at the Villa Diodati on Lake Geneva with Byron, Percy Shelley, and Shelley's bride-to-be, Mary Wollstonecraft, Polidori took part in a story-writing contest, devised to pass the time during a stretch of bad weather. While Shelley and Byron produced almost nothing, Mary conceived the germ of *Frankenstein* (1818) and Polidori wrote "The Vampyre," one of the most imitated and influential horror stories ever published.

The first, unauthorized appearance of "The Vampyre," in the *New Monthly Magazine* in April 1819, proved a scandal when its publisher, evidently to increase sales, falsely attributed the tale to Byron rather than Polidori. Polidori (who had had a tense, quarrelsome relationship with Byron during the period of the story's composition) had taken inspiration from a novel fragment by Byron, in which a traveller somewhat like Byron himself plots a method to return from the grave. The publisher seized on this

connection to cast doubt on Polidori's authorship—generating more publicity and sales, of course.

Polidori was outraged at the violation of his copyright, especially because he considered the story "imperfect and unfinished." Although Polidori ultimately intended to call his vampire Lord Strongmore, in the published version the character was called Lord Ruthven, apparently in reference to the rakish character, also named Ruthven, in Caroline Lamb's novel *Glenarvon* (1814), who was transparently based on Byron. Lamb was only one of Byron's many sexual conquests, but one of the few able to take her revenge in print.

Although Byron disavowed authorship ("I have a personal dislike to Vampires, and the little acquaintance I have with them would by no means induce me to reveal their secrets"), the story clung stubbornly to him, and even appeared in a volume of his collected works. No less a literary figure than Goethe was taken in, and called "The Vampyre" one of Byron's finest accomplishments.

Despite much legal wrangling, Polidori was finally able to collect only £ 30 for "The Vampyre," much less than its fair value. He published only one other work of fiction, the novella *Ernestus Berchtold* (1819). The circumstances of his premature death in 1821 at the age of 26 have not been completely verified, although he may have committed suicide by taking prussic acid following a debilitating head injury.

The most complete recounting of Polidori's life and the "Vampyre" affair can be found in *Poor Polidori* by D. L. MacDonald (University of Toronto Press, 1991).

of great wealth, by parents who died while he was yet in childhood. Left also to himself by guardians, who thought it their duty merely to take care of his fortune, while they relinquished the more important charge of his mind to the care of mercenary subalterns, he cultivated more his imagination than his judgment. He had, hence, that high romantic feeling of honour and candour, which daily ruins so many milliners' apprentices. He believed all to sympathise with virtue, and thought that vice was thrown in by Providence merely for the picturesque effect of the scene, as we see in romances: he thought that the misery of a cottage merely consisted in the vesting of clothes, which were as warm, but which were better adapted to the painter's eye by their irregular folds and various coloured patches. He thought, in fine, that the dreams of poets were the realities of life. He was handsome, frank, and rich: for these reasons, upon his entering into the gay circles, many mothers surrounded him, striving which should describe with least truth their languishing or romping favourites: the daughters at the same time, by their brightening countenances when he approached, and by their sparkling eyes, when he opened his lips, soon led him into false notions of his talents and his merit. Attached as he was to the romance of his solitary hours, he was startled at finding, that, except in the tallow and wax candles that flickered, not from the presence of a ghost, but from want of snuffing, there was no foundation in real life for any of that congeries of pleasing pictures and descriptions contained in those volumes, from which he had formed his study. Finding, however, some compensation in his gratified vanity, he was about to relinquish his dreams, when the extraordinary being we have above described, crossed him in his career.

He watched him; and the very impossibility of forming an idea of the character of a man entirely absorbed in himself, who gave few other signs of his observation of external objects, than the tacit assent to their existence, implied by the avoidance of their contact: allowing his imagination to picture every thing that flattered its propensity to extravagant ideas, he soon formed this object into the hero of a romance, and determined to observe the offspring of his fancy, rather than the person before him. He became acquainted with him, paid him attentions, and so far advanced upon his notice, that his presence was always recognised. He gradually learnt that Lord Ruthven's affairs were embarrassed, and soon found, from the notes of preparation in —— Street, that he was about to travel. Desirous of gaining some information respecting this singular character, who, till now, had only whetted his curiosity, he hinted to his guardians, that it was time for him to perform the tour, which for many generations has been thought necessary to enable the young to take some rapid steps in the career of vice towards putting themselves upon an equality with the aged, and not allowing them to appear as if fallen from the skies, whenever scandalous intrigues are mentioned as the subjects of pleasantry or of praise, according to the degree of skill shewn in carrying them on.

They consented: and Aubrey immediately mentioning his intentions to Lord Ruthven, was surprised to receive from him a proposal to join him. Flattered by such a mark of esteem from him, who, apparently, had nothing in common with other men, he gladly accepted it, and in a few days they had passed the circling waters.

Hitherto, Aubrey had had no opportunity of studying Lord Ruthven's character, and now he found, that, though many more of his actions were exposed to his view, the results offered different conclusions from the apparent motives to his conduct. His companion was profuse in his liberality; — the idle, the vagabond, and the beggar, received from his hand more than enough to relieve their immediate wants. But Aubrey could not avoid remarking, that it was not upon the virtuous, reduced to indigence by the misfortunes attendant even upon virtue, that he bestowed his alms; — these were sent from the door with hardly suppressed sneers; but when the profligate came to ask something, not to relieve his wants, but to allow him to wallow in his lust, to sink him still deeper in his iniquity, he was sent away with rich charity. This was, however, attributed by him to the greater importunity of the vicious, which generally prevails over the retiring bashfulness of the virtuous indigent. There was one circumstance about the charity of his Lordship, which was still more impressed upon his mind: all those upon whom it was bestowed, inevitably found that there was a curse upon it, for they were all either led to the scaffold, or sunk to the lowest and the most abject misery. At Brussels and other towns through which they passed, Aubrey was surprised at the apparent eagerness with which his companion sought for the centres of all fashionable vice; there he entered into all the spirit of the faro table: he betted and always gambled with success, except where the known sharper was his antagonist, and then he lost even more than he gained; but it was always with the same unchanging face, with which he generally watched the society around: it was not, however, so when he encountered the rash youthful novice, or the luckless father of a numerous family; then his very wish seemed fortune's law — this apparent abstractedness of mind was laid aside, and his eyes sparkled with more fire than that of the cat whilst dallying with the half-dead mouse. In every town, he left the formerly affluent youth, torn from the circle he adorned, cursing, in the solitude of a dungeon, the fate that had drawn him within the reach of this fiend; whilst many a father sat frantic, amidst the speaking looks of mute hungry children, without a single farthing of his late immense wealth, wherewith to buy even sufficient to satisfy their present craving. Yet he took no money from the gambling table; but immediately lost, to the ruiner of many, the last gilder he had just snatched from the convulsive grasp of the innocent: this might but be the result of a certain degree of knowledge, which was not, however, capable of combating the cunning of the more experienced. Aubrey often wished to represent this to his friend, and beg him to resign that charity and pleasure which proved the ruin of all, and did not tend to his own profit; but he delayed it — for each day he hoped his friend would

John Polidori

give him some opportunity of speaking frankly and openly to him; however, this never occurred. Lord Ruthven in his carriage, and amidst the various wild and rich scenes of nature, was always the same: his eye spoke less than his lip; and though Aubrey was near the object of his curiosity, he obtained no greater gratification from it than the constant excitement of vainly wishing to break that mystery, which to his exalted imagination began to assume the appearance of something supernatural.

They soon arrived at Rome, and Aubrey for a time lost sight of his companion; he left him in daily attendance upon the morning circle of an Italian countess, whilst he went in search of the memorials of another almost deserted city. Whilst he was thus engaged, letters arrived from England, which he opened with eager impatience; the first was from his sister, breathing nothing but affection; the others were from his guardians, the latter astonished him; if it had before entered into his imagination that there was an evil power resident in his companion these seemed to give him almost sufficient reason for the belief. His guardians insisted upon his immediately leaving his friend, and urged that his character was dreadfully vicious, for that the possession of irresistible powers of seduction, rendered his licentious habits more dangerous to society. It had been discovered, that his contempt for the adultress had not originated in hatred of her character; but that he had required, to enhance his gratification, that his victim, the partner of his guilt, should be hurled from the pinnacle of unsullied virtue, down to the lowest abyss of infamy and degradation: in fine, that all those females whom he had sought, apparently on account of their virtue, had, since his departure, thrown even the mask aside, and had not scrupled to expose the whole deformity of their vices to the public gaze.

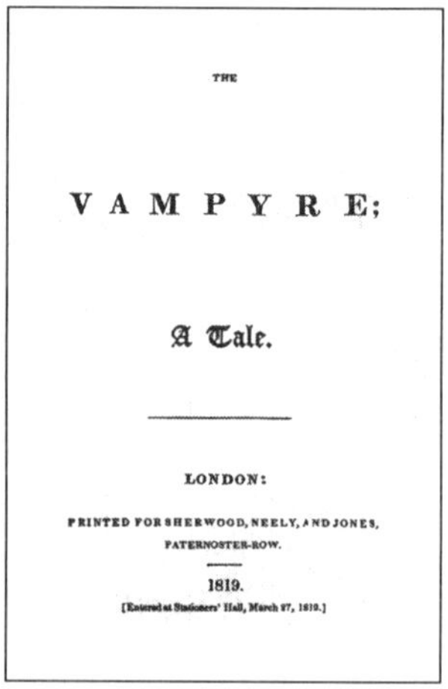
THE

VAMPYRE;

A Tale.

LONDON:

PRINTED FOR SHERWOOD, NEELY, AND JONES,
PATERNOSTER-ROW.

1819.

[Entered at Stationers' Hall, March 27, 1819.]

Original title page for "The Vampyre."

Aubrey determined upon leaving one, whose character had not shown a single bright point on which to rest the eye. He resolved to invent some plausible pretext for abandoning him altogether, purposing, in the meanwhile, to watch him more closely, and to let no slight circumstances pass by unnoticed. He entered into the same circle, and soon perceived, that his Lordship was endeavouring to work upon the inexperience of the daughter of the lady whose house he chiefly frequented. In Italy, it is seldom that an unmarried female is met with in society; he was therefore obliged to carry on his plans in secret; but Aubrey's eye followed him in all his windings, and soon discovered that an assignation had been appointed, which would most likely end in the ruin of an innocent, though thoughtless girl. Losing no time, he entered the apartment of Lord Ruthven, and abruptly asked him his intentions with respect to the lady, informing him at the same time that he was aware of his being about to meet her that very night. Lord Ruthven answered, that his intentions were such as he supposed all would have upon such an occasion; and upon being pressed whether he intended to marry her, merely laughed. Aubrey retired; and, immediately writing a note, to say, that from that moment he must decline accompanying his Lordship in the remainder of their proposed tour, he ordered his servant to seek other apartments, and calling upon the mother of the lady informed her of all he knew, not only with regard to her daughter, but also concerning the character of his Lordship. The assignation was prevented. Lord Ruthven next day merely sent his servant to notify his complete assent to a separation; but did not hint any suspicion of his plans having been foiled by Aubrey's interposition.

Having left Rome, Aubrey directed his steps towards Greece, and crossing the Peninsula, soon found himself at Athens. He then fixed residence in the house of a Greek; and soon occupied himself in tracing the faded records of ancient glory upon monuments that apparently, ashamed of chronicling the deeds of freemen only before slaves, had hidden themselves beneath the sheltering soil or many coloured lichen. Under the same roof as himself, existed a being, so beautiful and delicate, that she might have formed the model for a painter, wishing to portray on canvass the promised hope of the faithful in Mahomet's paradise, save that her eyes spoke too much mind for any one to think she could belong to those who had no souls. As she danced upon the plain, or tripped along the mountain's side, one would have thought the gazelle a poor type of her beauties; for who would have exchanged her eye, apparently the eye of animated nature, for that sleepy luxurious look of the animal suited but to the taste of an epicure. The light step of Ianthe often accompanied Aubrey in his search after antiquities, and often would the unconscious girl, engaged in the pursuit of a Kashmere butterfly, show the whole beauty of her form, boating as it were upon the wind, to the eager gaze of him, who forgot the letters he had just decyphered upon an almost effaced tablet, in the contemplation of her sylph-like figure. Often would her tresses falling, as she flitted around, exhibit in the sun's ray such delicately brilliant and swiftly fading hues, as might well excuse the forgetfulness of the antiquary, who let escape from his mind the very object he had before thought of vital importance to the proper interpretation of a passage in Pausanias. But why attempt to describe charms which all feel, but none can appreciate? — It was innocence, youth, and beauty, unaffected by crowded drawing-rooms and stifling balls. Whilst he drew those remains of which he wished to preserve a memorial for his future hours, she would stand by, and watch the magic effects of his pencil, in tracing the scenes of her native place; she would then describe to him the circling dance upon the open plain, would paint to him in all the glowing colours of youthful memory, the marriage pomp she remembered viewing in her infancy; and then, turning to subjects that had evidently made a greater impression upon her mind, would tell him all the supernatural tales of her nurse. Her earnestness and apparent belief of what she narrated, excited the interest even of Aubrey; and often as she told him the tale of the living vampyre, who had passed years amidst his friends, and dearest ties, forced every year, by feeding upon the life of a lovely female to prolong his existence for the ensuing months, his blood would run cold, whilst he attempted to laugh her out of such idle and

INTRODUCTION TO "THE VAMPYRE"

Following are excerpts from the unattributed preamble to the 1819 edition of "The Vampyre."

The superstition upon which this tale is founded is very general in the East. Among the Arabians it appears to be common: it did not, however, extend itself to the Greeks until after the establishment of Christianity; and it has only assumed its present form since the division of the Latin and Greek churches; at which time, the idea becoming prevalent, that a Latin body could not corrupt if buried in their territory, it gradually increased, and formed the subject of many wonderful stories, still extant, of the dead rising from their graves and feeding upon the blood of the young and beautiful. In the West it spread, with some slight variation, all over Hungary, Poland, Austria, and Lorraine, where the belief existed, that vampires nightly imbibed a certain portion of the blood of their victims, who became emaciated, lost their strength, and speedily died of consumptions; whilst these human blood-suckers fattened — and their veins distended to such a state of repletion, as to cause the blood to flow from all the passages of their bodies, and even from the very pores of their skins.

In many parts of Greece it is considered as a sort of punishment after death, for some heinous crime committed whilst in existence, that the deceased is not only doomed to vampyrise, but compelled to confine his infernal visitations solely to those beings he loved most while on earth—those to whom he was bound by ties of kindred and affection.

horrible fantasies; but Ianthe cited to him the names of old men, who had at last detected one living among themselves, after several of their near relatives and children had been found marked with the stamp of the fiend's appetite; and when she found him so incredulous, she begged of him to believe her, for it had been remarked, that those who had dared to question their existence, always had some proof given, which obliged them, with grief and heartbreaking, to confess it was true. She detailed to him the traditional appearance of these monsters, and his horror was increased by hearing a pretty accurate description of Lord Ruthven; he, however, still persisted in persuading her, that there could be no truth in her fears, though at the same time he wondered at the many coincidences which had all tended to excite a belief in the supernatural power of Lord Ruthven.

Many curious and interesting notices on this singularly horrible superstition might be added; though the present may suffice for the limits of a note, necessarily devoted to explanation, and which now may be concluded by merely remarking, that though the term Vampyre is the one in most general acceptation, there are several others synonymous with it, made use of in various parts of the world: as Vroucolocha, Vardoulacha, Goul, Broucoloka, &c.

Aubrey began to attach himself more and more to Ianthe; her innocence, so contrasted with all the affected virtues of the women among whom he had sought for his vision of romance, won his heart and while he ridiculed the idea of a young man of English habits, marrying an uneducated Greek girl, still he found himself more and more attached to the almost fairy form before him. He would tear himself at times from her, and, forming a plan for some antiquarian research, would depart, determined not to return until his object was attained; but he always found it impossible to fix his attention upon the ruins around him, whilst in his mind he retained an image that seemed alone the rightful possessor of his thoughts. Ianthe was unconscious of his love, and was ever the same frank infantile being he had first known. She always seemed to part from him with reluctance; but it was because she had no longer any one with whom she could visit her favourite haunts, whilst her guardian was occupied in sketching or uncovering some fragment which had yet escaped the destructive hand of time. She had appealed to her parents on the subject of Vampyres, and they both, with several present, affirmed their existence, pale with horror at the very name. Soon after, Aubrey determined to proceed upon one of his excursions, which was to detain him for a few hours; when they heard the name of the place, they all at once begged of him not to return at night, as he must necessarily pass through a wood, where no Greek would ever remain, after the day had closed, upon any consideration. They described it as the resort of the vampyres in their nocturnal orgies and denounced the most heavy evils as impending upon him who dared to cross their path. Aubrey made light of their representations, and tried to laugh them out of the idea; but when he saw them shudder at his daring thus to mock a superior, infernal power, the very name of which apparently made their blood freeze, he was silent.

POLIDORI'S COMPLAINT

The following letter was printed in the New Monthly Magazine *in the issue following the publication of "The Vampyre."*

Mr. Editor,

As the person referred to in the Letter from Geneva, prefixed to the Tale of the Vampyre, in your last Number, I beg leave to state, that your correspondent has been mistaken in attributing that tale, *in its present form*, to Lord Byron. The fact is, that though *the groundwork* is certainly Lord Byron's, its development is mine, produced at the request of a lady, who denied the possibility of any thing being drawn from the materials which Lord Byron had said he intended to have employed in the formation of his Ghost story.

I am, &c. JOHN W. POLIDORI

Next morning Aubrey set off upon his excursion unattended; he was surprised to observe the melancholy face of his host, and was concerned to find that his words, mocking the belief of those horrible fiends, had inspired them with such terror. When he was about to depart, Ianthe came to the side of his horse, and earnestly begged of him to return, ere night allowed the power of these beings to be put in action; — he promised. He was, however, so occupied in his research, that he did not perceive that daylight would soon end, and that in the horizon there was one of those specks which, in the warmer climates, so rapidly gather into a tremendous mass, and pour all their rage upon the devoted country. — He at last, however, mounted his horse, determined to make up by speed for his delay: but it was too late. Twilight, in these southern climates, is almost unknown; immediately the sun sets, night begins: and ere he had advanced far, the power of the storm was above — its echoing thunders had scarcely an interval of rest; — its thick heavy rain forced its way through the canopying foliage, whilst the blue forked lightning seemed to fall and radiate at his very feet. Suddenly his horse took fright, and he was carried with dreadful rapidity through the entangled forest.

The animal at last, through fatigue, stopped, and he found, by the glare of lightning, that he was in the neighbourhood of a hovel that hardly lifted itself up from the masses of dead leaves and brushwood which surrounded it. Dismounting, he approached, hoping to find some one to guide him to the town, or at least trusting to obtain shelter from the pelting of the storm. As he approached, the thunders, for a moment silent, allowed him to hear the dreadful shrieks of a woman mingling with the stifled, exultant mockery of a laugh, continued in one almost unbroken sound; — he was startled: but, roused by the thunder which again rolled over his head, he, with a sudden effort, forced open the door of the hut. He found himself in utter darkness: the sound, however, guided him. He was apparently unperceived; for, though he called, still the sounds continued, and no notice was taken of him. He found himself in contact with some one, whom he immediately seized; when a voice cried, "Again baffled!" to which a loud laugh succeeded; and he felt himself grappled by one whose strength seemed superhuman: determined to sell his life as dearly as he could, he struggled; but it was in vain: he was lifted from his feet and hurled with enormous force against the ground: — his enemy threw himself upon him, and kneeling upon his breast, had placed his hands upon his throat when the glare of many torches penetrating through the hole that gave light in the day, disturbed him; — he instantly rose, and, leaving his prey, rushed through the door, and in a moment the crashing of branches, as he broke through the wood, was no longer heard. The storm was now still; and Aubrey, incapable of moving, was soon heard by those without. They entered; the light of their torches fell upon mud walls, and the thatch loaded on every individual straw with heavy flakes of soot.

Thomas Potter Cooke played Lord Ruthuen onstage in The Vampire: or, The Bride of the Isles *(1820).*

At the desire of Aubrey they searched for her who had attracted him by her cries; he was again left in darkness; but what was his horror, when the light of the torches once more burst upon him, to perceive the airy form of his fair conductress brought in a lifeless corpse.

He shut his eyes, hoping that it was but a vision arising from his disturbed imagination; but he again saw the same form, when he unclosed them, stretched by his side. There was no colour upon her cheek, not even upon her lip; yet there was a stillness about her face that seemed almost as attaching as the life that once dwelt there: — upon her neck and breast was blood, and upon her throat were the marks of teeth having opened the vein: — to this the men pointed, crying, simultaneously struck with horror, "A Vampyre! a Vampyre!" A litter was quickly formed, and Aubrey was laid by the side of her who had lately been to him the object of so many bright and fairy visions, now fallen; with the flower of life that had died within her. He knew not what his thoughts were — his mind was benumbed and seemed to shun reflection and take refuge in vacancy; — he held almost unconsciously in his hand a naked dagger of a particular construction, which had been found in the hut. They were soon met by different parties who had been engaged in the search of her whom a mother had missed. Their lamentable cries as they approached the city, forewarned the parents of some dreadful catastrophe. — To describe their grief would be impossible; but when they ascertained the cause of their child's death, they looked at Aubrey and pointed to the corpse. They were inconsolable; both died brokenhearted.

Aubrey, being put to bed was seized with a most violent fever, and was often delirious; in these intervals he would call upon Lord Ruthven and upon Ianthe — by some unaccountable combination he seemed to beg of his former companion to spare the being he loved. At other times he would imprecate maledictions upon his head, and curse him as her destroyer. Lord Ruthven chanced at this time to arrive at Athens, and from whatever motive, upon hearing of the state of Aubrey, immediately placed himself in the same house, and became his constant attendant. When the latter recovered from his delirium, he was horrified and startled at the sight of him whose image he had now combined with that of a Vampyre; but Lord Ruthven, by his kind words, implying almost repentance for the fault that had caused their separation, and still more by the attention, anxiety, and care which he showed, soon reconciled him to his presence. His lordship seemed quite changed; he no longer appeared that apathetic being who had so astonished Aubrey; but as soon as his convalescence began to be rapid, he again gradually retired into the same state of mind, and Aubrey perceived no difference from the former man, except that at times he was surprised to meet his gaze fixed intently upon him, with a smile of malicious exultation playing upon his lips: he knew not why, but this smile haunted him. During the last stage of the invalid's recovery, Lord Ruthven was apparently engaged in watching the tideless waves raised by the cooling breeze, or in marking the progress of those orbs, circling, like our world, the moveless sun; — indeed, he appeared to wish to avoid the eyes of all.

Aubrey's mind, by this shock, was much weakened, and that elasticity of spirit which had once so distinguished him now seemed to have fled for ever. He was now as much a lover of solitude and silence as Lord Ruthven;

THE FATAL MAN

George Gordon, Lord Byron (1788-1824), was the leading figure of English Romanticism, whose legendary reputation as a seducer and rake forever changed the popular image of the vampire through his identification with Polidori's story. Previously, vampires were the stuff of peasant culture. Now, suddenly, they could be glamorous, brooding, sexy aristocrats. Being undead would never be the same.

Byron touched on the vampire theme only once in his own work. In the 1813 epic poem "The Giaour," a Muslim curse includes the scourge of vampirism:

But first on earth, as Vampyre sent,
Thy corse shall from its tomb be rent;
Then ghastly haunt thy native place,
And suck the blood of all thy race;
There from thy daughter, sister, wife,
At midnight drain the stream of life;
Yet loathe the banquet which perforce
Must feed thy livid living corse.
Thy victims, ere they yet expire,
Shall know the demon for their sire;
As cursing thee, thou cursing them,
Thy flowers are withered on the stem...
Yet with thine own best blood shall drip
Thy gnashing tooth and haggard lip;
Then stalking to thy sullen grave,
Go - and with Gouls and Afrits rave;
Till these in horror shrink away
From spectre more accursed than they!

In Tom Holland's 1995 novel *Vampyre* (American title: *Lord of the Dead*), Byron himself is presented as a literal Lord Ruthven, cutting an undead swath into the twentieth century (and bringing Polidori with him — also as a vampire).

but much as he wished for solitude, his mind could not find it in the neighbourhood of Athens; if he sought it amidst the ruins he had formerly frequented, Ianthe's form stood by his side; — if he sought it in the woods, her light step would appear wandering amidst the underwood, in quest of the modest violet; then suddenly turning round, would show, to his wild imagination, her pale face and wounded throat, with a meek smile upon her lips. He determined to fly scenes, every feature of which created such bitter associations in his mind. He proposed to Lord Ruthven, to whom he held himself bound by the tender care he had taken of him during his illness, that they should visit those parts of Greece neither had yet seen. They travelled in every direction, and sought every spot to which a recollection could be attached: but though they thus hastened from place to place, yet they seemed not to heed what they gazed upon. They heard much of robbers, but they gradually began to slight these reports, which they imagined were only the invention of individuals, whose interest it was to excite the generosity of those whom they defended from pretended dangers. In consequence of thus neglecting the advice of the inhabitants, on one occasion they travelled with only a few guards, more to serve as guides than as a defence. Upon entering, however, a narrow defile, at the bottom of which was the bed of a torrent, with large masses of rock brought down from the neighbouring precipices, they had reason to repent their negligence; for scarcely were the whole of the party engaged in the narrow pass, when they were startled by the whistling of bullets close to their heads, and by the echoed report of several guns. In an instant their guards had left them, and, placing themselves behind rocks, had begun to fire in the direction whence the report came. Lord Ruthven and Aubrey, imitating their example, retired for a moment behind the sheltering turn of the defile: but ashamed of being thus detained by a foe, who with insulting shouts bade them advance, and being exposed to unresisting slaughter, if any of the robbers should climb above and take them in the rear, they determined at once to rush forward in search of the enemy. Hardly had they lost the shelter of rock, when Lord Ruthven received a shot in the shoulder, which brought him to the ground. Aubrey hastened to his assistance; and, no longer heeding the contest or his own peril, was soon surprised by seeing the robbers' faces around him — his guards having, upon Lord Ruthven's being wounded, immediately thrown up their arms and surrendered.

By promises of great reward, Aubrey soon induced them to convey his wounded friend to a neighbouring cabin; and having agreed upon a ransom, he was no more disturbed by their presence — they being content merely to guard the entrance till their comrade should return with the promised sum, for which he had an order. Lord Ruthven's strength rapidly decreased; in two days mortification ensued, and death seemed advancing with hasty steps. His conduct and appearance had not changed; he seemed as unconscious of pain as he had been of the objects about him: but towards the close of the last evening, his mind became apparently uneasy, and his

Lord Byron

ROMANTICISM AND GOTHIC LITERATURE

The Gothic sensibility—that is, the taste in literature for morbid melodrama and supernatural sensation—was an offshoot of the Romantic revolution of the late 1700s and early 1800s, as a reaction against the Age of Reason took hold among European writers and artists.

Romanticism emphasized the subjective, emotional and irrational aspects of life over the neoclassical ideals that then dominated the arts. A particular fascination with death, decay, and the supernatural found its expression in the Gothic novel, so-called for the genre's favored images: crumbling castles, medieval monasteries, tombs and cemeteries, etc.

Lord Byron's entourage at the Villa Diodati in 1816 well embodied the Gothic/Romantic sensibility: free-thinking, melancholy, antiauthoritarian, preoccupied with sex, death and mysticism. The literary parlor game that inspired "The Vampyre" was the result of the group's reading of a collection of sensational European ghost stories, the *Fantasmagoriana*.

eye often fixed upon Aubrey, who was induced to offer his assistance with more than usual earnestness — "Assist me! you may save me — you may do more than that — I mean not life, I heed the death of my existence as little as that of the passing day; but you may save my honour, your friend's honour." — "How? tell me how? I would do any thing," replied Aubrey. — "I need but little, my life ebbs apace — I cannot explain the whole — but if you would conceal all you know of me, my honour were free from stain in the world's mouth — and if my death were unknown for some time in England —I — I — but life." — "It shall not be known." — "Swear!" cried the dying man raising himself with exultant violence. "Swear by all your soul reveres, by all your nature fears, swear that for a year and a day you will not impart your knowledge of my crimes or death to any living being in any way, whatever may happen, or whatever you may see." — His eyes seemed bursting from their sockets; "I swear!" said Aubrey; he sunk laughing upon his pillow, and breathed no more.

Aubrey retired to rest, but did not sleep; the many circumstances attending his acquaintance with this man rose upon his mind, and he knew not why; when he remembered his oath a cold shivering came over him, as if from the presentiment of something horrible awaiting him. Rising early in the morning, he was about to enter the hovel in which he had left the corpse, when a robber met him, and informed him that it was no longer there, having been conveyed by himself and comrades, upon his retiring, to the pinnacle of a neighbouring mount, according to a promise they had given his lordship, that it should be exposed to the first cold ray of the moon that rose after his death. Aubrey was astonished, and taking several of the men, determined to go and bury it upon the spot where it lay. But, when he had mounted to the summit he found no trace of either the corpse or the clothes, though the robbers swore they pointed out the identical rock on which they had laid the body. For a time his mind was bewildered in conjectures, but he at last returned, convinced that they had buried the corpse for the sake of the clothes.

Weary of a country in which he had met with such terrible misfortunes, and in which all apparently conspired to heighten that superstitious melancholy that had seized upon his mind, he resolved to leave it, and soon arrived at Smyrna. While waiting for a vessel to convey him to Otranto, or to Naples, he occupied himself in arranging those effects he had with him belonging to Lord Ruthven. Amongst other things there was a case containing several weapons of offence, more or less adapted to ensure the death of the victim. There were several daggers and ataghans. Whilst turning them over, and examining their curious forms, what was his surprise at finding a sheath apparently ornamented in the same style as the dagger discovered in the fatal hut; — he shuddered; hastening to gain further proof, he found the weapon, and his horror may be imagined when he discovered that it fitted, though peculiarly shaped, the sheath he held in his hand. His eyes seemed to need no further certainty — they seemed

Following are just a few of the Gothic/Romantic literary precursors to Polidori's "The Vampyre," Mary Shelley's *Frankenstein* and modern horror fiction in general:

GOETHE
A leading influence on the Romantics and one of Germany's greatest writers, Johann Wolfgang von Goethe penned the ballad "The Bride of Corinth" (1797), which concerned a malignant female revenant.

E.T.A. HOFFMAN
Ernst Theodor Amadeus Hoffman (1776-1822) was Germany's foremost proponent of the supernatural grotesque, whose *Tales* created a popular literary precedent for stories like "The Vampyre."

MRS. ANN RADCLIFFE
Radcliffe (1764-1823) popularized the Gothic novel with books like her bestselling *The Mysteries of Udolpho* (1794). While her stories avoided truly supernatural elements, she stabilized many of the familiar elements of later vampire fiction: ancient castles, cruel aristocrats, and a general atmosphere of the uncanny. Radcliffe's fiction followed the mold set by Horace Walpole (1717-1797) whose *The Castle of Otranto* (1763) is generally credited as the first Gothic novel.

JOHANN LUDWIG TIECK
Although the German Romantic Tieck (1773-1853) has never been proven to be the author of the story "Wake Not the Dead" (c. 1800), it is usually attributed to him, and is one of the earliest examples of the genre.

gazing to be bound to the dagger, yet still he wished to disbelieve; but the particular form, the same varying tints upon the haft and sheath were alike in splendour on both, and left no room for doubt; there were also drops of blood on each.

He left Smyrna, and on his way home, at Rome, his first inquiries were concerning the lady he had attempted to snatch from Lord Ruthven's seductive arts. Her parents were in distress, their fortune ruined, and she had not been heard of since the departure of his lordship. Aubrey's mind became almost broken under so many repeated horrors; he was afraid that this lady had fallen a victim to the destroyer of Ianthe. He became morose and silent; and his only occupation consisted in urging the speed of the postilions, as if he were going to save the life of some one he held dear. He arrived at Calais; a breeze, which seemed obedient to his will, soon wafted him to the English shores; and he hastened to the mansion of his fathers, and there, for a moment, appeared to lose, in the embraces and caresses of his sister, all memory of the past. If she before, by her infantine caresses, had gained his affection, now that the woman began to appear, she was still more attaching as a companion.

Miss Aubrey had not that winning grace which gains the gaze and applause of the drawing-room assemblies. There was none of that light brilliancy which only exists in the heated atmosphere of a crowded apartment. Her blue eye was never lit up by the levity of the mind beneath. There was a melancholy charm about it which did not seem to arise from misfortune, but from some feeling within, that appeared to indicate a soul conscious of a brighter realm. Her step was not that light footing, which strays where'er a butterfly or a colour may attract — it was sedate and pensive. When alone, her face was never brightened by the smile of joy; but when her brother breathed to her his affection, and would in her presence forget those griefs she knew destroyed his rest, who would have exchanged her smile for that of the voluptuary? It seemed as if those eyes, that face were then playing in the light of their own native sphere. She was yet only eighteen, and had not been presented to the world, it having been thought by her guardians more fit that her presentation should be delayed until her brother's return from the continent, when he might be her protector. It was now, therefore, resolved that the next drawing-room, which was fast approaching, should be the epoch of her entry into the "busy scene." Aubrey would rather have remained in the mansion of his fathers, and feed upon the melancholy which overpowered him. He could not feel interest about the frivolities of fashionable strangers, when his mind had been so torn by the events he had witnessed; but he determined to sacrifice his own comfort to the protection of his sister. They soon arrived in town, and prepared for the next day, which had been announced as a drawing-room.

The crowd was excessive — a drawing-room had not been held for long time, and all who were anxious to bask in the smile of royalty, hastened

THEATRE OF BLOOD: "THE VAMPYRE" ON STAGE

Polidori's tale inspired at least seven dramatic adaptations during the 1800s. The first and most influential was *Le Vampire* (1820), by the French Romantic writer Charles Nodier, which was reworked in English by James Robinson Planché as *The Vampire, or the Bride of the Isles* (1820). Two Parisian burlesques inspired by Nodier, one by Martinet and the other by Scribe and Melesville, also appeared in 1820, creating a virtual vampire mania on the Parisian stage. Lord Ruthven also appeared in the unauthorized two-volume novel *Lord Ruthwen ou les Vampires* by Cyprian Bérard (1920).

Modern readers of Anne Rice will recognize, in this period of Parisian vampire chic, the seeds of one of Rice's most memorable settings, Le Théâtre des Vampires, where the undead present themselves semi-openly to postrevolutionary audiences.

BOUCICAULT'S BLOOD FEAST

Irish-born actor and playwright Dion Boucicault (1820-1890) made his London debut in 1852 with *The Vampire:A Phantasm*, yet another theatrical adaptation of Polidori, by way of Nodier. Boucicault substituted a new blood drinker, Sir Alan Raby, for Lord Ruthven, and rather audaciously set its action in Scotland over three centuries — past, present, and future. The vampire, resurrected by "moonbames," forges his own will to reclaim his former castle. Boucicault later trimmed the third act and presented the play in America as *The Phantom*.

thither. Aubrey was there with his sister. While he was standing in a corner by himself, heedless of all around him, engaged in the remembrance that the first time he had seen Lord Ruthven was in that very place — he felt himself suddenly seized by the arm, and a voice he recognized too well, sounded in his ear—"Remember your oath." He had hardly courage to turn, fearful of seeing a spectre that would blast him, when he perceived, at a little distance, the same figure which had attracted his notice on this spot upon his first entry into society. He gazed till his limbs almost refusing to bear their weight, he was obliged to take the arm of a friend, and forcing a passage through the crowd, he threw himself into his carriage, and was driven home. He paced the room with hurried steps, and fixed his hands upon his head, as if he were afraid his thoughts were bursting from his brain. Lord Ruthven again before him — circumstances started up in dreadful array — the dagger — his oath. — He roused himself, he could not believe it possible — the dead rise again! — He thought his imagination had conjured up the image his mind was resting upon. It was impossible that it could be real — he determined, therefore, to go again into society; for though he attempted to ask concerning Lord Ruthven, the name hung upon his lips and he could not succeed in gaining information. He went a few nights after with his sister to the assembly of a near relation. Leaving her under the protection of a matron, he retired into a recess, and there gave himself up to his own devouring thoughts. Perceiving, at last, that many were leaving, he roused himself, and entering another room, found his sister surrounded by several, apparently in earnest conversation; he attempted to pass and get near her, when one, whom he requested to move, turned round, and revealed to him those features he most abhorred. He sprang forward, seized his sister's arm, and, with hurried step, forced her towards the street: at the door he found himself impeded by the crowd of servants who were waiting for their lords; and while he was engaged in passing them, he again heard that voice whisper close to him — "Remember your oath!" — He did not dare to turn, but, hurrying his sister, soon reached home.

Boucicault himself played the vampire, when the great nineteenth-century actor Charles Kean rejected the melodramatic role as being beneath him. The London *Examiner* noted that Boucicault "enacted the 'monster' with due paleness of visage, stealthiness of pace and solemnity of tone." The play itself was not as well-received. The *Illustrated London News* complained that "much ingenuity has been thrown away on a subject barren of interest, and, to some extent, disgusting." Queen Victoria herself, though quite haunted by Boucicault's performance, came back for a second look, but finally dismissed the play in her diary. "It does not bear seeing a second time," she wrote, "and is, in fact, very trashy."

One interesting aspect of Boucicault's conception of the vampire is that he may have been the first actor to incorporate the appearance of a bat into his performance. Odell's *Annals of the New York Stage* cites Mrs. M.E.W. Sherwood's 1875 recollection of Boucicault's characterization, "a dreadful and weird thing played with immortal genius. That great playwright would not have died unknown had he never done anything but flap his bat-like wings in that dream-disturbing piece."

The most recent theatrical version of Polidori is Tim Kelly's *The Vampyre* (1988), a very loose adaptation presented as a "penny-dreadful" stage thriller in two acts.

Aubrey became almost distracted. If before his mind had been absorbed by one subject, how much more completely was it engrossed now that the certainty of the monster's living again pressed upon his thoughts. His sister's attentions were now unheeded, and it was in vain that she intreated him to explain to her what had caused his abrupt conduct. He only uttered a few words, and those terrified her. The more he thought, the more he was bewildered. His oath startled him; — was he then to allow this monster to roam, bearing ruin upon his breath, amidst all he held dear, and not avert its progress? His very sister might have been touched by him. But even if he were to break his oath, and disclose his suspicions, who would believe him? He thought of employing his own hand to free the world from such a wretch; but death, he remembered, had been already mocked. For days he remained in state; shut up in his room, he saw no one, and ate only when his sister came, who, with eyes streaming with tears, besought him, for her sake, to support nature. At last, no longer capable of bearing stillness and solitude, he left his house, roamed from street to street, anxious to fly that image which haunted him. His dress became neglected, and he wandered, as often exposed to the noon-day sun as to the mid-night damps. He was no longer to be recognized; at first he returned with evening to the house; but at last he laid him down to rest wherever fatigue overtook him. His sister, anxious for his safety, employed people to follow him; but they were soon distanced by him who fled from a pursuer swifter than any — from thought. His conduct, however, suddenly changed. Struck with the idea that he left by his absence the whole of his friends, with a fiend amongst them, of whose presence they were unconscious, he determined to enter again into society, and watch him closely, anxious to forewarn, in spite of his oath, all whom Lord Ruthven approached with intimacy. But when he entered into a room, his haggard and suspicious looks were so striking, his inward shuddering so visible, that his sister was at last obliged to beg of him to abstain from seeking, for her sake, a society which affected him so strongly. When, however, remonstrance proved unavailing, the guardians thought proper to interpose, and, fearing that his mind was becoming alienated, they thought it high time to resume again that trust which had been before imposed upon them by Aubrey's parents.

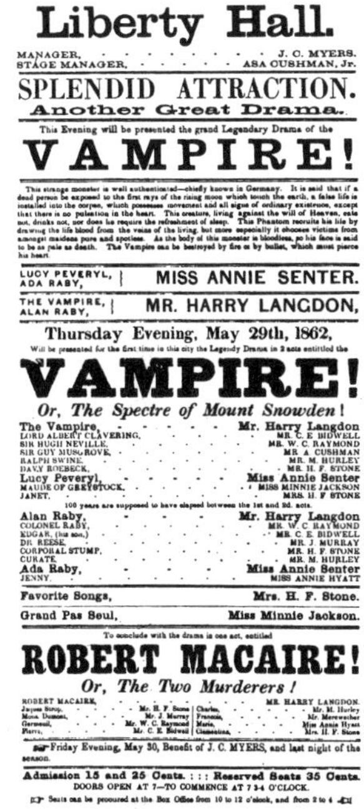

A poster for an 1862 revival of Boucicault's play.

Desirous of saving him from the injuries and sufferings he had daily encountered in his wanderings, and of preventing him from exposing to the general eye those marks of what they considered folly, they engaged a physician to reside in the house, and take constant care of him. He hardly appeared to notice it, so completely was his mind absorbed by one terrible subject. His incoherence became at last so great that he was confined to his chamber. There he would often lie for days, incapable of being roused. He had become emaciated, his eyes had attained a glassy lustre; — the only sign of affection and recollection remaining displayed itself upon the entry of his sister; then he would sometimes start, and, seizing her hands, with looks that severely afflicted her, he would desire her not to touch him. "Oh, do not touch him — if your love for me is aught, do not go near him!"

When, however, she inquired to whom he referred, his only answer was, "True! true!" and again he sank into a state, whence not even she could rouse him. This lasted many months: gradually, however, as the year was passing, his incoherences became less frequent, and his mind threw off a portion of its gloom, whilst his guardians observed, that several times in the day he would count upon his fingers a definite number, and then smile.

The time had nearly elapsed, when, upon the last day of the year, one of his guardians entering his room, began to converse with his physician upon the melancholy circumstance of Aubrey's being in so awful a situation, when his sister was going next day to be married. Instantly Aubrey's attention was attracted; he asked anxiously to whom. Glad of this mark of returning intellect, of which they feared he had been deprived, they mentioned the name of the Earl of Marsden. Thinking this was a young Earl whom he had met with in society, Aubrey seemed pleased, and astonished them still more by his expressing his intention to be present at the nuptials, and desiring to see his sister. They answered not, but in a few minutes his sister was with him. He was apparently again capable of being affected by the influence of her lovely smile; for he pressed her to his breast, and kissed her cheek, wet with tears, flowing at the thought of her brother's being once more alive to the feelings of affection. He began to speak with all his wonted warmth, and to congratulate her upon her marriage with a person so distinguished for rank and every accomplishment; when he suddenly perceived a locket upon her breast; opening it, what was his surprise at beholding the features of the monster who had so long influenced his life. He seized the portrait in a paroxysm of rage, and trampled it under foot. Upon her asking him why he thus destroyed the resemblance of her future husband, he looked as if he did not understand her; — then seizing her hands, and gazing on her with a frantic expression of countenance, he bade her swear that she would never wed this monster, for he — But he could not advance — it seemed as if that voice again bade him remember his oath - he turned suddenly round, thinking Lord Ruthven was near him but saw no one. In the meantime the guardians and physician, who had heard the whole, and thought this was but a return of his disorder, entered, and forcing him from Miss Aubrey, desired her to leave him. He fell upon his knees to them, he implored, he begged of them to delay but for one day. They, attributing this to the insanity they imagined had taken possession of his mind endeavoured to pacify him, and retired.

Lord Ruthven had called the morning after the drawing-room, and had been refused with every one else. When he heard of Aubrey's ill health, he readily understood himself to be the cause of it; but when he learned that he was deemed insane, his exultation and pleasure could hardly be concealed from those among whom he had gained this information. He hastened to the house of his former companion, and, by constant attendance, and the pretence of great affection for the brother and interest in his fate, he gradually won the ear of Miss Aubrey. Who could resist his power? His tongue had dangers and toils to recount — could speak of himself as of an individual having no sympathy with any being on the crowded earth, save with her to whom he addressed himself; — could tell how, since he knew her, his existence had begun to seem worthy of preservation, if it were merely that he might listen her soothing accents; — in fine, he knew so well how to use the serpent's art, or such was the will of fate, that he gained her affections. The title of the elder branch falling at length to him, he obtained an important embassy, which served as an excuse for hastening the marriage (in spite of her brother's deranged

THE OPERA ISN'T OVER TILL "THE VAMPYRE" SINGS

Based on Polidori's story, Heinrich Marschner's opera *Der Vampyr*, with a libretto by W.A. Wohlbruck, was first presented in Leipzig in 1828. The pleasantly melodic work was a standard fixture of the German repertoire for years and credited by musicologists as owing a distinct musical debt to Weber, while anticipating Wagner. Wagner, in fact, saw the premiere production during his student days, conducted the work himself a few years later, and may well have taken from *Der Vampyr* some measure of inspiration for his own opera on the theme of eternal life as damnation, *The Flying Dutchman* (1843).

A second 1828 opera, also called *Der Vampyr* and also based on the Polidori tale and its derivative melodramas, was composed by Peter Josef von Lindpainter with a libretto by Casar Max Heigel, and premiered in Stuttgart six month's after the Marschner work's debut. While it had a good run for its time, the Lindpainter/ Heigel piece had a confused story line, less musical interest, and has never been considered worthy of a full-scale revival.

The Marschner opera, however, was revived frequently through the turn of the twentieth century and is still produced occasionally in Germany. The work received its American premiere in a 1980 New York production with an English libretto by Michael Feingold. In 1992, the Marschner music was the basis for a cleverly modernized BBC television production that renamed Lord Ruthven "Ripley," a playboy investment banker. Called *The Vampyr: A Soap Opera*, the new libretto was written by Charles Hart, from a story by Janet Street-Porter and Nigel Finch (who also directed).

state), which was to take place the very day before his departure for the continent.

Aubrey, when he was left by the physician and his guardians, attempted to bribe the servants, but in vain. He asked for pen and paper; it was given him; he wrote a letter to his sister, conjuring her, as she valued her own happiness, her own honour, and the honour of those now in the grave, who once held her in their arms as their hope and the hope of their house, to delay but for a few hours that marriage, on which he denounced the most heavy curses. The servants promised they would deliver it; but giving it to the physician, he thought it better not to harass any more the mind of Miss Aubrey by, what he considered, the ravings of a maniac. Night passed on without rest to the busy inmates of the house; and Aubrey heard, with a horror that may more easily be conceived than described, the notes of busy preparation. Morning came, and the sound of carriages broke upon his ear. Aubrey grew almost frantic. The curiosity of the servants at last overcame their vigilance; they gradually stole away, leaving him in the custody of a helpless old woman. He seized the opportunity, with one bound was out of the room, and in a moment found himself in the apartment where all were nearly assembled. Lord Ruthven was the first to perceive him: he immediately approached, and, taking his arm by force, hurried him from the room, speechless with rage. When on the staircase, Lord Ruthven whispered in his ear — "Remember your oath, and know, if not my bride to day, your sister is dishonoured. Women are frail!" So saying, he pushed him towards his attendants, who, roused by the old woman, had come in search of him. Aubrey could no longer support himself; his rage not finding vent, had broken a blood-vessel, and he was conveyed to bed. This was not mentioned to his sister, who was not present when he entered, as the physician was afraid of agitating her. The marriage was solemnized, and the bride and bridegroom left London.

Title page to the score of Marschner's Der Vampyr.

Aubrey's weakness increased; the effusion of blood produced symptoms of the near approach of death. He desired his sister's guardians might be called, and when the midnight hour had struck, he related composedly what the reader has perused — he died immediately after.

The guardians hastened to protect Miss Aubrey; but when they arrived, it was too late. Lord Ruthven had disappeared, and Aubrey's sister had glutted the thirst of a VAMPYRE!

"I shall not die! I shall not die!" Illustration for the 1900 American edition of Gautier's "La morte amoureuse."

Loving Lady Death

(*La morte amoureuse*)

THEOPHILE GAUTIER
(1843)
translated by F. C. de Sumichrast

ABOUT THE AUTHOR

Theophile Gautier (1811-1872) was a leading literary figure of French Romanticism. Much influenced by E.T.A. Hoffman, Gautier's early works display a predilection for the grotesque and supernatural. Though greatly respected, Gautier struggled financially and supported himself primarily as a journalist and critic. In addition to "La morte amoureuse,"Gautier also utilized quasi-vampiric themes in his novel *Romance of a Mummy* (1863) and the story "Sprite" (1866). The translation of "La morte amoureuse" reprinted here is the work of the distinguished French literature authority F. C. de Sumichrast, and had its first American publication in 1900, under the title "The Vampire."

Theophile Gautier

YOU ASK ME, brother, if I have ever loved. I have. It is a strange story, and though I am sixty, I scarce venture to stir the ashes of that remembrance. I mean to refuse you nothing, but to no soul less tried than yours would I tell the story. The events are so strange that I can hardly believe they did happen. I was for more than three years the plaything of a singular and diabolical illusion. I, a poor priest, I led in my dreams every night — God grant they were dreams only!—the life of the damned, the life of the worldly, the life of Sardanapalus. A single glance, too full of approval, cast upon a woman, nearly cost me the loss of my soul. But at last, by the help of God and of my holy patron, I was able to drive away the evil spirit which had possessed me. My life was complicated by an entirely different nocturnal life. During the day I was a priest of God, chaste, busied with prayers and holy things; at night, as soon as I had closed my eyes, I became a young nobleman, a connoisseur of women, of horses and dogs, gambling, drinking, and cursing, and when at dawn I awoke, it seemed to me rather that I was going to sleep and dreaming of being a priest. Of that somnambulistic life there have remained in my remembrance things and words I cannot put away, and although I have never left the walls of my presbytery, you will be apt to think, on hearing me, that I am a man who, having worn out everything and having given up the world and entered religion, means to end in the bosom of God days too greatly agitated, rather than a humble student in a seminary, who has grown old in a forgotten parish in the depths of a forest, and who has never had anything to do with the things of the day.

Yes, I have loved, as no one on earth ever loved, with an insensate and furious love, so violent that I wonder it did not break my heart. Ah! what nights what nights I have had! From my youngest childhood I felt the vocation to the priesthood and all my studies were therefore bent in that direction. My life until the age of twenty-four was nothing but one long novitiate. Having finished my theological studies, I passed successfully through the minor orders, and my superiors considered me worthy, in spite of my youth, of crossing the last dread limit. The day of my ordination was fixed for Easter week. I had never gone into the world. The world, to me, lay within the walls of the college and of the seminary. I knew vaguely that there was something called a woman, but my thoughts never dwelt upon it; I was utterly innocent. I saw my old, infirm mother but twice a year; she was the only connection I had with the outer world. I regretted nothing; I felt not the least hesitation in the presence of the irrevocable engagement I was about to enter into; nay, I was joyous and full of impatience. Never did a young bridegroom count the hours with more feverish ardour. I could not sleep; I dreamed that I was saying Mass; I saw nothing more glorious in the world than to be a priest. I would have refused, had I been offered a kingdom, to be a king or a poet instead, for my ambition conceived nothing finer.

What I am telling you is to show you that what happened to me ought not to have happened, and that I was the victim of the most inexplicable fascination.

The great day having come, I walked to the church with so light a step that it seemed to me that I was borne in the air, or that I had wings on my shoulders; I thought myself an angel, and I was amazed at the sombre and preoccupied expression of my companions, — for there were several of us. I had spent the night in prayer, and was in a state bordering on ecstasy. The bishop, a venerable old man, seemed to me like God the Father bending from eternity, and I beheld the heavens through the vault of the dome.

You are acquainted with the details of the ceremony: the benediction, the Communion in both kinds, the anointing of the palms of the hands with the oil of the catechumens, and finally the sacred sacrifice offered in conjunction with the bishop. I will not dwell on these things. Oh! how right was job, "Imprudent is he who has not made a covenant with his eyes"! I happened to raise my head, which until then I had kept bent down, and I saw before me, so close that I might have touched her, although in reality she was a long way off, on the other side of the railing, a young woman of wondrous beauty dressed with regal magnificence. It was as though scales had fallen from my eyes. I felt like a blind man suddenly recovering his sight. The bishop, so radiant but now, was suddenly dimmed, the flame of the tapers on their golden candlesticks turned pale like stars in the morning light, and the whole church was shrouded in deep obscurity. The lovely creature stood out against this shadow like an angelic revelation. She seemed illumined from within, and to give forth light rather than to receive it. I cast down my eyes, determined not to look up again, so as to avoid the influence of external objects, for I was becoming more and more inattentive and I scarcely knew what I was about. Yet a moment later I opened my eyes again, for through my eyelids I saw her dazzling with the prismatic colours in a radiant penumbra, just as when one has gazed upon the sun.

Oh, how beautiful she was! The greatest painters had never approached this fabulous reality, even when, pursuing ideal beauty in the heavens, they brought back to earth the divine portrait of the Madonna. Neither the verse of the poet nor the palette of the painter can give you an idea of her. She was rather tall, with the figure and the port of a goddess. Her hair,

BAUDELAIRE AND GAUTIER: FLOWERS OF EVIL

One of Gautier's greatest admirers was the symbolist poet Charles Baudelaire (1821-1867), whose *Les fleurs du mal* (*Flowers of Evil*, 1857) was dedicated to Gautier. Teeming with ripe decadence and controversial from the very beginning, *Les fleurs du mal* included two vampire poems, "The Vampire" and "Metamorphoses of the Vampire," which were excised from the original collection due to their alleged obscenity, for which Baudelaire was tried and convicted.

Charles Baudelaire

of a pale gold, was parted on her brow and flowed down her temples like two golden streams; she looked like a crowned queen. Her forehead, of a bluish whiteness, spread out broad and serene over the almost brown eye-brows, a singularity which added to the effect of the sea-green eyes, the brilliancy and fire of which were unbearable. Oh, what eyes! With one flash they settled a man's fate. They were filled with a life, a limpidity, an ardour, a moist glow, which I have never seen in any other human eyes. From them flashed glances like arrows, which I distinctly saw striking my heart. I know not whether the flame that illumined them came from heaven or hell, but undoubtedly it came from one or the other place. That woman was an angel or a demon, perhaps both. She certainly did not come from the womb of Eve, our common mother. Teeth of the loveliest pearl sparkled through her rosy smile, and little dimples marked each inflection of her mouth in the rosy satin of her adorable cheeks. As to her nose, it was of regal delicacy and pride, and betrayed the noblest origin. An agate polish played upon the smooth, lustrous skin of her half-uncovered shoulders, and strings of great fair pearls, almost similar in tone to her neck, fell upon her bosom. From time to time she drew up her head with the undulating movement of an adder or of a peacock, and made the tall embroidered ruff that surrounded her like a silver trellis tremble slightly. She wore a dress of orange-red velvet, and out of the broad, ermine-lined sleeves issued won-drously delicate patrician hands, with long, plump fingers, so ideally trans-parent that the light passed through them as through the fingers of Dawn.

All these details are still as vivid to me as if I had seen her but yesterday, and although I was a prey to the greatest agitation, nothing escaped me; the faintest tint, the smallest dark spot on the corner of the chin, the scarcely perceptible down at the corners of the lips, the velvety brow, the trembling shadow of the eyelashes on her cheeks, — I noted all with astonishing lucidity.

As I gazed at her, I felt open within me doors hitherto fast-closed; passages obstructed until now were cleared away in every direction and revealed unsuspected prospects; life appeared in a new guise; I had just been born into a new order of ideas. Frightful anguish clutched my heart, and every minute that passed seemed to me a second and an age. Yet the ceremony was proceeding, and I was being carried farther from the world, the entrance to which was fiercely besieged by my nascent desires. I said "Yes," however, when I meant to say "no," when everything in me was revolting and protesting against the violence my vow was doing to my will. An occult force dragged the words from my mouth in spite of myself.

It is perhaps just what so many young girls do when they go to the altar with a firm resolve to boldly refuse the husband forced upon them. Not one carries out her intention. It is no doubt the same thing which makes so many poor novices take the veil, although they are quite determined to tear it to pieces at the moment of speaking their vows. No one dares to cause such a scandal before everybody, nor to deceive the

THE VAMPIRE
by Charles Baudelaire

You who, with the stroke of a sword
Has entered my plaintive heart,
You who, strong as a horde
Of demon women come apart,

Humiliate my soul
To make your bed and domain
— I am bound to your squalid whole
Like a convict to a chain.

Like a gambler to his cards,
Like worms to a decaying corpse,
To your bottle I'm a drunkard
I wish you damned, forever cursed!

For your swift sword I often pray
To defeat my liberty,
And I've spoken the poison of betrayal
To feel less cowardly.

Alas! The poison and the blade,
Only say, with disdain, to me:
"You're not worthy to be lifted away
From your cursed slavery.

Fool! If we could fate
Your delivery from her empire,
Your kisses would merely resuscitate
The cadaver of your vampire."

expectations of so many present. The numerous wills, the numerous glances, seem to weigh down on one like a leaden cloak. And then, every precaution is so carefully taken, everything is so well settled beforehand in a fashion so evidently irrevocable that thought yields to the weight of fact and completely gives way.

The expression of the fair unknown changed as the ceremony progressed. Her glance, tender and caressing at first, became disdainful and dissatisfied as if to reproach me with dullness of perception. I made an effort, mighty enough to have overthrown a mountain, to cry out that I would not be a priest, but I could not manage it; my tongue clove to the roof of my mouth and it was impossible for me to express my will by the smallest negative sign. I was, although wide-awake, in a state similar to that of nightmare, when one seeks to call out a word on which one's life depends, and yet is unable to do so.

She seemed to understand the martyrdom I was suffering, and as if to encourage me, she cast upon me a look full of divine promise. Her eyes were a poem, her every glance was a canto; she was saying to me, "If you will come with me, I will make you more happy than God Himself in Paradise. The angels will be jealous of you. Tear away the funeral shroud in which you are about to wrap yourself. I am beauty and youth and love; come to me, and together we shall be Love. What can Jehovah offer you in compensation? Our life shall pass like a dream, and will be but one eternal kiss. Pour out the wine in that cup and you are free. We will go away to unknown isles and you shall sleep on my bosom on a bed of massive gold under a pavilion of silver. For I love you and mean to take you from your God, before whom so many youthful hearts pour out floods of love that never reach Him."

It seemed to me that I heard these words on a rhythm of infinite sweetness, for her glance was almost sonorous, and the phrases her eyes sent me sounded within my heart as if invisible lips had breathed them. I felt myself ready to renounce God, but my hand was mechanically accomplishing the formalities of the ceremony. The beauty cast upon me a second glance so beseeching, so despairing that sharp blades pierced my heart, and I felt more swords enter my breast than did the Mother of Sorrows.

Never did any human face exhibit more poignant anguish. The maiden who sees her betrothed fall suddenly dead by her side, the mother by the empty cradle of her child, Eve seated on the threshold of the gate of Paradise, the miser who finds a stone in place of his treasure, the poet who has accidentally dropped into the fire the only manuscript of his favourite work, — not one of them could look more inconsolable, more stricken to the heart. The blood left her lovely face and she turned pale as marble. Her beautiful arms hung limp by her body as if the muscles had been unknotted, and she leaned against a pillar, for her limbs were giving way under her. As for me, livid, my brow covered with a sweat more bloody than that of Calvary, I staggered towards the church door. I was stifling; the vaulting

THE BEAUTIFUL VAMPIRE: AN APPRECIATION

Following is an abridgement of an essay introducing an early twentieth-century translation of Gautier by Oxford University's distinguished authority on French literature, Gustave Rudler. From The Beautiful Vampire *(New York: Robert M. McBride & Company, 1926).*

This exquisite prose poem, which ranks with *Albertus* as the most perfect specimen of the author's manner during his first period, shows, as might be expected, the riotous imagination of youth; but it is, nonetheless, not devoid of significance. The love story of the humble priest and the splendid courtesan sometimes reminds the reader of a fairy story, with its airy flights, its dream-like unreality and its gorgeous settings, Oriental rather than Italian; sometimes it terrifies, compresses the throat like a nightmare, so bitter is its emotion, so desperate its lament.

seemed to press down on me and my hand to upbear alone the weight of the cupola.

As I was about to cross the threshold, a woman's hand suddenly touched mine. I had never touched one before. It was cold like the skin of a serpent, yet it burned me like the print of a red-hot iron. It was she. "Oh, unfortunate man! Unfortunate man! What have you done?" she whispered; then disappeared in the crowd.

The old bishop passed by. He looked severely at me. My appearance was startlingly strange. I turned pale, blushed red, and flames passed before my eyes. One of my comrades took pity on me and led me away; I was incapable of finding alone the road to the seminary. At the corner of a street, while the young priest happened to look in another direction, a quaintly dressed negro page approached me and without staying his steps handed me a small pocket-book with chased gold corners, signing to me to conceal it. I slipped it into my sleeve and kept it there until I was alone in my cell. I opened it. It contained but two leaves with these words: "Clarimonda, at the Palazzo Concini." I was then so ignorant of life that I did not know of Clarimonda, in spite of her fame, and I was absolutely ignorant where the Palazzo Concini was situated. I made innumerable conjectures of the most extravagant kind, but the truth is that, provided I could see her again, I cared little what she might be, whether a great lady or a courtesan.

This new-born love of mine was hopelessly rooted within me. I did not even attempt to expel it from my heart, for I felt that that was an impossibility. The woman had wholly seized upon me; a single glance of hers had been sufficient to change me; she had breathed her soul into me, and I no longer lived but in her and through her. I indulged in countless extravagant fancies; I kissed on my hand the spot she had touched, and I repeated her name for hours at a time. All I needed to do to see her as plainly as if she had been actually present was to close my eyes; I repeated the words which she had spoken to me, "Unfortunate man! Unfortunate man! What have you done ?" I grasped the full horror of my situation, and the dread, sombre aspects of the state which I had embraced were plainly revealed to me. To be a priest; that is, to remain chaste, never to love, never to notice sex or age; to turn aside from beauty, to voluntarily blind myself, to crawl in the icy shadows of a cloister or a church, to see none but the dying, to watch by strangers' beds, to wear mourning for myself in the form of the black cassock, a robe that may readily be used to line your coffin.

Meanwhile I felt life rising within me like an internal lake, swelling and overflowing; my blood surged in my veins; my youth, so long suppressed, burst out suddenly like the aloe that blooms but once in a hundred years, and then like a thunder-clap. How could I manage to see Clarimonda again? I could find no pretext to leave the seminary, for I knew no one in town. Indeed, my stay in it was to be very short, for I was

The originality of *La Morte Amoureuse* does not consist in its ideas, or, as we say, its themes; for they are not new. Speaking generally, the satanism in the story is Byronic; the element of fantasy comes from Hoffmann; the priest, the courtesan, the very vague Italian setting—Gautier did not see Italy until 1850—from the English and French romantic tradition; the night rides, dream steeds, phantom forests and cloud castles, from the Rhine legends. If all that Gautier had done had been to weld all these elements into a cunning combination, we should only have to admire the brilliant ingenuity of the mosaic, the richness of his colours, his power of evocation, the vehemence of his desires and emotions, the glow or the delicacy of his style—and that would be all. But he has utilised all this

merely waiting to be appointed to a parish. I tried to loosen the bars of the window, but it was at a terrific height from the ground, and having no ladder, I had to give up that plan. Besides, I could go out at night only, and how should I ever find my way through the labyrinth of streets? All these difficulties, which would have been slight to other men, were tremendous for me, a poor seminarist, in love since yesterday, without experience, without money, and without clothes.

"Ah, if only I had not been a priest, I might have seen her every day; I might have been her lover, her husband," I said to myself in my blindness. Instead of being wrapped in my gloomy shroud, I should have worn silk and velvet, chains of gold, a sword and a plume, like handsome young cavaliers. My hair, instead of being dishonoured by a broad tonsure, would have fallen in ringlets around my neck; I should have worn a handsome waxed moustache; I should have been a valiant man. A single hour spent before an altar, a few words scarcely breathed, had cut me off forever from the living; I had myself sealed the stone of my tomb; I had pushed with my own hand the bolts of my prison door.

material in a highly individual manner. One notices again and again in this novelette the effort he makes to extract a human significance from what seem to be absurd freaks of the imagination. He deliberately examines and interprets them in a modern, one might even say up-to-date manner; the result is a fascinating blend of past and future, a skillful reconciliation of the most unbridled romanticism with the realities of life and of the human heart. We should add that throughout there is a tendency towards symbolism, and that presentiments of symbolist technique are evident.

Let us examine the main features of the story and see exactly in what Gautier's originality consists.

I looked out of the window. The heavens were wondrously blue, the trees had assumed their springtime livery, nature exhibited ironical joy. The square was full of people coming and going. Young dandies and young beauties in couples were going towards the gardens and the arbours; workmen passed by, singing drinking songs; there was an animation, a life, a rush, a gaiety, which contrasted all the more painfully with my mourning and my solitude. A young mother was playing with her child on the threshold of a door. She kissed its little rosy lips still pearly with drops of milk, and indulged, as she teased it, in those many divine puerilities which mothers alone can invent. The father, who stood a little way off, was smiling gently at the charming group, and his crossed arms pressed his joy to his heart. I could not bear the sight. I closed the window and threw myself on my bed, my heart filled with frightful hatred and jealousy, and I bit my fingers and my coverlet as if I had been a tiger starving for three days.

I know not how long I remained in this condition, but in turning over in a furious spasm, I perceived Father Serapion standing in the middle of the room gazing attentively at me. I was ashamed of myself, and letting fall my head upon my breast, I covered my face with my hands.

"Romualdo, my friend, something extraordinary is taking place in you," said Serapion after a few moments' silence. "Your conduct is absolutely inexplicable. You, so pious, so calm, and so gentle, you have been raging in your cell like a wild beast. Beware, my brother, and do not listen to the suggestions of the devil. The evil spirit, angered at your having devoted yourself to the Lord, prowls around you like a ravening wolf, and is making a last effort to draw you to himself. Instead of allowing yourself to be cast down, dear Romualdo, put on the breastplate of prayer, take up the shield of mortification, and valiantly fight the enemy. You will overcome him. Trial is indispensable to virtue, and gold emerges finer from the crucible.

Be not dismayed nor discouraged; the best guarded and the strongest souls have passed through just such moments. Pray, fast, meditate, and the evil one will flee from you."

The father's discourse brought me back to myself, and I became somewhat calmer. "I was coming," he said, "to inform you that you are appointed to the parish of C—. The priest who occupied it has just died, and his lordship the Bishop has charged me to install you there. Be ready to-morrow."

I signed that I would be ready, and the father withdrew.

I opened my breviary and began to read my prayers, but the lines soon became confused; I lost the thread of my thoughts, and the book slipped from my hands without my noticing it.

To leave to-morrow without having seen her again.

To add one more impossibility to all those that already existed between us! To lose forever the hope of meeting her unless a miracle occurred! Even if I were to write to her, how could I send my letter? Considering the sacred functions which I had assumed, to whom could I confide, in whom could I trust? I felt terrible anxiety. Then what Father Serapion had just said to me of the wiles of the devil recurred to my memory. The strangeness of the adventure, the supernatural beauty of Clarimonda, the phosphorescent gleam of her glance, the burning touch of her hand, the trouble into which she had thrown me, the sudden change which had occurred in me, my piety vanished in an instant, —everything went to prove plainly the presence of the devil, and that satin-like hand could only be the glove that covered his claws. These thoughts caused me much terror. I picked up the breviary that had fallen to the ground from my knees, and I again began to pray.

The next day Serapion came for me. Two mules were waiting for us at the door, carrying our small valises. He got on one and I on the other as well as I could. While traversing the streets of the town, I looked at every window and every balcony in the hope of seeing Clarimonda, but it was too early; and the town was not yet awake. My glance tried to pierce through the blinds and curtains of all the palaces in front of which we were passing. No doubt Serapion thought my curiosity was due to the admiration caused in me by the beauty of the architecture, for he slackened his mule's speed to give me time to look. Finally we reached the city gate and began to ascend the hill. When we reached the top, I turned around once again to gaze at the spot where lived Clarimonda. The shadow of a cloud covered the whole town; the blue and red roofs were harmonized in one uniform half-tint, over which showed, like flecks of foam, the morning smoke. By a singular optical effect there stood out bright under a single beam of light a building that rose far above the neighbouring houses, wholly lost in the mist. Although it was certainly three miles away, it seemed quite close; the smallest detail could be made out, — the turrets, the platforms, the windows, even the swallow-tailed vanes.

The central idea of *La Morte Amoureuse* is the all-powerful appeal made by youth, life and beauty to the heart of a theological student at the very moment when, by his ordination, he is renouncing them all; or, differently expressed, the impossibility of renunciation and the sovereignty of passion; in other words, the desperate regret a man feels for happiness denied to him, his inability to attain it except in dreams, his folly in destroying it when dreams have given it to him, his urge to escape from reality and his eagerness to return to it, the conflict between a man's life in the profession he has chosen, or had forced upon him, and his dreams . . . There we have the romantic idea *par excellence*, but at the same time a universal idea; it is applicable, in one form or another, to the whole of life, and to age even more than youth; it has a general symbolic value, and Gautier makes us feel it.

It is remarkable how he has concealed the anti-social aspect of his story; the novelette appears to have a conservative tendency. But in reality he has only avoided emphasising the conflict between the forces of nature and the constraints imposed by morality and civil society by separating them and allotting to each its own sphere—night and day, dream and reality. He does nothing to solve the problem—if there is a problem to be solved, for we all separate dream from reality, though in a less violent and definite form.

The French title of the book, deliberately enigmatic, is bound to puzzle a modern reader. It mysteriously indicates some triumph of Love over Death, but, in fact, refers to one of the most sinister of the legends which made their appearance at the dawn of romanticism—the vampire legend.

"What is that palace yonder lighted by a sunbeam?" I asked Serapion.

He shaded his eyes with his hand, and after having looked, answered: "That is the old palazzo which Prince Concini gave to Clarimonda the courtesan. Fearful things take place there."

At that moment, —I have never known whether it was a reality or an illusion, — I thought I saw on the terrace a slender white form that gleamed for a second and vanished. It was Clarimonda. Oh! Did she know that at that very moment, from the top of the rough road which was taking me away from her, ardent and restless, I was watching the palace she dwelt in, and which a derisive effect of light seemed to draw near to me as if to invite me to enter it as its master? No doubt she knew it, for her soul was too much in sympathy with mine not to have felt its every emotion, and it was that feeling which had urged her, still wearing her night-dress, to ascend to the terrace in the icy-cold dew of morning.

The shadow reached the palace, and all turned into a motionless ocean of roofs and attics in which nothing was to be distinguished save swelling undulations. Serapion urged on his mule; mine immediately started too, and a turn in the road concealed forever from me the town of S, for I was never to return there. After three days' travelling through a monotonous country, we saw rising above the trees the weathercock of the steeple of the church to which I had been appointed; and after having traversed some tortuous streets bordered by huts and small gardens, we arrived before the facade, which was not very magnificent. A porch adorned with a few mouldings and two or three sandstone pillars roughly cut, a tiled roof, and buttresses of the same sandstone as the pillars,—that was all. On the left, the cemetery overgrown with grass, with a tall iron cross in the centre; to the right, in the shadow of the church, the presbytery, a very plain, poor, but clean house. We entered. A few hens were picking up scattered grain. Accustomed, apparently, to the black dress of ecclesiastics, they were not frightened by our presence, and scarcely moved out of the way. A hoarse bark was heard, and an old dog ran up to us; it was my predecessor's dog. Its eye was dim, its coat was gray, and it exhibited every symptom of the greatest age a dog can reach. I patted it gently with my hand, and it immediately walked beside me with an air of inexpressible satisfaction. An old woman, who had been housekeeper to the former priest, also came to meet us, and after having shown us into the lower room, asked me if I intended to keep her. I told her that I should do so, and the dog and the hens also, and whatever furniture her master had left her at his death, which caused her a transport of joy, Father Serapion having at once paid her the price she had set upon it.

Having thus installed me, Father Serapion returned to the seminary. I therefore remained alone and without any other help than my own. The thought of Clarimonda again began to haunt me, and in spite of the efforts I made to drive it away, I was not always successful. One evening as I was walking through the box-edged walks of my little garden, I thought I saw through the shrubbery a female form watching my movements, and two sea-green eyes flashing amid the foliage, but it was merely an illusion. Having passed on the other side of the walk, I found only the imprint of a foot on the sand, so small that it looked like a child's foot. The garden was shut in by very high walls. I visited every nook and corner of it, but found no one. I have never been able to explain the fact, which, for the matter of that, was nothing by comparison with the strange things that were to happen to me.

I had been living in this way for a year, carefully fulfilling all the duties of my profession, praying, fasting, exhorting, and succouring the sick, giving alms even to the extent of depriving myself of the most indispensable necessaries; but I felt within me extreme aridity, and the sources of grace were closed to me. I did not enjoy the happiness which comes of fulfilling a holy mission; my thoughts were elsewhere, and Clarimonda's words often recurred to me. O my brother, ponder this carefully. Because I had a single time looked at a woman, because I had committed a fault apparently so slight, I suffered for several years the most dreadful agitation and my life was troubled forever.

From immemorial times and in different countries — especially in the East, Greece, Hungary and Illyria — the popular imagination has accepted "vampires" as dead persons who come out of their graves at night to suck the blood of the living while asleep, and so preserve themselves from dissolution. They belong to the weird fraternity of ogres, wizards, ghouls, gnomes, sprites, elves, etc., in which humanity symbolizes its fear of night, mystery and death, or, at a more advanced stage, its taste for agreeable terrors.

In April, 1819, a dull story entitled "The Vampire" was published by the New Monthly Magazine and later by the Librarie Galignani; it was by Polidori, but the name of Lord Byron was appended to it. Faber translated it immediately (1820), and a new

I shall not dwell longer upon these inward defeats and victories which were always followed by greater falls, but I shall pass at once to a decisive circumstance. One night there was a violent ringing at my door. The housekeeper went to open it, and a dark complexioned man, richly dressed in a foreign fashion, wearing a long dagger, showed under the rays of Barbara's lantern. Her first movement was one of terror, but the man reassured her, and told her that he must see me at once on a matter concerning my ministry. Barbara brought him upstairs. I was just about to go to bed. The man told me that his mistress, a very great lady, was dying and asking for a priest. I replied that I was ready to follow him, took what was needed for extreme unction, and descended quickly. At the door were impatiently pawing and stamping two horses black as night, breathing out long jets of smoke. He held the stirrup for me and helped me to mount one, then sprang on the other, merely resting his hand upon the pommel of the saddle. He pressed in his knees and gave his horse its head, when it went off like an arrow. My own, of which he held the bridle, also started at a gallop and kept up easily with the other. We rushed over the ground, which flashed by us gray and streaked, and the black silhouettes of the trees fled like the rout of an army. We traversed a forest, the darkness of which was so dense and icy that I felt a shudder of superstitious terror. The sparks which our horses' hoofs struck from the stones formed a trail of fire, and if any one had seen us at that time of night, he would have taken us for two spectres bestriding nightmares. From time to time will-o'-the-wisps flashed across the road, and the jackdaws croaked sadly in the thickness of the wood, in which shone here and there the phosphorescent eyes of wildcats. Our horses' manes streamed out wildly, sweat poured down their sides, and their breath came short and quick through their

translation by Eusèbe de Salles appeared in 1825. The literary vogue of vampirism came to France, therefore, from England, the paradise of sensationalism and emotionalism. A character in a French burlesque was made to say: "Vampires . . . they come to us from England . . . That's another kindness the English have done to us . . . . They make us such charming presents."

Strange to say, "The Vampire" did as much to make Byron popular in France as his greatest works had done, even after it had been shown not to be by him at all. Pichot was begged by his readers not to expunge it from his translation of Byron's works, and had to comply with the request. It is a mediocre story, and leaves the modern reader perfectly cold; our forefathers were given the "creeps" very easily.

nostrils; but when the equerry saw them slackening speed, he excited them by a guttural cry which had nothing of human in it, and the race began again madder than ever. At last our whirlwind stopped. A black mass dotted with brilliant points suddenly rose before us. The steps of our steeds sounded louder upon the ironbound flooring, and we entered under an archway the sombre mouth of which yawned between two huge towers. Great excitement reigned in the chateau. Servants with torches in their hands were traversing the courts in every direction, and lights were ascending and descending from story to story. I caught a confused glimpse of vast architecture,— columns, arcades, steps, stairs, a perfectly regal and fairy-like splendour of construction. A negro page, the same who had handed me Clarimonda's tablets, and whom I at once recognised, helped me to descend, and a majordomo, dressed in black velvet, with a gold chain around his neck and an ivory cane, advanced towards me. Great tears fell from his eyes and flowed down his cheeks upon his white beard. "Too late," he said, shaking his head. "Too late, my lord priest. But if you have not been able to save the soul, come and pray for the poor body." He took me by the arm and led me to the room of death. I wept as bitterly as he did, for I had understood that the dead woman was none else than Clarimonda, whom I had loved so deeply and madly. A *prie-dieu* was placed by the bedside; a bluish flame rising from a bronze cup cast through the room a faint, vague light, and here and there brought out of the shadow the corner of a piece of furniture or of a cornice. On a table, in a chased urn, was a faded white rose, the petals of which, with a single exception, had all fallen at the foot of the vase like perfumed tears. A broken black mask, a fan, and disguises of all kinds lay about on the armchairs, showing that death had entered this sumptuous dwelling unexpectedly and without warning, I knelt, not daring to cast my eyes on the bed, and began to recite the psalms with great fervour, thanking God for having put the tomb between the thought of that woman and myself, so that I might add to my prayers her name, henceforth sanctified. Little by little, however, my fervour diminished, and I fell into a reverie. The room had in no wise the aspect of a chamber of death. Instead of the fetid and cadaverous air which I was accustomed to breathe during my funeral watches, a languorous vapour of Oriental incense, a strange, amorous odour of woman, floated softly in the warm air. The pale light resembled less the yellow flame of the night-light that flickers by the side of the dead than the soft illumination of voluptuousness. I thought of the strange chance which made me meet Clarimonda at the very moment when I had lost her forever, and a sigh of regret escaped from my breast. I thought I heard some one sigh behind me, and I turned involuntarily. It was the echo. As I turned, my eyes fell upon the state-bed which until then I had avoided looking at. The red damask curtains with great flowered pattern, held back by golden cords, allowed the dead woman to be seen, lying full length, her hands crossed on her breast. She was covered with a linen veil of dazzling whiteness, made still more brilliant by the dark purple of the hangings; it was so tenuous that it concealed nothing of the charming form of her body, and allowed me to note the lovely lines, undulating like the neck of a swan, which even death itself had been unable to stiffen. She looked like an alabaster statue, the work of some clever sculptor, intended to be placed on a queen's tomb, or a young sleeping girl on whom snow had fallen.

I was losing my self-mastery. The sensuous air intoxicated me, the feverish scent of the half-faded rose went to my brain, and I strode up and down the room, stopping every time before the dais to gaze at the lovely dead woman through her transparent shroud. Strange thoughts came into my mind; I imagined that she was not really dead, that this was but a feint she had employed to draw me to her chateau and to tell me of her love. Once indeed I thought I saw her foot move under the white veil, disarranging the straight folds of the shroud.

Then I said to myself, "But is it Clarimonda? How do I know? The black page may have passed into some other woman's service. I am mad to grieve and worry as I am doing." But my heart replied, as it beat loud, "It is she, — it is none but she." I drew nearer the bed and gazed with increased attention at the object of my uncertainty. Shall I confess it? The perfection of her form, though refined and sanctified by the shadow of death, troubled me more voluptuously than was right, and her repose was so like sleep that any one might have been deceived by it. I forgot that I had come there to perform the funeral offices, and I imagined that I was a young husband entering the room of his bride who hides her face through modesty and will not allow herself to be seen. Sunk in grief, mad with joy, shivering with fear and pleasure, I bent towards her and took up the corner of the shroud; I raised it slowly, holding in my breath for fear of waking her. My arteries palpitated with such force that I felt the blood surging in my temples and my brow was covered with sweat as if I had been lifting a marble slab. It was indeed Clarimonda, such as I had seen her in the church on the day of my ordination. She was as lovely as then, and death seemed to be but a new coquetry of hers. The pallor of her cheeks, the paler rose of her lips, the long closed eyelashes showing their brown fringes against the whiteness, gave her an inexpressibly seductive expression of melancholy chastity and of pensive suffering. Her long hair, undone, in which were still a few little blue flowers, formed a pillow for her head and protected with its curls the nudity of her shoulders. Her lovely hands, purer and more diaphanous than the Host, were crossed in an attitude of pious repose and of silent prayer that softened the too great seduction, even in death, of the exquisite roundness and the ivory polish of her bare arms from which the pearl bracelets had not been removed. I remained long absorbed in mute contemplation. The longer I looked at her, the less I could believe that life had forever forsaken that lovely frame. I know not whether it was an illusion or a reflection of the lamp, but it seemed to me that the blood was beginning to course again under the mat pallor; yet she still remained perfectly

A French author served up the same dish, piping hot and highly flavoured, when in February 1820, Cyprian Bérard published his *Lord Ruthven ou les Vampires*, under the patronage of Charles Nodier. A second edition, with notes on vampirism, appeared in July. It must be said, in justice to French common-sense, that it appealed to French humour; numerous burlesques appeared, including one by Nodier himself. At the same time books continued to be written in which vampires were treated, outwardly at least, in a serious manner, such as Nodier's *Infernalia* (1822) and Mérimée's *Guzla* (1827).

***La Morte Amoureuse* was the lineal descendant of these works.**

It may seem to us now that it required some courage, in 1836, to make use of a theme which had been subjected to so much ridicule. But Gautier treated it most discreetly. He skimmed repulsive and morbid legends of their light and fantastic element, and flung aside the rest. His Clarimonde is infinitely more charming than terrible, more woman than vampire; her thirst for blood is excused by her love, and she drinks so little—only a few drops, enough to satisfy a bird's thirst. I do not know whether Gautier was endeavoring to furnish a symbolic, human, one might almost say scientific justification for the blood-drinking; but he suggests it. What man, Romuald asks himself, would not gladly give a little, nay, much of his blood to rescue his beloved from the claws of death? In our time, when transfusions of blood are frequently carried out, this then purely moral truth has acquired a literal significance. Gautier may almost be hailed as a prophet!

motionless. I gently touched her arm; it was cold, yet no colder than her hand on the day it touched me under the porch of the church. I resumed my position, bending my face over hers, and let fall upon her cheeks the warm dew of my tears. Oh, what a bitter despair and powerlessness I felt! Oh, what agony I underwent during that watch! I wished I could take my whole life in order to give it to her, and breathe upon her icy remains the flame that devoured me. Night was passing, and feeling the moment of eternal separation approaching, I was unable to refuse myself the sad and supreme sweetness of putting one kiss upon the dead lips of her who had had all my love. But, oh, wonder! A faint breath mingled with mine, and Clarimonda's lips answered to the pressure of mine. Her eyes opened, became somewhat brighter, she sighed, and moving her arms, placed them around my neck with an air of ineffable delight. "Oh, it is you, Romualdo!" she said in a voice as languishing and soft as the last faint vibrations of a harp. "I waited for you so long that I am dead. But now we are betrothed; I shall be able to see you and to come to you. Farewell, Romualdo, farewell! I love you; that is all I wish to say to you, and I give you back the life which you have recalled to me for one moment with your kiss. Good-bye, but not for long."

Her head fell back, but her arms were still around me as if to hold me. A wild gust of wind burst in the window and rushed into the room; the last leaf of the white rose fluttered for a moment like a wing at the top of the stem, then broke away and flew out of the casement, bearing Clarimonda's soul. The lamp went out and I swooned away on the bosom of the lovely dead.

When I recovered my senses, I was lying on my bed in my little room in my house, and the old dog of the former priest was licking my hand that was hanging out from under the blanket. Barbara, shaky with old age, was busy opening and closing drawers and mixing powders in glasses. On seeing me open my eyes, the old woman uttered a cry of joy, while the dog yelped and wagged his tail; but I was so weak that I could neither move nor speak. I learned later that I had remained for three days in that condition, giving no other sign of life than faint breathing. These three days are cut out of my life. I do not know where my mind was during that time, having absolutely no remembrance of it. Barbara told me that the same copper-complexioned man who had come to fetch me during the night, had brought me back the next morning in a closed litter and had immediately departed. As soon as I could collect my thoughts, I went over in my own mind all the circumstances of that fatal night. At first I thought I had been the dupe of some magical illusion, but real and palpable circumstances soon shattered that supposition. I could not believe I had been dreaming, since Barbara had seen, just as I had, the man with two black horses, and described his dress and appearance accurately. Yet no one knew of any chateau in the neighbourhood answering to the description of that in which I had again met Clarimonda.

One morning I saw Father Serapion enter. Barbara had sent him word that I was ill, and he had hastened to come to me. Although this eagerness proved affection for and interest in me, his visit did not give me the pleasure I should have felt. The penetration and the inquisitiveness of his glance troubled me; I felt embarrassed and guilty in his presence. He had been the first to notice my inward trouble, and I was annoyed by his clear-sightedness. While asking news of my health in a hypocritically honeyed tone he fixed upon me his two yellow, lion-like eyes, and plunged his glance into my soul like a sounding-rod. Then he asked me a few questions as to the way in which I was working my parish, if I enjoyed my position, how I spent the time which my duties left me, if I had made any acquaintances among the inhabitants of the place, what was my favourite reading, and many other details of the same kind. I answered as briefly as possible, and he himself, without waiting for me to finish, passed on to something else. The conversation evidently had nothing to do with what he meant to say to me. Then, without any preparation, as if it were a piece of news which he had just recollected and which he was afraid to again forget, he said, in a clear, vibrant voice that sounded in my ear like the trump of the Last judgment: —

The author has also, in harmony with the general tendency of his story, given a modified version of the method which the living used to employ to free themselves from the attentions of vampires. What was actually done was this: the coffin was opened, the corpse was pierced through the heart with a stake or the head cut off—it was considered advisable to do both to make quite sure!—then the remains were burned and the ashes flung to the winds. In *La Morte Amoureuse*, a single sprinkling with holy water suffices to make fair Clarimonde's body crumble into dust, nor does she even utter the fearful shriek which a vampire always emitted when the stake was driven through the heart.

"The great courtesan Clarimonda died recently, after an orgy that lasted eight days and nights. It was infernally splendid. They renewed the abominations of the feasts of Belshazzar and Cleopatra. What an age we are living in! The guests were served by dark slaves speaking an unknown language, who, I think, must have been fiends; the livery of the meanest of them might have served for the gala dress of an emperor. There have always been very strange stories about this Clarimonda; all her lovers have died a wretched and violent death. It is said that she was a ghoul, a female vampire, but I am of opinion that she was Beelzebub in person."

He was silent and watched me more attentively than ever to see the effect his words produced upon me. I had been unable to repress a start on hearing the name of Clarimonda, and the news of her death, besides the grief it caused me, through the strange coincidence with the nocturnal scene of which I had been a witness, filled me with a trouble and terror that showed in my face in spite of the efforts I made to master myself. Serapion looked at me anxiously and severely; then he said, "My son, I am bound to warn you that you have one foot over the abyss. Beware lest you fall in. Satan has a long arm, and tombs are not always faithful. The stone over Clarimonda should be sealed with a triple seal, for it is not, I am told, the first time that she has died. May God watch over you, Romualdo!"

With these words he walked slowly towards the door, and I did not see him again, for he left for S almost immediately.

I had at last entirely recovered, and had resumed my usual duties. The remembrance of Clarimonda and the words of the old priest were ever present to my mind; yet no extraordinary event had confirmed Serapion's gloomy predictions. I therefore began to believe that his fears and my terrors were exaggerated; but one night I dreamed a dream. I had scarcely fallen asleep when I heard the curtains of my bed open and the rings sliding over the bars with a rattling sound. I sat up abruptly, leaning on my elbow, and saw the shadow of a woman standing before me. I at once recognised Clarimonda. In her hand she bore a small lamp, of the shape of those put into tombs, the light of which gave to her slender fingers a rosy transparency that melted by insensible gradations into the opaque milky whiteness of her bare arm. Her sole vestment was the linen shroud that had covered her upon her state bed, and the folds of which she drew over her bosom as if she were ashamed of being so little clothed, but her small hand could not manage it. It was so white that the colour of the drapery was confounded with that of the flesh under the pale light of the lamp. Enveloped in the delicate tissue which revealed all the contours of her body, she resembled an antique marble statue of a bather rather than a woman filled with life. Dead or living, statue or woman, shadow or body, her beauty was still the same; only the green gleam of her eyes was somewhat dulled, and her mouth, so purple of yore, had now only a pale, tender rose-tint almost like that of her cheeks. The little blue flowers which I had noticed in her hair were dried up and had lost most of their leaves. And yet she was charming, so charming that in spite of the strangeness of the adventure and the inexplicable manner in which she had entered the room, I did not experience a single thrill of terror.

At the same time, Gautier has given us a glimpse of the diabolical vehemence of passion, the terrific forces of darkness and crime which lie at the root of these legends. But his Clarimonde, a complex but scarcely enigmatic creation, very little ghoul, almost all woman, with something of the demon in her and more of the angel, and withal only a dream figure, is a being instinct with charm and beauty, passion and tender love, and, in spite of her sinful life, with innocence. She transfigures the story, which might have been merely a low, coarse tale, and makes it a marvel of delicacy and refinement.

She placed the lamp on the table and sat down on the foot of my bed. Then bending towards me, she said in the silvery, velvety voice which I had heard from no one but her: —

"I have made you wait a long time, dear Romualdo, and you must have thought I had forgotten you. But I have come from a very long distance, from a bourne whence no traveller has yet returned. There is neither moon nor sun in the country whence I have come; neither road nor path; naught but space and shadow; no ground for the foot, no air for the wing; and yet I am here, for love is stronger than death and overcomes it. Ah, what worn faces, what terrible things I have seen on my way! What difficulty my soul, which returned to this world by the power of will, experienced before it could find its own body and re-enter it! What efforts I had to make before I could push up the tombstone with which they had covered me! See! The palms of my poor hands are all bruised. Kiss them and cure them, my dear love." And one after the other, she put the cold palms of her hands upon my lips. I did kiss them many a time, and she watched me with a smile of ineffable satisfaction.

I confess it to my shame, — I had wholly forgotten the counsels of Father Serapion and my own profession; I had fallen without resisting and at the first blow; I had not even endeavoured to drive away the tempter. The freshness of Clarimonda's skin penetrated mine, and I felt voluptuous thrills running through my body. Poor child! In spite of all that I have seen of her, I find it difficult to believe that she was a demon; she certainly did not look like one, and never did Satan better conceal his claws and horns. She had pulled her feet up under her, and was curled up on the edge of my bed in an attitude full of nonchalant coquetry. From time to time she passed her little hand through my hair and rolled it into ringlets as if to try how different ways of dressing it would suit my face. I allowed her to go on with the most guilty complaisance, and while she toyed with me she chatted brightly. The remarkable thing is that I experienced no astonishment at so extraordinary an adventure, and with the facility we enjoy in dreams of admitting as quite simple the most amazing events, it seemed to me that everything that was happening was quite natural.

"I loved you long before I had seen you, dear Romualdo, and I had looked for you everywhere. You were my dream, and when I saw you in church at that fatal moment, I at once said, 'It is he!' I cast on you a glance in which I put all the love which I had had, which I had, and which I was to have for you; a glance that would have damned a cardinal and made a king kneel before my feet in the presence of his whole court. But you remained impassible; you preferred your God to me. Oh, I am jealous of God, whom you loved, and whom you still love more than me! Unfortunate that I am, — oh, most unfortunate! Your heart will never be wholly mine, though you brought me back to life with a kiss, though I am Clarimonda, who was dead and who for your sake burst the cerements of the tomb, and has come to devote to you a life which she has resumed only to make you happy!"

"Oh, what eyes! With one flash they settled a man's fate."

With these words she mingled intoxicating caresses which penetrated my senses and my reason to such a degree that I did not hesitate, in order to console her, to utter frightful blasphemies and to tell her that I loved her as much as I did God.

Her eyes brightened and shone like chrysoprase. "True? Quite true? As much as God ?" she said, clasping me in her lovely arms. "Since that is so, you will go with me, you will follow me where I will. You shall cast off your ugly black clothes, you shall be the proudest and most envied of men, you shall be my lover. Oh, the lovely, happy life we shall lead! When shall we start?"

"To-morrow ! To-morrow!" I cried in my delirium.

"To-morrow be it," she replied. "I shall have time to change my dress, for this one is rather scanty and not of much use for travelling. Then I must also warn my people, who think me really dead, and who are mourning as hard as they can. Money, clothes, and carriage, — everything shall be ready, and I shall call for you at this same hour. Good-bye, dear heart," and she touched my brow with her lips.

The lamp went out, the windows were closed, and I saw no more. A leaden, dreamless sleep, overcame me and held me fast until the next morning. I awoke later than usual, and the remembrance of the strange vision agitated me the livelong day. At last I managed to persuade myself that it was a mere fever of my heated brain. Yet the sensation had been so intense that it was difficult to believe it was not real, and it was not without some apprehension of what might happen that I went to bed, after having prayed God to drive away from me evil thoughts and to protect the chastity of my sleep.

I soon fell fast asleep and my dream continued. The curtains were opened, and I saw Clarimonda, not as the first time, wan in her pale shroud, and the violets of death upon her cheeks, but gay, bright, and dainty, in a splendid travelling-dress of green velvet with gold braid, caught up on the side and showing a satin under-skirt. Her fair hair escaped in great curls from below her broad black felt hat with capriciously twisted white-feathers. She held in her hand a small riding-whip ending in a golden whistle. She touched me lightly with it and said: "Well, handsome sleeper, is that the way you get ready? I expected to find you up. Rise quickly, we have no time to lose."

The priest whom Gautier makes his hero, and with whom the heroine, by a rather strange freak of predestination, falls in love, is a familiar figure. From Lewis' *Le Moine* onwards, he has constantly appeared in romantic literature. One of his latest incarnations was Claude Frollo in *Notre-Dame de Paris*, not to mention the awful curé Mingrat. Gautier needed only to call in at the second-hand shop of romanticism to obtain this character.

He could not, however, have chosen a character more appropriate to his subject, and his priest enabled him to particularise his general theme with peculiar aptness. The idea of the struggle against and the victory of passion in the heart of a priest is more arresting, more vivid, than if the hero were an

I sprang from my bed.

"Come, put on your clothes and let us go," she said, pointing to a small parcel which she had brought. "The horses are impatiently champing their bits at the door. We ought to be thirty miles away by now."

I dressed hastily, and she herself passed me the clothes, laughing at my awkwardness and telling me what they were when I made a mistake. She arranged my hair for me, and when it was done, she held out a small pocket-mirror of Venice crystal framed with silver filigree and said to me, "What do you think of yourself ? Will you take me as your valet?"

I was no longer the same man and did not recognise myself. I was no more like myself than a finished statue is like a block of stone. My former face seemed to me but a coarse sketch of the one reflected in the mirror. I was handsome, and my vanity was sensibly tickled by the metamorphosis. The elegant clothes, the rich embroidered jacket, made me quite a different person, and I admired the power of transformation possessed by a few yards of stuff cut in a certain way. The spirit of my costume entered into me, and in ten minutes I was passably conceited. I walked up and down the room a few times to feel more at my ease in my new garments. Clarimonda looked at me with an air of maternal complaisance and appeared well satisfied with her work.

"Now, that is childishness enough. Let us be off, dear Romualdo; we are going a long way and we shall never get there." As she touched the doors they opened, and we passed by the dog without waking it.

At the door we found Margheritone, the equerry who had already conducted me. He held three horses, black like the first, one for me, one for himself, and one for Clarimonda. The horses must have been Spanish jennets, sired by the gale, for they went as fast as the wind, and the moon, which had risen to light us at our departure, rolled in the heavens like a wheel detached from its car. We saw it on our right spring from tree to tree, breathlessly trying to keep up with us. We soon reached a plain where by a clump of trees waited a carriage drawn by four horses. We got into it and the horses started off at a mad gallop. I had one arm around Clarimonda's waist and one of her hands in mine; she leaned her head on my shoulder, and I felt her half-bare bosom against my arm. I had never enjoyed such lively happiness. I forgot everything at that moment. I no more remembered having been a priest, so great was the fascination which the evil spirit exercised over me. From that night my nature became in some sort double. There were in me two men unknown to each other. Sometimes I fancied myself a priest who dreamed every night he was a nobleman; sometimes I fancied I was a nobleman who dreamed he was a priest. I was unable to distinguish between the vision and the waking, and I knew not where reality began and illusion ended. The conceited libertine rallied the priest; the priest hated the excesses of the young nobleman. Two spirals, twisted one within the other and confounded without ever touching, very aptly represent this bicephalous life of mine. Yet, in spite of the strangeness of this position, I do not think that for one instant I was mad. I always preserved very clearly the perception of my double life. Only there was an absurd fact which I could not explain: it was that the feeling of the same self should exist in two men so utterly different. That was an anomaly which I did not understand, whether I believed myself to be the parish priest of the little village of —— or il Signor Romualdo, the declared lover of Clarimonda.

What is certain is that I was, or at least believed that I was, in Venice. I have never yet been able to make out what was true and what was imaginary in that strange adventure. We dwelt in a great marble palace on the Canaleio, full of frescoes and statues, with two paintings in Titian's best manner in Clarimonda's bedroom. It was a palace worthy of a king. Each of us had his own gondola and gondoliers, his own livery, music-room, and poet. Clarimonda liked to live in great style, and she had something of Cleopatra in her nature. As for me, I lived like a prince's son, and acted as if I belonged to the family of the twelve Apostles or the four Evangelists of the Most Serene Republic; I would not have got out of my way to let the Doge pass, and I do not think that since Satan fell from heaven there was any one so proud and so insolent as I. I used to go to the Ridotto and gamble fearfully. I met the best society in the world, ruined eldest sons, swindlers, parasites, and swashbucklers, yet in spite of this dissipated life, I remained faithful to Clarimonda. I loved her madly. She would have awakened satiety itself and fixed inconstancy. I should have been perfectly happy but for the accursed nightmare which returned every night, and in which I thought myself a parish priest living an ascetic life and doing penance for his excesses of the daytime. Reassured by the habit of being with her, I scarcely ever thought of the strange manner in which I had made her acquaintance. However, what Father Serapion had told me about her occasionally occurred to my mind and caused me some uneasiness.

For some time past Clarimonda's health had been failing. Her complexion was becoming paler and paler every day. The doctors, when

ordinary man; the reactions are more violent; the ordination scene, for example, is gripping in its realism. The antithesis of a priest and courtesan followed naturally—too naturally; as in Hugo's dramas, the contrasts are rather crude. But Gautier, at any rate, had no blasphemous or irreverent intention. An almost unbroken tradition impelled him to make his Romuald a pleasure-loving, hypocritical priest; on the contrary, he made him pious, clean-living, sincere, modest, generous, devoted to his duties and anxious for his spiritual security—the victim only of a diabolical obsession which he detests, but from which he cannot free himself.

It is, in my judgement, from this study of obsession that the book derives its novelty and its intellectual significance, and Gautier knew very well what he was doing. One is faintly reminded, in reading the book, of Flaubert's *Tentation*.

called in, failed to understand her disease and knew not how to treat it. They prescribed insignificant remedies, and did not return. Meanwhile she became plainly paler, and colder and colder. She was almost as white and as dead as on that famous night in the unknown chateau. I was bitterly grieved to see her thus slowly pining away. She, touched by my sorrow, smiled gently and sadly at me with the smile of one who knows she is dying.

I do not mean only that he has striven with all the power of his art to blend reality with imagination, waking with dreaming, so that the frontier between them is wiped out and Romuald does not know whether he is really living or in a state of hallucination. This might be nothing more than the device of a skillful writer, at pains to surround a fantastic theme with the atmosphere best calculated to lend it probability and make the illusion effective. But it is much more than this. Gautier intended this double life led by Romuald to represent a case of double personality. He expresses a curiosity and uneasiness as to the nature of the ego. Romuald is astonished at the strength of the feeling of identity which both his personalities possess; he speculates as to the real nature of a somnambulistic life. A passage

One morning I was seated by her bed breakfasting at a small table, in order not to leave her a minute. As I pared a fruit I happened to cut my finger rather deeply. The blood immediately flowed in a purple stream, and a few drops fell upon Clarimonda. Her eyes lighted up, her face assumed an expression of fierce and savage joy which I had never before beheld. She sprang from her bed with the agility of an animal, of a monkey or of a cat, and sprang at my wound, which she began to suck with an air of inexpressible delight. She sipped the blood slowly and carefully like a gourmand who enjoys a glass of sherry or Syracuse wine; she winked her eyes, the green pupils of which had become oblong instead of round. From time to time she broke off to kiss my hand, then she again pressed the wound with her lips so as to draw out a few more red drops. When she saw that the blood had ceased to flow, she rose up, rosier than a May morn, her face full, her eyes moist and shining, her hand soft and warm; in a word, more beautiful than ever and in a perfect state of health.

"I shall not die! I shall not die!" she said, half mad with joy, as she hung around my neck. "I shall be able to love you a long time yet. My life is in yours, and all that I am comes from you. A few drops of your rich, noble blood, more precious and more efficacious than all the elixirs in the world, have restored my life."

The scene preoccupied me a long time and filled me with strange doubts concerning Clarimonda. That very evening, when sleep took me back to the presbytery, I saw Father Serapion, graver and more care-worn than ever. He looked at me attentively, and said to me: "Not satisfied with losing your soul, you want to lose your body also. Unfortunate youth, what a trap you have fallen into!" The tone in which he said these few words struck me greatly, but in spite of its vivacity, the impression was soon dispelled and numerous other thoughts effaced it from my mind. However, one evening I saw in my mirror, the perfidious position of which she had not taken into account, Clarimonda pouring a powder into the cup of spiced wine she was accustomed to prepare for me after the meal. I took the cup, feigned to carry it to my lips, and put it away as if to finish it later at leisure, but I profited by a moment when my beauty had turned her back, to throw the contents under the table, after which I withdrew to my room and went to bed, thoroughly determined not to sleep, and to see what she would do. I had not long to wait. Clarimonda entered in her night-dress, and having thrown it off, stretched herself in the bed by me. When she was quite certain that I was asleep, she bared my arm, drew a golden pin from her hair, and whispered, "One drop, nothing but a little red drop, a ruby at the end of my needle! Since you still love me, I must not die. Oh, my dear love! I shall drink your beautiful, brilliant, purple blood. Sleep, my sole treasure, my god and my child. I shall not hurt you, I shall only take as much of your life as I need not to lose my own. If I did not love you so much, I might make up my mind to have other lovers whose veins I would drain; but since I have known you, I have a horror of every one else. Oh, what a lovely arm how round and white it is! I shall never dare to prick that pretty blue vein." And as she spoke, she wept, and I felt her tears upon my arm which she held in her hands. At last she made up her mind, pricked me with the needle, and began to suck the blood that flowed. Though she had scarcely imbibed a few drops, she feared to exhaust me. She tied my arm with a narrow band, after having rubbed my wound with an unguent which healed it immediately.

I could no longer doubt; Father Serapion was right. However, in spite of the certainty, I could not help loving Clarimonda, and I would willingly have given her all the blood she needed in order to support her factitious existence. Besides, I was not much afraid, for the woman guarded me against the vampire; what I had heard and seen completely reassured me. At that time I had full-blooded veins which would not be very speedily exhausted, and I did not care whether my life went drop by drop. I would have opened my arm myself and said to her, "Drink, and let my life enter your body with my blood." I avoided alluding in the least to the narcotic which she had poured out for me and the scene of the pin,—and we lived in the most perfect harmony.

Yet my priestly scruples tormented me more than ever, and I knew not what new penance to invent to tame and mortify my flesh. Although all these visions were involuntary and I in no wise took part in them, I dared not touch the crucifix with hands so impure and a mind so soiled by such debauch, whether real or imaginary. After falling into these fatiguing hallucinations, I tried to keep from sleeping. I kept my eyes open with my fingers, and remained standing by the wall struggling against slumber with all my strength; but soon it would force itself into my eyes, and seeing that the struggle was useless, I let fall my arms with discouragement and weariness, while the current carried me again to the perfidious shores. Serapion exhorted me most vehemently, and harshly reproached me with weakness and lack of fervour. One day, when he had been more agitated than usual, he said to me: —

"There is but one way of ridding you of this obsession, and although it is extreme, we must make use of it. Great evils require great remedies. I know where Clarimonda is buried. We must dig her up, and you shall see in what a pitiful condition is the object of your love. You will no longer be tempted to lose your soul for a loathsome body devoured by worms and about to fall into dust. It will assuredly bring you back to your senses."

For myself, I was so wearied of my double life that I accepted, wishing

which seems to me most significant is that in which Romuald relates how, on waking up, he believes he is falling asleep; he has lost his sense of values, and his dream life is more real to him than his life on earth and his daily round of duties. This seems to me even more true than it is romantic and artistic; a prolonged, regular, rhythmic hallucination should have this effect.

The magnetism of love, the blending of souls, the effect on another of will-power, and some form of thought-transmission were clearly in Gautier's mind.

The attempt which Gautier makes to reduce the fantastic to terms of reason and science seems to me most remarkable, but I do not think it was supported by much real observation; most probably a philosopher or a doctor would not find utilisable material in the

to know once for all whether it was the priest or the nobleman who was the dupe of an illusion. I was determined to kill, for the benefit of the one or the other, one of the two men who were in me, or to kill them both, for such a life as I had been leading was unendurable. Father Serapion provided a pick, a crowbar, and a lantern, and at midnight we repaired to the cemetery of the place of which he knew accurately, as well as the disposition of the graves. Having cast the light of our lantern upon the inscriptions on several tombs, we at last reached a stone half hidden by tall grass and covered with moss and parasitical plants, on which we made out this partial inscription: "Here lies Clarimonda, who in her lifetime was the most beautiful woman in the world . . ."

"This is the spot," said Serapion, and putting down the lantern, he introduced the crowbar in the joints of the stone and began to raise it. The stone yielded, and he set to work with the pick. I watched him, darker and more silent than the night itself. As for him, bending over this funereal work, he perspired heavily and his quick breath sounded like the rattle in a dying man's throat. It was a strange spectacle, and any one who might have seen us would have taken us rather for men profaning the tomb and robbing the shrouds than for priests of God. Serapion's zeal had something harsh and savage which made him resemble a demon rather than an apostle or an angel, and his face, with its austere features sharply brought out by the light of the lantern, was in no wise reassuring. I felt an icy sweat break out on my limbs, my hair rose upon my head. Within myself I considered the action of the severe Serapion an abominable sacrilege, and I wished that from the sombre clouds that passed heavily over our heads might flash a bolt that would reduce him to powder. The owls, perched on the cypresses, troubled by the light of the lantern, struck the glass with their dusty wings and uttered plaintive cries. The foxes yelped in the distance, and innumerable sinister noises rose in the silence.

At last Serapion's pick struck the coffin, which gave out the dull, sonorous sound which nothingness gives out when it is touched. He pulled off the cover, and I saw Clarimonda, pale as marble, her hands clasped, her white shroud forming but one line from her head to her feet. A little red drop shone like a rose at the corner of her discoloured lips. Serapion at the sight of it became furious.

"Ah! There you are, you demon, you shameless courtesan! You who drink blood and gold!" and he cast on the body and the coffin quantities of holy water, tracing with the sprinkler a cross upon the coffin.

The holy dew no sooner touched poor Clarimonda than her lovely body fell into dust and became only a hideous mass of ashes and half-calcined bones.

"There is your mistress, my lord Romualdo," said the inexorable priest, as he pointed to the remains. "Are you now still tempted to go to the Lido and Fusino with your beauty?"

I bowed my head. Something had been shattered within me. I re-

story. Nevertheless, it reveals a train of thought in the author's mind of which it would be interesting to ascertain the source; it is not common among the writers of the period, except in Balzac. But Gautier ranks much higher as an *artist* than Balzac; his concentrated method has more affinity with that of Mérimée, though the romanticism of the latter is much less flamboyant.

I should be extremely sorry if these introductory remarks, possibly a trifle sceptical in places, were to keep readers from appreciating the fantasy of *La Morte Amoureuse*; the magic of the story remains, and if people are still able to shiver to the marrow at the thought of Gautier's charming vampire, I would not wish for a moment that they should be deprived of that enjoyment.

turned to my presbytery, and lord Romualdo, the lover of Clarimonda, left the poor priest with whom he had so long kept such strange company. Only the next night I saw Clarimonda. She said to me, as the first time under the porch of the church, "Unfortunate man! Unfortunate man! What have you done? Why did you listen to that foolish priest? Were you not happy? What have I done to you, that you should go and violate my poor tomb and lay bare the wretchedness of my nothingness? All communion between our souls and bodies is henceforth broken. Farewell; you will regret me."

She vanished in air like a vapour, and I never saw her again. Alas! She spoke the truth. I have regretted her more than once, and I still regret her. I purchased the peace of my soul very dearly. The love of God was not too much to replace her love.

Such, brother, is the story of my youth. Never look upon a woman, and walk always with your eyes cast on the ground, for chaste and calm though you may be, a single minute may make you lose eternity.

The haunting atmosphere of the Cemetery Pere-Lachaise in Paris was well-known to the French Romantics. (Photo by David J. Skal)

In one of the last great roles of his career, Boris Karloff played the vampire Gorcha in "The Wurdalak" segment of Mario Bava' Black Sabbath *(1963). (Photofest)*

The Family of the Vourdalak

ALEXIS TOLSTOY
(1843)

A New Translation by
DAVID J. SKAL

THE CONGRESS OF VIENNA,1815: an assembly of the most distinguished European intellectuals, glamorous society figures, and diplomats of the highest standing.[1]

When eventually the official Congress concluded, the royalist emigrées prepared to return permanently to their chateaux, the Russian soldiers to their abandoned homes, while some Polish malcontents returned to Cracow still nurturing their true love of freedom, only tenuously realized under the partitioned governance of Prince Metternich, Prince Hardenberg, and Count Nesselrode.

As at the end of a lively ball, there remained from the gathering a small number of persons still disposed to pleasure, and who, fascinated by the charms of the Austrian ladies, delayed packing their bags and delayed their departure.

This happy group, of which I was part, assembled twice a week at the chateau of the dowager Princess Schwarzenberg, some miles from the city in a little village called Hitzing. The grand manner of the chateau's mistress, heightened by her gracious amiability, rendered the sojourn extremely pleasant.

Our mornings were dedicated to leisurely walks; we took our lunch together at the chateau or on its grounds, and, in the evenings, seated by a warming hearth, we amused ourselves with conversation and storytelling. Political talk, of course, was strictly forbidden — we had all had enough of that. The stories we told were based upon legends of our respective homelands, or drawn from our personal memories.

One night, after each of us had told a story, and we found our minds in that intensified state of apprehension which darkness and silence only serve to increase, the Marquise d'Urfé, an elderly emigré whom we all liked for his youthful good cheer, not to mention the piquant recollections of his early life and good fortunes, broke the silence and offered the following:

"Your stories, messieurs, are quite amazing, without doubt. But it seems to me that each lacks an essential element of authenticity. I doubt that any

ABOUT THE AUTHOR

Count Alexis Konstantinovich Tolstoy (1817-1875) was an older, distant cousin of the far more celebrated Leo Tolstoy, but a significant literary figure in his own right. A precocious writer, he began writing verse at the age of six, and his diverse output included poetry, ballads, plays, satire and fiction. Best-known for his historical novel *The Silver Knight* (1862) and the dramatic trilogy *The Death of Ivan the Terrible*, *Czar Theodore*, and *Czar Boris* (1866-70), he also wrote a series of supernatural stories in the early 1840s, of which "The Family of the Vourdalak" (first published in Russian in 1884) is the outstanding example, richly reflecting his interest in Russian folk culture. Despite his estimable talent, he did not manage his affairs well and his last years were marked by isolation, bankruptcy and depression. He finally killed himself with an overdose of morphine at the age of fifty-eight.

The present translation of "The Family of the Vourdalak" is based on Tolstoy's original manuscript in French, which only surfaced following World War II, seventy-five years after the author's death.

1. The Congress of Vienna reassembled Europe after the Napoleonic wars, but declined to give Poland true independence, instead establishing a shaky tripartate government engineered by Metternich (Austria), Hardenburg (Prussia), and Nesselrode (Russia).

of you has seen with your own eyes such fantastic things as you have recounted, nor are able to affirm the truth with your pledge of honor as gentlemen."

We were obliged to agree, and the old man continued, smoothing the ruffle of his coat.

"As for me, messieurs, I know of only one case of the kind, but it is a case so strange, so horrible and so true as to strike fear into even the most incredulous imaginations. Unfortunately, I was at the same time both witness and participant, and though I ordinarily do not like to recall the circumstances, I will do so this one time, if mesdames will permit."

Alexis Tolstoy

The assent was unamimous, though some of the group glanced down nervously at the illuminated squares the candle light had begun to trace on the parquet, but soon the small circle pulled together, the better to hear the Marquise's story. He took a pinch of snuff, inhaled it slowly, and thus began:

First and foremost, mesdames, I beg your pardon if, during the course of my story, I come to speak of affairs of the heart more often than might be seemly for a man of my age. But I must include them for your full understanding of the tale. Besides, it is forgiveable in one's old age to have momentary lapses, and it will be your fault, ladies, if seeing such beautiful women before me, I am tempted to fancy myself a young man once more. I tell you then, without further preliminaries, that in the year 1759 I was passionately in love with the beautiful Duchess de Gramont. This passion, which I then believed to be profound and enduring, gave me no rest by day or by night, and the duchess, as pretty women often do, pleased herself with coquetry that only added to my torment. So much so, that in a moment of pure spite, I solicited and obtained a diplomatic mission to the Hospadar of Moldavia, where negotiations were underway with the cabinet of Versailles on matters too boring and pointless to bear repeating. The day before my departure, I visited the duchess at home. She received me in a less teasing manner than was her habit, and spoke to me in a voice suffused with real emotion.

"D'Urfé, you are making a serious mistake. But I know you, and I understand you will never go back on a decision made. Therefore, I only ask of you one thing: accept this little cross as a sign of my good will and carry it with you until you return. It is a family heirloom to which we attach great value." With a perhaps displaced gallantry at such a moment, I kissed not only the relic, but also the charming hand that presented it to me. I put the cross around my neck, and have never removed it since.

I will not tire you, mesdames, with the details of my trip, nor with my observations about the Hungarians and Serbs — poor, ignorant, yet brave and honorable people. Though subjugated by the Turks, they had forgotten neither their dignity or their former independence. Suffice it to say that I learned a bit of Polish during a sojourn in Warsaw, and quickly familiarized myself with Serbian, because these two languages, like Russian and

Bohemian, as you are no doubt are aware, are just so many branches of the same tongue, called Slavonic.

Therefore, I was able to make myself understood when I arrived in a village, which shall remain nameless. I found the inhabitants of the house where I intended to stay in a state of dismay, which seemed to me all the more strange because it was a Sunday, the day when the Serbian people customarily abandon themselves to various pleasures like dancing, rifle shooting, wrestling, and so on. I attributed the mood of my hosts to some recent calamity, and I was going to take my leave when a man of about thirty years, tall and imposing, approached me and took my hand.

"Enter, enter, stranger," he said to me. "Don't be put off by our sadness; you will understand when you know the cause."

He told me that his old father, whose name was Gorcha — a restless, rigid man — rose from his bed one day and took down from the wall his long Turkish rifle.

"Children," he said to his two sons, one named Georges and the other Pierre, "I am going to the mountains to join the brave men who are giving chase to that dog Alibek" (the name of a brigand Turk who, for some time, had devastated the land). "Wait for me for ten days, and, if I do not return on the tenth, have said for me a funeral mass, for I will have been killed." "But," old Gorcha added in his most serious manner, "if (may God protect you from this) I return after ten days have passed, for your salvation do not let me enter. In that case I order you to forget that I was your father and you must pierce me with a stake of aspen, regardless of what I may say or do, because I would then be nothing but an accursed *vourdalak*, coming to suck your blood."

It is fitting to tell you, mesdames, that the *vourdalaks*, or vampires of the Slavic people, are, according to local belief, dead bodies risen from their graves to suck the blood of the living. To this extent, their habits are those of all vampires, but they have another trait that renders them even more fearsome. The *vourdalaks*, mesdames, prefer to suck the blood of their closest relatives and most intimate friends, who, once dead, become vampires in turn — such that people claim to have seen entire villages in Bosnia and Hungary transfomed into *vourdalaks*. The abbé Augustin Calmet, in his curious book on apparitions, cites some deadful examples. The Germanic emperors named many commissions to throw light on cases of vampirism. They held inquests and exhumed bodies, which they found were gorged with blood, and burned the corpses in public after piercing their hearts with wooden stakes. Certain witnessing magistrates confirmed that they heard the cadavers let out shrieks at the moment the stakes were driven; their formal depositions were solemnized by their oaths and their signatures.

With this information, it should be easy for you to understand, mesdames, the effect Old Gorcha's words had upon his sons. They both leapt to their feet and begged to go in his stead. But his only response was to turn

MORE TALES OF RUSSIAN VAMPIRES

According to vampirologist Montague Summers, "It is no matter for surprise that in so sad and sick a country as Russia the tradition of the vampire should assume, if it be possible, an even intenser darkness. We find, indeed, a note of something deformed, as it were, something cariously diseased and unclean, a rank wealth of grotesque and fetid details which but serve to intensify the loathliness and horror."

Alexis Tolstoy was one of the few writers of his time to incorporate the gritty folklore of Russian peasants into vampire literature during a time when the fictional undead were elsewhere rapidly metamorphosing into genteel aristocrats.

In *The Vampire in Europe* (1929), Summers quotes the nineteenth-century folklore expert W.R.S. Ralston on the special characteristics of the Russian vampire:

"The districts of the Russian Empire in which a belief in vampires mostly prevails are White Russia and the Ukraine. But the ghastly blood-sucker, the *Upir*, whose name has been naturalized in so many alien lands under forms resembling our 'Vampire,' disturbs the peasant-mind in many other parts of Russia, though not perhaps with the same intense fear which it

away from the pair, and began singing the refrain of an old ballad. The day I arrived in the village was precisely the expiration of the time specified by Gorcha, and I had no difficulty comprehending his children's unease.

This was a good and honest family. Georges, the older of the two brothers, virile and rugged, appeared a serious and determined man. He was married, and the father of two children. His brother Pierre, a handsome young man of eighteen, had a more gentle appearance, and seemed to be the favorite of his younger sister, Sdenka, who had all the qualities of a classic Slavic beauty. In addition to her own incontestable loveliness, a distant resemblance to the Duchess de Gramont immediately struck me. The distinctive shape of her forehead was a trait I have never seen in my life, except in these two persons. Nothing special upon a first glance, it nonetheless became irresistible after one had seen it several times. Whether I was just very young then, or merely naive and eccentric, I could not be in Sdenka's company for two minutes without this sudden fondness threatening to evolve into a deeper attraction if I prolonged my visit to the village.

We were all gathered before the house around a table laden with cheese and bowls of milk. Sdenka was sewing; her sister-in-law prepared supper for the children, who were playing in the sand, and Pierre, with an affected indifference, whistled while cleaning a yatagan, a long Turkish knife. Georges, his brow furrowed, his elbows on the table and his head in his hands, directed his gaze down the main road and said not a word.

As for me, depressed by the general melancholy, I sadly regarded the clouds that framed the evening sky, as well as the silhouette of a monastery, partially hidden by a black forest of pines.

This monastery, I was later to learn, had once been celebrated for a miraculous icon of the Virgin, which had been transported by angels and set in an oak tree. But at the beginning of the last century, the Turks had invaded the region. They cut the throats of the monks and sacked the monastery. Nothing remained but the walls, and a chapel attended by a lone hermit, who did the honor of showing the ruins to the curious and gave hospitality to pilgrims, who, making the rounds on foot from one place of devotion to another, especially liked to rest at the Monastery of Our Lady of the Oak. As I have said, I wasn't aware of any of this at the time, for the last thing on my mind that night was the archeology of Serbia. As it often happens when one lets the imagination run wild, I dreamed of times past, of the idyllic days of my childhood, of my beautiful France, which I had left for a distant and savage land.

I dreamed of the Duchess de Gramont, and — why not admit it — I dreamed of several other contemporaries of your grandmothers. Their images had stolen into my heart, unconsciously echoing my feelings for the charming duchess.

Soon I had forgotten my hosts and their distress. All at once, Georges broke the silence.

spreads through the inhabitants of the above-named districts, or of some other Slavonic lands. The numerous traditions which have gathered around the original idea vary to some extent according to their locality, but they are never radically inconsistent.
"Some of the details are curious. The Little-Russians hold that if a vampire's hands have grown numb from remaining long crossed in the grave, he makes use of his teeth, which are like steel. When he has gnawed his way with these through all obstacles, he first destroys the babes he finds in a house, and then the older inmates. If fine salt be scattered on the floor of a room, the vampire's footsteps may be traced to his grave, in which he will be found resting with rosy cheek and gory mouth.

"The Kashoubes say that when a *Vieszcy*, as they call the Vampire, wakes from his sleep within the grave, he begins to gnaw his hands and feet: and as he gnaws, one after another, first his relations, then his other neighbours, sicken and die. When he has finished his own store of flesh, he rises at midnight and destroys cattle, or climbs a belfry and sounds the bell. All who hear the ill-omened tones will soon die. But generally he sucks the blood of sleepers. Those on whom he has operated will be found next morning dead, with a very small wound on the left side of the breast, directly over the heart. The Lusatian Wends hold that when a corpse chews its shroud or sucks its own breast, all its kin will soon follow it to the grave."

"Wife," he said, "at what hour did the old man leave?"

"At eight o'clock," she responded. "I clearly heard the monastery bell."

"Well, good," replied Georges. "It can't be more than half past seven." And he once again fixed his gaze down the road that faded into the dark forest.

I have forgotten to tell you, ladies, that when the Serbians suspect someone of vampirism, they avoid referring to him by name, or indicating him directly, because they believe this will summon a vampire from the grave. Therefore, Georges, when speaking of his father, called him nothing but "the old man."

There passed several minutes of silence. Suddenly, one of the children pulled Sdenka's apron string and said:

"Aunt Sdenka, when will grand-papa return home?"

A slap in the face from Georges was his answer to the untimely question.

The child began to cry, and his little brother reacted with astonishment and fear.

"But why, father, do you forbid us to speak of grand-papa?"

Another slap shut his mouth. Both children began to bawl, and the whole family made the sign of the cross.

Such was the scene when the monastery bell slowly sounded eight o'clock. The first sound had scarcely reached our ears, when a human shape stood out in relief from the woods and advanced toward us.

"It's him! God be praised!" cried Sdenka, Pierre, and their sister-in-law, all at once.

"God and the saints protect us!" said Georges solemnly. "How are we to know whether the ten days are passed or not?"

Everyone looked at him in fright. Meanwhile, the human shape still advanced. It was a tall old man with a silver moustache, a pale, severe figure who dragged himself painfully with the aid of a stick. Step by step he advanced, and Georges became more gloomy still. At last the newcomer stopped before us, and regarded his family with dark, sunken eyes that seemed not to see.

"Well," he said in a hollow voice, "will no one rise to greet me? What is the meaning of this silence? Can't you see that I am wounded?"

I could see that the old man's side was covered with blood.

"Go to your father's aid," I said to Georges, "and you, Sdenka, you must give him some something warm — he's ready to fall down from exhaustion!"

"My father," said Georges, approaching Gorcha, "show me your injury. I know how to treat it — "

He made as if to open Gorcha's cloak, but the old man rudely pushed him away and covered his side with both hands.

"Get away, you clumsy oaf! You are hurting me!"

"But you are wounded in the heart!" cried Georges, all pale. "Come, remove your cloak. It is necessary — you must do it, I tell you!"

These type of folk beliefs were quite familiar to Alexis Tolstoy as he began writing his cycle of supernatural fiction. The following traditional story, translated by Ralston in 1875, has a certain resonance with an incident in "The Family of the Vourdalak":

The Dog and the Corpse

A moujik went out in pursuit of game one day, and took a favorite dog with him. He walked and walked through the woods and bogs, but got nothing for his pains. At last the darkness of night surprised him. At an uncanny hour he passed by a graveyard, and there, at a place where two roads met, he saw standing a corpse in a white shroud. The moujik was horrified, and knew not which way to go — whether to keep on or to turn back.

"Well, whatever happens, I'll go on," he thought; and on he went, his dog running at his heels. When the corpse perceived him, it came to meet him; not touching the earth with its feet, but keeping about a foot above it — the shroud fluttering after it. When it had come up with the sportsman, it made a rush at him; but the dog seized hold of it by its bare calves, and began a tussle with it. When the moujik saw his dog and the corpse grappling with each other, he was delighted that things had turned out so well for himself, and he set off running home with all his might. The dog kept up the struggle until cock-crow, when

"Be careful," he said in a muted voice. "If you touch me, I will curse you."

Pierre stepped between Georges and his father.

"Let go," he said, "or you'll be the one who suffers!"

"Don't cross him," added his wife. "You know he will never tolerate it."

At that moment, we saw a herd of cattle returning from the pasture under a cloud of dust. But the dog which accompanied them either did not recognize his old master, or sensed something else amiss. As soon as he saw Gorcha, he stopped, his hair bristled and he started to howl as if he had seen a ghost.

"What has gotten into this dog?" said the old man, more and more displeased. "Have I become a stranger in my own home? Have ten days in the mountains changed me to the point that my own dogs don't recognize me?"

"Do you hear?" said Georges to his wife.

"What?"

"He admits that ten days are passed."

"But no — he has returned at exactly the appointed time!"

"Yes, and it is good that I know what is to be done."

As the dog continued to howl, Gorcha cried: "I want it killed! Do you understand me?"

Georges didn't budge, but Pierre rose, tears in his eyes, and seized his father's Turkish rifle. He tugged at the dog, which was rolling in the dust.

"He was always my favorite dog," he said in a low voice. "I do not know why the old man wants it killed!"

"Because he deserves it," said Gorcha. "Go on, it is cold and I want to go inside."

Meanwhile, Sdenka had prepared for the old man a hot tea made from clear brandy boiled with pears, honey and raisins, but her father pushed it away in disgust. He showed the same aversion to the plate of mutton and rice he was offered. Georges went to sit down in the corner by the fireplace, murmuring unintelligible words through clenched teeth.

A pinewood fire crackled in the hearth, its flickering light animating the pale, weak figure of the old man. Without this effect of the light, one would have taken him for dead. Sdenka sat down beside him.

"My father," she said, "you don't want to eat or sleep. Do you want to tell us of your adventures in the mountains?"

In saying this, the young girl knew she touched a sympathetic chord, because the old man loved to talk about warfare and battles. In response, a kind of smile parted his discolored lips, although his eyes remained vacant, and he passed a hand through her beautiful blonde hair.

"Yes, my daughter, yes, Sdenka. I do want to tell of my trip to the mountains, but that must be another time, for I am tired today. But I will tell you in the meantime that Alibek is dead by my hand. If anyone doubts it," continued the old man, casting his gaze over his family, "here is the proof!"

the corpse fell motionless to the ground. Then the dog ran off in pursuit of its master, caught him up just as he reached home, and rushed at him, furiously trying to bite and to rend him. So savage was it, and so persistent, that it was as much as the people of the house could do to beat it off.

"Whatever has come over that dog?" asked the moujik's old mother. "Why should it hate its master so?"

The moujik told her all that had happened.

"A bad piece of work, my son!" said the old woman.

"The dog was disgusted at your not helping it. There it was fighting with the corpse — and you deserted it, and thought only of saving yourself! Now it will owe you a grudge for ever so long."

Next morning, while the family were going about the farmyard, the dog was perfectly quiet. But the moment its master made its appearance, it began to growl like anything.

They fastened it to a chain; for a whole year they kept it chained up. But in spite of that, it never forgot how its master had offended it. One day it got loose, flew straight at him, and began to throttle him.

So they had to kill it.

He untied a kind of beggar's bag which hung behind his back, and pulled out a livid, bloody head which rivaled Gorcha's own in its ghastly pallor. We turned away in horror while Gorcha gave it to Pierre.

"Now then," he said, "hang it above the door, so that all who pass will know that Alibek is killed and the roads are purged of brigands — all except the sultan's guards!"

Pierre obeyed, with disgust.

"I understand everything now," he said. "The poor dog I killed only howled because it smelled dead flesh."

"Yes, he smelled dead flesh," responded Georges, darkly. He had gone out for a moment without anyone noticing, and returned with an object in his hand which he placed in a corner, and which I thought to be a wooden stake.

"Georges," said his wife in a soft voice, "You don't want, I hope…"

"My brother," added his sister, "what are you intending? But no, you won't do anything, will you?"

"Leave me alone," responded Georges. "I know what I have to do, and I won't do anything that isn't necessary."

In the meantime, night had fallen, and the family went to bed in a part of the house that was separated from my room only by a very thin partition. I must admit that what I had seen that night had made a strong impression on my mind. My candle was extinguished, and the moon shone brightly through a small, low window very close to my bed, and threw on the floor and walls a faint glow — a bit like the room we are in now, mesdames. I wanted to sleep but could not. I attributed my insomnia to the brightness of the moon; I searched for something which could serve as a curtain, but could find nothing. Then, hearing a confusion of voices behind the partition, I made myself listen.

"Go to sleep, woman," said Georges, "and you, Pierre, and you, Sdenka. Don't worry about anything, I will keep watch for you."

"But Georges," replied his wife, "it is my place to stay up. You have worked the night through, you must be very tired. Besides, I must look after our eldest. You know he has not been well since yesterday!"

"Be quiet and lie down," said Georges. "I am doing it for us both."

"But, my brother," said Sdenka in a very soft voice, "it seems to me that there's no point in anyone keeping watch. Our father is already asleep, and look at how calm and peaceful he seems."

"Neither one of you understands anything," said Georges in a tone that did not allow for a reply. "I'm telling you to go to sleep and to let me keep watch."

There followed a profound silence. Soon I felt my eyelids grow heavy, and sleep overtook my senses.

Dreaming, I thought I saw my door open slowly, with old Gorcha at the threshold. Or, I sensed his form more than I saw it, because the room behind him was very dark. It seemed to me that his blank eyes were straining

(The next story mixes Russian witchcraft with blood-stealing, and makes a special point about the anti-vampire powers of aspen wood, which Tolstoy also invokes in "The Family of the Vourdalak." Aspen was traditionally believed to be the wood of the True Cross.)

The Soldier and the Vampire

A certain soldier was allowed to go home on furlough. Well, he walked and walked, and after a time he began to draw near his native village. Not far off from that village lived a miller in his mill. In old times the Soldier had been very intimate with him; why shouldn't he go and see his friend? He went. The Miller received him cordially, and at once brought out liquor; and the two began drinking and chattering about their ways and doings. All this took place towards nightfall, and the Soldier stopped so long at the Miller's that it grew quite dark.

When he proposed to start for his village, his host exclaimed:

"Spend the night here, trooper! It's very late now, and perhaps you might run into mischief."

"How so?"

"God is punishing us! A terrible warlock has died among us, and by night he rises from his grave, and does such things as bring fear upon the very boldest! How could you even help being afraid of him?"

to read my thoughts, and followed the movement of my breathing. He put one foot forward, then another. Then, with extreme stealth, he began to approach me like a wolf. He jumped, landing next to my bed. I felt an inexpressible anguish, but an overpowering force kept me still. The old man bent over me, and brought his face so close to mine that I thought I sensed his cadaverous breath. Then, with an almost supernatural effort, I woke myself up, bathed in sweat. There was no one in my room, but, glancing at the window, I distinctly saw old Gorcha, who, outside, had pressed his face against the glass and fixed his fearsome eyes upon me. I had the strength not to scream and the presence of mind to stay in bed, as if I had seen nothing. Nonetheless, the old man seemed only to have come to make sure that I was sleeping, because he did not attempt to enter, but, after having thoroughly scrutinized me, he moved back from the window and I heard him walking in the adjacent room. Georges had fallen asleep and was snoring loudly enough to shake the walls. At that moment, the child coughed, and I discerned Gorcha's voice.

"You're not sleeping, little one?"

"No, grand-papa," replied the child, "and I would like to talk with you."

"Ah, you'd like to talk with me. And what will we talk about?"

"I would like you to tell me how you battled the Turks, because I would also gladly fight against them!"

"I thought about that, child, and I brought you back a little yatagan that I will give you tomorrow."

"Ah, grand-papa, since you're not sleeping, give it to me now instead."

"But why, little one, didn't you talk with me during the day?"

"Because papa said it was forbidden!"

"He is careful, your papa. And so, you would like to have your little yatagan."

"Oh, yes, I would like it, but not here, because papa might wake up."

"But where, then?"

"If we went out, I promise to be good and not to make the slightest sound!"

I thought I could discern Gorcha's sardonic laugh and heard the child get up. I did not believe in vampires, but the nightmare that I just endured had shaken my nerves, and, not wanting to have anything to reproach myself for later, I got up, and banged on the partition. The banging was loud enough to wake the seven sleepers, but there was no indication that the family heard it. I threw myself toward the door, firmly resolved to save the child, but I found that it was locked from the outside, and the bolts resisted all my efforts. While I was trying to break through, I saw the old man with the child in his arms pass in front of my window.

"Get up! Get up!" I shouted with all my might, and I broke down the partition with my blows. Only Georges had awakened.

"Where is the old man?" he said.

"Not a bit of it! A soldier is a man who belongs to the crown, and 'crown property cannot be drowned in water nor burnt in fire.' I'll be off: I'm tremendously anxious to see my people as soon as possible."

Off he set. His road lay in front of a graveyard. On one of the graves he saw a great fire blazing. "What's that?" thinks he. "Let's have a look." When he drew near, he saw that the Warlock was sitting by the fire, sewing boots.

"Hail, brother!" calls out the Soldier.

The Warlock looked up and said:

"What have you come here for?"

"Why, I wanted to see what you're doing."

The Warlock threw his work aside and invited the Soldier to a wedding.

"Come along, brother," says he, "let's enjoy ourselves. There's a wedding going on in the village."

"Come along!" says the Soldier.

They came to where the wedding was; there they were given drink, and treated with the utmost hospitality. The Warlock drank and drank, revelled and revelled, and then grew angry. He chased all the guests and relatives out of the house, threw the wedded pair into a slumber, took out two

"Go quickly!" I cried. "The old man has taken your child!"

With a single kick, Georges broke down his own door, which, like mine, had been locked from without, and began to run in the direction of the woods. I finally managed to wake Pierre, his sister-in-law, and Sdenka. We gathered in front of the house, and, after a wait of several minutes, we saw Georges return with his son. He had found him passed out on the main road, but soon he had revived and didn't appear any worse than before. Pounded with questions, he replied that his grandfather had done him no harm, that they had gone out together to talk more freely, but once outside, he had fainted without remembering how. As for Gorcha, he had disappeared.

The rest of the night, as you can well imagine, was sleepless.

The next day, I learned that the Danube, which cut the main road a quarter mile from the village, was full of drift ice, which always happens in these regions toward the end of autumn and the beginning of spring. The passage was blocked for several days, and I couldn't even dream of leaving. Besides, even if I could have, curiosity, along with a more powerful attraction, would have kept me. The more I saw Sdenka, the more I was drawn to love her. I am not, mesdames, one of those who believe in sudden and irresistible passions that novels are full of. But I think that there are cases where love develops more rapidly than usual. Sdenka's unique beauty, her striking resemblance to the Duchess of Gramont, from whom I had fled in Paris and whom I rediscovered here in picturesque dress, speaking in a foreign and euphonious tongue, the special character of her face — the kind over which I had wanted to kill myself twenty times in France — all of that, coupled with the peculiarity of my situation and the mysteries surrounding me, must have combined to ripen in me an emotion, which, in other circumstances, would have perhaps only manifested itself in a vague and passing fashion.

In the course of the day I heard Sdenka speaking with her younger brother.

"What do you make of all of this?" she asked. "Do you also have doubts about our father?"

"I don't dare suspect him," replied Pierre, "especially since the child said he did him no harm. And as far as his disappearance goes, you know he never accounted for his absences."

"I know," said Sdenka, "but in that case we have to save him, because you know Georges. . . ."

"Yes, yes, I know him. Talking to him would be pointless, but we will hide the stake, and he won't find another, because on this side of the mountains there is not a single aspen tree."

"Yes, let's hide the stake, but don't tell the children, because they might chatter about it in front of Georges."

"We'll be careful," said Pierre. And they parted.

Night fell without our having learned anything about old Gorcha.

phials and an awl, and began drawing off their blood. Having done this, he said to the Soldier:

"Now, lets be off!"

Well, off they went. On the way the Soldier said:

"Tell me; why did you draw off their blood in those phials?"

"Why, in order that the bride and bridegroom might die. To-morrow morning no one will be able to wake them. I alone know how to bring them back to life."

"How's that managed?"

"The bride and bridegroom must have cuts made in their heels, and some of their own blood must then be poured back into those wounds. I've got the bridegroom's blood stowed away in my right-hand pocket, and the bride's in my left."

The Soldier listened to this without letting a single word escape him. Then the Warlock began boasting again.

"Whatever I wish," says he, "that I can do!"

"I suppose it's quite impossible to get the better of you?" says the Soldier.

"Why impossible? If anyone were to make a pyre of aspen boughs, a hundred loads of them, and were to burn me on that pyre, then he'd be able to get the better of me. Only he'd have to look out

Like the previous night, I was sprawled on my bed, the moon shining brightly into my room. When sleep began to confuse my thoughts, I felt, as if by instinct, the approach of the old man. I opened my eyes and saw his livid face pressed against my window.

This time I wanted to get up, but it was impossible for me to do so. It seemed to me that all of my limbs were paralyzed. After looking me over intently, the old man went away. I heard him walk through the house and tap softly on the window of the room where Georges and his wife were sleeping. The child went back to bed, groaning in troubled dreams. Several minutes of silence passed, and then I heard once more a knocking at the window. The child moaned again and woke up.

"Is that you, grand-papa?" he asked.

"It is me," replied a hollow voice, "and I have your little yatagan."

"But I don't dare go outside — papa has forbidden it."

"You don't have to go out. Just open the window and come hug me!"

The child got up and I heard the window open. Drawing on all of my energy, I leapt from the bottom of my bed and began rapping on the partition. In a minute Georges was up. I heard him swear; his wife let out a loud scream, and soon the whole house had gathered around the motionless child. Gorcha had disappeared like the night before. Through our efforts, we managed to bring the child back to consciousness, but he was extremely weak and breathed with difficulty. The poor child did not know what caused him to faint. His mother and Sdenka attributed it to the fear of having been caught speaking to his grandfather. I said nothing. In the meantime, the child calmed down, and everyone except Georges went to bed.

Towards dawn I heard him wake up his wife, and they spoke to each other in hushed tones. Sdenka joined them; she and her sister-in-law were sobbing.

The child was dead.

I will omit any description here of the family's profound despair. Nonetheless, no one attributed the cause of death to old Gorcha. At least, no one said so openly.

George sat silently, but his expression, always somber, had become something terrible indeed. For two days, the old man did not reappear. On the night of the third day (the day the child was buried) I thought I heard footsteps around the house and the voice of an old man calling out the name of the dead child's little brother. For a moment, I thought I saw the face of Gorcha pressed again against my window, but I could not tell whether it was real, or merely an effect of my imagination, for it was quite dark that night, the moon veiled with clouds. In any event, I felt obliged to talk to Georges about the matter. He questioned he child, who reported that he had indeed heard his grandfather calling, and had seen him staring at the window.

All these circumstances did nothing to stop my attraction to Sdenka, which intensified from day to day.

sharp in burning me; for snakes and worms and different kinds of reptiles would creep out of my inside, and crows and magpies and jackdaws would come flying up. All these must be caught and flung on the pyre. If so much as a single maggot were to escape, then there'd be no help for it; in that maggot I should slip away!"

The Soldier listened to all this and did not forget it. He and the Warlock talked and talked, and at last they arrived at the grave.

"Well, brother," said the Warlock, "now I'll tear you to pieces. Otherwise you'd be telling all this."

"What are you talking about? Don't deceive yourself; I serve God and the Emperor."

The Warlock gnashed his teeth, howled aloud, and sprang at the Soldier, who drew his sword and began laying about him with sweeping blows. They struggled and struggled; the Soldier was all but at the end of his strength. "Ah!" thinks he, "I'm a lost man — and all for nothing!" Suddenly the cocks begin to crow. The Warlock fell lifeless to the ground.

The Soldier took the phials of the blood out of the Warlock's pocket, and went on to the house of his own people. When he had got there, and had exchanged greetings with his relatives, they said:

"Did you see any disturbance, Soldier?"

By day, I had no way to speak to her unobserved. Come the night, the very thought of my approaching departure grieved my heart. Sdenka's room was separated from mine by a kind of hallway leading to the road on one side and to the back yard on the other.

When my host family had gone to bed, I had the idea to take a walk through the countryside to lessen my distress. In the hallway I saw that Sdenka's door was open.

Almost involuntarily, I stopped. The familiar rustling of her dress set my heart pounding. Then, I heard some lyrics, sung in a soft voice. She was singing about a Serbian king setting off to war, and bidding farewell to his lady love.

"Oh, my young poplar, said the old king, I am going to war and you will forget me.

"The trees that grow at the foot of the mountain are slender and pliable, but your waistline is even more so!

"The fruits of the rowan tree, which the wind sets swaying, are red, but your lips are redder than the fruits!

"And I am like an old oak, barren of leaves, and my beard is whiter than the Danube's foam!

"And you will forget me, oh my soul, and I will die of sorrow, for the enemy won't dare to kill the old king!

"Then the beautiful lady replied: I swear to be faithful and not to forget you. If I break my oath, come, after your death, to suck all the blood from my heart!

"And the old king said: So be it! And he went off to war. And soon the beautiful lady forgot him!"

Here, Sdenka stopped, as if she was afraid to complete the ballad. I could contain myself no longer. That voice... so sweet, so expressive... it was the voice of the Duchess of Gramont. Without even thinking, I pushed open the door and entered. Sdenka had taken off her outer garment, a kind of large-sleeved bodice worn by the women of the region. All she wore was a blouse made of gold-embroidered red silk, cinched around her waist by a simple checkered skirt. Her beautiful blonde braids had been let down and their dishabille only heightened their attraction. Without irritation at my brusque entrance, she was nonetheless confused and blushed slightly.

"Oh," she said to me. "Why have you come? What will they think of me if they catch us like this?"

"Sdenka, my soul, be calm. Everyone is asleep but us — only the cricket in the grass and the mayfly in the air can hear what I have to tell you."

"Oh, my friend, get out, get out! If my brother discovers us, I am lost!"

"Sdenka, I won't leave until you make me a promise to love me always, like the girl promised the king in your song. I am leaving soon, Sdenka, and who knows when we will see each other again? Sdenka, I love you more than my soul, more than my salvation... my life and my blood are yours... won't you give me one hour in exchange?"

"No, I saw none."

"There now! Why we've had a terrible piece of work going on in the village. A Warlock has taken to haunting it!"

After talking a while, they lay down to sleep. Next morning the Soldier awoke, and began asking:

"I'm told you've got a wedding going on somewhere here?"

"There was a wedding in the house of a rich moujik," replied his relatives, "but the bride and bridegroom have died this very night — what from, nobody knows."

"Where does this moujik live?"

They showed him the house. Thither he went without speaking a word. When he got there, he found the whole family in tears.

"What are you mourning about?" says he.

"Such and such is the state of things, Soldier," say they.

"I can bring your young people to life again. What will you give me if I do?"

"Take what you like, even were it half of what we've got!"

The Soldier did as the Warlock had instructed him, and brought the young people back to life. Instead of weeping there began to be

"A good many things can happen in an hour," said Sdenka, looking pensive, but she left her hand in mine. "You do not know my brother," she continued, shuddering. "I have a feeling that he will come."

"Calm yourself, my Sdenka," I told her. "Your brother is tired from so many sleepless nights. The wind playing in the trees has made him drowsy; his sleep is very heavy, the night is very long, and I am only asking for one hour! And then, adieu... perhaps forever."

"Oh, no, no — not forever! Sdenka blurted; then she recoiled, as if frightened by her own voice.

"Sdenka," I cried, "I see only you, I hear only you, I'm no longer in control of myself. I am obeying a superior force. Forgive me, Sdenka!" And like a madman I clasped her to my heart.

"You're not my friend," she said, disengaging herself and seeking refuge in the back of her room. I do not know how I responded, because I was myself shocked by my effrontery — not because such boldness hadn't served me well on similar occasions — but because, in spite of my passion, I could not overcome my deep respect for Sdenka's innocence

It is true that at the start of our conversation I had tossed out some gallant phrases that were not displeasing to young women of the time, but soon I was ashamed of my words and I renounced them, seeing that the young lady's simplicity prevented her from understanding what — I can tell by your half smile — you, mesdames, have already guessed.

I was there, in front of her, not knowing what to say, when suddenly I saw her shudder and fix a look of terror on the window. I followed the direction of her eyes and distinctly saw the rigid face of Gorcha, who was watching us from outside.

At the same moment, I felt a heavy hand on my shoulder. I turned around. It was Georges.

"What are you doing here?" he demanded.

Disconcerted by this sudden interruption, I pointed out his father watching at the window, who disappeared just at the moment Georges turned to see.

"I had heard the old man and had come to warn your sister," I told him.

Georges stared at me as if he was able to read the bottom of my soul. Then he took me by the arm, conducted me to my room, and left without a word.

The next day the family had regrouped in front of the house around a table laden with milk and cheese.

"Where is the child?" asked George.

"He is in the yard," replied his mother. "He plays his favorite game all by himself, pretending that he is fighting the Turks."

As soon as she had pronounced these words, it was to our extreme surprise that we saw, advancing from the dark woods, the imposing figure of Gorcha, who walked slowly toward us and sat at the table as he had

happiness and rejoicing; the Soldier was hospitably treated and well rewarded. Then — left about face! Off he marched to the Starosta, and told him to call the peasants together and to get ready a hundred loads of aspen wood. Well, they took the wood into the graveyard, dragged the Warlock out of his grave, placed him on ohe pyre, and set it alight — the people all standing round in a circle with brooms, shovels, and fire-irons. The pyre became wrapped in flames, the Warlock began to burn. His corpse burst, and out of it crept snakes, worms and all sorts of reptiles, and up came flying crows, magpies and jackdaws. The peasants knocked them down and flung them into the fire, not allowing so much as a single maggot to creep away! And so the Warlock was thoroughly consumed, and the Soldier collected his ashes and strewed them to the winds. From that time forth there was peace in the village.

The Soldier received the thanks of the whole community. He stayed at home some time, enjoying himself thoroughly. Then he went back to the Tsar's service with money in his pocket. When he had served his time, he retired from the army, and began to live at his ease.

done the day of my arrival.

"Welcome, father," murmured his beautiful daughter-in-law in an almost inaudible voice.

"Welcome, father," Sdenka and Pierre echoed softly.

"My father," said Georges firmly, but with a change in his color, "we are waiting for you to say grace!"

The old man turned away, furrowing his brow.

"Grace, at this very minute!" repeated Georges, "and make the sign of the cross, or, by St. George —"

Sedanka and her sister-in-law leaned toward the old man and beseeched him to say grace.

"No, no, no," said the old man, "he has no right to give me orders, and if he insists, I will curse him!"

Georges rose and ran into the house. Soon he returned, fury in his eyes.

"Where is the stake?" he cried. "Where have you hidden the stake?"

Sdenka and Pierre exchanged a glance.

"Corpse!" Georges then shouted at the old man, "What have you done with my oldest son? Why have you killed my child? Give me back my son, you cadaver!"

In speaking thus, he grew more and more pale, his eyes even wilder.

The old man gave him an ugly look and didn't move.

"Oh, the stake, the stake!" cried Georges. "May whoever hid it answer for the evils that await us!"

At that moment we heard the joyous bursts of laughter of the youngest child and we saw him arrive riding the large stake like a hobby horse, rearing and prancing while shouting in his little voice the war cry of the Serbs.

At this sight, Georges's eyes flashed. He tore the stake from the child and hurled himself at his father. Gorcha let out a yell and began to run in the direction of the woods with a speed so out of keeping with his age that it appeared supernatural.

Georges followed him across the fields and soon we lost sight of them.

Sun had fallen by the time Georges came back to the house, pale as a dead man, his hair standing on end. He sat down near the fire and I thought I heard his teeth chatter. No one dared question him. Toward the hour when the family usually took their leave of one another for the night, he seemed to recover some of his energy and, taking me aside, told me in the most natural way:

"My dear guest, I just saw the river. There are no more ice drifts, the road is clear, nothing prevents your departure. There is no point," he added, glancing at Sdenka, "in formal good byes. My family, through me, wishes you all the earthly happiness one could desire, and I hope that you, too, will remember us fondly. Tomorrow, at first light, you will find your horse saddled and your guide ready to serve. Adieu. Think from time to time of

(The following tale shares some features with "The Soldier and the Vampire," and includes the classic dispatching of a vampire by wooden stake.)

The Coffin Lid

A moujik was driving along one night with a load of pots. His horse grew tired, and all of a sudden it came to a stand-still alongside of a graveyard. The moujik unharnessed his horse and set it free to graze; meanwhile he laid himself down on one of the graves. But somehow he didn't go to sleep.

He remained lying there some time. Suddenly the grave began to open beneath him: he felt the movement and sprang to his feet. The grave opened, and out of it came a corpse — wrapped in a white shroud, and holding a coffin-lid — came out and ran to the church, laid the coffin-lid at the door, and then set off for the village.

The moujik was a daring fellow. He picked up the coffin-lid and remained standing beside his cart, waiting to see what would happen. After a short delay the dead man came back, and was going to snatch up his coffin-lid — but it was not to be seen. Then the corpse began to track it out, traced it up to the moujik, and said:

"Give me my lid: if you don't, I'll tear you to bits!"

your host and forgive him if your stay here was not as free of tribulations as he would have wished."

Georges's features had, at that moment, almost a friendly expression. He led me to my room and shook my hand a last time. Then he shuddered and his teeth chattered as if he was shivering from the cold.

Alone, I could not even dream of going to bed, as you can well imagine. Other ideas preoccupied me. I had loved several times in my life. I had experienced passion, jealousy, but never, even when I fled the Duchess of Gramont, I never felt a sadness to equal to that which then tore at my heart. Before the sun was gone, I put on my travel clothing, and hoped to have one last meeting with Sdenka. But Georges was waiting for me in the hallway. Any possibility of a meeting was dashed.

I mounted my horse and spurred it on. I made a promise to myself to pass through this village upon my return from Jassy. Even though this might be some time in the future, the idea eased my sadness a bit. I was already imagining, with satisfaction, all the details in advance, when the horse suddenly lurched, and I almost lost my hold on the pommel. The animal stopped completely, dug in its legs and gave a snort of alarm, as if danger was near. I looked around warily, and saw, about a hundred feet ahead, a wolf digging in the earth. Hearing the horse, the wolf quickly fled. I spurred the horse again and got him to continue forward. At the place the wolf had been was a freshly-dug pit. I thought I also saw the end of a wooden stake protruding a few inches from the ground the wolf had been pawing. However, I cannot confirm this point, since I rode by very quickly.

Here, the marquis stopped his story and took a pinch of snuff.

"Is that all?" asked the ladies.

"Sadly, no," replied d'Urfé. "That which I have left to tell you is a very painful, and I would give much to rid myself of the memory."

He continued:

The affairs that engaged me in Jassy kept me there much longer than I had intended, about six months. What can I tell you? It is sad but true that there are few lasting emotions in this life. The success of my negotiations, the praise I received from the cabinet of Versailles — in a word, politics, nasty politics, of which we have become so very bored lately — did nothing to diminish my memory of Sdenka. Then, the wife of a *hospodar*, a beautiful woman who spoke French perfectly, honored me upon my arrival by singling me out from all the other young foreigners in Jassy. I had been raised according to the principles of French gallantry, and my Gallic blood would have revolted at the idea of repaying the beautiful woman's kind attentions with ingratitude. So I responded courteously to her advances, and, to enable me to press forward the interests and rights of France, you might say I began by identifying myself with the interests and rights of her husband.

Returning home, I followed the same road I had taken to Jassy.

"And my hatchet, how about that?" answers the moujik. "Why, it's I who'll be chopping you into small pieces!"

"Do give it back to me, good man!" begs the corpse.

"I'll give it when you tell me where you've been and what you've done."

"Well, I've been in the village, and there I've killed a couple of youngsters."

"Well then, now tell me how they can be brought back to life."

The corpse reluctantly made answer:

"Cut off the left skirt of my shroud, and take it with you. When you come into the house where the youngsters were killed, pour some live coals into a pot and put the piece of the shroud in with them, and then lock the door. The lads will be revived by the smoke immediately."

The moujik cut off the left skirt of the shroud, and gave up the coffin-lid. The corpse went to its grave — the grave opened. But just as the dead man was descending into it, all of a sudden the cocks began to crow, and it hadn't time to get properly covered over. One end of the coffin-lid remained sticking out of the ground.

The moujik saw all this and made a note of it. The day began to

I didn't think at all about Sdenka, nor her family, until one night, as I rode through the countryside, I head a bell sounding the hour of eight. It sounded familiar to me, and my guide told me it came from a monastery nearby. I asked the name of the monastery, and learned that it was Our Lady of the Oak. I raced ahead on my horse and soon was knocking at the monastery door. The hermit caretaker opened the door, welcomed us and showed us the guest quarters, which I found to be too crowded. I asked if I could find a place to stay in the village.

"You will find more than one," he replied with a deep sigh. "Thanks to that wretched Gorcha, there's no lack of empty houses!"

"What are you saying?" I asked. "Is old Gorcha still alive?"

"Oh, no, he is surely buried, with a stake in his heart! But he sucked the blood of Georges's son. The child returned one night, crying at the door that he was cold and wanted to come inside. His stupid mother, even though she had buried him herself, didn't have the courage to send him back to the cemetery, and opened the door for him. Then he threw himself at her, and sucked her to death. Buried in her turn, she came back to suck the blood of her second son, and then that of her husband, and then that of her brother-in-law. All of them went that way."

"And Sdenka?"

"Oh, she went mad from grief. Poor child, don't talk to me about it any more!"

The hermit had not fully answered my question, but I did not have the courage to ask again.

"Vampirism is contagious," the hermit continued, crossing himself. "Many of the families in the village have suffered from it, many families are dead to the last member, and if you want my advice, stay here at the monastery tonight, for even if you are not devoured by the *vourdalaks*, it is still possible that the fear that they will inspire in you will suffice to turn your hair white before I have finished ringing the morning bells. I am only a poor monk," he continued, "but the generosity of travelers has enabled me to accommodate their needs. I have exquisite cheeses, raisins that will make your mouth water just by looking at them, and some bottles of Tokay that is the equal of any wine served to His Holiness."

It appeared to me that the holy man was turning into an innkeeper. I thought that he had purposely told a tall tale to induce me to please heaven by imitating the other travelers' financial generosity.

But the word "fear" has always had an effect on me like that of a bugle on a war horse. I would have been ashamed of myelf if I had not set out immediately. My trembling guide asked my permission to stay, which I gladly gave.

It took about half an hour for me to reach the village. I found it deserted. Not a single light shone through any window, not a single song could be heard. I passed in silence in front of all of these houses, most of which were known to me, until I finally arrived at Georges's. Due either to

dawn; he harnessed his horse and drove into the village. In one of the houses he heard cries and wailing. In he went — there lay two dead lads.

"Don't cry," says he, "I can bring them to life!"

"Do bring them to life, kinsman," say their relatives. "We'll give you half of all we possess."

The moujik did everything as the corpse had instructed him, and the lads came back to life. Their relatives were delighted, but they immediately seized the moujik and bound him with cords, saying:

"No, no, trickster! We'll hand you over to the authorities. Since you knew how to bring them back to life, maybe it was you who killed them!"

"What are you thinking about, true believers! Have the fear of God before your eyes!" cried the moujik.

Then he told them everything that had happened to him during the night. Well, they spread the news through the village; the whole population assembled and swarmed into the graveyard. They found out the grave from which the dead man had come out, they tore it open, and they drove an aspen stake right into the heart of the corpse, so it might no more rise up and slay. But they rewarded the moujik richly, and sent him away home with great honour.

sentimental memory, or youthful temerity, it was there I decided to stay the night.

I got down from my horse and knocked at the carriage gate. No one responded. I pushed on the gate, which groaned on its hinges as it opened, and entered the yard.

I tethered my horse, fully saddled, in an outdoor structure where I found a provision of oats sufficient for one night, and walked resolutely toward the house.

None of the doors was locked, although the house appeared to be completely uninhabited. Sdenka's room seemed to have been abandoned only the night before. Some clothes were strewn across the bed. Several jewelry pieces I had given her, among which I recognized the little enamel cross I had bought when passing through Pesth, sparkled on a table in the moonlight. I could not help feeling a pang in my heart, even though my love for her was no more. Nevertheless I wrapped myself in my coat and streched out on the bed. Soon sleep overtook me. I do not recall the details of my dream, but I know that I saw Sdenka once again, as beautiful, naive and loving as she was in the past. Seeing her, I reproached myself for my egotism and my inconstancy. How could I have abandoned this poor child who loved me, I wondered, how could I have forgotten her? Then my memory of her blended with that of the Duchess of Gramont and I saw in these two images only one and the same person. I threw myself at Sdenka's feet and implored her forgiveness. My entire being, my entire soul were infused with an ineffable feeling of melancholy and happiness.

At that point in my dream, I was half-awoken by a melodious sound, similar to the rustling of a wheat field animated by a light breeze. It seemed to me that I heard the stalks melodiously brushing against each other, and the singing of birds mix with the rushing sound of a waterfall and the whispering of the trees. Then I realized that this confusion of sounds was only the rustle of a woman's dress. I opened my eyes and saw Sdenka next to my bed. The moon shone so intensely that I could see in the finest detail her adorable features, which had been so dear to me in the past, a treasure my dream had just made me realize anew. I found Sdenka more beautiful and more womanly than ever. She wore the same nightclothes as the last time I saw her alone: a simple silk blouse embroidered in gold, with a skirt tightly cinched above her hips.

"Sdenka!" I said, sitting up. "Is it truly you, Sdenka?"

"Yes, it is me," she replied in a soft, sad voice, "your Sdenka, whom you have forgotten. Ah, why didn't you return sooner? Everything is finished now, you must leave; one moment more and you are lost! Adieu, my friend, adieu forever!"

"Sdenka," I said, "I have been told that you have suffered many misfortunes. Come, we will talk and that will ease your pain."

"Oh, my friend," she said, "you shouldn't believe everything that is said about us. But leave, leave as fast as you can, because, if you stay here,

ANOTHER TOLSTOY VAMPIRE TALE

Alexis Tolstoy's first published story, "Vampires" (1841) might be best considered as a half-successful finger-exercise for "The Family of the Vourdalak." Part supernatural thriller and part social satire, the story still makes a fascinating read, even if its eclectic youthful ambitions are not quite achieved in its execution.

In the following excerpt from its only published English translation, by Fedor Nikanov (included in *Vampires: Stories of the Supernatural*; New York: Hawthorn Books, 1969), Tolstoy casts a satiric eye at the brazen society *oupyrs* who present themselves openly, even before those who already know they are dead:

One character points out another at a society ball. "You see? She was Madame Sugobrina many years ago, but now she's nothing but a vampire waiting for the opportunity to glut herself on human blood. Look how she glances at her granddaughter. Listen to what she says. She's flattering and urging her to visit the estate. I guarantee you that in less than three days the poor young thing will be dead. Doctors will call it fever and pneumonia. Don't believe them."

"Their impertinence is astounding," he says. "But you asked how I recognized *oupyrs*.

your ruin is certain."

"But Sdenka, what is this danger that threatens me? Can't you give me one hour, just one hour to talk with you?"

Sdenka shivered, a strange transformation overtook her entire being.

"Yes," she said, "an hour, yes, like when I was singing the ballad of the old king and you came into this room? Is that what you mean? Well, so be it, I will give you an hour."

But then, recovering herself, she said, "No, no, leave, go! Quickly, I tell you, flee! Flee while you still can!"

A savage energy animated her features.

I did not understand the reason she talked like this, but she was so beautiful that I resolved to stay, no matter what. Finally, at my insistance, she sat next to me, and spoke to me of times past, admitting while blushing, that she had loved me from the day of my arrival. Nonetheless, little by little, I noticed an overwhelming change in Sdenka. Her former reserve had given way to a strange informality. Her look, previously so timid, now had something brazen about it. Finally, I saw with surprise that her manner with me was far from the modesty that formerly distinguished it.

Notice that upon meeting one another they click their tongues. Actually, it's not a click so much as the sound one makes when sucking an orange. Such is the clandestine signal by which they greet and identify one another ..."

Alexis Tolstoy

Would it be possible, I wondered, that Sdenka was not the pure and innocent young girl that she had once seemed to be? Had she only feigned chastity for fear of her brother? Had I been grossly deceived by her counterfeit virtue? But then, why would she implore me to leave? Could it be the refined stratagem of a coquette? And I thought I knew her! But no matter. If Sdenka is not the Diana I thought her to be, I can still compare her to another divinity, no less lovable, and, by God, I prefer the role of Adonis to that of Acteon!

If this classical reference, which I addressed to myself, seems out of place to you, mesdames, please remember that the story that I have the honor of telling you took place in the year of our Lord 1758. Mythology then was much in fashion, and I did not fancy myself to be ahead of my time. Much has changed since then and it is not so long ago that the Revolution, which simultaneously overthrew the vestiges of paganism and Christianity, put the goddess Reason in their place. This goddess, mesdames, has never been my patron when in the presence of beautiful women, and, at the time of which I speak, I was even less disposed than ever to offer sacrifices in her name. I abandoned myself to the inclination that led me toward Sdenka and I delighted in her new, wanton manner. After some time had passed in a sweet intimacy, I amused myself by decorating Sdenka with her jewels and wished to place around her neck the little enamel cross that I had found on the table. Shivering, Sdenka recoiled from my gesture.

"Enough of this childishness, my friend," she said. "Leave these baubles — let's talk about you, and what you want!"

Sdenka's troubled demeanor caused me to think. Upon a closer look, I noted that she no longer wore around her neck, as she had before, a necklace of little icons, religious relics, and sachets of incense which the Serbians are typically given in childhood, and which they do not remove until death.

"Sdenka," I said. "Where are those icons which you wore around your neck?"

"I have lost them," she replied, with an air of impatience, quickly changing the topic of conversation.

I don't know what vague premonition overtook me. I wanted to leave, but Sdenka restrained me back.

"What? You asked me for an hour, and now you want to leave after the passing of a few minutes!"

"Sdenka," I said, "you were right to tell me to leave. I think I heard a noise, and I fear someone will discover us!"

"Be calm, my friend, everyone is asleep except us. Only the cricket in the grass and the mayfly in the air can hear what I have to tell you!"

"No, no, Sdenka, it is necessary for me to leave!"

"Stop, stop," said Sdenka. "I love you more than my soul, more than my salvation, you told me that your life and your blood were mine..."

"But your brother, your brother, Sdenka, I have a premonition that he will find us!"

"Don't worry, my soul, my brother is drowsy from the wind playing in the trees; his sleep is very heavy, the night is very long, and I am only asking for one hour!"

In saying this, Sdenka was so beautiful, that the vague terror that was gripping me began to give way to the desire to remain near her. A mixture of fear and lust impossible to describe filled my entire being. As I weakened, Sdenka became more affectionate, to the point that I decided to give in, even as I promised myself to keep on my guard. Nonetheless, as I said a few moments ago, I have never done entirely what I should, and when Sdenka, noticing my reserve, suggested that we chase the cold of the night with a few generous glasses of wine that she told me came from the good hermit, I accepted her proposal with an urgency that made her smile. The wine produced its effect. Upon the second glass, the ugly impression that the episode with the cross and the icons had made on me was completely erased; I found Sdenka irresistible in the disorder of her toilette, with her beautiful, half-braided hair and moonlit jewels. I contained myself no longer and took her in my arms.

Then, mesdames, occurred one of those mysterious revelations that I will never be able to explain, but whose existence experience has forced me to believe in, even though, up to that point, I had been little inclined to accept.

The force with which I wrapped my arms around Sdenka caused one of the points of the cross that you just saw and which the Duchess of Gramont had given me upon my departure, to prick my chest. The sharp pain that I experienced was like a ray of light that penetrated my entire being. I looked at Sdenka and I saw that her features, while still beautiful, had shrunken in death, that her eyes did not see and that her smile was an

VOURDALAK CINEMA

"The Family of the Vourdalak" was effectively adapted for the screen as part of Mario Bava's stylish anthology film *I tre volti della paura (The Three Faces of Fear*, 1963, released in the United States the following year as *Black Sabbath*).

As Gorcha, Boris Karloff essayed the only vampire role of his career in a performance that is generally regarded as one of the best of his final years. Ironically, the actor had been slated to play Dracula in a proposed 1958 adaptation of Stoker's novel that ultimately failed to materialize. It is not difficult to imagine glimpses of this "lost" Dracula in Karloff's nuanced portrayal of an ancient, Slavic blood-drinker, even though Gorcha is a peasant rather than an aristocrat.

The screenplay (called "The Wurdalak") takes only minimal liberties with Tolstoy's original story, eliminating the framing narration and, in a twist, making the character of d'Urfé a vampire at the end. This was a reversal of the filmmaker's original intention; production notes reveal a regrettably "happy" ending in which d'Urfé saves Sdenka from the family curse.

A second, modernized "The Family of the Vourdalak" was lensed by Italian filmmaker in 1971, and released under the title *Night of the Devils*.

agonized convulsion imprinted on the face of a cadaver. At the same time, I sensed in the room the nauseating odor ordinarily emitted by poorly-sealed tombs. The awful truth rose up before me in all of its ugliness, and I remembered, too late, the hermit's warnings. I understood how precarious my position was, and I felt that everything depended on my courage and clearheadedness. I turned away from Sdenka to hide from her the horror that my features must have been expressing. Then my gaze fell on the window and I saw the infamous Gorcha, leaning on a bloody wooden stake and fixing upon me the eyes of a hyena. The other window was taken by the pale visage of Georges, who, at this moment had a frightening resemblance to his father. Both seemed to spy on my movements and I did not doubt that they would pounce on me if I made the least effort to flee. Therefore, I pretended not to see them, but with a violent internal effort, I continued — yes, mesdames — I continued to lavish upon Sdenka the same caresses that I enjoyed giving her before my terrible discovery. All the while, I pondered with anguish the means of my escape. I noticed that Gorcha and Georges exchanged knowing glances with Sdenka and that they were growing impatient. I also heard outside a woman's voice and the cries of children, but sounds so awful that one would have taken them for the howls of wild cats.

It was time to get out, I told myself, and the sooner the better.

I told Sdenka, in a voice loud enough to be heard by her hideous family, "I am very tired, my child, I would like to go to bed and sleep for a few hours, but first I have to see if my horse has eaten his provisions. Please do not leave, but wait for my return."

I pressed a kiss against her cold and discolored lips, and left. I found my horse foaming at the mouth and struggling in the tether-shed. He had not touched the oats, but his whinnying upon seeing me gave me gooseflesh, because I was afraid that he would betray my intentions. Meanwhile, the vampires, whom had probably heard my conversation with Sdenka, did not suspect they might have cause for alarm. I made certain that the gate was open, and, jumping on the saddle, I dug my spurs into my horse's flanks.

Advertising art for Black Sabbath.

I had the time to notice, in passing through the gate, that outside the house a mob had assembled, comprised for the most part of individuals pressing their faces against the windows. I think that my sudden exit baffled them at first, because for a certain time I discerned only the steady gallop of my horse in the silence of the night. I was already thinking about congratulating myself for my ruse, when suddenly I heard a noise behind me, similar to a windstorm breaking in the mountains. A thousand confused voices shrieked, howled and seemed to be fighting among themselves. Then all the voices went silent, and it sounded as if a troop of soldiers was rapidly approaching on foot.

I pressed my mount hard enough to rip his flanks. My blood pounded with a burning fever, and, while I was exhausting myself in an unheard-of effort to keep my wits, I heard behind me a voice that cried:

"Stop, stop, my friend! I love you more than my soul, I love you more than my salvation! Stop, stop, your blood is mine!"

At the same time, I felt a cold breath in my ear and I knew that Sdenka had leapt up behind me and rode on the horse's rump.

"My heart, my soul!" she said to me, "I see only you, I feel only you, I am not the mistress of myself, I am obeying a superior force, forgive me, my friend, forgive me!"

And wrapping me in her arms, she tried to pull me back and bite me in the throat. A terrible struggle ensued between us. For a long time I defended myself with difficulty, but finally I managed to seize Sdenka with one hand at her belt, and the other on her hair, and standing in the stirrups, I threw her to the ground!

Immediately my strength abandoned me and delirium took hold. I was pursued by a thousand terrible, mad, grimacing images. First Georges and his brother Pierre ran alongside the road and tried to block my path. They didn't succeed, and I was about to rejoice in their failure when I turned and saw old Gorcha, who was using his stake to vault himself forward, rather in the manner of Tyrolean mountain men traversing a ravine. But Gorcha, too, fell behind. Then his daughter-in-law, who was dragging her children behind her, threw one of them to the old man, who caught the child on the end of his stake. Employing it like a catapult, he threw the child after me with all his might. I avoided the hit, but, with the true instinct of a bull-dog, the odious brat attached himself to my horse's neck, and I had difficulty pulling him off. The other child was delivered to me in the same fashion, but he fell in front of the horse and was crushed. I do not know any more of what I saw, but when I regained my senses, it was daylight and found myself lying on the road next to my dying horse.

Thus ends, mesdames, a little love affair which should have cured me forever of the desire to seek out new ones. Several contemporaries of your grandmothers would be able to tell you if I was wiser in the future.

Regardless, I still tremble at the idea that if I had succumbed to my enemies, I would have become a vampire in my turn. But heaven did not permit things to come to this point, and, far from thirsting for your blood, mesdames, I do not ask for better, as old as I am, than to spill my own in your service.

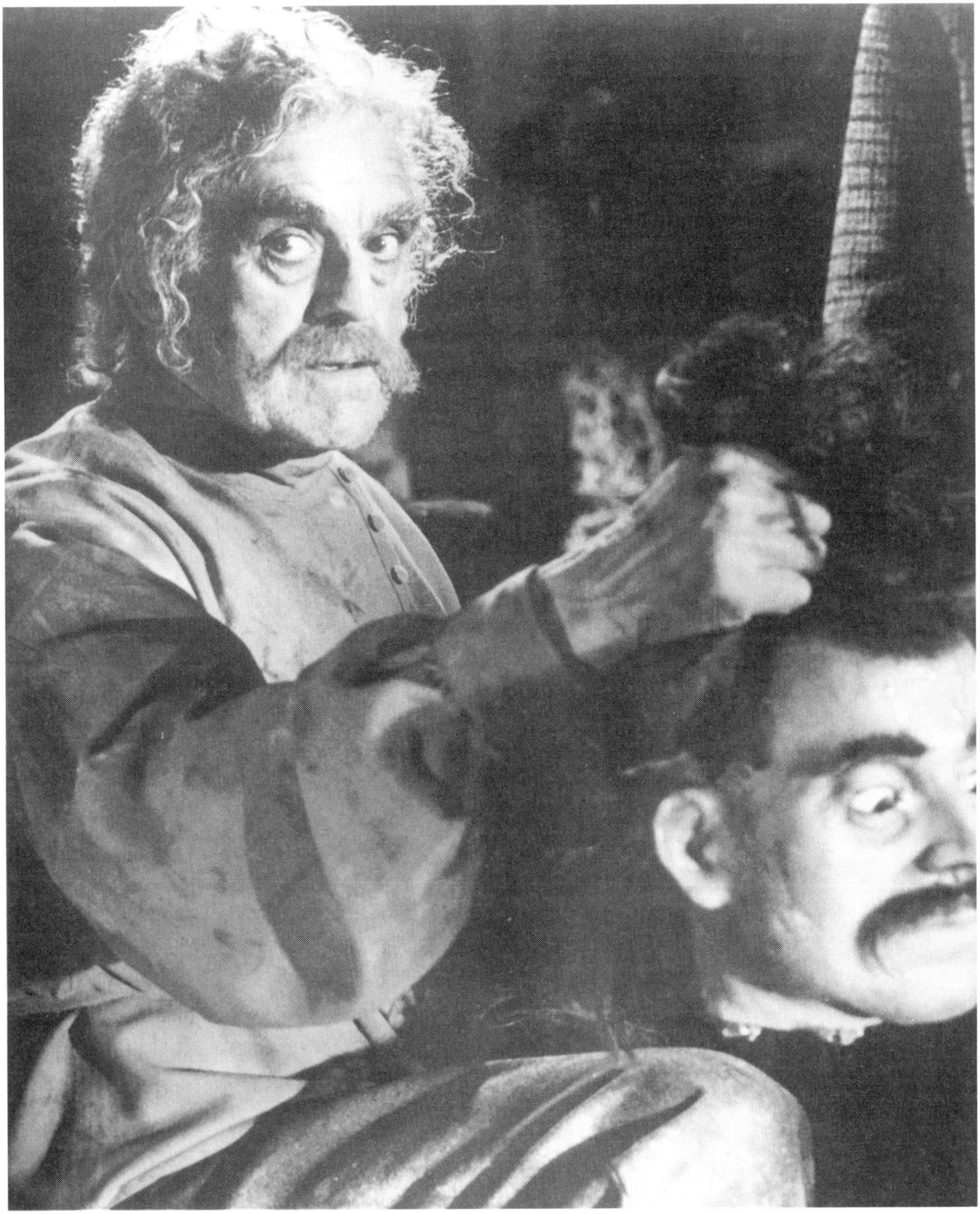

Black Sabbath*: Boris Karloff as Gorcha with the head of Alibek, the Turkish brigand.*

Anonymous woodcut illustration for Varney the Vampire *(1847).*

Varney the Vampyre
Or, The Feast of Blood

JAMES MALCOLM RYMER
(1847)

CHAPTER I.

— "How graves give up their dead,
And how the night air hideous grows
With shrieks!"

MIDNIGHT. — THE HAIL-STORM.—
THE DREADFUL VISITOR.— THE VAMPYRE.

THE SOLEMN TONES of an old cathedral clock have announced midnight — the air is thick and heavy — a strange, death-like stillness pervades all nature. Like the ominous calm which precedes some more than usually terrific outbreak of the elements, they seem to have paused even in their ordinary fluctuations, to gather a terrific strength for the great effort. A faint peal of thunder now comes from far off. Like a signal gun for the battle of the winds to begin, it appeared to awaken them from their lethargy, and one awful, warring hurricane swept over a whole city, producing more devastation in the four or five minutes it lasted, than would a half century of ordinary phenomena.

It was as if some giant had blown upon some toy town, and scattered many of the buildings before the hot blast of his terrific breath; for as suddenly as that blast of wind had come did it cease, and all was as still and calm as before.

Sleepers awakened, and thought that what they had heard must be the confused chimera of a dream. They trembled and turned to sleep again.

All is still — still as the very grave. Not a sound breaks the magic of repose. What is that — a strange pattering noise, as of a million fairy feet? It is hail — yes, a hail-storm has burst over the city. Leaves are dashed from the trees, mingled with small boughs; windows that lie most opposed to the direct fury of the pelting particles of ice are broken, and the rapt repose that before was so remarkable in its intensity, is exchanged for a noise which, in its accumulation, drowns every cry of surprise or consternation which here and there arose from persons who found their houses invaded by the storm.

Now and then, too, there would come a sudden gust of wind that in its strength, as it blew laterally, would, for a moment, hold millions of the

ABOUT THE AUTHOR

***Varney the Vampyre* was long thought to be the work of Victorian hack writer Thomas Preskett Prest (1810-1859), but modern scholarship has fairly well proved that the real culprit was James Malcolm Rymer (1804-1884), a Scottish-born writer and editor who also toiled in the trenches of penny-dreadful fiction. Penny-dreadfuls were serial novels published in cheap installments; *Varney* originally appeared between 1845 and 1847, and was a tremendous popular success, stabilizing the conventions of literary vampirism for all future horror writers.**

hailstones suspended in mid air, but it was only to dash them with redoubled force in some new direction, where more mischief was to be done.

Oh, how the storm raged! Hail — rain — wind. It was, in very truth, an awful night.

There was an antique chamber in an ancient house. Curious and quaint carvings adorn the walls, and the large chimneypiece is a curiosity of itself. The ceiling is low, and a large bay window, from roof to floor, looks to the west. The window is latticed, and filled with curiously painted glass and rich stained pieces, which send in a strange, yet beautiful light, when sun or moon shines into the apartment. There is but one portrait in that room, although the walls seem paneled for the express purpose of containing a series of pictures. That portrait is of a young man, with a pale face, a stately brow, and a strange expression about the eyes, which no one cared to look on twice.

There is a stately bed in that chamber, of carved walnut-wood is it made, rich in design and elaborate in execution; one of those works which owe their existence to the Elizabethan era. It is hung with heavy silken and damask furnishing; nodding feathers are at its corners — covered with dust are they, and they lend a funereal aspect to the room. The floor is of polished oak.

God! How the hail dashes on the old bay window! Like an occasional discharge of mimic musketry, it comes clashing, beating, and cracking upon the small panes; but they resist it — their small size saves them; the wind, the hail, the rain, expend their fury in vain.

The bed in that old chamber is occupied. A creature formed in all fashions of loveliness lies in a half sleep upon that ancient couch — a girl young and beautiful as a spring morning. Her long hair has escaped from its confinement and streams over the blackened coverings of the bedstead; she has been restless in her sleep, for the clothing of the bed is in much confusion. One arm is over her head, the other hangs nearly off the side of the bed near to which she lies. A neck and bosom that would have formed a study for the rarest sculptor that ever Providence gave genius to, were half disclosed. She moaned slightly in her sleep, and once or twice the lips moved as if in prayer — at least one might judge so, for the name of Him who suffered for all came once faintly from them.

She had endured much fatigue, and the storm does not awaken her; but it can disturb the slumbers it does not possess the power to destroy entirely. The turmoil of the elements wakes the senses, although it cannot entirely break the repose they have lapsed into.

Oh, what a world of witchery was in that mouth, slightly parted, and exhibiting within the pearly teeth that glistened even in the faint light that came from that bay window. How sweetly the long silken eyelashes lay upon the cheek. Now she moves, and one shoulder is entirely visible — whiter, fairer than the spotless clothing of the bed on which she lies, is the smooth skin of that fair creature, just budding into womanhood, and in that transition state which presents to us all the charms of the girl — almost of the child, with the more matured beauty and gentleness of advancing years.

Was that lightning? Yes — an awful, vivid, terrifying flash — then a roaring peal of thunder, as if a thousand mountains were rolling one over the other in the blue vault of Heaven! Who sleeps now in that ancient city? Not one living soul. The dread trumpet of eternity could not more effectually have awakened any one.

The hail continues. The wind continues. The uproar of the elements seems at its height. Now she awakens — that beautiful girl on the antique bed; she opens those eyes of celestial blue, and a faint cry of alarm bursts from her lips. At least it is a cry which, amid the noise and turmoil without, sounds but faint and weak. She sits upon the bed and presses her hands upon her eyes. Heavens! What a wild torrent of wind, and rain, and hail! The thunder likewise seems intent upon awakening sufficient echoes to last until the next flash of forked lightning should again produce the wild concussion of the air. She murmurs a prayer — a prayer for those she loves best; the names of those dear to her gentle heart come from her lips; she weeps and prays; she thinks then of what devastation the storm must surely produce, and to the great God of Heaven she prays for all living things. Another flash — a wild, blue, bewildering flash of lightning streams across that bay window, for an instant bringing out every colour in it with terrible distinctness. A shriek bursts from the lips of the young girl, and then, with eyes fixed upon that window, which, in another moment, is all darkness, and with such an expression of terror upon her face as it had never before known, she trembled, and the perspiration of intense fear stood upon her brow.

"What— what was it?" she gasped, "real or delusion? Oh, God, what was it? A figure tall and gaunt, endeavouring from the outside to unclasp the window. I saw it. That flash of lightning revealed it to me. It stood the whole length of the window."

There was a lull of the wind. The hail was not falling so thickly — moreover, it now fell, what there was of it, straight, and yet a strange clattering sound came upon the glass of that long window. It could not be a delusion — she is awake, and she hears it. What can produce it? Another flash of lightning — another shriek — there could be now no delusion.

A tall figure is standing on the ledge immediately outside the long window. It is its finger-nails upon the glass that produces the sound so like the hail, now that the hail has ceased. Intense fear paralysed the limbs of the beautiful girl. That one shriek is all she can utter — with hand clasped, a face of marble, a heart beating so wildly in her bosom, that each moment it seems as if it would break its confines, eyes distended and fixed upon the window, she waits, froze with horror. The pattering and clattering of the nails continue. No word is spoken, and now she fancies she can trace the darker form of that figure against the window, and she can see the long

THE END OF LORD VARNEY

Although it is not possible in the space available to recount the full adventures of Varney the Vampyre, suffice it to say that Varney was in many ways the antecedent of television's Barnabas Collins, a vampire who grew to regret his curse over endless installments. Following is the denouement of Varney's saga, after more than 800 melodramatically attenuated chapters.

The manuscript which the clergyman had read with so much interest, here abruptly terminated. He was left to conclude that Varney after that had been resuscitated; and he was more perplexed than ever to come to any opinion concerning the truth of the narration which he had now concluded.

It was one week after he had finished the perusal of Varney's papers that the clergyman read in an English newspaper the following statement.

"We extract from the *Algemeine Zeitung* the following most curious story, the accuracy of which of course we cannot vouch for, but still there is a sufficient air of probability about it to induce us to present it to our readers.

arms moving to and fro, feeling for some mode of entrance. What strange light is that which now gradually creeps up into the air? Red and terrible —brighter and brighter it grows. The lightning has set fire to a mill, and the reflection of the rapidly consuming building falls upon that long window. There can be no mistake. The figure is there, still feeling for an entrance, and clattering against the glass with its long nails, that appear as if the growth of many years had been untouched. She tries to scream again but a choking sensation comes over her, and she cannot. It is too dreadful — she tries to move — each limb seems weighted down by tons of lead — she can but in a hoarse faint whisper cry, —

"Help— help— help— help!"

And that one word she repeats like a person in a dream. The red glare of the fire continues. It throws up the tall gaunt figure in hideous relief against the long window. It shows, too, upon the one portrait that is in the chamber, and the portrait appears to fix its eyes upon the attempting intruder, while the flickering light from the fire makes it look fearfully lifelike. A small pane of glass is broken, and the form from without introduces a long gaunt hand, which seems utterly destitute of flesh. The fastening is removed, and one-half of the window, which opens like folding doors, is swung wide open upon its hinges.

And yet now she could not scream — she could not move. "Help! — help! — help!" was all she could say. But, oh, that look of terror that sat upon her face, it was dreadful — a look to haunt the memory for a lifetime — a look to obtrude itself upon the happiest moments, and turn them to bitterness.

The figure turns half round, and the light falls upon its face. It is perfectly white — perfectly bloodless. The eyes look like polished tin; the lips are drawn back, and the principal feature next to those dreadful eyes is the teeth — the fearful looking teeth — projecting like those of some wild animal, hideously, glaringly white, and fang-like. It approaches the bed with a strange, gliding movement. It clashes together the long nails that literally appear to hang from the finger ends. No sound comes from its lips. Is she going mad — that young and beautiful girl exposed to so much terror? She has drawn up all her limbs; she cannot even now say help. The power of articulation is gone, but the power of movement has returned to her; she can draw herself slowly along to the other side of the bed from that towards which the hideous appearance is coming.

But her eyes are fascinated. The glance of a serpent could not have produced a greater effect upon her than did the fixed gaze of those awful, metallic-looking eyes that were bent down on her face. Crouching down so that the gigantic height was lost, and the horrible, protruding white face was the most prominent object, came on the figure. What was it? — What did it want there? — What made it look so hideous — So unlike an inhabitant of the earth, and yet be on it?

Now she has got to the verge of the bed, and the figure pauses. It

"Late in the evening, about four days since, a tall and melancholy-looking stranger arrived, and put up at one of the principal hotels at Naples. He was a most peculiar looking man, and considered by the persons of the establishment as about the ugliest guest they had ever had within the walls of their place.

"In short time he summoned the landlord, and the following conversation ensued between him and the strange guest.

"'I want,' said the stranger, 'to see all the curiosities of Naples, and among the rest Mount Vesuvius. Is there any difficulty?'

"'None,' replied the landlord, 'with a proper guide.'

"A guide was secured, who set out with the adventurous Englishman to make the ascent of the burning mountain.

"They went on then until the guide did not think it quite prudent to go any further, as there was a great fissure in the side of the mountain, out of which a stream of lava was slowly issuing and spreading itself in rather an alarming manner.

"The ugly Englishman, however, pointed to a secure mode of getting higher still, and they proceeded until they were very near the edge of the crater itself.

seemed as if when it paused she lost the power to proceed. The clothing of the bed was now clutched in her hands with unconscious power. She drew her breath short and thick. Her bosom heaves, and her limbs tremble, yet she cannot withdraw her eyes from that marble-looking face. He holds her with his glittering eye.

The storm has ceased — all is still. The winds are hushed; the church clock proclaims the hour of one: a hissing sound comes from the throat of the hideous being, and he raises his long, gaunt arms — the lips move. He advances. The girl places one small foot on to the floor. She is unconsciously dragging the clothing with her. The door of the room is in that direction — can she reach it? Has she power to walk? — can she withdraw her eyes from the face of the intruder, and so break the hideous charm? God of Heaven! Is it real, or some dream so like reality as to nearly overturn judgment forever?

The figure has paused again, and half on the bed and half out of it that young girl lies trembling. Her long hair streams across the entire width of the bed. As she has slowly moved along she has left it streaming across the pillows. The pause lasted about a minute — oh, what an age of agony. That minute was, indeed, enough for madness to do its full work in.

With a sudden rush that could not be foreseen — with a strange howling cry that was enough to awaken terror in every breast, the figure seized the long tresses of her hair, and twining them round his bony hands he held her to the bed. Then she screamed — Heaven granted her then power to scream. Shriek followed shriek in rapid succession. The bed-clothes fell in a heap by the side of the bed — she was dragged by her long silken hair completely on to it again. Her beautifully rounded limbs quivered with the agony of her soul. The glassy, horrible eyes of the figure ran over that angelic form with a hideous satisfaction — horrible profanation. He drags her head to the bed's edge. He forces it back by the long hair still entwined in his grasp. With a plunge he seizes her neck in his fang-like teeth — a gush of blood, and a hideous sucking noise follows. *The girl has swooned, and the vampyre is at his hideous repast!*

The stranger then took his purse from his pocket and flung it at the guide saying, —

" 'You can keep that for your pains, and for coming into some danger with me. But the fact was, that I wanted a witness to an act which I have set my mind upon performing.'

"The guide says that these words were spoken with so much calmness, that he verily believed the act mentioned as about to be done was some scientific experiment of which he knew the English were very fond, and he replied, —

" 'Sir, I am only too proud to serve so generous and so distinguished a gentleman. In what way can I be useful?'

" 'You will make what haste you can,' said the stranger, 'from the mountain, inasmuch as it is covered with sulfurous vapours, inimical to human life, and when you reach the city you will cause to be published an account of my proceedings, and what I say. You will say that you accompanied Varney the Vampyre to the crater of Mount Vesuvius, and that, tired and disgusted with a life of horror, he flung himself in to prevent the possibility of a reanimation of his remains.' "

Before then the guide could utter anything but a shriek, Varney took one tremendous leap, and disappeared into the burning mouth of the mountain.

The best-known screen adaptation of Le Fanu's "Carmilla": The Vampire Lovers *(1970).*
Pictured: Peter Cushing and Ingrid Pitt. (Photofest)

Carmilla

J. SHERIDAN LE FANU
(1872)

Prologue

UPON A PAPER attached to the Narrative which follows, Doctor Hesselius has written a rather elaborate note, which he accompanies with a reference to his Essay on the strange subject which the MS. illuminates.

This mysterious subject he treats, in that Essay, with his usual learning and acumen, and with remarkable directness and condensation. It will form but one volume of the series of that extraordinary man's collected papers.

As I publish the case, in this volume, simply to interest the "laity," I shall forestall the intelligent lady, who relates it, in nothing; and after due consideration, I have determined, therefore, to abstain from presenting any precis of the learned Doctor's reasoning, or extract from his statement on a subject which he describes as "involving, not improbably, some of the profoundest arcana of our dual existence, and its intermediates."

I was anxious on discovering this paper, to reopen the correspondence commenced by Doctor Hesselius, so many years before, with a person so clever and careful as his informant seems to have been. Much to my regret, however, I found that she had died in the interval.

She, probably, could have added little to the Narrative which she communicates in the following pages, with, so far as I can pronounce, such conscientious particularity.

ABOUT THE AUTHOR

Joseph Thomas Sheridan Le Fanu (1814-1873) was a Dublin-born writer of elegant ghost stories and gothic novels, of which "Carmilla" is the most famous, and certainly the most influential; among vampire tales, it ranks second only to Bram Stoker's *Dracula* for the sheer number of screen adaptations it has inspired. Stoker, a fellow Irishman, was, in fact, powerfully influenced by Le Fanu, and originally planned to set his story in "Carmilla"'s semi-imaginary country of "Styria" instead of Transylvania.

Le Fanu had an instinctive, pre-Freudian grasp of the underlying psychodynamics of vampirism, and "Carmilla" is usually acknowledged as the first literary conflation of same-sex love and the living dead.; no other work has had such a powerful shaping influence on the now-familiar lesbian vampire genre.

"Carmilla" originally appeared in Le Fanu's 1864 collection, *In a Glass Darkly*, and is generally regarded as his masterwork.

I
An Early Fright

In Styria, we, though by no means magnificent people, inhabit a castle, or schloss. A small income, in that part of the world, goes a great way. Eight or nine hundred a year does wonders. Scantily enough ours would have answered among wealthy people at home. My father is English, and I bear an English name, although I never saw England.

But here, in this lonely and primitive place, where everything is so marvellously cheap, I really don't see how ever so much more money would at all materially add to our comforts, or even luxuries.

My father was in the Austrian service, and retired upon a pension and his patrimony, and purchased this feudal residence, and the small estate on which it stands, a bargain.

Nothing can be more picturesque or solitary. It stands on a slight eminence in a forest. The road, very old and narrow, passes in front of its drawbridge, never raised in my time, and its moat, stocked with perch, and sailed over by many swans, and floating on its surface white fleets of water-lilies.

Over all this the schloss shows its many-windowed front; its towers, and its Gothic chapel.

The forest opens in an irregular and very picturesque glade before its gate, and at the right a steep Gothic bridge carries the road over a stream that winds in deep shadow through the wood.

J. Sheridan Le Fanu

I have said that this is a very lonely place. Judge whether I say truth. Looking from the hall door towards the road, the forest in which our castle stands extends fifteen miles to the right, and twelve to the left. The nearest inhabited village is about seven of your English miles to the left. The nearest inhabited schloss of any historic associations, is that of old General Spielsdorf, nearly twenty miles away to the right.

I have said "the nearest *inhabited* village," because there is, only three miles westward, that is to say in the direction of General Spielsdorf's schloss, a ruined village, with its quaint little church, now roofless, in the aisle of which are the mouldering tombs of the proud family of Karnstein, now extinct, who once owned the equally desolate chateau which, in the thick of the forest, overlooks the silent ruins of the town.

Respecting the cause of the desertion of this striking and melancholy spot, there is a legend which I shall relate to you another time.

I must tell you now, how very small is the party who constitute the inhabitants of our castle. I don't include servants, or those dependents who occupy rooms in the buildings attached to the schloss. Listen, and wonder! My father, who is the kindest man on earth, but growing old; and I, at the date of my story, only nineteen. Eight years have passed since then. I and my father constituted the family at the schloss. My mother, a Styrian lady, died in my infancy, but I had a good-natured governess, who had been with me from, I might almost say, my infancy. I could not remember the time time when her fat, benignant face was not a familiar picture in my memory. This was Madame Perrodon, a native of Berne, whose care and good nature now in part supplied to me the loss of my mother, whom I do not even remember, so early I lost her. She made a third at our little dinner party. There was a fourth, Mademoiselle De La Fontaine, a lady such as you term, I believe, a "finishing governess." She spoke French and German, Madame Perrodon French and broken English, to which my father and I added English, which, partly to prevent its becoming a lost language among us, and partly from patriotic motives, we spoke every day. The consequence was a Babel, at which strangers used to laugh, and which I shall make no attempt to reproduce in this narrative. And there were two or three young lady friends besides, pretty nearly of my own age, who were occasional visitors, for longer or shorter terms; and these visits I sometimes returned.

These were our regular social resources; but of course there were chance visits from "neighbours" of only five or six leagues distance. My life was, notwithstanding, rather a solitary one, I can assure you.

My gouvernantes had just so much control over me as you might conjecture such sage persons would have in the case of a rather spoiled girl, whose only parent allowed her pretty nearly her own way in everything.

The first occurrence in my existence, which produced a terrible impression upon my mind, which, in fact, never has been effaced, was one of the very earliest incidents of my life which I can recollect. Some people will think it so trifling that it should not be recorded here. You will see, however, by-and-by, why I mention it. The nursery, as it was called, though I had it all to myself, was a large room in the upper story of the castle, with a steep oak roof. I can't have been more than six years old, when one night I awoke, and looking round the room from my bed, failed to see the nursery-maid. Neither was my nurse there; and I thought myself alone. I was not frightened, for I was one of those happy children who are studiously kept in ignorance of ghost stories, of fairy tales, and of all such lore as makes us cover up our heads when the door cracks suddenly, or the flicker of an expiring candle makes the shadow of a bed-post dance upon the wall, nearer to our faces. I was vexed and insulted at finding myself, as I conceived, neglected, and I began to whimper, preparatory to a hearty bout of roaring; when to my surprise, I saw a solemn, but very pretty face looking at me from the side of the bed. It was that of a young lady who was kneeling, with her hands under the coverlet. I looked at her with a kind of pleased wonder, and ceased whimpering. She caressed me with her hands, and lay down beside me on the bed, and drew me towards her, smiling; I felt immediately delightfully soothed, and fell asleep again. I was wakened by a sensation as if two needles ran into my breast very deep at the same moment, and I cried loudly. The lady started back, with her eyes fixed on me, and then slipped down upon the floor, and, as I thought, hid herself under the bed.

I was now for the first time frightened, and I yelled with all my might and main. Nurse, nursery-maid, housekeeper, all came running in, and hearing my story, they made light of it, soothing me all they could meanwhile. But, child as I was, I could perceive that their faces were pale with an unwonted look of anxiety, and I saw them look under the bed, and about the room, and peep under tables and pluck open cupboards; and the housekeeper whispered to the nurse: "Lay your hand along that hollow in the bed; some one *did* lie there, so sure as you did not; the place is still warm."

I remember the nursery-maid petting me, and all three examining my chest, where I told them I felt the puncture, and pronouncing that there was no sign visible that any such thing had happened to me.

The housekeeper and the two other servants who were in charge of the

THE FATAL FEMALE

In his provocative book *Femme Fatale: Images of Evil and Fascinating Women* (New York: Mayflower Books, 1979), art historian Patrick Bade comments "In the second half of the nineteenth century there was an extraordinary proliferation of femmes fatales in European art and culture. Wherever he went, the exhibition visitor of the 1890s found walls crowded with malefactors of the female sex. They also appeared with alarming frequency in poetry, plays, novels and operas . . . This preoccupation with evil and destructive women is one of the most striking features of late nineteenth-century culture."

Supernatural vampire women were only a small subset of this popular obsession, the roots of which could be traced to antiquity. The *lamia* of classical mythology was a vampiric female serpent, shape-shifting and deadly. Another classical antecedent is Lilith, in Hebrew tradition the first wife of Adam, who abandoned her mate and the Garden of Eden to become the malevolent Queen of the Night. The figure of Lilith has her roots in Babylonian legends and is echoed in various guises throughout world mythology, often as a kind

nursery, remained sitting up all night; and from that time a servant always sat up in the nursery until I was about fourteen.

I was very nervous for a long time after this. A doctor was called in, he was pallid and elderly. How well I remember his long saturnine face, slightly pitted with smallpox, and his chestnut wig. For a good while, every second day, he came and gave me medicine, which of course I hated.

The morning after I saw this apparition I was in a state of terror, and could not bear to be left alone, daylight though it was, for a moment.

I remember my father coming up and standing at the bedside, and talking cheerfully, and asking the nurse a number of questions, and laughing very heartily at one of the answers; and patting me on the shoulder, and kissing me, and telling me not to be frightened, that it was nothing but a dream and could not hurt me.

But I was not comforted, for I knew the visit of the strange woman was *not* a dream; and I was *awfully* frightened.

I was a little consoled by the nursery-maid's assuring me that it was she who had come and looked at me, and lain down beside me in the bed, and that I must have been half-dreaming not to have known her face. But this, though supported by the nurse, did not quite satisfy me.

I remembered, in the course of that day, a venerable old man, in a black cassock, coming into the room with the nurse and housekeeper, and talking a little to them, and very kindly to me; his face was very sweet and gentle, and he told me they were going to pray, and joined my hands together, and desired me to say, softly, while they were praying, "Lord hear all good prayers for us, for Jesus' sake." I think these were the very words, for I often repeated them to myself, and my nurse used for years to make me say them in my prayers.

I remembered so well the thoughtful sweet face of that white-haired old man, in his black cassock, as he stood in that rude, lofty, brown room, with the clumsy furniture of a fashion three hundred years old about him, and the scanty light entering its shadowy atmosphere through the small lattice. He kneeled, and the three women with him, and he prayed aloud with an earnest quavering voice for, what appeared to me, a long time. I forget all my life preceding that event, and for some time after it is all obscure also, but the scenes I have just described stand out vivid as the isolated pictures of the phantasmagoria surrounded by darkness.

of sperm-vampire or succubus (the female equivalent of the incubus, or sexually-draining nightmare or demon).

In literature, the powerful allure of the female vampire had already memorably characterized by Theophile Gautier in "La morte amoureuse" (presented elsewhere in this volume as "Loving Lady Death"), a story with which Le Fanu was no doubt quite familiar. But the complex, obsessive, and langorous sexuality of "Wake Not the Dead" (attributed to, but never proven to be the work of German romanticist Johann Ludwig Tieck), even more distinctly anticipates "Carmilla." The story's vampire, Brunhilda, ultimately assumes the serpent-like form of the classical lamia, but her voracious appetite first builds with an understated dramatic intensity, as demonstrated in the following excerpt wherein her lover Walter resurrects her with the help of a wizard, and sets in motion a bloody reign of terror.

Carmilla and her new friend. An 1872 illustration.

II
A Guest

I am now going to tell you something so strange that it will require all your faith in my veracity to believe my story. It is not only true, nevertheless, but truth of which I have been an eye-witness.

It was a sweet summer evening, and my father asked me, as he sometimes did, to take a little ramble with him along that beautiful forest vista which I have mentioned as lying in front of the schloss.

"General Spielsdorf cannot come to us so soon as I had hoped," said my father, as we pursued our walk.

He was to have paid us a visit of some weeks, and we had expected his arrival next day. He was to have brought with him a young lady, his niece and ward, Mademoiselle Rheinfeldt, whom I had never seen, but whom I had heard described as a very charming girl, and in whose society I had promised myself many happy days. I was more disappointed than a young lady living in a town, or a bustling neighbourhood can possibly imagine. This visit, and the new acquaintance it promised, had furnished my day dream for many weeks.

"And how soon does he come?" I asked.

"Not till autumn. Not for two months, I dare say," he answered. "And I am very glad now, dear, that you never knew Mademoiselle Rheinfeldt."

"And why?" I asked, both mortified and curious.

"Because the poor young lady is dead," he replied. "I quite forgot I had not told you, but you were not in the room when I received the General's letter this evening."

I was very much shocked. General Spielsdorf had mentioned in his first letter, six or seven weeks before, that she was not so well as he would wish her, but there was nothing to suggest the remotest suspicion of danger.

"Here is the General's letter," he said, handing it to me. "I am afraid he is in great affliction; the letter appears to me to have been written very nearly in distraction."

We sat down on a rude bench, under a group of magnificent lime trees. The sun was setting with all its melancholy splendour behind the sylvan horizon, and the stream that flows beside our home, and passes under the steep old bridge I have mentioned, wound through many a group of noble trees, almost at our feet, reflecting in its current the fading crimson of the sky. General Spielsdorf's letter was so extraordinary, so vehement, and in some places so self-contradictory, that I read it twice over-the second time aloud to my father-and was still unable to account for it, except by supposing that grief had unsettled his mind.

It said, "I have lost my darling daughter, for as such I loved her. During the last days of dear Bertha's illness I was not able to write to you. Before then I had no idea of her danger. I have lost her, and now learn *all*, too late. She died in the peace of innocence, and in the glorious hope of a blessed futurity. The fiend who betrayed our infatuated hospitality has done it all. I thought I was receiving into my house innocence, gaiety, a charming companion for my lost Bertha. Heavens! What a fool have I been! I thank God my child died without a suspicion of the cause of her sufferings. She is gone without so much as conjecturing the nature of her illness, and the accursed passion of the agent of all this misery. I devote my remaining days to tracking and extinguishing a monster. I am told I may hope to accomplish my righteous and merciful purpose. At present there

An unexpected night-visitor. Illustration for "Carmilla" (1872).

From
WAKE NOT THE DEAD
(English translation, 1823)

The old man now drew a circle round the grave, all the while muttering words of enchantment. Immediately the storm began to howl among the tops of the trees; owls flapped their wings, and uttered their low voice of omen; the stars hid their mild, beaming aspect, that they might not behold so unholy and impious a spectacle; the stone then rolled from the grave with a hollow sound, leaving a free passage for the inhabitant of that dreadful tenement. The sorcerer scattered into the yawning earth, roots and

is scarcely a gleam of light to guide me. I curse my conceited incredulity, my despicable affectation of superiority, my blindness, my obstinacy — all — too late. I cannot write or talk collectedly now. I am distracted. So soon as I shall have a little recovered, I mean to devote myself for a time to enquiry, which may possibly lead me as far as Vienna. Some time in the autumn, two months hence, or earlier if I live, I will see you — that is, if you permit me; I will then tell you all that I scarce dare put upon paper now. Farewell. Pray for me, dear friend."

In these terms ended this strange letter. Though I had never seen Bertha Rheinfeldt my eyes filled with tears at the sudden intelligence; I was startled, as well as profoundly disappointed.

The sun had now set, and it was twilight by the time I had returned the General's letter to my father.

It was a soft clear evening, and we loitered, speculating upon the possible meanings of the violent and incoherent sentences which I had just been reading. We had nearly a mile to walk before reaching the road that passes the schloss in front, and by that time the moon was shining brilliantly. At the drawbridge we met Madame Perrodon and Mademoiselle De La Fontaine, who had come out, without their bonnets, to enjoy the exquisite moonlight.

We heard their voices gabbling in animated dialogue as we approached. We joined them at the drawbridge, and turned about to admire with them the beautiful scene.

The glade through which we had just walked lay before us. At our left the narrow road wound away under clumps of lordly trees, and was lost to sight amid the thickening forest. At the right the same road crosses the steep and picturesque bridge, near which stands a ruined tower which once guarded that pass; and beyond the bridge an abrupt eminence rises, covered with trees, and showing in the shadows some grey ivy-clustered rocks.

Over the sward and low grounds a thin film of mist was stealing like smoke, marking the distances with a transparent veil; and here and there we could see the river faintly flashing in the moonlight.

No softer, sweeter scene could be imagined. The news I had just heard made it melancholy; but nothing could disturb its character of profound serenity, and the enchanted glory and vagueness of the prospect.

My father, who enjoyed the picturesque, and I, stood looking in silence over the expanse beneath us. The two good governesses, standing a little way behind us, discoursed upon the scene, and were eloquent upon the moon.

Madame Perrodon was fat, middle-aged, and romantic, and talked and sighed poetically. Mademoiselle De La Fontaine — in right of her father who was a German, assumed to be psychological, metaphysical, and something of a mystic — now declared that when the moon shone with a light so intense it was well known that it indicated a special spiritual

herbs of most magic power, and of most penetrating odour. so that the worms crawling forth from the earth congregated together, and raised themselves in a fiery column over the grave: while rushing wind burst from the earth, scattering the mould before it, until at length the coffin lay uncovered. The moonbeams fell on it, and the lid burst open with a tremendous sound. Upon this the sorcerer poured upon it some blood from out of a human skull, exclaiming at the same time, "Drink, sleeper, of this warm stream, that thy heart may again beat within thy bosom." And, after a short pause, shedding on her some other mystic liquid, he cried aloud with the voice of one inspired: "Yes, thy heart beats once more with the flood of life: thine eye is again opened to sight. Arise, therefore, from the tomb."

As an island suddenly springs forth from the dark waves of the ocean, raised upwards from the deep by the force of subterraneous fires, so did Brunhilda start from her earthy couch, borne forward by some invisible power. Taking her by the hand, the sorcerer led her towards Walter, who stood at some little distance, rooted to the ground with amazement.

"Receive again," said he, "the object of thy passionate sighs:

activity. The effect of the full moon in such a state of brilliancy was manifold. It acted on dreams, it acted on lunacy, it acted on nervous people, it had marvelous physical influences connected with life. Mademoiselle related that her cousin, who was mate of a merchant ship, having taken a nap on deck on such a night, lying on his back, with his face full in the light on the moon, had wakened, after a dream of an old woman clawing him by the cheek, with his features horribly drawn to one side; and his countenance had never quite recovered its equilibrium.

"The moon, this night," she said, "is full of idyllic and magnetic influence — and see, when you look behind you at the front of the schloss how all its windows flash and twinkle with that silvery splendour, as if unseen hands had lighted up the rooms to receive fairy guests."

There are indolent styles of the spirits in which, indisposed to talk ourselves, the talk of others is pleasant to our listless ears; and I gazed on, pleased with the tinkle of the ladies' conversation.

"I have got into one of my moping moods to-night," said my father, after a silence, and quoting Shakespeare, whom, by way of keeping up our English, he used to read aloud, he said:

"'In truth I know not why I am so sad.
It wearies me: you say it wearies you;
But how I got it — came by it.'

"I forget the rest. But I feel as if some great misfortune were hanging over us. I suppose the poor General's afflicted letter has had something to do with it."

At this moment the unwonted sound of carriage wheels and many hoofs upon the road, arrested our attention.

They seemed to be approaching from the high ground overlooking the bridge, and very soon the equipage emerged from that point. Two horsemen first crossed the bridge, then came a carriage drawn by four horses, and two men rode behind.

It seemed to be the travelling carriage of a person of rank; and we were all immediately absorbed in watching that very unusual spectacle. It became, in a few moments, greatly more interesting, for just as the carriage had passed the summit of the steep bridge, one of the leaders, taking fright, communicated his panic to the rest, and after a plunge or two, the whole team broke into a wild gallop together, and dashing between the horsemen who rode in front, came thundering along the road towards us with the speed of a hurricane.

The excitement of the scene was made more painful by the clear, long-drawn screams of a female voice from the carriage window.

We all advanced in curiosity and horror; me rather in silence, the rest with various ejaculations of terror.

Our suspense did not last long. Just before you reach the castle drawbridge, on the route they were coming, there stands by the roadside a magnificent lime tree, on the other stands an ancient stone cross, at sight

mayest thou never more require my aid; should that, however, happen, so wilt thou find me, during the full of the moon, upon the mountains in that spot and where the three roads meet."

Instantly did Walter recognize in the form that stood before him, her whom he so ardently loved; and a sudden glow shot through his frame at finding her thus restored to him: yet the night-frost had chilled his limbs and palsied his tongue. For a while he gazed upon her without either motion or speech, and during this pause, all was again become hushed and serene; and the stars shone brightly in the clear heavens.

"Walter!" exclaimed the figure; and at once the well-known sound, thrilling to his heart, broke the spell by which he was bound.

"Is it reality? Is it truth?" cried he, "or a cheating delusion?"

"No, it is no imposture; I am really living: — conduct me quickly to thy castle in the mountains."

Walter looked around: the old man had disappeared, but he perceived close by his side, a coal-black steed of fiery eye, ready equipped to conduct him thence; and on his back lay all proper

of which the horses, now going at a pace that was perfectly frightful, swerved so as to bring the wheel over the projecting roots of the tree.

I knew what was coming. I covered my eyes, unable to see it out, and turned my head away; at the same moment I heard a cry from my lady-friends, who had gone on a little.

Curiosity opened my eyes, and I saw a scene of utter confusion. Two of the horses were on the ground, the carriage lay upon its side with two wheels in the air; the men were busy removing the traces, and a lady, with a commanding air and figure had got out, and stood with clasped hands, raising the handkerchief that was in them every now and then to her eyes. Through the carriage door was now lifted a young lady, who appeared to be lifeless. My dear old father was already beside the elder lady, with his hat in his hand, evidently tendering his aid and the resources of his schloss. The lady did not appear to hear him, or to have eyes for anything but the slender girl who was being placed against the slope of the bank.

I approached; the young lady was apparently stunned, but she was certainly not dead. My father, who piqued himself on being something of a physician, had just had his fingers on her wrist and assured the lady, who declared herself her mother, that her pulse, though faint and irregular, was undoubtedly still distinguishable. The lady clasped her hands and looked upward, as if in a momentary transport of gratitude; but immediately she broke out again in that theatrical way which is, I believe, natural to some people.

She was what is called a fine looking woman for her time of life, and must have been handsome; she was tall, but not thin, and dressed in black velvet, and looked rather pale, but with a proud and commanding countenance, though now agitated strangely.

"Who was ever being so born to calamity?" I heard her say, with clasped hands, as I came up. "Here am I, on a journey of life and death, in prosecuting which to lose an hour is possibly to lose all. My child will not have recovered sufficiently to resume her route for who can say how long. I must leave her: I cannot, dare not, delay. How far on, sir, can you tell, is the nearest village? I must leave her there; and shall not see my darling, or even hear of her till my return, three months hence."

I plucked my father by the coat, and whispered earnestly in his ear: "Oh! Papa, pray ask her to let her stay with us — it would be so delightful. Do, pray."

"If Madame will entrust her child to the care of my daughter, and of her good gouvernante, Madame Perrodon, and permit her to remain as our guest, under my charge, until her return, it will confer a distinction and an obligation upon us, and we shall treat her with all the care and devotion which so sacred a trust deserves."

"I cannot do that, sir, it would be to task your kindness and chivalry too cruelly," said the lady, distractedly.

"It would, on the contrary, be to confer on us a very great kindness at

attire for Brunhilda, who lost no time in arraying herself. This being done, she cried, "Haste, let us away ere the dawn breaks, for my eye is yet too weak to endure the light of day." Fully recovered from his stupor, Walter leaped into his saddle, and catching up, with a mingled feeling of delight and awe, the beloved being thus mysteriously restored from the power of the grave, he spurred on across the wild, towards the mountains, as furiously as if pursued by the shadows of the dead, hastening to recover from him their sister.

The castle to which Walter conducted his Brunhilda, was situated on a rock between other rocks rising up above it. Here they arrived, unseen by any save one aged domestic, on whom Walter imposed secrecy by the severest threats.

"Here will we tarry," said Brunhilda, "until I can endure the light, and until thou canst look upon me without trembling as if struck with a cold chill." They accordingly continued to make that place their abode: yet no one knew that Brunhilda existed, save only that aged attendant, who provided their meals. During seven entire days they had no light except that of tapers: during the next seven, the light was admitted through the lofty

the moment when we most need it. My daughter has just been disappointed by a cruel misfortune, in a visit from which she had long anticipated a great deal of happiness. If you confide this young lady to our care it will be her best consolation. The nearest village on your route is distant, and affords no such inn as you could think of placing your daughter at; you cannot allow her to continue her journey for any considerable distance without danger. If, as you say, you cannot suspend your journey, you must part with her to-night, and nowhere could you do so with more honest assurances of care and tenderness than here."

There was something in this lady's air and appearance so distinguished and even imposing, and in her manner so engaging, as to impress one, quite apart from the dignity of her equipage, with a conviction that she was a person of consequence.

By this time the carriage was replaced in its upright position, and the horses, quite tractable, in the traces again.

The lady threw on her daughter a glance which I fancied was not quite so affectionate as one might have anticipated from the beginning of the scene; then she beckoned slightly to my father, and withdrew two or three steps with him out of hearing; and talked to him with a fixed and stern countenance, not at all like that with which she had hitherto spoken.

I was filled with wonder that my father did not seem to perceive the change, and also unspeakably curious to learn what it could be that she was speaking, almost in his ear, with so much earnestness and rapidity.

Two or three minutes at most I think she remained thus employed, then she turned, and a few steps brought her to where her daughter lay, supported by Madame Perrodon. She kneeled beside her for a moment and whispered, as Madame supposed, a little benediction in her ear; then hastily kissing her she stepped into her carriage, the door was closed, the footmen in stately liveries jumped up behind, the outriders spurred on, the postillions cracked their whips, the horses plunged and broke suddenly into a furious canter that threatened soon again to become a gallop, and the carriage whirled away, followed at the same rapid pace by the two horsemen in the rear.

III
We Compare Notes

We followed the *cortege* with our eyes until it was swiftly lost to sight in the misty wood; and the very sound of the hoofs and the wheels died away in the silent night air.

Nothing remained to assure us that the adventure had not been an illusion of a moment but the young lady, who just at that moment opened her eyes. I could not see, for her face was turned from me, but she raised her head, evidently looking about her, and I heard a very sweet voice ask complainingly, "Where is mamma?"

casements only while the rising or setting-sun faintly illumined the mountain-tops, the valley being still enveloped in shade.

Seldom did Walter quit Brunhilda's side: a nameless spell seemed to attach him to her; even the shudder which he felt in her presence, and which would not permit him to touch her, was not unmixed with pleasure, like that thrilling awful emotion felt when strains of sacred music float under the vault of some temple; he rather sought, therefore, than avoided this feeling. Often too as he had indulged in calling to mind the beauties of Brunhilda, she had never appeared so fair, so fascinating, so admirable when depicted by his imagination, as when now beheld in reality. Never till now had her voice sounded with such tones of sweetness; never before did her language possess such eloquence as it now did, when she conversed with him on the subject of the past. And this was the magic fairy-land towards which her words constantly conducted him. Ever did she dwell upon the days of their first love, those hours of delight which they had participated together when the one derived all enjoyment from the other: and so rapturous, so enchanting, so full of life did she recall to his imagination that blissful season, that he even

Our good Madame Perrodon answered tenderly, and added some comfortable assurances.

I then heard her ask:

"Where am I? What is this place?" and after that she said, "I don't see the carriage; and Matska, where is she?"

Madame answered all her questions in so far as she understood them; and gradually the young lady remembered how the misadventure came about, and was glad to hear that no one in, or in attendance on, the carriage was hurt; and on learning that her mamma had left her here, till her return in about three months, she wept.

I was going to add my consolations to those of Madame Perrodon when Mademoiselle De Lafontaine placed her hand upon my arm, saying:

"Don't approach, one at a time is as much as she can at present converse with; a very little excitement would possibly overpower her now."

As soon as she is comfortably in bed, I thought, I will run up to her room and see her.

My father in the meantime had sent a servant on horseback for the physician, who lived about two leagues away; and a bedroom was being prepared for the young lady's reception.

The stranger now rose, and leaning on Madame's arm, walked slowly over the drawbridge and into the castle gate.

In the hall, servants waited to receive her, and she was conducted forthwith to her room. The room we usually sat in as our drawing-room is long, having four windows, that looked over the moat and drawbridge, upon the forest scene I have just described.

It is furnished in old carved oak, with large carved cabinets, and the chairs are cushioned with crimson Utrecht velvet. The walls are covered with tapestry, and surrounded with great gold frames, the figures being as large as life, in ancient and very curious costume, and the subjects represented are hunting, hawking, and generally festive. It is not too stately to be extremely comfortable; and here we had our tea, for with his usual patriotic leanings he insisted that the national beverage should make its appearance regularly with our coffee and chocolate.

We sat here this night, and with candles lighted, were talking over the adventure of the evening.

Madame Perrodon and Mademoiselle De Lafontaine were both of our party. The young stranger had hardly lain down in her bed when she sank into a deep sleep; and those ladies had left her in the care of a servant.

"How do you like our guest?" I asked, as soon as Madame entered. "Tell me all about her?"

"I like her extremely," answered Madame, "she is, I almost think, the prettiest creature I ever saw; about your age, and so gentle and nice."

"She is absolutely beautiful," threw in Mademoiselle, who had peeped for a moment into the stranger's room.

"And such a sweet voice!" added Madame Perrodon.

doubted whether he had ever experienced with her so much felicity, or had been so truly happy. And, while she thus vividly portrayed their hours of past delight, she delineated in still more glowing, more enchanting colours, those hours of approaching bliss which now awaited them, richer in enjoyment than any preceding ones. In this manner did she charm her attentive auditor with enrapturing hopes for the future, and lull him into dreams of more than mortal ecstasy; so that while he listened to her siren strain, he entirely forgot how little blissful was the latter period of their union, when he had often sighed at her imperiousness, and at her harshness both to himself and all his household. Yet even had he recalled this to mind would it have disturbed him in his present delirious trance? Had she not now left behind in the grave all the frailty of mortality? Was not her whole being refined and purified by that long sleep in which neither passion nor sin had approached her even in dreams? How different now was the subject of her discourse! Only when speaking of her affection for him, did she betray anything of earthly feeling: at other times, she uniformly dwelt upon themes relating to the invisible and future world; when in descanting and declaring the mysteries of

"Did you remark a woman in the carriage, after it was set up again, who did not get out," inquired Mademoiselle, "but only looked from the window?"

"No, we had not seen her."

Then she described a hideous black woman, with a sort of coloured turban on her head. and who was gazing all the time from the carriage window, nodding and grinning derisively towards the ladies, with gleaming eyes and large white eye-balls, and her teeth set as if in fury.

"Did you remark what an ill-looking pack of men the servants were?" asked Madame.

"Yes," said my father, who had just come in, "ugly, hang-dog looking fellows. as ever I beheld in my life. I hope they mayn't rob the poor lady in the forest. They are clever rogues, however; they got everything to rights in a minute."

"I dare say they are worn out with too long travelling — said Madame. "Besides looking wicked, their faces were so strangely lean, and dark, and sullen. I am very curious, I own; but I dare say the young lady will tell you all about it to-morrow, if she is sufficiently recovered."

"I don't think she will," said my father, with a mysterious smile, and a little nod of his head, as if he knew more about it than he cared to tell us.

This made us all the more inquisitive as to what had passed between him and the lady in the black velvet, in the brief but earnest interview that had immediately preceded her departure.

We were scarcely alone, when I entreated him to tell me. He did not need much pressing.

"There is no particular reason why I should not tell you. She expressed a reluctance to trouble us with the care of her daughter, saying she was in delicate health, and nervous, but not subject to any kind of seizure — she volunteered that — nor to any illusion; being, in fact, perfectly sane."

"How very odd to say all that!" I interpolated. "It was so unnecessary."

"At all events it was said," he laughed, "and as you wish to know all that passed, which was indeed very little, I tell you. She then said, 'I am making a long journey of *vital* importance — she emphasized the word — rapid and secret; I shall return for my child in three months; in the meantime, she will be silent as to who we are, whence we come, and whither we are travelling.' That is all she said. She spoke very pure French. When she said the word 'secret,' she paused for a few seconds, looking sternly, her eyes fixed on mine. I fancy she makes a great point of that. You saw how quickly she was gone. I hope I have not done a very foolish thing, in taking charge of the young lady."

For my part, I was delighted. I was longing to see and talk to her; and only waiting till the doctor should give me leave. You, who live in towns, can have no idea how great an event the introduction of a new friend is, in such a solitude as surrounded us.

The doctor did not arrive till nearly one o'clock; but I could no more

eternity, a stream of prophetic eloquence would burst from her lips.

In this manner had twice seven days elapsed, and, for the first time, Walter beheld the being now dearer to him than ever, in the full light of day. Every trace of the grave had disappeared from her countenance; a roseate tinge like the ruddy streaks of dawn again beamed on her pallid cheek; the faint, mouldering taint of the grave was changed into a delightful violet scent; the only sign of earth that never disappeared. He no longer felt either apprehension or awe, as he gazed upon her in the sunny light of day: it was not until now, that he seemed to have recovered her completely; and, glowing with all his former passion towards her, he would have pressed her to his bosom, but she gently repulsed him, saying: — "Not yet — spare your caresses until the moon has again filled her horn."

Spite of his impatience, Walter was obliged to await the lapse of another period of seven days: but, on the night when the moon was arrived at the full, he hastened to Brunhilda, whom he found more lovely than she had ever appeared before. Fearing no obstacles to his transports, he embraced with all the fervour of a deeply enamoured and successful lover.

have gone to my bed and slept, than I could have overtaken, on foot, the carriage in which the princess in black velvet had driven away.

When the physician came down to the drawing-room, it was to report very favourably upon his patient. She was now sitting up, her pulse quite regular, apparently perfectly well. She had sustained no injury, and the little shock to her nerves had passed away quite harmlessly. There could be no harm certainly in my seeing her, if we both wished it; and, with this permission I sent, forthwith, to know whether she would allow me to visit her for a few minutes in her room.

The servant returned immediately to say that she desired nothing more.

You may be sure I was not long in availing myself of this permission.

Our visitor lay in one of the handsomest rooms in the schloss. It was, perhaps, a little stately. There was a sombre piece of tapestry opposite the foot of the bed, representing Cleopatra with the asps to her bosom; and other solemn classic scenes were displayed, a little faded, upon the other walls. But there was gold carving, and rich and varied colour enough in the other decorations of the room, to more than redeem the gloom of the old tapestry.

There were candles at the bed-side. She was sitting up; her slender pretty figure enveloped in the soft silk dressing-gown, embroidered with flowers, and lined with thick quilted silk, which her mother had thrown over her feet as she lay upon the ground.

What was it that, as I reached the bed-side and had just begun my little greeting, struck me dumb in a moment, and made me recoil a step or two from before her? I will tell you.

I saw the very face which had visited me in my childhood at night, which remained so fixed in my memory, and on which I had for so many years so often ruminated with horror, when no one suspected of what I was thinking.

It was pretty, even beautiful; and when I first beheld it, wore the same melancholy expression.

But this almost instantly lighted into a strange fixed smile of recognition.

There was a silence of fully a minute, and then at length *she* spoke; *I* could not.

"How wonderful!" she exclaimed. "Twelve years ago, I saw your face in a dream, and it has haunted me ever since."

"Wonderful indeed!" I repeated, overcoming with an effort the horror that had for a time suspended my utterances. "Twelve years ago, in vision or reality, I certainly saw you. I could not forget your face. It has remained before my eyes ever since."

Her smile had softened. Whatever I had fancied strange in it, was gone, and it and her dimpling cheeks were now delightfully pretty and intelligent.

I felt reassured, and continued more in the vein which hospitality indicated, to bid her welcome, and to tell her how much pleasure her

Brunhilda, however, still refused to yield to his passion. "What!" exclaimed she, "is it fitting that I who have been purified by death from the frailty of mortality, should become thy concubine, while a mere daughter of the earth bears the title of thy wife: never shall it be. No, it must be within the walls of thy palace, within that chamber where I once reigned as queen, that thou obtainest the end of thy wishes, — and of mine also," added she, imprinting a glowing kiss on the lips, and immediately disappeared.

Heated with passion, and determined to sacrifice everything to the accomplishment of his desires, Walter hastily quitted the apartment, and shortly after the castle itself. He travelled over mountain and across heath, with the rapidity of a storm, so that the turf was flung up by his horse's hoofs; nor once stopped until he arrived home.

Here, however, neither the affectionate caresses of Swanhilda, or those of his children could touch his heart, or induce him to restrain his furious desires. Alas! Is the impetuous torrent to be checked in its devastating course by the beauteous flowers over which it rushes, when they exclaim: — "Destroyer, commiserate our helpless innocence and

accidental arrival had given us all, and especially what a happiness it was to me.

I took her hand as I spoke. I was a little shy, as lonely people are, but the situation made me eloquent, and even bold. She pressed my hand, she laid hers upon it, and her eyes glowed, as, looking hastily into mine, she smiled again, and blushed.

She answered my welcome very prettily. I sat down beside her, still wondering; and she said:

"I must tell you my vision about you; it is so very strange that you and I should have had, each of the other so vivid a dream, that each should have seen, I you and you me, looking as we do now, when of course we both were mere children. I was a child, about six years old, and I awoke from a confused and troubled dream, and found myself in a room, unlike my nursery, wainscoted clumsily in some dark wood, and with cupboards and bedsteads, and chairs, and benches placed about it. The beds were, I thought, all empty, and the room itself without anyone but myself in it; and I, after looking about me for some time, and admiring especially an iron candlestick with two branches, which I should certainly know again, crept under one of the beds to reach the window; but as I got from under the bed, I heard someone crying; and looking up, while I was still upon my knees, I saw you — most assuredly you — as I see you now; a beautiful young lady, with golden hair and large blue eyes, and lips — your lips — you as you are here. Your looks won me; I climbed on the bed and put my arms about you, and I think we both fell asleep. I was aroused by a scream; you were sitting up screaming. I was frightened, and slipped down upon the ground, and, it seemed to me, lost consciousness for a moment; and when I came to myself, I was again in my nursery at home. Your face I have never forgotten since. I could not be misled by mere resemblance. You are the lady whom I saw then."

It was now my turn to relate my corresponding vision, which I did, to the undisguised wonder of my new acquaintance.

"I don't know which should be most afraid of the other," she said, again smiling —"If you were less pretty I think I should be very much afraid of you, but being as you are, and you and I both so young, I feel only that I have made your acquaintance twelve years ago, and have already a right to your intimacy; at all events it does seem as if we were destined, from our earliest childhood, to be friends. I wonder whether you feel as strangely drawn towards me as I do to you; I have never had a friend — shall I find one now?" She sighed, and her fine dark eyes gazed passionately on me.

Now the truth is, I felt rather unaccountably towards the beautiful stranger. I did feel, as she said, "drawn towards her," but there was also something of repulsion. In this ambiguous feeling, however, the sense of attraction immensely prevailed. She interested and won me; she was so beautiful and so indescribably engaging.

beauty, nor lay us waste?" – the stream sweeps over them unregarding, and a single moment annihilates the pride of a whole summer.

Shortly afterwards did Walter begin to hint to Swanhilda that they were ill-suited to each other; that he was anxious to taste that wild, tumultuous life, so well according with the spirit of his sex, while she, on the contrary, was satisfied with the monotonous circle of household enjoyments: — that he was eager for whatever promised novelty, while she felt most attached to what was familiarized to her by habit: and lastly, that her cold disposition, bordering upon indifference, but ill assorted with his ardent temperament: it was therefore more prudent that they should seek apart from each other that happiness which they could not find together. A sigh, and a brief acquiescence in his wishes was all the reply that Swanhilda made: and, on the following morning, upon his presenting her with a paper of separation, informing her that she was at liberty to return home to her father, she received it most submissively: yet, ere she departed, she gave him the following warning: "Too well do I conjecture to whom I am indebted for this our separation. Often have I seen thee at Brunhilda's grave, and beheld

I perceived now something of languor and exhaustion stealing over her, and hastened to bid her good night.

"The doctor thinks," I added, "that you ought to have a maid to sit up with you to-night; one of ours is waiting, and you will find her a very useful and quiet creature."

"How kind of you, but I could not sleep, I never could with an attendant in the room. I shan't require any assistance — and, shall I confess my weakness, I am haunted with a terror of robbers. Our house was robbed once, and two servants murdered, so I always lock my door. It has become a habit — and you look so kind I know you will forgive me. I see there is a key in the lock."

She held me close in her pretty arms for a moment and whispered in my ear, "Good night, darling, it is very hard to part with you, but good night; to-morrow, but not early, I shall see you again."

She sank back on the pillow with a sigh, and her fine eyes followed me with a fond and melancholy gaze, and she murmured again, "Good night, dear friend."

Young people like, and even love, on impulse. I was flattered by the evident, though as yet undeserved, fondness she showed me. I liked the confidence with which she at once received me. She was determined that we should be very dear friends.

Next day came and we met again. I was delighted with my companion; that is to say, in many respects.

Her looks lost nothing in daylight — she was certainly the most beautiful creature I had ever seen, and the unpleasant remembrance of the face presented in my early dream, had lost the effect of the first unexpected recognition.

She confessed that she had experienced a similar shock on seeing me, and precisely the same faint antipathy that had mingled with my admiration of her. We now laughed together over our momentary horrors.

IV
Her Habits — A Saunter

I told you that I was charmed with her in most particulars.

There were some that did not please me so well.

She was above the middle height of women. I shall begin by describing her. She was slender, and wonderfully graceful. Except that her movements were languid — very languid — indeed, there was nothing in her appearance to indicate an invalid. Her complexion was rich and brilliant; her features were small and beautifully formed; her eyes large, dark, and lustrous; her hair was quite wonderful, I never saw hair so magnificently thick and long when it was down about her shoulders; I have often placed my hands under it, and laughed with wonder at its weight. It was exquisitely fine and soft, and in colour a rich very dark brown, with something of

thee there even on that night when the face of the heavens was suddenly enveloped in a veil of clouds. Hast thou rashly dared to tear aside the awful veil that separates the mortality that dreams, from that which dreameth not? Oh! then woe to thee, thou wretched man, for thou hast attached to thyself that which will prove thy destruction."

She ceased: nor did Walter attempt any reply, for the similar admonition uttered by the sorcerer flashed upon his mind, all obscured as it was by passion, just as the lightning glares momentarily through the gloom of night without dispersing the obscurity.

Swanhilda then departed, in order to pronounce to her children, a bitter farewell, for they, according to national custom, belonged to the father; and, having bathed them in her tears, and consecrated them with the holy water of maternal love, she quitted her husband's residence, and departed to the home of her father's.

Thus was the kind and benevolent Swanhilda driven an exile from those halls where she had presided with grace; — from halls which were now newly decorated to receive another mistress. The day at length arrived on which Walter, for the second time,

gold. I loved to let it down, tumbling with its own weight, as, in her room, she lay back in her chair talking in her sweet low voice, I used to fold and braid it, and spread it out and play with it. Heavens! If I had but known all!

I said there were particulars which did not please me. I have told you that her confidence won me the first night I saw her; but I found that she exercised with respect to herself, her mother, her history, everything in fact connected with her life, plans, and people, an ever wakeful reserve. I dare say I was unreasonable, perhaps I was wrong; I dare say I ought to have respected the solemn injunction laid upon my father by the stately lady in black velvet. But curiosity is a restless and unscrupulous passion, and no one girl can endure, with patience, that hers should be baffled by another. What harm could it do anyone to tell me what I so ardently desired to know? Had she no trust in my good sense or honour? Why would she not believe me when I assured her, so solemnly, that I would not divulge one syllable of what she told me to any mortal breathing.

There was a coldness, it seemed to me, beyond her years, in her smiling melancholy persistent refusal to afford me the least ray of light.

I cannot say we quarrelled upon this point, for she would not quarrel upon any. It was, of course, very unfair of me to press her, very ill-bred, but I really could not help it; and I might just as well have let it alone.

What she did tell me amounted, in my unconscionable estimation — to nothing.

It was all summed up in three very vague disclosures:

First — Her name was Carmilla.

Second — Her family was very ancient and noble.

Third — Her home lay in the direction of the west.

She would not tell me the name of her family, nor their armorial bearings, nor the name of their estate, nor even that of the country they lived in.

You are not to suppose that I worried her incessantly on these subjects. I watched opportunity, and rather insinuated than urged my inquiries. Once or twice, indeed, I did attack her more directly. But no matter what my tactics, utter failure was invariably the result. Reproaches and caresses were all lost upon her. But I must add this, that her evasion was conducted with so pretty a melancholy and deprecation, with so many, and even passionate declarations of her liking for me, and trust in my honour, and with so many promises that I should at last know all, that I could not find it in my heart long to be offended with her.

She used to place her pretty arms about my neck, draw me to her, and laying her cheek to mine, murmur with her lips near my ear, "Dearest, your little heart is wounded; think me not cruel because I obey the irresistible law of my strength and weakness; if your dear heart is wounded, my wild heart bleeds with yours. In the rapture of my enormous humiliation I live in your warm life, and you shall die — die, sweetly die — into mine. I cannot help it; as I draw near to you, you, in your turn, will draw near to

conducted Brunhilda home as a newly made bride. And he caused it to be reported among his domestics that his new consort had gained his affections by her extraordinary likeness to Brunhilda, their former mistress. How ineffably happy did he deem himself as he conducted his beloved once more into the chamber which had often witnessed their former joys, and which was now newly gilded and adorned in a most costly style: among the other decorations were figures of angels scattering roses, which served to support the purple draperies whose ample folds o'ershadowed the nuptial couch. With what impatience did he await the hour that was to put him in possession of those beauties for which he had already paid so high a price, but, whose enjoyment was to cost him most dearly yet! Unfortunate Walter! Revelling in bliss, thou beholdest not the abyss that yawns beneath thy feet, intoxicated with the luscious perfume of the flower thou hast plucked, thou little deemest how deadly is the venom with which it is fraught, although, for a short season, its potent fragrance bestows new energy on all thy feelings.

Happy, however, as Walter was now, his household were far from being equally so. The strange resemblance between their new

others, and learn the rapture of that cruelty, which yet is love; so, for a while, seek to know no more of me and mine, but trust me with all your loving spirit."

And when she had spoken such a rhapsody, she would press me more closely in her trembling embrace, and her lips in soft kisses gently glow upon my cheek.

Her agitations and her language were unintelligible to me.

From these foolish embraces, which were not of very frequent occurrence, I must allow, I used to wish to extricate myself; but my energies seemed to fail me. Her murmured words sounded like a lullaby in my ear, and soothed my resistance into a trance, from which I only seemed to recover myself when she withdrew her arms.

In these mysterious moods I did not like her. I experienced a strange tumultuous excitement that was pleasurable, ever and anon, mingled with a vague sense of fear and disgust. I had no distinct thoughts about her while such scenes lasted, but I was conscious of a love growing into adoration, and also of abhorrence. This I know is paradox, but I can make no other attempt to explain the feeling.

I now write, after an interval of more than ten years, with a trembling hand, with a confused and horrible recollection of certain occurrences and situations, in the ordeal through which I was unconsciously passing; though with a vivid and very sharp remembrance of the main current of my story. But, I suspect, in all lives there are certain emotional scenes, those in which our passions have been most wildly and terribly roused, that are of all others the most vaguely and dimly remembered.

Sometimes after an hour of apathy, my strange and beautiful companion would take my hand and hold it with a fond pressure, renewed again and again; blushing softly, gazing in my face with languid and burning eyes, and breathing so fast that her dress rose and fell with the tumultuous respiration. It was like the ardour of a lover; it embarrassed me; it was hateful and yet over-powering; and with gloating eyes she drew me to her, and her hot lips travelled along my cheek in kisses; and she would whisper, almost in sobs, "You are mine, you shall be mine, you and I are one for ever." Then she has thrown herself back in her chair, with her small hands over her eyes, leaving me trembling.

"Are we related," I used to ask; "what can you mean by all this? I remind you perhaps of some one whom you love; but you must not, I hate it; I don't know you — I don't know myself when you look so and talk so."

She used to sigh at my vehemence, then turn away and drop my hand.

Respecting these very extraordinary manifestations I strove in vain to form any satisfactory theory — I could not refer them to affectation or trick. It was unmistakably the momentary breaking out of suppressed instinct and emotion. Was she, notwithstanding her mother's volunteered denial, subject to brief visitations of insanity; or was there here a disguise and a romance? I had read in old story books of such things. What if a

lady and the deceased Brunhilda filled them with a secret dismay, — an undefinable horror; for there was not a single difference of feature, of tone of voice, or of gesture. To add too to these mysterious circumstances, her female attendants discovered a particular mark on her back, exactly like one which Brunhilda had. A report was now soon circulated, that their lady was no other than Brunhilda herself, who had been recalled to life by the power of necromancy. How truly horrible was the idea of living under the same roof with one who had been an inhabitant of the tomb, and of being obliged to attend upon her, and acknowledge her as mistress! There was also in Brunhilda much to increase this aversion, and favour their superstition: no ornaments of gold ever decked her person; all that others were wont to wear of this metal, she had formed of silver: no richly coloured and sparkling jewels glittered upon her; pearls alone, lent their pale lustre to adorn her bosom. Most carefully did she always avoid the cheerful light of the sun, and was wont to spend the brightest days in the most retired and gloomy apartments: only during the twilight of the commencing or declining day did she ever walk abroad, but her favourite hour was when the phantom light of the moon bestowed on all objects a shadowy appearance and a

boyish lover had found his way into the house, and sought to prosecute his suit in masquerade, with the assistance of a clever old adventuress. But there were many things against this hypothesis, highly interesting as it was to my vanity.

I could boast of no little attentions such as masculine gallantry delights to offer. Between these passionate moments there were long intervals of common-place, of gaiety, of brooding melancholy, during which, except that I detected her eyes so full of melancholy fire, following me, at times I might have been as nothing to her. Except in these brief periods of mysterious excitement her ways were girlish; and there was always a languor about her, quite incompatible with a masculine system in a state of health.

In some respects her habits were odd. Perhaps not so singular in the opinion of a town lady like you, as they appeared to us rustic people. She used to come down very late, generally not till one o'clock, she would then take a cup of chocolate, but eat nothing; we then went out for a walk, which was a mere saunter, and she seemed, almost immediately, exhausted, and either returned to the schloss or sat on one of the benches that were placed, here and there, among the trees. This was a bodily languor in which her mind did not sympathise. She was always an animated talker, and very intelligent.

She sometimes alluded for a moment to her own home, or mentioned an adventure or situation, or an early recollection, which indicated a people of strange manners, and described customs of which we knew nothing. I gathered from these chance hints that her native country was much more remote than I had at first fancied.

As we sat thus one afternoon under the trees a funeral passed us by. It was that of a pretty young girl, whom I had often seen, the daughter of one of the rangers of the forest. The poor man was walking behind the coffin of his darling; she was his only child, and he looked quite heartbroken. Peasants walking two-and-two came behind, they were singing a funeral hymn.

I rose to mark my respect as they passed, and joined in the hymn they were very sweetly singing.

My companion shook me a little roughly, and I turned surprised.

She said brusquely, "Don't you perceive how discordant that is?"

"I think it very sweet, on the contrary," I answered, vexed at the interruption, and very uncomfortable, lest the people who composed the little procession should observe and resent what was passing.

I resumed, therefore, instantly, and was again interrupted. "You pierce my ears," said Carmilla, almost angrily, and stopping her ears with her tiny fingers. "Besides, how can you tell that your religion and mine are the same; your forms wound me, and I hate funerals. What a fuss! Why you must die — everyone must die; and all are happier when they do. Come home."

"My father has gone on with the clergyman to the churchyard. I thought you knew she was to be buried to-day."

sombre hue; always too at the crowing of the cock an involuntary shudder was observed to seize her limbs. Imperious as before her death, she quickly imposed her iron yoke on every one around her, while she seemed even far more terrible than ever, since a dread of some supernatural power attached to her, appalled all who approached her. A malignant withering glance seemed to shoot from her eye on the unhappy object of her wrath, as if it would annihilate its victim. In short, those halls which, in the time of Swanhilda were the residence of cheerfulness and mirth, now resembled an extensive desert tomb. With fear imprinted on their pale countenances, the domestics glided through the apartments of the castle; and in this abode of terror, the crowing of the cock caused the living to tremble, as if they were the spirits of the departed; for the sound always reminded them of their mysterious mistress. There was no one but who shuddered at meeting her in a lonely place, in the dusk of evening, or by the light of the moon, a circumstance that was deemed to be ominous of some evil: so great was the apprehension of her female attendants, they pined in continual disquietude, and, by degrees, all quitted her. In the course of time even others of the domestics fled, for an insupportal

"She? I don't trouble my head about peasants. I don't know who she is," answered Carmilla, with a flash from her fine eyes.

"She is the poor girl who fancied she saw a ghost a fortnight ago, and has been dying ever since, till yesterday, when she expired."

"Tell me nothing about ghosts. I shan't sleep to-night if you do."

"I hope there is no plague or fever coming; all this looks very like it," I continued. "The swineherd's young wife died only a week ago, and she thought something seized her by the throat as she lay in her bed, and nearly strangled her. Papa says such horrible fancies do accompany some forms of fever. She was quite well the day before. She sank afterwards, and died before a week."

"Well, her funeral is over, I hope, and her hymn sung; and our ears shan't be tortured with that discord and jargon. It has made me nervous. Sit down here, beside me; sit close; hold my hand; press it hard-hard-harder."

We had moved a little back, and had come to another seat.

She sat down. Her face underwent a change that alarmed and even terrified me for a moment. It darkened, and became horribly livid; her teeth and hands were clenched, and she frowned and compressed her lips, while she stared down upon the ground at her feet, and trembled all over with a continued shudder as irrepressible as ague. All her energies seemed strained to suppress a fit, with which she was then breathlessly tugging; and at length a low convulsive cry of suffering broke from her, and gradually the hysteria subsided. "There! That comes of strangling people with hymns!" she said at last. "Hold me, hold me still. It is passing away."

And so gradually it did; and perhaps to dissipate the sombre impression which the spectacle had left upon me, she became unusually animated and chatty; and so we got home.

This was the first time I had seen her exhibit any definable symptoms of that delicacy of health which her mother had spoken of. It was the first time, also, I had seen her exhibit anything like temper.

Both passed away like a summer cloud; and never but once afterwards did I witness on her part a momentary sign of anger. I will tell you how it happened.

She and I were looking out of one of the long drawing-room windows, when there entered the courtyard, over the drawbridge, a figure of a wanderer whom I knew very well. He used to visit the schloss generally twice a year.

It was the figure of a hunchback, with the sharp lean features that generally accompany deformity. He wore a pointed black beard, and he was smiling from ear to ear, showing his white fangs. He was dressed in buff, black, and scarlet, and crossed with more straps and belts than I could count, from which hung all manner of things. Behind, he carried a magic-lantern, and two boxes, which I well knew, in one of which was a salamander, and in the other a mandrake. These monsters used to make my father laugh. They were compounded of parts of monkeys, parrots

horror had seized them.

The art of the sorcerer had indeed bestowed upon Brunhilda an artificial life, and due nourishment had continued to support the restored body: yet this body was not able of itself to keep up the genial glow of vitality, and to nourish the flame whence springs all the affections and passions, whether of love or hate; for death had for ever destroyed and withered it: all that Brunhilda now possessed was a chilled existence, colder than that of the snake. It was nevertheless necessary that she should love, and return with equal ardour the warm caresses of her spell-enthralled husband, to whose passion alone she was indebted for her renewed existence. It was necessary that a magic draught should animate the dull current in her veins and awaken her to the glow of life and the flame of love — a potion of abomination — one not even to be named without a curse — human blood, imbibed whilst yet warm, from the veins of youth. This was the hellish drink for which she thirsted: possessing no sympathy with the purer feelings of humanity; deriving no enjoyment from aught that interests in life and occupies its varied hours; her existence was a mere blank, unless when in the arms of her paramour husband, and therefore was it that she

squirrels, fish, and hedgehogs, dried and stitched together with great neatness and startling effect. He had a fiddle, a box of conjuring apparatus, a pair of foils and masks attached to his belt, several other mysterious cases dangling about him, and a black staff with copper ferrules in his hand. His companion was a rough spare dog, that followed at his heels, but stopped short, suspiciously at the drawbridge, and in a little while began to howl dismally.

In the meantime, the mountebank, standing in the midst of the courtyard, raised his grotesque hat, and made us a very ceremonious bow, paying his compliments very volubly in execrable French, and German not much better. Then, disengaging his fiddle, he began to scrape a lively air to which he sang with a merry discord, dancing with ludicrous airs and activity, that made me laugh, in spite of the dog's howling.

Then he advanced to the window with many smiles and salutations, and his hat in his left hand, his fiddle under his arm, and with a fluency that never took breath, he gabbled a long advertisement of all his accomplishments, and the resources of the various arts which he placed at our service, and the curiosities and entertainments which it was in his power, at our bidding, to display.

"Will your ladyships be pleased to buy an amulet against the oupire, which is going like the wolf, I hear, through these woods," he said dropping his hat on the pavement. "They are dying of it right and left and here is a charm that never fails; only pinned to the pillow, and you may laugh in his face."

These charms consisted of oblong slips of vellum, with cabalistic ciphers and diagrams upon them.

Carmilla instantly purchased one, and so did I.

He was looking up, and we were smiling down upon him, amused; at least, I can answer for myself. His piercing black eye, as he looked up in our faces, seemed to detect something that fixed for a moment his curiosity.

In an instant he unrolled a leather case, full of all manner of odd little steel instruments.

"See here, my lady," he said, displaying it, and addressing me, "I profess, among other things less useful, the art of dentistry. Plague take the dog!" he interpolated. "Silence, beast! He howls so that your ladyships can scarcely hear a word. Your noble friend, the young lady at your right, has the sharpest tooth, — long, thin, pointed, like an awl, like a needle; ha, ha! With my sharp and long sight, as I look up, I have seen it distinctly; now if it happens to hurt the young lady, and I think it must, here am I, here are my file, my punch, my nippers; I will make it round and blunt, if her ladyship pleases; no longer the tooth of a fish, but of a beautiful young lady as she is. Hey? Is the young lady displeased? Have I been too bold? Have I offended her?"

The young lady, indeed, looked very angry as she drew back from the window.

craved incessantly after the horrible draught. It was even with the utmost effort that she could forbear sucking even the blood of Walter himself, reclined beside her.

Whenever she beheld some innocent child whose lovely face denoted the exuberance of infantine health and vigour, she would entice it by soothing words and fond caresses into her most secret apartment, where, lulling it to sleep in her arms, she would suck from its bosom the warm, purple tide of life. Nor were youths of either sex safe from her horrid attack: having first breathed upon her unhappy victim, who never failed immediately to sink into a lengthened sleep, she would then in a similar manner drain his veins of the vital juice. Thus children, youths, and maidens quickly faded away, as flowers gnawn by the cankering worm: the fullness of their limbs disappeared; a sallow line succeeded to the rosy freshness of their cheeks, the liquid lustre of the eye was deadened, even as the sparkling stream when arrested by the touch of frost; and their locks became thin and grey, as if already ravaged by the storm of life. Parents beheld with horror this desolating pestilence devouring their offspring; nor could simple or charm, potion or amulet avail aught against it. The grave swallowed up one after the

"How dares that mountebank insult us so? Where is your father? I shall demand redress from him. My father would have had the wretch tied up to the pump, and flogged with a cart-whip, and burnt to the bones with the castle brand!"

She retired from the window a step or two, and sat down, and had hardly lost sight of the offender, when her wrath subsided as suddenly as it had risen, and she gradually recovered her usual tone, and seemed to forget the little hunchback and his follies.

My father was out of spirits that evening. On coming in he told us that there had been another case very similar to the two fatal ones which had lately occurred. The sister of a young peasant on his estate, only a mile away, was very ill, had been, as she described it, attacked very nearly in the same way, and was now slowly but steadily sinking.

"All this," said my father, "is strictly referable to natural causes. These poor people infect one another with their superstitions, and so repeat in imagination the images of terror that have infested their neighbours."

"But that very circumstance frightens one horribly," said Carmilla.

"How so?" inquired my father.

"I am so afraid of fancying I see such things; I think it would be as bad as reality."

"We are in God's hands: nothing can happen without his permission, and all will end well for those who love him. He is our faithful creator; He has made us all, and will take care of us."

"Creator! Nature!" said the young lady in answer to my gentle father. "And this disease that invades the country is natural. Nature. All things proceed from Nature — don't they? All things in the heaven, in the earth, and under the earth, act and live as Nature ordains? I think so."

"The doctor said he would come here to-day," said my father, after a silence. "I want to know what he thinks about it, and what he thinks we had better do."

"Doctors never did me any good," said Carmilla.

"Then you have been ill?" I asked.

"More ill than ever you were," she answered.

"Long ago?"

"Yes, a long time. I suffered from this very illness; but I forget all but my pain and weakness, and they were not so bad as are suffered in other diseases."

"You were very young then?"

"I dare say; let us talk no more of it. You would not wound a friend?"

She looked languidly in my eyes, and passed her arm round my waist lovingly, and led me out of the room. My father was busy over some papers near the window.

"Why does your papa like to frighten us?" said the pretty girl with a sigh and a little shudder.

"He doesn't, dear Carmilla, it is the very furthest thing from his mind."

other; or did the miserable victim survive, he became cadaverous and wrinkled even in the very morn of existence. Parents observed with horror this devastating pestilence snatch away their offspring — a pestilence which, nor herb however potent, nor charm, nor holy taper, nor exorcism could avert. They either beheld their children sink one after the other into the grave, or their youthful forms, withered by the unholy, vampire embrace of Brunhilda, assume the decrepitude of sudden age.

At length strange surmises and reports began to prevail; it was whispered that Brunhilda herself was the cause of all these horrors; although no one could pretend to tell in what manner she destroyed her victims, since no marks of violence were discernible. Yet when young children confessed that she had frequently lulled them asleep in her arms, and elder ones said that a sudden slumber had come upon them whenever she began to converse with them, suspicion became converted into certainty, and those whose offspring had hitherto escaped unharmed, quitted their hearths and home — all their little possessions — the dwellings of their fathers and the inheritance of their children, in order to rescue from so horrible

"Are you afraid, dearest?"

"I should be very much if I fancied there was any real danger of my being attacked as those poor people were."

"You are afraid to die?"

"Yes, every one is."

"But to die as lovers may — to die together, so that they may live together. Girls are caterpillars while they live in the world, to be finally butterflies when the summer comes; but in the meantime there are grubs and larvae, don't you see — each with their peculiar propensities, necessities and structure. So says Monsieur Buffon, in his big book, in the next room."

Later in the day the doctor came, and was closeted with papa for some time. He was a skillful man, of sixty and upwards, he wore powder, and shaved his pale face as smooth as a pumpkin. He and papa emerged from the room together, and I heard papa laugh, and say as they came out:

"Well, I do wonder at a wise man like you. What do you say to hippogriffs and dragons?"

The doctor was smiling, and made answer, shaking his head —

"Nevertheless life and death are mysterious states, and we know little of the resources of either."

And so they walked on, and I heard no more. I did not then know what the doctor had been broaching, but I think I guess it now.

V
A Wonderful Likeness

This evening there arrived from Gratz the grave, dark-faced son of the picture cleaner, with a horse and cart laden with two large packing cases, having many pictures in each. It was a journey of ten leagues, and whenever a messenger arrived at the schloss from our little capital of Gratz, we used to crowd about him in the hall, to hear the news.

This arrival created in our secluded quarters quite a sensation. The cases remained in the hall, and the messenger was taken charge of by the servants till he had eaten his supper. Then with assistants, and armed with hammer, ripping-chisel, and turnscrew, he met us in the hall where we had assembled to witness the unpacking of the cases.

Carmilla sat looking listlessly on, while one after the other the old pictures, nearly all portraits, which had undergone the process of renovation, were brought to light. My mother was of an old Hungarian family, and most of these pictures, which were about to be restored to their places, had come to us through her.

My father had a list in his hand, from which he read, as the artist rummaged out the corresponding numbers. I don't know that the pictures were very good, but they were, undoubtedly, very old, and some of them very curious also. They had, for the most part, the merit of being now seen

a fate those who were dearer to their simple affections than aught else the world could give.

Thus daily did the castle assume a more desolate appearance; daily did its environs become more deserted; none but a few aged decrepit old women and grey-headed menials were to be seen remaining of the once numerous retinue. Such will in the latter days of the earth be the last generation of mortals, when childbearing shall have ceased, when youth shall no more be seen, nor any arise to replace those who shall await their fate in silence.

Walter alone noticed not, or heeded not, the desolation around him; he apprehended not death, lapped as he was in a glowing elysium of love. Far more happy than formerly did he now seem in the possession of Brunhilda. All those caprices and frowns which had been wont to overcloud their former union had now entirely disappeared. She even seemed to dote on him with a warmth of passion that she had never exhibited even during the happy season of bridal love; for the flame of that youthful blood, of which she drained the veins of others, rioted in her own. At night, as soon as he closed his eyes, she would breathe on him till he sank into delicious dreams, from which he awoke only to

by me, I may say, for the first time; for the smoke and dust of time had all but obliterated them.

"There is a picture that I have not seen yet," said my father. "In one corner, at the top of it, is the name, as well as I could read, 'Marcia Karnstein,' and the date '1698'; and I am curious to see how it has turned out."

I remembered it; it was a small picture, about a foot and a half high, and nearly square, without a frame; but it was so blackened by age that I could not make it out.

The artist now produced it, with evident pride. It was quite beautiful; it was startling; it seemed to live. It was the effigy of Carmilla!

"Carmilla, dear, here is an absolute miracle. Here you are, living, smiling, ready to speak, in this picture. Isn't it beautiful, Papa? And see, even the little mole on her throat."

My father laughed, and said, "Certainly it is a wonderful likeness," but he looked away, and to my surprise seemed but little struck by it, and went on talking to the picture cleaner, who was also something of an artist, and discoursed with intelligence about the portraits or other works, which his art had just brought into light and colour, while I was more and more lost in wonder the more I looked at the picture.

"Will you let me hang this picture in my room, papa?" I asked.

"Certainly, dear," said he, smiling, "I'm very glad you think it so like. It must be prettier even than I thought it, if it is.

The young lady did not acknowledge this pretty speech, did not seem to hear it. She was leaning back in her seat, her fine eyes under their long lashes gazing on me in contemplation, and she smiled in a kind of rapture.

"And now you can read quite plainly the name that is written in the corner. It is not Marcia; it looks as if it was done in gold. The name is Mircalla, Countess Karnstein, and this is a little coronet over and underneath A.D. 1698. I am descended from the Karnsteins; that is, mamma was."

"Ah!" said the lady, languidly, "so am I, I think, a very long descent, very ancient. Are there any Karnsteins living now?"

"None who bear the name, I believe. The family were ruined, I believe, in some civil wars, long ago, but the ruins of the castle are only about three miles away."

"How interesting!" she said, languidly. "But see what beautiful moonlight!" She glanced through the hall-door, which stood a little open. "Suppose you take a little ramble round the court, and look down at the road and river."

"It is so like the night you came to us," I said.

She sighed; smiling.

She rose, and each with her arm about the other's waist, we walked out upon the pavement.

In silence, slowly we walked down to the drawbridge, where the beautiful landscape opened before us.

experience more rapturous enjoyments. By day she would continually discourse with him on the bliss experienced by happy spirits beyond the grave, assuring him that, as his affection had recalled her from the tomb, they were now irrevocably united. Thus fascinated by a continual spell, it was not possible that he should perceive what was taking place around him. Brunhilda, however, foresaw with savage grief that the source of her youthful ardour was daily decreasing, for, in a short time, there remained nothing gifted with youth, save Walter and his children, and these latter she resolved should be her next victims.

On her first return to the castle, she had felt an aversion towards the offspring of another, and therefore abandoned them entirely to the attendants appointed by Swanhilda. Now, however, she began to pay considerable attention to them, and caused them to be frequently admitted into her presence. The aged nurses were filled with dread at perceiving these marks of regard from her towards their young charges, yet dared they not to oppose the will of their terrible and imperious mistress. Soon did Brunhilda gain the affection of the children, who were too unsuspecting of guile to apprehend any danger from her; on the contrary,

"And so you were thinking of the night I came here?" she almost whispered. "Are you glad I came?"

"Delighted, dear Carmilla," I answered.

"And you asked for the picture you think like me, to hang in your room," she murmured with a sigh, as she drew her arm closer about my waist, and let her pretty head sink upon my shoulder. "How romantic you are, Carmilla," I said. "Whenever you tell me your story, it will be made up chiefly of some one great romance."

She kissed me silently.

"I am sure, Carmilla, you have been in love; that there is, at this moment, an affair of the heart going on."

"I have been in love with no one, and never shall," she whispered, "unless it should be with you."

How beautiful she looked in the moonlight!

Shy and strange was the look with which she quickly hid her face in my neck and hair, with tumultuous sighs, that seemed almost to sob, and pressed in mine a hand that trembled.

Her soft cheek was glowing against mine. "Darling, darling," she murmured, "I live in you; and you would die for me, I love you so."

I started from her.

She was gazing on me with eyes from which all fire, all meaning had flown, and a face colourless and apathetic.

"Is there a chill in the air, dear?" she said drowsily. "I almost shiver; have I been dreaming? Let us come in. Come; come; come in."

"You look ill, Carmilla; a little faint. You certainly must take some wine," I said.

"Yes. I will. I'm better now. I shall be quite well in a few minutes. Yes, do give me a little wine," answered Carmilla, as we approached the door. "Let us look again for a moment; it is the last time, perhaps, I shall see the moonlight with you."

"How do you feel now, dear Carmilla? Are you really better?" I asked.

I was beginning to take alarm, lest she should have been stricken with the strange epidemic that they said had invaded the country about us.

"Papa would be grieved beyond measure," I added, "if he thought you were ever so little ill, without immediately letting us know. We have a very skillful doctor near this, the physician who was with papa to-day."

"I'm sure he is. I know how kind you all are; but, dear child, I am quite well again. There is nothing ever wrong with me, but a little weakness. People say I am languid; I am incapable of exertion; I can scarcely walk as far as a child of three years old: and every now and then the little strength I have falters, and I become as you have just seen me. But after all I am very easily set up again; in a moment I am perfectly myself. See how I have recovered."

So, indeed, she had; and she and I talked a great deal, and very animated she was; and the remainder of that evening passed without any

her caresses won them completely to her. Instead of ever checking their mirthful gambols, she would rather instruct them in new sports: often too did she recite to them tales of such strange and wild interest as to exceed all the stories of their nurses. Were they wearied either with play or with listening to her narratives, she would take them on her knees and lull them to slumber. Then did visions of the most surpassing magnificence attend their dreams: they would fancy themselves in some garden where flowers of every hue rose in rows one above the other, from the humble violet to the tall sunflower, forming a parti-coloured broidery of every hue, sloping upwards towards the golden clouds where little angels whose wings sparkled with azure and gold descended to bring them delicious cakes or splendid jewels; or sung to them soothing melodious hymns. So delightful did these dreams in short time become to the children that they longered for nothing so eagerly as to slumber on Brunhilda's lap, for never did they else enjoy such visions of heavenly forms. They were then most anxious for that which was to prove their destruction:— yet do we not all aspire after that which conducts us to the grave — after the enjoyment of life? These innocents stretched out their arms to approaching death because it assumed the

recurrence of what I called her infatuations. I mean her crazy talk and looks, which embarassed, and even frightened me.

But there occurred that night an event which gave my thoughts quite a new turn, and seemed to startle even Carmilla's languid nature into momentary energy.

VI
A Very Strange Agony

When we got into the drawing-room, and had sat down to our coffee and chocolate, although Carmilla did not take any, she seemed quite herself again, and Madame, and Mademoiselle De Lafontaine, joined us, and made a little card party, in the course of which papa came in for what he called his "dish of tea."

When the game was over he sat down beside Carmilla on the sofa, and asked her, a little anxiously, whether she had heard from her mother since her arrival.

She answered "No."

He then asked whether she knew where a letter would reach her at present.

"I cannot tell," she answered ambiguously, "but I have been thinking of leaving you; you have been already too hospitable and too kind to me. I have given you an infinity of trouble, and I should wish to take a carriage to-morrow, and post in pursuit of her; I know where I shall ultimately find her, although I dare not yet tell you."

"But you must not dream of any such thing," exclaimed my father, to my great relief. "We can't afford to lose you so, and I won't consent to your leaving us, except under the care of your mother, who was so good as to consent to your remaining with us till she should herself return. I should be quite happy if I knew that you heard from her: but this evening the accounts of the progress of the mysterious disease that has invaded our neighbourhood, grow even more alarming; and my beautiful guest, I do feel the responsibility, unaided by advice from your mother, very much. But I shall do my best; and one thing is certain, that you must not think of leaving us without her distinct direction to that effect. We should suffer too much in parting from you to consent to it easily."

"Thank you, sir, a thousand times for your hospitality," she answered, smiling bashfully. "You have all been too kind to me; I have seldom been so happy in all my life before, as in your beautiful chateau, under your care, and in the society of your dear daughter."

So he gallantly, in his old-fashioned way, kissed her hand, smiling and pleased at her little speech.

I accompanied Carmilla as usual to her room, and sat and chatted with her while she was preparing for bed.

"Do you think," I said at length, "that you will ever confide fully in me?"

mask of pleasure; for, which they were lapped in these ecstatic slumbers, Brunhilda sucked the life-stream from their bosoms. On waking, indeed, they felt themselves faint and exhausted, yet did no pain nor any mark betray the cause. Shortly, however, did their strength entirely fail, even as the summer brook is gradually dried up: their sports became less and less noisy; their loud, frolicsome laughter was converted into a faint smile; the full tones of their voices died away into a mere whisper. Their attendants were filled with horror and despair; too well did they conjecture the horrible truth, yet dared not to impart their suspicions to Walter, who was so devotedly attached to his horrible partner. Death had already smote his prey: the children were but the mere shadows of their former selves, and even this shadow quickly disappeared.

The anguished father deeply bemoaned their loss, for, notwithstanding his apparent neglect, he was strongly attached to them, nor until he had experienced their loss was he aware that his love was so great. His affliction could not fail to excite the displeasure of Brunhilda: "Why dost thou lament so fondly," said she, "for these little ones? What satisfaction could such unformed beings

She turned round smiling, but made no answer, only continued to smile on me.

"You won't answer that?" I said. "You can't answer pleasantly; I ought not to have asked you."

"You were quite right to ask me that, or anything. You do not know how dear you are to me, or you could not think any confidence too great to look for. But I am under vows, no nun half so awfully, and I dare not tell my story yet, even to you. The time is very near when you shall know everything. You will think me cruel, very selfish, but love is always selfish; the more ardent the more selfish. How jealous I am you cannot know. You must come with me, loving me, to death; or else hate me and still come with me and *hating* me through death and after. There is no such word as indifference in my apathetic nature."

"Now, Carmilla, you are going to talk your wild nonsense again," I said hastily.

"Not I, silly little fool as I am, and full of whims and fancies; for your sake I'll talk like a sage. Were you ever at a ball?"

"No; how you do run on. What is it like? How charming it must be."

"I almost forget, it is years ago."

I laughed.

"You are not so old. Your first ball can hardly be forgotten yet."

"I remember everything about it — with an effort. I see it all, as divers see what is going on above them, through a medium, dense, rippling, but transparent. There occurred that night what has confused the picture, and made its colours faint. I was all but assassinated in my bed, wounded here," she touched her breast, "and never was the same since."

"Were you near dying?"

"Yes, very — a cruel love — strange love, that would have taken my life. Love will have its sacrifices. No sacrifice without blood. Let us go to sleep now; I feel so lazy. How can I get up just now and lock my door?"

She was lying with her tiny hands buried in her rich wavy hair, under her cheek, her little head upon the pillow, and her glittering eyes followed me wherever I moved, with a kind of shy smile that I could not decipher.

I bid her good night, and crept from the room with an uncomfortable sensation.

I often wondered whether our pretty guest ever said her prayers. I certainly had never seen her upon her knees. In the morning she never came down until long after our family prayers were over, and at night she never left the drawing-room to attend our brief evening prayers in the hall.

If it had not been that it had casually come out in one of our careless talks that she had been baptised, I should have doubted her being a Christian. Religion was a subject on which I had never heard her speak a word. If I had known the world better, this particular neglect or antipathy would not have so much surprised me.

The precautions of nervous people are infectious, and persons of a like

yield to thee unless thou wert still attached to their mother? Thy heart then is still hers? Or dost thou now regret her and them because thou art satiated with my fondness and weary of my endearments? Had these young ones grown up, would they not have attached thee, thy spirit and thy affections more closely to this earth of clay — to this dust and have alienated thee from that sphere to which I, who have already passed the grave, endeavour to raise thee? Say is thy spirit so heavy, or thy love so weak, or thy faith so hollow, that the hope of being mine for ever is unable to touch thee?" Thus did Brunhilda express her indignation at her consort's grief, and forbade him her presence. The fear of offending her beyond forgiveness and his anxiety to appease her soon dried up his tears; and he again abandoned himself to his fatal passion, until approaching destruction at length awakened him from his delusion.

Neither maiden, nor youth, was any longer to be seen, either within the dreary walls of the castle, or the adjoining territory: — all had disappeared; for those whom the grave had not swallowed up had fled from the region of death. Who, therefore, now remained to quench the horrible thirst of the female vampire save Walter himself? And his death she

temperament are pretty sure, after a time, to imitate them. I had adopted Carmilla's habit of locking her bedroom door, having taken into my head all her whimsical alarms about midnight invaders and prowling assassins. I had also adopted her precaution of making a brief search through her room, to satisfy herself that no lurking assassin or robber was "ensconced."

These wise measures taken, I got into my bed and fell asleep. A light was burning in my room. This was an old habit, of very early date, and which nothing could have tempted me to dispense with.

Thus fortifed I might take my rest in peace. But dreams come through stone walls, light up dark rooms, or darken light ones, and their persons make their exits and their entrances as they please, and laugh at locksmiths.

I had a dream that night that was the beginning of a very strange agony.

I cannot call it a nightmare, for I was quite conscious of being asleep. But I was equally conscious of being in my room, and lying in bed, precisely as I actually was. I saw, or fancied I saw, the room and its furniture just as I had seen it last, except that it was very dark, and I saw something moving round the foot of the bed, which at first I could not accurately distinguish. But I soon saw that it was a sooty-black animal that resembled a monstrous cat. It appeared to me about four or five feet long for it measured fully the length of the hearthrug as it passed over it; and it continued to-ing and fro-ing with the lithe, sinister restlessness of a beast in a cage. I could not cry out, although as you may suppose, I was terrified. Its pace was growing faster, and the room rapidly darker and darker, and at length so dark that I could no longer see anything of it but its eyes. I felt it spring lightly on the bed. The two broad eyes approached my face, and suddenly I felt a stinging pain as if two large needles darted, an inch or two apart, deep into my breast. I waked with a scream. The room was lighted by the candle that burnt there all through the night, and I saw a female figure standing at the foot of the bed, a little at the right side. It was in a dark loose dress, and its hair was down and covered its shoulders. A block of stone could not have been more still. There was not the slightest stir of respiration. As I stared at it, the figure appeared to have changed its place, and was now nearer the door; then, close to it, the door opened, and it passed out.

I was now relieved, and able to breathe and move. My first thought was that Carmilla had been playing me a trick, and that I had forgotten to secure my door. I hastened to it, and found it locked as usual on the inside. I was afraid to open it — I was horrified. I sprang into my bed and covered my head up in the bedclothes, and lay there more dead than alive till morning.

dared to contemplate unmoved; for that divine sentiment that unites two beings in one joy and one sorrow was unknown to her bosom. Was he in his tomb, so was she free to search out other victims and glut herself with destruction, until she herself should, at the last day, be consumed with the earth itself, such is the fatal law to which the dead are subject when awoke by the arts of necromancy from the sleep of the grave.

•

Struggling with the madness that was beginning to seize him, and brooding incessantly on the ghastly visions that presented themselves to his horror-stricken mind, he lay motionless in the gloomiest recesses of the woods, even from the rise of sun till the shades of eve. But, no sooner was the light of day extinguished in the west, and the woods buried in impenetrable darkness, than the apprehension of resigning himself to sleep drove him forth among the mountains. The storm played wildly with the fantastic clouds, and with the rattling leaves, as they were caught up into the air, as if some dread spirit was sporting with these images of transitoriness and decay: it roared among the summits of the oaks as if uttering a voice of fury, while its hollow sound rebounding among

VII
Descending

It would be vain my attempting to tell you the horror with which, even now, I recall the occurrence of that night. It was no such transitory terror as a dream leaves behind it. It seemed to deepen by time, and communicated itself to the room and the very furniture that had encompassed the apparition.

I could not bear next day to be alone for a moment. I should have told papa, but for two opposite reasons. At one time I thought he would laugh at my story, and I could not bear its being treated as a jest; and at another I thought he might fancy that I had been attacked by the mysterious complaint which had invaded our neighbourhood. I had myself no misgiving of the kind, and as he had been rather an invalid for some time, I was afraid of alarming him.

I was comfortable enough with my good-natured companions, Madame Perrodon, and the vivacious Mademoiselle La Fontaine. They both perceived that I was out of spirits and nervous, and at length I told them what lay so heavy at my heart.

Mademoiselle laughed, but I fancied that Madame Perrodon looked anxious.

"By-the-by," said Mademoiselle, laughing, "the long lime-tree walk, behind Carmilla's bedroom-window, is haunted!"

"Nonsense!" exclaimed Madame, who probably thought the theme rather inopportune, "and who tells that story, my dear?"

"Martin says that he came up twice, when the old yard-gate was being repaired, before sunrise, and twice saw the same female figure walking down the lime-tree avenue."

"So he well might, as long as there are cows to milk in the river fields," said Madame.

"I daresay; but Martin chooses to be frightened, and never did I see a fool more frightened."

"You must not say a word about it to Carmilla, because she can see down that walk from her room window," I interposed, "and she is, if possible, a greater coward than I."

Carmilla came down rather later than usual that day.

"I was so frightened last night," she said, so soon as we were together, "and I am sure I should have seen something dreadful if it had not been for that charm I bought from the poor little hunchback whom I called such hard names. I had a dream of something black coming round my bed, and I awoke in a perfect horror, and I really thought, for some seconds, I saw a dark figure near the chimney-piece, but I felt under my pillow for my charm, and the moment my fingers touched it, the figure disappeared, and I felt quite certain, only that I had it by me, that something frightful would have made its appearance, and, perhaps, throttled me, as it did those poor people we heard of."

"Well, listen to me," I began, and recounted my adventure, at the recital of which she appeared horrified.

"And had you the charm near you?" she asked, earnestly.

"No, I had dropped it into a china vase in the drawing-room, but I

the distant hills, seemed as the moans of a departing sinner, or as the faint cry of some wretch expiring under the murderer's hand: the owl too, uttered its ghastly cry as if foreboding the wreck of nature. Walter's hair flew disorderly in the wind, like black snakes wreathing around his temples and shoulders; while each sense was awake to catch fresh horror. In the clouds he seemed to behold the forms of the murdered; in the howling wind to hear their laments and groans; in the chilling blast itself he felt the dire kiss of Brunhilda; in the cry of the screeching bird he heard her voice; in the mouldering leaves he scented the charnel-bed out of which he had awakened her. "Murderer of thy own offspring," exclaimed he in a voice making night, and the conflict of the element still more hideous, "paramour of a blood-thirsty vampire, reveller with the corruption of the tomb!" while in his despair he rent the wild locks from his head. Just then the full moon darted from beneath the bursting clouds; and the sight recalled to his remembrance the advice of the sorcerer, when he trembled at the first apparition of Brunhilda rising from her sleep of death; — namely, to seek him at the season of the full moon in the mountains, where three roads met. Scarcely had this gleam of hope broke in on his bewildered

shall certainly take it with me to-night, as you have so much faith in it."

At this distance of time I cannot tell you, or even understand, how I overcame my horror so effectually as to lie alone in my room that night. I remember distinctly that I pinned the charm to my pillow. I fell asleep almost immediately, and slept even more soundly than usual all night.

Next night I passed as well. My sleep was delightfully deep and dreamless. But I wakened with a sense of lassitude and melancholy, which, however, did not exceed a degree that was almost luxurious.

"Well, I told you so," said Carmilla, when I described my quiet sleep, "I had such delightful sleep myself last night; I pinned the charm to the breast of my nightdress. It was too far away the night before. I am quite sure it was all fancy, except the dreams. I used to think that evil spirits made dreams, but our doctor told me it is no such thing. Only a fever passing by, or some other malady, as they often do, he said, knocks at the door, and not being able to get in, passes on, with that alarm."

"And what do you think the charm is?" said I.

"It has been fumigated or immersed in some drug, and is an antidote against the malaria," she answered.

"Then it acts only on the body?"

"Certainly; you don't suppose that evil spirits are frightened by bits of ribbon, or the perfumes of a druggist's shop? No, these complaints, wandering in the air, begin by trying the nerves, and so infect the brain, but before they can seize upon you, the antidote repels them. That I am sure is what the charm has done for us. It is nothing magical, it is simply natural.

I should have been happier if I could have quite agreed with Carmilla, but I did my best, and the impression was a little losing its force.

For some nights I slept profoundly; but still every morning I felt the same lassitude, and a languor weighed upon me all day. I felt myself a changed girl. A strange melancholy was stealing over me, a melancholy that I would not have interrupted. Dim thoughts of death began to open, and an idea that I was slowly sinking took gentle, and, somehow, not unwelcome, possession of me. If it was sad, the tone of mind which this induced was also sweet. Whatever it might be, my soul acquiesced in it.

I would not admit that I was ill, I would not consent to tell my papa, or to have the doctor sent for.

Carmilla became more devoted to me than ever, and her strange paroxysms of languid adoration more frequent. She used to gloat on me with increasing ardour the more my strength and spirits waned. This always shocked me like a momentary glare of insanity.

Without knowing it, I was now in a pretty advanced stage of the strangest illness under which a mortal ever suffered. There was an unaccountable fascination in its earlier symptoms that more than reconciled me to the incapacitating effect of that stage of the malady. This fascination increased for a time, until it reached a certain point, when gradually a sense of the horrible mingled itself with it, deepening, as you shall hear,

mind than he flew to the appointed spot.

On his arrival, Walter found the old man seated there upon a stone as calmly as though it had been a bright sunny day and completely regardless of the uproar around. "Art thou come then?" exclaimed he to the breathless wretch, who, flinging himself at his feet, cried in a tone of anguish: — "Oh save me — succour me — rescue me from the monster that scattereth death and desolation around her.

"Wherefore a mysterious warning? why didst thou not rather disclose to me at once all the horrors that awaited my sacrilegious profanation of the grave?"

"And wherefore a mysterious warning? why didst thou not perceivest how wholesome was the advice — 'Wake not the dead.'

"Wert thou able to listen to another voice than that of thy impetuous passions? Did not thy eager impatience shut my mouth at the very moment I would have cautioned thee?"

"True, true: — thy reproof is just: but what does it avail now; — I need the promptest aid."

"Well," replied the old man,

until it discoloured and perverted the whole state of my life.

The first change I experienced was rather agreeable. It was very near the turning point from which began the descent of Avernus.

Certain vague and strange sensations visited me in my sleep. The prevailing one was of that pleasant, peculiar cold thrill which we feel in bathing, when we move against the current of a river. This was soon accompanied by dreams that seemed interminable, and were so vague that I could never recollect their scenery and persons, or any one connected portion of their action. But they left an awful impression, and a sense of exhaustion, as if I had passed through a long period of great mental exertion and danger. After all these dreams there remained on waking a remembrance of having been in a place very nearly dark, and of having spoken to people whom I could not see; and especially of one clear voice, of a female's, very deep, that spoke as if at a distance, slowly, and producing always the same sensation of indescribable solemnity and fear. Sometime there came a sensation as if a hand was drawn softly along my cheek and neck. Sometimes it was as if warm lips kissed me, and longer and longer and more lovingly as they reached my throat, but there the caress fixed itself. My heart beat faster, my breathing rose and fell rapidly and full drawn; a sobbing, that rose into a sense of strangulation, supervened, and turned into a dreadful convulsion, in which my senses left me and I became unconscious.

It was now three weeks since the commencement of this unaccountable state. My sufferings had, during the last week, told upon my appearance. I had grown pale, my eyes were dilated and darkened underneath, and the languor which I had long felt began to display itself in my countenance.

My father asked me often whether I was ill; but, with an obstinacy which now seems to me unaccountable, I persisted in assuring him that I was quite well.

In a sense this was true. I had no pain, I could complain of no bodily derangement. My complaint seemed to be one of the imagination, or the nerves, and, horrible as my sufferings were, I kept them, with a morbid reserve, very nearly to myself.

It could not be that terrible complaint which the peasants called the oupire, for I had now been suffering for three weeks, and they were seldom ill for much more than three days, when death put an end to their miseries.

Carmilla complained of dreams and feverish sensations, but by no means of so alarming a kind as mine. I say that mine were extremely alarming. Had I been capable of comprehending my condition, I would have invoked aid and advice on my knees. The narcotic of an unsuspected influence was acting upon me, and my perceptions were benumbed.

I am going to tell you now of a dream that led immediately to an odd discovery.

One night, instead of the voice I was accustomed to hear in the dark, I heard one, sweet and tender, and at the same time terrible, which said,

"there remains even yet a means of rescuing thyself, but it is fraught with horror and demands all thy resolution."

"Utter it then, utter it; for what can be more appalling, more hideous than the misery I now endure?"

"Know then," continued the sorcerer, "that only on the night of the new moon does she sleep the sleep of mortals; and then all the supernaturural power which she inherits from the grave totally fails her. 'Tis then that thou must murder her."

"How! murder her!" echoed Walter.

"Aye," returned the old man calmly, "pierce her bosom with a sharpened dagger, which I will furnish thee with; at the same time renounce her memory for ever, swearing never to think of her intentionally, and that, if thou dost involuntarily, thou wilt repeat the curse."

"Most horrible! yet what can be more horrible than she herself is? — I'll do it."

"Keep then this resolution until the next new moon."

"What, must I wait until then?" cried Walter, "alas ere then.

"Your mother warns you to beware of the assassin."

At the same time a light unexpectedly sprang up, and I saw Carmilla, standing, near the foot of my bed, in her white nightdress, bathed, from her chin to her feet, in one great stain of blood.

I wakened with a shriek, possessed with the one idea that Carmilla was being murdered. I remember springing from my bed, and my next recollection is that of standing on the lobby, crying for help.

Madame and Mademoiselle came scurrying out of their rooms in alarm; a lamp burned always on the lobby, and seeing me, they soon learned the cause of my terror.

I insisted on our knocking at Carmilla's door. Our knocking was unanswered. It soon became a pounding and an uproar. We shrieked her name, but all was vain.

We all grew frightened, for the door was locked. We hurried back, in panic, to my room. There we rang the bell long and furiously. If my father's room had been at that side of the house, we would have called him up at once to our aid. But, alas! He was quite out of hearing, and to reach him involved an excursion for which we none of us had courage.

Servants, however, soon came running up the stairs; I had got on my dressing-gown and slippers meanwhile, and my companions were already similarly furnished. Recognising the voices of the servants on the lobby, we sallied out together; and having renewed, as fruitlessly, our summons at Carmilla's door, I ordered the men to force the lock. They did so, and we stood, holding our lights aloft, in the doorway, and so stared into the room.

We called her by name; but there was still no reply. We looked round the room. Everything was undisturbed. It was exactly in the state in which I had left it on bidding her good night. But Carmilla was gone.

VIII
Search

At sight of the room, perfectly undisturbed except for our violent entrance, we began to cool a little, and soon recovered our senses sufficiently to dismiss the men. It had struck Mademoiselle that possibly Carmilla had been wakened by the uproar at her door, and in her first panic had jumped from her bed, and hid herself in a press, or behind a curtain, from which she could not, of course, emerge until the majordomo and his myrmidons had withdrawn. We now recommenced our search, and began to call her name again.

It was all to no purpose. Our perplexity and agitation increased. We examined the windows, but they were secured. I implored of Carmilla, if she had concealed herself, to play this cruel trick no longer — to come out and to end our anxieties. It was all useless. I was by this time convinced that she was not in the room, nor in the dressing-room, the door of which

either her savage thirst for blood will have forced me into the night of the tomb, or horror will have driven me into the night of madness."

"Nay," replied the sorcerer, "that I can prevent;" and, so saying, he conducted him to a cavern further among the mountains. "Abide here twice seven days," said he; "so long can I protect thee against her deadly caresses. Here wilt thou find all due provision for thy wants; but take heed that nothing tempt thee to quit this place. Farewell, when the moon renews itself, then do I repair hither again." So saying, the sorcerer drew a magic circle around the cave, and then immediately disappeared.

Twice seven days did Walter continue in this solitude, where his companions were his own terrifying thoughts, and his bitter repentance. The present was all desolation and dread; the future presented the image of a horrible deed which he must perforce commit; while the past was empoisoned by the memory of his guilt. Did he think on his former happy union with Brunhilda, her horrible image presented itself to his imagination with her lips defiled with dropping blood: or, did he call to mind the peaceful days he had passed with Swanhilda, he beheld her

was still locked on this side. She could not have passed it. I was utterly puzzled. Had Carmilla discovered one of those secret passages which the old housekeeper said were known to exist in the schloss, although the tradition of their exact situation had been lost? A little time would, no doubt, explain all—utterly perplexed as, for the present, we were.

It was past four o'clock, and I preferred passing the remaining hours of darkness in Madame's room. Daylight brought no solution of the difficulty.

The whole household, with my father at its head, was in a state of agitation next morning. Every part of the chateau was searched. The grounds were explored. No trace of the missing lady could be discovered. The stream was about to be dragged; my father was in distraction; what a tale to have to tell the poor girl's mother on her return. I, too, was almost beside myself, though my grief was quite of a different kind.

The morning was passed in alarm and excitement. It was now one o'clock, and still no tidings. I ran up to Carmilla's room, and found her standing at her dressing-table. I was astounded. I could not believe my eyes. She beckoned me to her with her pretty finger, in silence. Her face expressed extreme fear.

I ran to her in an ecstasy of joy; I kissed and embraced her again and again. I ran to the bell and rang it vehemently, to bring others to the spot who might at once relieve my father's anxiety.

"Dear Carmilla, what has become of you all this time? We have been in agonies of anxiety about you," I exclaimed. "Where have you been? How did you come back?"

"Last night has been a night of wonders," she said.

"For mercy's sake, explain all you can."

"It was past two last night," she said, "when I went to sleep as usual in my bed, with my doors locked, that of the dressing-room, and that opening upon the gallery. My sleep was uninterrupted, and, so far as I know, dreamless; but I woke just now on the sofa in the dressing-room there, and I found the door between the rooms open, and the other door forced. How could all this have happened without my being wakened? It must have been accompanied with a great deal of noise, and I am particularly easily wakened; and how could I have been carried out of my bed without my sleep having been interrupted, I whom the slightest stir startles?"

By this time, Madame, Mademoiselle, my father, and a number of the servants were in the room. Carmilla was, of course, overwhelmed with inquiries, congratulations, and welcomes. She had but one story to tell, and seemed the least able of all the party to suggest any way of accounting for what had happened.

My father took a turn up and down the room, thinking. I saw Carmilla's eye follow him for a moment with a sly, dark glance.

When my father had sent the servants away, Mademoiselle having gone in search of a little bottle of valerian and salvolatile, and there being no one now in the room with Carmilla, except my father, Madame, and myself,

sorrowful spirit with the shadows of her murdered children. Such were the horrors that attended him by day: those of night were still more dreadful, for then he beheld Brunhilda herself, who, wandering round the magic circle which she could not pass, called upon his name till the cavern reechoed the horrible sound. "Walter, my beloved," cried she, "wherefore dost thou avoid me? art thou not mine? for ever mine — mine here, and mine hereafter? And dost thou seek to murder me? — ah! commit not a deed which hurls us both to perdition — thyself as well as me." In this manner did the horrible visitant torment him each night, and, even when she departed, robbed him of all repose.

The night of the new moon at length arrived, dark as the deed it was doomed to bring forth. The sorcerer entered the cavern; "Come," said he to Walter, "let us depart hence, the hour is now arrived:" and he forthwith conducted him in silence from the cave to a coal-black steed, the sight of which recalled to Walter's remembrance the fatal night. He then related to the old man Brunhilda's nocturnal visits and anxiously inquired whether her apprehensions of eternal perdition would be fulfilled or not. "Mortal eye," exclaimed the sorcerer, "may not pierce the

he came to her thoughtfully, took her hand very kindly, led her to the sofa, and sat down beside her.

"Will you forgive me, my dear, if I risk a conjecture, and ask a question?"

"Who can have a better right?" she said. "Ask what you please, and I will tell you everything. But my story is simply one of bewilderment and darkness. I know absolutely nothing. Put any question you please, but you know, of course, the limitations mamma has placed me under."

"Perfectly, my dear child. I need not approach the topics on which she desires our silence. Now, the marvel of last night consists in your having been removed from your bed and your room, without being wakened, and this removal having occurred apparently while the windows were still secured, and the two doors locked upon the inside. I will tell you my theory and ask you a question."

Carmilla was leaning on her hand dejectedly; Madame and I were listening breathlessly.

"Now, my question is this. Have you ever been suspected of walking in your sleep?"

"Never, since I was very young indeed."

"But you did walk in your sleep when you were young?"

"Yes; I know I did. I have been told so often by my old nurse."

My father smiled and nodded.

"Well, what has happened is this. You got up in your sleep, unlocked the door, not leaving the key, as usual, in the lock, but taking it out and locking it on the outside; you again took the key out, and carried it away with you to some one of the five-and-twenty rooms on this floor, or perhaps upstairs or downstairs. There are so many rooms and closets, so much heavy furniture, and such accumulations of lumber, that it would require a week to search this old house thoroughly. Do you see, now, what I mean?"

"I do, but not all," she answered.

"And how, papa, do you account for her finding herself on the sofa in the dressing-room, which we had searched so carefully?"

"She came there after you had searched it, still in her sleep, and at last awoke spontaneously, and was as much surprised to find herself where she was as any one else. I wish all mysteries were as easily and innocently explained as yours, Carmilla," he said, laughing. "And so we may congratulate ourselves on the certainty that the most natural explanation of the occurrence is one that involves no drugging, no tampering with locks, no burglars, or poisoners, or witches — nothing that need alarm Carmilla, or anyone else, for our safety."

Carmilla was looking charmingly. Nothing could be more beautiful than her tints. Her beauty was, I think, enhanced by that graceful languor that was peculiar to her. I think my father was silently contrasting her looks with mine, for he said:

"I wish my poor Laura was looking more like herself," and he sighed.

So our alarms were happily ended, and Carmilla restored to her friends.

dark secrets of another world, or penetrate the deep abyss that separates earth from heaven." Walter hesitated to mount the steed. "Be resolute," exclaimed his companion, "but this once is it granted to thee to make the trial, and, should thou fail now, nought can rescue thee from her power."

"What can be more horrible than she herself? — I am determined:" and he leaped on the horse, the sorcerer mounting also behind him.

Carried with a rapidity equal to that of the storm that sweeps across the plain they in brief space arrived at Walter's castle. All the doors flew open at the bidding of his companion, and they speedily reached Brunhilda's chamber, and stood beside her couch. Reclining in a tranquil slumber; she reposed in all her native loveliness, every trace of horror had disappeared from her countenance; she looked so pure, meek and innocent that all the sweet hours of their endearments rushed to Walter's memory, like interceding angels pleading in her behalf. His unnerved hand could not take the dagger which the sorcerer presented to him."The blow must be struck even now:" said the latter, "shouldst thou delay but an hour, she will lie at daybreak on thy bosom, sucking the warm life drops from thy heart."

IX
The Doctor

As Carmilla would not hear of an attendant sleeping in her room, my father arranged that a servant should sleep outside her door, so that she would not attempt to make another such excursion without being arrested at her own door.

That night passed quietly; and next morning early, the doctor, whom my father had sent for without telling me a word about it, arrived to see me.

Madame accompanied me to the library; and there the grave little doctor, with white hair and spectacles, whom I mentioned before, was waiting to receive me.

I told him my story, and as I proceeded he grew graver and graver.

We were standing, he and I, in the recess of one of the windows, facing one another. When my statement was over, he leaned with his shoulders against the wall, and with his eyes fixed on me earnestly, with an interest in which was a dash of horror.

After a minute's reflection, he asked Madame if he could see my father.

He was sent for accordingly, and as he entered, smiling, he said:

"I dare say, doctor, you are going to tell me that I am an old fool for having brought you here; I hope I am."

But his smile faded into shadow as the doctor, with a very grave face, beckoned him to him.

He and the doctor talked for some time in the same recess where I had just conferred with the physician. It seemed an earnest and argumentative conversation. The room is very large, and I and Madame stood together, burning with curiosity, at the farther end. Not a word could we hear, however, for they spoke in a very low tone, and the deep recess of the window quite concealed the doctor from view, and very nearly my father, whose foot, arm, and shoulder only could we see; and the voices were, I suppose, all the less audible for the sort of closet which the thick wall and window formed.

After a time my father's face looked into the room; it was pale, thoughtful, and, I fancied, agitated.

"Laura, dear, come here for a moment. Madame, we shan't trouble you, the doctor says, at present."

Accordingly I approached, for the first time a little alarmed; for, although I felt very weak, I did not feel ill; and strength, one always fancies, is a thing that may be picked up when we please.

My father held out his hand to me, as I drew near, but he was looking at the doctor, and he said:

"It certainly is very odd; I don't understand it quite. Laura, come here, dear; now attend to Doctor Spielsberg, and recollect yourself."

"You mentioned a sensation like that of two needles piercing the skin,

"Horrible! most horrible!" faltered the trembling Walter, and turning away his face, he thrust the dagger into her bosom, exclaiming — "I curse thee for ever! — and the cold blood gushed upon his hand. Opening her eyes once more, she cast a look of ghastly horror on her husband, and, in a hollow dying accent said — "Thou too art doomed to perdition."

somewhere about your neck, on the night when you experienced your first horrible dream. Is there still any soreness?"

"None at all," I answered.

"Can you indicate with your finger about the point at which you think this occurred?"

"Very little below my throat — *here*," I answered.

I wore a morning dress, which covered the place I pointed to.

"Now you can satisfy yourself," said the doctor. "You won't mind your papa's lowering your dress a very little. It is necessary, to detect a symptom of the complaint under which you have been suffering."

I acquiesced. It was only an inch or two below the edge of my collar.

"God bless me! — so it is," exclaimed my father, growing pale.

"You see it now with your own eyes," said the doctor, with a gloomy triumph.

"What is is?" I exclaimed, beginning to be frightened.

"Nothing, my dear young lady, but a small blue spot, about the size of the tip of your little finger; and now," he continued, turning to papa, "the question is what is best to be done?"

"Is there any danger?" I urged, in great trepidation.

"I trust not, my dear," answered the doctor. "I don't see why you should not recover. I don't see why you should not begin immediately to get better. That is the point at which the sense of strangulation begins?"

"Yes," I answered.

"And — recollect as well as you can — the same point was a kind of centre of that thrill which you described just now, like the current of a cold stream running against you?"

"It may have been; I think it was."

"Ay, you see?" he added, turning to my father. "Shall I say a word to Madame?"

"Certainly," said my father.

He called Madame to him, and said:

"I find my young friend here far from well. It won't be of any great consequence, I hope; but it will be necessary that some steps be taken, which I will explain by-and-by; but in the meantime, Madame, you will be so good as not to let Miss Laura be alone for one moment. That is the only direction I need give for the present. It is indispensable."

"We may rely upon your kindness, Madame, I know," added my father.

Madame satisfied him eagerly.

"And you, dear Laura, I know you will observe the doctor's direction."

"I shall have to ask your opinion upon another patient, whose symptoms slightly resemble those of my daughter, that have just been detailed to you — very much milder in degree, but I believe quite of the same sort. She is a young lady — our guest; but as you say you will be passing this way again this evening, you can't do better than take your supper here, and you can then see her. She does not come down till the afternoon."

THE BLOODY COUNTESS

Just as Dracula would have a bloodthirsty, though nonsupernatural, historical antecedent in the person of Vlad the Impaler (see commentary for "Dracula's Guest"), so too did Carmilla have a model, of sorts, in a similarly bloodthirsty Hungarian noble named Erzebet (Elizabeth) Bathory (1540-1614). Unlike Vlad, Bathory's story had already taken on a mythic/vampiric coloration, her bloodlust explained as a means to uncanny rejuvenation.

Sabine Baring-Gould, in *The Book of Were-Wolves* (1865), quotes the following popular account:

"Elizabeth —— was wont to dress well in order to please her husband, and she spent half the day over her toilet. On one occasion, a lady's maid saw something wrong in her head-dress, and as a recompense for observing it, received such a severe box on the ears that the blood gushed from her nose, and spurted on to her mistress's face. When the blood drops were washed off her face, her skin appeared much more beautiful — whiter and more transparent on the spots where the blood had been.

"Elizabeth formed the resolution to bathe her face and her whole body in human blood so as to enhance her beauty. Two old women and a certain Fitzko assisted her in her undertaking. This monster used to ill the luckless victim, and the old women caught the blood, in which Elizabeth was wont to bathe at

"I thank you," said the doctor. "I shall be with you, then, at about seven this evening."

And then they repeated their directions to me and to Madame, and with this parting charge my father left us, and walked out with the doctor; and I saw them pacing together up and down between the road and the moat, on the grassy platform in front of the castle, evidently absorbed in earnest conversation.

The doctor did not return. I saw him mount his horse there, take his leave, and ride away eastward through the forest.

Nearly at the same time I saw the man arrive from Dranfield with the letters, and dismount and hand the bag to my father.

In the meantime, Madame and I were both busy, lost in conjecture as to the reasons of the singular and earnest direction which the doctor and my father had concurred in imposing. Madame, as she afterwards told me, was afraid the doctor apprehended a sudden seizure, and that, without prompt assistance, I might either lose my life in a fit, or at least be seriously hurt.

Erzebet Bathory

The interpretation did not strike me; and I fancied, perhaps luckily for my nerves, that the arrangement was prescribed simply to secure a companion, who would prevent my taking too much exercise, or eating unripe fruit, or doing any of the fifty foolish things to which young people are supposed to be prone.

About half an hour after my father came in — he had a letter in his hand — and said:

"This letter had been delayed; it is from General Spielsdorf. He might have been here yesterday, he may not come till to-morrow or he may be here to-day."

He put the open letter into my hand; but he did not look pleased, as he used to when a guest, especially one so much loved as the General, was coming. On the contrary, he looked as if he wished him at the bottom of the Red Sea. There was plainly something on his mind which he did not choose to divulge.

"Papa, darling, will you tell me this?" said I, suddenly laying my hand on his arm, and looking, I am sure, imploringly in his face.

"Perhaps," he answered, smoothing my hair caressingly over my eyes.

"Does the doctor think me very ill?"

"No, dear; he thinks, if right steps are taken, you will be quite well again, at least, on the high road to a complete recovery, in a day or two," he answered, a little dryly. "I wish our good friend, the General, had chosen any other time; that is, I wish you had been perfectly well to receive him."

"But do tell me, papa," I insisted, "what does he think is the matter with me?"

"Nothing; you must not plague me with questions," he answered, with more irritation than I ever remember him to have displayed before; and seeing that I looked wounded, I suppose, he kissed me, and added, "You shall know all about it in a day or two; that is, all that I know. In the meantime you are not to trouble your head about it."

He turned and left the room, but came back before I had done wondering and puzzling over the oddity of all this; it was merely to say that he was going to Karnstein, and had ordered the carriage to be ready at twelve, and that I and Madame should accompany him; he was going to see a priest who lived near those picturesque grounds, upon business, and as Carmilla had never seen them, she could follow, when she came down, with Mademoiselle, who would bring materials for what you call a picnic, which might be laid for us in the ruined castle.

At twelve o'clock, accordingly, I was ready, and not long after, my father, Madame and I set out upon our projected drive.

Passing the drawbridge we turn to the right, and follow the road over the steep Gothic bridge, westward, to reach the deserted village and ruined castle of Karnstein.

No sylvan drive can be fancied prettier. The ground breaks into gentle hills and hollows, all clothed with beautiful wood, totally destitute of the comparative formality which artificial planting and early culture and pruning impart.

The irregularities of the ground often lead the road out of its course, and cause it to wind beautifully round the sides of broken hollows and the steeper sides of the hills, among varieties of ground almost inexhaustible.

Turning one of these points, we suddenly encountered our old friend, the General, riding towards us, attended by a mounted servant. His portmanteaus were following in a hired wagon, such as we term a cart.

The General dismounted as we pulled up, and, after the usual greetings, was easily persuaded to accept the vacant seat in the carriage and send his horse on with his servant to the schloss.

X
Bereaved

It was about ten months since we had last seen him: but that time had sufficed to make an alteration of years in his appearance. He had grown thinner; something of gloom and anxiety had taken the place of that cordial serenity which used to characterise his features. His dark blue eyes, always penetrating, now gleamed with a sterner light from under his shaggy grey eyebrows. It was not such a change as grief alone usually induces, and angrier passions seemed to have had their share in bringing it about.

We had not long resumed our drive, when the General began to talk, with his usual soldierly directness, of the bereavement, as he termed it, which he had sustained in the death of his beloved niece and ward; and he then broke out in a tone of intense bitterness and fury, inveighing against the "hellish arts" to which she had fallen a victim, and expressing, with more exasperation than piety, his wonder that Heaven should tolerate so

the hour of four in the morning. After the bath she appeared more beautiful than before.

"She continued this habit after the death of her husband (1604) in the hopes of gaining new suitors. The unhappy girls who were allured to the castle, under the plea that they were to be taken into service there, were locked up in a cellar. Here they were beaten till their bodies were swollen. Elizabeth not infrequently tortured the victims herself; often she changed their clothes which dripped with blood, and then renewed her cruelties. Their swollen bodies were then cut up with razors.

"Occasionally she had the girls burned, and then cut up, but the great majority were beaten to death.

"At last her cruelty became so great, that she would stick needles into those who sat with her in a carriage, especially if they were of her own sex. One of her servant girls she stripped naked, smeared her with honey, and so drove her out of the house.

"When she was ill, and could not indulge her cruelty, she bit a person who came near her sick bed as though she were a wild beast.

monstrous an indulgence of the lusts and malignity of hell.

My father, who saw at once that something very extraordinary had befallen, asked him, if not too painful to him, to detail the circumstances which he thought justified the strong terms in which he expressed himself.

"I should tell you all with pleasure," said the General, "but you would not believe me."

"Why should I not?" he asked.

"Because," he answered testily, "you believe in nothing but what consists with your own prejudices and illusions. I remember when I was like you, but I have learned better."

"Try me," said my father; "I am not such a dogmatist as you suppose. Besides which, I very well know that you generally require proof for what you believe, and am, therefore, very strongly predisposed to respect your conclusions."

"You are right in supposing that I have not been led lightly into a belief in the marvellous — for what I have experienced is marvellous — and I have been forced by extraordinary evidence to credit that which ran counter, diametrically, to all my theories. I have been made the dupe of a preternatural conspiracy."

Notwithstanding his professions of confidence in the General's penetration, I saw my father, at this point, glance at the General, with, as I thought, a marked suspicion of his sanity.

The General did not see it, luckily. He was looking gloomily and curiously into the glades and vistas of the woods that were opening before us.

"You are going to the Ruins of Karnstein?" he said. "Yes, it is a lucky coincidence; do you know I was going to ask you to bring me there to inspect them. I have a special object in exploring. There is a ruined chapel, ain't there, with a great many tombs of that extinct family?"

"So there are — highly interesting," said my father. "I hope you are thinking of claiming the title and estates?"

My father said this gaily, but the General did not recollect the laugh, or even the smile, which courtesy exacts for a friend's joke; on the contrary, he looked grave and even fierce, ruminating on a matter that stirred his anger and horror.

"Something very different," he said, gruffly. "I mean to unearth some of those fine people. I hope, by God's blessing, to accomplish a pious sacrilege here, which will relieve our earth of certain monsters, and enable honest people to sleep in their beds without being assailed by murderers. I have strange things to tell you, my dear friend, such as I myself would have scouted as incredible a few months since."

My father looked at him again, but this time not with a glance of suspicion — with an eye, rather, of keen intelligence and alarm.

"The house of Karnstein," he said, "has been long extinct: a hundred years at least. My dear wife was maternally descended from the Karnsteins. But the name and title have long ceased to exist. The castle is a ruin; the

"She caused, in all, the death of 650 girls, some in Tscheita, on the neutral ground, where she had a cellar constructed for the purpose; others in different localities; for murder and bloodshed became with her a necessity.

"When at last the parents of the lost children could no longer be cajoled, the castle was seized, and the traces of the murders were discovered. Her accomplices were executed, and she was imprisoned for life."

In fact, Bathory was walled up in her own castle bedroom and kept in solitary confinement until her death. But much of the rest of the story is fanciful, and the celebrated bloody beauty treatments may never have taken place — although the murders certainly did, and are horrifyingly chronicled in Bathory's trial transcripts.

Film treatments have included *Countess Dracula* (1970), a Hammer Films production starring Ingrid Pitt, who the same year also played Carmilla in Hammer's *The Vampire Lovers*, followed in 1971 by *Daughters of Darkness*, in which Bathory is depicted as a soignée vampire slinking around European resort hotels, looking for young lives to

very village is deserted; it is fifty years since the smoke of a chimney was seen there; not a roof left."

"Quite true. I have heard a great deal about that since I last saw you; a great deal that will astonish you. But I had better relate everything in the order in which it occurred," said the General. "You saw my dear ward — my child, I may call her. No creature could have been more beautiful, and only three months ago none more blooming."

"Yes, poor thing! When I saw her last she certainly was quite lovely," said my father. "I was grieved and shocked more than I can tell you, my dear friend; I knew what a blow it was to you."

He took the General's hand, and they exchanged a kind pressure. Tears gathered in the old soldier's eyes. He did not seek to conceal them. He said:

"We have been very old friends; I knew you would feel for me, childless as I am. She had become an object of very near interest to me, and repaid my care by an affection that cheered my home and made my life happy. That is all gone. The years that remain to me on earth may not be very long; but by God's mercy I hope to accomplish a service to mankind before I die, and to subserve the vengeance of Heaven upon the fiends who have murdered my poor child in the spring of her hopes and beauty!"

"You said, just now, that you intended relating everything as it occurred," said my father. "Pray do; I assure you that it is not mere curiosity that prompts me."

By this time we had reached the point at which the Drunstall road, by which the General had come, diverges from the road which we were travelling to Karnstein.

"How far is it to the ruins?" inquired the General, looking anxiously forward.

"About half a league," answered my father. "Pray let us hear the story you were so good as to promise."

corrupt and destroy. *The Legend of Blood Castle* (1972), a Spanish/Italian coproduction, played up the sensationalism while retaining much of the original story. Pablo Picasso's daughter Paloma played Erzebet as a lesbian monster in *Three Immoral Women*, an anthology film released in 1974. The story gave way to laughs in *Mama Dracula* (1979), starring Louise Fletcher. Perhaps the low point of the Bathory film legacy was *the Craving* (1980), a film that pitted her against a revenant werewolf, and bombed.

The Bathory legend also provided a loose inspiration for *I Vampiri* (1956), photographed and codirected by Mario Bava, and *The Devil's Wedding Night* (1973). *The Mysterious Death of Nina Chereau* (1988) seems to be a standard psychological thriller until its climax, when a female murder suspect is revealed to be the rejuvenated Bathory. A novel based on Bathory's legend, *The Blood Countess* by Andrei Codrescu, was published in 1995.

XI
The Story

"With all my heart," said the General, with an effort; and after a short pause in which to arrange his subject, he commenced one of the strangest narratives I ever heard.

"My dear child was looking forward with great pleasure to the visit you had been so good as to arrange for her to your charming daughter." Here he made me a gallant but melancholy bow. "In the meantime we had an invitation to my old friend the Count Carlsfeld, whose schloss is about six leagues to the other side of Karnstein. It was to attend the series of fetes which, you remember, were given by him in honour of his illustrious visitor, the Grand Duke Charles."

"Yes; and very splendid, I believe, they were," said my father.

"Princely! But then his hospitalities are quite regal. He has Aladdin's lamp. The night from which my sorrow dates was devoted to a magnificent masquerade. The grounds were thrown open, the trees hung with coloured lamps. There was such a display of fireworks as Paris itself had never witnessed. And such music — music, you know, is my weakness — such ravishing music! The finest instrumental band, perhaps, in the world, and the finest singers who could be collected from all the great operas in Europe. As you wandered through these fantastically illuminated grounds, the moon-lighted chateau throwing a rosy light from its long rows of windows, you would suddenly hear these ravishing voices stealing from the silence of some grove, or rising from boats upon the lake. I felt myself, as I looked and listened, carried back into the romance and poetry of my early youth.

Before and after: Ingrid Pitt in Countess Dracula *(1970).*

"When the fireworks were ended, and the ball beginning, we returned to the noble suite of rooms that were thrown open to the dancers. A masked ball, you know, is a beautiful sight; but so brilliant a spectacle of the kind I never saw before.

"It was a very aristocratic assembly. I was myself almost the only 'nobody' present.

"My dear child was looking quite beautiful. She wore no mask. Her excitement and delight added an unspeakable charm to her features, always lovely. I remarked a young lady, dressed magnificently, but wearing a mask, who appeared to me to be observing my ward with extraordinary interest. I had seen her, earlier in the evening, in the great hall, and again, for a few minutes, walking near us, on the terrace under the castle windows, similarly employed. A lady, also masked, richly and gravely dressed, and with a stately air, like a person of rank, accompanied her as a chaperon. Had the young lady not worn a mask, I could, of course, have been much more certain upon the question whether she was really watching my poor darling. I am now well assured that she was.

"We were now in one of the salons. My poor dear child had been dancing, and was resting a little in one of the chairs near the door; I was standing near. The two ladies I have mentioned had approached and the younger took the chair next my ward; while her companion stood beside me, and for a little time addressed herself, in a low tone, to her charge.

"Availing herself of the privilege of her mask, she turned to me, and in the tone of an old friend, and calling me by my name, opened a conversation with me, which piqued my curiosity a good deal. She referred to many scenes where she had met me — at Court, and at distinguished houses. She alluded to little incidents which I had long ceased to think of, but which, I found, had only lain in abeyance in my memory, for they instantly started into life at her touch.

"I became more and more curious to ascertain who she was, every moment. She parried my attempts to discover very adroitly and pleasantly. The knowledge she showed of many passages in my life seemed to me all but unaccountable; and she appeared to take a not unnatural pleasure in foiling my curiosity, and in seeing me flounder in my eager perplexity, from one conjecture to another.

"In the meantime the young lady, whom her mother called by the odd name of Millarca, when she once or twice addressed her, had, with the same ease and grace, got into conversation with my ward.

"She introduced herself by saying that her mother was a very old acquaintance of mine. She spoke of the agreeable audacity which a mask rendered practicable; she talked like a friend; she admired her dress, and insinuated very prettily her admiration of her beauty. She amused her with laughing criticisms upon the people who crowded the ballroom, and laughed at my poor child's fun. She was very witty and lively when she pleased, and after a time they had grown very good friends, and the young stranger lowered her mask, displaying a remarkably beautiful face. I had never seen it before, neither had my dear child. But though it was new to us, the features were so engaging, as well as lovely, that it was impossible not to feel the attraction powerfully. My poor girl did so. I never saw anyone more taken with another at first sight, unless, indeed, it was the stranger herself, who seemed quite to have lost her heart to her.

"In the meantime, availing myself of the licence of a masquerade, I put not a few questions to the elder lady.

" 'You have puzzled me utterly,' I said, laughing. 'Is that not enough? Won't you, now, consent to stand on equal terms, and do me the kindness to remove your mask?'

" 'Can any request be more unreasonable?' she replied. 'Ask a lady to yield an advantage! Beside, how do you know you should recognise me? Years make changes.'

" 'As you see,' I said, with a bow, and, I suppose, a rather melancholy little laugh.

" 'As philosophers tell us,' she said; 'and how do you know that a sight of my face would help you?'

" 'I should take chance for that,' I answered. 'It is vain trying to make yourself out an old woman; your figure betrays you.'

" 'Years, nevertheless, have passed since I saw you, rather since you saw me, for that is what I am considering. Millarca, there, is my daughter; I cannot then be young, even in the opinion of people whom time has taught to be indulgent, and I may not like to be compared with what you remember me. You have no mask to remove. You can offer me nothing in exchange.'

" 'My petition is to your pity, to remove it.'

" 'And mine to yours, to let it stay where it is,' she replied.

" 'Well, then, at least you will tell me whether you are French or German; you speak both languages so perfectly.'

" 'I don't think I shall tell you that, General; you intend a surprise, and are meditating the particular point of attack.'

" 'At all events, you won't deny this,' I said, 'that being honoured by

VAMPYR (1931)

The first film which claimed "Carmilla" as an inspiration was Carl Theodor Dreyer's *Vampyr*, a French/German production of 1931. Although the film finally draws nothing from Le Fanu except for the central idea of a female vampire (in the film, a crone reminiscent of "Good Lady Ducayne"), it remains one of the most evocatively dreamlike films ever made. *Vampyr* was financed by a rich German baron, Nicolas de Gunzberg, who also acted in the film under the name "Julian West" as a young man drawn into a mysterious vampire realm.

The film's most famous sequence features Gunzberg watching his own funeral (the camera assuming his place in the coffin) and the sudden, unsettling appearance of the ancient female vampire who peers down at him, and us. The film was shot silent, with a sparse dialogue track added later in French, German, and English.

Director Dreyer had previously received high critical praise for *The Passion of Joan of Arc* (1928), but *Vampyr* (like *Nosferatu* at the time of its first American release) was roundly snubbed. *New York Times* critic C. Hooper Trask was typical in his dismissal of the film, which he called "peculiarly irritating" and "one of the worst pictures I have ever attended," though admitting that "there were some scenes that gripped with a brutal directness." Trask attributed this not to Dreyer but the actors, whom, he speculated, grew into their roles over an uncommonly long production period — more than a year.

your permission to converse, I ought to know how to address you. Shall I say Madame la Comtesse?'

"She laughed, and she would, no doubt, have met me with another evasion — if, indeed, I can treat any occurrence in an interview every circumstance of which was pre-arranged, as I now believe, with the profoundest cunning, as liable to be modified by accident.

" 'As to that,' she began; but she was interrupted, almost as she opened her lips, by a gentleman, dressed in black, who looked particularly elegant and distinguished, with this drawback, that his face was the most deadly pale I ever saw, except in death. He was in no masquerade — in the plain evening dress of a gentleman; and he said, without a smile, but with a courtly and unusually low bow: —

" 'Will Madame la Comtesse permit me to say a very few words which may interest her?'

"The lady turned quickly to him, and touched her lip in token of silence; she then said to me, 'Keep my place for me, General; I shall return when I have said a few words.'

"And with this injunction, playfully given, she walked a little aside with the gentleman in black, and talked for some minutes, apparently very earnestly. They then walked away slowly together in the crowd, and I lost them for some minutes.

"I spent the interval in cudgelling my brains for a conjecture as to the identity of the lady who seemed to remember me so kindly, and I was thinking of turning about and joining in the conversation between my pretty ward and the Countess's daughter, and trying whether, by the time she returned, I might not have a surprise in store for her, by having her name, title, chateau, and estates at my fingers' ends. But at this moment she returned, accompanied by the pale man in black, who said:

" 'I shall return and inform Madame la Comtesse when her carriage is at the door.'

"He withdrew with a bow."

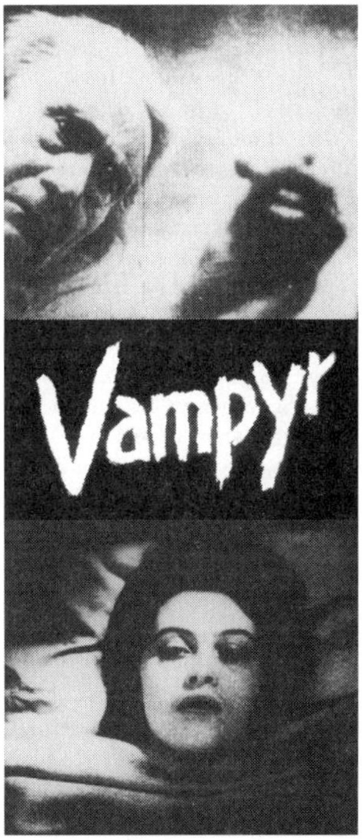

Two frame enlargements from Vampyr, *Carl Theodor Dreyer's moody meditation on the vampire theme. (Courtesy of Scott MacQueen)*

XII
A Petition

" 'Then we are to lose Madame la Comtesse, but I hope only for a few hours,' I said, with a low bow.

" 'It may be that only, or it may be a few weeks. It was very unlucky his speaking to me just now as he did. Do you now know me?'

"I assured her I did not.

" 'You shall know me,' she said, 'but not at present. We are older and better friends than, perhaps, you suspect. I cannot yet declare myself. I shall in three weeks pass your beautiful schloss, about which I have been making enquiries. I shall then look in upon you for an hour or two, and renew a friendship which I never think of without a thousand pleasant recollections. This moment a piece of news has reached me like a thunderbolt. I must set out now, and travel by a devious route, nearly a hundred miles, with all the dispatch I can possibly make. My perplexities multiply. I am only deterred by the compulsory reserve I practise as to my name from making a very singular request of you. My poor child has not quite recovered her strength. Her horse fell with her, at a hunt which she had ridden out to witness, her nerves have not yet recovered the shock, and our physician says that she must on no account exert herself for some time to come. We came here, in consequence, by very easy stages — hardly six leagues a day. I must now travel day and night, on a mission of life and death — a mission the critical and momentous nature of which I shall be able to explain to you when we meet, as I hope we shall, in a few weeks, without the necessity of any concealment.'

"She went on to make her petition, and it was in the tone of a person from whom such a request amounted to conferring, rather than seeking a favour. This was only in manner, and, as it seemed, quite unconsciously. Than the terms in which it was expressed, nothing could be more deprecatory. It was simply that I would consent to take charge of her daughter during her absence.

"This was, all things considered, a strange, not to say, an audacious request. She in some sort disarmed me, by stating and admitting everything that could be urged against it, and throwing herself entirely upon my chivalry. At the same moment, by a fatality that seems to have predetermined all that happened, my poor child came to my side, and, in an undertone, besought me to invite her new friend, Millarca, to pay us a visit. She had just been sounding her, and thought, if her mamma would allow her, she would like it extremely.

"At another time I should have told her to wait a little, until, at least, we knew who they were. But I had not a moment to think in. The two ladies assailed me together, and I must confess the refined and beautiful face of the young lady, about which there was something extremely engaging, as well as the elegance and fire of high birth, determined me; and, quite overpowered, I submitted, and undertook, too easily, the care of the young lady, whom her mother called Millarca.

"The Countess beckoned to her daughter, who listened with grave attention while she told her, in general terms, how suddenly and peremptorily she had been summoned, and also of the arrangement she had made for her under my care, adding that I was one of her earliest and most valued friends.

"I made, of course, such speeches as the case seemed to call for, and found myself, on reflection, in a position which I did not half like.

"The gentleman in black returned, and very ceremoniously conducted the lady from the room.

"The demeanour of this gentleman was such as to impress me with the conviction that the Countess was a lady of very much more importance

BLOOD AND ROSES

In 1961, Roger Vadim's loose update of "Carmilla," originally titled *Et mourir de plaisir* (*And Die of Pleasure*), was released in America only after its lesbian eroticism was significantly cut. The film has a somewhat better reputation than it probably deserves, perhaps in part because of Claude Renoir's accomplished cinematography. *Film Quarterly* called it "the most elegant and intelligent vampire film in decades, despite a few lines such as, 'What do you make of these marks on her throat, Doctor?' "

Brendan Gill, writing in *The New Yorker*, begged Vadim that *Blood and Roses* "be his last crack at a supernatural thriller. Vampires just aren't what they used to be; they seem to lack the old — well, I guess you'd have to call it spirit. The setting for this preposterous farrago is, of all places, Hadrian's Villa. It is beautiful an will survive." The film starred Mel Ferrer, Elsa Martinelli, and Annette Vadim (as the vampire). (Paramount)

than her modest title alone might have led me to assume.

"Her last charge to me was that no attempt was to be made to learn more about her than I might have already guessed, until her return. Our distinguished host, whose guest she was, knew her reasons.

" 'But here,' she said, 'neither I nor my daughter could safely remain for more than a day. I removed my mask imprudently for a moment, about an hour ago, and, too late, I fancied you saw me. So I resolved to seek an opportunity of talking a little to you. Had I found that you had seen me, I would have thrown myself on your high sense of honour to keep my secret some weeks. As it is, I am satisfied that you did not see me; but if you now suspect, or, on reflection, should suspect, who I am, I commit myself, in like manner, entirely to your honour. My daughter will observe the same secrecy, and I well know that you will, from time to time, remind her, lest she should thoughtlessly disclose it.'

"She whispered a few words to her daughter, kissed her hurriedly twice, and went away, accompanied by the pale gentleman in black, and disappeared in the crowd.

" 'In the next room,' said Millarca, 'there is a window that looks upon the hall door. I should like to see the last of mamma, and to kiss my hand to her.'

"We assented, of course, and accompanied her to the window. We looked out, and saw a handsome old-fashioned carriage, with a troop of couriers and footmen. We saw the slim figure of the pale gentleman in black, as he held a thick velvet cloak, and placed it about her shoulders and threw the hood over her head. She nodded to him, and just touched his hand with hers. He bowed low repeatedly as the door closed, and the carriage began to move.

A vampire masquerade costume covers the genuine article in Roger Vadim's Blood and Roses. *(Photofest)*

" 'She is gone,' said Millarca, with a sigh.

" 'She is gone,' I repeated to myself, for the first time — in the hurried moments that had elapsed since my consent — reflecting upon the folly of my act.

" 'She did not look up,' said the young lady, plaintively.

" 'The Countess had taken off her mask, perhaps, and did not care to show her face,' I said; 'and she could not know that you were in the window.'

"She sighed, and looked in my face. She was so beautiful that I relented. I was sorry I had for a moment repented of my hospitality, and I determined to make her amends for the unavowed churlishness of my reception.

"The young lady, replacing her mask, joined my ward in persuading me to return to the grounds, where the concert was soon to be renewed. We did so, and walked up and down the terrace that lies under the castle windows. Millarca became very intimate with us, and amused us with lively descriptions and stories of most of the great people whom we saw upon the terrace. I liked her more and more every minute. Her gossip without being ill-natured, was extremely diverting to me, who had been so long out of the great world. I thought what life she would give to our sometimes lonely evenings at home.

"This ball was not over until the morning sun had almost reached the horizon. It pleased the Grand Duke to dance till then, so loyal people could not go away, or think of bed.

"We had just got through a crowded saloon, when my ward asked me what had become of Millarca. I thought she had been by her side, and she fancied she was by mine. The fact was, we had lost her.

"All my efforts to find her were vain. I feared that she had mistaken, in the confusion of a momentary separation from us, other people for her new friends, and had, possibly, pursued and lost them in the extensive grounds which were thrown open to us.

"Now, in its full force, I recognised a new folly in my having undertaken the charge of a young lady without so much as knowing her name; and fettered as I was by promises, of the reasons for imposing which I knew nothing, I could not even point my inquiries by saying that the missing young lady was the daughter of the Countess who had taken her departure a few hours before.

"Morning broke. It was clear daylight before I gave up my search. It was not till near two o'clock next day that we heard anything of my missing charge.

"At about that time a servant knocked at my niece's door, to say that he had been earnestly requested by a young lady, who appeared to be in great distress, to make out where she could find the General Baron Spielsdorf and the young lady his daughter, in whose charge she had been left by her mother.

"There could be no doubt, notwithstanding the slight inaccuracy, that our young friend had turned up; and so she had. Would to heaven we had lost her!

"She told my poor child a story to account for her having failed to recover us for so long. Very late, she said, she had got to the housekeeper's bedroom in despair of finding us, and had then fallen into a deep sleep which, long as it was, had hardly sufficed to recruit her strength after the fatigues of the ball.

"That day Millarca came home with us. I was only too happy, after all, to have secured so charming a companion for my dear girl."

THE VAMPIRE LOVERS (1970)

The first of three "Carmilla"-inspired features produced by Hammer Films, *The Vampire Lovers* (1970) was a landmark move by Hammer away from sexual insinuation into overt eroticism. Polish-born beauty Ingrid Pitt is particularly striking as a topless, deathless lesbian-vamp. The *New York Times* took note of the film's unblinking Sapphism: "Sure, she entices and destroys a couple of guys, but only because they're in her way en route to those gorgeous girls. Miss Pitt, specifically, is a luscious brunette, who is exposed in a nude scene and prowls about the rest of the time in a diaphonous shift that leaves little to the imagination. And her willing victims. . . are just as nobly endowed. Vampirism, which has become a silly business on the screen, is, at least, easy on the eyes in this case." *New York* magazine added, "Cultists may be interested to know that lesbian-oriented vampires bite their men victims in the neck and their lady victims in the bosom (just a bit above the nipple, accounting for the film's R rating, since those fang marks do, after all, have to be examined)." *The Vampire Lovers* was directed by Roy Ward Baker from a screenplay by Tudor Gates. Also starring are Peter Cushing, Madeleine Smith, Pippa Steele, George Cole and Dawn Addams. (American International)

XIII
The Woodman

"There soon, however, appeared some drawbacks. In the first place, Millarca complained of extreme languor — the weakness that remained after her late illness — and she never emerged from her room till the afternoon was pretty far advanced. In the next place, it was accidentally discovered, although she always locked her door on the inside, and never disturbed the key from its place till she admitted the maid to assist at her toilet, that she was undoubtedly sometimes absent from her room in the very early morning, and at various times later in the day, before she wished it to be understood that she was stirring. She was repeatedly seen from the windows of the schloss, in the first faint grey of the morning, walking through the trees, in an easterly direction, and looking like a person in a trance. This convinced me that she walked in her sleep. But this hypothesis did not solve the puzzle. How did she pass out from her room, leaving the door locked on the inside? How did she escape from the house without unbarring door or window?

"In the midst of my perplexities, an anxiety of a far more urgent kind presented itself.

"My dear child began to lose her looks and health, and that in a manner so mysterious, and even horrible, that I became thoroughly frightened.

"She was at first visited by appalling dreams; then, as she fancied, by a spectre, sometimes resembling Millarca, sometimes in the shape of a beast, indistinctly seen, walking round the foot of her bed, from side to side. Lastly came sensations. One, not unpleasant, but very peculiar, she said, resembled the flow of an icy stream against her breast. At a later time, she felt something like a pair of large needles pierce her, a little below the throat, with a very sharp pain. A few nights after, followed a gradual and convulsive sense of strangulation; then came unconsciousness."

I could hear distinctly every word the kind old General was saying, because by this time we were driving upon the short grass that spreads on either side of the road as you approach the roofless village which had not shown the smoke of a chimney for more than half a century.

You may guess how strangely I felt as I heard my own symptoms so exactly described in those which had been experienced by the poor girl who, but for the catastrophe which followed, would have been at that moment a visitor at my father's chateau. You may suppose, also, how I felt as I heard him detail habits and mysterious peculiarities which were, in fact, those of our beautiful guest, Carmilla!

Advertising art for The Vampire Lovers.

A vista opened in the forest; we were on a sudden under the chimneys and gables of the ruined village, and the towers and battlements of the dismantled castle, round which gigantic trees are grouped, overhung us from a slight eminence.

In a frightened dream I got down from the carriage, and in silence, for we had each abundant matter for thinking; we soon mounted the ascent, and were among the spacious chambers, winding stairs, and dark corridors of the castle.

"And this was once the palatial residence of the Karnsteins!" said the old General at length, as from a great window he looked out across the village, and saw the wide, undulating expanse of forest. "It was a bad family, and here its blood-stained annals were written," he continued. "It is hard that they should, after death, continue to plague the human race with their atrocious lusts. That is the chapel of the Karnsteins, down there."

He pointed down to the grey walls of the Gothic building partly visible through the foliage, a little way down the steep. "And I hear the axe of a woodman," he added, "busy among the trees that surround it; he possibly may give us the information of which I am in search, and point out the grave of Mircalla, Countess of Karnstein. These rustics preserve the local traditions of great families, whose stories die out among the rich and titled so soon as the families themselves become extinct."

"We have a portrait, at home, of Mircalla, the Countess Karnstein; should you like to see it?" asked my father.

"Time enough, dear friend," replied the General. "I believe that I have seen the original; and one motive which has led me to you earlier than I at first intended, was to explore the chapel which we are now approaching."

"What! See the Countess Mircalla," exclaimed my father; "why, she has been dead more than a century!"

"Not so dead as you fancy, I am told," answered the General.

"I confess, General, you puzzle me utterly," replied my father, looking at him, I fancied, for a moment with a return of the suspicion I detected before. But although there was anger and detestation, at times, in the old General's manner, there was nothing flighty.

Ingrid Pitt seduces a victim in The Vampire Lovers. *(Photofest)*

"There remains to me," he said, as we passed under the heavy arch of the Gothic church — for its dimensions would have justified its being so styled —"but one object which can interest me during the few years that remain to me on earth, and that is to wreak on her the vengeance which, I thank God, may still be accomplished by a mortal arm."

"What vengeance can you mean?" asked my father, in increasing amazement.

"I mean, to decapitate the monster," he answered, with a fierce flush, and a stamp that echoed mournfully through the hollow ruin, and his clenched hand was at the same moment raised, as if it grasped the handle of an axe, while he shook it ferociously in the air.

"What?" exclaimed my father, more than ever bewildered.

"To strike her head off."

"Cut her head off!"

"Aye, with a hatchet, with a spade, or with anything that can cleave through her murderous throat. You shall hear," he answered, trembling with rage. And hurrying forward he said:

"That beam will answer for a seat; your dear child is fatigued; let her be seated, and I will, in a few sentences, close my dreadful story."

The squared block of wood, which lay on the grass-grown pavement of the chapel, formed a bench on which I was very glad to seat myself, and in the meantime the General called to the woodman, who had been removing some boughs which leaned upon the old walls; and, axe in hand, the hardy old fellow stood before us.

He could not tell us anything of these monuments; but there was an old man, he said, a ranger of this forest, at present sojourning in the house of the priest, about two miles away, who could point out every monument of the old Karnstein family; and, for a trifle, he undertook to bring him back with him, if we would lend him one of our horses, in little more than half an hour. "Have you been long employed about this forest?" asked my father of the old man.

"I have been a woodman here," he answered in his patois, "under the forester, all my days; so has my father before me, and so on, as many generations as I can count up. I could show you the very house in the village here, in which my ancestors lived."

"How came the village to be deserted?" asked the General.

"It was troubled by revenants, sir; several were tracked to their graves, there detected by the usual tests, and extinguished in the usual way, by decapitation, by the stake, and by burning; but not until many of the villagers were killed.

"But after all these proceedings according to law," he continued — "so many graves opened, and so many vampires deprived of their horrible animation — the village was not relieved. But a Moravian nobleman, who happened to be travelling this way, heard how matters were, and being skilled — as many people are in his country — in such affairs, he offered to deliver the village from its tormentor. He did so thus: There being a bright moon that night, he ascended, shortly after sunset, the towers of the chapel here, from whence he could distinctly see the churchyard beneath him; you can see it from that window. From this point he watched until he saw the vampire come out of his grave, and place near it the linen clothes in which he had been folded, and then glide away towards the village to plague its inhabitants.

"The stranger, having seen all this, came down from the steeple, took the linen wrappings of the vampire, and carried them up to the top of the tower, which he again mounted. When the vampire returned from his prowlings and missed his clothes, he cried furiously to the Moravian, whom he saw at the summit of the tower, and who, in reply, beckoned him to ascend and take them. Whereupon the vampire, accepting his invitation, began to climb the steeple, and so soon as he had reached the battlements, the Moravian, with a stroke of his sword, clove his skull in twain, hurling him down to the churchyard, whither, descending by the winding stairs, the stranger followed and cut his head off, and next day delivered it and the body to the villagers, who duly impaled and burnt them.

"This Moravian nobleman had authority from the then head of the family to remove the tomb of Mircalla, Countess Karnstein, which he did effectually, so that in a little while its site was quite forgotten."

"Can you point out where it stood?" asked the General, eagerly.

The forester shook his head, and smiled.

"Not a soul living could tell you that now," he said; "besides, they say her body was removed; but no one is sure of that either."

Having thus spoken, as time pressed, he dropped his axe and departed, leaving us to hear the remainder of the General's strange story.

LUST FOR A VAMPIRE (1971)

Jimmy Sangster directed Hammer's second "Carmilla"-derived flesh-fest cum blood feast, set in a nineteenth-century girls' finishing school (the equivalent of a convenience store for the undead, who un-live in a castle nearby). The lesbian aspects went further than anything the studio had previously attempted, and the film was heavily censored for its 1971 American release.

The script was again the work of Tudor Gates, and the cast included Ralph Bates, Suzanna Leigh, Michael Johnson, Yutte Stensgaard (as Carmilla/Mircalla) and Mike Raven (as the vampire Count Karnstein).

Poster for Lust for a Vampire.

TWINS OF EVIL (1971)

Hammer's final installment of its three-part exploration of the "Carmilla" family saga quickly followed *Lust for a Vampire* into theaters, continuing Hammer's profitable exploitation of supernatural lesbianism, mixing vampire conventions with the trappings of Puritan witch hysteria, not to mention the old movie chestnut about good and evil twins who cannot be told apart.

Peter Cushing is memorable as the hypocritical witch-finder, Gustav Weil, whose pious fanaticism thinly covers his own, nonsupernatural brand of bloodlust. John Hough directed, from Anthony Tudor's screenplay. In addition to Cushing, the cast included Madeline and Mary Collinson (as the twins), Kathleen Bryon, and Dennis Price. (Hammer/Universal)

XIV
The Meeting

"My beloved child," he resumed, "was now growing rapidly worse. The physician who attended her had failed to produce the slightest impression on her disease, for such I then supposed it to be. He saw my alarm, and suggested a consultation. I called in an abler physician, from Gratz. Several days elapsed before he arrived. He was a good and pious, as well as a learned man. Having seen my poor ward together, they withdrew to my library to confer and discuss. I, from the adjoining room, where I awaited their summons, heard these two gentlemen's voices raised in something sharper than a strictly philosophical discussion. I knocked at the door and entered. I found the old physician from Gratz maintaining his theory. His rival was combating it with undisguised ridicule, accompanied with bursts of laughter. This unseemly manifestation subsided and the altercation ended on my entrance.

" 'Sir,' said my first physician, 'my learned brother seems to think that you want a conjuror, and not a doctor.'

" 'Pardon me,' said the old physician from Gratz, looking displeased, 'I shall state my own view of the case in my own way another time. I grieve, Monsieur le General, that by my skill and science I can be of no use. Before I go I shall do myself the honour to suggest something to you.'

"He seemed thoughtful, and sat down at a table and began to write. Profoundly disappointed, I made my bow, and as I turned to go, the other doctor pointed over his shoulder to his companion who was writing, and then, with a shrug, significantly touched his forehead.

"This consultation, then, left me precisely where I was. I walked out into the grounds, all but distracted. The doctor from Gratz, in ten or fifteen minutes, overtook me. He apologised for having followed me, but said that he could not conscientiously take his leave without a few words more. He told me that he could not be mistaken; no natural disease exhibited the same symptoms; and that death was already very near. There remained, however, a day, or possibly two, of life. If the fatal seizure were at once arrested, with great care and skill her strength might possibly return. But all hung now upon the confines of the irrevocable. One more assault might extinguish the last spark of vitality which is, every moment, ready to die.

" 'And what is the nature of the seizure you speak of?' I entreated.

" 'I have stated all fully in this note, which I place in your hands upon the distinct condition that you send for the nearest clergyman, and open my letter in his presence, and on no account read it till he is with you; you would despise it else, and it is a matter of life and death. Should the priest fail you, then, indeed, you may read it.'

"He asked me, before taking his leave finally, whether I would wish to see a man curiously learned upon the very subject, which, after I had read his letter, would probably interest me above all others, and he urged me earnestly to invite him to visit him there; and so took his leave.

"The ecclesiastic was absent, and I read the letter by myself. At another time, or in another case, it might have excited my ridicule. But into what quackeries will not people rush for a last chance, where all accustomed means have failed, and the life of a beloved object is at stake?

"Nothing, you will say, could be more absurd than the learned man's letter. It was monstrous enough to have consigned him to a madhouse. He said that the patient was suffering from the visits of a vampire! The punctures which she described as having occurred near the throat, were, he insisted, the insertion of those two long, thin, and sharp teeth which, it is well known, are peculiar to vampires; and there could be no doubt, he added, as to the well-defined presence of the small livid mark which all concurred in describing as that induced by the demon's lips, and every symptom described by the sufferer was in exact conformity with those recorded in every case of a similar visitation.

Twins of Evil (Photofest)

"Being myself wholly skeptical as to the existence of any such portent as the vampire, the supernatural theory of the good doctor furnished, in my opinion, but another instance of learning and intelligence oddly associated with some one hallucination. I was so miserable, however, that, rather than try nothing, I acted upon the instructions of the letter.

"I concealed myself in the dark dressing-room, that opened upon the poor patient's room, in which a candle was burning, and watched there till she was fast asleep. I stood at the door, peeping through the small crevice, my sword laid on the table beside me, as my directions prescribed, until, a little after one, I saw a large black object, very ill-defined, crawl, as it seemed to me, over the foot of the bed, and swiftly spread itself up to the poor girl's throat, where it swelled, in a moment, into a great, palpitating mass.

"For a few moments I had stood petrified. I now sprang forward, with my sword in my hand. The black creature suddenly contracted towards the foot of the bed, glided over it, and, standing on the floor about a yard below the foot of the bed, with a glare of skulking ferocity and horror fixed on me, I saw Millarca. Speculating I know not what, I struck at her instantly with my sword; but I saw her standing near the door, unscathed. Horrified, I pursued, and struck again. She was gone; and my sword flew to shivers against the door.

"I can't describe to you all that passed on that horrible night. The whole house was up and stirring. The spectre Millarca was gone. But her victim was sinking fast, and before the morning dawned, she died."

The old General was agitated. We did not speak to him. My father walked to some little distance, and began reading the inscriptions on the tombstones; and thus occupied, he strolled into the door of a side-chapel to prosecute his researches. The General leaned against the wall, dried his eyes, and sighed heavily. I was relieved on hearing the voices of

Carmilla and Madame, who were at that moment approaching. The voices died away.

In this solitude, having just listened to so strange a story, connected, as it was, with the great and titled dead, whose monuments were mouldering among the dust and ivy round us, and every incident of which bore so awfully upon my own mysterious case — in this haunted spot, darkened by the towering foliage that rose on every side, dense and high above its noiseless walls — a horror began to steal over me, and my heart sank as I thought that my friends were, after all, not about to enter and disturb this triste and ominous scene.

The old General's eyes were fixed on the ground, as he leaned with his hand upon the basement of a shattered monument.

THE BLOOD-SPATTERED BRIDE (1972)

This Spanish soft-porn spinoff on "Carmilla" features a ménage-à-trois theme that evokes D.H. Lawrence's "The Fox" more than it does J. Sheridan Le Fanu. The film contains one of the most bizarre images of a sleeping vampire in cinema history, when the young husband finds the creature who will seduce his wife supine beneath the beach, her breasts and scuba goggles peeking through the sand.

Directed by Vicente Aranda. With Alexandra Bastedo, Mirabel Martin, and Simon Andrev. (Morgana Films)

Under a narrow, arched doorway, surmounted by one of those demoniacal grotesques in which the cynical and ghastly fancy of old Gothic carving delights, I saw very gladly the beautiful face and figure of Carmilla enter the shadowy chapel.

I was just about to rise and speak, and nodded smiling, in answer to her peculiarly engaging smile; when with a cry, the old man by my side caught up the woodman's hatchet, and started forward. On seeing him a brutalised change came over her features. It was an instantaneous and horrible transformation, as she made a crouching step backwards. Before I could utter a scream, he struck at her with all his force, but she dived under his blow, and unscathed, caught him in her tiny grasp by the wrist. He struggled for a moment to release his arm, but his hand opened, the axe fell to the ground, and the girl was gone.

He staggered against the wall. His grey hair stood upon his head, and a moisture shone over his face, as if he were at the point of death.

The frightful scene had passed in a moment. The first thing I recollect after, is Madame standing before me, and impatiently repeating again and again, the question, "Where is Mademoiselle Carmilla?"

I answered at length, "I don't know — I can't tell — she went there," and I pointed to the door through which Madame had just entered; "only a minute or two since."

"But I have been standing there, in the passage, ever since Mademoiselle Carmilla entered; and she did not return."

She then began to call, "Carmilla," through every door and passage and from the windows, but no answer came.

"She called herself Carmilla?" asked the General, still agitated.

"Carmilla, yes," I answered.

"Aye," he said; "that is Millarca. That is the same person who long ago was called Mircalla, Countess Karnstein. Depart from this accursed ground, my poor child, as quickly as you can. Drive to the clergyman's house, and stay there till we come. Begone! May you never behold Carmilla more; you will not find her here."

XV
Ordeal and Execution

As he spoke one of the strangest looking men I ever beheld entered the chapel at the door through which Carmilla had made her entrance and her exit. He was tall, narrow-chested, stooping, with high shoulders, and dressed in black. His face was brown and dried in with deep furrows; he wore an oddly-shaped hat with a broad leaf. His hair, long and grizzled, hung on his shoulders. He wore a pair of gold spectacles, and walked slowly, with an odd shambling gait, with his face sometimes turned up to the sky, and sometimes bowed down towards the ground, seemed to wear a perpetual smile; his long thin arms were swinging, and his lank hands, in old black gloves ever so much too wide for them, waving and gesticulating in utter abstraction.

"The very man!" exclaimed the General, advancing with manifest delight. "My dear Baron, how happy I am to see you, I had no hope of meeting you so soon." He signed to my father, who had by this time returned, and leading the fantastic old gentleman, whom he called the Baron to meet him. He introduced him formally, and they at once entered into earnest conversation. The stranger took a roll of paper from his pocket, and spread it on the worn surface of a tomb that stood by. He had a pencil case in his fingers, with which he traced imaginary lines from point to point on the paper, which from their often glancing from it, together, at certain points of the building, I concluded to be a plan of the chapel. He accompanied, what I may term, his lecture, with occasional readings from a dirty little book, whose yellow leaves were closely written over.

Advertising art for The Blood-Spattered Bride.

They sauntered together down the side aisle, opposite to the spot where I was standing, conversing as they went; then they began measuring distances by paces, and finally they all stood together, facing a piece of the side-wall, which they began to examine with great minuteness; pulling off the ivy that clung over it, and rapping the plaster with the ends of their sticks, scraping here, and knocking there. At length they ascertained the existence of a broad marble tablet, with letters carved in relief upon it.

With the assistance of the woodman, who soon returned, a monumental inscription, and carved escutcheon, were disclosed. They proved to be those of the long lost monument of Mircalla, Countess Karnstein.

The old General, though not I fear given to the praying mood, raised his hands and eyes to heaven, in mute thanksgiving for some moments.

"To-morrow," I heard him say; "the commissioner will be here, and the Inquisition will be held according to law."

Then turning to the old man with the gold spectacles, whom I have described, he shook him warmly by both hands and said:

"Baron, how can I thank you? How can we all thank you? You will have delivered this region from a plague that has scourged its inhabitants for more than a century. The horrible enemy, thank God, is at last tracked."

My father led the stranger aside, and the General followed. I know that he had led them out of hearing, that he might relate my case, and I saw them glance often quickly at me, as the discussion proceeded.

My father came to me, kissed me again and again, and leading me from the chapel, said:

"It is time to return, but before we go home, we must add to our party the good priest, who lives but a little way from this; and persuade him to accompany us to the schloss."

In this quest we were successful: and I was glad, being unspeakably fatigued when we reached home. But my satisfaction was changed to dismay, on discovering that there were no tidings of Carmilla. Of the scene that had occurred in the ruined chapel, no explanation was offered to me, and it was clear that it was a secret which my father for the present determined to keep from me.

CARMILLA (1989)

This television adaptation, part of the "Nightmare Theatre" series, is the only version of the Le Fanu story that retains the original title, while jettisoning the European location in favor of the antebellum South. Gabrielle Beaumont directed a cast including Meg Tilley as Carmilla, Ione Skye as her victim, Roy Dotrice and Roddy McDowall. (Showtime)

The sinister absence of Carmilla made the remembrance of the scene more horrible to me. The arrangements for the night were singular. Two servants, and Madame were to sit up in my room that night; and the ecclesiastic with my father kept watch in the adjoining dressing-room.

The priest had performed certain solemn rites that night, the purport of which I did not understand any more than I comprehended the reason of this extraordinary precaution taken for my safety during sleep.

I saw all clearly a few days later.

The disappearance of Carmilla was followed by the discontinuance of my nightly sufferings.

You have heard, no doubt, of the appalling superstition that prevails in Upper and Lower Styria, in Moravia, Silesia, in Turkish Servia, in Poland, even in Russia; the superstition, so we must call it, of the Vampire.

If human testimony, taken with every care and solemnity, judicially, before commissions innumerable, each consisting of many members, all chosen for integrity and intelligence, and constituting reports more voluminous perhaps than exist upon any one other class of cases, is worth anything, it is difficult to deny, or even to doubt the existence of such a phenomenon as the Vampire.

For my part I have heard no theory by which to explain what I myself have witnessed and experienced, other than that supplied by the ancient and well-attested belief of the country.

The next day the formal proceedings took place in the Chapel of Karnstein. The grave of the Countess Mircalla was opened; and the General and my father recognised each his perfidious and beautiful guest, in the face now disclosed to view. The features, though a hundred and fifty years had passed since her funeral, were tinted with the warmth of life. Her eyes were open; no cadaverous smell exhaled from the coffin. The two medical men, one officially present, the other on the part of the promoter of the inquiry, attested the marvellous fact that there was a faint but appreciable respiration, and a corresponding action of the heart. The limbs were perfectly flexible, the flesh elastic; and the leaden coffin floated with blood, in which to a depth of seven inches, the body lay immersed. Here then, were all the admitted signs and proofs of vampirism. The body, therefore, in accordance with the ancient practice, was raised, and a sharp stake driven through the heart of the vampire, who uttered a piercing shriek at the moment, in all respects such as might escape from a living person in the last agony. Then the head was struck off, and a torrent of blood flowed from the severed neck. The body and head was next placed on a pile of wood, and reduced to ashes, which were thrown upon the river and borne away, and that territory has never since been plagued by the visits of a vampire.

My father has a copy of the report of the Imperial Commission, with the signatures of all who were present at these proceedings, attached in verification of the statement. It is from this official paper that I have summarized my account of this last shocking scene.

The cast of television's "Carmilla." (Photofest)

XVI
Conclusion

I write all this you suppose with composure. But far from it; I cannot think of it without agitation. Nothing but your earnest desire so repeatedly expressed, could have induced me to sit down to a task that has unstrung my nerves for months to come, and reinduced a shadow of the unspeakable horror which years after my deliverance continued to make my days and nights dreadful, and solitude insupportably terrific.

Let me add a word or two about that quaint Baron Vordenburg, to whose curious lore we were indebted for the discovery of the Countess Mircalla's grave.

He had taken up his abode in Gratz, where, living upon a mere pittance, which was all that remained to him of the once princely estates of his family, in Upper Styria, he devoted himself to the minute and laborious investigation of the marvellously authenticated tradition of Vampirism. He had at his fingers' ends all the great and little works upon the subject. "Magia Posthuma," "Phlegon de Mirabilibus," "Augustinus de cura pro Mortuis," "Philosophicae et Christianae Cogitationes de Vampiris," by John Christofer Herenberg; and a thousand others, among which I remember only a few of those which he lent to my father. He had a voluminous digest of all the judicial cases, from which he had extracted a system of principles that appear to govern — some always, and others occasionally only — the condition of the vampire. I may mention, in passing, that the deadly pallor attributed to that sort of *revenants*, is a mere melodramatic fiction. They present, in the grave, and when they show themselves in human society, the appearance of healthy life. When disclosed to light in their coffins, they exhibit all the symptoms that are enumerated as those which proved the vampire-life of the long-dead Countess Karnstein.

How they escape from their graves and return to them for certain hours every day, without displacing the clay or leaving any trace of disturbance in the state of the coffin or the cerements, has always been admitted to be utterly inexplicable. The amphibious existence of the vampire is sustained by daily renewed slumber in the grave. Its horrible lust for living blood supplies the vigour of its waking existence. The vampire is prone to be fascinated with an engrossing vehemence, resembling the passion of love, by particular persons. In pursuit of these it will exercise inexhaustible patience and stratagem, for access to a particular object may be obstructed in a hundred ways. It will never desist until it has satiated its passion, and drained the very life of its coveted victim. But it will, in these cases, husband and protract its murderous enjoyment with the refinement of an epicure, and heighten it by the gradual approaches of an artful courtship. In these cases it seems to yearn for something like sympathy and consent. In ordinary ones it goes direct to its object, overpowers with violence, and strangles and exhausts often at a single feast.

The vampire is, apparently, subject, in certain situations, to special conditions. In the particular instance of which I have given you a relation, Mircalla seemed to be limited to a name which, if not her real one, should at least reproduce, without the omission or addition of a single letter, those, as we say, anagrammatically, which compose it. Carmilla did this; so did Millarca.

My father related to the Baron Vordenburg, who remained with us for two or three weeks after the expulsion of Carmilla, the story about the Moravian nobleman and the vampire at Karnstein churchyard, and then he asked the Baron how he had discovered the exact position of the long-concealed tomb of the Countess Mircalla? The Baron's grotesque features puckered up into a mysterious smile; he looked down, still smiling on his worn spectacle-case and fumbled with it. Then looking up, he said:

"I have many journals, and other papers, written by that remarkable man; the most curious among them is one treating of the visit of which you speak, to Karnstein. The tradition, of course, discolours and distorts a little. He might have been termed a Moravian nobleman, for he had changed his abode to that territory, and was, beside, a noble. But he was, in truth, a native of Upper Styria. It is enough to say that in very early youth he had been a passionate and favoured lover of the beautiful Mircalla, Countess Karnstein. Her early death plunged him into inconsolable grief. It is the nature of vampires to increase and multiply, but according to an ascertained and ghostly law.

"Assume, at starting, a territory perfectly free from that pest. How does it begin, and how does it multiply itself? I will tell you. A person, more or less wicked, puts an end to himself. A suicide, under certain circumstances, becomes a vampire. That spectre visits living people in their slumbers; they die, and almost invariably, in the grave, develop into vampires. This happened in the case of the beautiful Mircalla, who was

"CARMILLA" ON STAGE

"Carmilla" very nearly had its stage debut in 1928, when Hamilton Deane, coauthor of the Broadway adaptation of *Dracula*, feuded with producer Horace Liveright over an American tour of the popular vampire play. In collaboration with his wife, actress Dora Mary Patrick, Deane wrote a theatrical adaptation of "Carmilla" very much in the style of *Dracula*, and threatened to tour it himself. Liveright relented, and the Deane/Patrick adaptation was never performed.

In 1937, the Earl of Longford presented a stage version in Dublin and London. Apropos the thick lesbian subtext, the *Sunday Times* admitted that it was prepared to spend an evening's playgoing in the company of such a fearful denizen of the charnel house as a vampire without embarassment. But this play... causes discomfort by apparently giving bald statement to a theme which, in the theatre at any rate, is usually treated with delicate insinuation."

New York's La Mama Experimental Theatre Company produced a curiously stylized chamber musical based on "Carmilla" written by Wilford Leach, in the fall of 1970. Owning to a knee operation undergone by the actress who played Carmilla, the staging was altered to allow the vampire to remain seated throughout the performance on a grandly carved Victorian sofa through which singers' faces protruded like animated masks.

haunted by one of those demons. My ancestor, Vordenburg, whose title I still bear, soon discovered this, and in the course of the studies to which he devoted himself, learned a great deal more.

"Among other things, he concluded that suspicion of vampirism would probably fall, sooner or later, upon the dead Countess, who in life had been his idol. He conceived a horror, be she what she might, of her remains being profaned by the outrage of a posthumous execution. He has left a curious paper to prove that the vampire, on its expulsion from its amphibious existence, is projected into a far more horrible life; and he resolved to save his once beloved Mircalla from this.

"He adopted the stratagem of a journey here, a pretended removal of her remains, and a real obliteration of her monument. When age had stolen upon him, and from the vale of years, he looked back on the scenes he was leaving, he considered, in a different spirit, what he had done, and a horror took possession of him. He made the tracings and notes which have guided me to the very spot, and drew up a confession of the deception that he had practised. If he had intended any further action in this matter, death prevented him; and the hand of a remote descendant has, too late for many, directed the pursuit to the lair of the beast."

We talked a little more, and among other things he said was this:

"One sign of the vampire is the power of the hand. The slender hand of Mircalla closed like a vice of steel on the General's wrist when he raised the hatchet to strike. But its power is not confined to its grasp; it leaves a numbness in the limb it seizes, which is slowly, if ever, recovered from."

The following Spring my father took me on a tour through Italy. We remained away for more than a year. It was long before the terror of recent events subsided; and to this hour the image of Carmilla returns to memory with ambiguous alternations — sometimes the playful, languid, beautiful girl; sometimes the writhing fiend I saw in the ruined church; and often from a reverie I have started, fancying I heard the light step of Carmilla at the drawing-room door.

The discovery of a female vampire's ritual destruction. From Varney the Vampyre *(1847)*

The Fate of Madame Cabanel

ELIZA LYNN LINTON

(1880)

PROGRESS HAD NOT invaded, science had not enlightened, the little hamlet of Pieuvrot, in Brittany. They were a simple, ignorant, superstitious set who lived there, and the luxuries of civilization were known to them as little as its learning. They toiled hard all the week on the ungrateful soil that yielded them but a bare subsistence in return; they went regularly to mass in the little rock-set chapel on Sundays and saints' days; believed implicitly all that Monsieur le curé said to them, and many things which he did not say; and they took all the unknown, not as magnificent, but as diabolical.

The sole link between them and the outside world of mind and progress was Monsieur Jules Cabanel, the proprietor, par excellence, of the place; *maire, juge de paix*, and all the public functionaries rolled into one. And he sometimes went to Paris whence he returned with a cargo of novelties that excited envy, admiration, or fear, according to the degree of intelligence in those who beheld them.

Monsieur Jules Cabanel was not the most charming man of his class in appearance, but he was generally held to be a good fellow at bottom. A short, thickset, low-browed man, with blue-black hair cropped close like a mat, as was his blue-black beard, inclined to obesity and fond of good living, he had need have some virtues behind the bush to compensate for his want of personal charms. He was not bad, however; he was only common and unlovely.

Up to fifty years of age he had remained the unmarried prize of the surrounding country; but hitherto he had resisted all the overtures made by maternal fowlers, and had kept his liberty and his bachelorhood intact. Perhaps his handsome housekeeper, Adèle, had something to do with his persistent celibacy. They said she had, under their breath as it were, down at *la Veuve Prieur's*; but no one dared to so much as hint the like to herself. She was a proud, reserved kind of woman; and had strange notions of her own dignity which no one cared to disturb. So, whatever the underhand gossip of the place might be, neither she nor her master got wind of it.

Presently and quite suddenly, Jules Cabanel, who had been for a longer time than usual in Paris, came home with a wife. Adèle had only twenty-four hours' notice to prepare for this strange home-coming; and the task seemed heavy. But she got through it in her old way of silent determination;

ABOUT THE AUTHOR

Although "The Fate of Madame Cabanel" is often cited as a terrifying feminist parable of female scapegoating, Eliza Lynn Linton (1822-1898) was no friend of the assertive and independent "new woman" of her time, but instead a staunch defender of the sexual status quo. Like so many social commentators of her time, she was also fervently anti-Zionist, and found much to fear in the perceived rise of a Jewish business class. The fear of females and foreigners was a constant subtext of late Victorian vampire narratives, reaching a delirious, near-paranoid climax in Bram Stoker's *Dracula* (1897).

The lasting power of "The Fate of Madame Cabanel" may be the result of the author's own cultural ambivalences about the individual and society, which imbues the story with a disturbing resonance.

arranged the rooms as she knew her master would wish them to be arranged; and even supplemented the usual nice adornments by a voluntary bunch of flowers on the salon table.

"Strange flowers for a bride," said to herself little Jeannette, the goose-girl who was sometimes brought into the house to work, as she noticed heliotrope — called in France *la fleur des veuves* — scarlet poppies, a bunch of belladonna, another of aconite — scarcely, as even ignorant little Jeannette said, flowers of bridal welcome or bridal significance. Nevertheless, they stood where Adèle had placed them; and if Monsieur Cabanel meant anything by the passionate expression of disgust with which he ordered them out of his sight, madame seemed to understand nothing, as she smiled with that vague, half-deprecating look of a person who is assisting at a scene of which the true bearing is not understood.

Madame Cabanel was a foreigner, and an Englishwoman; young, pretty and fair as an angel.

"*La beauté du diable*," said the Pieuvrotines, with something between a sneer and a shudder; for the words meant with them more than they mean in ordinary use. Swarthy, ill-nourished, low of stature and meagre in frame as they were themselves, they could not understand the plump form, tall figure and fresh complexion of the Englishwoman. Unlike their own experience, it was therefore more likely to be evil than good. The feeling which had sprung up against her at first sight deepened when it was observed that, although she went to mass with praiseworthy punctuality, she did not know her missal and signed herself *a tràvers*. *La beauté du diable*, in faith!

"*Pouf!*" said Martin Briolic, the old gravedigger of the little cemetery; "with those red lips of hers, her rose cheeks and her plump shoulders, she looks like a vampire and as if she lived on blood."

He said this one evening down at *la Veuve Prieur's*; and he said it with an air of conviction that had its weight. For Martin Briolic was reputed the wisest man of the district; not even excepting Monsieur le curé who was wise in his own way, which was not Martin's — nor Monsieur Cabanel who was wise in his, which was neither Martins's nor the cure's. He knew all about the weather and the stars, the wild herbs that grew on the plains and the wild shy beasts that eat them; and he had the power of divination and could find where the hidden springs of water lay far down in the earth when he held the baguette in his hand. He knew too, where treasures could be had on Christmas Eve if only you were quick and brave enough to enter the cleft in the rock at the right moment and come out again before too late; and he had seen with his own eyes the White Ladies dancing in the moonlight; and the little imps, the Infins, playing their prankish gambols by the pit at the edge of the wood. And he had a shrewd suspicion as to who, among those blackhearted men of La Crèche-en-bois — the rival hamlet — was a loup-garou, if ever there was one on the face of the earth and no one had doubted that! He had other powers of a yet

Eliza Lynn Linton may have been primarily a fiction writer, but she lived to see a macabre echo of her story enacted in the real world. A decade and a half following the publication of "The Fate of Madame Cabanel," a woman named Bridget Cleary met an eerily similar fate in Clonmel, Ireland. Cleary was not accused of vampirism herself, but rather of being the victim of a vampirish possessing spirit. E.F. Benson, soon to be one of the greatest purveyors of English ghost stories (see "The Room in the Tower" in this collection), wrote an essay analyzing the case for the June 1895 issue of* The Nineteenth Century, *which is excerpted below.

THE RECENT "WITCH BURNING" AT CLONMEL
by E. F. Benson

There seems to be no doubt, if we examine the motives which appear to have led to this crime, that the ten persons, nine of whom are to be tried on the capital charge of wilfully murdering Bridget Cleary, in the recent case of witch-burning at

more mystic kind; so that Martin Briolic's bad word went for something, if, with the illogical injustice of ill-nature his good went for nothing.

Fanny Campbell, or, as she was now Madame Cabanel, would have excited no special attention in England, or indeed anywhere but at such a dead-alive, ignorant, and consequently gossiping place as Pieuvrot. She had no romantic secret as her background; and what history she had was commonplace enough, if sorrowful too in its own way. She was simply an orphan and a governess; very young and very poor; whose employers had quarrelled with her and left her stranded in Paris, alone and almost moneyless; and who had married Monsieur Jules Cabanel as the best thing she could do for herself. Loving no one else, she was not difficult to be won by the first man who showed her kindness in her hour of trouble and destitution; and she accepted her middle-aged suitor, who was fitter to be her father than her husband, with a clear conscience and a determination to do her duty cheerfully and faithfully—all without considering herself as a martyr or an interesting victim sacrificed to the cruelty of circumstances. She did not know, however, of the handsome housekeeper Adèle, nor of the housekeeper's little nephew — to whom the master was so kind that he allowed him to live at the Maison Cabanel and had him well taught by the cure. Perhaps if she had she would have thought twice before she put herself under the same roof with a woman who for a bridal bouquet offered her poppies, heliotrope and poison-flowers.

If one had to name the predominant characteristic of Madame Cabanel it would be easiness of temper. You saw it in the round, soft, indolent lines of her face and figure; in her mild blue eyes and placid, unvarying smile; which irritated the more petulant French temperament and especially disgusted Adèle. It seemed impossible to make madame angry or even to make her understand when she was insulted, the housekeeper used to say with profound disdain; and, to do the woman justice, she did not spare her endeavours to enlighten her. But madame accepted all Adèle's haughty reticence and defiant continuance of mistress-hood with unwearied sweetness; indeed, she expressed herself gratified that so much trouble was taken off her hands, and that Adèle so kindly took her duties on herself.

The consequences of this placid lazy life, where all her faculties were in a manner asleep, and where she was enjoying the reaction from her late years of privation and anxiety, was, as might be expected, an increase in physical beauty that made her freshness and good condition still more remarkable. Her lips were redder, her cheeks rosier, her shoulders plumper than ever; but as she waxed, the health of the little hamlet waned, and not the oldest inhabitant remembered so sickly a season, or so many deaths. The master too, suffered slightly; the little Adolphe desperately.

This failure of general health in undrained hamlets is not uncommon in France or in England; neither is the steady and pitiable decline of French children; but Adèle treated it as something out of all the lines of normal experience; and, breaking her habits of reticence spoke to everyone quite

Clonmel, acted, if we may use such a word in connection with so ghastly a tragedy, honestly, and, as they undoubtedly appear to have believed, for the best.

This cruel and unnatural murder, as at first sight the crime seems to be, was committed deliberately, in cold blood, and throughout the preliminary examination no evidence, however slight, tended to indicate that either the woman's father, her husband, or any of those concerned in it were actuated by anger, malice, or motives of gain. The account given on all sides of the reasons for inflicting the tortures which eventually caused the death of this woman, with whom, apparently, all parties concerned were on the best of terms, are both too fantastic on the one hand, and, on the other, too authentically based on what was at one time a very widely spread superstition, to have been invented. That such a superstition should still be so deeply ingrained in the minds of these peasants as to lead in practice to so horrible a deed seems surprising enough on first sight, and becomes doubly surprising when we consider to

fiercely of the strange sickliness that had fallen on Pieuvrot and the Maison Cabanel; and how she believed it was something more than common; while as to her little nephew, she could give neither a name nor find a remedy for the mysterious disease that had attacked him. There were strange things among them, she used to say; and Pieuvrot had never done well since the old, times were changed. Jeannette used to notice how she would sit gazing at the English lady, with such a deadly look on her handsome face when she turned from the foreigner's fresh complexion and grand physique to the pale face of the stunted, meagre, fading child. It was a look, she said afterwards, that used to make her flesh get like ice and creep like worms.

how primitive a stratum of belief it belongs.

It appears from evidence given at the committal that Bridget Cleary, in consequence of certain nervous, excited symptoms which she had exhibited, was forced to swallow certain herbs, was beaten and burned, and was repeatedly asked the question, "In the name of God, are you Bridget Cleary?" During this treatment she received the injuries of which she died, but it is, I think, quite clear that the men neither wished nor meant to kill her. All the witnesses, who on important points entirely corroborated each other, agreed in saying that the object of this treatment was to drive out from her body the fairy which had taken possession of it, and which exhibited its presence by her nervous disorder. The woman's own soul would then be able to return.

This is not, then, a case of witch-burning at all. Bridget Cleary, they believed, was possessed, and they tried by violent means to make the invading spirit come out of her. But what is more convincing than the testimony of all the

One night Adèle, as if she could bear it no longer, dashed down to where old Martin Briolic lived, to ask him to tell her how it had all come about — and the remedy.

"Hold, Ma'am Adèle," said Martin, as he shuffled his greasy tarot cards and laid them out in triplets on the table; 'there is more in this than one sees. One sees only a poor little child become suddenly sick; that may be, is it not so? And no harm done by man? God sends sickness to us all and makes my trade profitable to me. But the little Adolphe has not been touched by the Good God. I see the will of a wicked woman in this. Rein!" Here he shuffled the cards and laid them out with a kind of eager distraction of manner, his withered hands trembling and his mouth uttering words that Adèle could not catch. "Saint Joseph and all the saints protect us!" he cried; "the foreigne — the Englishwoman — she whom they call Madame Cabanel — no rightful madame she! — Ah, misery!"

"Speak, Father Martin! What do you mean!" cried Adèle, grasping his arm. Her black eyes were wild; her arched nostrils dilated; her lips, thin, sinuous, flexible, were pressed tight over her small square teeth.

"Tell me in plain words what you would say!"

"Broucolaque!" said Martin in a low voice.

"It is what I believed!" cried Adèle. "It is what I knew. Ah, my Adolphe! Woe on the day when the master brought that fair-skinned devil home!"

"Those red lips don't come by nothing, Ma'am Adèle," cried Martin nodding his head. "Look at them — they glisten with blood! I said so from the beginning; and the cards, they said so too. I drew 'blood' and a 'bad fair woman' on the evening when the master brought her home, and I said to myself, 'Ha, ha, Martin! you are on the track, my boy — on the track. Martin!' — and, Ma'am Adèle, I have never left it! Broucolaque! That's what the cards say, Ma'am Adèle. Vampire. Watch and see; watch and see; and you'll find that the cards have spoken true."

"And when we have found, Martin?" said Adèle in a hoarse whisper.

The old man shuffled his cards again. "When we have found, Ma'am Adèle?" he said slowly. "You know the old pit out there by the forest? — The old pit where the lutins run in and out, and where the White Ladies wring the necks of those who come upon them in the moonlight? Perhaps the White Ladies will do as much for the English wife of Monsieur Cabanel; who knows?"

"They may," said Adèle, gloomily.

"Courage, brave woman!" said Martin. "They will."

The only really pretty place about Pieuvrot was the cemetery. To be sure there was the dark gloomy forest which was grand in its own mysterious way; and there was the broad wide plain where you might wander for a long summer's day and not come to the end of it; but these were scarcely places where a young woman would care to go by herself; and for the rest, the miserable little patches of cultivated ground, which the peasants had snatched from the surrounding waste and where they had raised poor crops, were not very lovely. So Madame Cabanel, who, for all the soft indolence that had invaded her, had the Englishwoman's inborn love for walking and fresh air, haunted the pretty little graveyard a good deal. She had no sentiment connected with it. Of all the dead who laid there in their narrow coffins, she knew none and cared for none; but she liked to see the pretty little flower-beds and the wreaths of immortelles, and the like; the distance too, from her own home was just enough for her; and the view over the plain to the dark belt of forest and the mountains beyond, was fine.

The Pieuvrotines did not understand this. It was inexplicable to them that any one, not out of her mind, should go continually to the cemetery — not on the day of the dead and not to adorn the grave of one she loved — only to sit there and wander among the tombs, looking out on to the plain and the mountains beyond when she was tired.

"It was just like —" The speaker, one Lesouëf, had got so far as this, when he stopped for a word.

He said this down at *la Veuve Prieur's* where the hamlet collected nightly to discuss the day's small doings, and where the main theme, ever since she had come among them, three months ago now, had been Madame Cabanel and her foreign ways and her wicked ignorance of her mass-book and her wrong-doings of a mysterious kind generally, interspersed with jesting queries, bandied from one to the other, of how Ma'am Adèle liked it? — and what would become of le petit Adolphe when the rightful heir appeared? — some adding that monsieur was a brave man to shut up two wild cats under the same roof together; and what would become of it in the end? Mischief of a surety.

"Wander about the tombs just like what, Jean Lesouëf?" said Martin Briolic. Rising, he added in a low but distinct voice, every word falling clear and clean: "I will tell you like what, Lesouëf — like a vampire! La femme Cabanel has red lips and red cheeks; and Ma'am Adèle's little nephew is perishing before your eyes. La femme Cabanel has red lips and red cheeks; and she sits for hours among the tombs. Can you read the riddle, my friends? For me it is as clear as the blessed sun."

"Ha, Father Martin, you have found the word — like a vampire!" said Lesouëf with a shudder.

witnesses is the fact that they were acting strictly in accordance with a primitive and savage superstition.

However, the woman died, and we notice that after her death, which I hope to show was entirely unpremeditated, and undesired, the prescription as it were having failed to act, these men fell back on another very common superstition. They do not seem ever to have been thoroughly convinced that they had succeeded in driving the spirit out of the woman's body, and when she died they still hoped that it was the possessing spirit they had killed, and believed that the real Bridget Cleary would return, sitting on a grey or white horse, the reins of which would have to be cut to enable her to come back. According to the latest accounts this is still believed, and men wait in the "fairy inhabitance" for the coming of the white horse that will bear Bridget Cleary . . .

Primitive man knows nothing about germs, microbes, nerves, or laws of nature, and when the simplest and most evident of

"Like a vampire!" they all echoed with a groan.

"And I said vampire the first," said Martin Briolic. "Call to mind I said it from the first."

"Faith! And you did," they answered; "and you said true."

So now the unfriendly feeling that had met and accompanied the young Englishwoman ever since she came to Pieuvrot had drawn to a focus. The seed which Martin and Adèle had dropped so sedulously, had at last taken root; and the Pieuvrotines would have been ready to accuse of atheism and immorality any one who had doubted their decision, and had declared that pretty Madame Cabanel was only a young woman with nothing special to do, a naturally fair complexion, superb health — and no vampire at all, sucking the blood of a living child or living among the tombs to make the newly buried her prey.

The little Adolphe grew paler and paler, thinner and thinner; the fierce summer sun told on the half-starved dwellers within those foul mud-huts surrounded by undrained marshes; and Monsieur Jules Cabanel's former solid health followed the law of the rest. The doctor, who lived at Crêche-en-bois, shook his head at the look of things; and said it was grave. When Adèle pressed him to tell her what was the matter with the child and with monsieur, he evaded the question; or gave her a word which she neither understood nor could pronounce. The truth was, he was a credulous and intensely suspicious man; a viewy man who made theories and then gave himself to the task of finding them true. He had made the theory that Fanny was secretly poisoning both her husband and the child; and though he would not give Adèle a hint of this, he would not set her mind at rest by a definite answer that went on any other line.

As for Monsieur Cabanel, he was a man without imagination and without suspicion; a man to take life easily and not distress himself too much for the fear of wounding others; a selfish man but not a cruel one; a man whose own pleasure was his supreme law and who could not imagine, still less brook, opposition or the want of love and respect for himself. Still, he loved his wife as he had never loved a woman before. Coarsely moulded, common-natured as he was, he loved her with what strength and passion of poetry nature had given him; and if the quantity was small, the quality was sincere. But that quality was sorely tried when — now Adèle, now the doctor — hinted mysteriously, the one at diabolical influences, the other at underhand proceedings of which it behooved him to be careful, especially careful what he ate and drank and how it was prepared and by whom; Adèle adding hints about the perfidiousness of English women and the share which the devil had in fair hair and brilliant complexions. Love his young wife as he might, this constant dropping of poison was not without some effect. It told much for his steadfastness and loyalty that it should have had only so small effect.

One evening, however, when Adèle, in an agony, was kneeling at his feet — madame had gone out for her usual walk — crying: "Why did you

natural phenomena, like disease, death, or storm, are brought before his notice, he forms an equally simple and natural theory about them when he refers to them, as he invariably does, to the work of some malignant spirits. To the savage, and to primitive religion generally, the idea of a purely beneficent god is altogether foreign. To him life, health, and the supplies of the simpler means of life are the normal conditions of his consciousness; sickness, famine, and death the abnormal, the work of some external agency, obviously malignant. And we therefore find that the earliest conceptions which he forms about the forces over which he has no control represent them as entirely evil, interfering in the normal and beneficent course of events.

That the idea of external agencies being malignant should belong to the earliest stages of religious belief is natural enough, for evil is more readily referred to as a cause than its opposite. The destruction of a crop by a thunderstorm, for instance, is easily traceable to the malignant spirit in the thunderstorm; whereas the slow ripening of

leave me for such as she is? — I, who loved you, who was faithful to you, and she, who walks among the graves, who sucks your blood and our child's — she who has only the devil's beauty for her portion and who loves you not?" — something seemed suddenly to touch him with electric force.

"Miserable fool that I was!" he said, resting his head on Adèle's shoulders and weeping. Her heart leapt with joy. Was her reign to be renewed? Was her rival to be dispossessed?

From that evening Monsieur Cabanel's manner changed to his young wife but she was too easy-tempered and unsuspicious to notice anything, or if she did, there was too little depth in her own love for him, — it was so much a matter of untroubled friendliness only — that she did not fret but accepted the coldness and brusqueness that had crept into his manner as good-naturedly as she accepted all things. It would have been wiser if she had cried and made a scene and come to an open fracas with Monsieur Cabanel. They would have understood each other better; and Frenchmen like the excitement of a quarrel and a reconciliation.

Naturally kind hearted, Madame Cabanel went much about the village, offering help of various kinds to the sick. But no one among them all, not the very poorest — indeed, the very poorest the least — received her civilly or accepted her aid. If she attempted to touch one of the dying children, the mother, shuddering, withdrew it hastily to her own arms; if she spoke to the adult sick, the wan eyes would look at her with a strange horror and the feeble voice would mutter words in a patois she could not understand. But always came the same word, "broucolaque!"

"How these people hate the English!" she used to think as she turned away, perhaps just a little depressed, but too phlegmatic to let herself be uncomfortable or troubled deeply.

It was the same at home. If she wanted to do any little act of kindness to the child, Adèle passionately refused her. Once she snatched him rudely from her arms, saying as she did so: "Infamous broucolaque! Before my very eyes?" And once, when Fanny was troubled about her husband and proposed to make him a cup of beef-tea a l'Anglaise, the doctor looked at her as if he would have looked through her; and Adèle upset the saucepan; saying insolently — but yet hot tears were in her eyes — "Is it not fast enough for you, madame? Not faster, unless you kill me first!"

To all of which Fanny replied nothing; thinking only that the doctor was very rude to stare so fixedly at her and that Adèle was horribly cross; and what an ill-tempered creature she was; and how unlike an English housekeeper!

But Monsieur Cabanel, when he was told of the little scene, called Fanny to him and said in a more caressing voice than he had used to her of late: "Thou wouldst not hurt me, little wife? It was love and kindness, not wrong, that thou wouldst do?"

"Wrong? What wrong could I do?"answered Fanny, opening her blue

wheat under the kindly influence of the sun and rain is a more subtle phenomenon, because it is a less evident process, the workings of which are altogether hidden from him, and a more natural one, since, on the whole, wheat ripens more often than it fails to do so. Health, again, is naturally regarded as normal, and when disease attacks a town it is obviously due to a demon of disease, an evil hostile power, whereas health cannot be definitively referred to any one cause. Similarly, recovery and convalescence are but a reversion to normal conditions, due to the cessation of the action of the demon who sent the disease.

This theory accounts for the undeniable fact that a purely healing cult finds no place in any system of early religious belief. The primary function of the spirit who is concerned with such matters is, not to remove, but to send disease, though, in a secondary manner, he is regarded as being able to remove it, inasmuch as he can stop sending it. But by degrees, as spirits pass from being considered the enemies of into becoming the

eyes wide. "What wrong should I do to my best and only friend?"

"And I am thy friend? Thy lover? Thy husband? Thou lovest me dear?" said Monsieur Cabanel.

"Dear Jules, who is so dear; who so near?" she said kissing him, while he said fervently:

"God bless thee!"

The next day Monsieur Cabanel was called away on urgent business. He might be absent for two days, he said, but he would try to lessen the time; and the young wife was left alone in the midst of her enemies, without even such slight guard as his presence might prove.

Adèle was out. It was a dark, hot summer's night, and the little Adolphe had been more feverish and restless than usual all the day. Towards evening he grew worse; and though Jeannette, the goose-girl, had strict commands not to allow madame to touch him, she grew frightened at the condition of the boy; and when madame came into the small parlour to offer her assistance, Jeannette gladly abandoned a charge that was too heavy for her and let the lady take him from her arms.

Sitting there with the child in her lap, cooing to him, soothing him by a low, soft nursery song, the paroxysm of his pain seemed to her to pass and it was as if he slept. But in that paroxysm he had bitten both his lip and tongue; and the blood was now oozing from his mouth. He was a pretty boy; and his mortal sickness made him at this moment pathetically lovely. Fanny bent her head and kissed the pale still face — and the blood that was on his lips was transferred to hers.

While she still bent over him — her woman's heart touched with a mysterious force and prevision of her own future motherhood — Adèle, followed by old Martin and some others of the village, rushed into the room.

"Behold her!" she cried, seizing Fanny by the arm and forcing her face upwards by the chin — "behold her in the act! Friends, look at my child — dead, dead in her arms; and she with his blood on her lips! Do you want more proofs? — Vampire that she is, can you deny the evidence of your own senses?"

"No! No!" roared the crowd hoarsely. "She is a vampire — a creature cursed by God and the enemy of man; away with her to the pit. She must die as she has made others to die!"

"Die, as she has made my boy to die!" said Adèle; and more than one who had lost a relative or child during the epidemic echoed her words, "Die, as she has made mine to die!"

"What is the meaning of all this?" said Madame Cabanel, rising and facing the crowd with the true courage of an Englishwoman. "What harm have I done to any of you that you should come about me, in the absence of my husband, with these angry looks and insolent words?"

"What harm hast thou done?" cried old Martin, coming close to her.

"Sorceress as thou art, thou hast bewitched our good master; and vampire as thou art, thou nourishest thyself on our blood! Have we not

friends of man, prayer and sacrifice begin to be offered to them, that they may confer directly prosperity and health, and remove evils not necessarily of their own sending. In other words, the secondary function of primitive gods becomes at a later stage their primary function. So, also, the sorcerer, who was the medium between man and the malignant spirit, and who induced him to send disease and death, while remaining his medium, becomes the priest of the god of health and life, and induces him to restore his suppliants to strength . . .

Now, all the details of the death of Bridget Cleary point to the earliest stage in such beliefs, when the possessing spirit was thought to be purely malignant, and had to be forcibly ejected. What caused these men to imagine that she had been taken possession of by a fairy was the exhibition of certain nervous and excited symptoms. This . . . is well paralleled from the folk-lore of other nations. To take a few instances: among the Zulus there are spirits, called *amatongo*, supposed to be the ghosts of

proof of that at this very moment? Look at thy mouth — cursed broucolaque; and here lies thy victim, who accuses thee in his death!"

Fanny laughed scornfully, "I cannot condescend to answer such folly," she said lifting her head. "Are you men or children?"

"We are men, madame," said Legros the miller; "and being men we must protect our weak ones. We have all had our doubts — and who more cause than I, with three little ones taken to heaven before their time — and now we are convinced."

"Because I have nursed a dying child and done my best to soothe him!" said Madame Cabanel with unconscious pathos.

"No more words!" cried Adèle, dragging her by the arm from which she had never loosed her hold. "To the pit with her, my friends, if you would not see all your children die as mine has died — as our good Legros's have died!"

A kind of shudder shook the crowd; and a groan that sounded in itself a curse burst from them.

"To the pit!" they cried. "Let the demons take their own!"

Quick as thought Adèle pinioned the strong white arms whose shape and beauty had so often maddened her with jealous pain; and before the poor girl could utter more than one cry Legros had placed his brawny hand over her mouth. Though this destruction of a monster was not the murder of a human being in his mind, or in the mind of any there, still they did not care to have their nerves disturbed by cries that sounded so human as Madame Cabanel's. Silent then, and gloomy, that dreadful cortege took its way to the forest, carrying its living load; gagged and helpless as if it had been a corpse among them. Save with Adèle and old Martin, it was not so much personal animosity as the instinctive self-defence of fear that animated them. They were executioners, not enemies; and the executioners of a more righteous law than that allowed by the national code. But one by one they all dropped off till their numbers were reduced to six; of whom Legros was one, and Lesouëf, who had lost his only sister, was also one.

The pit was not more than an English mile from the Maison Cabanel. It was a dark and lonesome spot, where not the bravest man of all that assembly would have dared to go alone after nightfall, not even if the cure had been with him; but a multitude gives courage, said old Martin Briolic; and half a dozen stalwart men, led by such a woman as Adèle, were not afraid of even lutins or the White Ladies.

As swiftly as they could for the burden they bore, and all in utter silence, the cortege strode over the moor; one or two of them carrying rude torches; for the night was black and the way was not without its physical dangers. Nearer and nearer they came to the fatal bourn; and heavier grew the weight of their victim. She had long ceased to struggle; and now lay as if dead in the hands of her bearers. But no one spoke of this or of aught else. Not a word was exchanged between them; and more than

ancestors. A man possessed by them exhibits hysterical symptoms, which are regarded as a sign of possession. Among Siberian tribes, children liable to convulsions are looked upon as being possessed by the divining spirit. So, too, the priestess at Delphi passed into an epileptic trance before the oracular spirit descended on her, and the oracle of Trophonios produced hysterical laughter in those who consulted it.

A further confirmation of this theory as applied to this case is shown in the fact that the door was left open, in order that the ejected spirit might pass out. Had these men wished to murder the woman, the very last thing they would have done would be to leave the door open; but such a proceeding is entirely in accordance with the idea they had in their minds. Like the Hottentots, who "jolt and pommel" a dying man in order to drive the spirit of disease out, so they beat and burned this wretched victim of superstition in order to expel the spirit that possessed her. Far from wishing to murder her, they wished to bring her back into the land of the

one, even of those left, began to doubt whether they had done wisely, and whether they had not better have trusted to the law. Adèle and Martin alone remained firm to the task they had undertaken; and Legros too was sure; but he was weakly and humanly sorrowful for the thing he felt obliged to do. As for Adèle, the woman's jealousy, the mother's anguish and the terror of superstition, had all wrought in her so that she would not have raised a finger to have lightened her victim of one of her pains, or have found her a woman like herself and no vampire after all.

living, and it seems that the remorse with which her husband was seized after her death was perfectly genuine. It is inconceivable that, if they had wished to kill her, they would have left the door open, that they should have allowed their shous to attract the neighbours, or that ten persons should have been admitted to witness the deed. Terrible and ghastly as the case is, we cannot call it wilful murder.

The way got darker; the distance between them and their place of execution shorter; and at last they reached the border of the pit where this fearful monster, this vampire — poor innocent Fanny Cabanel — was to be thrown. As they lowered her, the light of their torches fell on her face.

"Grand Dieu!" cried Legros, taking off his cap; "she is dead!"

"A vampire cannot die," said Adèle, "it is only an appearance. Ask Father Martin."

"A vampire cannot die unless the evil spirits take her, or she is buried with a stake thrust through her body," said Martin Briolic sententiously.

"I don't like the look of it," said Legros; and so said some others.

They had taken the bandage from the mouth of the poor girl; and as she lay in the flickering light, her blue eyes half open; and her pale face white with the whiteness of death, a little return of human feeling among them shook them as if the wind had passed over them.

French funerary sculpture
(Photo: David J. Skal)

Suddenly they heard the sound of horses' hoofs thundering across the plain. They counted two, four, six; and they were now only four unarmed men, with Martin and Adèle to make up the number. Between the vengeance of man and the power and malice of the wood-demons, their courage faded and their presence of mind deserted them. Legros rushed frantically into the vague darkness of the forest; Lesouëf, followed him; the other two fled over the plain while the horsemen came nearer and nearer. Only Adèle held the torch high above her head, to show more clearly both herself in her swarthy passion and revenge and the dead body of her victim. She wanted no concealment; she had done her work, and she gloried in it. Then the horsemen came plunging to them — Jules Cabanel the first, followed by the doctor and four gardes champetres.

"Wretches! Murderers!" was all he said, as he flung himself from his horse and raised the pale face to his lips.

"Master," said Adèle, "she deserved to die. She is a vampire and she has killed our child."

"Fool!" cried Jules Cabanel, flinging off her hand. "Oh, my loved wife! Thou who did no harm to man or beast, to be murdered now by men who are worse than beasts!"

"She was killing thee," said Adèle. "Ask monsieur le docteur. What ailed the master, monsieur?"

"Do not bring me into this infamy," said the doctor looking up from the dead. "Whatever ailed monsieur, she ought not to be here. You have made yourself her judge and executioner, Adèle, and you must answer: for it to the law."

"You say this too, master?" said Adèle.

"I say so too," returned Monsieur Cabanel. "To the law you must answer for the innocent life you have so cruelly taken — you and all the fools and murderers you have joined to you."

"And is there to be no vengeance for our child?"

"Would you revenge yourself on God, woman?" said Monsieur Cabanel sternly.

"And our past years of love, master?"

"Are memories of hate, Adèle," said Monsieur Cabanel, as he turned again to the pale face of his dead wife.

"Then my place is vacant," said Adèle, with a bitter cry. "Ah, my little Adolphe, it is well you went before!"

"Hold, Ma'am Adèle!" cried Martin.

But before a hand could be stretched out, with one bound, one shriek, she had flung herself into the pit where she had hoped to bury Madame Cabanel; and they heard her body strike the water at the bottom with a dull splash, as of something falling from a great distance.

"They can prove nothing against me, Jean," said old Martin to the garde who held him. "I neither bandaged her mouth nor carried her on my shoulders. I am the gravedigger of Pieuvrot, and, *ma foi*, you would all do badly, you poor creatures, when you die, without me! I shall have the honour of digging madame's grave, never doubt it; and, Jean," he whispered, "they may talk as they like, those rich aristos who know nothing. She is a vampire, and she shall have a slatte through her body yet! Who knows better than I? If we do not tie her down like this, she will come out of her grave and suck our blood; it is a way these vampires have."

"Silence there!" said the garde, commanding the little escort. "To prison with the assassins; and keep their tongues from wagging."

"To prison with martyrs and the public benefactors," retorted old Martin. "So the world rewards its best!"

And in this faith he lived and died, as a forçat at Toulon, maintaining to the last that he had done the world a good service by ridding it of a monster who else would not have left one man in Pieuvrot to perpetuate his name and race. But Legros and also Lesouëf, his companion, doubted gravely of the righteousness of that act of theirs on that dark summer's night in the forest; and though they always maintained that they should not have been punished, because of their good motives, yet they grew in time to disbelieve old Martin Briolic and his wisdom, and to wish that they had let the law take its own course unhelped by them — reserving their strength for the grinding of the hamlet's flour and the mending of the hamlet's sabots — and the leading of a good life according to the teaching of Monsieur le curé and the exhortations of their own wives.

Original dust jacket art for Studies in Death.

A True Story of A Vampire

ERIC, COUNT STENBOCK
(1894)

ABOUT THE AUTHOR

Eric Magnus Andreas Harry Stanislaus Stenbock was born March 12, 1860 to an aristocratic Estonian family. His parents, the Count and Countess Stenbock, were both eccentric characters, the mother "odd," and the father a terminal alcoholic who died shortly after the birth of their son.

According to biographer John Adlard in *Stenbock, Yeats and the Nineties* (London: Cecil & Amelia Woolf, 1969), one of the child's earliest memories was a macabre Punch and Judy show which "filled me with intense horror." Perhaps appropriately, the rest of Stenbock's life would be consecrated to the pursuit of the grotesque.

Privately educated, enrolled at Oxford in 1879, he may have spent some part of his youth in Russia. Arthur Symons recalled him as "one of these extraordinary Slav creatures, who, coming to settle down in London after half a lifetime spent in travelling, live in a bizarre, fantastic, feverish, eccentric, extravagant, morbid and perverse fashion... he was one of the most inhuman beings I have ever encountered; inhuman and abnormal; a degenerate, who had I know not how many vices."

VAMPIRE STORIES ARE generally located in Styria; mine is also. Styria is by no means the romantic kind of place described by those who have certainly never been there. It is a flat, uninteresting country, only celebrated by its turkeys, its capons, and the stupidity of its inhabitants. Vampires generally arrive by night, in carriages drawn by two black horses.

Our Vampire arrived by the commonplace means of the railway train, and in the afternoon. You must think that I am joking, or perhaps that by the word "Vampire" I mean a financial vampire. No, I am quite serious. The Vampire of whom I am speaking, who laid waste our hearth and home, was a real vampire.

Vampires are generally described as dark, sinister-looking, and singularly handsome. Our Vampire was, on the contrary, rather fair, and certainly not at first sight sinister-looking, and though decidedly attractive in appearance, not what one would call singularly handsome.

Yes, he desolated our home, killed my brother — the one object of my adoration — also my dear father. Yet, at the same time, I must say that I myself came under the spell of his fascination, and, in spite of all, have no ill-will towards him now.

Doubtless you have read in the papers *passim* of "the Baroness and her beasts." It is to tell how I came to spend most of my useless wealth on an asylum for stray animals that I am writing this.

I am old now; what happened then was when I was a little girl of about thirteen. I will begin by describing our household. We were Poles; our name was Wronski: we lived in Styria, where we had a castle. Our household was very limited. It consisted, with the exclusion of domestics, of only my father, my governess — a worthy Belgian named mademoiselle Vonnaert — my brother, and myself. Let me begin with my father: he was old, and both my brother and I were children of his old age. Of my mother I remember nothing: she died in giving birth to my brother, who is only one year, or not as much, younger than myself. Our father was studious, continually occupied in reading books, chiefly on recondite subjects and in all kinds of unknown languages. He had a long white beard, and wore habitually a black velvet skull-cap.

How kind he was to us! It was more than I could tell. Still it was not I who was the favourite. His whole heart went out to Gabriel — Gabryel as we spelt it in Polish. He was always called by the Russian abbreviation Gavril — I mean, of course, my brother, who had a resemblance to the only portrait of my mother, a slight chalk sketch which hung in my father's study. But I was by no means jealous: my brother was and has been the only love of my life. It is for his sake that I am now keeping in Westbourne Park a home for stray cats and dogs.

I was at that time, as I said before, a little girl; my name was Carmela. My long tangled hair was always all over the place, and never would be combed straight. I was not pretty — at least, looking at a photograph of me at that time, I do not think I could describe myself as such. Yet at the same time, when I look at my photograph, I think my expression may have been pleasing to some people: irregular features, large mouth, and large wild eyes.

I was by way of being naughty — not so naughty as Gabriel in the opinion of Mlle. Vonnaert. Mlle. Vonnaert, I may intercalate, was a wholly excellent person, middle-aged, who really did speak good French, although she was a Belgian, and could make herself understood in German, which, as you may or may not know, is the current language of Styria.

I find it difficult to describe my brother Gabriel; there was something about him strange and superhuman, or perhaps I should rather say praeter-human, something between the animal and the divine. Perhaps the Greek idea of the Faun might illustrate what I mean; but that will not do either. He had large, wild, gazelle-like eyes: his hair, like mine, was in a perpetual tangle — that point he had in common with me, and indeed, as I afterwards heard, our mother having been of gypsy race, it will account for much of the innate wildness there was in our natures. I was wild enough, but Gabriel was much wilder. Nothing would induce him to put on shoes and socks, except on Sundays — when he also allowed his hair to be combed, but only by me. How shall I describe the grace of that lovely mouth, shaped verily "en arc d'amour." I always think of the text in the Psalm, "Grace is shed forth on thy lips, therefore has God blessed thee eternally" — lips that seemed to exhale the very breath of life. Then that beautiful, lithe, living, elastic form!

He could run faster than any deer: spring like a squirrel to the topmost branch of a tree: he might have stood for the sign and symbol of vitality itself. But seldom could he be induced by Mlle. Vonnaert to learn lessons, but when he did so, he learned with extraordinary quickness. He would play upon every conceivable instrument, holding a violin here, there, and everywhere except the right place: manufacturing instruments for himself out of reeds — even sticks. Mlle. Vonnaert made futile efforts to induce him to learn to play the piano. I suppose he was what was called spoilt, though merely in the superficial sense of the word. Our father allowed him to indulge in every caprice.

One of his peculiarities, when quite a little child, was horror at the sight of meat. Nothing on earth would induce him to taste it. Another

Symons compared Stenbock to the lines in Wilde's poem "The Sphinx" (1894):

Your eyes are like fantastic moons that
Shiver in some stagnant lake,

Your tongue is like a scarlet snake that
Dances to fantastic tunes.

Get hence, you loathsome mystery!
Hideous
Animal, get hence!

You wake in me each bestial sense,
You make me what I would not be.

To all accounts, Stenbock was an extremely strange young man. A consummate dilletante, he would ultimately publish only two slim volumes of his poetry and fiction; his real creative achievement was the transformation of his life itself into bizarre performance art, recalling the decadent excesses of J. K. Huysman's peerlessly decadent French novel *A Rebours* (1884), with which Stenbock was, no doubt, thoroughly acquainted.

Stenbock's life was consecrated in equal parts to to artifice and animalism. Like the doomed Gabriel of "A True Story of a

thing which was particularly remarkable about him was his extraordinary power over animals. Everything seemed to come tame to his hand. Birds would sit on his shoulder. Then sometimes Mlle. Vonnaert and I would lose him in the woods — he would suddenly dart away. Then we would find him singing softly or whistling to himself, with all manner of woodland creatures around him, — hedgehogs, little foxes, wild rabbits, marmots, squirrels, and such like. He would frequently bring these things home with him and insist on keeping them. This strange menagerie was the terror of poor Mlle. Vonnaert's heart. He chose to live in a little room at the top of a turret; but which, instead of going upstairs, he chose to reach by means of a very tall chestnut tree, through the window. But in contradiction to all this, it was his custom to serve every Sunday Mass in the parish church, with hair nicely combed and with white surplice and red cassock. He looked as demure and tamed as possible. Then came the element of the divine. What an expression of ecstasy there was in those glorious eyes!

Thus far I have not been speaking about the Vampire. However, let me begin with my narrative at last. One day my father had to go to the neighboring town — as he frequently had. This time he returned accompanied by a guest. The gentleman, he said, had missed his train, through the late arrival of another at our station, which was a junction, and he would therefore, as trains were not frequent in our parts, have had to wait there all night. He had joined in conversation with my father in the too-late-arriving train from the town: and had consequently accepted my father's invitation to stay the night at our house. But of course, you know, in those out-of-the-way parts we are almost patriarchal in our hospitality.

He was announced under the name of Count Vardalek — the name being Hungarian. But he spoke German well enough: not with the monotonous accentuation of Hungarians, but rather, if anything, with a slight Slavonic intonation. His voice was particularly soft and insinuating. We soon afterwards found out he could talk Polish, and Mlle. Vonnaert vouched for his good French. Indeed he seemed to know all languages. But let me give my first impressions. He was rather tall, with fair wavy hair, rather long, which accentuated a certain effeminacy about his smooth face. His figure had something — I cannot say what — serpentine about it. The features were refined; and he had long, slender, magnetic-looking hands, a somewhat long sinuous nose, a graceful mouth, and an attractive smile, which belied the intense sadness of the expression of the eyes. When he arrived his eyes were half closed — indeed they were habitually so — so that I could not decide their colour. He looked worn and wearied. I could not possibly guess his age.

Suddenly Gabriel burst into the room: a yellow butterfly was clinging to his hair. He was carrying in his arms a little squirrel. Of course he was bare-legged as usual. The stranger looked up at his approach; then I noticed his eyes. They were green: they seemed to dilate and grow larger. Gabriel

Vampire," Stenbock himself had a special affinity for undomesticated animals, especially monkeys and reptiles, with which he shared close living quarters. According to biographer Adlard, "Often in the morning he would appear in his marvellous Japanese dressing-gown, with the snake coiled around his neck, and let it slither its way down his sleeve and out at the cuff, to the horror of those present." Once, he served his dinner guests from a centerpiece in the shape of a coffin, figuratively transforming those assembled into a party of ravenous ghouls.

Stenbock emblemized a hot-button topic of Victorian culture, the fear of evolutionary degeneracy, an issue demonized and popularized by the Austrian social critic Max Nordau, whose book *Degeneration* appeared in English translation in 1895. Nordau's goal was nothing short of the total medicalization of art and literature. "Degenerates are not always criminals," he wrote. "They are often authors and artists [who] manifest the same mental characteristics, and for the most part the same somatic

stood stock-still, with a startled look, like that of a bird fascinated with a serpent. But nevertheless he held out his hand to the newcomer Vardalek, taking his hand — I don't know why I noticed this trivial thing, — pressed the pulse with his forefinger. Suddenly Gabriel darted from the room and upstairs, going to his turret-room this time by the staircase instead of the tree. I was in terror of what the Count might think of him. Great was my relief when he came down in his velvet Sunday suit, and shoes and stockings. I combed his hair, and set him generally right.

When the stranger came to dinner his appearance had somewhat altered; he looked much younger. There was an elasticity of the skin, combined with a delicate complexion, rarely to be found in a man. Before, he had struck me as very pale.

Well, at dinner we were all charmed with him, especially my father. He seemed to be thoroughly acquainted with all my father's peculiar hobbies. Once, when my father was relating some of his military experiences, he said something about a drummer-boy who was wounded in battle. His eyes opened completely again and dilated: this time with a particularly disagreeable expression, dull and dead, yet at the same time animated by some horrible excitement. But this was only momentary.

Max Nordau, author of Degeneration *and leading critic of literary "decadence."*

The chief subject of his conversation with my father was about certain curious mystical books which my father had just lately picked up, and which he could not make out, but Vardalek seemed completely to understand. At dessert-time my father asked him if he were in a great hurry to reach his destination: if not, would he not stay on with us a little while: though our place was out of the way, he would find much that would interest him in his library.

He answered, "I am in no hurry. I have no particular reason for going to that place at all, and if I can be of service to you in deciphering these books, I shall be only too glad." He added with a smile that was bitter, very very bitter: "You see, I am a cosmopolitan, a wanderer on the face of the earth."

After dinner my father asked him if he played the piano. He said, "Yes, I can do a little," and he sat down at the piano. Then he played a Hungarian *csardas* — wild, rhapsodic, wonderful.

That is the music which makes men mad. He went on in the same strain.

Gabriel stood stock still by the piano, his eyes dilated and fixed, his form quivering. At last he said very slowly, at one particular motive — for want of a better word you may call it the *relâche* of a *casardas*, by which I mean that point where the original quasi-slow movement begins again, — "Yes, I think I could play that."

Then he quickly fetched his fiddle and self-made xylophone, and did actually, alternating the instruments, render the same very well indeed.

Vardalek looked at him, and said in a very sad voice, "Poor child! You have the soul of music within you."

I could not understand why he should seem to commiserate instead of congratulate Gabriel on what certainly showed an extraordinary talent.

Gabriel was shy even as the wild animals who were tame to him. Never before had he taken to a stranger. Indeed, as a rule, if any stranger came to the house by any chance, he would hide himself, and I had to bring him up his food to the turret chamber. You may imagine what was my surprise when I saw him walking about hand in hand with Vardalek the next morning, in the garden, talking livelily with him, and showing his collection of pet animals, which he had gathered from the woods, and for which we had had to fit up a regular zoological gardens. He seemed utterly under the domination of Vardalek. What surprised us was (for otherwise we liked the stranger, especially for being kind to him) that he seemed, though not noticeably at first — except to me, who noticed everything with regard to him — to be gradually losing his health and vitality. He did not become pale as yet; but there was a certain languor about his movements which certainly there was by no means before.

My father got more and more devoted to Count Vardalek. He helped him in his studies: and my father would hardly allow him to go away, which he did sometimes — to Trieste, he said — he always came back, bringing us presents of strange Oriental jewellery or textures.

I knew all kinds of people came to Trieste, Orientals included. Still, there was a strangeness and magnificence about these things which I was sure even then could not have possibly come from such a place as Trieste, memorable to me chiefly for its necktie shops.

When Vardalek was away, Gabriel was continually asking for him and talking about him. Then at the same time he seemed to regain his old vitality and spirits. Vardalek always returned looking much older, wan, and weary. Gabriel would rush to meet him, and kiss him on the mouth. Then he gave a slight shiver, and after a little while began to look quite young again.

Things continued like this for some time. My father would not hear of Vardalek's going away permanently. He came to be an inmate of our house. I indeed, and Mlle. Vonnaert also, could not help noticing what a difference there was altogether about Gabriel. But my father seemed totally blind to it.

One night I had gone downstairs to fetch something which I had left in the drawing room. As I was going up again I passed Vardalek's room. He was playing on a piano, which had been specially put there for him, one of Chopin's nocturnes, very beautifully; I stopped, leaning on the banisters to listen.

Something white appeared on the dark staircase. We believed in ghosts in our part. I was transfixed with terror, and clung to the banisters. What was my astonishment to see Gabriel walking slowly down the staircase, his eyes fixed as though in a trance! This terrified me even more than a ghost would. Could I believe my senses? Could that be Gabriel?

features as assassins and anarchists."

"Zoophilia," or an excessive attraction to animals, was a surefire giveaway of degeneracy, according to Nordau, whose theories were enthusiastically embraced by Bram Stoker, who employed pronounced animal characteristics in his portrait of the eponymous *Dracula* (1897).

Stenbock's work was not well-received by critics. Of his collection *The Shadow of Death*, the *Pall Mall Gazette* opined "... it must be a parody — an elaborate and screaming parody of that latterday literary abortion, the youthful decadent. The slipshod versification, the maudlin sentiment, the affected preciousness, the sham mysticism and sham aestheticism, the ridiculous medley of Neo-Paganism and Neo-Catholicism... all the nauseating characterstics of the type, in short, are here reproduced in lively burlesque, and the result is in its way quite one of the most amusing books we have ever seen."

I simply could not move. Gabriel, clad in his long white night-shirt, came downstairs and opened the door. He left it open. Vardalek still continued playing, but talked as he played.

He said — this time speaking in Polish — *Nie umiem wyrazic jak ciebie kocham,*—"My darling, I fain would spare thee; but thy life is my life, and I must live, I who would rather die. Will God not have *any* mercy on me? Oh! oh! life; oh, the torture of life!" Here he struck one agonised and strange chord, then continued playing softly, "Oh, Gabriel, my beloved! My life, yes *life* — oh, why life? I am sure this is but a little of what I demand of thee. Surely thy superabundance of life can spare a little to one who is already dead. No, stay," he said now almost harshly, "what must be, must be!"

Arthur Symons wrote of his work in general, "... there are in his pages too much play made with skulls and crossbones and macabre images and symbols drawn from charnel houses... Morbidly fascinated by a fantastic attraction toward the violent delights of horror and the nervous exasperation of his sensations, he became a kind of Judas Iscariot to himself."

Ravaged by opium and alcohol, Stenbock died on April 26, 1896, coincidentally the first day of Oscar Wilde's trial for acts of gross indecency.

Symons remarked that "A True Story of a Vampire" held an "unholy fascination," but of its author, "Satan and the Senses, the Seven Deadly Sins, were at his beck and call; so he imagined. Only, he had none of the magic of an Exorciser, he was no actual evoker of dead ghosts; he was, in one word, one of those conspicuous failures in life and in art which leave no traces behind them, save some faint drift in one's memory."

Gabriel stood there quite still, with the same fixed vacant expression, in the room. He was evidently walking in his sleep. Vardalek played on: then said, "Ah!" with a sigh of terrible agony. Then, very gently, "Go now, Gabriel; it is enough." And Gabriel went out of the room and ascended the staircase at the same slow pace, with the same unconscious stare. Vardalek struck the piano, and although he did not play loudly, it seemed as though the strings would break. You never heard music so strange and so heart-rending!

I only know I was found by Mlle. Vonnaert in the morning, in an unconscious state, at the foot of the stairs. Was it a dream after all? I am sure now that it was not. I thought then it might be, and said nothing to any one about it. Indeed, what could I say?

Well, to let me cut a long story short, Gabriel, who had never known a moment's sickness in his life, grew ill; and we had to send to Gratz for a doctor, who could give no explanation of Gabriel's strange illness. Gradual wasting away, he said: absolutely no organic complaint. What could this mean?

My father at last became conscious of the fact that Gabriel was ill. His anxiety was fearful. The last trace of grey faded from his hair, and it became quite white. We sent to Vienna for doctors. But all with the same result.

Gabriel was generally unconscious, and when conscious, only seemed to recognize Vardalek, who sat continually by his bedside, nursing him with the utmost tenderness.

One day I was alone in the room: and Vardalek cried suddenly, almost fiercely, "Send for a priest at once, at once," he repeated. "It is now almost too late!"

Gabriel stretched out his arms spasmodically, and put them around Vardalek's neck. This was the only movement he had made for some time. Vardalek bent down and kissed him on the lips. I rushed downstairs: and the priest was sent for. When I came back Vardalek was not there. The priest administered extreme unction. I think Gabriel was already dead, although we did not think so at the time.

Vardalek had utterly disappeared; and when we looked for him he was nowhere to be found; nor have I seen or heard of him since.

My father died very soon afterwards: suddenly aged, and bent down with grief. And so the whole of the Wronski property came into my sole possession. And here I am, an old woman, generally laughed at for keeping, in memory of Gabriel, an asylum for stray animals — and — people do not, as a rule, believe in Vampires!

Title illustration by Gordon Browne for the Strand Magazine, *February 1896.*

Good Lady Ducayne

MARY ELIZABETH BRADDON
(1896)

I

BELLA ROLLERSTON HAD made up her mind that her only chance of earning her bread and helping her mother to an occasional crust was by going out into the great unknown world as companion to a lady. She was willing to go to any lady rich enough to pay her a salary and so eccentric as to wish for a hired companion. Five shillings told off reluctantly from one of those sovereigns which were so rare with the mother and daughter, and which melted away so quickly, five solid shillings, had been handed to a smartly-dressed lady in an office in Harbeck Street, London, W., in the hope that this very Superior Person would find a situation and a salary for Miss Rolleston. The Superior Person glanced at the two half-crowns as they lay on the table where Bella's hand had placed them, to make sure they were neither of them florins, before she wrote a description of Bella's qualifications and requirements in a formidable-looking ledger.

"Age?" she asked, curtly.

"Eighteen, last July."

"Any accomplishments?"

"No; I am not at all accomplished. If I were I should want to be a governess — a companion seems the lowest stage."

"We have some highly accomplished ladies on our books as companions, or chaperon companions."

"Oh, I know!" babbled Bella, loquacious in her youthful candor. "But that is quite a different thing. Mother hasn't been able to afford a piano since I was twelve years old, so I'm afraid I've forgotten how to play. And I have had to help mother with her needlework, so there hasn't been much time to study."

"Please don't waste time upon explaining what you can't do, but kindly tell me anything you can do," said the Superior Person, crushingly, with her pen poised between delicate fingers waiting to write. "Can you read aloud for two or three hours at a stretch? Are you active and handy, an early riser, a good walker, sweet tempered, and obliging?"

"I can say yes to all those questions except about the sweetness. I think I have a pretty good temper, and I should be anxious to oblige anybody — who paid for my services. I should want them to feel that I was really earning my salary."

ABOUT THE AUTHOR

Mary Elizabeth Braddon (1835-1915) was a highly prolific fiction writer (the author of more than sixty books) whose most popular work was the novel *Lady Audley's Secret* (1862). "Good Lady Ducayne" is one of the first stories to posit a scientific rationalization for its vampire, and its plucky, if clueless, protagonist quaintly prefigures the heroines of modern medical thrillers, who invariably discover the equivalent of a black cape beneath the white coat.

"Good Lady Ducayne" has been reprinted many times, but rarely with the full set of original pen and ink illustrations by Gordon Browne that accompanied its first publication in *The Strand* magazine.

"The kind of ladies who come to me would not care for a talkative companion," said the Person, severely, having finished writing in her book. "My connection lies chiefly among the aristocracy, and in that class considerable deference is expected."

"Oh, of course," said Bella; "but it's quite different when I'm talking to you. I want to tell you all about myself once and forever."

"I am glad it is to be only once!" said the Person, with the edges of her lips.

The Person was of uncertain age, tightly laced in a black silk gown. She had a powdery complexion and a handsome clump of somebody else's hair on the top of her head. It may be that Bella's girlish freshness and vivacity had an irritating effect upon nerves weakened by an eight hour day in that overheated second floor in Harbeck Street. To Bella the official apartment, with its Brussels carpet, velvet curtains and velvet chairs, and French clock, ticking loud on the marble chimney-piece, suggested the luxury of a palace, as compared with another second floor in Walworth where Mrs. Rolleston and her daughter had managed to exist for the last six years.

"Not a love affair, I hope?"

"Do you think you have anything on your books that would suit me?" faltered Bella, after a pause.

"Oh, dear, no; I have nothing in view at present," answered the Person, who had swept Bella's half-crowns into a drawer, absentmindedly, with the tips of her fingers. "You see, you are so very unformed — so much too young to be companion to a lady of position. It is a pity you have not enough education for a nursery governess; that would be more in your line."

"And do you think it will be very long before you can get me a situation?" asked Bella, doubtfully.

"I really cannot say. Have you any particular reason for being so impatient — not a love affair, I hope?"

"A love affair!" cried Bella, with flaming cheeks. "What utter nonsense. I want a situation because mother is poor, and I hate being a burden to her. I want a salary that I can share with her."

"There won't be much margin for sharing in the salary you are likely to get at your age — and with your — very — unformed manners," said the Person, who found Bella's peony cheeks, bright eyes, and unbridled vivacity more and more oppressive.

"Perhaps if you'd be kind enough to give me back the fee I could take it to an agency where the connection isn't quite so aristocratic," said Bella, who — as she told her mother in her recital of the interview — was determined not to be sat upon.

"You will find no agency that can do more for you than mine," replied the Person, whose harpy fingers never relinquished coin. "You will have to wait for your opportunity. Yours is an exceptional case: but I will bear you in mind, and if anything suitable offers I will write to you. I cannot say more than that."

The half-contemptuous bend of the stately head, weighted with borrowed hair, indicated the end of the interview. Bella went back to Walworth — tramped sturdily every inch of the way in the September afternoon — and "took off" the Superior Person for the amusement of her mother and the landlady, who lingered in the shabby little sitting-room after bringing in the tea-tray, to applaud Miss Rolleston's "taking off."

"Dear, dear, what a mimic she isl" said the landlady. "You ought to have let her go on the stage, mum. She might have made her fortune as an actress."

II

Bella waited, hoped, and listened for the postman's knocks which brought such store of letters for the parlors and the first floor, and so few for that humble second floor, where mother and daughter sat sewing with hand and with wheel and treadle, for the greater part of the day. Mrs. Rolleston was a lady by birth and education; but it had been her bad fortune to marry a scoundrel; for the last half-dozen years she had been that worst of widows, a wife whose husband had deserted her. Happily, she was courageous, industrious, and a clever needlewoman; and she had been able just to earn a living for herself and her only child, by making mantles and cloaks for a West-end house. It was not a luxurious living. Cheap lodgings in a shabby street off the Walworth Road, scanty dinners, homely food, well-worn raiment, had been the portion of mother and daughter; but they loved each other so dearly, and Nature had made them both so light-hearted, that they had contrived somehow to be happy.

"Good Lady Ducayne."

But now this idea of going out into the world as companion to some fine lady had rooted itself into Bella's mind, and although she idolized her mother, and although the parting of mother and daughter must needs tear two loving hearts into shreds, the girl longed for enterprise and change and excitement, as the pages of old longed to be knights, and to start for the Holy Land to break a lance with the infidel.

She grew tired of racing downstairs every time the postman knocked, only to be told "nothing for you, miss," by the smudgy-faced drudge who picked up the letters from the passage floor. "Nothing for you, miss," grinned the lodging-house drudge, till at last Bella took heart of grace and walked up to Harbeck Street, and asked the Superior Person how it was that no situation had been found for her.

"You are too young," said the Person, "and you want a salary."

"Of course I do," answered Bella; "don't other people want salaries?"

"Young ladies of your age generally want a comfortable home."

"I don't," snapped Bella: "I want to help mother."

"You can call again this day week," said the Person; "or, if I hear of anything in the meantime, I will write to you."

No letter came from the Person, and in exactly a week Bella put on her neatest hat, the one that had been seldomest caught in the rain, and trudged off to Harbeck Street.

It was a dull October afternoon, and there was a greyness in the air which might turn to fog before night. The Walworth Road shops gleamed brightly through that grey atmosphere, and though to a young lady reared in Mayfair or Belgravia such shop-windows would have been unworthy of a glance, they were a snare and temptation for Bella. There were so many things that she longed for, and would never be able to buy.

Harbeck Street is apt to be empty at this dead season of the year, a long, long street, an endless perspective of eminently respectable houses. The Person's office was at the further end, and Bella looked down that long, grey vista almost despairingly, more tired than usual with the trudge from Walworth. As she looked, a carriage passed her, an old-fashioned, yellow chariot, on cee springs, drawn by a pair of high grey horses, with the stateliest of coachmen driving them, and a tall footman sitting by his side.

"In the olive woods."

"It looks like the fairy godmother's coach," thought Bella. "I shouldn't wonder if it began by being a pumpkin."

It was a surprise when she reached the Person's door to find the yellow chariot standing before it, and the tall footman waiting near the doorstep. She was almost afraid to go in and meet the owner of that splendid carriage. She had caught only a glimpse of its occupant as the chariot rolled by, a plumed bonnet, a patch of ermine.

The Person's smart page ushered her upstairs and knocked at the official door. "Miss Rolleston," he announced, apologetically, while Bella waited outside.

"Show her in," said the Person, quickly; and then Bella heard her murmuring something in a low voice to her client.

Bella went in fresh, blooming, a living image of youth and hope, and before she looked at the Person her gaze was riveted by the owner of the chariot.

Never had she seen anyone as old as the old lady sitting by the Person's fire: a little old figure, wrapped from chin to feet in an ermine mantle; a withered, old face under a plumed bonnet — a face so wasted by age that it seemed only a pair of eyes and a peaked chin. The nose was peaked, too, but between the sharply pointed chin and the great, shining eyes, the small, aquiline nose was hardly visible.

"This is Miss Rolleston, Lady Ducayne."

Claw-like fingers, flashing with jewels, lifted a double eyeglass to Lady Ducayne's shining black eyes, and through the glasses Bella saw those unnaturally bright eyes magnified to a gigantic size, and glaring at her awfully.

"Miss Torpinter has told me all about you," said the old voice that belonged to the eyes. "Have you good health? Are you strong and active, able to eat well, sleep well, walk well, able to enjoy all that there is good in life?"

"I have never known what it is to be ill, or idle," answered Bella.

"Then I think you will do for me."

"Of course, in the event of references being perfectly satisfactory," put in the Person.

"I don't want references. The young woman looks frank and innocent. I'll take her on trust."

"So like you, dear Lady Ducayne," murmured Miss Torpinter.

"I want a strong young woman whose health will give me no trouble."

"You have been so unfortunate in that respect," cooed the Person, whose voice and manner were subdued to a melting sweetness by the old woman's presence.

"Yes, I've been rather unlucky," grunted Lady Ducayne.

"But I am sure Miss Rolleston will not disappoint you, though certainly after your unpleasant experience with Miss Tomson, who looked the picture of health — and Miss Blandy, who said she had never seen a doctor since she was vaccinated — "

"Lies, no doubt," muttered Lady Ducayne, and then turning to Bella, she asked, curtly, "You don't mind spending the winter in Italy, I suppose?"

In Italy! The very word was magical. Bella's fair young face flushed crimson.

"It has been the dream of my life to see Italy," she gasped.

From Walworth to Italy! How far, how impossible such a journey had seemed to that romantic dreamer.

"With yearning eyes looking Westward."

"Well, your dream will be realized. Get yourself ready to leave Charing Cross by the train deluxe this day week at eleven. Be sure you are at the station a quarter before the hour. My people will look after you and your luggage."

Lady Ducayne rose from her chair, assisted by her crutch-stick, and Miss Torpinter escorted her to the door.

"And with regard to salary?" questioned the Person on the way.

"Salary, oh, the same as usual — and if the young woman wants a quarter's pay in advance you can write to me for a check," Lady Ducayne answered, carelessly.

Miss Torpinter went all the way downstairs with her client, and waited to see her seated in the yellow chariot. When she came upstairs again she was slightly out of breath, and she had resumed that superior manner which Bella had found so crushing.

"You may think yourself uncommonly lucky, Miss Rolleston," she said. "I have dozens of young ladies on my books whom I might have recommended for this situation — but I remembered having told you to call this afternoon — and I thought I would give you a chance. Old Lady Ducayne is one of the best people on my books. She gives her companion a hundred a year, and pays all travelling expenses. You will live in the lap of luxury."

"A hundred a year! How too lovely! Shall I have to dress very grandly? Does Lady Ducayne keep much company?"

"At her age! No, she lives in seclusion — in her own apartments — her French maid, her footman, her medical attendant, her courier."

"Why did those other companions leave her?" asked Bella.

"Their health broke down!"

"Poor things, and so they had to leave?"

"Yes, they had to leave. I suppose you would like a quarter's salary in advance?"

"Oh, yes, please. I shall have things to buy."

"Very well, I will write for Lady Ducayne's check, and I will send you the balance after deducting my commission for the year."

"To be sure, I had forgotten the commission."

"You don't suppose I keep this office for pleasure."

"Of course not," murmured Bella, remembering the five shillings entrance fee; but nobody could expect a hundred a year and a winter in Italy for five shillings.

"What a vampire!"

III

"From Miss Rolleston, at Cap Ferrino, to Mrs. Rolleston, in Beresford Street, Walworth, London.

"How I wish you could see this place, dearest; the blue sky, the olive woods, the orange and lemon orchards between the cliffs and the sea — sheltering in the hollow of the great hill — and with summer waves dancing up to the narrow ridge of pebbles and weeds which is the Italian idea of a beach! Oh, how I wish you could see it all, mother dear, and bask in this sunshine, that makes it so difficult to believe the date at the head of this paper. November! The air is like an English June — the sun is so hot that I can't walk a few yards without an umbrella. And to think of you at Walworth while I am here! I could cry at the thought that perhaps you will never see this lovely coast, this wonderful sea, these summer flowers that bloom in winter. There is a hedge of pink geraniums under my window, mother — a thick, rank hedge, as if the flowers grew wild — and there are Dijon roses climbing over arches and palisades all along the terrace — a rose garden full of bloom in November! Just picture it all! You could never imagine the luxury of this hotel. It is nearly new, and has been built and decorated regardless of expense. Our rooms are upholstered in pale blue satin, which shows up Lady Ducayne's parchment complexion; but as she sits all day in a corner of the balcony basking in the sun, except when she is in her carriage, and all the evening in her armchair close to the fire, and never sees anyone but her own people, her complexion matters very little.

"She has the handsomest suite of rooms in the hotel. My bedroom is inside hers, the sweetest room — all blue satin and white lace — white enamelled furniture, looking-glasses on every wall, till I know my pert little profile as I never knew it before. The room was really meant for Lady Ducayne's dressing-room, but she ordered one of the blue satin couches to be arranged as a bed for me — the prettiest little bed, which I can wheel near the window on sunny mornings, as it is on castors and easily moved about. I feel as if Lady Ducayne were a funny old grandmother, who had suddenly appeared in my life, very, very rich, and very, very kind.

"She is not at all exacting. I read aloud to her a good deal, and she dozes and nods while I read. Sometimes I hear her moaning in her sleep — as if she had troublesome dreams. When she is tired of my reading she orders Francine, her maid, to read a French novel to her, and I hear her chuckle and groan now and then, as if she were more interested in those books than in Dickens or Scott. My French is not good enough to follow Francine, who reads very quickly. I have a great deal of liberty, for Lady Ducayne often tells me to run away and amuse myself; I roam about the hills for hours. Everything is so lovely. I lose myself in olive woods, always climbing up and up towards the pine woods above — and above the pines there are the snow mountains that just show their white peaks above the dark hills. Oh, you poor dear, how can I ever make you understand what this place is like — you, whose poor, tired eyes have only the opposite side of Beresford Street? Sometimes I go no farther than the terrace in front of the hotel, which is a favorite lounging-place with everybody. The gardens lie below, and the tennis courts where I sometimes play with a very nice girl, the only person in the hotel with whom I have made friends. She is a year older than I, and has come to Cap Ferrino with her brother, a doctor — or a medical student, who is going to be a doctor. He passed his M.B. exam at Edinburgh, just before they left home, Lotta told me. He came to Italy entirely on his sister's account. She had a troublesome chest attack last summer and was ordered to winter abroad. They are orphans, quite alone in the world, and so fond of each other. It is very nice for me to have such a friend as Lotta. She is so thoroughly respectable. I can't help using that word, for some of the girls in this hotel go on in a way that I know you would shudder at. Lotta was brought up by an aunt, deep down in the country, and knows hardly anything about life. Her brother won't allow her to read a novel, French or English, that he has not read and approved.

"'He treats me like a child' she told me, 'but I don't mind, for it's nice to know somebody loves me, and cares about what I do, and even about my thoughts.'"

THE TERRORS OF TRANSFUSION

Before the age of modern medicine, a great number of erroneous beliefs about the need to remove and/or replace blood were a standard part of a physician's training. Bleeding was a very common treatment for a wide variety of ills, and while it usually posed little risk to healthy patients, a related procedure — transfusion — was fraught with terrible risks.

The earliest experiments in blood transfusion were conducted over three hundred years ago — disastrous attempts to introduce animal blood into humans, which resulted in shock, kidney damage and death. But later doctors knew that transfusions from one human being to another *could* be beneficial in certain cases of traumatic blood loss or anemia, but they also knew the procedure could just as easily prove fatal. During the nineteenth century it was simply not understood why certain bloods were compatible and others weren't.

> "Perhaps this is what makes some girls so eager to marry — the want of someone strong and brave and honest and true to care for them and order them about. I want no one, mother darling, for I have you, and you are all the world to me. No husband could ever come between us two. If I ever were to marry he would have only the second place in my heart. But I don't suppose I ever shall marry, or even know what it is like to have an offer of marriage. No young man can afford to marry a penniless girl nowadays. Life is too expensive.
>
> "Mr. Stafford, Lotta's brother, is very clever, and very kind. He thinks it is rather hard for me to have to live with such an old woman as Lady Ducayne, but then he does not know how poor we are — you and I — and what a wonderful life this seems to me in this lovely place. I feel a selfish wretch for enjoying all my luxuries, while you, who want them so much more than I, have none of them — hardly know what they are like — do you, dearest? — for my scamp of a father began to go to the dogs soon after you were married and since then life has been all trouble and care and struggle for you."

It was not until the early twentieth century that human blood groups were properly identified, and transfusion techniques were perfected, largely in the trenches of battlefield medicine during World War I.

Mary Elizabeth Braddon seems to be the first fiction writer to make an association between the dark scientific history of blood transfusion and the equally dark history of the vampire. Bram Stoker would make transfusion a key plot element in *Dracula* (1897), which he was completing around the time Braddon, a personal friend, published "Good Lady Ducayne."

This letter was written when Bella had been less than a month at Cap Ferrino, before the novelty had worn off the landscape, and before the pleasure of luxurious surroundings had begun to cloy. She wrote to her mother every week, such long letters as girls who have lived in closest companionship with a mother alone can write; letters that are like a diary of heart and mind. She wrote gaily always; but when the new year began Mrs. Rolleston thought she detected a note of melancholy under all those lively details about the place and the people.

"My poor girl is getting homesick," she thought. "Her heart is in Beresford Street."

It might be that she missed her new friend and companion, Lotta Stafford, who had gone with her brother for a little tour to Genoa and Spezia, and as far as Pisa. They were to return before February; but in the meantime Bella might naturally feel very solitary among all those strangers, whose manners and doings she described so well.

The mother's instinct had been true. Bella was not so happy as she had been in that first flush of wonder and delight which followed the change from Walworth to the Riviera. Somehow, she knew not how, lassitude had crept upon her. She no longer loved to climb the hills, no longer flourished her orange stick in sheer gladness of heart as her light feet skipped over the rough ground and the coarse grass on the mountain side. The odor of rosemary and thyme, the fresh breath of the sea, no longer filled her with rapture. She thought of Beresford Street and her mother's face with a sick longing. They were so far — so far away! And then she thought of Lady Ducayne, sitting by the heaped-up olive logs in the overheated salon — thought of that wizened-nutcracker profile, and those gleaming eyes, with an invincible horror.

Visitors at the hotel had told her that the air of Cap Feffino was relaxing — better suited to age than to youth, to sickness than to health. No doubt it was so. She was not so well as she had been at Walworth; but she told herself that she was suffering only from the pain of separation from the dear companion of her girlhood, the mother who had been nurse, sister, friend, flatterer, all things in this world to her. She had shed many tears over that parting, had spent many a melancholy hour on the marble terrace with yearning eyes looking westward, and with her heart's desire a thousand miles away.

She was sitting in her favorite spot, an angle at the eastern end of the terrace, a quiet little nook sheltered by orange trees, when she heard a couple of Riviera habitués talking in the garden-below. They were sitting on a bench against the terrace wall.

She had no idea of listening to their talk, till the sound of Lady Ducayne's name attracted her, and then she listened without any thought of wrong-doing. They were talking no secrets — just casually discussing a hotel acquaintance.

They were two elderly people whom Bella only knew by sight. An English clergyman who had wintered abroad for half his lifetime; a stout, comfortable, well-to-do spinster, whose chronic bronchitis obliged her to migrate annually.

"I have met her about Italy for the last ten years," said the lady; "but have never found out her real age."

"I put her down at a hundred — not a year less," replied the parson. "Her reminiscences all go back to the Regency. She was evidently then in her zenith; and I have heard her say things that showed she was in Parisian society when the First Empire was at its best — before Josephine was divorced."

"She doesn't talk much now."

"No; there's not much life left in her. She is wise in keeping herself secluded. I only wonder that wicked old quack, her Italian doctor, didn't finish her off years ago."

"I should think it must be the other way, and that he keeps her alive."

"My dear Miss Manders, do you think foreign quackery ever kept anybody alive?"

"Well, there she is — and she never goes anywhere without him. He certainly has an unpleasant countenance."

"Unpleasant," echoed the parson, "I don't believe the foul fiend himself can beat him in ugliness. I pity that poor young woman who has to live between old Lady Ducayne and Dr. Parravicini."

"But the old lady — is very good to her companions."

"No doubt. She is very free with her cash; the servants call her good Lady Ducayne. She is a withered old female Croesus, and knows she'll

VICTORIAN VAMPIRES: OTHER WOMEN WRITERS

Vampire stories attracted several Victorian women writers; beyond Mary Elizabeth Braddon and Eliza Lynn Linton there was also Anne Crawford, the older sister of F. Marion Crawford, whose classic story "For the Blood is the Life" can be found in this volume.

Following is an excerpt from the blood-curdling climax of Crawford's memorable exercise in continental vampirism.

From
A MYSTERY OF THE CAMPAGNA
Anne Crawford
(1886)

Yes; Marcello was there. He was lying stretched upon the floor, staring at the ceiling, dead, and already stiff, as I could see at a glance. We stood over him, saying not a word, then I knelt down and felt him, for mere form's sake, and said, as though I had not known it before, "He has been dead for some hours."

never be able to get through her money, and doesn't relish the idea of other people enjoying it when she's in her coffin. People who live to be as old as she is become slavishly attached to life. I daresay she's generous to those poor girls — but she can't make them happy. They die in her service."

"Don't say they, Mr. Carton; I know that one poor girl died at Mentone last spring."

"Yes, and another poor girl died in Rome three years ago. I was there at the time. Good Lady Ducayne left her there in an English family. The girl had every comfort. The old woman was very liberal to her — but she died. I tell you, Miss Manders, it is not good for any young woman to live with two such horrors as Lady Ducayne and Parravicini."

They talked of other things — but Bella hardly heard them. She sat motionless, and a cold wind seemed to come down upon her from the mountains and to creep up to her from the sea, till she shivered as she sat there in the sunshine, in the shelter of the orange trees in the midst of all that beauty and brightness.

Yes, they were uncanny, certainly, the pair of them — she so like an aristocratic witch in her withered old age; he of no particular age, with a face that was more like a waxen mask than any human countenance Bella had ever seen. What did it matter? Old age is venerable, and worthy of all reverence; and Lady Ducayne had been very kind to her. Dr. Parravicini was a harmless, inoffensive student, who seldom looked up from the book he was reading. He had his private sitting-room, where he made experiments in chemistry and natural science — perhaps in alchemy. What could it matter to Bella? He had always been polite to her, in his far-off way. She could not be more happily placed than she was — in this palatial hotel, with this rich old lady.

No doubt she missed the young English girl who had been so friendly, and it might be that she missed the girl's brother, for Mr. Stafford had talked to her a good deal — had interested himself in the books she was reading, and her manner of amusing herself when she was not on duty.

"You must come to our little salon when you are 'off,' as the hospital nurses call it, and we can have some music. No doubt you play and sing?" Upon which Bella had to own with a blush of shame that she had forgotten how to play the piano ages ago.

"Mother and I used to sing duets sometimes between the lights, without accompaniment," she said, and the tears came into her eyes as she thought of the humble room, the half-hour's respite from work, the sewing machine standing where a piano ought to have been, and her mother's plaintive voice, so sweet, so true, so dear.

Sometimes she found herself wondering whether she would ever see that beloved mother again. Strange forebodings came into her mind. She was angry with herself for giving way to melancholy thoughts.

One day she questioned Lady Ducayne's French maid about those two companions who had died within three years.

"Since yesterday evening," said Magnin, in a horror-stricken voice, yet with a certain satisfaction in it, as though to say, "You see, I was right."

Marcello was lying with his head slightly thrown back, no contortions in his handsome features; rather the look of a person who has quietly died of exhaustion — who has slipped unconsciously from life to death. His collar was thrown open and a part of his breast, of a ghastly white, was visible. Just over the heart was a small spot.

"Give me the lantern," I whispered, as I stooped over it. It was a very little spot, of a faint purplish-brown, and must have changed colour within the night.

I examined it intently, and should say that the blood had been sucked to the surface, and then a small prick or incision made. The slight subcutaneous effusion led me to this conclusion. One tiny drop of coagulated blood closed the almost imperceptible wound. I probed it with the end of one of

"They were poor, feeble creatures," Francine told her. "They looked fresh and bright enough when they came to Miladi; but they ate too much, and they were lazy. They died of luxury and idleness. Miladi was too kind to them. They had nothing to do; and so they took to fancying things; fancying the air didn't suit them, that they couldn't sleep."

"I sleep well enough, but I have had a strange dream several times since I have been in Italy."

"Ah, you had better not begin to think about dreams, or you will be like those other girls. They were dreamers — and they dreamt themselves into the cemetery."

The dream troubled her a little, not because it was a ghastly or frightening dream, but on account of sensations which she had never felt before in sleep — a whirring of wheels that went round in her brain, a great noise like a whirlwind, but rhythmical like the ticking of a gigantic clock: and then in the midst of this uproar as of winds and waves she seemed to sink into a gulf of unconsciousness, out of sleep into far deeper sleep — total extinction. And then, after that black interval, there had come the sound of voices, and then again the whirr of wheels, louder and louder — and again the black—and then she awoke, feeling languid and oppressed.

She told Dr. Parravicini of her dream one day, on the only occasion when she wanted his professional advice. She had suffered rather severely from the mosquitoes before Christmas — and had been almost frightened at finding a wound upon her arm which she could only attribute to the venomous sting of one of these torturers. Parravicini put on his glasses, and scrutinized the angry mark on the round, white arm, as Bella stood before him and Lady Ducayne with her sleeve rolled up above her elbow. "Yes, that's rather more than a joke," he said; "he has caught you on the top of a vein. What a vampire! But there's no harm done, signorina, nothing that a little dressing of mine won't heal. You must always show me any bite of this nature. It might be dangerous if neglected. These creatures feed on poison and disseminate it."

"And to think that such tiny creatures can bite like this," said Bella; "my arm looks as if it had been cut by a knife."

"If I were to show you a mosquito's sting under my microscope you wouldn't be surprised at that," replied Parravicini.

Bella had to put up with the mosquito bites, even when they came on the top of a vein, and produced that ugly wound. The wound recurred now and then at longish intervals, and Bella found Dr. Parravicini's dressing a speedy cure. If he were the quack his enemies called him, he had at least a light hand and a delicate touch in performing this small operation.

"Bella Rolleston to Mrs. Rolleston — April 14th.

"EVER DEAREST,

Behold the check for my second quarter's salary — five and twenty pounds. There is no one to pinch off a whole tenner for a

Magnin's matches. It was scarcely more than skin deep, so it could not be the stab of a stiletto, however slender, or the track of a bullet. Still, it was strange, and with one impulse we turned to see if no one were concealed there, or if there were no second exit. It would be madness to suppose that the murderer, if there was one, would remain by his victim. Had Marcello been making love to a pretty rontadina, and was this some jealous lover's vengeance? But it was not a stab. Had one drop of poison in the little wound done this deadly work?

We peered about the place, and I saw that Magnin's eyes were blinded by tears and his face as pale as that upturned one on the floor whose lids I had vainly tried to close. The chamber was low, and beautifully ornamented with stucco bas-reliefs, in the manner of the well-known one not far from there upon the same road. Winged genii, griffins, and arabesques, modelled with marvellous lightness, covered the walls and ceiling. There was no other door than the one we had

year's commission as there was last time, so it is all for you, mother, dear. I have plenty of pocket-money in hand from the cash I brought away with me, when you insisted on my keeping more than I wanted. It isn't possible to spend money here — except on occasional tips to servants, or sous to beggars and children — unless one had lots to spend, for everything one would like to buy — tortoise-shell, coral, lace — is so ridiculously dear that only a millionaire ought to look at it. Italy is a dream of beauty: but for shopping, give me Newington Causeway.

"You ask me so earnestly if I am quite well that I fear my letters must have been very dull lately. Yes, dear, I am well — but I am not quite so strong as I was when I used to trudge to the West-end to buy half a pound of tea — just for a constitutional walk — to Dulwich to look at the pictures. Italy is relaxing; and I feel what the people here call 'slack.' But I fancy I can see your dear face looking worried as you read this. Indeed, and indeed, I am not ill. I am only a little tired of this lovely scene — as I suppose one might get tired of looking at one of Turner's pictures if it hung on a wall that was always opposite one. I think of you every hour in every day — think of you and our homely little room — our dear little shabby parlor, with the armchairs from the wreck of your old home, and Dick singing in his cage over the sewing machine. Dear, shrill, maddening Dick, who, we flattered ourselves, was so passionately fond of us. Do tell me in your next letter that he is well.

"My friend Lotta and her brother never came back after all. They went from Pisa to Rome. Happy mortals! And the Italian lakes in May; which lake was not decided when Lotta last wrote to me. She has been a charming correspondent, and has confided all her little flirtations to me. We are all to go to Bellaggio next week — by Genoa and Milan. Isn't that lovely? Lady Ducayne travels by the easiest stages — except when she is bottled up in the train deluxe. We shall stop two days at Genoa and one at Milan. What a bore I shall be to you with my talk about Italy when I come home.

"Love and love — and ever more love from your adoring, BELLA."

IV

Herbert Stafford and his sister had often talked of the pretty English girl with her fresh complexion, which made such a pleasant touch of rosy color among all those sallow faces at the Grand Hotel. The young doctor thought of her with a compassionate tenderness — her utter loneliness in that great hotel where there were so many people, her bondage to that old, old woman, where everybody else was free to think of nothing but enjoying

entered by. In the centre stood a marble sarcophagus, with the usual subjects sculptured upon it, on the one side Hercules conducting a veiled figure, on the other a dance of nymphs and fauns. A space in the middle contained the following inscription, deeply cut in the stone, and still partially filled with red pigment:

D.M.
VESPERTILIAE • THC • AIMA-
TOΠΩTΔOC • Q • FLAVIVS •
VIX • IPSE • SOSPES • MON •
POSVIT

"What is this?" whispered Magnin. It was only a pickaxe and a long crowbar, such as the country people use in hewing out their blocks of "tufa," and his foot had struck against them. Who could have brought them here? They must belong to the guardiano above, but he said that he had never come here, and I believed him, knowing the Italian horror of darkness and lonely places; but what had Marcello wanted with them? It did not occur to us that

life. It was a hard fate; and the poor child was evidently devoted to her mother, and felt the pain of separation — "only two of them, and very poor, and all the world to each other," he thought.

Lotta told him one morning that they were to meet again at Bellaggio. "The old thing and her court are to be there before we are," she said. "I shall be charmed to have Bella again. She is so bright and gay in spite of an occasional touch of homesickness. I never took to a girl on a short acquaintance as I did to her."

"I like her best when she is homesick," said Herbert; "for then I am sure she has a heart."

"What have you to do with hearts, except for dissection? Don't forget that Bella is an absolute pauper. She told me in confidence that her mother makes mantles for a West-end shop. You can hardly have a lower depth than that."

"I shouldn't think any less of her if her mother made matchboxes."

"Not in the abstract — of course not. Matchboxes are honest labor. But you couldn't marry a girl whose mother makes mantles."

"We haven't come to the consideration of that question yet," answered Herbert, who liked to provoke his sister.

In two years' hospital practice he had seen too much of the grim realities of life to retain any prejudices about rank. Cancer, phthisis, gangrene, leave a man with little respect for the humanity. The kernel is always the same — fearfully and wonderfully made — a subject for pity and terror.

Mr. Stafford and his sister arrived at Bellaggio in a fair May evening. The sun was going down as the steamer approached the pier; and all that glory of purple bloom which curtains every wall at this season of the year flushed and deepened in the glowing light. A group of ladies were standing on the pier watching the arrivals, and among them Herbert saw a pale face that startled him out of his wonted composure.

"There she is," murmured Lotta, at his elbow, "but how dreadfully changed. She looks a wreck."

They were shaking hands with her a few minutes later, and a flush had lighted up her poor pinched face in the pleasure of meeting.

"I thought you might come this evening," she said. "We have been here a week."

She did not add that she had been there every evening to watch the boat in, and a good many times during the day. The Grand Bretagne was close by, and it had been easy for her to creep to the pier when the boat bell rang. She felt a joy in meeting these people again; a sense of being with friends; a confidence which Lady Ducayne's goodness had never inspired in her.

"Oh, you poor darling, how awfully ill you must have been," exclaimed Lotta, as the two girls embraced.

Bella tried to answer, but her voice was choked with tears.

"What has been the matter, dear? That horrid influenza, I suppose?"

archaeological curiosity could have led him to attempt to open the sarcophagus, the lid of which had evidently never been raised, thus justifying the expression, "piously preserved."

As I rose from examining the tools my eyes fell upon the line of mortar where the cover joined to the stone below, and I noticed that some of it had been removed, perhaps with the pickaxe which lay at my feet. I tried it with my nails and found that it was very crumbly. Without a word I took the tool in my hand, Magnin instinctively following my movements with the lantern. What impelled us I do not know. I had myself no thought, only an irresistible desire to see what was within. I saw that much of the mortar had been broken away, and lay in small fragments upon the ground, which I had not noticed before. It did not take long to complete the work. I snatched the lantern from Magnin's hand and set it upon the ground, where it shone full upon Marcello's dead face, and by its light I found a little break

"No, no, I have not been ill — I have only felt a little weaker than I used to be. I don't think the air of Cap Ferrino quite agreed with me."

"It must have disagreed with you abominably. I never saw such a change in anyone. Do let Herbert doctor you. He is fully qualified, you know. He prescribed for ever so many influenza patients at the Londres. They were glad to get advice from an English doctor in a friendly way."

"I am sure he must be very clever," faltered Bella, "but there is really nothing the matter. I am not ill, and if I were ill, Lady Ducayne's physician —"

"That dreadful man with the yellow face? I would as soon one of the Borgias prescribed for me. I hope you haven't been taking any of his medicines."

"No, dear, I have taken nothing. I have never complained of being ill."

This was said while they were all three walking to the hotel. The Staffords' rooms had been secured in advance, pretty ground-floor rooms, opening into the garden. Lady Ducayne's statelier apartments were on the floor above.

"I believe these rooms are just under ours," said Bella.

"Then it will be all the easier for you to run down to us," replied Lotta, which was not really the case, as the grand staircase was in the center of the hotel.

"Oh, I shall find it easy enough," said Bella. "I'm afraid you'll have too much of my society. Lady Ducayne sleeps away half the day in this warm weather, so I have a good deal of idle time; and I get awfully moped thinking of mother and home."

Her voice broke upon the last word. She could not have thought of that poor lodging which went by the name of home more tenderly had it been the most beautiful that art and wealth ever created. She moped and pined in this lovely garden, with the sunlit lake and the romantic hills spreading out their beauty before her. She was homesick and she had dreams; or, rather, an occasional recurrence of that one bad dream with all its strange sensations — it was more like a hallucination than dreaming — the whirring of wheels, the sinking into an abyss, the struggling back to consciousness. She had the dream shortly before she left Cap Ferrino, but not since she had come to Bellaggio, and she began to hope the air in this lake district suited her better, and that those strange sensations would never return.

Mr. Stafford wrote a prescription and had it made up at the chemist's near the hotel. It was a powerful tonic, and after two bottles, and a row or two on the lake, and some rambling over the hills and in the meadows where the spring flowers made earth seem paradise, Bella's spirits and looks improved as if by magic.

"It is a wonderful tonic," she said, but perhaps in her heart of hearts she knew that the doctor's kind voice, and the friendly hand that helped her in and out of the boat, and the lake, had something to do with her cure.

between the two masses of stone and managed to insert the end of my crowbar, driving it in with a blow of the pickaxe. The stone chipped and then cracked a little. Magnin was shivering.

"What are you going to do?" he said, looking around at where Marcello lay.

"Help me!" I cried, and we two bore with all our might upon the crowbar. I am a strong man, and I felt a sort of blind fury as the stone refused to yield. What if the bar should snap? With another blow I drove it in still further, then using it as a lever, we weighed upon it with our outstretched arms until every muscle was at its highest tension. The stone moved a little, and almost fainting we stopped to rest.

From the ceiling hung the rusty remnant of an iron chain which must once have held a lamp. To this, by scrambling upon the sarcophagus, I contrived to make fast the lantern.

"I hope you don't forget that her mother makes mantles," Lotta said warningly.

"Or matchboxes; it is just the same thing, so far as I am concerned."

"You mean that in no circumstances could you think of marrying her?"

"I mean that if ever I love a woman well enough to think of marrying her, riches or rank will count for nothing with me. But I fear — I fear your poor friend may not live to be any man's wife."

"Do you think her so very ill?"

He sighed, and left the question unanswered.

One day, while they were gathering wild hyacinths in an upland meadow, Bella told Mr. Stafford about her bad dream.

"It is curious only because it is hardly like a dream," she said. "I daresay you could find some commonsense reason for it. The position of my head on my pillow, or the atmosphere, or something."

And then she described her sensations; how in the midst of sleep there came a sudden sense of suffocation; and then those whirring wheels, so loud, so terrible; and then a blank, and then a coming back to waking consciousness.

"Have you ever had chloroform given you — by a dentist, for instance?"

"Never — Dr. Parravicini asked me that question one day."

"Lately?"

"No, long ago, when we were in the train deluxe."

"Has Dr. Parravicini prescribed for you since you began to feel weak and ill?"

"Oh, he has given me a tonic from time to time, but I hate medicine, and took very little of the stuff. And then I am not ill, only weaker than I used to be. I was ridiculously strong and well when I lived at Walworth, and used to take long walks every day. Mother made me take those tramps to Dulwich or Norwood, for fear I should suffer from too much sewing machine; sometimes — but very seldom — she went with me. She was generally toiling at home while I was enjoying fresh air and exercise. And she was very careful about our food — that, however plain it was, it should be always nourishing and ample. I owe it to her care that I grew up such a great, strong creature."

"You don't look great or strong now, you poor dear," said Lotta.

"I'm afraid Italy doesn't agree with me."

"Perhaps it is not Italy, but being cooped up with Lady Ducayne that has made you ill."

"But I am never cooped up. Lady Ducayne is absurdly kind, and lets me roam about or sit in the balcony all day if I like. I have read more novels since I have been with her than in all the rest of my life."

"Then she is very different from the average old lady, who is usually a slave driver," said Stafford. "I wonder why she carries a companion about with her if she has so little need of society."

"Oh, I am only part of her state. She is inordinately rich — and the

"Now!" said I, and we heaved again at the lid. It rose, and we alternately heaved and pushed until it lost its balance and fell with a thundering crash upon the other side; such a crash that the walls seemed to shake, and I was for a moment utterly deafened, while little pieces of stucco rained upon us from the ceiling. When we had paused to recover from the shock we leaned over the sarcophagus and looked in.

The light shone full upon it, and we saw — how is it possible to tell? We saw lying there, amidst folds of mouldering rags, the body of a woman, perfect as in life, with faintly rosy face, soft crimson lips, and a breast of living pearl, which seemed to heave as though stirred by some delicious dream. The rotten stuff swathed about her was in ghastly contrast to this lovely form, fresh as the morning! Her hands lay stretched at her side, the pink palms were turned a little outwards, her eyes were closed as peacefully as those of a sleeping child, and her long hair, which shone red-gold in the dim light from above, was wound

salary she gives me doesn't count. Apropos of Dr. Parravicini, I know he is a clever doctor, for he cures my horrid mosquito bites."

"A little ammonia would do that, in the early stage of the mischief. But there are no mosquitoes to trouble you now."

"Oh, yes, there are; I had a bite just before we left Cap Ferrino." She pushed up her loose lawn sleeve, and exhibited a scar, which he scrutinized intently, with a surprised and puzzled look.

"This is no mosquito bite," he said.

"Oh, yes it is — unless there are snakes or adders at Cap Ferrino."

"It is not a bite at all, You are trifling with me. Miss Rolleston — you have allowed that wretched Italian quack to bleed you. They killed the greatest man in modern Europe that way, remember. How very foolish of you."

"I was never bled in my life, Mr. Stafford."

"Nonsense! Let me look at your other arm. Are there any more mosquito bites?"

"Yes; Dr. Parravicini says I have a bad skin for healing, and that the poison acts more virulently with me than with most people."

Stafford examined both her arms in the broad sunlight, scars new and old.

"You have been very badly bitten, Miss Rolleston," he said, "and if ever I find the mosquito I shall make him smart. But, now tell me, my dear girl, on your word of honor, tell me as you would tell a friend who is sincerely anxious for your health and happiness — as you would tell your mother if she were here to question you — have you no knowledge of any cause for these scars except mosquito bites — no suspicion even?"

"No, indeed! No, upon my honor! I have never seen a mosquito biting my arm. One never does see the horrid little fiends. But I have heard them trumpeting under the curtains and I know that I have often had one of the pestilent wretches buzzing about me."

Later in the day Bella and her friends were sitting at tea in the garden, while Lady Ducayne took her afternoon drive with her doctor.

"How long do you mean to stop with Lady Ducayne, Miss Rolleston?" Herbert Stafford asked, after a thoughtful silence, breaking suddenly upon the trivial talk of the two girls.

"As long as she will go on paying me twenty-five pounds a quarter."

"Even if you feel your health breaking down in her service?"

"It is not the service that has injured my health. You can see that I have really nothing to do — to read aloud for an hour or so once or twice a week; to write a letter once in a while to a London tradesman. I shall never have such an easy time with anybody. And nobody else would give me a hundred a year."

"Then you mean to go on till you break down; to die at your post?"

"Like the other two companions? No! If ever I feel seriously ill — really ill — I shall put myself in a train and go back to Walworth

around her head in numberless finely plaited tresses, beneath which little locks escaped in rings upon her brow. I could have sworn that the blue veins on that divinely perfect bosom held living blood!

We were absolutely paralyzed, and Magnin leaned gasping over the edge as pale as death, paler by far than this living, almost smiling face to which his eyes were glued. I do not doubt that I was as pale as he at this inexplicable vision. As I looked the red lips seemed to grow redder. They were redder! The little pearly teeth showed between them. I had not seen them before, and now a clear ruby drop trickled down to her rounded chin and from there slipped sideways and fell upon her neck. Horror-struck I gazed upon the living corpse, till my eyes could not bear the sight any longer. As I looked away my glance fell once more upon the inscription, but now I could see — and read — it all. "To Vespertilia"— that was in Latin, and even the Latin name of the woman suggested a thing of evil

without stopping."

"What about the other two companions?"

"They both died. It was very unlucky for Lady Ducayne. That's why she engaged me; she chose me because I was ruddy and robust. She must feel rather disgusted at my having grown white and weak. By-the-bye, when I told her about the good your tonic had done me, she said she would like to see you and have a little talk with you about her own case."

"And I should like to see Lady Ducayne. When did she say this?"

"The day before yesterday."

"Will you ask her if she will see me this evening?"

"With pleasure! I wonder what you will think of her? She looks rather terrible to a stranger; but Dr. Parravicini says she was once a famous beauty."

It was nearly ten o'clock when Mr. Stafford was summoned by message from Lady Ducayne, whose courier came to conduct him to her ladyship's salon. Bella was reading aloud when the visitor was admitted; and he noticed the languor in the low, sweet tones, the evident effort.

"Shut up the book," said the querulous old voice. "You are beginning to drawl like Miss Blandy."

Stafford saw a small, bent figure crouching over the piled-up olive logs; a shrunken old figure in a gorgeous garment of black and crimson brocade, a skinny throat emerging from a mass of old Venetian lace, clasped with diamonds that flashed like fireflies as the trembling old head turned towards him.

The eyes that looked at him out of the face were almost as bright as the diamonds — the only living feature in that narrow parchment mask. He had seen terrible faces in the hospital — faces on which disease had set dreadful marks — but he had never seen a face that impressed him so painfully as this withered countenance, with its indescribable horror of death outlived, a face that should have been hidden under a coffin-lid years and years ago. The Italian physician was standing on the other side of the fireplace, smoking a cigarette, and looking down at the little old woman brooding over the hearth as if he were proud of her. "Good evening, Mr. Stafford; you can go to your room, Bella, and write your everlasting letter to your mother at Walworth," said Lady Ducayne. "I believe she writes a page about every wildflower she discovers in the woods and meadows. I don't know what else she can find to write about," she added, as Bella quietly withdrew to the pretty little bedroom opening out of Lady Ducayne's spacious apartment. Here, as at Cap Ferrino, she slept in a room adjoining the old lady's.

"You are a medical man, I understand, Mr. Stafford."

"I am a qualified practitioner, but I have not begun to practice."

"You have begun upon my companion, she tells me."

"I have prescribed for her, certainly, and I am happy to find my prescription has done her good; but I look upon that improvement as temporary. Her case will require more drastic treatment."

"Never mind her case. There is nothing the matter with the girl —

flitting in the dusk. But the full horror of the nature of that thing had been veiled to Roman eyes under the Greek [τηζ αιματοπωτιδοζ] "The blood-drinker, the vampire woman." And Flavius —her lover — *vix ipse sospes*, "himself hardly saved" from that deadly embrace, had buried her here, and set a seal upon her sepulchre, trusting to the weight of stone and the strength of clinging mortar to imprison for ever the beautiful monster he had loved.

"Infamous murderess!" I cried, "you have killed Marcello!" and a sudden, vengeful calm came over me.

"Give me the pickaxe," I said to Magnin; I can hear myself saying it still. He picked it up and handed it to me as in a dream; he seemed little better than an idiot, and the beads of sweat were shining on his forehead. I took my knife, and from the long wooden handle of the pickaxe I cut a fine, sharp stake. Then I clambered, scarcely feeling any repugnance, over the side of the sarcophagus, my feet

absolutely nothing — except girlish nonsense; too much liberty and not enough work."

"I understand that two of your ladyship's previous companions died of the same disease," said Stafford, looking first at Lady Ducayne, who gave her tremulous old head an impatient jerk, and then at Parravicini, whose yellow complexion had paled a little under Stafford's scrutiny.

"Don't bother me about my companions, sir," said Lady Ducayne. "I sent for you to consult you about myself — not about a parcel of anemic girls. You are young, and medicine is a progressive science, the newspapers tell me. Where have you studied?"

amongst the folds of Vespertilia's decaying winding-sheet, which crushed like ashes beneath my boot.

I looked for one moment at that white breast, but only to choose the loveliest spot, where the network of azure veins shimmered like veiled turquoises, and then with one blow I drove the pointed stake deep down through the breathing snow and stamped it in with my heel.

An awful shriek, so ringing and horrible that I thought my ears must have burst; but even then I felt neither fear nor horror. There are times when these cannot touch us. I stopped and gazed once again at the face, now undergoing a fearful change — fearful and final!

"Foul vampire!" I said quietly in my concentrated rage. "You will do no more harm now!" And then, without looking back upon her cursed face, I clambered out of the horrible tomb.

"In Edinburgh — and in Paris."

"Two good schools. And know all the new-fangled theories, the modern discoveries — that remind one of the medieval witchcraft, of Albertus Magnus, and George Ripley; you have studied hypnotism — electricity?"

"And the transfusion of blood," said Stafford, very slowly, looking at Parravicini.

"Have you made any discovery that teaches you to prolong human life — any elixir — any mode of treatment? I want my life prolonged, young man. That man there has been my physician for thirty years. He does all he can to keep me alive—after his lights. He studies all the new theories of all the scientists — but he is old; he gets older every day — his brain-power is going — he is bigoted — prejudiced — can't receive new ideas — can't grapple with new systems. He will let me die if I am not on my guard against him."

"You are of an unbelievable ingratitude, Ecclenza," said Parravicini.

"Oh, you needn't complain. I have paid you thousands to keep me alive. Every year of my life has swollen your hoards; you know there is nothing to come to you when I am gone. My whole fortune is left to endow a home for indigent women of quality who have reached their ninetieth year. Come, Mr. Stafford, I am a rich woman. Give me a few years more in the sunshine, a few years more above ground, and I will give you the price of a fashionable London practice — I will set you up at the West-end."

"How old are you, Lady Ducayne?"

"I was born the day Louis XVI was guillotined."

"Then I think you have had your share of the sunshine and the pleasures of the earth, and that you should spend your few remaining days in repenting your sins and trying to make atonement for the young lives that have been sacrificed to your love of life."

"What do you mean by that, sir?"

"Oh, Lady Ducayne, need I put your wickedness and your physician's still greater wickedness in plain words? The poor girl who is now in your employment has been reduced from robust health to a condition of absolute danger by Dr. Parravicini's experimental surgery; and I have no doubt those other two young women who broke down in your service were treated by him in the same manner. I could take upon myself to demonstrate — by most convincing evidence, to a jury of medical men — that Dr. Parravicini has been bleeding Miss Rolleston after putting her under chloroform, at intervals, ever since she has been in your service. The deterioration in the girl's health speaks for itself; the lancet marks upon the girl's arms are unmistakable; and her description of a series of sensations, which she calls a dream, points unmistakably to the administration of chloroform while she was sleeping. A practice so nefarious, so murderous, it must, if exposed, result in a sentence only less severe than the punishment of murder."

"I laugh," said Parravicini, with an airy motion of his skinny fingers; "I laugh at once at your theories and at your threats. I, Parravicini Leopold, have no fear that the law can question anything I have done."

"Take the girl away and let me hear no more of her," cried Lady Ducayne, in the thin, old voice, which so poorly matched the energy and fire of the wicked old brain that guided its utterances. "Let her go back to her mother — I want no more girls to die in my service. There are girls enough and to spare in the world, God knows."

"Her brain power is going."

"If you ever engage another companion — or take another English girl into your service, Lady Ducayne, I will make all England ring with the story of your wickedness."

"I want no more girls. I don't believe in his experiments. They have been full of danger for me as well as for the girl — an air bubble, and I should be gone. I'll have no more of his dangerous quackery. I'll find some new man — a better man than you, sir, a discoverer like Pasteur, or Virchow, a genius — to keep me alive. Take your girl away, young man. Marry her if you like. I'll write a check for a thousand pounds, and let her go and live on beef and beer, and get strong and plump again. I'll have no more such experiments. Do you hear, Parravicini?" she screamed, vindictively, the yellow, wrinkled face distorted with fury, the eyes glaring at him.

The Staffords carried Bella Rolleston off to Varese next day, she very loath to leave Lady Ducayne, whose liberal salary afforded such help for the dear mother. Herbert Stafford insisted, however, treating Bella as coolly as if he had been the family physician, and she had been given over wholly to his care.

"Do you suppose your mother would let you stop here to die?" he asked. "If Mrs. Rolleston knew how ill you are, she would come post haste to fetch you."

"I shall never be well again till I get back to Walworth," answered Bella, who was low-spirited and inclined to tears this morning, a reaction after her good spirits of yesterday.

"We'll try a week or two at Varese first," said Stafford. "When you can walk halfway up Monte Generoso without palpitation of the heart, you shall go back to Walworth."

"Poor mother, how glad she will be to see me, and how sorry that I've lost such a good place."

This conversation took place on the boat when they were leaving Bellaggio. Lotta had gone to her friend's room at seven o'clock that morning, long before Lady Ducayne's withered eyelids had opened to the daylight, before even Francine, the French maid, was astir, and had helped to pack a Gladstone bag with essentials, and hustled Bella downtairs and out of doors before she could make any strenuous resistance.

"It's all right," Lotta assured her. "Herbert had a good talk with Lady Ducayne last night, and it was settled for you to leave this morning. She doesn't like invalids, you see."

"No," sighed Bella, "she doesn't like invalids. It was very unlucky that I should break down, just like Miss Tomson and Miss Blandy."

"At any rate, you are not dead, like them," answered Lotta, "and my brother says you are not going to die."

"A check for a thousand."

It seemed rather a dreadful thing to be dismissed in that offhand way, without a word of farewell from her employer.

"I wonder what Miss Torpinter will say when I go to her for another situation," Bella speculated, ruefully, while she and her friends were breakfasting on board the steamer.

"Perhaps you may never want another situation," said Stafford.

"You mean that I may never be well enough to be useful to anybody?"

"No, I don't mean anything of the kind."

It was after dinner at Varese, when Bella had been induced to take a whole glass of Chianti, and quite sparkled after that unaccustomed stimulant, that Mr. Stafford produced a letter from his pocket.

"I forgot to give you Lady Ducayne's letter of adieu!" he said.

"What, did she write to me? I am so glad — I hated to leave her in such a cool way; for after all she was very kind to me, and if I didn't like her it was only because she was too dreadfully old."

She tore open the envelope. The letter was short and to the point: —

> "Goodbye, child. Go and marry your doctor. I enclose a farewell gift for your trousseau.
> —ADELINE DUCAYNE

"A hundred pounds, a whole year's salary — no — why, it's for a — a check for a thousand!" cried Bella. "What a generous old soul! She really is the dearest old thing."

"She just missed being very dear to you, Bella," said Stafford.

He had dropped into the use of her Christian name while they were on board the boat. It seemed natural now that she was to be in his charge till they all three went back to England.

"I shall take upon myself the privileges of an elder brother till we land at Dover," he said; "after that — well, it must be as you please."

The question of their future relations must have been satisfactorily settled before they crossed the Channel, for Bella's next letter to her mother communicated three startling facts.

First, that the enclosed check for £1,000 was to be invested in debenture stock in Mrs. Rolleston's name, and was to be her very own, income and principal, for the rest of her life.

Next, that Bella was going home to Walworth immediately.

And last, that she was going to be married to Mr. Herbert Stafford in the following autumn.

"And I am sure you will adore him, mother, as much as I do," wrote Bella.

"It is all good Lady Ducayne's doing. I never could have married if I had not secured that little nest-egg for you. Herbert says we shall be able to add to it as the years go by, and that wherever we live there shall be always a room in our house for you. The word 'mother-in-law' has no terrors for him."

Cover of an early British edition of Dracula's Guest.
(Courtesy of Ronald V. Borst / Hollywood Movie Posters)

Dracula's Guest

BRAM STOKER
(c.1895, published in 1914)

WHEN WE STARTED for our drive the sun was shining brightly on Munich, and the air was full of the joyousness of early summer. Just as we were about to depart, Herr Delbruck (the maître d'hotel of the Quatre Saisons, where I was staying) came down bareheaded to the carriage and, after wishing me a pleasant drive, said to the coachman, still holding his hand on the handle of the carriage door, "Remember you are back by nightfall. The sky looks bright but there is a shiver in the north wind that says there may be a sudden storm. But I am sure you will not be late." Here he smiled and added, "for you know what night it is."

Johann answered with an emphatic, "Ja, mein Herr," and, touching his hat, drove off quickly. When we had cleared the town, I said, after signalling to him to stop:

"Tell me, Johann, what is tonight?"

He crossed himself, as he answered laconically: "Walpurgis nacht." Then he took out his watch, a great, old-fashioned German silver thing as big as a turnip and looked at it, with his eyebrows gathered together and a little impatient shrug of his shoulders. I realized that this was his way of respectfully protesting against the unnecessary delay and sank back in the carriage, merely motioning him to proceed. He started off rapidly, as if to make up for lost time. Every now and then the horses seemed to throw up their heads and sniff the air suspiciously. On such occasions I often looked round in alarm. The road was pretty bleak, for we were traversing a sort of high windswept plateau. As we drove, I saw a road that looked but little used and which seemed to dip through a little winding valley. It looked so inviting that, even at the risk of offending him, I called Johann to stop — and when he had pulled up, I told him I would like to drive down that road. He made all sorts of excuses and frequently crossed himself as he spoke. This somewhat piqued my curiosity, so I asked him various questions. He answered fencingly and repeatedly looked at his watch in protest.

Finally I said, "Well, Johann, I want to go down this road. I shall not ask you to come unless you like; but tell me why you do not like to go, that is all I ask." For answer he seemed to throw himself off the box, so quickly did he reach the ground. Then he stretched out his hands appealingly to me and implored me not to go. There was just enough of English mixed with the German for me to understand the drift of his talk. He seemed always just about to tell me something — the very idea of which

ABOUT THE AUTHOR

The world-famous author of *Dracula* remains a tantalizing enigma for literary commentators. While Abraham "Bram" Stoker (1847-1912) wrote fiction voluminously, he did not keep diaries or journals that might have illuminated his creative process. As a result, Stoker has ended up as a kind of blank canvas on to which commentators are free to project all manner of their own obsessions.

Stoker himself seems to have been obsessed with *Dracula*—at least to the degree to which he extensively researched and reworked the manuscript. The story "Dracula's Guest" is not, as his widow later claimed, a deleted chapter from the finished novel, but instead a kind of discarded finger exercise for the novel's opening, which Stoker completely rewrote. Nevertheless, "Dracula's Guest" remains a polished and self-contained mood piece with a central, striking image of a female vampire that inspired the motion picture *Dracula's Daughter* (1936).

evidently frightened him; but each time he pulled himself up saying, "Walpurgis nacht!"

I tried to argue with him, but it was difficult to argue with a man when I did not know his language. The advantage certainly rested with him, for although he began to speak in English, of a very crude and broken kind, he always got excited and broke into his native tongue — and every time he did so, he looked at his watch. Then the horses became restless and sniffed the air. At this he grew very pale, and, looking around in a frightened way, he suddenly jumped forward, took them by the bridles, and led them on some twenty feet. I followed and asked why he had done this. For an answer he crossed himself, pointed to the spot we had left, and drew his carriage in the direction of the other road, indicating a cross, and said, first in German, then in English, "Buried him — him what killed themselves."

As for the facts of Stoker's life, he was born in Dublin, the third of seven children, and suffered from a long, peculiar, and possibly even hysterical paralysis which, by Stoker's account, left him completely bedridden until the age of seven (but somehow did nothing to prevent him from later becoming an accomplished athlete). Stoker was educated at Trinity College, was active in the Trinity Philosophical Society (where he sponsored the membership of his friend Oscar Wilde), and became a devoted, public partisan of Walt Whitman, to whom he wrote long, passionate missives: "How sweet a thing it is for a strong healthy man who can be if he wishes father, and brother and wife to his soul."

I remembered the old custom of burying suicides at crossroads: "Ah! I see, a suicide. How interesting!" But for the life of me I could not make out why the horses were frightened.

Whilst we were talking, we heard a sort of sound between a yelp and a bark. It was far away; but the horses got very restless, and it took Johann all his time to quiet them. He was pale and said, "It sounds like a wolf — but yet there are no wolves here now."

"No?" I said, questioning him. "Isn't it long since the wolves were so near the city?"

"Long, long," he answered, "in the spring and summer; but with the snow the wolves have been here not so long."

Whilst he was petting the horses and trying to quiet them, dark clouds drifted rapidly across the sky. The sunshine passed away, and a breath of cold wind seemed to drift over us. It was only a breath, however, and more of a warning than a fact, for the sun came out brightly again.

Bram Stoker

Johann looked under his lifted hand at the horizon and said, "The storm of snow, he comes before long time." Then he looked at his watch again, and, straightway holding his reins firmly — for the horses were still pawing the ground restlessly and shaking their heads — he climbed to his box as though the time had come for proceeding on our journey.

I felt a little obstinate and did not at once get into the carriage.

"Tell me," I said, " about this place where the road leads," and I pointed down.

Again he crossed himself and mumbled a prayer before he answered, "It is unholy."

"What is unholy?" I enquired.

"The village."

"Then there is a village?"

"No, no. No one lives there hundreds of years."

My curiosity was piqued, "But you said there was a village."

"There was."

"Where is it now?"

Whereupon he burst out into a long story in German and English, so mixed up that I could not quite understand exactly what he said. Roughly I gathered that long ago, hundreds of years, men had died there and been buried in their graves; but sounds were heard under the clay, and when the graves were opened, men and women were found rosy with life and their mouths red with blood. And so, in haste to save their lives (aye, and their souls! — and here he crossed himself) those who were left fled away to other places, where the living lived and the dead were dead and not — not something. He was evidently afraid to speak the last words. As he proceeded with his narration, he grew more and more excited. It seemed as if his imagination had got hold of him, and he ended in a perfect paroxysm of fear —white-faced, perspiring, trembling, and looking round him as if expecting that some dreadful presence would manifest itself there in the bright sunshine on the open plain.

Finally, in an agony of desperation, he cried, "Walpurgis nacht!" and pointed to the carriage for me to get in.

All my English blood rose at this, and standing back I said, "You are afraid, Johann — you are afraid. Go home, I shall return alone, the walk will do me good." The carriage door was open. I took from the seat my oak walking stick—which I always carry on my holiday excursions — and closed the door, pointing back to Munich, and said, "Go home, Johann — Walpurgis nacht doesn't concern Englishmen."

The horses were now more restive than ever, and Johann was trying to hold them in, while excitedly imploring me not to do anything so foolish. I pitied the poor fellow, he was so deeply in earnest; but all the same I could not help laughing.

His English was quite gone now. In his anxiety he had forgotten that his only means of making me understand was to talk my language, so he jabbered away in his native German. It began to be a little tedious. After giving the direction, "Home!" I turned to go down the cross road into the valley.

With a despairing gesture, Johann turned his horses towards Munich. I leaned on my stick and looked after him. He went slowly along the road for a while, then there came over the crest of the hill a man tall and thin. I could see so much in the distance. When he drew near the horses, they began to jump and kick about, then to scream with terror. Johann could not hold them in; they bolted down the road, running away madly. I watched them out of sight, then looked for the stranger; but I found that he, too, was gone.

With a light heart I turned down the side road through the deepening valley to which Johann had objected. There was not the slightest reason, that I could see, for his objection; and I daresay I tramped for a couple of hours without thinking of time or distance and certainly without seeing a person or a house. So far as the place was concerned, it was desolation

Stoker's susceptibility to male charisma reached a climax in his professional association with the actor Henry Irving, who rescued him from the life of a Dublin petty-clerk and part-time theatre critic, elevating him as second-in-command of his company at the Royal Lyceum Theatre in London.

The critical moment between the men seems to have occurred in 1876 when Stoker, then twenty-eight years old, had a hysterical fit following one of Irving's intensely emotional dramatic recitations. The actor shrewdly sensed that Stoker's rapt response to his art, coupled with his love of the theatre and general business acumen, would be useful in establishing his theatrical dominion over London.

Stoker devoted his life to Irving for three decades, writing melodramatic fiction as a sideline. His friend, the novelist Hall Caine, would later write that "I say without hesitation that I have never seen, never do I expect to see, such absorption of one man's life in the life of another...with Irving's life, poor Bram's had really ended."

itself. But I did not notice this particularly till, on turning a bend in the road, I came upon a scattered fringe of wood; then I recognized that I had been impressed unconsciously by the desolation of the region through which I had passed.

I sat down to rest myself and began to look around. It struck me that it was considerably colder than it had been at the commencement of my walk — a sort of sighing sound seemed to be around me with, now and then, high overhead, a sort of muffled roar. Looking upwards I noticed that great thick clouds were drafting rapidly across the sky from north to south at a great height. There were signs of a coming storm in some lofty stratum of the air. I was a little chilly, and, thinking that it was the sitting still after the exercise of walking, I resumed my journey.

The vampire-like dynamics of Stoker's relationship with Irving have been widely acknowledged as having influenced the composition of *Dracula*; indeed, Stoker himself explicitly told at least one American theatre critic that he conceived the title character as a composite of Irving's villainous portrayals, and vainly tried to persuade him to play the master vampire on stage.

Even though Irving snubbed it ("Dreadful!" he is said to have declaimed), *Dracula* found a wide popular audience and is the only one of his books to have remained steadily in print. But his royalties could not really sustain him after Irving's death in 1906 and the ensuing dissolution of the Lyceum company. Stoker died, semi-impoverished, the same week as the sinking of the *Titanic* in 1912. He never lived to see the extraordinary life *Dracula* would enjoy through dramatic and film adaptations in the twentieth century.

Henry Irving

The ground I passed over was now much more picturesque. There were no striking objects that the eye might single out, but in all there was a charm of beauty. I took little heed of time, and it was only when the deepening twilight forced itself upon me that I began to think of how I should find my way home. The air was cold, and the drifting of clouds high overhead was more marked. They were accompanied by a sort of far away rushing sound, through which seemed to come at intervals that mysterious cry which the driver had said came from a wolf. For a while I hesitated. I had said I would see the deserted village, so on I went and presently came on a wide stretch of open country, shut in by hills all around. Their sides were covered with trees which spread down to the plain, dotting in clumps the gentler slopes and hollows which showed here and there. I followed with my eye the winding of the road and saw that it curved close to one of the densest of these clumps and was lost behind it.

As I looked there came a cold shiver in the air, and the snow began to fall. I thought of the miles and miles of bleak country I had passed, and then hurried on to seek shelter of the wood in front. Darker and darker grew the sky, and faster and heavier fell the snow, till the earth before and around me was a glistening white carpet the further edge of which was lost in misty vagueness. The road was here but crude, and when on the level its boundaries were not so marked as when it passed through the cuttings; and in a little while I found that I must have strayed from it, for I missed underfoot the hard surface, and my feet sank deeper in the grass and moss. Then the wind grew stronger and blew with ever increasing force, till I was fain to run before it. The air became icy-cold, and in spite of my exercise I began to suffer. The snow was now falling so thickly and whirling around me in such rapid eddies that I could hardly keep my eyes open. Every now and then the heavens were torn asunder by vivid lightning, and in the flashes I could see ahead of me a great mass of trees, chiefly yew and cypress all heavily coated with snow.

I was soon amongst the shelter of the trees, and there in comparative silence I could hear the rush of the wind high overhead. Presently the blackness of the storm had become merged in the darkness of the night.

By-and-by the storm seemed to be passing away, it now only came in fierce puffs or blasts. At such moments the weird sound of the wolf appeared to be echoed by many similar sounds around me.

Now and again, through the black mass of drifting cloud, came a straggling ray of moonlight which lit up the expanse and showed me that I was at the edge of a dense mass of cypress and yew trees. As the snow had ceased to fall, I walked out from the shelter and began to investigate more closely. It appeared to me that, amongst so many old foundations as I had passed, there might be still standing a house in which, though in ruins, I could find some sort of shelter for a while.

As I skirted the edge of the copse, I found that a low wall encircled it, and following this I presently found an opening. Here the cypresses formed an alley leading up to a square mass of some kind of building. Just as I caught sight of this, however, the drifting clouds obscured the moon, and I passed up the path in darkness. The wind must have grown colder, for I felt myself shiver as I walked; but there was hope of shelter, and I groped my way blindly on.

I stopped, for there was a sudden stillness. The storm had passed; and, perhaps in sympathy with nature's silence, my heart seemed to cease to beat. But this was only momentarily; for suddenly the moonlight broke through the clouds showing me that I was in a graveyard and that the square object before me was a great massive tomb of marble, as white as the snow that lay on and all around it. With the moonlight there came a fierce sigh of the storm which appeared to resume its course with a long, low howl, as of many dogs or wolves. I was awed and shocked, and I felt the cold perceptibly grow upon me till it seemed to grip me by the heart. Then while the flood of moonlight still fell on the marble tomb, the storm gave further evidence of renewing, as though it were returning on its track. Impelled by some sort of fascination, I approached the sepulchre to see what it was and why such a thing stood alone in such a place. I walked around it and read, over the Doric door, in German —

COUNTESS DOLINGEN OF GRATZ
IN STYRIA
SOUGHT AND FOUND DEAD
1801

On the top of the tomb, seemingly driven through the solid marble—for the structure was composed of a few vast blocks of stone—was a great iron spike or stake. On going to the back I saw, graven in great Russian letters: "The dead travel fast."

There was something so weird and uncanny about the whole thing that it gave me a turn and made me feel quite faint. I began to wish, for the first time, that I had taken Johann's advice. Here a thought struck me, which came under almost mysterious circumstances and with a terrible shock. This was Walpurgis Night!

Walpurgis Night was when, according to the belief of millions of people,

Emily Gerard (*Courtesy of Lokke Heiss*)

THE MOTHER OF DRACULA

Bram Stoker never visited Transylvania, but instead made significant use of the work of the Scottish-born writer and folklorist Emily Gerard (1849-1905). Gerard married a Polish/Austrian cavalry officer, Miecislaus Laszowski, and spent two years with him in Transylvania, where she amassed a prodigious amount of research, used for an 1885 article in the influential British journal *The Nineteenth Century* and for a full-length book, *The Land Beyond the Forest* (1888).

From Gerard, Stoker gleaned some near-verbatim descriptions of Transylvanian folk beliefs, and, most important, her citation of the controversial term *nosferatu* as a Romanian word for vampire, which Stoker happily appropriated, despite the fact that *nosferatu* appears in no Romanian dictionary (or any other dictionary, for that matter). Nonetheless, the word *nosferatu* remains Gerard's indelible gift to the realm of the undead.

the devil was abroad—when the graves were opened and the dead came forth and walked. When all evil things of earth and air and water held revel. This very place the driver had specially shunned. This was the depopulated village of centuries ago. This was where the suicide lay; and this was the place where I was alone — unmanned, shivering with cold in a shroud of snow with a wild storm gathering again upon me! It took all my philosophy, all the religion I had been taught, all my courage, not to collapse in a paroxysm of fright.

And now a perfect tornado burst upon me. The ground shook as though thousands of horses thundered across it; and this time the storm bore on its icy wings, not snow, but great hailstones which drove with such violence that they might have come from the thongs of Balearic slingers — hailstones that beat down leaf and branch and made the shelter of the cypresses of no more avail than though their stems were standing corn. At the first I had rushed to the nearest tree; but I was soon fain to leave it and seek the only spot that seemed to afford refuge, the deep Doric doorway of the marble tomb. There, crouching against the massive bronze door, I gained a certain amount of protection from the beating of the hailstones, for now they only drove against me as they ricochetted from the ground and the side of the marble.

As I leaned against the door, it moved slightly and opened inwards. The shelter of even a tomb was welcome in that pitiless tempest and I was about to enter it when there came a flash of forked lightning that lit up the whole expanse of the heavens. In the instant, as I am a living man, I saw, as my eyes turned into the darkness of the tomb, a beautiful woman with rounded cheeks and red lips, seemingly sleeping on a bier. As the thunder broke overhead, I was grasped as by the hand of a giant and hurled out into the storm. The whole thing was so sudden that, before I could realize the shock, moral as well as physical, I found the hailstones beating me down. At the same time I had a strange, dominating feeling that I was not alone. I looked towards the tomb. Just then there came another blinding flash which seemed to strike the iron stake that surmounted the tomb and to pour through to the earth, blasting and crumbling the marble, as in a burst of flame. The dead woman rose for a moment of agony while she was lapped in the flame, and her bitter scream of pain was drowned in the thundercrash. The last thing I heard was this mingling of dreadful sound, as again I was seized in the giant grasp and dragged away, while the hailstones beat on me and the air around seemed reverberant with the howling of wolves. The last sight that I remembered was a vague, white, moving mass, as if all the graves around me had sent out the phantoms of their sheeted dead, and that they were closing in on me through the white cloudiness of the driving hail. Gradually there came a sort of vague beginning of consciousness, then a sense of weariness that was dreadful. For a time I remembered nothing, but slowly my senses returned. My feet seemed positively racked with pain, yet I could not move them. They seemed to

From
TRANSYLVANIAN SUPERSTITIONS
Emily de Laszowska Gerard
(1885)

Transylvania might well be termed the land of superstition, for nowhere else does this curious crooked plant of delusion flourish as persistently and in such bewildering variety. It would almost seem as though the whole species of demons, pixies, witches, and hobgoblins, driven from the rest of Europe by the wand of science, had taken refuge within this mountain rampart, well aware that here they would find secure lurking-places, whence they might defy their persecutors yet awhile.

There are many reasons why these fabulous beings should retain an abnormally firm hold on the soul of these parts; and looking at the matter closely we find here no less than three separate sources of superstition.

First, there is what might be called the indigenous superstition of the country, the scenery of which is peculiarly adapted to serve as background to all sorts of supernatural beings and monsters. There are innumerable caverns, whose mysterious depths seem made to harbour whole legions of evil spirits: forest glades fit only for fairy folk on moonlight nights, solitary lakes which instinctively call up visions of water sprites, golden treasures lying hidden in mountain chasms, all of which have gradually

be numbed. There was an icy feeling at the back of my neck and all down my spine, and my ears, like my feet, were dead yet in torment; but there was in my breast a sense of warmth which was by comparison delicious. It was as a nightmare — a physical nightmare, if one may use such an expression; for some heavy weight on my chest made it difficult for me to breathe.

This period of semilethargy seemed to remain a long time, and as it faded away I must have slept or swooned. Then came a sort of loathing, like the first stage of seasickness, and a wild desire to be free of something — I knew not what. A vast stillness enveloped me, as though all the world were asleep or dead — only broken by the low panting as of some animal close to me. I felt a warm rasping at my throat, then came a consciousness of the awful truth which chilled me to the heart and sent the blood surging up through my brain. Some great animal was lying on me and now licking my throat. I feared to stir, for some instinct of prudence bade me lie still; but the brute seemed to realize that there was now some change in me, for it raised its head. Through my eyelashes I saw above me the two great flaming eyes of a gigantic wolf. Its sharp white teeth gleamed in the gaping red mouth, and I could feel its hot breath fierce and acrid upon me.

For another spell of time I remembered no more. Then I became conscious of a low growl, followed by a yelp, renewed again and again. Then seemingly very far away, I heard a "Holloa! Holloa!" as of many voices calling in unison. Cautiously I raised my head and looked in the direction whence the sound came, but the cemetery blocked my view. The wolf still continued to yelp in a strange way, and a red glare began to move round the grove of cypresses, as though following the sound.

As the voices drew closer, the wolf yelped faster and louder. I feared to make either sound or motion. Nearer came the red glow over the white pall which stretched into the darkness around me. Then all at once from beyond the trees there came at a trot a troop of horsemen bearing torches. The wolf rose from my breast and made for the cemetery. I saw one of the horsemen (soldiers by their caps and their long military cloaks) raise his carbine and take aim. A companion knocked up his arm, and I heard the ball whiz over my head. He had evidently taken my body for that of the wolf. Another sighted the animal as it slunk away, and a shot followed. Then, at a gallop, the troop rode forward — some towards me, others following the wolf as it disappeared amongst the snow-clad cypresses.

As they drew nearer I tried to move but was powerless, although I could see and hear all that went on around me. Two or three of the soldiers jumped from their horses and knelt beside me. One of them raised my head and placed his hand over my heart.

"Good news, comrades!" he cried. "His heart still beats!"

Then some brandy was poured down my throat; it put vigor into me, and I was able to open my eyes fully and look around. Lights and shadows

insinuated themselves into the minds of the oldest inhabitants, the Roumenians, and influenced their way of thinking, so that these people, by nature imaginative and poetically inclined, have built up for themselves out of the surrounding materials a whole code of fanciful superstition, to which they adhere as closely as to their religion itself.

Secondly, there is here the imported superstition! That is to say, the old German customs and beliefs brought hither seven hundred years ago by the Saxon colonists from their native land, and like many other things, preserved here in greater perfection than in the original country.

Thirdly, there is the wandering superstition of the gypsy tribes, themselves a race of fortune-tellers and witches, whose ambulating caravans cover the country as with a network, and whose less vagrant members fill up the suburbs of towns and villages.

Of course all these various sorts of superstition have twined and intermingled, acted and reacted upon one another, until in many cases it is a difficult matter to determine the exact parentage of some particular belief or custom; but in a general way the three sources I have named may be admitted as a rough sort of classification in dealing with the principal superstitions afloat in Transylvania.

were moving among the trees, and I heard men call to one another. They drew together, uttering frightened exclamations; and the lights flashed as the others came pouring out of the cemetery pell-mell, like men possessed. When the further ones came close to us, those who were around me asked them eagerly, "Well, have you found him?"

The reply rang out hurriedly, "No! No! Come away quick — quick! This is no place to stay, and on this of all nights!"

"What was it?" was the question, asked in all manner of keys. The answer came variously and all indefinitely as though the men were moved by some common impulse to speak yet were restrained by some common fear from giving their thoughts.

"It — it — indeed!" gibbered one, whose wits had plainly given out for the moment.

"A wolf — and yet not a wolf!" another put in shudderingly.

"No use trying for him without the sacred bullet," a third remarked in a more ordinary manner.

"Serve us right for coming out on this night! Truly we have earned our thousand marks!" were the ejaculations of a fourth.

"There was blood on the broken marble," another said after a pause, "the lightning never brought that there. And for him — is he safe? Look at his throat! See comrades, the wolf has been lying on him and keeping his blood warm."

The officer looked at my throat and replied, "He is all right, the skin is not pierced. What does it all mean? We should never have found him but for the yelping of the wolf."

"What became of it?" asked the man who was holding up my head and who seemed the least panic-stricken of the party, for his hands were steady and without tremor. On his sleeve was the chevron of a petty officer.

"It went home," answered the man, whose long face was pallid and who actually shook with terror as he glanced around him fearfully. "There are graves enough there in which it may lie. Come, comrades — come quickly! Let us leave this cursed spot."

The officer raised me to a sitting posture, as he uttered a word of command; then several men placed me upon a horse. He sprang to the saddle behind me, took me in his arms, gave the word to advance; and, turning our faces away from the cypresses, we rode away in swift military order.

As yet my tongue refused its office, and I was perforce silent. I must have fallen asleep; for the next thing I remembered was finding myself standing up, supported by a soldier on each side of me. It was almost broad daylight, and to the north a red streak of sunlight was reflected like a path of blood over the waste of snow. The officer was telling the men to say nothing of what they had seen, except that they found an English stranger, guarded by a large dog.

"Dog! That was no dog," cut in the man who had exhibited such fear.

THE "HISTORICAL" DRACULA: VLAD THE IMPALER

"Dracula," the sobriquet given to the Wallachian warlord Vlad Tepes (1431-1476) means "son of the devil" (or "dragon") in Romanian — Tepes' father had been called *Dracul.* Vlad's nickname "The Impaler" derives from his favorite method of dispatching enemies — by impalement on wooden stakes. On one atrocious occasion, 20,000 Turkish captives were exterminated in this manner and displayed in a mile-long semicircle outside Dracula's capital city, Tirgoviste, to ward off oncoming enemy troops. (It worked.) By all accounts, Vlad reveled in the death agony of his victims and often dined in the shadow of their writhing, rotting bodies. Other victims seem to have been boiled, or hacked apart "like cabbages," according to one account.

While ruthless, sadistic, and undeniably psychopathic, Vlad Tepes is nonetheless a hero of Romanian history, who successfully protected the country against foreign incursions.

Sometime in the early 1890s, Bram Stoker came across a reference to the Voivode (Prince) Dracula in a historical book and decided to use the name for his vampire villain (his original choice for the character's name was "Count Wampyr"). In other words, Vlad the Impaler did not inspire Stoker to write his vampire novel, but rather provided a modicum of historical

"I think I know a wolf when I see one."

The young officer answered calmly, "I said a dog."

"Dog!" reiterated the other ironically. It was evident that his courage was rising with the sun; and, pointing to me, he said, "Look at his throat. Is that the work of a dog, master?"

Instinctively I raised my hand to my throat, and as I touched it I cried out in pain. The men crowded round to look, some stooping down from their saddles; and again there came the calm voice of the young officer, "A dog, as I said. If aught else were said we should only be laughed at."

I was then mounted behind a trooper, and we rode on into the suburbs of Munich. Here we came across a stray carriage into which I was lifted, and it was driven off to the Quatre Saisons — the young officer accompanying me, whilst a trooper followed with his horse, and the others rode off to their barracks.

When we arrived, Herr Delbruck rushed so quickly down the steps to meet me, that it was apparent he had been watching within. Taking me by both hands he solicitously led me in. The officer saluted me and was turning to withdraw, when I recognized his purpose and insisted that he should come to my rooms. Over a glass of wine I warmly thanked him and his brave comrades for saving me. He replied simply that he was more than glad, and that Herr Delbruck had at the first taken steps to make all the searching party pleased; at which ambiguous utterance the maître d'hotel smiled, while the officer plead duty and withdrew.

"But Herr Delbruck," I enquired, "how and why was it that the soldiers searched for me?"

He shrugged his shoulders, as if in depreciation of his own deed, as he replied, "I was so fortunate as to obtain leave from the commander of the regiment in which I serve, to ask for volunteers."

"But how did you know I was lost?" I asked.

"The driver came hither with the remains of his carriage, which had been upset when the horses ran away."

"But surely you would not send a search party of soldiers merely on this account?"

"Oh, no!" he answered, "but even before the coachman arrived, I had this telegram from the Boyar whose guest you are," and he took from his pocket a telegram which he handed to me, and I read:

> Bistritz.
>
> Be careful of my guest — his safety is most precious to me. Should aught happen to him, or if he be missed, spare nothing to find him and ensure his safety. He is English and therefore adventurous. There are often dangers from snow and wolves and night. Lose not a moment if you suspect harm to him. I answer your zeal with my fortune.
>
> — *Dracula.*

verisimilitude to a fictional story Stoker had already conceived. Other than an account of Vlad dipping bread into the blood of a victim, and the coincidence of the wooden stake motif, there is no historical correspondence between the bloody voivode and traditional vampire folklore.

The life and times of Vlad the Impaler have been extensively documented by Historians Raymond T. McNally and Radu R. Florescu in two books, *In Search of Dracula* (1972) and *Dracula: Prince of Many Faces* (1989).

Vlad the Impaler enjoys a hideous repast. From a sixteenth-century woodcut.

As I held the telegram in my hand, the room seemed to whirl around me, and if the attentive mâitre d'hotel had not caught me, I think I should have fallen. There was something so strange in all this, something so weird and impossible to imagine, that there grew on me a sense of my being in some way the sport of opposite forces — the mere vague idea of which seemed in a way to paralyze me. I was certainly under some form of mysterious protection. From a distant country had come, in the very nick of time, a message that took me out of the danger of the snow sleep and the jaws of the wolf.

"Dracula's Guest" inspired the 1936 film Dracula's Daughter *with Gloria Holden in the title role. (Photofest)*

FROM

Dracula

BRAM STOKER
(1897)

Chapter II

JONATHAN HARKER'S JOURNAL.

5 MAY.— I must have been asleep, for certainly if I had been fully awake I must have noticed the approach of such a remarkable place. In the gloom the courtyard looked of considerable size, and as several dark ways led from it under great round arches it perhaps seemed bigger than it really is. I have not yet been able to see it by daylight.

When the caleche stopped the driver jumped down, and held out his hand to assist me to alight. Again I could not but notice his prodigious strength. His hand actually seemed like a steel vice that could have crushed mine if he had chosen. Then he took out my traps, and placed them on the ground beside me as I stood close to a great door, old and studded with large iron nails, and set in a projecting doorway of massive stone. I could see even in the dim light that the stone was massively carved, but that the carving had been much worn by time and weather. As I stood, the driver jumped again into his seat and shook the reins; the horses started forward, and trap and all disappeared down one of the dark openings.

I stood in silence where I was, for I did not know what to do. Of bell or knocker there was no sign; through these frowning walls and dark window openings it was not likely that my voice could penetrate. The time I waited seemed endless, and I felt doubts and fears crowding upon me. What sort of place had I come to, and among what kind of people? What sort of grim adventure was it on which I had embarked? Was this a customary incident in the life of a solicitor's clerk sent out to explain the purchase of a London estate to a foreigner? Solicitor's clerk! Mina would not like that. Solicitor,— for just before leaving London I got word that my examination was successful; and I am now a full-blown solicitor! I began to rub my eyes and pinch myself to see if I were awake. It all seemed like a horrible nightmare to me, and I expected that I should suddenly awake, and find myself at home, with the dawn struggling in through the windows, as I had now and again felt in the morning after a day of overwork. But my flesh answered the pinching test, and my eyes were not to be deceived. I was indeed awake and among the Carpathians. All I could do now was to be patient, and to wait the coming of the morning.

Just as I had come to this conclusion I heard a heavy step approaching

ABOUT THE NOVEL

Bram Stoker's *Dracula* (originally entitled *The Un-Dead*), was first published in London by Constable in 1897 and has never been out of print since. Inarguably the most influential vampire novel of all time, *Dracula* is nonetheless a book unknown to many modern readers, who are familiar with Count Dracula primarily through his romanticized and sanitized portrayals in the theatre and cinema.

As this excerpt from the novel demonstrates, the original Dracula was not a particularly charming character, but rather a repulsive old man who reflected the popular Victorian stereotype of evolutionary degeneracy: a pointy-eared, hook-nosed, hairy-palmed satyr.

Following is a recently-rediscovered review of Stoker's novel from *The Stage*, June 17, 1897:

"Mr. Bram Stoker has already made his mark as a writer of romances, but in his latest book, *Dracula*, just published by Archibald Constable & Co., he has done more ambitious work. Grim legends in which strange beings such as Were-Wolf and the Vampire are represented as preying upon human life have for ages found a place in European folklore, and a theme of this weird and eerie kind Mr. Bram Stoker has worked out with the zeal and ingenuity of a Wilkie Collins, telling his story, we should add, entirely by means of letters,

behind the great door, and saw through the chinks the gleam of a coming light. Then there was the sound of rattling chains and the clanking of massive bolts drawn back. A key was turned with the loud grating noise of long disuse, and the great door swung back.

Within, stood a tall old man, clean shaven save for a long white moustache, and clad in black from head to foot, without a single speck of colour about him anywhere. He held in his hand an antique silver lamp, in which the flame burned without chimney or globe of any kind, throwing long quivering shadows as it flickered in the draught of the open door. The old man motioned me in with his right hand with a courtly gesture, saying in excellent English, but with a strange intonation:—

"Welcome to my house! Enter freely and of your own will!" He made no motion of stepping to meet me, but stood like a statue, as though his gesture of welcome had fixed him into stone. The instant, however, that I had stepped over the threshold, he moved impulsively forward, and holding out his hand grasped mine with a strength which made me wince, an effect which was not lessened by the fact that it seemed as cold as ice — more like the hand of a dead than a living man. Again he said:—

"Welcome to my house. Come freely. Go safely; and leave something of the happiness you bring!" The strength of the handshake was so much akin to that which I had noticed in the driver, whose face I had not seen, that for a moment I doubted if it were not the same person to whom I was speaking; so to make sure, I said interrogatively:—

"Count Dracula?" He bowed in a courtly way as he replied:—

"I am Dracula; and I bid you welcome, Mr. Harker, to my house. Come in; the night air is chill, and you must need to eat and rest." As he was speaking he put the lamp on a bracket on the wall, and stepping out, took my luggage; he had carried it in before I could forestall him. I protested but he insisted:—

"Nay, sir, you are my guest. It is late, and my people are not available. Let me see to your comfort myself." He insisted on carrying my traps along the passage, and then up a great winding stair, and along another great passage, on whose stone floor our steps rang heavily. At the end of this he threw open a heavy door, and I rejoiced to see within a well-lit room in which a table was spread for supper, and on whose mighty hearth a great fire of logs, freshly replenished, flamed and flared.

The Count halted, putting down my bags, closed the door, and crossing the room, opened another door, which led into a small octagonal room lit by a single lamp, and seemingly without a window of any sort. Passing through this, he opened another door, and motioned me to enter. It was a welcome sight; for here was a great bedroom well lighted and warmed with another log fire,— also added to but lately for the top logs were fresh — which sent a hollow roar up the wide chimney. The Count himself left my luggage inside and withdrew, saying, before he closed the door:—

"You will need, after your journey, to refresh yourself by making your

diaries and memoranda written by various personages, and afterwards typewritten, with all the sets of facts set in chronological order. In surrounding his gruesome and fantastically supernatural root idea with a framework plainly matter-of-fact and purely of 19th Century structure, Mr. Stoker has, we think, gone too far in the introduction of complicated details. As the book also contains much about hypnotism medically employed, semi-medieval philosophy, and applications of the latest information concerning the workings of the abnormal brain, it must be conned very carefully indeed if the reader wishes to grasp all the threads in the author's elaborately constructed argument. The opening chapters of Mr. Stoker's brilliant tour de force describing, almost as it were by way of prologue, the first journey of the young English solicitor to the lonely Transylvanian castle haunted by the lonely Vampire or Un Dead, Count Dracula, and his three ghostly sisters who share his avidity for sucking human blood, give the reader at once the proper "atmosphere," and this section of the book has been more artistically done than the somewhat involved and disjointed conclusion that deals with the pursuit of the dreadful Count and his capture and mercilessly merciful true death just as he is regaining his ancestral lair. Mr. Stoker has a keen eye for the picturesque and appropriate in his choice of epithets and in his

toilet. I trust you will find all you wish. When you are ready come into the other room, where you will find your supper prepared."

The light and warmth and the Count's courteous welcome seemed to have dissipated all my doubts and fears. Having then reached my normal state, I discovered that I was half famished with hunger; so making a hasty toilet, I went into the other room.

I found supper already laid out. My host, who stood on one side of the great fireplace, leaning against the stonework, made a graceful wave of his hand to the table, and said:—

"I pray you, be seated and sup how you please. You will, I trust, excuse me that I do not join you; but I have dined already, and I do not sup."

I handed to him the sealed letter which Mr. Hawkins had entrusted to me. He opened it and read it gravely; then, with a charming smile, he handed it to me to read. One passage of it, at least, gave me a thrill of pleasure:

"I much regret that an attack of gout, from which malady I am a constant sufferer, forbids absolutely any travelling on my part for some time to come; but I am happy to say I can send a sufficient substitute, one in whom I have every possible confidence. He is a young man, full of energy and talent in his own way, and of a very faithful disposition. He is discreet and silent, and has grown into manhood in my service. He shall be ready to attend on you when you will during his stay, and shall take your instructions in all matters."

The Count himself came forward and took off the cover of a dish, and I fell to at once on an excellent roast chicken. This, with some cheese and a salad and a bottle of old Tokay, of which I had two glasses, was my supper. During the time I was eating it the Count asked me many questions as to my journey, and I told him by degrees all I had experienced.

By this time I had finished my supper, and by my host's desire had drawn up a chair by the fire and begun to smoke a cigar which he offered me, at the same time excusing himself that he did not smoke. I had now an opportunity of observing him, and found him of a very marked physiognomy.

His face was a strong — a very strong — aquiline, with high bridge of the thin nose and peculiarly arched nostrils; with lofty domed forehead, and hair growing scantily round the temples but profusely elsewhere. His eyebrows were very massive, almost meeting over the nose, and with bushy hair that seemed to curl in its own profusion. The mouth, so far as I could see it under the heavy moustache, was fixed and rather cruel-looking, with peculiarly sharp white teeth; these protruded over the lips, whose remarkable ruddiness showed astonishing vitality in a man of his years. For the rest, his ears were pale and at the tops extremely pointed; the chin was broad and strong, and the cheeks firm though thin. The general effect was one of extraordinary pallor.

Hitherto I had noticed the backs of his hands as they lay on his knees

word-painting, and many passages of his story are indeed remarkably written. Those who know the Rev. S. Baring Gould's little volume on the Were-Wolf, a theme also touched on here and there by Kipling, may possibly not be repelled by the grisly details of two beautiful and virtuous women having the veins in their throats sucked by the red lips, and lacerated by the gleaming white teeth of this centuries-old Transylvanian warrior and statesman, who often appars as a gaunt wolf and a huge bat. Still more horrible is the scene where the solicitor's brave wife is actually forced by the Vampire to quaff his own nauseating blood, and Mr. Stoker's treatment of the semi-spiritual connection, even at a great distance, thus established between the Count and his second English victim, recalls one of Bulwer Lytton's novels.

"A careful study of the zoophagous maniac who, after devouring flies and spiders, is tempted by the Count to taste human blood, is one of the most interesting things in a volume full of excellently drawn character sketches; the old Amsterdam professor, for instance, with his curious blend of ancient and modern science and Catholic superstitions, the self-sacrificing young American, and the asylum doctor, being admirably depicted. A white mist, sea-fog, specks of dust floating in the air, are among the elemental machinery employed by Mr. Stoker, who also lays stress upon the development of canine teeth in Dracula's prey,

in the firelight, and they had seemed rather white and fine; but seeing them now close to me, I could not but notice that they were rather coarse-broad, with squat fingers. Strange to say, there were hairs in the centre of the palm. The nails were long and fine, and cut to a sharp point. As the Count leaned over me and his hands touched me, I could not repress a shudder. It may have been that his breath was rank, but a horrible feeling of nausea came over me, which, do what I would, I could not conceal. The Count, evidently noticing it, drew back; and with a grim sort of smile, which showed more than he had yet done his protuberant teeth, sat himself down again on his own side of the fireplace. We were both silent for a while; and as I looked towards the window I saw the first dim streak of the coming dawn. There seemed a strange stillness over everything; but as I listened I heard as if from down below in the valley the howling of many wolves. The Count's eyes gleamed, and he said:—

and brings in, mutatis mutandis, the stabbing of women recently notorious in London. The author has, perhaps, knocked the nail too often upon the head in his constant allusions to the exact periods of the day during which the Un Dead may arise from their mouldy earth-filled coffins, but yet all who are attracted by the supernatural in literature will find fascination enough in Mr. Stoker's *Dracula*. We must not omit to mention that one of the most effective elements of the horror rests in the fact that the Vampire's victims, unless purified by a terrible process we need not describe, become, even after their natural death, of the corruption-spreading family of the Un Dead."

"Listen to them — the children of the night. What music they make!"

Seeing, I suppose, some expression in my face strange to him, he added:—

"Ah, sir, you dwellers in the city cannot enter into the feelings of the hunter." Then he rose and said:—

"But you must be tired. Your bedroom is all ready, and to-morrow you shall sleep as late as you will. I have to be away till the afternoon; so sleep well and dream well!" With a courteous bow, he opened for me himself the door to the octagonal room, and I entered my bedroom...

I am all in a sea of wonders. I doubt; I fear, I think strange things which I dare not confess to my own soul. God keep me, if only for the sake of those dear to me!

Following pages: various editions of Bram Stoker's Dracula, *never out of print after more than a century. (Courtesy of Ronald V. Borst / Hollywood Movie Posters and the Count Dracula Fan Club)*

7 May.— it is again early morning, but I have rested and enjoyed the last twenty-four hours. I slept till late in the day, and awoke of my own accord. When I had dressed myself I went into the room where we had supped, and found a cold breakfast laid out, with coffee kept hot by the pot being placed on the hearth. There was a card on the table, on which was written:—

"I have to be absent for a while. Do not wait for me.— D." I set to and enjoyed a hearty meal. When I had done, I looked for a bell, so that I might let the servants know I had finished; but I could not find one. There are certainly odd deficiencies in the house, considering the extraordinary evidences of wealth which are round me. The table service is of gold, and so beautifully wrought that it must be of immense value. The curtains and upholstery of the chairs and sofas and the hangings of my bed are of the costliest and most beautiful fabrics, and must have been of fabulous value when they were made, for they are centuries old, though in excellent order. I saw something like them in Hampton Court, but there they were worn and frayed and moth-eaten. But still in none of the rooms is there a mirror. There is not even a toilet glass on my table and I had to get the little shaving glass from my bag before I could either shave or brush my hair. I have not yet seen a servant anywhere, or heard a sound near the castle except the howling of wolves. Some time after I had finished my meal — I do not know whether to call it breakfast or dinner, for it was between five and six o'clock when I had it — I looked about for something to read, for I did not like to go about the castle until I had asked the Count's permission. There was absolutely nothing in the room, book, newspaper, or even writing materials; so I opened another door in the room and found a sort of library. The door opposite mine I tried, but found it locked.

In the library I found, to my great delight, a vast number of English books, whole shelves full of them, and bound volumes of magazines and newspapers. A table in the centre was littered with English magazines and newspapers, though none of them were of very recent date. The books were of the most varied kind — history, geography, politics, political economy, botany, geology, law — all relating to England and English life and customs and manners. There were even such books of reference as the London Directory, the "Red" and "Blue" books, Whitaker's Almanac, the Army and Navy Lists, and — it somehow gladdened my heart to see it — the Law List.

Whilst I was looking at the books, the door opened, and the Count entered. He saluted me in a hearty way, and hoped that I had had a good night's rest. Then he went on:—

"I am glad you found your way in here, for I am sure there is much that will interest you. These companions" — and he laid his hand on some of the books — "have been good friends to me, and for some years past, ever since I had the idea of going to London, have given me many, many hours of pleasure. Through them I have come to know your great England; and to know her is to love her. I long to go through the crowded streets of your mighty London, to be in the midst of the whirl and rush of humanity, to share its life, its change, its death, and all that makes it what it is. But alas! As yet I only know your tongue through books. To you, my friend, I look that I know it to speak."

"But, Count," I said, "you know and speak English thoroughly!" He bowed gravely.

"I thank you, my friend, for your all too flattering estimate, but yet I fear that I am but a little way on the road I would travel. True, I know the grammar and the words, but yet I know not how to speak them."

"Indeed," I said, "you speak excellently."

"Not so," he answered. "Well I know that, did I move and speak in your London, none there are who would not know me for a stranger. That is not enough for me. Here I am noble; I am boyar; the common people know me, and I am master. But a stranger in a strange land, he is no one; men know him not — and to know not is to care not for. I am content if I am like the rest, so that no man stops if he see me, or pause in his speaking if he hear my words, 'Ha, ha! A stranger!' I have been so long master that

I would be master still — or at least that none other should be master of me. You come to me not alone as agent of my friend Peter Hawkins, of Exeter, to tell me all about my new estate in London. You shall, I trust, rest here with me a while, so that by our talking I may learn the English intonation; and I would that you tell me when I make error, even of the smallest, in my speaking. I am sorry that I had to be away so long to-day; but you will, I know, forgive one who has so many important affairs in hand."

Of course I said all I could about being willing, and asked if I might come into that room when I chose. He answered: "Yes, certainly," and added: —

"You may go anywhere you wish in the castle, except where the doors are locked, where of course you will not wish to go. There is reason that all things are as they are, and did you see with my eyes and know with my knowledge, you would perhaps better understand." I said I was sure of this, and then he went on: —

"We are in Transylvania; and Transylvania is not England. Our ways are not your ways, and there shall be to you many strange things. Nay, from what you have told me of your experiences already, you know something of what strange things there may be."

This led to much conversation; and as it was evident that he wanted to talk, if only for talking's sake, I asked him many questions regarding things that had already happened to me or come within my notice. Sometimes he sheered off the subject, or turned the conversation by pretending not to understand; but generally he answered all I asked most frankly. Then as time went on, and I had got somewhat bolder, I asked him of some of the strange things of the preceding night, as, for instance, why the coachman went to the places where he had seen the blue flames. He then explained to me that it was commonly believed that on a certain night of the year — last night, in fact, when all evil spirits are supposed to have unchecked sway — a blue flame is seen over any place where treasure has been concealed. "That treasure has been hidden," he went on, "in the region through which you came last night, there can be but little doubt; for it was the ground fought over for centuries by the Wallachian, the Saxon, and the Turk. Why, there is hardly a foot of soil in all this region that has not been enriched by the blood of men, patriots or invaders. In old days there were stirring times, when the Austrian and the Hungarian came up in hordes, and the patriots went out to meet them — men and women, the aged and the children too — and waited their coming on the rocks above the passes, that they might sweep destruction on them with their artificial avalanches. When the invader was triumphant he found but little, for whatever there was had been sheltered in the friendly soil."

"But how," said I, "can it have remained so long undiscovered, when there is a sure index to it if men will but take the trouble to look?" The Count smiled, and as his lips ran back over his gums, the long, sharp, canine teeth showed out strangely; he answered: —

"Because your peasant is at heart a coward and a fool! Those names only appear on one night; and on that night no man of this land will, if he can help it, stir without his doors. And, dear sir, even if he did he would not know what to do. Why, even the peasant that you tell me of who marked the place of the flame would not know where to look in daylight even for his own work. Even you would not, I dare be sworn, be able to find these places again?"

"There you are right," I said. "I know no more than the dead where even to look for them." Then we drifted into other matters.

"Come," he said at last, "tell me of London and of the house which you have procured for me." With an apology for my remissness, I went into my own room to get the papers from my bag. Whilst I was placing them in order I heard a rattling of china and silver in the next room, and as I passed through, noticed that the table had been cleared and the lamp lit, for it was by this time deep into the dark. The lamps were also lit in the study or library, and I found the Count lying on the sofa, reading, of all things in the world, an English Bradshaw's Guide. When I came in he cleared the books and papers from the table; and with him I went into plans and deeds and figures of all sorts. He was interested in everything, and asked me a myriad questions about the place and its surroundings. He clearly had studied beforehand all he could get on the subject of the neighborhood, for he evidently at the end knew very much more than I did. When I remarked this, he answered:—

"Well, but, my friend, is it not needful that I should? When I go there I shall be all alone, and my friend Harker Jonathan — nay, pardon me, I fall into my country's habit of putting your patronymic first — my friend Jonathan Harker will not be by my side to correct and aid me. He will be in Exeter, miles away, probably working at papers of the law with my other friend, Peter Hawkins. So!"

We went thoroughly into the business of the purchase of the estate at Purfleet. When I had told him the facts and got his signature to the necessary papers, and had written a letter with them ready to post to Mr. Hawkins, he began to ask me how I had come across so suitable a place. I read to him the notes which I had made at the time, and which I inscribe here: —

"At Purfleet, on a by-road, I came across just such a place as seemed to be required, and where was displayed a dilapidated notice that the place was for sale. It is surrounded by a high wall, of ancient structure, built of heavy stones, and has not been repaired for a large number of years. The closed gates are of heavy old oak and iron, all eaten with rust.

"The estate is called Carfax, no doubt a corruption of the old Quatre Face, as the house is four-sided, agreeing with the cardinal points of the compass. It contains in all some twenty acres, quite surrounded by the solid stone wall above mentioned. There are many trees on it, which make it in places gloomy, and there is a deep, dark-looking pond or small lake, evidently fed by some springs, as the water is clear and flows away in a fair-sized stream. The house is very large and of all periods back, I should say, to mediaeval times, for one part is of stone immensely thick, with only a few windows high up and heavily barred with iron. It looks like part of a keep, and is close to an old chapel or church. I could not enter it, as I had not the key of the door leading to it from the house, but I have taken with my kodak views of it from various points. The house has been added to, but in a very straggling way, and I can only guess at the amount of ground it covers, which must be very great. There are but few houses close at hand, one being a very large house only recently added to and formed into a private lunatic asylum. It is not, however, visible from the grounds."

When I had finished, he said —

"I am glad that it is old and big. I myself am of an old family, and to live in a new house would kill me. A house cannot be made habitable in a day; and, after all, how few days go to make up a century. I rejoice also that there is a chapel of old times. We Transylvanian nobles love not to think that our bones may lie amongst the common dead. I seek not gaiety nor mirth, not the bright voluptuousness of much sunshine and sparkling waters which please the young and gay. I am no longer young; and my heart, through weary years of mourning over the dead, is not attuned to mirth. Moreover, the walls of my castle are broken; the shadows are many, and the wind breathes cold through the broken battlements and casements. I love the shade and the shadow, and would be alone with my thoughts when I may."

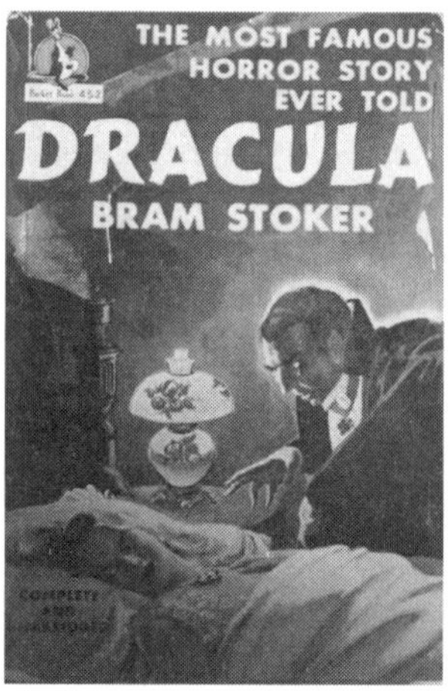

Somehow his words and his look did not seem to accord, or else it was that his cast of face made his smile look malignant and saturnine.

Presently, with an excuse, he left me, asking me to put all my papers together. He was some little time away, and I began to look at some of the books around me. One was an atlas, which I found opened naturally at England, as if that map had been much used. On looking at it I found in certain places little rings marked, and on examining these I noticed that one was near London on the east side, manifestly where his new estate was situated; the other two were Exeter, and Whitby on the Yorkshire coast.

It was the better part of an hour when the Count returned. "Aha!" he said; "still at your books? Good! But you must not work always. Come; I am informed that your supper is ready." He took my arm, and we went into the next room, where I found an excellent supper ready on the table. The Count again excused himself, as he had dined out on his being away from home. But he sat as on the previous night, and chatted whilst I ate. After supper I smoked, as on the last evening, and the Count stayed with me, chatting and asking questions on every conceivable subject, hour after hour. I felt that it was getting very late indeed, but I did not say anything, for I felt under obligation to meet my host's wishes in every way. I was not sleepy as the long sleep yesterday had fortified me; but I could not help experiencing that chill which comes over one at the coming of the dawn, which is like, in its way, the turn of the tide. They say that people who are near death die generally at the change to the dawn or at the turn of the tide; any one who has when tired, and tied as it were to his post, experienced this change in the atmosphere can well believe it. All at once we heard the crow of a cock coming up with preternatural shrillness through the clear morning air, Count Dracula, jumping to his feet, said: —

"Why, there is the morning again! How remiss I am to let you stay up so long. You must make your conversation regarding my dear new country of England, less interesting, so that I may not forget how time flies by us," and, with courtly bow, he quickly left me.

I went into my own room and drew the curtains, but there was little to notice; my window opened into the courtyard, all I could see was the warm grey of quickening sky. So I pulled the curtains again, and have written of this day.

8 MAY. — I began to fear as I wrote in this book that I was getting too diffuse; but now I am glad that I went into detail from the first, for there is something so strange about this place and all in it that I cannot but feel uneasy. I wish I were safe out of it, or that I had never come. It may be that this strange night-existence is telling on me; but would that that were all! If there were any one to talk to I could bear it, but there is no one. I have only the Count to speak with, and he!— I fear I am myself the only living soul within the place. Let me be prosaic so far as facts can be; it will help me to bear up, and imagination must not run riot with me. If it does I am lost. Let me say at once how I stand — or seem to.

I only slept a few hours when I went to bed, and feeling that I could not sleep any more, got up. I had hung my shaving glass by the window, and was just beginning to shave. Suddenly I felt a hand on my shoulder, and heard the Count's voice saying to me, "Good-morning." I started, for it amazed me that I had not seen him, since the reflection of the glass covered the whole room behind me. In starting I had cut myself slightly, but did not notice it at the moment. Having answered the Count's salutation, I turned to the glass again to see how I had been mistaken. This time there could be no error, for the man was close to me, and I could see him over my shoulder. But there was no reflection of him in the mirror! The whole room behind me was displayed; but there was no sign of a man in it, except myself. This was startling, and, coming on the top of so many strange things, was beginning to increase that vague feeling of uneasiness which I always have when the Count is near; but at the instant I saw that the cut had bled a little, and the blood was trickling over my chin. I laid down the razor, turning as I did so half round to look for some sticking plaster. When the Count saw my face, his eyes blazed with a sort of demoniac fury, and he suddenly made a grab at my throat. I drew away, and his hand touched the string of beads which held the crucifix. It made an instant

Whitby Abbey, North Yorkshire, one of many real-life locations Stoker employed in Dracula.
(Photo by David J. Skal)

change in him, for the fury passed so quickly that I could hardly believe that it was ever there.

"Take care," he said, "take care how you cut yourself. It is more dangerous than you think in this country." Then seizing the shaving glass, he went on: "And this is the wretched thing that has done the mischief. It is a foul bauble of man's vanity. Away with it!" and opening the heavy window with one wrench of his terrible hand, he flung out the glass, which was shattered into a thousand pieces on the stones of the courtyard far below. Then he withdrew without a word. It is very annoying, for I do not see how I am to shave, unless in my watch-case or the bottom of the shaving-pot, which is fortunately of metal.

When I went into the dining-room, breakfast was prepared; but I could not find the Count anywhere. So I breakfasted alone. It is strange that as yet I have not seen the Count eat or drink. He must be a very peculiar man! After breakfast I did a little exploring in the castle. I went out on the stairs and found a room looking towards the South. The view was magnificent, and from where I stood there was every opportunity of seeing it. The castle is on the very edge of a terrible precipice. A stone falling from the window would fall a thousand feet without touching anything! As far as the eye can reach is a sea of green tree tops, with occasionally a deep rift where there is a chasm. Here and there are silver threads where the rivers wind in deep gorges through the forests.

But I am not in heart to describe beauty, for when I had seen the view I explored further, doors, doors, doors everywhere, and all locked and bolted. In no place save from the windows in the castle walls is there an available exit.

The castle is a veritable prison, and I am a prisoner!

The Vampire *by Philip Burne-Jones*

FOR THE PICTURE

A fool there was and he made his prayer
(Even as you and I!)
To a rag and a bone and a hank of hair
(We called her the woman who did not care)
But the fool he called her his lady fair--
(Even as you and I!)

Oh, the years we waste and the tears we waste
And the work of our head and hand,
Belong to the woman who did not know
(And now we know that she never could know)
And did not understand!

A fool there was and his goods he spent
(Even as you and I!)
Honour and faith and a sure intent
(And it wasn't the least what the lady meant)
But a fool must follow his natural bent
(Even as you and I!)

Oh, the toil we lost and the spoil we lost
And the excellent things we planned
Belong to the woman who didn't know why
(And now we know that she never knew why)
And did not understand!

The fool was stripped to his foolish hide
(Even as you and I!)
Which she might have seen when she
threw him aside
(But it isn't on record the lady tried)
So some of him lived but the most
of him died—
(Even as you and I!)

And it isn't the shame and it isn't the blame
That stings like a white hot brand.
It's coming to know that she never knew why
(Seeing at last she could never know why)
And never could understand.

RUDYARD KIPLING
(1897)

Fatal Image
The Artist, the Actress, and "The Vampire"

DAVID J. SKAL

Then he leapt up, and went to the door. "Yes," he cried, "you have killed my love. You used to stir my imagination. Now you don't even stir my curiosity.... I loved you because you were marvelous, because you had genius and intellect, because you realized the dreams of great poets and gave shape and substance to the shadows of art. You have thrown it all away. You are shallow and stupid. My God! How mad I was to love you! What a fool I have been!

Oscar Wilde
The Picture of Dorian Gray
(1891)

WHEN SHE SAW the painting at last, it was said that her heart nearly stopped. The resemblance was a strong one, as she had been warned. "There were the long arms, the same supple figure, the high shoulders, the big black eyes and the mass of dark hair that had become so well known," one newspaper would write of the affair, "there was the strange, exotic, sorceress-like air that had so often held thousands spell-bound in the theatre..."[1] But nothing could have prepared her for the brutality of the canvas. She knew the artist — she knew him well. She had befriended him and his illustrious family. He had painted her, idealized her. But that was...before. This time she was painted not even as human, but instead as a foul, predatory thing of the night stretched over an ashen victim on a theatrically curtained bed. A red wound over the dead man's heart marked the spot from which the monster had taken its nourishment. A printed poem accompanied the painting: *A fool there was* ... it began. A look of triumph suffused the vampire's face — a face instantly recognizable to all of theatrical London. It was too much. *A rag, and a bone, and a hank of hair* . . . She reached out to her entourage for support. *The fool, he called her his lady fair* . . .

And then she fainted dead away.

Of course, it didn't happen that way at all. Mrs. Patrick Campbell, the tempestuous toast of the West End in the spring of 1897, did not collapse in a fashionable Regent Street gallery before a painting called *The Vampire*. She did, however, little to scotch the rumors. As for the author of the libel on canvas, the playboy painter Philip (later Sir Philip) Burne-Jones

1. "Gossip of 'Vampire' Painting Revived by Actress' Visit Here," *Brooklyn Daily Eagle*, February 9, 1902.

("always an unstable personality," according to *The Dictionary of Victorian Painters*) did not, in fact, commit suicide — entries in more than one standard reference book notwithstanding.

What actually happened was infinitely more interesting.

Philip Burne-Jones was borne in October, 1861, the son of Edward Coley Burne-Jones and Georgiana (née MacDonald) Burne-Jones. His father was the celebrated decorative painter and protegé of Dante Gabriel Rossetti and the Pre-Raphaelites; his mother was one of four sisters from humble origins who married into illustrious circumstances; Georgie's sister Alice was the mother of Rudyard Kipling; Agnes MacDonald married the painter Edward Poynter; and the fourth sister, Louisa, would be the mother of British Prime Minister Stanley Baldwin.

Phil's birth had been the cause of some apprehension; Burne-Jones confided to a correspondent about the impending event, adding "don't tell, I keep it quiet for fear it should be a monster."[2] Victorian England was fascinated with freaks; annual festivals like Bartholomew Fair and permanent exhibition halls in the West End allowed Londoners of all classes to catch a glimpse of themselves in the distorted mirror of Darwinism — human anomalies were often advertised as half-animal. In the grinding age of industrial expansionism and imperial "progress," images of evolutionary failure provided a cautionary, cathartic shudder. The most spectacular of these Victorian sideshow attractions was Joseph Carey Merrick, the celebrated "Elephant Man." Born within a year of each other, Merrick and Philip Burne-Jones would have contemporaneous, though hardly parallel, childhoods.

A doctor and "wise woman" had been engaged but could not assist at Phil's delivery; Burne-Jones himself acted as physician-midwife. "The arrival of our child," Georgie recalled, "while not a 'monster,' brought us face to face with some strange experiences. No one had told us any details connected with it essential for our guidance . . . By his own energy, however, [Edward] guided the disjointed time and set it straight, for with him intellect was a manageable force applicable to everything."[3] Analytical control had its limits, however, and Burne-Jones suffered a nervous collapse after the nurse arrived.

Both parents were disturbed by the new, intrusive force in their lives. Burne-Jones feared the baby was usurping his place in Georgie's affections; Georgie, for her part, was ambivalent about motherhood, and experienced a sense of "exile" from her husband's world. "I sat with my little son on my knee and dropped selfish tears upon him," she wrote, wondering if the child was "the separator of companions and terminator of delights."[4]

Less than six months old, Phil took ill and nearly died; but no sooner had the crisis passed than Burne-Jones and Georgie left him with his maternal grandparents for an extended trip to Italy. After a few months of homesickness —"nothing would serve her but her husband must draw the baby for her on the sand,"[5] wrote John Ruskin — they returned, but Phil's

2. Georgiana Burne-Jones, *Memorials of Edward Burne-Jones*, Vol. 1 (New York: The MacMillan Co., 1904), p. 229.

3. *Ibid.*, p. 230.

4. *Ibid.*, p. 235-36.

5. Ina Taylor, *Victorian Sisters: The Remarkable MacDonald Women and the Great Men They Inspired* (Bethesda, Maryland: Adler & Adler Publishers, Inc., 1987), p. 75.

childhood would continue to be marked by parental ambivalence and anxiety. In October 1864, while in the throes of scarlet fever — contracted from Phil — Georgie gave birth to another son, Christopher. Georgie nearly died; the baby did die. Grief-stricken, Georgie took Phil and spent the Christmas holidays with her family. Burne-Jones hated Christmas, for reasons unexplained; the holiday separation, in any event, reflected a growing estrangement between Georgie and her husband.

When Phil was five, the family moved to The Grange, a rambling house in Fulham that had once been the residence of Samuel Richardson. There, Phil grew up in a household where the schism between the conspicuous display of courtly love in his father's artwork and the deteriorated state of the Burne-Jones marriage could not have been more pronounced. As a boy Phil was withdrawn and did not make friends easily; his father found Rudyard Kipling, Phil's rambunctious younger cousin, a better playmate than his own son. Arthur Baldwin, the Prime Minister's son, wrote that, "As a child, Phil must have been one of the queerest and most fidgety in all Kensington."[6] Relatives commented on his restlessness and endless chatter.

Edward Burne-Jones

That year, 1869, must have been a strange one indeed for young Phil. His father left the family to explore the charms of a rich, red-headed siren named Maria Cassavetti Zambaco, who had been one of his models. His affair with the volatile Greek beauty became common gossip. Phil's mother grew increasingly despondent, and finally unable to put on a brave face for the world, much less to her children. On the very verge of fleeing to the continent with Mrs. Zambaco (who was threatening suicide, and wanted Phil's father to join her) Burne-Jones instead returned home in a guilty panic, prostrated by an apparently psychosomatic "brain-fever." "For two or three days he was so ill we kept his being at home a secret, that the house might be quite quiet," Georgie wrote.[7] Sexual panic was nothing new to Edward Burne-Jones; he had suffered a similar attack on his wedding night. "Lust," he once confided to his studio assistant, "does frighten me."[8] Nonetheless, his affair with Zambaco continued for years.

In the world Philip Burne-Jones inhabited as a child, sex was scary stuff, bound up as it was with parental brain-fevers, the threat of invading temptresses, abandonment and depression. For better or worse, children of visual artists often have an inescapable access to the interior lives of their parents. The elder Burne-Jones's sexual imagination was on constant, conspicuous display. His oversized, tapestry-like canvases brimmed with images of fantastic, fatal women. His mentors of the Pre-Raphaelite Brotherhood sought an escape from the aesthetic outrages of industrialism by cultivating an alternate, "medieval" universe heavily imbued with Arthurian themes and characters, and Burne-Jones carried on the quest. One of the most compelling archetypes of this fantasyland was Arthur's fairy/witch sister, Morgan Le Fay, a.k.a. Fata Morgana. Alternately a hag or an irresistible siren, the protean pagan goddess exerted a considerable influence on

6. A.W. Baldwin, *The MacDonald Sisters* (London: Peter Davies, 1960), p. 150.

7. *Victorian Sisters*, p. 82.

8. *Ibid.*, p. 70.

Victorian painters and writers, and provided an imaginative substructure for the burgeoning concept of the female vampire. The nightmarish power of his themes — sorceresses, medusae, inhuman-below-the-waist mermaids dragging their resigned lovers to watery graves — was considerably offset by his tasteful decorative treatment and the languid, resigned expressions worn by nearly everyone in his ouevre.

Water witch: Edward Burne-Jones's The Depths of the Sea (1885). *From A.L. Baldry's* Burne-Jones *(New York, 1909).*

The chasm between the courtly ideals (rigid behavior codes that held sexual terror at bay) embodied by Pre-Raphaelite art, and male-female relationships as they were practiced in the Burne-Jones family was deep indeed. Phil's two godfathers by proxy, John Ruskin and Dante Gabriel Rossetti, each in their own way provide illustrations of the tensions and contradictions inherent in placing women on pedestals of art. Ruskin had been unable to consummate his first marriage, possibly due to the discovery that his wife had pubic hair ("He had been raised on classical nudes, you see, and so he never suspected," the late critic Martin Esslin memorably told this writer in 1973. "I think the incident amply demonstrated the dangers of over-idealization.") A preeminent arbiter of artistic taste in the 19th century, Ruskin reveals a strong antifeminine streak in his criticism, especially in his interpretation of myth. (Ruskin, who often dreamed about snakes, once recalled "the most horrible serpent dream I ever had yet in my life. The deadliest came out into the room under a door. It rose up like a Cobra — with horrible round eyes and had woman's, or at least Medusa's, breasts.")[9]

Later in the same year as the Zambaco affair, Rossetti made a ghoulish mockery of the concept of undying love by exhuming the corpse of his wife, Elizabeth Siddal, to retrieve a notebook of his poetry he had buried with her seven years earlier. Legend had it that Siddall's body remained in an astonishing state of preservation, her red hair still growing and filling the coffin. Graveworms had only nibbled at the book of poems — or so it was said. Beyond its singular sensationalism, the Siddal exhumation captured the public imagination by literalizing the prevailing Victorian concept of the "sleeping dead" — death was held to be a restful hiatus, the prelude to resurrection. Lizzie Siddal was probably not so well preserved as the gossips claimed, but to admit this would be to go against the extraordinary grain of popular delusion.

Siddal was, and is, buried in Highgate Cemetery in Hampstead, once of several burial grounds that opened in suburban London in the mid 1800s as a reaction to the revolting and unsanitary conditions that plagued the overcrowded city churchyards. At Highgate, the physical face of death was largely obliterated through gracious landscaping, sculpture, and an overall feeling of sanctuary. The distinction between life and death became blurred. And if the dead were not "really" dead, then they could exist in an idealized, even eroticized state of un-life.

These themes, of course, were to be made grotesquely literal in Bram Stoker's 1897 novel *Dracula*. In describing the soon-to-awaken corpse of

9. John Ruskin, *Diaries*; quoted in Stephen Brook, ed., *The Oxford Book of Dreams* (Oxford and New York: Oxford University Press, 1983), p. 92.

Miss Lucy Westenra, the vampire's first conquest in England, Stoker's Dr. Seward notes in his diary that "All Lucy's loveliness had come back to her in death, and the hours that had passed, instead of leaving traces of 'decay's effacing fingers,' had but restored the beauty of life, till positively I could not believe my eyes that I was looking at a corpse."[10] Stoker's victim is interred in a crypt at a cemetery very much like Highgate, and the ensuing scene in which her coffin is opened and a stake driven through her heart strongly recalls the violation of the Siddal grave. Stoker may have made use of the other part of the Siddal myth — the luxuriant, still-growing hair that was said to fill her coffin — in his short story "The Secret of the Growing Gold," in which a murdered woman's hair grows up through the floor of a castle as retribution upon the killer.

Dead, sick and collapsing women were favorite subjects of Victorian art. It was rather as if Poe's dictum that "there is no more poetical topic in the world than the death of a beautiful woman" had been taken up as an ideological creed by an army of painters who conducted a final aesthetic solution. In *Idols of Perversity*, Bram Dijkstra's exhaustive study of the phenomenon, a case is convincingly made that men of the nineteenth century were obsessed with subjugating woman by controlling her image in art.[11] Sleeping, swooning, sacrificial, tubercular, broken-backed, drowned — these were the favored, favorable poses. Women who were not victimized in art were alternately presented as victimizers — sphinxes, gorgons, vampire-whores. In any event, women tended to be explicitly associated with death — either as eager recipients, or enthusiastic dispensers, *sans merci*. Sarah Bernhardt's famous "coffin portrait" — marketed as a best-selling postcard in its day — is a clear demonstration of how even a still-living woman could capitalize on the necrophile craze in art. The image also concretized the idea of the actress as a kind of vampire, a shape-changing creature who rose each night to feed on the energy of her audience.

Sarah Bernhardt: the coffin portrait.

Philip Burne-Jones became acquainted with the world of the theatre, if not yet its female vampires, at an early age. By the age of eleven his mother had taken him to see the stage spectacle *The Last Days of Pompeii* at the Queen's Theatre, Long Acre, and a few months later he was treated to Madame Tussauds. The boy developed an early flair for dramatics, and was sometimes enlisted by other cousins to participate in amateur theatricals for the entertainment of adult visitors to the Grange. Burne-Jones's biographer David Cecil recounted the pleasure the artist took in drawing for his children, "either comic pictures, often of pigs, or horrific scenes of ghosts and monsters, with titles like 'The Mist Walkers' and 'The Heath Horror.' These last were alarming enough for other people to tell him that they were unfit for children. The children themselves disagreed."[12]

At age twelve Phil was sent off to Marlborough. Burne-Jones later regretted the decision. His son was plagued by bullies and loneliness. Throughout his boarding school days, Phil would send his father drawings for his approval. By the time Phil graduated from Marlborough and

10. Bram Stoker, *Dracula* (Oxford and New York: Oxford University Press, 1988), p. 164.

11. Bram Dijkstra, *Idols of Perversity: Fantasies of Feminine Evil in Fin-de-Siecle Culture* (New York and Oxford: Oxford University Press, 1986).

12. David Cecil, *Visionary and Dreamer: Two Poetic Painters: Samuel Palmer and Edward Burne-Jones* (Princeton, New Jersey: Princeton University Press, 1969), p. 135.

matriculated at Oxford, Burne-Jones declared that his son was a better draftsman at age sixteen than he had been at twenty-two. But in fairness, he tended to overidealize his son; surviving family sketches depicting Phil present him in an almost ethereal light at odds with actual photographs of the boy, which show him to be rather homely.

Phil and Oxford did not mix well; he showed an aptitude for mathematics, but had little enthusiasm for French, and left Oxford without earning a degree. One talent he had developed, however, was his distinct gift for comic illustration, but to his father's horror; the elder Burne Jones implored him to give up cartoons and train in his studio as a proper painter. Phil consented, though he might have done better to pursue his artistic bent a bit further from home. As Angela Thirkell recalled in her memoirs, "Uncle Phil . . . could have been a distinguished painter and would have been one under a luckier star, but two things told fatally against him. He never needed to work, and he was cursed with a sense of diffidence and a feeling that whatever he did would be contrasted unfavorably with his father's work."[13]

Rudyard Kipling

Perhaps with a certain calculated rebellion, Phil began to pursue the kind of flashy social life his parents abhorred. Nonetheless, his father conceded to hyphenate his last name in1885, partly to assist his son's society pretensions. "His ambition was to cut a dash in the Prince of Wales set," wrote Edward Burne-Jones's biographer Penelope Fitzgerald.[14]

Phil often acted or spoke impulsively, much to his later regret. Once he sold an indiscreet article to a publishing syndicate. Then, having realized the implications of his actions, he enlisted his cousin Rudyard Kipling to help him retrieve his manuscript. The incident later found its way into Kipling's novel *The Light that Failed* (1891), Phil's article transformed into an artist's sketches.

Though Ruddy could be counted on in a pinch, it was Phil's family that more often bailed him out. His parents indulged him even when he ran up huge expenses, and endured a procession of his insufferable celebrity "friends." Phil's cousin Alice "Trix" Kipling recalled a visitor in 1882 in a letter to her brother Rudyard: ". . . I went to the Grange for half-term holiday and Phil's new adoration, Oscar Wilde, was there...he is like a bad copy of a bust of a very decadent Roman Emperor, roughly modelled in suet pudding. I sat opposite him and could not make out what his lips reminded me of — they are exactly like the big brown slugs we used to hate so in the garden..." At each infrequent pause in Wilde's incessant chatter, Trix remembered, Phil would set him off again with a breathless "Oscar, tell us so and so."[15]

In *The Picture of Dorian Gray* (1891), Wilde produced one of the essential texts of fin-de-siecle vampirism, a fantastic allegory of energy transference and moral corruption that had a discernible influence on his friend Bram Stoker (who had married Wilde's one-time Dublin heartthrob Florence Balcombe), then beginning the composition of *Dracula*. Stoker

13. Angela Thirkell, *Three Houses* (London & Humphrey Milford: Oxford University Press, 1931), p. 66.

14. Penelope Fitzgerald, *Edward Burne-Jones: A Biography* (London: Michael Joseph, 1975), p. 207.

15. Alice Kipling, letter to Rudyard Kipling, March 18, 1882, quoted in Taylor, *op. cit.*, pp. 136-37.

16. Ernest Dudley, *The Gilded Lily: The Life and Loves of the Fabulous Lillie Langtry* (London: Odhams Press, Ltd., 1958), p. 15.

17. Barbara Belford, *Violet: The Story of the Irrepressible Violet Hunt and her Circle of Lovers and Friends — Ford Madox Ford, H.G. Wells, Somerset Maugham, and Henry James* (New York: Simon and Schuster, 1990), p. 48.

originally considered an art-world subplot for his vampire story, in which a painter's attempts to create a portrait of the vampire would be frustrated: the image would always look like someone else. Many who beheld Oscar Wilde in the flesh wished they were looking at someone else — the writer's appearance seems to have set off Dracula-style shudders in many who made his acquaintance; Trix Kipling was hardly alone in her visceral reaction.

Wilde, for all his intellectual brilliance, seemed to blur distinctions between the human and lower animal realms, much like Dracula himself. Again and again, Wilde is described by detractors, critics, and even friends, as pale and bloodless, a kind of fleshy, engulfing amoeba. Though Wilde's homosexuality didn't register explicitly in the public consciousness until his trial in 1895, he was widely perceived as the embodiment of decadence and transgression; much of the furor over *The Picture of Dorian Gray* stemmed from a public confusion of Wilde's own persona, and that of his beautiful monster. A thinly veiled fear of homosexuality has been part and parcel of the vampire literary formula ever since Sheridan Le Fanu's *Carmilla* in 1871. And the whole idea of "decadence" suggested a kind of evolution-in-reverse, a prominent theme in horror fiction and vampire stories in an age still wrestling with the disturbing theories of Darwin.

Philip Burne-Jones never painted Wilde, but he did cast more than an artist's eye in the direction of some of his father's female models. One was the celebrated actress Lillie Langtry, whom Phil's father had used as the pitiless female figure of Fate in his painting *The Wheel of Fortune*. To persuade Langtry to pose, Burne-Jones began serenading her beneath her bedroom window at the break of day, loudly declaring her to be a heartless enemy of art. The actress relented, and soon found herself cast as a tall figure in grey presiding over an endless cycle of men being crushed on an allegorical wheel. "It's a cruel picture," she told the artist, "but horribly true."[16] Phil created something of an embarrassment for the family by pursuing the actress to Monte Carlo and making a nuisance of himself.

Another of Phil's crushes was for Violet Hunt (later the biographer of Elizabeth Siddal) who modeled for both Phil and his father, most notably as the female half of Burne-Jones's *King Cophetua and the Beggar Maid*. Hunt, however, found Phil not to her liking. "[He] obliges me to take an interest in him, which his own manner and appearance do not warrant," she wrote in her diary in June 1882.[17] He also seems to have had an unrequited passion for Edith Balfour, later the second Mrs. Alfred Lyttleton.

But it was another face who would capture Phil's fancy, and, for better or worse, inspire his most famous production on canvas.

The woman who would eventually be known to the world as Mrs. Patrick Campbell was born Beatrice Stella Tanner on February 9, 1865 in Kensington. Her father was a manufacturer of military accoutrements whose business could fluctuate wildly; her mother was the daughter of an artistocratic, exiled Italian family. Stella (as she called herself) was a restless child with strong, but undefined, interest in the arts artistic. She had a

OSCAR WILDE: VAMPIRE?

Oscar Wilde (1854-1900) is often overlooked in histories of vampirism, but *The Picture of Dorian Gray* is a brilliant and influential update on the theme. Its title character, a decadent Victorian dandy, makes a devilish pact with his own portrait, and remains preternaturally young and handsome while the painting reflects his increasing debauchery, ultimately coming to resemble an ancient, leering satyr (which, interesting enough, was very much Bram Stoker's conception of Dracula). Dorian has, in effect, been undead since he sold his soul, and when he finally stabs the painting, he dies — instantly aging in the manner of a traditional vampire whose heart has been ritually pierced.

When Wilde died, a bizarre hoax was launched on the pages of a literary journal, insinuating that Wilde was not dead, but like Dracula (or Jesus) had transcended death and still walked the earth. The hoax's perpetrator, George Sylvester Viereck, later wrote a bizarre novel, *House of the Vampire* (1907) about a malignant artist who absorbed the talent and creativity of others. The face that appeared on the original German edition was unmistakably that of Oscar Wilde.

strong aptitude for music, but became bored with the piano. At the age of seventeen she met Patrick Campbell, a nondescript bank manager who shared none of her artistic leanings. Nonetheless, according to her biographer Margot Peters, "Seeing that she could fascinate and rule him, Stella was immediately attracted." [18]

Pregnant at nineteen, Stella was forced into a marriage she didn't really want, and which she was sure would be the end of all her aspirations. Miserable, she wrote to her sister, "A poor man's wife is seldom fit for anything but a superior sort of servant .... My life is sure to be a short one. I wish it to be."[19] The first years of her marriage proved a nightmare of money worries as Patrick faced repeated demotions. Following the birth of her second child in 1886, a crisis was reached. Patrick decided to seek employment in Australia, and Stella decided to pursue a career in the theatre, based on the encouragement and positive notices she had received for her work in amateur theatricals.

Mrs. Patrick Campbell

She made her professional debut in Liverpool in 1888, and over the next few years worked with a variety of provincial companies. Having no formal training, she drew immediate attention to herself onstage through her unaffected, almost offhand performances. Augustin Filon, in *The English Stage* (1897) gave the following analysis of the spell that Mrs. Pat cast over her public:

> She is said to have Italian blood in her veins; hence, no doubt, that nervous delicacy of hers, that *morbidezza* which shades, veils, tempers, refines her talent no less than her beauty. She has neither the originality, nor the knowledge, nor the voice of Sarah Bernhardt, but she possesses that magnetic personality . . . with which there is no such thing as a bad part. If this personality must be described, I would say that Mrs. Campbell's province as an actress is more particularly that of dangerous love. That voice of hers, though it has but little sonorousness, power, or richness, produces in one a sense of disquiet and distress, straitens the heart with a kind of fascinating delicious fear that I would describe as the *curiosite de souffrir*. You feel that if you love her you are lost, but once you have seen her it is too late to attempt resistance.[20]

Mrs. Pat's major break came in 1893, when Arthur Wing Pinero cast her, somewhat reluctantly, as the lead in his grim society drama *The Second Mrs. Tanqueray*, in which a beautiful, middleclass woman facing blackmail over her sordid, earlier life commits suicide to spare her husband and daughter from scandal. Her *morbidezza* paid off. The play was a sensation, and the actress became the toast of theatrical London. Oscar Wilde himself, fresh from the success of *A Woman of No Importance*, came to her dressing room, the tubercular Aubrey Beardsley in tow. Beardsley penned a morbid, wraith-like portrait of her for *The Yellow Book*, notorious for its

18. Margot Peters, *Mrs. Pat: The Life of Mrs. Patrick Campbell* (New York: Alfred A. Knopf, 1984), p. 20.

19. *Ibid.*, p. 20.

20. Augustin Filon, *The English Stage; being an account of the Victorian drama* (1897; reprint, New York: Benjamin Blom, 1969).

"decadent"content and contributors.

Stella was indeed a woman of some importance now, a fact that bewildered her husband, returned from abroad. He had left a young wife with two small children eking out a living by teaching piano. Now she was one of the most glamorous theatrical figures in London, hobnobbing with the likes of Max Beerbohm and J. M. Barrie.

At some point during the run of *Mrs. Tanqueray*, Mrs. Pat received the following note:

> Dear Mrs Campbell,
> We think the play should end at the finish of the third act — except that you appear again.
> We also think that you are the greatest living actress.
> LOUIS N. PARKER
> PHILIP BURNE-JONES[21]

Phil's friend Louis Napoleon Parker (1852-1944) was a rising young dramatist and composer who had composed the incidental music for Mrs. Pat's first professional production in London in 1890. Shortly thereafter, when the actress appeared in an alfresco production of his pastoral play *Love-in-a-Mist*, Parker found himself mesmerized by the actress, who "had a fantastic part; and she wore beautiful, shimmering gowns; and she wandered through glades like an exquisite queen of dryads; and she spoke a sonnet in a way which persuaded me I was a poet; and, oh, climax! she sang a song of my setting (and a very pretty setting, too!) and her golden voice fetched my heart out of my body and brought it rolling to her feet; and my one ambition was to . . . follow the company round the world, ever making my poor heart Mrs. Campbell's footstool."[22]

But if the actress could evoke heaven, she also had the talent to summon hell, and Parker recalled that she "frightened me nearly out of my wits" when, following an alfresco performance in a rainstorm, he escorted Mrs. Pat to her lodgings, several miles from the soggy stage.

> The rain had ceased, but it was a wild, weird, night, with a watery moon chasing through clouds of sinister shape. "Look!" cried Mrs. Campbell, clutching my arm, "the dead are hurrying to overwhelm us! The dead in their millions! They are coming! They are coming!" All the way she kept up this cheerful talk. Murderers were lurking behind corners; corpses were gibbering behind trees; drowned men were floating in with the tide, beckoning to us with fleshless hands — O! the artistic temperament![23]

Phil was more than happy to have Mrs. Pat take his arm. He escorted her to the 1894 Royal Academy exhibit, where her own portrait as Paula Tanqueray was one of the most talked-about paintings of the season. A

21. Mrs. Patrick Campbell, *My Life and Some Letters* (New York: Dodd, Mead and Company, 1922), p. 102.

22. Louis N. Parker, *Several of My Lives* (London: Chapman and Hall, 1928), pp. 152-153.

23. *Ibid.*

mob swelled around her in the Burlington House galleries and Phil had the chivalrous pleasure of engineering her escape through a side door. She became a regular guest at The Grange, surrounded by images of courtly supernatural fantasy. She remembered, in her memoirs, the "wonderful day" when Phil took her to his father's studio. "I suppose we all have a period in art which appeals to us in an intimate way," she wrote. "Perhaps, because of my Italian blood, the Pre-Raphaelite School spoke to me in my own language: my very first visit to the Grange seemed a visit to my home. I wanted to stretch my arms in welcome to all that rich colour, pure design, and loveliness."[24] Mrs. Pat took to the house immediately. Upon arrival, she would sometimes ask to lie down in a darkened room, the better to commune with the peace and the beauty of the place.

Phil's mother did not take to Mrs. Pat — one imagines she had a natural wariness of sirens with artistic leanings. The actress conceded that Georgiana "made me doubt myself,"[25] and Edward Burne-Jones apparently could deal with her only in small doses. On one occasion, to break the tension around the dinner table, Phil's father excused himself, making a solemn re-entrance cowled as a cloistered monk, mumbling medieval mumbo-jumbo. All seriousness immediately evaporated.

Burne-Jones was now Sir Edward, having reluctantly accepted a baronetcy in 1894 at Phil's tearful pleading. The title was, of course, inheritable. "I am almost in a hurry to be gone that he might light a long cigar and march down Bond Street—will a rich widow want him now?" Burne-Jones wrote.[26]

The short, froggy-looking Phil must have made an odd escort indeed for the willowy, unwidowed Mrs. Pat. "[U]nlike his father, he is the most awkward and prosaic looking creature imaginable," writer Henry P. Marston told the readers of the *New England Home Magazine* a few years later. "However, he laid siege to the heart of Mrs. Pat and, to the surprise of his scoffers, succeeded marvelously. He became her 'angel.' The money that he earned by painting he spent in diamonds, carriages, furs and other little trifles that a popular actress considers needful to make life tolerable, and she rewarded him with smiles — for a while."[27]

Naturally, Phil painted her. One extravagant canvas, entitled *Stella*, depicted the actress in a literally celestial light, standing on a rock and gazing up into a star-filled sky. She hung it in her dressing room. Another painting, *By a Summer Sea*, depicted her in a pensive mode, standing on the beach holding a book. He also caricatured her good-naturedly in pen-and-ink, an inflated fashion-balloon with enormous hat and mutton sleeves, being pulled along by a tiny leashed dog.

Phil's flirtation with Stella Campbell continued for some time. She may, indeed, have been using him as a social stepping-stone, as well as for material favors. ("Clothes began to matter," she recalled, "and to fuss me. To feel dressed up was misery, and to be dowdy — impossible.")[28] She was famous, though not rich, and her money problems were chronic. Wealthy

24. *My Life and Some Letters*, pp. 113-114.

25. *Ibid.*, p. 106.

26. *Edward Burne-Jones*, p. 251.

27. Henry P. Marston, "An Artist's Revenge: the Romance of the Famous 'Vampire,'" *New England Home Magazine*, October 9, 1898, pp. 56-59.

28. *My Life and Some Letters*, p. 106.

admirers came in handy. "Men made love to me," she admitted, "and I was accused of being a wicked flirt. I deny that. In more than one case I cared: but my first love had taught me love's true face."[29]

In fact, her first love was pretty much dashed on the rocks of her theatrical success. While Phil squired the actress around in a showy, platonic courtship, another, a less lofty liaison was brewing in Stella's life. Phil's parents lent her the use of one of their houses for a second honeymoon with Pat Campbell, but the marriage was crumbling irretrievably. Unfortunately for Phil, the breakup provided him no opportunity. Johnston Forbes-Robertson, her leading man in *Romeo and Juliet* at the Lyceum, had fallen in love with her, and she with him.

"Burne-Jones did not appear to take it too much to heart," Marston wrote. "Outwardly, he preserved his usual cheerfulness of demeanor. But the shaft rankled sore. He locked himself in his studio and labored diligently. What he did he showed to no one. Even his servants were never allowed to see what was growing beneath his brush on the easel. Model he had none. It was not needed."[30]

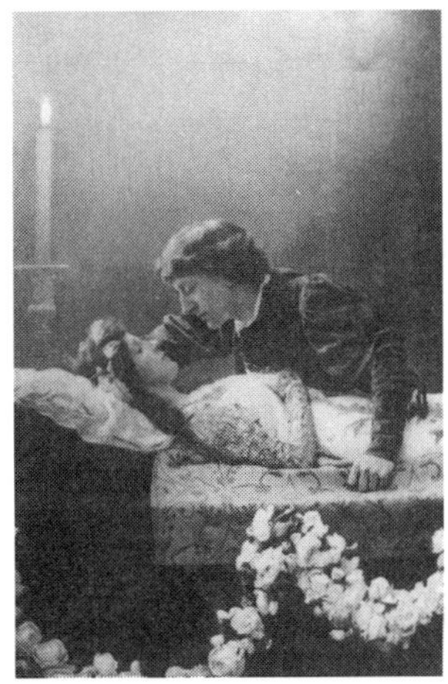

Playing dead: Mrs. Patrick Campbell and Johnston Forbes-Robertson in Romeo and Juliet (1895).

Just how and when Phil struck on the idea of portraying Beatrice Stella Campbell as the queen of vampires is not known, but images of the living dead were becoming ever more prominent in fin de siècle literature and art, and Phil could have been introduced to the theme in any number of ways. The previous year, Arthur Symons had published a poem entitled "The Vampire" which strikingly anticipates both the mood and pose of Phil's masterpiece:

> Intolerable woman, where's the name
> For your insane complexity of shame?
> Vampire! white bloodless creature of the night,
> Whose lust of blood has blanched her chill veins white,
> Veins fed with moonlight over dead men's tombs;
> Whose eyes remember many martyrdoms,
> So that their depths, whose depth cannot be found,
> Are shadowed pools in which a soul lies drowned;
> Who would fain have pity, but she may not rest
> Till she have sucked a man's heart from his breast,
> And drained his life-blood from him, vein, by vein,
> And seen his eyes grow brighter for the pain,
> And his lips sigh her name with his last breath,
> As the man swoons ecstatically on death.

The Victorians were obsessed with blood-weakness and blood contamination; at the time Phil was working on *The Vampire*, "anemic" women (and "effeminate" men) were actually being advised by some doctors to visit slaughterhouses to quaff cups of fresh blood from freshly killed oxen. The ghoulish phenomenon was recorded by the French painter

29. *Ibid.*, p. 107.

30. "An Artist's Revenge" *op. cit.*

31. A reproduction of the Gueldry painting appears in Dijkstra, *op. cit.*, p. 338.

Joseph-Ferdinand Gueldry in his 1898 canvas *The Blood Drinkers.*[31] Patent medicines offering to restore and purify the bloodstream were widely advertised; one, Clark's Blood Mixture, promised an end to sanguinary "mischief."

The obsession with unclean blood was not without some basis in fact. Just as the AIDS epidemic today has fueled the current fascination with vampire themes, so too was the Victorian imagination morbidly stimulated by another incurable, blood-borne disease associated with sexual license. Syphilis was the AIDS of the 1890s, a disease that widely afflicted all classes, yet it was a disease that dared not speak its name — at least not in polite society. It is no wonder, then, that syphilis found a symbolic representation in countless images of wanton, destructive, and vampiric women that were staple fixtures of the Victorian art world. In J. K. Huysman's peerlessly decadent novel, *A Rebours* (1884), syphilis was presented as a monstrously literal nightmare-hag. Prostitutes were especially to blame, in Victorian eyes, for the spread of the disease—forget that it was men who chose to pay for their services, and who infected them in the first place. Numerous artists of the time depicted streetwalkers as living, livid corpses who preyed upon both the pocketbooks and health of their customers. In vampire fiction, epitomized by *Dracula*, the theme of blood-contamination runs riot, the vampire-disease being associated with the appearance of telltale marks on the skin, and voluptuous wantoness on the part of afflicted women.

A vampiric Victorian encounter between the sexes. Artist unkown.

It was common for the prostitute to be blatantly demonized along vampire lines. The Rev. William Bevan, in his 1843 tract *Prostitution in the Borough of Liverpool,* effectively conveyed the odor of graveyard and brimstone. "Their name is Legion!" Bevan wrote, evoking the biblical story of Christ's encounter with a spirit possessed man living in a tomb. "They have their lurking places in the sepulchres! They hold their orgies over the entombing of wealth, and strength, and honesty, and virtue! They fix the death spot on all they touch!" Another writer likened whores to "a multitudinous amazonian army the devil keeps in constant field service."[32]

The connection in the public mind between actresses and prostitutes has deep roots in theatre history, dating to antiquity. The actress-courtesan and showgirl-whore are persistent, pesky archetypes, even today. Actresses, of course, typically work at night, and, heavily painted, accept payment from the public to assume a false persona for the payer's pleasure, to perform certain "acts" that end in "climaxes," and so on. Since Mrs. Patrick Campbell frequently played fallen women, the public had little trouble in drawing unsavory conclusions. It was for Philip Burne-Jones, drawing in part on a lifetime's exposure to misogynistic sex allegories in art, and his own immediate displeasure with Mrs. Pat, to make complete the actress/gold-digger/vampire fantasy and put it on public display.

Phil dropped some of his customary working methods for *The Vampire.* His usual canvas was a twenty-by-thirty inch panel that he claimed to

32. Rev. William Bevan, *Prostitution in the Borough of Liverpool* (Liverpool, 1843), p. 12. Cited in Nead, p. 118.

Philip Burne-Jones in his studio with The Vampire.

favor for its portability, but one must wonder if there was some reluctance on Phil's part to go one-on-one with his father's vast extravanganzas. The odd size resulted in three-quarter-sized portraits, a format "peculiarly his own," in the recollection of his niece, Angela Thirkell.[33] But at more than twice the usual size, *The Vampire* was to be the largest painting Phil would ever attempt. And whatever conflicted emotions may have propelled him in its execution, it is clear that the picture was carefully planned. The overall composition owed much — it might even be considered positively an homage — to Henry Fuseli's 1781 painting *The Nightmare*, in which a Rubenesque woman lies swooning on a curtained bed, a gnome-like creature perched oppressively atop her. This, of course, was Fuseli's concept of the incubus of antiquity; the nightmare made concrete as a kind of breath-stealing gremlin.

John Henry Fuseli, The Nightmare *(1781).*

While modern stories specify the jugular vein as the locus of a vampire's attention, Phil chose the victim's breast as the site of the fatal incision. The result was a more allegorical statement, one more in keeping with Arthur Symons's verses about the creature who "sucked a man's heart from his breast." And he may well have been familiar with the catlike vampire in J. Sheridan Le Fanu's *Carmilla* who spread over her victims like a huge furry coverlet, taking nourishment from the region of the heart. The site of the blood-pumping heart has in many traditions also been held to be the seat of the emotions, if not the soul itself. The following description of a typical vampire visitation was quoted in a newspaper article accompanying a discussion of Phil's painting:

> You are lying in your bed at night, when you see, by the faint light, a shape entering at the door and gliding toward you with a long sigh. The thing moves along the air as if by the mere act of volition.
>
> You lie still — like one under the influence of a nightmare — and the thing floats slowly over you. Presently you fall into a deep sleep or a swoon, returning, up to the last moment of consciousness the fixed stare of the phantom. When you awake in the morning you think it is all a dream, until you perceive a small, blue, deadly looking spot on your chest near the heart; and the truth flashes on you . . .
>
> Day after day you grow paler and more languid; your face becomes livid, your eyes leaden, your cheeks hollow. Your friends advise you to seek medical aid, to take a change of air, but you are aware that it is all in vain. You therefore keep your fearful secret to yourself, and pine, and droop, and languish, till you die.[34]

In contrast to the Phil's idea of a mystical-allegorical theft of blood directly from the symbolical heart, Bram Stoker's *Dracula* had scuttled a lot of previous romantic nonsense about vampires and made the whole process clinical and up-to-date. One can easily imagine that Stoker moved

33. *Three Houses*, op. cit.

34. "Vampires," *Chicago Tribune*, January 25, 1903.

the whole blood-business to the jugular vein because it was logical to do so — it was a major blood vessel, after all, readily accessible for quick and copious quenching. Dracula, in Stoker's characterization, was a tactician, not a symbolist.

As an added metaphor, merging style with subject, Phil decided to drain the color from his palette. He painted *The Vampire* in monochromatic tones — chalky greens predominating, with a weird hint of bluish moonlight.[35]

Phil arranged to have *The Vampire* included as part of The New Gallery's 1897 summer exhibition. Founded a decade earlier by Charles Halle and Comyns Carr to promote the tastes formed by the Pre-Raphaelites and the Aesthetic Movement, the New Gallery was located in Regent Street in spacious rooms surrounding a fountained center court. The annual exhibit was a major social event, rivaling that of the stodgier Royal Academy. Edward Burne-Jones's work was a mainstay at the New Gallery, as were the paintings of John Singer Sargent. For the new exhibit, the elder Burne-Jones was unveiling what would become one of his most celebrated paintings, *The Pilgrim of Love*.

Phil, on his father's coattails, had exhibited paintings in previous New Gallery shows, but *The Vampire* was to be his first real splash. He took the unusual step of adding a set of verses titled "For the Picture" by his cousin Rudyard Kipling. Both Phil and Ruddy had an early literary exposure to *Fatal Women*; Nora Crook, in Kipling's *Myths of Love and Death*, notes that Kipling, at the age of twelve, first came under the spell of the Wilhelm Meinhold novel *Sidonia the Sorceress*, translated by Lady Wilde in 1849. The book was a cult favorite of the Pre-Raphaelite circle; Phil's father made it the subject of two watercolors, and Phil himself illustrated a similarly-themed Meinhold book, *The Amber Witch*, in 1899.

Kipling's poem was published under the title "The Vampire" in the *Daily Mail* as a rather shameless advance advertisement. It also appeared in the printed exhibition catalog, taking up a fair amount of space in the pocket-sized pamphlet. As the painting itself was an example of the fashion known as "the problem picture" — an ambiguous scene which the viewer must complete with an interpretation — the Kipling ballad, itself ambiguous, was calculated to tease the reader with the allegory of an everyman-fool who succumbed to the attractions of a heartless woman who took, and took, and took, only to throw the fool aside, "stripped to his foolish hide."

The verses amounted to a kind of boudoir variation on the military ballads of which he was an indisputed master — a music hall or drinking song of male camaraderie ("Even as you and I!") in the endless, bloody war between the sexes. Richard Le Galliene, who appraised the poem at the turn of the century, called it "something like a great achievement in satire, as it is surely the bitterest thing ever written by a man against woman. The touch of hysteria in it, as of personal pain, will save women from taking it

35. Although more than one published description of *The Vampire* maintains that the canvas was painted in vivid shades of crimson, green, and white, these details seem drawn from a brightly colored postcard published around the turn of the century. No accurate color reproduction ever seems to have been struck, but direct critical appraisals of the painting in English and American publications consistently describe the leeched-out quality of the color scheme, a pale greyish-green predominating.

too much to heart; and, of course, like all recrimination between the sexes, it is necessarily one-sided. But all that discounts in no way from its murderous force."[36]

And thus, Rudyard Kipling, the greatest literary celebrant of Empire, also penned one of the most memorable evocations of Vampire.

The private viewing of the New Gallery's Tenth Annual Summer Exhibition was held on Saturday, April 24, 1897. Owing to winter-like weather and the coinciding Easter holiday, attendance was not as strong as usual. Georgiana Burne-Jones attended, wearing a dark blue cloth gown and cape, and a bonnet with light flowers. As per custom at the New Gallery, all paintings were exhibited in gilt frames, without exception. An ominous chord was struck from the beginning, when a young girl knocked over an expensive piece of statuary, damaging it beyond repair. The *Sunday Times* previewed the exhibition on April 25. "The chief picture in the first or South Room is 'The Vampire' by Philip Burne-Jones. What will strike the public is the weird and powerful fancy. What strikes the jaded critic is the great technical advance. Mr. Philip Burne-Jones is hardly the same man that he was only three years since. Even from last year the improvement is surprising. His figures are now alive, and they have anatomy beneath their outward show. The woman leaning over the dead youth whose life-blood has nourished her is painted with a power that arrests attention. The lighting of the whole scene, though by no means true, is a great advance toward naturalism as compared with previous work. The poem by Rudyard Kipling — Mr. Burne-Jones's cousin — is in its strongest manner."[37]

Cover decoration for an early edition of Kipling's poem.

The *Daily Mail*, which had already run Kipling's poem without a context or explanation, noted that a smug portrait of a clergyman, "seems curiously out of place" by its proximity to *The Vampire*. "In a moonlit room an eerie figure clad in the night costume that is usually worn in the security of locked doors, sits on a couch and gloats over the body of her victim, upon whose bare chest is an ominous crimson stain. Mr. Kipling has called her 'a rag and a bone and a hank of hair,' which is a singularly happy description. One may cavil at Mr. Burne-Jones's predeliction for the gruesome; it is, however, evident that his style and draftsmanship have improved steadily . . . 'The Vampire' will be much talked about during the coming season."[38]

The *Pall Mall Gazette* called the painting "a piece that cannot but attract attention . . . because its subject is sensational and the model chosen engaging and morbid, [it] may please the public. The painting seems coarse and stringy, but who cares when there is emotion in the model?"[39] Where the *Pall Mall Gazette* only hinted at the celebrity of the subject, the *Westminster Gazette* was explicit. "Mr. Philip Burne-Jones' disagreeable picture of Mrs. Patrick Campbell as a 'Vampire' in a dirty nightgown and a weird light, is uncommonly clever in many respects, but it is hardly the kind of thing one would want to live with. Its main

36. Richard Le Galliene, *Rudyard Kipling: A Criticism* (London and New York: John Lane: The Bodley Head, 1900), p. 63.

37. "Art and Artists — The New Gallery," *Sunday Times*, April 25, 1897, p. 2.

38. Gilbert Burgess, "The New Gallery: Some Impressions," *Daily Mail*, April 26, 1897.

39. R.A.M.S., "The New Gallery: Preliminary Note," *Pall Mall Gazette*, April 24, 1897.

fault lies in an utter absence of any idealization . . . all the more marked because as a realistic study it is both unsuccessful and unpleasing. His portrait of Lady Bett Balfour is more satisfactory; he has skill and a certain originality, but he must do something better than 'The Vampire' if he is to prove worthy of a great name."[40]

The *St. James Gazette* called *The Vampire* one of the most striking figure paintings in the exhibition. But Phil, the *Gazette* noted, did not rest content with his own talents: " . . . he has applied to his cousin, Mr. Rudyard Kipling, for some illustrative verses; and Mr. Kipling has replied with a sequence of stanzas so penetrating, so energetic, so grimly humorous, that the picture (which is by no means tame in itself) appears a little tame in comparison to them."[41] The society weekly *Black and White* called the painting "sufficiently sensational," but opined that the real reason to see it was because "it has inspired Mr. Rudyard Kipling to write one of the best sets of verses he has ever done."[42]

The Athenaeum published the most detailed appraisal:

> The subject of The Vampire has taxed the resources and invention of Mr. P. Burne-Jones to the utmost, and, indeed, the result of his studies does not quite equal our hopes. The scene is a moonlit room, where the victim of the vampire, a stalwart young man, who is either dead, dying, or in a trance of terror, lies supine on his bed; upon his uncovered breast is the red mark of the monster's fatal caress, and she, in the shape of a wan, demon-like woman, sits at his side, and looks as if her lips had just parted from their horrid work. So far this is an easily understood illustration of what are, after all, but the externals of the ghastly legend, and there is nothing in the design or the picture as such which adds to its horror. What we see is doubtless suitable enough to the tale, but we are unable to discover signs of deeper insight at work to raise the picture far above the level of an illustration in the common sense of that term.[43]

Nevertheless, the reviewer conceded to praise

> the good judgement which selected this cold and wan moonlight as the effect best suited to the theme, the skill which enabled Mr. Burne-Jones to paint it so well; also let us commend to students the simplicity of the design, which contains no extraneous and unnecessary features nor any excess of details, incidents and circumstances. The livid greyness and the olive pallor of the woman are exactly what they ought to be. The artist's technique is evidently rapidly improving, so that it is easy for us to see how much better 'The Vampire' is painted than any of his pictures which have preceded it. Mr. Rudyard Kipling's lines, "For the

40. "The New Gallery and What Is In It," *Westminster Gazette*, April 24, 1897, p. 2.

41."The New Gallery," *St. James Gazette*, April 26, 1897.

42. Review of The New Gallery summer exhibition, *Black and White*, May 1, 1897.

43. "The New Gallery. (First Notice.)," *Athenaeum*, May 1. 1897, p. 585.

44. *Ibid.*

> Picture," which are printed in the catalogue under its name, may have a very remote and quite indirect connexion with the vampire legend, and to be a sort of allegory unattached, but that is as much as can be said of them.[44]

But the most interesting appraisals of the picture, of course, were the ones that didn't appear in print. According to Henry P. Marston, the *New England Home Magazine* journalist, "there is a story behind it all, and now the London gossips are rolling the sweet morsel under their tongues with the most pleasurable sensations."

In Marston's retelling of the tale, "It is whispered that spite and revenge were the motives that prompted the artist to represent the features of the actress in such an unpleasant guise, and that Mr. Philip Burne-Jones had been a passionate admirer of Mrs. Campbell (quite platonic, of course, because there is understood to be a Mr. Campbell packed away somewhere in the Egyptian civil service [sic]), and had been thrown overboard in a way that hurt his feelings and lacerated his susceptible artistic tempermanent."[45]

The original edition of Dracula *(1897).*

A few weeks after the New Gallery exhibition opened, Constable and Co. published *Dracula*, which, for all its subsequent appeal, stirred much less comment in the London press than did the Burne-Jones painting. The *Weekly Sun* reviewer noted the coincidental appearance of two major vampires in London in the spring of 1897, observing that "It was not till I had read Mr. Stoker's book that I grasped the full meaning and weirdness of that striking picture."[46]

Mrs. Patrick Campbell did not attend the New Gallery exhibition — although, in a sense, she was a central fixture of the show. A play she had just opened with Forbes-Robertson closed abruptly in March, a disastrous flop. Her nerves were shattered from overwork. She pushed on with her activities anyway, and her mental state worsened. Phil's sister Margaret tried to comfort her, to no avail. Unable to sleep, prone to hysterical outbursts, she was finally persuaded to check into a nursing home. Her doctor blamed it on acting. Gossip about *The Vampire* may or may not have reached her, although there are hints that some letters may have been exchanged over the matter between her and a possibly embarrassed Edward Burne-Jones. But, then as now, almost any public notice or comment was better than none for those who toiled in the arts. Mrs. Pat spent eight weeks convalescing, but, thanks to *The Vampire* and the throngs it attracted, she never really left the spotlight that summer at all.

Kipling neglected to copyright his poem, and it appeared in numerous editions on both sides of the Atlantic over the next few years. Parodies were inevitable, and one of the best, "from woman's point of view," was penned by Mary C. Low, and published in both New York and London in 1899:

45. "An Artist's Revenge," *op. cit.*

46. "Books & Bookmen," *Weekly Sun*, June 6, 1897.

A woman there was who heard a prayer,
(Even as you and I!)
From flesh and bones and a lock of hair
(He called her the woman beyond compare),
But he only used her to lighten his care,
(Even as you and I!)
Oh, the walks we had and the talks we had,
And the best of our heart and hand,
Were sought by the man who pretended to care,
He didn't — but why he pretended to care,
We cannot understand.

A woman received the flowers he sent,
(Even as you and I!)
Honour and faith she thought his intent,
(But God only knows what the gentleman meant),
Yet a man must follow his natural bent,
(Even as you and I!)

Oh, the vows we spoke and the vows we broke,
And the various things we planned,
Belonged to the man who said he was true,
(But now we know that he never was true)
And we cannot understand.

One favor she asked — but it was denied,
(Even as you and I!)
In some or other he might have replied,
(But it isn't on record the gentleman tried),
Her faith in him faltered and finally died,
(Even as you and I!)

And it isn't the shame and it isn't the blame,
That stings like a white hot brand,
It's coming to know he would never say why,
Seeing at last she could never know why,
And never could understand.[47]

Mrs. Patrick Campbell in Magda *(1894).*

Like most things macabre and English, *The Vampire* eventually found its way to America. Phil, who became bored and idle in the years following the splash of the New Gallery exhibition, found himself "spellbound" by an ennui, "describing a tiresome circle round and round an invisible centre." His father's death, in 1899, disturbed him greatly ("I failed him at every turn.") He yearned to be "among new faces, and under new conditions"

47. Mary C. Low, "The Vampire," *The Academy* (London), April 1, 1899.

48. Philip Burne-Jones, *Dollars and Democracy* (New York: D. Appleton and Company, 1904), p. 4. Phil was an engaging writer when he set himself to the task, and this book of his observations.

where limy own personality would not be such a terribly open book to my companions; where there would be some novelty and mystery about them for me. In short, to get to some place where I wasn't known myself, and didn't know others."[48]

Phil had never been to America, but he envisioned "a vast continent full of the most bewitching girls, clad in the daintiest costumes, delighted to see me, and ready to extend their hands in a natural and unaffected camaraderie, only possible in America." He imagined their brothers and husbands ("strong, manly, simple") as "constantly at work somewhere out of sight, chiefly in 'Wall Street,' wherever that was, leaving their wives and sisters free to entertain me, and glad to think they were doing so. I had heard much of the unselfishness of American men."[49]

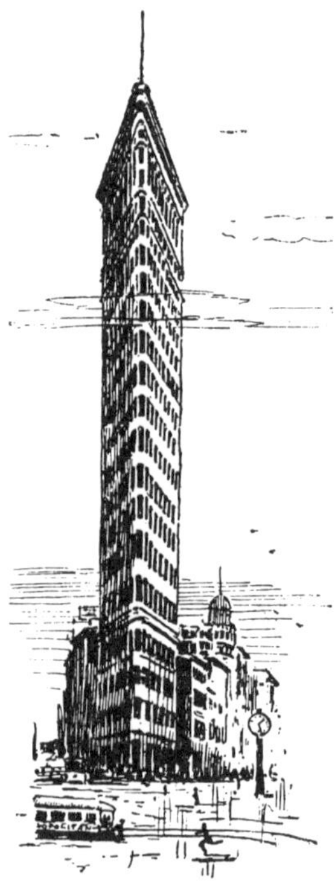

New York's Flatiron Building, as drawn by Philip Burne-Jones. From Dollars and Democracy.

Whatever his fantasies about sunnier arrangements between the sexes in America, Phil fully intended to take *The Vampire*, along with other recent canvases. His sister saw him off in Euston in February aboard the *Oceanic*, and expressed doubts about his company. "He starts for New York today: he appears to be travelling with a quite particularly shady set of people, most of them looking as if their only reason for going was to escape the strong arm of the law."[50] *The Vampire* traveled separately aboard the *S.S. Philadelphia.* Like the sea-crossing of Dracula, the trip proved tumultuous, not because of storms, but due to a protracted struggle with the customs authorities. Phil arrived on February 26 and, while searching for his scattered luggage ("half of it had got stuck under 'B' and half under 'J'") was set upon by "the little band of reporters who hang about the quay on the arrival of the European steamers, eager for 'copy,' however trivial — something, anything that can be twisted or turned into a paragraph that will bring them a few cents." Unfamiliar with their ways at this early point in his visit, Phil talked to them. A few days later "I read in a newspaper that I had met with a terrible accident, and had been badly burned — set on fire, in fact, the evening before, in a friend's rooms which I had never visited, and under graphically described circumstances which never had the remotest existence in fact."[51]

While waiting for *The Vampire* to be released from its official quarantine, he found a studio and apartments in a building directly opposite Bryant Park. "During my first weeks in my new home," he wrote, "I remember undergoing one or two acute attacks of nostagia, as I sat alone in the cafe, eating my solitary dinner to the strain of a heart-breaking orchestra. One can have some of the saddest moments of one's life, I think, dining alone at restaurants and listening to the band."[52]

"After a fortnight's struggle with the custom-house officials over my poor paintings, which I finally got released after appealing for help to Washington,"[53] Phil agreed to file a bond, which the *New York Daily Tribune* reported as being $20,000, with the sworn intention of sending the paintings and their frames back to England. Phil finally arranged to have his unchained monster — as well as twenty-five other pictures in oil and

49. *Ibid.*, p. 5

50. *Victorian Sisters*, p. 187.

51. *Dollars and Democracy*, pp. 9-10.

52. *Ibid.*, pp. 10-11.

53. *Ibid.*, p. 12.

watercolor — shown at the galleries of M. Knoedler & Co. at 355 Fifth Avenue, at Thirty-Fourth Street, for two weeks beginning March 17.

The event, somewhat tartly reported by the *Evening Post* on March 22, was "duly and fully announced by his advance agent or some one who was willing Sir Philip's work should be discussed." The *Post* found that "The woman who plays the title role in the tableau is queer in anatomy, to be sure, and the lighting of her figure is false; but in its own unwholesome way the work is bound to command respectful consideration. The dead fool by her side, upon whom she has preyed, is dead, indeed, and the expression of the vampire conveys much . . . but a great work of art this is not.'"

The *New York Daily Tribune* made note of some fantastic treatments in the exhibition other than *The Vampire: The Phantom Ship*, suggested by Coleridge's *Rime of the Ancient Mariner* and depicting Life and Death casting lots for the sailor's soul; and Earthrise from the Moon, (a curiously morbid science-fiction illustration, possibly inspired by the moon-inhabiting "Selenites" of H.G. Wells's *First Men in the Moon* (1901). Phil painted his moonscape with the skeletal remains of an Lunarian prominent in the moondust. The *Tribune* called it "bleak and ghastly.") Other weird pictures in the exhibit included The *Shadow of the Saint*, in which two girls returning from a feast through the moonlit streets of a city in ancient Syria, are startled by the shadow of St. Simeon Stylites upon a wall. An *Unpainted Masterpiece* was based on a story by Henry James, depicting the "grim ghost" of an artist's studio. As James described the scene, "Before him, upon the easel, stood the picture, presumably the painted Madonna — a mere blank canvas, cracked and discolored by time. This was his immortal work!" In a similar vein, *The Shadowless Man* illustrated the famous story of Peter Schlemel. The major works of the exhibition, in other words, involved not only vampirism, but the related themes of loss of shadow/soul, and a strikingly plaintive mood of artistic impotence, death and desolation.

Philip Burne-Jones

As to the artist's technique, the *Tribune* called Phil's use of color "tasteless, deplorably lacking in quality. He has none of that sense of decorative beauty which enabled his father to triumph over his shortcomings as a painter."[54]

The *New York Times* was especially nasty. "Sir Philip Burne-Jones, Bart., has inherited from his distinguished father a title, but also the awkward bequest of a comparison." The *Times* also attacked Kipling's verses ("...hack work in verse in the style of Bulwer Lytton, with a coarse touch to make them seem modern"). The figure of the vampire herself was an "overrated enigma," according to the unsigned appraisal of March 20. "At best the man and woman here are acting 'living pictures,' but whether the lady intends to rouse the man from a nightmare or is waiting till he wakes in order to give him a curtain lecture, the painter has forgotten to say." In a final dismissal, the *Times* found it "unfortunate that so much bother has been raised in the papers about Sir Philip and his 'Vampire,' for expectations

54. *New York Daily Tribune*, March 27, 1902.

55. *New York Times*, March 20, 1902.

56. *Dollars and Democracy*, pp. 12-13.

naturally rose mountains high, and all that appears is a little mouse of a talent, which seems to have lost its way."[55]

The bad reviews did nothing, however, to keep the curious public away from *The Vampire*, and, though Phil "retired the poorer" from the exhibition, he recalled that "crowds flocked in all day, and I had the satisfaction of realizing, as I paid the dealer his bill, that I had at least helped to advertise his gallery for him."[56]

Mrs. Pat played opposite Sarah Bernhardt in Pelléas and Mélisande *(1898).*

Mrs. Patrick Campbell, long recovered from her nervous prostration and now touring the states, had inadvertently spurred public interest in *The Vampire* during a New York engagement the season prior to Sir Philip's exhibition. She had engaged a particularly aggressive press agent, a portly Dane with the singular name of A. Toxen Worm, who had no qualms about playing up the vampire business for all it was worth. A lengthy unsigned article that appeared in the *Brooklyn Daily Eagle* was likely Worm's doing. "That Mrs. Patrick Campbell was the model from which the artist drew his tragic conception of this popular legend is a well defined theory in literary and artistic circles," Brooklyn readers were told, "a rumor that has now become so well established that it will require more than a few words of bare denial to set it at rest, and the general opinion in London is complimentary neither to young Burne-Jones, nor to his more distinguished relative, Rudyard Kipling." The article continued:

> As the story was first told by those who pretended to know, Mrs. Campbell had often used her influence to obtain commissions for Burne-Jones and befriended him in other ways. For some reason that the gossips did not pretend to discover the friendship was broken off. Soon after this picture was placed on exhibition...It may not be impossible to imagine why Mrs. Campbell objects to being regarded as the vampire woman.[57]

Mrs. Pat was not crazy about A. Toxen Worm, who instead of protecting her reputation seemed intent on just spreading more gossip. But, as Philip Burne-Jones was discovering for himself, the American press operated on terms quite different than that of the staid London papers. The American tabloids had already been full of stories that she had "ruined" the young Gerald Du Maurier, who had played her boy lover in the Oscar Wilde-inspired play *Mr. & Mrs. Daventry* by Frank Harris. Du Maurier was, apparently, quite smitten with the actress, and finally withdrew from the production. To the sensational press, however, the vampire was at it again. The *Chicago North American* ran the headline MRS. PAT IS A PALE, POSTER-LIKE LADY over an article that called her "the high priestess of the Decadent, the Saint of the Order of the Degenerates." When she appeared at the Harlem Opera House to display her "morbid heroines," one especially brief — and bitchy — item ran in a New York paper: "Up above the dead line of sporting and theatrical New York, where Mrs. Campbell is to display

57. "Gossip of 'Vampire' Painting Revived by Actress' Visit Here," *op. cit.*

58. Unsourced New York newspaper clipping; Robinson Locke Theatrical Scrapbooks, Vol. 97, Billy Rose Theatre Collection, New York Public Library at Lincoln Center.

59. "Notes of the Stage," *New York Daily Tribune*, November 17, 1902. A. Toxen Worm became a legend in his time and was an effective fixture of the Shubert's operation in the teens and twenties. On the occasion of the 1915 amputation of Sarah Bernhardt's leg, he proposed a special engagement "because of the news interest attached to seeing a woman of her age...appearing with a cork leg. The newspapers will eat this up, with descriptions of her legs, how it feels to play on a cork leg, etc., because it is something no woman did before." He also suggested that Bernhardt, figuratively battle-scarred, might be profitably presented as the "undaunted spirit" of wartime France. For more, see Brooks McNamara, *The Shuberts of Broadway* (New York and Oxford: Oxford University Press, 1990), pp. 66-67.

her weird studies of women who don't exist, the unappreciative natives cannot understand the merits of her acting, so that she has had to come down to their level by reducing her prices..."[58]

The impolite treatment at the hands of journalists "was a new experience for the actress," reported the *New York Daily Tribune*, which noted that the most frequent way she referred to her press agent was as "that — that — creature!" (To call him "Worm," the paper explained, would have amounted to an anticlimax.)[59]

With Phil, *The Vampire*, and Mrs. Pat simultaneously touring America and attracting press attention wherever they went, a collision was almost inevitable. It happened in January 1903. In an incident worthy of Noel Coward, all three had an inadvertent rendezvous in Chicago. There is no evidence that Phil timed his trip to coincide with Mrs. Pat's engagement, or intended to stay in the same hotel as the actress. But it happened nonetheless. The press immediately picked up the nuances and had a field day letting the story unfold. The *Chicago Chronicle* got the ball rolling — or the bat flying:

Chicago newspaper illustrations that accompanied a discussion of the controversy.

> "The Vampire" is in Chicago. It came yesterday and direct from Sir Philip Burne-Jones' New York studio. It will be placed on exhibition tomorrow if the painter succeeds in finding a place to hang it.
>
> If as you stand before the picture of "a rag and a bone and a hank of hair" you should see a woman, tall and stately, with large, dark eyes and raven hair, advance, watch her closely, for it will be Mrs. Patrick Campbell, who does not deny that she is the female figure represented. A strange meeting may occur if Sir Philip, Mrs. Campbell and "The Vampire" should find themselves in the same room.[60]

The *Chicago Daily News* reported on January 17 that "the two principals in the story are both occupying apartments at the Auditorium Annex and it is said that when the coming of Sir Philip was announced Mrs. Campbell expressed satisfaction at the news and told many interesting stories about the English artist."[61] The *Chronicle* even reported that Mrs. Pat, sitting in her hotel room, would occasionally assume the famous catlike pose. No doubt, she knew how to give reporters what they wanted, with or without A. Toxen Worm.

Phil, flustered, told the *Chicago Record Herald* that "I do not know Mrs. Campbell at all. Furthermore, I do not want — that is, I have no particular desire to meet the lady, you know." Phil told the *Chronicle* that the painting's exhibition "will not be an invitation affair." Instead, "an admission will be charged. It will be open to the public."

"And to Mrs. Campbell?" asked the reporter.

"If she has the price," Phil snorted.[62]

Mrs. Pat's "interesting stories" dried up instantly. She dictated a statement to her secretary, who typed it up on hotel stationery:

60. "'The Vampire' in the City," *Chicago Chronicle*, January 20, 1903.

61. "Actress Replies to Artist," *Chicago Daily News*, January 16, 1903.

62. "Mrs. Pat No Vampire," *Chicago Record Herald*, January 16, 1903.

January 16, 1903

To the Editor of The Daily News:

I knew Sir Philip Burne-Jones nine years ago. I also had the privilege of knowing intimately his great and most beloved father, Sir Edward Burne-Jones, and his sister, Mrs. J. W. Mackail who became then, and is now, my dearest friend.

At the time the "Vampire" picture was exhibited, I was not in London, and since, I have had no opportunity of seeing it. I always understood the female face in the picture resembled mine, and that this fact has added to the sensational success of the picture. Sir Philip Burne-Jones was at Mrs. Caton's delightful party last night, but we did not meet.

BEATRICE STELLA CAMPBELL[63]

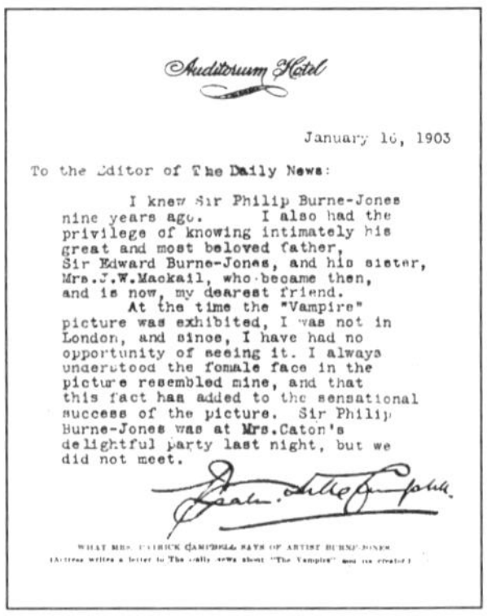
Auditorium Hotel

January 16, 1903

To the Editor of The Daily News:

I knew Sir Philip Burne-Jones nine years ago. I also had the privilege of knowing intimately his great and most beloved father, Sir Edward Burne-Jones, and his sister, Mrs. J. W. Mackail, who became then, and is now, my dearest friend.

At the time the "Vampire" picture was exhibited, I was not in London, and since, I have had no opportunity of seeing it. I always understood the female face in the picture resembled mine, and that this fact has added to the sensational success of the picture. Sir Philip Burne-Jones was at Mrs. Caton's delightful party last night, but we did not meet.

Mrs. Pat's letter to the press.

Phil stuck to his denials, adding embellishments to the story. "The woman who did the posing was a paid model in Brussels," he said, according to the *Record Herald*. "I was staying in the Belgian capital for the summer, and saw among the professional models there a young woman who seemed to be just the one I wanted for such a picture. I hired her at so much a day, and that's all there is to it. It seems I made a good choice of model."[64]

The paper asked him what he thought of the unfavorable criticism of *The Vampire* by a certain Frenchman named H. Pene du Bois. "Frenchman?" Phil was said to have retorted. "He looks like a mulatto, and he knows nothing of art."[65] (Phil, to be frank, was not very fond of people of color. "When they were transported in ship-loads to America as slaves, of course they hadn't quite a fair chance; but since freedom was given to them, have they shown any particular aptitude for distinguishing themselves in any way?" he asked, adding that "the best that can be said for them is that they make tolerably good servants." In truth, black people often frightened him. "I can imagine, if one were at all feverish or delirious, seeing their faces in the dark would produce a fit," he wrote, finally declaring that the black race could never be assimilated in America, "for nothing can eradicate the inherent antipathy which the black man arouses in the white, and they cannot work in double harness. One must go to the wall, and it will not be the white man.")[66]

The Chicago press loved the idea of a malicious Sir Phil wielding a rapier brush or pen, and did its best to encourage the image. One day, while sitting in the hotel's court, a reporter begged him to scribble some cartoons, particularly one of "a woman yawning." Heedless to the strange specificity of the request, Phil complied, adding a cartoon of himself, and a grotesque male face taken completely from his imagination. The reporter's purpose became clear the following morning, when the drawings were reproduced in his newspaper with the following caption:

63. "Actress Replies to Artist," *op. cit.*

64. "Mrs. Pat No Vampire," *op. cit.*

65. *Ibid.*

66. *Dollars and Democracy*, pp. 166-167.

67. *Ibid.*, p. 227.

> Directing the eye to these interesting outlines, in the upper left corner will be recognized a distinguished "leader of the 400." Beneath are depicted the features of a no less prominent society matron of the present day. Sir Philip Burne-Jones has drawn his own figure in the centre — while on the right, undoubtably, may be found his concrete idea of the Chicago man. [67]

The following week, a New York society journal accused him of amusing friends in Chicago by carcicaturing hostesses who had entertained him, and, as Phil wrote, "it was not ashamed to mention by name one of the best known ladies in New York as having been one of my victims."[68]

Phil felt it necessary to reply publicly, which he did in the letters column of the *New York Daily Tribune*. "The result of this protest was that the aforesaid scurrilous journal in New York 'went for' me the following week with renewed fury, pouring out nearly a column of venomous abuse interspersed with lies, and repeating the libel of the week before, with circumstantial details (all false) and mentioning another lady, well-known in Chicago by name, as having suffered in a similar way at my hands."[69]

As to *The Vampire*, the *Chicago Tribune* decided that the painting was "not calculated to inspire terror, she looks too much like a fashion plate or a stuffed scarecrow for that. The man, her victim, has the air of having 'died easy.'" In a story headlined SIR PHILIP BURNE-JONES' 'VAMPIRE' IS VERY CREEPY, the *Chicago Journal* captured the ambiance of the painting's exhibition at Russell's Gallery at 40 Madison Street. "The show opened at 10 o'clock," the paper reported dutifully. "A small boy in buttons barred the way to the great treasure until the ante of 25 cents was produced." The monochromatic painting was displayed before a plush red backdrop, and illuminated by electrical light. A visit to the morgue, the paper opined, could produce "just about the same feeling" as Phil's traveling horror show.[70]

On January 22, acccording to a gossip column in the *Chronicle*,[71] an attractive young woman approached Phil in a corridor of the Auditorium Annex and introduced herself as a palmist who was interested in giving him a reading. Charmed, he consented, and they sat down near the entrance to the hotel's Pompeiian room. The woman took his left hand and examined it. "Not used to labor," she said, and Phil laughed. A couple passed and the male half cracked, "Great stunt, that palmistry." Phil blushed instantly and drew back his hand. The reader coaxed it back from him. His head and heart lines were entirely too close, she said. And his finger bulbs — well, the size of finger bulbs indicated sensitivity and Phil's were enormous — the biggest, in fact, that she'd ever seen.

And finally, the woman came to the mocking heart of the encounter. "Now, let's trace your love line," she said, examining the evidence. "Sir Philip, you have been disappointed. Now, don't deny it. You have been crossed. Now haven't you?"

Phil snatched his hand away. The charming hand-holding with a pretty

68. *Ibid.*

69. *Ibid.*, p 229.

70. "In the Field of Art," *Chicago Tribune*, January 25, 1903.

71. "Odd Tales of the Town," *Chicago Chronicle*, January 23, 1903.

American girl had been reduced to a cruel joke. "I don't believe you know anything about palmistry," Phil said. "I think you are just poking fun at me." Was this an example of the open-handed American camaraderie he had once anticipated? Even the pretty American girls were using him for sport.

On February 1, 1902, the *Tribune* reported that "Mr. Burne-Jones returned to New York early this week, having failed to receive sufficient encouragement for his portraits. His painting of 'The Vampire' attracted over 1200 visitors last week. Chicago has therefore made up for the losses that he incurred in New York by exhibiting his picture without charging for admission as he did here."

Porter Emerson Browne

Phil left *The Vampire* in Chicago, "for continued exhibition here and possible sale to a Chicago admirer," according to The *Chicago Chronicle*. This, despite the reported $20,000 bond in New York. He refused to discuss reports that he had received "an off" for the canvas. "That is a matter that I must decline to talk about for publication," he told a reporter. "If my picture is sold the purchaser is the person who should announce the fact that he has gained possession. When I part with the painting I also part with the right to advertise the plans in which it is involved."[72]

His final public words in Chicago: "I repeat," he told a reporter as they ascended the steps of his train, "that I do not know Mrs. Patrick Campbell."

And there is no evidence that he ever acknowledged her again.

It was inevitable that a painting as intrinsically dramatic as *The Vampire* — and one that additionally took as its model one of the most celebrated personalities of the theatre world —would eventually attract a playwright eager to explain the tantalizing enigmas of the image and the poem.

At the age of twenty in 1899, Porter Emerson Browne had resigned himself to giving up writing altogether. He had been a cub reporter for a Boston newspaper when a diagnosis of tuberculosis prompted him to seek outdoor employment in the rarified air of Colorado. But his new job — supervising a coal tipple — only succeeded in coating his lungs in black dust, worsening his fiction. Following a period of rejection slips, began to sell short stories regularly to such magazines as *Collier's*, *McClure's*, *Everybody's*, and the *Saturday Evening Post*.

Browne's work came to the attention of the noted actor Robert Hilliard, whose stage career had stalled in vaudeville and who was eager for a vehicle to re-enter the legitimate theatre. He had carried around a reproduction of the Burne-Jones painting and the Kipling poem for some years; one day around 1907 he noticed a story of Browne's called *The Proof* in *The Red Book* and invited the young writer to his rooms to make a proposition. He produced the Burne-Jones picture and pointed at it. "I believe," he said, "there is a play in that. Write it for me, Porter."

Since neither the painting nor the poem provided anything but the

72. *Chicago Chronicle*, undated clipping.

vaguest sense of a narrative line, Browne used his imagination to concoct a plot. He Americanized the characters and settings, and opened the play on a cross-Atlantic passenger ship about to New York. A young man is seen groveling at the feet of a cold, beautiful Woman (all the characters have allegorical, Albee-esque names — the Man, the Wife, the Woman, etc.) who holds a thorny bouquet of blood-red roses. The young man begs — for a kiss, for a kind word, for acknowledgment — anything. But the Woman stares him down until he can do nothing else but remove a gun from his pocket and blow out his brains. With his death, the Woman becomes gleefully animated, laughing as she sprinkles red rose petals over the corpse.

Enter the Man, a diplomat devoted to his Wife and Child, about to leave on an important overseas mission. But the Woman catches his eye, and transfixes him. By the time he returns home, he is the Woman's slave — addicted equally to her, it would seem, as to the whiskey bottle. His marriage crumbles. Even the President of the United States denounces him. A concerned Friend tries to intervene, literally throwing cold water in his face. He seems to understand what has happened to him and swears to reform. But the Woman returns, mocking him. "Kiss me, my fool!" she commands. Instead he tries to strangle her, but it is a quixotic effort — his heart weakened by fatal dissipation, he falls lifeless to the floor. More rose petals flutter down, and the Woman laughs as the curtain falls. Moments later, it rises again in a *tableau vivant* — the actor and actress brilliantly illuminated by calcium lights in the predatory pose immortalized by Philip Burne-Jones.

Katherine Kaelred as the Vampire Woman in A Fool There Was *(1909).*

Robert Hilliard loved the play and immediately arranged for it to be produced at the Liberty Theatre by Frederic Thompson, the impresario who had created Coney Island's Luna Park. Hilliard, of course, took the lead, and the role of the vampire was essayed by a dark-haired Australian actress named Katherine Kaelred.

"Mr. Porter Emerson Browne is the daring young author who has undertaken to prove Mr. Kipling's poem in seven scenes with slow music," The *New York Times* reported on March 25, 1909. "By the time he is through there is not the least little doubt in the world about the rag and the bone theory, though it would be ungenerous to allude to the splendid masses of woman's crowning glory exhibited by Miss Katherine Kaelred as anything like a hank." The *Times* called Hilliard's performance "a carefully delivered study in the enervating effects of rum and rose leaves," and noted that the final tableau resembled nothing so much as Little Eva on her way to glory. "Only, of course, this pair was bound in the other direction, a fact frequently emphasized in the course of the evening."

Despite some condescending reviews, *A Fool There Was* was an unqualified hit, an unvarnished morality play that played cannily to popular tastes. Though Carrie Nation was nearly at the end of her life in 1909, the temperance movements she symbolized were gathering momentum, and soon would merge into the disastrous juggernaut of Prohibition. On another level, the vampire woman of *A Fool There Was* embodied anxieties about women's growing independence. As in the hit film *Fatal Attraction*, released eight decades later, the unattached ("loose") woman was represented as a mortal threat to the middle-class family, its security and property. It also reflected a barely-concealed backlash (by men and women) against the swelling movement for female suffrage.

Porter Browne simultaneously published a novelization of the drama. His vampire's first shipboard entrance is described exactly as it appeared on stage: "Coming down the deck was a woman, a woman darkly beautiful, tall, lithe, sinuous. Great masses of dead black hair were coiled about her head. Her cheeks were white; her lips very red. Eyes heavy lidded looked out in cold, inscrutable hauteur . . . She wore a gown that clung to her perfectly modelled figure — that seemed almost a part of her being. She carried, in her left arm, a great cluster of crimson roses."[73] But Browne went even further, embellishing the story with a dramatic explanation of the vampire's origins. She came from France, it seemed, a feral child who first sought revenge on the father who raped and abandoned her mother. Having once tasted blood, she kept on revenging. The book, predictably, became a best seller and the play toured the country to packed houses.

"Are there, in your opinion," asked the *Chicago Tribune* of Robert Hilliard on Halloween, 1909, "such women in the world as are depicted in Kipling's poem and Burne-Jones' painting?"

"Yes," Hilliard replied. "Many of them. Strange women who slink through the world, bringing corruption to everything they touch. Women who are not beautiful — many of them are homely — but who possess a power of attraction that never fails to enchant when and wherever directed." Hilliard called this power "impossible to analyze," but capable of imparting in even the most emotionally detached male a "thrill which takes him ... as a terrier does a rat, never again leaving him alone until all the life and sense have been shaken from him..."[74]

The *Atlanta Georgian* called *A Fool There Was* less a play than "an intimate view of a dissecting room. It is brutal, sometimes almost sickening; one who saw it will not get the bad taste from his mouth or the picture from his memory for many a day." The *Georgian* critic, who evidently had his own considerable store of morbid imagery to draw upon, called upon his readers to enhance the occasion with memories of unrelated horrors:

> You who have read Dickens remember the nauseous chapter describing the death of the man upstairs by spontaneous combustion, the greasy ashes which descend from the room of death and settled on the hands and faces of the watchers below. The closing act of *A Fool There Was* is no less brutal in its realism. Playwright and scenic artist and actor have combined to make it a picture which haunts the memory. You grip the arms of your seat

73. Porter Emerson Browne, *A Fool There Was* (New York: The H.K. Fly Company, Publishers, 1909), p. 111.

74. "Are There Real Vampires? Lots of 'Em, Says Hilliard," *Chicago Tribune*, October 31, 1909.

75. Dudley Glass, "Robert Hilliard's 'Fool' Terrible in Its Power," *Atlanta Georgian*, November 24, 1910.

and wait until the last piercing shriek of the vampire woman echoes through the wings and then you hurry out and give thanks for the fresh air and the stars overhead.[75]

Just as England had made a gift of *The Vampire* to American consciousness, the favor was returned when *A Fool There Was* crossed the Atlantic, regurgitating Burne-Jones's and Kipling's fantasy on its home shore. "It is staggering," wrote the *London Clarion*, "to reflect on the stage reached in the evolution of American theatrical art if *A Fool There Was* by Porter Emerson Browne, produced at the Queen's last week, is a fair specimen of trans-Atlantic taste. The play is based on Rudyard Kipling's well-known verse called 'The Vampire,' and depicts a lurid lady of the genus ... this one has the suction power of a 40 h.p. vacuum cleaner."[76]

Theda Bara

The *London Evening News* noted that "It was said that Mr. Kipling was last night in the theatre. If so, he must have been astonished to find how little he had to do with it all. Only the title was his and honestly it was the only part of the play worth having."[77]

A Fool There Was evidently came within Philip Burne-Jones's purview, and judging from the following clipping, he was not amused. "Did you notice on the hoardings," asked the *London Opinion*, "that striking reproduction of the Burne-Jones 'Vampire' picture? It was used to announce Mr. Sleath's production of *A Fool There Was*, but has had to be relinquished, some copyright trouble having arisen."[78]

The play generated an enormous amount of ancillary discussion in the press about the menace of "vampire wives" in the real world. Canon Horsley, the rector of St. Peter's, Walworth, told the *Daily Express* that vampire wives "are mainly West End products. The majority of our West End ladies have been brought up to a life of frivolity and pleasure." The only answer, in the rector's opinion, was for the Church and State to be given "autocratic power" over private lives.[79]

Meanwhile, in America, the burgeoning motion picture industry was moving to take its own brand of control power over American life. In the post-nickelodeon era, the vampire, as imagined by Burne-Jones and Kipling and transmogrified by Emerson Browne, would provide the template for an even larger cultural icon — the movie star. In particular, "the vamp."

The screen rights to *A Fool There Was* were acquired by William Fox, who cast a screen unknown, Theodosia Goodman, as the female monster. Goodman's professional name was Theda Bara, a studio-concocted anagram for "Arab Death." For five years, Theda Bara was filmdom's reigning vampire, her real identity hidden beneath mountains of publicity stunts and shameless misinformation. Both the press and the public were more than willing to go along with the gag. Bara held press conferences in darkened rooms amid billowing incense; she posed for publicity photographs rearing over the stripped-clean skeletons of her male victims. In a 1950 reminiscence, journalist Adela Rogers St. John recalled that "audiences

76. Review of *A Fool There Was*; *The Clarion* (London), March 31, 1910.

77. Undated clipping, *London Evening News*, probably March 22, 1911.

78. *London Opinion*, April 15, 1910.

79. "Canon's Condemnation," *Daily Express* (London), April 18, 1911.

80. Adela Rogers St. John, "The Hollywood Story," *Los Angeles Examiner*, October 15, 1950.

"KISS ME, MY FOOL"

Although Theda Bara was one of the silent screen's most indelible icons, appearing in over forty films, only four (including *A Fool There Was*) are known to survive. Most of the other films that shaped her reputation as film's leading "vamp" — *Sin* (1915), *Carmen* (1915), *The Serpent* (1916) and *Cleopatra* (1917) — have all been lost to neglect, nitrate decomposition, and studio vault fires. Still photographs, such as those shown here, remain the primary record of her work.

Bara's film career ended almost as abruptly as it had begun, when the vampire formula lost its appeal following World War I. In the 1920s, she tried the theatre, with devastating results. In his memoirs, Hollywood veteran Budd Schulberg recalled the premiere of her stage melodrama *The Blue Flame*. "Her opening on Broadway drew a sold-out audience laced with all the reigning celebrities," Schulberg wrote. "The first time she opened her mouth, they laughed. *This* was the irresistible vampire against whom the Church and an organized group of outraged wives had fulminated as a threat to the established order? This was the Serpent Woman? Cleopatra and Salome incarnate? At the first sound of her childlike piping, cruel laughter ended Theodosia Goodman's career."

Following her retirement from the stage and screen, Bara and her husband, director Charles Brabin, became Los Angeles society fixtures. She died of abdominal cancer in Hollywood on April 7, 1955.

RECOMMENDED READING

***Vamp: The Rise and Fall of Theda Bara* by Eve Golden (Emprise Publishing, Inc., 1996.)**

were torn between a fear of the Vampire and a wild desire to have some of her strange power rub off on them. The head of a New York department store pleaded with her, 'Please don't come in, Miss Bara. We'll send gowns to your hotel, but we can't stand any more of these riots.' Mobs of women had broken plate glass windows to grab a hat Theda Bara had touched, in hope that they, too, might be able to make men grovel."[80]

Even as the idea of "the vamp" swept the world in the first two decades of the twentieth century, the two personalities whose embattled relationship was the start of it all increasingly became anachronisms, further and further outside the mainstream of theatre and art. Phil maintained his wit, but he failed to see the significance of the emerging energies that were changing the visual arts forever. Where were the myths, the parables, the allegories? "A stroll through the Louvre or the National Gallery would convince the most superficial observer that it is to the 'anecdote' and the 'literary ideal' that the world owes its mightiest masterpieces," Phil told a 1912 assembly at the Author's Club, London, in an address called "The Cult of the Hideous and the Modern Iconoclasts."[81] The world as envisioned by the Post-Impressionists, the Futurists, the Cubists seemed "the untutored scribbles of the nursery enlivened by the delirium of the madhouse."[82] Phil worried that "There seems to have grown up, in certain quarters, a sort of reaction against Beauty — a fear of it, as though it were, in some way, an insidious form of weakness, and a set determination to hold no parley with the enemy."[83]

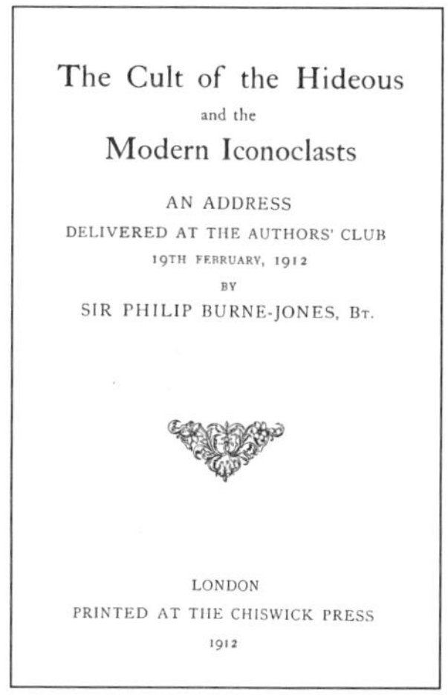
The Cult of the Hideous
and the
Modern Iconoclasts
AN ADDRESS
DELIVERED AT THE AUTHORS' CLUB
19TH FEBRUARY, 1912
BY
SIR PHILIP BURNE-JONES, BT.
LONDON
PRINTED AT THE CHISWICK PRESS
1912

Phil's screed against modern art.

> The House of Art has many mansions — and room to spare for all who can justify their presence by singleness of purpose and competent achievement.
>
> The Classic and Romantic and Realistic Schools — however wide apart in aim and spirit — live peacefully, in mutual respect, within those all-embracing walls. Only for hysterical and epileptic excresences, as exemplified by your "Cubistes" "Pointillistes" "Futuristes" etc., there should be no harbourage higher than the cellar...and may it be a dark one!
>
> How long the Spirit of Beauty will withdraw herself from us it would be idle to surmise. She seems, at present, to have gone into retreat, waiting, as it were, for this wave of universal Ugliness to pass. That she will one day return to us, I have no sort of doubt — but it will not be in our time.[84]

Two years earlier, the New Gallery in Regent Street, that stronghold of Pre-Raphaelite and Neo-Classical energies, had given up the ghost for good; its fixtures were auctioned off and the galleries were converted into a restaurant. The *New York Daily Tribune*, noting its passing, commented, "There are sentimentalists who cannot enjoy melodrama at the Lyceum Theatre, popular as the prices may be and excellent as the acting and

81. Sir Philip Burne Jones, Bt., "The Cult of the Hideous and the Modern Iconoclasts," text of address delivered at the Author's Club, London, February 19, 1912 (London: Chiswick Press, 1912), p. 8.

82. *Ibid.*, p. 7.

83. *Ibid.*

84. *Ibid.*, pp. 13-14.

85. "Art in London," *New York Daily Tribune* (supplement), March 6, 1910.

management undoubtably are. They cannot forget the high drama once enacted there, when Sir Henry Irving was the leader of the English stage and the glamour of a unique personality fascinated sympathetic audiences. People of equal sensibility are not likely to enjoy their luncheons and dinners at the new restaurant…"[85]

When, for reasons of financial necessity, Mrs. Patrick Campbell decided to publish her memoirs in 1921, she sounded out George Bernard Shaw on her rights, if any, to include Edward Burne-Jones's correspondence with her over the Vampire affair. (Shaw was simultaneously making her miserable with his refusal to allow her to publish any of his own platonic love letters to her, which she knew would make a sensational and lucrative public splash.) "I think," Shaw responded, "you must obtain Ph. B.J.'s permission to publish his father's letters. Otherwise he might possibly have the edition confiscated and ruin you and Constables." Shaw also suggested that she ask permission to reproduce the painting itself (if she felt "mischievously disposed.") "If he refuses, you can paraphrase the letters (copyright is only in B.J.'s exact words, not in the meaning) and explain that Ph. has made himself disagreeable. But it would be hard for him to justify a refusal…"[86]

Mrs. Pat as Eliza Doolittle in Shaw's Pygmalion *(1914) — a role especially written for her.*

Evidently, it wasn't hard at all. Mrs. Patrick Campbell's *My Life and Some Letters* was published in 1922 with no mention whatsoever of *The Vampire*, or Edward Burne-Jones's evident apologies on behalf of his son. Instead, Mrs. Pat included the following passage, as if to heal an old wound for once and for all:

> Philip Burne-Jones was among the many new acquaintances my success brought me.
>
> We soon became warm friends, and what unforgettable kindness he showed me. His talent for painting and drawing, his keen appreciation of the comedy of life, his interest in the theatre, and his genuine love of children made him a delightful companion. All friends of Phil will remember, as I do, the almost exaggerated devotion and service he offered them.[87]

In his final years, Phil drifted "miserably" among wealthy former patrons of his father's.[88] His mother died in 1920, deepening his chronic depression. A few years later, he wrote to family friend Sydney Cockerell, then the director of the Fitzwilliam Museum at Cambridge, asking for a copy of an extract from one his mother's letters to Cockerell, in which she had written of Phil as having been a decent son. He had mislaid an earlier copy, and missed the comfort it had provided.[89]

Phil shortly thereafter sent Cockerell an undated manuscript entitled *The Allotted Span,* that revolved around the Victorian picture books known as "Quarles' Emblems." At the end of the book were a series of illustrations in which the duration of human life was represented by a burning candle.

86. Alan Dent, ed., *Bernard Shaw and Mrs. Patrick Campbell: Their Correspondence* (New York: Alfred A. Knopf, 1952), pp. 256-257.

87. *My Life and Some Letters*, p. 113.

88. *Victorian Sisters*, p. 187.

89. Philip Burne-Jones to Sydney Cockerell, February 11, 1923, British Library Department of Manuscripts, Cockerell collection.

His own candle, he wrote, was now burning dismally.

On March 9, 1926, he wrote Cockerell from the Golf Hotel Beauvallon, where he was confined to his room, attended by a doctor and a nurse. He could hardly walk without the aid of an unbrella and a supporting arm. He considered himself a "deplorable and degraded object."

On April 8, 1926, he wrote that he was entering a rest-cure facility called St. Michaels, at Ascot.

In the following weeks, his handwriting deteriorated badly. On May 5, he complained of the noisy housekeepers and "horrible" food that was almost more difficult to eat than to eliminate.[90]

A card from Phil's sister Margaret reached Sydney Cockerell in June from Ascot, the first part of the message informing him that Phil had taken a turn for the worse. After improving temporarily, he developed a serious heart condition, barely recognizing his sister, who made plans to set up hospital care for him at home in Egerton Terrace. But before Margaret Mackail could finish writing the card, her brother had peacefully departed the world.[91]

Because Phil never married, the baronetcy died with him. He left dozens of paintings, countless letters and holiday cards, illuminated with whimsical drawings, and several books that he wrote or illustrated, or both. One of his most revealing works he left unsigned, in the pseudonymously-published children's book *Fables by Fal*.[92] "Fal" is unmistakably Phil, and his poem "The Wingless Frog" is rather like Kipling's "The Vampire," writ small and sad and wistful, with Phil in the title role of the doomed, lovesick amphibian:

THE WINGLESS FROG
On the brink of P'uddmore's Marshdom
Lived a green-backed,
good, young frog, With a stomach far more yellow
Than the yellow of a fog.
Ah! his eyes were bright and lustrous,
All lit up with love's bright ray,
For he saw a young lark soaring,
Singing sweet her soul's pure lay.
Oh! my reachless, unknown, loved one,
Would that I were with you skied!
Cupid, scornful, overhearing,
Straight poor Froggie Heav'nward shied.
Upward sent, first green, then yellow,
Saw the earth as up he sped,
Neared the lark, ecstatic moment,
Kissed ! but wingless, reached earth dead!

90. *Ibid.*, March 9, 1926.

91. Cockerell collection, letter from Margaret Mackail, June 1926.

92. *Fables by Fal* (London: Duckworth & Co., 1898)

Mrs. Patrick Campbell pushed on thirteen years longer than the man who had epitomized her as the vampire of a generation's dreams. By the time she reached Hollywood in the early 1930s, the movies had abandoned the notion of Theda Bara as vampire for the new, Bela Lugosi conception. But even films like Tod Browning's *Dracula* (1931) and *Mark of the Vampire* (1935) continued to feature images of pale, dark-eyed women in nightgown-shrouds — the very image of Mrs. Pat in the 1897 painting.

The vampire dowager in Hollywood: Mrs. Pat as she appeared in James Whale's One More River *(1934).*

"There's something about me that creates legends,"[93] she told a reporter for the *Los Angeles Times* in 1933, when she was both trying to start a new career in films and doing her best to sabotage her efforts with her legendary, acid wit. Withering lines like "I'm no Hedda Garbo" did little to endear her to the film colony, which was remarkably resistant to the affectations of a faded grande dame. According to John Gielgud, who acted with her in Ibsen's *Ghosts* near the end of her stage career, "why she had to be so ill-behaved, sometimes even common and rude, it was difficult to tell. A kind of demon seized her and she could not resist being unkind to people, making cheap jokes at their expenses."[94]

She also directed many of her jokes at herself. Cecil Beaton recalled photographing her near the end of her life. "Why didn't you appear forty years ago?" she asked him, producing faded photographs of herself as Melisande and Paula Tanqueray. "I was a beauty then. You could have taken wonderful pictures then. But what can you do with all these dewlaps? I look like an old paper bag which has burst."[95] Such observations, Beaton recalled, were followed by explosions of earthy, self-deprecating laughter.

In comparison with the old photographs (one of which was the very role in which Philip Burne-Jones and his friend decided she was "the greatest living actress"), Beaton wrote that "it was almost alarming to compare them with her now and to witness the transformation of that exquisite, mysterious-looking creature. Although Mrs. Pat still possessed a majestic quality and an innate grandeur, it must be conceded that she had become a fat old woman with few relics of past beauty." The woman whom Oscar Wilde had once introduced to Aubrey Beardsley for the purpose of a portrait had, in the intervening decades, become her own Picture of Dorian Gray.

> "Look at the glory of that neck, at that line of cheek, and look at me now, sagging like an old cow! Look at that jaw, and now look at me! All wind and water. Oh God! How can he be so unkind to do this to me? Why must we suffer this terrible change? Why must we all become ugly? I don't know how some women stand it. I don't know how they don't commit suicide!" And then her eyes would twinkle and her pursed lips quiver into a raucous Hogarthian laugh.[96]

Beaton's portraits survived, as did the portrait of Paula Tanqueray and Beardsley's curious, anorectic tribute to a theatrical legend. But *The Vampire*

93. Alma Whitaker, "Mrs. Pat Campbell, Age 70, Sets Filmdom on Ear Again," *Los Angeles Times*, December 10, 1933.

94. John Gielgud, John Miller, and John Powell, *Gielgud: An Actor and His Time* (New York: Clarkson N. Potter, Inc., 1979).

95. Cecil Beaton, *Photobiography* (Garden City, New York: Doubleday & Company, Inc., 1951), p. 114.

96. *Ibid.*

97. *Mrs. Pat,* p. 486

disappeared, as if into thin air. Attempts to locate it over the years have been fruitless. In the late 1940s, Columbia Pictures asked the New York Public Library's fine arts division to help locate the canvas, presumably in connection with the Theda Bara biography they were planning to (but never actually did) film. The library's researchers turned up a blank. Margot Peters, the most recent biographer of Mrs. Patrick Campbell, was told that the painting was at the Tate Gallery in London, but upon making inquiries she found that they had no record of the thing whatsoever.[97]

Did Phil finally sell the painting? Or did he destroy it, fed up at last with the grief and public ridicule it had caused him? It is perhaps appropriate that we may never know the truth. For whatever happened to the painting itself, *The Vampire* has survived, in its own way, through its almost incalculable impact on popular culture. The painting and poem became not only a major repository for turn-of-the century anti-feminist backlash and a rallying point for the temperance movement, as well as the prototype inspiration for the twentieth century's most powerful female icon — the movie star. Arguably, even the seed of Madonna was sown in Theda Bara — the "material girl" being an image known all too well to Rudyard Kipling and Philip Burne-Jones.

While Phil's painting may be "lost," it is missing only in the manner of a real vampire's reflection. The mirror may indeed seem empty as we gaze into the frame. But behind us, the phantom fills the room.

PART THREE

Vampires in the Twentieth Century

An early New England tombstone rubbing.

Luella Miller

MARY E. WILKINS-FREEMAN
(1903)

CLOSE TO THE village street stood the one-story house in which Luella Miller, who had an evil name in the village, had dwelt. She had been dead for years, yet there were those in the village who, in spite of the clearer light which comes on a vantage-point from a long-past danger, half believed in the tale which they had heard from their childhood. In their hearts, although they scarcely would have owned it, was a survival of the wild horror and frenzied fear of their ancestors who had dwelt in the same age with Luella Miller. Young people even would stare with a shudder at the old house as they passed, and children never played around it as was their wont around an untenanted building. Not a window in the old Miller house was broken: the panes reflected the morning sunlight in patches of emerald and blue, and the latch of the sagging front door was never lifted, although no bolt secured it. Since Luella Miller had been carried out of it, the house had had no tenant except one friendless old soul who had no choice between that and the far-off shelter of the open sky. This old woman, who had survived her kindred and friends, lived in the house one week, then one morning no smoke came out of the chimney, and a body of neighbours, a score strong, entered and found her dead in her bed. There were dark whispers as to the cause of her death, and there were those who testified to an expression of fear so exalted that it showed forth the state of the departing soul upon the dead face. The old woman had been hale and hearty when she entered the house, and in seven days she was dead; it seemed that she had fallen a victim to some uncanny power. The minister talked in the pulpit with covert severity against the sin of superstition; still the belief prevailed. Not a soul in the village but would have chosen the almshouse rather than that dwelling. No vagrant, if he heard the tale, would seek shelter beneath that old roof, unhallowed by nearly half a century of superstitious fear.

There was only one person in the village who had actually known Luella Miller. That person was a woman well over eighty, but a marvel of vitality and unextinct youth. Straight as an arrow, with the spring of one recently let loose from the bow of life, she moved about the streets, and she always went to church, rain or shine. She had never married, and had lived alone for years in a house across the road from Luella Miller's.

This woman had none of the garrulousness of age, but never in all her life had she ever held her tongue for any will save her own, and she never spared the truth when she essayed to present it. She it was who bore

ABOUT THE AUTHOR

Mary Eleanor Wilkins-Freeman (1852-1930) was a critically-acclaimed proponent of the "local color" movement in American fiction, using regional settings and dialects in her carefully observed character studies of quiet desperation in small-town New England.

As Mary Eleanor Wilkins, she published many collections of her fiction, most notably *A Humble Romance and Other Stories* (1887) and *A New England Nun and Other Stories* (1891) as well as the 1894 novel *Pembroke*. In 1902, the year "Luella Miller" was published, she married Charles M. Freeman, and lived in Metuchen, New Jersey for the rest of her life. Long neglected, "Luella Miller" is now frequently reprinted as a prime example of the psychological vampire story.

testimony to the life, evil, though possibly wittingly or designedly so, of Luella Miller, and to her personal appearance. When this old woman spoke — and she had the gift of description, although her thoughts were clothed in the rude vernacular of her native village — one could seem to see Luella Miller as she had really looked. According to this woman, Lydia Anderson by name, Luella Miller had been a beauty of a type rather unusual in New England. She had been a slight, pliant sort of creature, as ready with a strong yielding to fate and as unbreakable as a willow. She had glimmering lengths of straight, fair hair, which she wore softly looped round a long, lovely face. She had blue eyes full of soft pleading, little slender, clinging hands, and a wonderful grace of motion and attitude.

"Luella Miller used to sit in a way nobody else could if they sat up and studied a week of Sundays,"said Lydia Anderson, "and it was a sight to see her walk. If one of them willows over there on the edge of the brook could start up and get its roots free of the ground, and move off, it would go just the way Luella Miller used to. She had a green shot silk she used to wear, too, and a hat with green ribbon streamers, and a lace veil blowing across her face and out sideways, and a green ribbon flyin' from her waist. That was what she came out bride in when she married Erastus Miller. Her name before she was married was Hill. There was always a sight of "l's"in her name, married or single. Erastus Miller was good lookin', too, better lookin' than Luella. Sometimes I used to think that Luella wa'n't so handsome after all. Erastus just about worshiped her. I used to know him pretty well. He lived next door to me, and we went to school together. Folks used to say he was waitin' on me, but he wa'n't. I never thought he was except once or twice when he said things that some girls might have suspected meant somethin'. That was before Luella came here to teach the district school. It was funny how she came to get it, for folks said she hadn't any education, and that one of the big girls, Lottie Henderson, used to do all the teachin' for her, while she sat back and did embroidery work on a cambric pocket-handkerchief. Lottie Henderson was a real smart girl, a splendid scholar, and she just set her eyes by Luella, as all the girls did. Lottie would have made a real smart woman, but she died when Luella had been here about a year — just faded away and died: nobody knew what aided her. She dragged herself to that schoolhouse and helped Luella teach till the very last minute. The committee all knew how Luella didn't do much of the work herself, but they winked at it. It wa'n't long after Lottie died that Erastus married her. I always thought he hurried it up because she wa'n't fit to teach. One of the big boys used to help her after Lottie died, but he hadn't much government, and the school didn't do very well, and Luella might have had to give it up, for the committee couldn't have shut their eyes to things much longer. The boy that helped her was a real honest, innocent sort of fellow, and he was a good scholar, too. Folks said he overstudied, and that was the reason he was took crazy the year after Luella married, but I don't know. And I don't know what

NEW VAMPIRES FOR A NEW CENTURY

When Mary Eleanor Wilkins Freeman wrote "Luella Miller," vampire fiction stood at a distinct crossroads, looking back to the age of the Romantic Agony and its cast of traditional, blood-drinking seducers, while simultaneously exploring new, scientifically-rationalized forms of vampirism, as well as the vampire as a psychological metaphor for unhealthy human relationships.

Take, for example, the Australian-born writer Hume Nisbet's tried-and-true approach to the small-town female vampire.

From
THE VAMPIRE MAID
Hume Nisbet
(1900)

Ariadne was not beautiful in the strictly classical sense, her complexion being too lividly white and her expression too set to be quite pleasant at first sight; yet, as her mother had informed me, she had been ill for some time, which accounted for that defect. Her features were not regular, her hair and eyes seemed too black with that strangely white skin, and her lips too red for any except the decadent harmonies of an Aubrey Beardsley.

Yet my fantastic dreams of the preceding night, with my morning walk, had prepared me to be enthralled by this modern poster-like invalid.

made Erastus Miller go into consumption of the blood the year after he was married: consumption wa'n't in his family. He just grew weaker and weaker, and went almost bent double when he tried to wait on Luella, and he spoke feeble, like an old man. He worked terrible hard till the last trying to save up a little to leave Luella. I've seen him out in the worst storms on a wood-sled — he used to cut and sell wood — and he was hunched up on top lookin' more dead than alive. Once I couldn't stand it: I went over and helped him pitch some wood on the cart — I was always strong in my arms. I wouldn't stop for all he told me to, and I guess he was glad enough for the help. That was only a week before he died. He fell on the kitchen floor while he was gettin' breakfast. He always got the breakfast and let Luella lay abed. He did all the sweepin' and the washin' and the ironin' and most of the cookin'. He couldn't bear to have Luella lift her finger, and she let him do for her. She lived like a queen for all the work she did. She didn't even do her sewin'. She said it made her shoulder ache to sew, and poor Erastus's sister Lily used to do all her sewin'. She wa'n't able to, either; she was never strong in her back, but she did it beautifully. She had to, to suit Luella, she was so dreadful particular. I never saw anythin' like the fagottin' and hemstitchin' that Lily Miller did for Luella. She made all Luella's weddin' outfit, and that green silk dress, after Maria Babbit cut it. Maria she cut it for nothin', and she did a lot more cuttin' and fittin' for nothin' for Luella, too. Lily Miller went to live with Luella after Erastus died. She gave up her home, though she was real attached to it and wa'n't a mite afraid to stay alone. She rented it and she went to live with Luella right away after the funeral."

Then this old woman, Lydia Anderson, who remembered Luella Miller, would go on to relate the story of Lily Miller. It seemed that on the removal of Lily Miller to the house of her dead brother, to live with his widow, the village people first began to talk. This Lily Miller had been hardly past her first youth, and a most robust and blooming woman, rosy-cheeked, with curls of strong, black hair overshadowing round, candid temples and bright dark eyes. It was not six months after she had taken up her residence with her sister-in-law that her rosy colour faded and her pretty curves became wan hollows. White shadows began to show in the black rings of her hair, and the light died out of her eyes, her features sharpened, and there were pathetic lines at her mouth, which yet wore always an expression of utter sweetness and even happiness. She was devoted to her sister; there was no doubt that she loved her with her whole heart, and was perfectly content in her service. It was her sole anxiety lest she should die and leave her alone.

"The way Lily Miller used to talk about Luella was enough to make you mad and enough to make you cry," said Lydia Anderson. "I've been in there sometimes toward the last when she was too feeble to cook and carried her some blanc-mange or custard — somethin' I thought she might relish, and she'd thank me, and when I asked her how she was, say she felt

The loneliness of the moor, with the singing of the ocean, had gripped my heart with a wistful longing. The incongruity of those flaunting and evanescent poppy flowers, dashing their giddy tints in the face of that sober heath, touched me with a shiver as I approached the cottage, and lastly that weird embodiment of startling contrasts completed my subjugation.

She rose from her chair as her mother introduced her, and smiled while she held out her hand. I clasped that soft snow-flake, and as I did so a faint thrill tingled over me and rested on my heart, stopping for the moment its beating.

This contact seemed to have affected her as it did me; a clear flush, like a white flame, lighted up her face, so that it glowed as if an alabaster lamp had been lit; her black eyes became softer and more humid as our glances crossed, and her scarlet lips grew moist. She was a living woman now, while before she had seemed half a corpse.

She permitted her white slender hand to remain in mine longer than most people do at an introduction, and then she slowly withdrew it, still regarding me with steadfast eyes for a second or two afterwards.

Fathomless velvety eyes these were, yet before they were shifted from mine they appeared to have absorbed all my willpower and made me her adject slave. They looked like deep dark pools of clear water, yet they filled me with fire and deprived me of strength. I sank into my chair

better than she did yesterday, and asked me if I didn't think she looked better, dreadful pitiful, and say poor Luella had an awful time takin' care of her and doin' the work — she wa'n't strong enough to do anythin' — when all the time Luella wa'n't liftin' her finger and poor Lily didn't get any care except what the neighbours gave her, and Luella eat up everythin' that was carried in for Lily. I had it real straight that she did. Luella used to just sit and cry and do nothin'. She did act real fond of Lily, and she pined away considerable, too. There was those that thought she'd go into a decline herself. But after Lily died, her Aunt Abby Mixter came, and then Luella picked up and grew as fat and rosy as ever. But poor Aunt Abby begun to droop just the way Lily had, and I guess somebody wrote to her married daughter, Mrs. Sam Abbot, who lived in Barre, for she wrote her mother that she must leave right away and come and make her a visit, but Aunt Abby wouldn't go. I can see her now. She was a real good-lookin' woman, tall and large, with a big, square face and a high forehead that looked of itself kind of benevolent and good. She just tended out on Luella as if she had been a baby, and when her married daughter sent for her she wouldn't stir one inch. She'd always thought a lot of her daughter, too, but she said Luella needed her and her married daughter didn't. Her daughter kept writin' and writin', but it didn't do any good. Finally she came, and when she saw how bad her mother looked, she broke down and cried and all but went on her knees to have her come away. She spoke her mind out to Luella, too. She told her that she'd killed her husband and everybody that had anythin' to do with her, and she'd thank her to leave her mother alone. Luella went into hysterics, and Aunt Abby was so frightened that she called me after her daughter went. Mrs. Sam Abbot she went away fairly cryin' out loud in the buggy, the neighbours heard her, and well she might, for she never saw her mother again alive. I went in that night when Aunt Abby called for me, standin' in the door with her little green-checked shawl over her head. I can see her now. 'Do come over here, Miss Anderson,' she sung out, kind of gasping for breath. I didn't stop for anythin'. I put over as fast as I could, and when I got there, there was Luella laughin' and cryin' all together, and Aunt Abby trying to hush her, and all the time she herself was white as a sheet and shakin' so she could hardly stand. 'For the land sakes, Mrs. Mixter,' says I, 'you look worse than she does. You ain't fit to be up out of your bed.

" 'Oh, there ain't anythin' the matter with me,' says she. Then she went on talkin' to Luella. 'There, there, don't, don't, poor little lamb,' says she. 'Aunt Abby is here. She ain't goin' away and leave you. Don't, poor little lamb.'

" 'Do leave her with me, Mrs. Mixter, and you get back to bed,' says I, for Aunt Abby had been layin' down considerable lately, though somehow she contrived to do the work.

" 'I'm well enough,' says she. 'Don't you think she had better have the doctor, Miss Anderson?'

almost as languidly as I had risen from my bed that morning.

Yet I made a good breakfast, and although she hardly tasted anything, this strange girl rose much refreshed with a slight glow of colour on her cheeks, which improved her so greatly that she appeared younger and almost beautiful.

I had come here seeking solitude, but since I had seen Ariadne it seemed as if I had come for her only. She was not very lively; indeed, thinking back, I cannot recall any spontaneous remark of hers; she answered my questions by monosyllables and left me to lead in words; yet she was insinuating and appeared to lead my thoughts in her direction and speak to me with her eyes. I cannot describe her minutely, I only know that from the first glance and touch she gave me I was bewitched and could think of nothing else.

It was a rapid, distracting, and devouring infatuation that possessed me; all day long I followed her about like a dog, every night I dreamed of that white glowing face, those steadfast black eyes, those moist scarlet lips, and each morning I rose more languid than I had been the day before. Sometimes I dreamt that she was kissing me with those red lips, while I shivered at the contact of her silky black tresses as they covered my throat; sometimes that we were floating in the air, her arms about me and her long hair enveloping us both like an inky cloud, while I lie supine and helpless.

" 'The doctor,' says I, 'I think *you* had better have the doctor. I think you need him much worse than some folks I could mention.' And I looked right straight at Luella Miller laughin' and cryin' and goin' on as if she was the centre of all creation. All the time she was actin' so — seemed as if she was too sick to sense anythin' — she was keepin' a sharp lookout as to how we took it out of the corner of one eye. I see her. You could never cheat me about Luella Miller. Finally I got real mad and I run home and I got a bottle of valerian I had, and I poured some boilin' hot water on a handful of catnip, and I mixed up that catnip tea with most half a wineglass of valerian, and I went with it over to Luella's. I marched right up to Luella, a-holdin' out of that cup, all smokin'. 'Now,' says I, 'Luella Miller, *you swaller this!'*

" 'What is — what is it, oh, what is it?' she sort of screeches out. Then she goes off a-laughin' enough to kill.

" 'Poor lamb, poor little lamb,' says Aunt Abby, standin' over her, all kind of tottery, and tryin' to bathe her head with camphor.

" *'You swaller this right down,'* says I. And I didn't waste any ceremony. I just took hold of Luella Miller's chin and I tipped her head back, and I caught her mouth open with laughin', and I clapped that cup to her lips, and I fairly hollered at her: 'Swaller, swaller, swaller!' and she gulped it right down. She had to, and I guess it did her good. Anyhow, she stopped cryin' and laughin' and let me put her to bed, and she went to sleep like a baby inside of half an hour. That was more than poor Aunt Abby did. She lay awake all that night and I stayed with her, though she tried not to have me; said she wa'n't sick enough for watchers. But I stayed, and I made some good cornmeal gruel and I fed her a teaspoon every little while all night long. It seemed to me as if she was jest dyin' from bein' all wore out. In the mornin' as soon as it was light I run over to the Bisbees and sent Johnny Bisbee for the doctor. I told him to tell the doctor to hurry, and he come pretty quick. Poor Aunt Abby didn't seem to know much of anythin' when he got there. You couldn't hardly tell she breathed, she was so used up. When the doctor had gone, Luella came into the room lookin' like a baby in her ruffled nightgown. I can see her now. Her eyes were as blue and her face all pink and white like a blossom, and she looked at Aunt Abby in the bed sort of innocent and surprised. 'Why,' says she, 'Aunt Abby ain't got up yet?'

" 'No, she ain't,' says I, pretty short.

" 'I thought I didn't smell the coffee,' says Luella.

" 'Coffee,' says I. 'I guess if you have coffee this mornin' you'll make it yourself.'

" 'I never made the coffee in all my life,' says she, dreadful astonished. 'Erastus always made the coffee as long as he lived, and then Lily she made it, and then Aunt Abby made it. I don't believe I can make the coffee, Miss Anderson.'

" 'You can make it or go without, jest as you please,' says I.

One night, about a couple of weeks after my coming to the cottage, I had returned after a delicious moonlight walk with Ariadne. The night was warm and the moon at the full, therefore I left my bedroom window open to let in what little air there was.

I was more than usually fagged out, so that I had only strength enough to remove my boots and coat before I flung myself wearily on the coverlet and fell almost instantly asleep without tasting the nightcap draught that was constantly placed on the table, and which I had always drained thirstily.

I had a ghastly dream this night. I thought I saw a monster bat, with the face and tresses of Ariadne, fly into the open window and fasten its white teeth and scarlet lips on my arm. I tried to beat the horror away, but could not, for I seemed chained down and thralled also with drowsy delight as the beast sucked my blood with a gruesome rapture.

I looked out dreamily and saw a line of dead bodies of young men lying on the floor, each with a red mark on their arms, on the same part where the vampire was then sucking me, and I remembered having seen and wondered at such a mark on my own forearm for the past fortnight. In a flash I understood the reason for my strange weakness, and at the same moment a sudden prick of pain roused me from my dreamy pleasure.

The vampire in her eagerness had bitten a little too deeply that night, unaware that I had not

" 'Ain't Aunt Abby goin' to get up?' says she.

" 'I guess she won't get up,' says I, 'sick as she is.' I was gettin' madder and madder. There was somethin' about that little pink-and-white thing standin' there and talkin' about coffee, when she had killed so many better folks than she was, and had jest killed another, that made me feel 'most as if I wished somebody would up and kill her before she had a chance to do any more harm.

" 'Is Aunt Abby sick?' says Luella, as if she was sort of aggrieved and injured.

" 'Yes,' says I, 'she's sick, and she's goin' to die, and then you'll be left alone, and you'll have to do for yourself and wait on yourself, or do without things.' I don't know but I was sort of hard, but it was the truth, and if I was any harder than Luella Miller had been I'll give up. I ain't never been sorry that I said it. Well, Luella, she up and had hysterics again at that, and I jest let her have 'em. All I did was to bundle her into the room on the other side of the entry where Aunt Abby couldn't hear her, if she wa'n't past it — I don't know but she was — and set her down hard in a chair and told her not to come back into the other room, and she minded. She had her hysterics in there till she got tired. When she found out that nobody was comin' to coddle her and do for her she stopped. At least I suppose she did. I had all I could do with poor Aunt Abby tryin' to keep the breath of life in her. The doctor had told me that she was dreadful low, and give me some very strong medicine to give to her in drops real often, and told me real particular about the nourishment. Well, I did as he told me real faithful till she wa'n't able to swaller any longer. Then I had her daughter sent for. I had begun to realize that she wouldn't last any time at all. I hadn't realized it before, though I spoke to Luella the way I did. The doctor he came, and Mrs. Sam Abbot, but when she got there it was too late; her mother was dead. Aunt Abby's daughter just give one look at her mother layin' there, then she turned sort of sharp and sudden and looked at me.

" 'Where is she?' says she, and I knew she meant Luella.

" 'She's out in the kitchen,' says I. 'She's too nervous to see folks die. She's afraid it will make her sick.'

"The Doctor he speaks up then. He was a young man. Old Doctor Park had died the year before, and this was a young fellow just out of college. 'Mrs. Miller is not strong,' says he, kind of severe, 'and she is quite right in not agitating herself.'

" 'You are another, young man; she's got her pretty claw on you,' thinks I, but I didn't say anythin' to him. I just said over to Mrs. Sam Abbot that Luella was in the kitchen, and Mrs. Sam Abbot she went out there, and I went, too, and I never heard anythin' like the way she talked to Luella Miller. I felt pretty hard to Luella myself, but this was more than I ever would have dared to say. Luella she was too scared to go into hysterics. She jest flopped. She seemed to jest shrink away to nothin' in that kitchen

tasted the drugged draught. As I woke I saw her fully revealed by the midnight moon, with her black tresses flowing loosely, and with her red lips glued to my arm. With a shriek of horror I dashed her backwards, getting one last glimpse of her savage eyes, glowing white face and blood-stained red lips; then I rushed out to the night, moved on by my fear and hatred, nor did I pause in my mad flight until I had left miles between me and that accursed Cottage on the Moor.

SIR ARTHUR CONAN DOYLE AND PSYCHIC VAMPIRISM

In* Dracula*, Bram Stoker had quasi-medicalized vampirism in order to make the tale more plausible to contemporary readers. In his short fiction of the same period, Sir Arthur Conan Doyle began to explore the relationship between vampirism and the pseudoscientific disciplines of animal magnetism, mesmerism, and thought projection. Doyle himself was a fierce believer in these movements, despite his association with that paradigm of rational skepticism, Sherlock Holmes.

In "John Barrington Cowles" (1891) Doyle presented the story of a harrowing love affair that is clearly vampiric, but in which the precise nature of the vampirism remains psychologically ambiguous.

chair, with Mrs. Sam Abbot standin' over her and talkin' and tellin' her the truth. I guess the truth was most too much for her and no mistake, because Luella presently actually did faint away, and there wa'n't any sham about it, the way I always suspected there was about them hysterics. She fainted dead away and we had to lay her flat on the floor, and the Doctor he came runnin' out and he said somethin' about a weak heart dreadful fierce to Mrs. Sam Abbot, but she wa'n't a mite scared. She faced him jest as white as even Luella was layin' there lookin' like death and the Doctor feelin' of her pulse.

" 'Weak heart,' says she, 'weak heart; weak fiddlesticks! There ain't nothin' weak about that woman. She's got strength enough to hang onto other folks till she kills 'em. Weak? It was my poor mother that was weak: this woman killed her as sure as if she had taken a knife to her.'

"But the Doctor he didn't pay much attention. He was bendin' over Luella layin' there with her yellow hair all streamin' and her pretty pink-and-white face all pale, and her blue eyes like stars gone out, and he was holdin' onto her hand and smoothin' her forehead, and tellin' me to get the brandy in Aunt Abby's room, and I was sure as I wanted to be that Luella had got somebody else to hang onto, now Aunt Abby was gone, and I thought of poor Erastus Miller, and I sort of pitied the poor young Doctor, led away by a pretty face, and I made up my mind I'd see what I could do.

"I waited till Aunt Abby had been dead and buried about a month, and the Doctor was goin' to see Luella steady and folks were beginnin' to talk; then one evenin', when I knew the Doctor had been called out of town and wouldn't be round, I went over to Luella's. I found her all dressed up in a blue muslin with white polka dots on it, and her hair curled jest as pretty, and there wa'n't a young girl in the place could compare with her. There was somethin' about Luella Miller seemed to draw the heart right out of you, but she didn't draw it out of me. She was settin' rocking in the chair by her sittin'-room window, and Maria Brown had gone home. Maria Brown had been in to help her, or rather to do the work, for Luella wa'n't helped when she didn't do anythin'. Maria Brown was real capable and she didn't have any ties; she wa'n't married, and lived alone, so she'd offered. I couldn't see why she should do the work any more than Luella; she wa'n't any too strong; but she seemed to think she could and Luella seemed to think so, too, so she went over and did all the work — washed, and ironed, and baked, while Luella sat and rocked. Maria didn't live long afterward. She began to fade away just the same fashion the others had. Well, she was warned, but she acted real mad when folks said anythin': said Luella was a poor, abused woman, too delicate to help herself, and they'd ought to be ashamed, and if she died helpin' them that couldn't help themselves she would — and she did.

" 'I s'pose Maria has gone home,' says I to Luella, when I had gone in and sat down opposite her.

From
JOHN BARRINGTON COWLES
Sir Arthur Conan Doyle
(1891)

As the door flew open I knew in a moment that my worst apprehensions had been fulfilled. Barrington Cowles was leaning against the railings outside with his face sunk upon his breast, and his whole attitude expressive of the most intense despondency. As he passed in he gave a stagger, and would have fallen had I not thrown my left arm around him. Supporting him with this, and holding the lamp in my other hand, I led him slowly upstairs into our sitting-room. He sank down upon the sofa without a word. Now that I could get a good view of him, I was horrified to see the change which had come over him. His face was deadly pale, and his very lips were bloodless. His cheeks and forehead were clammy, his eyes glazed, and his whole expression altered. He looked like a man who had gone through some terrible ordeal, and was thoroughly unnerved.

"My dear fellow, what is the matter?" I asked, breaking the silence. "Nothing amiss, I trust? Are you unwell?"

"Brandy!" he gasped. "Give me some brandy!"

I took out the decanter, and was about to help him, when he snatched it from me with a trembling hand, and poured out nearly half a tumbler of the spirit. He was usually a most abstemious man, but he took this off at a gulp without adding any water to it.

" 'Yes, Maria went half an hour ago, after she had got supper and washed the dishes,' says Luella, in her pretty way.

" 'I suppose she has got a lot of work to do in her own house tonight,' says I, kind of bitter, but that was all thrown away on Luella Miller. It seemed to her right that other folks that wa'n't any better able than she was herself should wait on her, and she couldn't get it through her head that anybody should think it wa'n't right.

" 'Yes,' says Luella, real sweet and pretty, 'yes, she said she had to do her washin' to-night. She has let it go for a fortnight along of comin' over here.'

" 'Why don't she stay home and do her washin' instead of comin' over here and doin' *your* work, when you are just as well able, and enough sight more so, than she is to do it?' says I.

"Then Luella she looked at me like a baby who has a rattle shook at it. She sort of laughed as innocent as you please. 'Oh, I can't do the work myself, Miss Anderson,' says she. 'I never did. Maria *has* to do it.'

"Then I spoke out: 'Has to do it!' says I. 'Has to do it!' She don't have to do it, either. Maria Brown has her own home and enough to live on. She ain't beholden to you to come over here and slave for you and kill herself.'

"Luella she jest set and stared at me for all the world like a doll-baby that was so abused that it was comin' to life.

" 'Yes,' says I, 'she's killin' herself. She's goin' to die just the way Erastus did, and Lily, and your Aunt Abby. You're killin' her jest as you did them. I don't know what there is about you, but you seem to bring a curse,' says I. 'You kill everybody that is fool enough to care anythin' about you and do for you.'

"She stared at me and she was pretty pale.

" 'And Maria ain't the only one you're goin' to kill,' says I. 'You're goin' to kill Doctor Malcom before you're done with him.'

"Then a red colour came flamin' all over her face. 'I ain't goin' to kill him, either,' says she, and she begun to cry.

" 'Yes, you *be*!' says I. Then I spoke as I had never spoke before. You see, I felt it on account of Erastus. I told her that she hadn't any business to think of another man after she'd been married to one that had died for her: that she was a dreadful woman; and she was, that's true enough, but sometimes I have wondered lately if she knew it — if she wa'n't like a baby with scissors in its hand cuttin' everybody without knowin' what it was doin'.

"Luella she kept gettin' paler and paler, and she never took her eyes off my face. There was somethin' awful about the way she looked at me and never spoke one word. After a while I quit talkin' and I went home. I watched that night, but her lamp went out before nine o'clock, and when Doctor Malcom came drivin' past and sort of slowed up he see there wa'n't any light and he drove along. I saw her sort of shy out of meetin' the next

It seemed to do him good, for the colour began to come back to his face, and he leaned upon his elbow.

"My engagement is off, Bob," he said, trying to speak calmly, but with a tremor in his voice which he could not conceal. "It is all over."

"Cheer up!" I answered, trying to encourage him.

"Don't get down on your luck. How was it? What was it all about?"

"About?" he groaned, covering his face with his hands. "If I did tell you, Bob, you would not believe it. It is too dreadful — too horrible — unutterably awful and incredible! O Kate, Kate!" and he rocked himself to and fro in his grief; "I pictured you an angel and I find you a ——"

"A what?" I asked, for he had paused.

He looked at me with a vacant stare, and then suddenly burst out, waving his arms: "A fiend!" he cried. "A ghoul from the pit! A vampire soul behind a lovely face! Now, God forgive me!" he went on in a lower tone, turning his face to the wall; "I have said more than I should. I have loved her too much to speak of her as she is. I love her too much now."

He lay still for some time, and I had hoped that the brandy had had the effect of sending him to sleep, when he suddenly turned his face towards me.

Sunday, too, so he shouldn't go home with her, and I begun to think mebbe she did have some conscience after all. It was only a week after that that Maria Brown died — sort of sudden at the last, though everybody had seen it was comin'. Well, then there was a good deal of feelin' and pretty dark whispers. Folks said the days of witchcraft had come again, and they were pretty shy of Luella. She acted sort of offish to the Doctor and he didn't go there, and there wa'n't anybody to do anythin' for her. I don't know how she *did* get along. I wouldn't go in there and offer to help her — not because I was afraid of dyin' like the rest, but I thought she was just as well able to do her own work as I was to do it for her, and I thought it was about time that she did it and stopped killin' other folks. But it wa'n't very long before folks began to say that Luella herself was goin' into a decline jest the way her husband, and Lily, and Aunt Abby and the others had, and I saw myself that she looked pretty bad. I used to see her goin' past from the store with a bundle as if she could hardly crawl, but I remembered how Erastus used to wait and 'tend when he couldn't hardly put one foot before the other, and I didn't go out to help her.

"But at last one afternoon I saw the Doctor come drivin' up like mad with his medicine chest, and Mrs. Babbit came in after supper and said that Luella was real sick.

" 'I'd offer to go in and nurse her,' says she, 'but I've got my children to consider, and mebbe it ain't true what they say, but it's queer how many folks that have done for her have died.'

"I didn't say anythin', but I considered how she had been Erastus's wife and how he had set his eyes by her, and I made up my mind to go in the next mornin', unless she was better, and see what I could do; but the next mornin' I see her at the window, and pretty soon she came steppin' out as spry as you please, and a little while afterward Mrs. Babbit came in and told me that the Doctor had got a girl from out of town, a Sarah Jones, to come there, and she said she was pretty sure that the Doctor was goin' to marry Luella.

"I saw him kiss her in the door that night myself, and I knew it was true. The woman came that afternoon, and the way she flew around was a caution. I don't believe Luella had swept since Maria died. She swept and dusted, and washed and ironed; wet clothes and dusters and carpets were flyin' over there all day, and every time Luella set her foot out when the Doctor wa'n't there there was that Sarah Jones helpin' of her up and down the steps, as if she hadn't learned to walk.

"Well, everybody knew that Luella and the Doctor were goin' to be married, but it wa'n't long before they began to talk about his lookin' so poorly, jest as they had about the others; and they talked about Sarah Jones, too.

"Well, the Doctor did die, and he wanted to be married first, so as to leave what little he had to Luella, but he died before the minister could get there, and Sarah Jones died a week afterward.

"Did you ever read of wehr-wolves?" he asked.

I answered that I had.

"There is a story," he said thoughtfully, "in one of Marryat's books, about a beautiful woman who took the form of a wolf at night and devoured her own children. I wonder what put that idea into Marryat's head?"

He pondered for some minutes, and then he cried out for some more brandy. There was a small bottle of laudanum upon the table, and I managed, by insisting upon helping him myself, to mix about half a drachm with the spirits. He drank it off, and sank his head once more upon the pillow. "Anything better than that," he groaned. "Death is better than that. Crime and cruelty; cruelty and crime. Anything is better than that," and so on, with the monotonous refrain, until at last the words became indistinct, his eyelids closed over his weary eyes, and he sank into a profound slumber. I carried him into his bedroom without arousing him; and making a couch for myself out of the chairs, I remained by his side all night.

•

In "The Parasite" (1894), Doyle expanded on the theme. A professor named Austin Gilroy consents to be mesmerized by Miss Helen Penclosa, a crippled woman whose physical frailty masks a terrible need to control the minds and souls of others. Once under her hypnotic

"Well, that wound up everything for Luella Miller. Not another soul in the whole town would lift a finger for her. There got to be a sort of panic. Then she began to droop in good earnest. She used to have to go to the store herself, for Mrs. Babbit was afraid to let Tommy go for her, and I've seen her goin' past and stoppin' every two or three steps to rest. Well, I stood it as long as I could, but one day I see her comin' with her arms full and stoppin' to lean against the Babbit fence, and I run out and took her bundles and carried them to her house. Then I went home and never spoke one word to her though she called after me dreadful kind of pitiful. Well, that night I was taken sick with a chill, and I was sick as I wanted to be for two weeks. Mrs. Babbit had seen me run out to help Luella and she came in and told me I was goin' to die on account of it. I didn't know whether I was or not, but I considered I had done right by Erastus's wife.

"That last two weeks Luella she had a dreadful hard time, I guess. She was pretty sick, and as near as I could make out nobody dared go near her. I don't know as she was really needin' anythin' very much, for there was enough to eat in her house and it was warm weather, and she made out to cook a little flour gruel every day, I know, but I guess she had a hard time, she that had been so petted and done for all her life.

"When I got so I could go out, I went over there one morning. Mrs. Babbit had just come in to say she hadn't seen any smoke and she didn't know but it was somebody's duty to go in, but she couldn't help thinkin' of her children, and I got right up, though I hadn't been out of the house for two weeks, and I went in there, and Luella she was layin' on the bed, and she was dyin'.

"She lasted all that day and into the night. But I sat there after the new doctor had gone away. Nobody else dared to go there. It was about midnight that I left her for a minute to run home and get some medicine I had been takin', for I begun to feel rather bad.

"It was a full moon that night, and just as I started out of my door to cross the street back to Luella's, I stopped short, for I saw something."

Lydia Anderson at this juncture always said with a certain defiance that she did not expect to be believed, and then proceeded in a hushed voice:

"I saw what I saw, and I know I saw it, and I will swear on my death bed that I saw it. I saw Luella Miller and Erastus Miller, and Lily, and Aunt Abby, and Maria, and the Doctor, and Sarah, all goin' out of her door, and all but Luella shone white in the moonlight, and they were all helpin' her along till she seemed to fairly fly in the midst of them. Then it all disappeared. I stood a minute with my heart poundin', then I went over there. I thought of goin' for Mrs. Babbit, but I thought she'd be afraid. So I went alone, though I knew what had happened. Luella was layin' real peaceful, dead on her bed."

This was the story that the old woman, Lydia Anderson, told, but the sequel was told by the people who survived her, and this is the tale which has become folklore in the village.

influence, Gilroy's brain is no longer under his control; Miss Penclosa can enter at will, rather like a crafty computer hacker of the present day, or like a traditional vampire that can enter a house as it pleases once a formal invitation has been made.

"Let me try to reason it out!" wrote the fictional Gilroy. "This woman, by her own explanation, can dominate my nervous organism. She can project herself into my body and take command of it. She has a parasite soul; yes, she is a parasite, a monstrous parasite. She creeps into my frame as the hermit crab does into the whelk's shell. I am powerless What can I do? I am dealing with forces of which I know nothing. And I can tell no one of my trouble. They would set me down as a madman."

Miss Penclosa, who uses mind-parasitism to control men she cannot otherwise attract due to her infirmity, sabotages Gilroy's lectures and ruins his academic career, and unsuccessfully sends him to his sleeping fiancée on a jealous mission of disfigurement; he awakes from his trance at the last moment to find himself standing over the girl, holding a bottle of acid. In the end, his own will proves the equal of Miss Penclosa's in this tightly-written, if misogynistic little thriller.

At the turn of the twentieth century, one thing was clear: vampires were here to stay, and blood was no longer the only thing they necessarily wanted.

Lydia Anderson died when she was eighty-seven. She had continued wonderfully hale and hearty for one of her years until about two weeks before her death.

One bright moonlight evening she was sitting beside a window in her parlour when she made a sudden exclamation, and was out of the house and across the street before the neighbour who was taking care of her could stop her. She followed as fast as possible and found Lydia Anderson stretched on the ground before the door of Luella Miller's deserted house, and she was quite dead.

The next night there was a red gleam of fire athwart the moonlight and the old house of Luella Miller was burned to the ground. Nothing is now left of it except a few old cellar stones and a lilac bush, and in summer a helpless trail of morning glories among the weeds, which might be considered emblematic of Luella herself.

A vampire's rest is disturbed. From Varney the Vampire *(1847).*

Count Magnus

M.R. JAMES
(1904)

BY WHAT MEANS the papers out of which I have made a connected story came into my hands is the last point which the reader will learn from these pages. But it is necessary to prefix to my extracts from them a statement of the form in which I possess them.

They consist, then, partly of a series of collections for a book of travels, such a volume as was a common product of the forties and fifties. Horace Marryat's *Journal of a Residence in Jutland and the Danish Isles* is a fair specimen of the class to which I allude. These books usually treated of some unfamiliar district on the Continent. They were illustrated with woodcuts or steel plates. They gave details of hotel accommodation, and of means of communication, such as we now expect to find in any well-regulated guidebook, and they dealt largely in reported conversations with intelligent foreigners, racy innkeepers and garrulous peasants. In a word, they were chatty.

Begun with the idea of furnishing material for such a book, my papers as they progressed assumed the character of a record of one single personal experience, and this record was continued up to the very eve, almost, of its termination.

The writer was a Mr. Wraxall. For my knowledge of him I have to depend entirely on the evidence his writings afford, and from these I deduce that he was a man past middle age, possessed of some private means, and very much alone in the world. He had, it seems, no settled abode in England, but was a denizen of hotels and boarding-houses. It is probable that he entertained the idea of settling down at some future time which never came; and I think it also likely that the Pantechnicon fire in the early seventies must have destroyed a great deal that would have thrown light on his antecedents, for he refers once or twice to property of his that was warehoused that establishment.

It is further apparent that Mr. Wraxall had published a book, and that it treated of a holiday he had once taken in Brittany. More than this I cannot say about his work, because a diligent search in bibliographical works has convinced me that it must have appeared either anonymously or under a pseudonym.

As to his character, it is not difficult to form some superficial opinion. He must have been an intelligent and cultivated man. It seems that he was near being a Fellow of his college at Oxford — Brasenose, as I judge from the Calendar. His besetting fault was pretty clearly that of

ABOUT THE AUTHOR

Montague Rhodes James (1862-1936) became fascinated with antiquarian books at an early age, and spent the rest of his life steeped in bible scholarship and medievalism. He was educated at Eton and King's College, Cambridge, where he was named provost in 1905. A brilliant scholar and bibliographer, his deep interest in the past eventually encompassed ghosts, and he is best known today for his elegant tales of the supernatural, which he was fond of reading aloud to his friends at Christmastide. No less an authority than H.P. Lovecraft praised James's "almost diabolic power of calling horror by gentle steps from the midst of prosaic daily life." "Count Magnus" originally appeared in his first collection, *Ghost Stories of an Antiquary* (1904); it was followed by *More Ghost Stories of an Antiquary* (1911), *A Thin Ghost and Others* (1919), and *A Warning to the Curious* (1925).

over-inquisitiveness, possibly a good fault in a traveller, certainly a fault for which this traveller paid dearly enough in the end.

On what proved to be his last expedition, he was plotting another book. Scandinavia, a region not widely known to Englishmen forty years ago, had struck him as an interesting field. He must have lighted on some old books of Swedish history, or memoirs, and the idea had struck him that there was room for a book descriptive of travel in Sweden, interspersed with episodes from the history of some of the great Swedish families. He procured letters of introduction, therefore, to some person of quality in Sweden, and set out thither in the early summer of 1863.

M.R. James's devilish story "Casting the Runes" was the inspiration for Jacques Tourneur's classic horror film Curse of the Demon *(1958).*

Of his travels in the North there is no need to speak, nor of his residence of some weeks in Stockholm. I need only mention that some *savant* resident there put him on the track of an important collection of family papers belonging to the proprietor of an ancient manor-house in Vestergothland, and obtained from him permission to examine them.

The manor-house, or *herrgård*, in question is to be called Råbäck (pronounced something like Roebeck), though that is not its name. It is one of the best buildings of its kind in the country, and the picture of it in Dahlenberg's *Suecia antiqua et moderna*, engraved in 1694, shows it very much as the tourist may see it today. It was built soon after 1600, and is, roughly speaking, very much like an English house of that period in respect of material — red-brick with stone facings — and style. The man who built it was a scion of the great house of De la Gardie, and his descendants possess it still. De la Gardie is the name by which I will designate them when mention of them becomes necessary.

They received Mr. Wraxall with great kindness and courtesy and pressed him to stay in the house as long as his researches lasted. But, preferring to be independent, and mistrusting his powers of conversing in Swedish, he settled himself in the village inn, which turned out quite sufficiently comfortable, at any rate during the summer months. This arrangement would entail a short walk daily to and from the manor-house of something under a mile. The house itself stood in a park, and was protected — we should say grown up — with large old timber. Near it you found the walled garden, and then entered a close wood fringing one of the small lakes with which the whole country is pitted. Then came the wall of the demesne, and you climbed a steep knoll — a knob of rock lightly covered with soil — and on the top of this stood the church, fenced in with tall dark trees. It was a curious building to English eyes. The nave and aisles were low, and filled with pews and galleries. In the western gallery stood the handsome old organ, gaily painted, and with silver pipes. The ceiling was flat, and had been adorned by a seventeenth-century artist with a strange and hideous "Last Judgment," full of lurid flames, falling cities, burning ships, crying souls, and brown and smiling demons. Handsome brass coronae hung from the roof; the pulpit was like a doll's-house, covered with little painted wooden cherubs and saints; a stand with three hour-glasses was hinged to the preacher's desk. Such sights as these may be seen in many a church in Sweden now, but what distinguished this one was an addition to the original building. At the eastern end of the north aisle the builder of the manor-house had erected a mausoleum for himself and his family. It was a largish eight-sided building, lighted by a series of oval windows, and had a domed roof, topped by a kind of pumpkin-shaped object rising into a spire, a form in which Swedish architects greatly delighted. The roof was of copper externally, and was painted black, while the walls in common with those of the church, were staringly white. To this mausoleum there was no access from the church. It had a portal and steps of its own on the northern side.

Past the churchyard the path to the village goes, and not more than three or four minutes bring you to the inn door.

On the first day of his stay at Råbäck, Mr. Wraxall found the church door open, and made those notes of the interior which I have epitomized. Into the mausoleum, however, he could not make his way. He could by looking through the keyhole just descry that there were fine marble effigies and sarcophagi of copper, and a wealth of armorial ornament, which made him very anxious to spend some time in investigation.

The papers he had come to examine at the manor-house proved to be of just the kind he wanted for his book. There were family correspondence, journals, and account-books of the earliest owners of the estate, very carefully kept and clearly written, full of amusing and picturesque detail. The first De la Gardie appeared in them as a strong and capable man. Shortly after the building of the mansion there had been a period of distress in the district, and the peasants had risen and attacked several châteaux and done some damage. The owner of Råbäck took a leading part in suppressing the trouble, and there was reference to executions of ringleaders and severe punishments inflicted with no sparing hand.

The portrait of this Magnus de la Gardie was one of the best in the house, and Mr. Wraxall studied it with no little interest after his day's work. He gives no detailed description of it, but I gather that the face impressed him rather by its power than by its beauty or goodness; in fact, he writes that Count Magnus was an almost phenomenally ugly man.

On this day Mr. Wraxall took his supper with the family and walked back in the late but still bright evening.

"I must remember," he writes, "to ask the sexton if he can let me into the mausoleum at the church. He evidently has access to it himself, for I saw him tonight standing on the steps, and as I thought, locking or unlocking the door."

I find that early on the following day Mr. Wraxall had some conversation with his landlord. His setting it down at such length as he does surprised me at first; but I soon realized that the papers I was reading were, at least in their beginning, the materials for the book he was meditating, and that it was to have been one of those quasi-journalistic

COFFINS, CRYPTS, AND CATACOMBS

The coffin/crypt/catacomb is the vampire's traditional daytime retreat, but aside from its obvious utilitarian aspects, the coffin also has a number of symbolic meanings. Boxes and other tight enclosures suggest secrets and the concomitant promise of revelations (this concealing/revealing characteristic is also true of cloaks and capes). As an enclosure of the human form, the coffin is additionally a womblike symbol of the mysterious transitions and interdependencies between life and death.

F.G. Loring's story "The Tomb of Sarah," excerpted on the following pages, is from the same period as "Count Magnus," and shares M.R. James's fascination with old chapels and crypts. It also distinctly anticipates a later James story, "An Episode of Cathedral History" (1919) in which a modern church renovation releases an undead creature from its tomb. However, where James's revenants are atmospherically evoked, Loring opts for a more explicit depiction of a traditional, fang-champing blood countess.

productions which admit of the introduction of an admixture of conversational matter.

His object, he says, was to find out whether any traditions of Count Magnus de la Gardie lingered on in the scenes of that gentleman's activity, and whether the popular estimate of him were favorable or not. He found that the Count was decidedly not a favorite. If his tenants came late to work on the days which they owed to him as Lord of the Manor, they were set on the wooden horse, or flogged and branded in the manor-house. One or two cases there were of men who had occupied lands which encroached on the lord's domain, and whose houses had been mysteriously burnt on a winter's night, with the whole family inside. But what seemed to dwell on the innkeeper's mind most — for he returned to the subject more than once — was that the Count had been on the Black Pilgrimage, and had brought something or someone back with him.

You will naturally inquire, as Mr. Wraxall did, what the Black Pilgrimage may have been. But your curiosity on the point must remain unsatisfied for the time being, just as his did. The landlord was evidently unwilling to give a full answer, or indeed any answer, on the point, and, being called out for a moment, trotted off with obvious alacrity, only putting his head in at the door a few minutes afterwards to say that he was called away to Skara, and should not be back till evening.

So Mr. Wraxall had to go unsatisfied to his day's work at the manor-house. The papers on which he was just then engaged soon put his thoughts into another channel, for he had to occupy himself with glancing over the correspondence between Sophia Albertina in Stockholm and her married cousin Ulrica Leonora at Råbäck in the years 1705-10. The letters were of exceptional interest from the light they threw upon the culture of that period in Sweden, as anyone can testify who has read the full edition of them in the publications of the Swedish Historical Manuscripts Commission.

In the afternoon he had done with these, and after returning the boxes in which they were kept to their places on the shelf, he proceeded, very naturally, to take down some of the volumes nearest to them, in order to determine which of them had best be his principal subject of investigation next day. The shelf he had hit upon was occupied mostly by a collection of account-books in the writing of the first Count Magnus. But one among them was not an account-book, but a book of alchemical and other tracts in another sixteenth-century hand. Not being very familiar with alchemical literature, Mr. Wraxall spends much space which he might have spared in setting out the names and beginnings of the various treatises: The book of the Phoenix, book of the Thirty Words, book of the Toad, book of Miriam, Turba philosophorum, and so forth; and then he announces with a good deal of circumstance his delight at finding, on a leaf originally left blank near the middle of the book, some writing of Count Magnus himself headed "Liber nigrae peregrinationis." It is true that only a few lines

From
THE TOMB OF SARAH
F.G. Loring
(1900)

Just before sunset last night the rector and I locked ourselves into the church, and took up our position in the pulpit. It was one of those pulpits, to be found in some churches, which is entered from the vestry, the preacher appearing at a good height through an arched opening in the wall. This gave us a sense of security, which we felt we needed, a good view of the interior, and direct access to the implements which I had concealed in the vestry.

The sun set and the twilight gradually deepened and faded. There was, so far, no sign of the usual fog, nor any howling of the dogs. At nine o'clock the moon rose, and her pale light gradually flooded the aisles, and still no sign of any kind from the "Sarah Tomb." The rector had asked me several times what he might expect, but I was determined that no words or thought of mine should influence him, and that he should be convinced by his own senses alone.

By half past ten we were both getting very tired, and I began to think that perhaps after all we should see nothing that night However, soon after eleven we observed a light mist rising from the "Sarah Tomb." It seemed to scintillate and sparkle as it rose, and curled in a sort of pillar or spiral.

were written, but there was quite enough to show that the landlord had that morning been referring to a belief at least as old as the time of Count Magnus, and probably shared by him. This is the English of what was written:

"If any man desires to obtain a long life, if he would obtain a faithful messenger and see the blood of his enemies, it is necessary that he should first go into the city of Chorazin, and there salute the prince..." Here there was an erasure of one word, not very thoroughly done, so that Mr. Wraxall felt pretty sure that he was right in reading it as *aëris* ("of the air"). But there was no more of the text copied, only a line in Latin: "Quaere reliqua hujus materiei inter secretiora" (See the rest of this matter among the more private things).

It could not be denied that this threw a rather lurid light upon the tastes and beliefs of the Count; but to Mr. Wraxall, separated from him by nearly three centuries, the thought that he might have added to his general forcefulness alchemy, and to alchemy something like magic, only made him a more picturesque figure; and when, after a rather prolonged contemplation of his picture in the hall, Mr. Wraxall set out on his homeward way, his mind was full of the thought of Count Magnus. He had no eyes for his surroundings, no perception of the evening scents of the woods or the evening light on the lake; and when all of a sudden he pulled up short, he was astonished to find himself already at the gate of the churchyard, and within a few minutes of his dinner. His eyes fell on the mausoleum.

"Ah," he said, "Count Magnus, there you are. I should dearly like to see you."

"Like many solitary men," he writes, "I have a habit of talkng to myself aloud; and, unlike some of the Greek and Latin particles, I do not expect an answer. Certainly, and perhaps fortunately in this case, there was neither voice nor any that regarded: only the woman who, I suppose, was cleaning up the church, dropped some metallic object on the floor, whose clang startled me. Count Magnus, I think, sleeps sound enough."

That same evening the landlord of the inn, who had heard Mr. Wraxall say that he wished to see the clerk or deacon (as he would be called in Sweden) of the parish, introduced him to that official in the inn parlour. A visit to the De la Gardie tomb-house was soon arranged for the next day, and a little general conversation ensued.

Mr. Wraxall, remembering that one function of Scandinavian deacons is to teach candidates for Confirmation, thought he would refresh his own memory on a Biblical point.

"Can you tell me," he said, "anything about Chorazin?"

The deacon seemed startled, but readily reminded him how that village had once been denounced.

"To be sure," said Mr. Wraxall; "it is, I suppose, quite a ruin now?"

"So I expect," replied the deacon. "I have heard some of our old priests say that Antichrist is to be born there; and there are tales —"

I said nothing, but heard the rector give a sort of gasp as he clutched my arm feverishly. "Great Heaven!" he whispered, "it is taking shape."

And, true enough, in a very few moments we saw standing erect by the tomb the ghastly figure of the Countess Sarah!

She looked thin and haggard still, and her face was deadly white; but the crimson lips looked like a hideous gash in the pale cheeks, and her eyes glared like red coals in the gloom of the church.

It was a fearful thing to watch as she stepped unsteadily down the aisle, staggering a little as if from weakness and exhaustion. This was perhaps natural, as her body must have suffered much physically from her long incarceration, in spite of the unholy forces which kept it fresh and well.

We watched her to the door, and wondered what would happen; but it appeared to present no difficulty, for she melted through it and disappeared.

"Now, Grant," I said, "do you believe?"

"Yes," he replied, "I must. Everything is in your hands, and I will obey your comands to the letter, if you can only instruct me how to rid my poor people of this unnameable terror."

"By God's help I will," said I; "but you shall be yet more convinced first, for we have a terrible work to do, and much to answer for in

"Ah! what tales are those?" Mr. Wraxall put in.

"Tales, I was going to say, which I have forgotten," said the deacon; and soon after that he said good night.

The landlord was now alone, and at Mr. Wraxall's mercy, and that inquirer was not inclined to spare him.

"Herr Nielsen," he said, "I have found out something about the Black Pilgrimage. You may as well tell me what you know. What did the Count bring back with him?"

Swedes are habitually slow, perhaps, in answering, or perhaps the landlord was an exception. I am not sure; but Mr. Wraxall notes that the landlord spent at least one minute in looking at him before he said anything at all. Then he came close up to his guest, and with a good deal of effort he spoke:

"Mr. Wraxall, I can tell you this one little tale, and no more — not any more. You must not ask anything when I have done. In my grandfather's time— that is, ninety-two years ago — there were two men who said: 'The Count is dead; we do not care for him. We will go tonight and have a free hunt in his wood' — the long wood on the hill that you have seen behind Råbäck. Well, those that heard them say this, they said: 'No, do not go; we are sure you will meet with persons walking who should not be walking. They should be resting, not walking.' These men laughed. There were no forest-men to keep the wood, because no one wished to hunt there. The family were not here at the house. These men could do what they wished.

"Very well, they go to the wood that night. My grandfather was sitting here in this room. It was the summer, and a light night. With the window open, he could see out to the wood, and hear.

"So he sat there, and two or three men with him, and they listened. At first they hear nothing at all; then they hear someone — you know how far away it is — they hear someone scream, just as if the most inside part of his soul was twisted out of him. All of them in the room caught hold of each other, and they sat so for three-quarters of an hour. Then they hear someone else, only about three hundred ells off. They hear him laugh out loud: it was not one of those two men that laughed, and, indeed, they have all of them said that it was not any man at all. After that they hear a great door shut.

"Then, when it was just light with the sun, they all went to the priest. They said to him:

"'Father, put on your gown and your ruff, and come to bury these men, Anders Bjornsen and Hans Thorbjorn.'

"You understand that they were sure these men were dead. So they went to the wood — my grandfather never forgot this. He said they were all like so many dead men themselves. The priest, too, he was in a white fear. He said when they came to him:

"'I heard one cry in the night, and I heard one laugh afterwards. If I cannot forget that, I shall not be able to sleep again.'

the future, before we leave the church again this morning. And now to work, for in its present weak state the vampire will not wander far, but may return at any time, and must not find us unprepared."

We stepped down from the pulpit, and taking dog-roses and garlic from the vestry, proceeded to the tomb. I arrived first, and throwing off the wooden cover cried: "Look! it's empty!" There was nothing there! Nothing except the impress of the body in the loose damp mold!

I took the flowers and laid them in a circle round the tomb for legend teaches us that vampires will not pass over these particular blossoms if they can avoid it.

Then, eight or ten feet away, I made a circle on the stone pavement, large enough for the rector and myself to stand in, and within the circle I placed the implements that I had brought into the church with me.

"Now," I said, "from this circle, which nothing unholy can step across, you shall see the vampire face to face, and see her afraid to cross that other circle of garlic and dog-roses to regain her unholy refuge. But on no account step beyond the holy place you stand in, for the vampire has a fearful strength not her own, and, like a snake, can draw her victim willingly to his own destruction." Now so far my work was done, and, calling the doctor, we stepped into the holy circle to await the vampire's return. Nor was this long delayed.

"So they went to the wood, and they found these men on the edge of the wood. Hans Thorbjorn was standing with his back against a tree, and all the time he was pushing with his hands — pushing something away from him which was not there. So he was not dead. And they led him away, and took him to the house at Nykjoping, and he died before the winter, but he went on pushing with his hands. Also Anders Bjornsen was there; but he was dead. And I tell you this about Anders Bjornsen, that he was once a beautiful man, but now his face was not there, because the flesh of it was sucked away off the bones. You understand that? My grandfather did not forget that. And they laid him on the bier which they brought, and they put a cloth over his head, and the priest walked before; and they began to sing the psalm for the dead as well as they could. So, as they were singing the end of the first verse, one fell down, who was carrying the head of the bier, and the others looked back, and they saw that the cloth had fallen off, and the eyes of Anders Bjornsen were looking up, because there was nothing to close over them. And this they could not bear. Therefore the priest laid the cloth upon him, and sent for a spade, and they buried him in that place."

The next day Mr. Wraxall records that the deacon called for him soon after his breakfast, and took him to the church, and mausoleum. He noticed that the key of the latter was hung on a nail just by the pulpit, and it occurred to him that, as the church door seemed to be left unlocked as a rule, it would not be difficult for him to pay a second and more private visit to the monuments if there proved to be more of interest among them than could be digested at first. The building, when he entered it, he found not unimposing. The monuments, mostly large erections of the seventeenth and eighteenth centuries were dignified if luxuriant, and the epitaphs and heraldry were copious. The central space of the dome room was occupied by three copper sarcophagi, covered with finely-engraved ornament. Two of them had, as is commonly the case in Denmark and Sweden, a large metal crucifix on the lid. The third, that of Count Magnus, as it appeared, had, instead of that, a full-length effigy engraved upon it, and round the edge were several bands of similar ornament representing various scenes. One was a battle, with cannon belching out smoke, and walled towns, and troops of pikemen. Another showed an execution. In a third, among trees, was a man running at full speed, with flying hair and outstretched hands. After him followed a strange form; it would be hard to say whether the artist had intended it for a man, and was unable to give the requisite similitude, or whether it was intentionally made as monstrous as it looked. In view of the skill with which the rest of the drawing was done, Mr. Wraxall felt inclined to adopt the latter idea. The figure was unduly short, and was for the most part muffled in a hooded garment which swept the ground. The only part of the form which projected from that shelter was not shaped like any hand or arm. Mr. Wraxall compares it to the tentacle of a devil-fish, and continues: "On seeing this, I said to myself, 'This,

Presently, a damp, cold odor seemed to pervade the church, which made our hair bristle and flesh creep. And then, down the aisle with noiseless feet, came that which we watched for.

I heard the rector mutter a prayer, and I held him tightly by the arm, for he was shivering violently.

Long before we could distinguish the features we saw the glowing eyes and crimson sensual mouth. She went straight to her tomb, but stopped short when she encountered my flowers. She walked right round the tomb seeking a place to enter, and as she walked she saw us. A spasm of diabolical hate and fury passed over her face, but it quickly vanished, and a smile of love, more devilish still, took its place. She stretched out her arms towards us. Then we saw that round her mouth gathered a bloody froth, and from under her lips long pointed teeth gleamed and champed.

She spoke: a soft soothing voice, a voice that carried a spell with it, and affected us both strangely, particularly the rector. I wished to test as far as possible, without endangering our lives, the vampire's power.

Her voice had a soporific effect, which I resisted easily enough, but which seemed to throw the rector into a sort of trance. More than this: it seemed to compel him to her in spite of his efforts to resist.

"Come!" she said, "come! I give sleep and peace — sleep and peace — sleep and peace."

then, which is evidently an allegorical representation of some kind — a fiend pursuing a hunted soul — may be the origin of the story of Count Magnus and his mysterious companion. Let us see how the huntsman is pictured: doubtless it will be a demon blowing his horn.' " But, as it turned out, there was no such sensational figure, only the semblance of a cloaked man on a hillock, who stood leaning on a stick, and watching the hunt with an interest which the engraver had tried to express in his attitude.

Mr. Wraxall noted the finely-worked and massive steel padlocks — three in number — which secured the sarcophagus. One of them, he saw, was detached, and lay on the pavement. And then, unwilling to delay the deacon longer or to waste his own working-time, he made his way onward to the manor-house.

"It is curious," he notes, "how on retracing a familiar path one's thoughts engross one to the absolute exclusion of surrounding objects. Tonight, for the second time, I had entirely failed to notice where I was going (I had planned a private visit to the tomb-house to copy the epitaphs), when I suddenly, as it were, awoke to consciousness, and found myself (as before) turning in at the churchyard gate, and, I believe, singing or chanting some such words as, 'Are you awake, Count Magnus? Are you asleep, Count Magnus?' and then something more which I have failed to recollect. It seemed to me that I must have been behaving in this nonsensical way for some time."

He found the key of the mausoleum where he had expected to find it, and copied the greater part of what he wanted; in fact, he stayed until the light began to fail him.

"I must have been wrong," he writes, "in saying that one of the padlocks of my Count's sarcophagus was unfastened; I see tonight that two are loose. I picked both up, and laid them carefully on the window-ledge, after trying unsuccessfully to close them. The remaining one is still firm, and, though I take it to be a spring lock, I cannot guess how it is opened. Had I succeeded in undoing it, I am almost afraid I should have taken the liberty of opening the sarcophagus. It is strange, the interest I feel in the personality of this, I fear, somewhat ferocious and grim old noble."

The day following was, as it turned out, the last of Mr. Wraxall's stay at Råbäck. He received letters connected with certain investments which made it desirable that he should return to England; his work among the papers was practically done, and travelling was slow. He decided, therefore, to make his farewells, put some finishing touches to his notes, and be off.

These finishing touches and farewells, as it turned out, took more time than he had expected. The hospitable family insisted on his staying to dine with them — they dined at three — and it was verging on half-past six before he was outside the iron gates of Råbäck. He dwelt on every step of his walk by the lake, determined to saturate himself, now that he trod it for the last time, in the sentiment of the place and hour. And when he reached the summit of the churchyard knoll, he lingered for many

She advanced a little towards us, but not far, for I noted that the sacred circle seemed to keep her back like an iron hand.

My companion seemed to become demoralized and spellbound. He tried to step forward and, finding me detaining him, whispered: "Harry, let go! I must go! She is calling me! I must! I must! Oh, help me! Help me!" And he began to struggle.

It was time to finish.

"Grant!" I cried, in a loud, firm voice, "in the name of all that you hold sacred, have done and play the man!" He shuddered violently and gasped: "Where am I?" Then he remembered, and clung to me convulsively for a moment.

At this a look of damnable hate changed the smiling face before us, and with a sort of shriek she staggered back.

"Back!" I cried: "back to your unholy tomb! No longer shall you molest the suffering world! Your end is near."

It was fear that now showed itself in her beautiful face (for it was beautiful in spite of its horror) as she shrank back, back and over the circlet of flowers, shivering as she did so. At last, with a low mournful cry, she appeared to melt back again into her tomb.

As she did so the first gleams of the rising sun lit up the world, and I knew all danger was over for the day.

minutes, gazing at the limitless prospect of woods near and distant, all dark beneath a sky of liquid green. When at last he turned to go, the thought struck him that surely he must bid farewell to Count Magnus as well as the rest of the De la Gardies. The church was but twenty yards away, and he knew where the key of the mausoleum hung. It was not long before he was standing over the great copper coffin, and, as usual, talking to himself aloud. "You may have been a bit of a rascal in your time, Magnus," he was saying, "but for all that I should like to see you, or, rather —"

"Just at that instant," he says, "I felt a blow on my foot. Hastily enough I drew it back, and something fell on the pavement with a clash. It was the third, the last of the three padlocks which had fastened the sarcophagus. I stooped to pick it up, and — Heaven is my witness that I am writing only the bare truth — before I had raised myself there was a sound of metal hinges creaking, and I distinctly saw the lid shifting upwards. I may have behaved like a coward, but I could not for my life stay for one moment. I was outside that dreadful building in less time than I can write — almost as quickly as I could have said — the words; and what frightens me yet more, I could not turn the key in the lock. As I sit here in my room noting these facts, I ask myself (it was not twenty minutes ago) whether that noise of creaking metal continued, and I cannot tell whether it did or not. I only know that there was something more than I have written that alarmed me, but whether it was sound or sight I am not able to remember. What is this that I have done?"

Poor Mr. Wraxall! He set out on his journey to England on the next day, as he had planned, and he reached England in safety; and yet, as I gather from his changed hand and inconsequent jottings, a broken man. One of several small notebooks that have come to me with his papers gives, not a key to, but a kind of inkling of, his experiences. Much of his journey was made by canal-boat, and I find not less than six painful attempts to enumerate and describe his fellow-passengers. The entries are of this kind:

"24. Pastor of village in Skane. Usual black coat and soft black hat.
"25. Commercial traveller from Stockholm going to Trollhattan. Black cloak, brown hat.
"26. Man in long black cloak, broad-leafed hat, very old-fashioned."

This entry is lined out, and a note added: "Perhaps identical with No. 13. Have not yet seen his face." On referring to No. 13, I find that he is a Roman priest in a cassock.

The net result of the reckoning is always the same. Twenty-eight people appear in the enumeration, one being always a man in a long black cloak and broad hat, and the other a "short figure in dark cloak and hood." On the other hand, it is always noted that only twenty-six passengers appear at

Taking Grant by the arm, I drew him with me out of the circle and led him to the tomb. There lay the vampire once more, still in her living death as we had a moment before seen her in her devilish life. But in the eyes remained that awful expression of hate, and cringing, appalling fear.

Grant was pulling himself together.

"Now," I said, "will you dare the last terrible act and rid the world forever of this horror?"

"By God!" he said solemnly, "I will. Tell me what to do."

"Help me lift her out of her tomb. She can harm us no more," I replied.

With averted faces we set to our terrible task, and laid her out upon the flags.

"Now," I said, "read the burial service over the poor body, and then let us give it its release from this living hell that holds it."

Reverently the rector read the beautiful words, and reverently I made the necessary responses. When it was over I took the stake and, without giving myself time to think, plunged it with all my strength through the heart.

As though really alive, the body for a moment writhed and kicked convulsively, and an awful heart-rending shriek rang through the silent church; then all was still.

Then we lifted the poor body back; and, thank God! The consolation that legend tells us is

meals, and that the man in the cloak is perhaps absent, and the short figure is certainly absent.

On reaching England, it appears that Mr. Wraxall landed at Harwich, and that he resolved at once to put himself out of the reach of some person or persons whom he never specifies, but whom he had evidently come to regard as his pursuers. Accordingly he took a vehicle — it was a closed fly — not trusting the railway, and drove across country to the village of Belchamp St. Paul. It was about nine o'clock on a moonlit August night when he neared the place. He was sitting forward, and looking out of the window at the fields and thickens — there was little else to be seen — racing past him. Suddenly, he came to a cross-road. At the corner two figures were standing motionless; both were in dark cloaks; the taller one wore a hat, the shorter a hood. He had no time to see their faces, nor did they make any motion that he could discern. Yet the horse shied violently and broke into a gallop, and Mr. Wraxall sank back into his seat in something like desperation. He had seen them before.

Arrived at Belchamp St. Paul, he was fortunate enough to find a decent furnished lodging, and for the next twenty-four hours he lived, comparatively speaking, in peace. His last notes were written on this day. They are too disjointed and ejaculatory be given here in full, but the substance of them is clear enough. He is expecting a visit from his pursuers — how or when he knows not — and his constant cry is "What has he done?" and "Is there no hope?" Doctors, he knows, would call him mad, policemen would laugh at him. The parson is away. What can he do but lock his door and cry to God?

People still remember last year at Belchamp St. Paul how a strange gentleman came one evening in August years back; and how the next morning but one he was found dead, there was an inquest; and the jury that viewed the body fainted, seven of 'em did, and none of 'em wouldn't speak to what they see, and the verdict was visitation of God; and how the people as kep' the 'ouse moved out that same week, and went away from that part. But they do not, I think, know any glimmer of light has ever been thrown, or could be thrown, on the mystery. It so happened that last year the house came into my hands as part of a legacy. It had stood empty since 1863, and there seemed no prospect of letting it; so I had it pulled down, and the papers of which I have given you an abstract were found in a forgotten cupboard under the window in the best bedroom.

never denied to those who have to do such awful work as ours came at last. Over the face stole a great and solemn peace; the lips lost their crimson hue, the prominent sharp teeth sank back into the mouth, and for a moment we saw before us the calm, pale face of a most beautiful woman, who smiled as she slept. A few minutes more, and she faded away to dust before our eyes as we watched. We set to work and cleaned up every trace of our work, and then departed for the rectory. Most thankful were we to step out of the church, with its horrible associations, into the rosy warmth of the summer morning. With the above end the notes in my father's diary, though a few days later this further entry occurs:

***15th July*. Since the 12th everything has been quiet and as usual. We replaced and sealed up the "Sarah Tomb" this morning. The workmen were surprised to find the body had disappeared, but took it to be the natural result of exposing it to the air.**

One odd thing came to my ears today. It appears that the child of one of the villagers strayed from home the night of the 11th inst., and was found asleep in a coppice near the church, very pale and quite exhausted. There were two small marks on her throat, which have since disappeared.

What does this mean? I have, however, kept it to myself, as, now the vampire is no more, no further danger either to that child or to any other is to be apprehended. It is only those who die of the vampire's embrace that become vampires at death in their turn.

The Singular Death of Morton

ALGERNON BLACKWOOD
(1910)

ABOUT THE AUTHOR

Algernon Blackwood (1869 – 1951) remains one of the most prolific and literate English authors of fantasy and horror, especially excelling at tales involving psychic and spiritual terror; a recurring character was John Silence, a psychic detective, who reflected Blackwood's own deep interest in the paranormal. Toward the end of his life he became a beloved television storyteller for the BBC and was made a Commander of the British Empire in 1949.

In his famous essay "Supernatural Horror in Literature," H.P. Lovecraft observed, "Of the quality of Mr. Blackwood's genius there can be no dispute, for no one has even approached the skill, seriousness, and minute fidelity with which he records the overtones of strangeness in ordinary things and experiences . . . he is the one absolute and unquestioned master or weird atmosphere; and can evoke what amounts to a story from a simple fragment of humorless psychological description. Above all others he understands how fully some sensitive minds dwell forever on the borderland of dream, and how relatively slight is the distinction betwixt those images formed from actual objects and those excited by the play of the imagination."

DUSK WAS MELTING into darkness as the two men slowly made their way through the dense forest of spruce and fir that clothed the flanks of the mountain. They were weary with the long climb, for neither was in his first youth, and the July day had been a hot one. Their little inn lay further in the valley among the orchards that separated the forest from the vineyards.

Neither of them talked much. The big man led the way, carrying the knapsack, and his companion, older, shorter, evidently the more fatigued of the two, followed with small footsteps. From time to time he stumbled among the loose rocks. An exceptionally observant mind would possibly have divined that his stumbling was not entirely due to fatigue, but to an absorption of spirit that made him careless how he walked.

"All right behind?" the big man would call from time to time, half glancing back.

"Eli? What?" the other would reply, startled out of a reverie.

"Pace too fast?"

"Not a bit. I'm coming." And once he added: "You might hurry on and see to supper, if you feel like it. I shan't be long behind you."

But his big friend did not adopt the suggestion. He kept the same distance between them. He called out the same question at intervals. Once or twice *he* stopped and looked back too.

In this way they came at length to the skirts of the wood. A deep hush covered all the valley; the limestone ridges they had climbed gleamed down white and ghostly upon them from the fading sky. Midway in its journeys, the evening wind dropped suddenly to watch the beauty of the moonlight to hold the branches still so that the light might slip between and weave its silver pattern on the moss below.

And, as they stood a moment to take it in, a step sounded behind them on the soft pine-needles, and the older man, still a little in the rear, turned with a start as though he had been suddenly called by name.

"There's that girl — again!" he said, and his voice expressed a curious mingling of pleasure, surprise and — apprehension.

Into a patch of moonlight passed the figure of a young girl, who looked at them as though about to stop yet thinking better of it, smiled softly, and moved on out of sight into the surrounding darkness. The moon just

Algernon Blackwood, England's master of dark fantasy and horror.

caught her eyes and teeth, so that they shone; the rest of her body stood in shadow; the effect was striking — almost as though head and shoulders hung alone in mid air, watching them with this shining smile, then fading away.

"Come on, for heaven's sake," the big man cried. There was patience in his manner, not unkindness. The other lingered a moment, peering closely into the gloom where the girl had vanished His friend repeated his injunction, and a moment later the two had emerged upon the high road with the village lights in sight beyond, and forest left behind them like a vast mantle that held the night within its folds.

For some minutes neither of them spoke; then the big man waited for his friend to draw up alongside.

"About all this valley of the Jura," he said presently, "there seems to me something — rather weird." He shifted the knapsack vigorously on his back. It was a gesture of unconscious protest. "Something uncanny," he added, as he set a good pace.

"But extraordinarily beautiful—"

"It attracts you more than it does me, I think," was the short reply.

"The picturesque superstitions still survive here," observed the older man. "They touch the imagination in spite of oneself."

A pause followed during which the other tried to increase the pace. The subject evidently made him impatient for some reason.

"Perhaps," he said presently. "Though I think myself it's due to the curious loneliness of the place. I mean, we're in the middle of tourist-Europe here, yet so utterly remote. It's such a neglected little corner of the world. The contradiction bewilders. Then, being so near the frontier, too, with the clock changing an hour a mile from the village, makes one think of time as unreal and imaginary." He laughed. He produced several other reasons as well. His friend admitted their value, and agreed half-heartedly. He still turned occasionally to look back. The mountain ridge where they had climbed was clearly visible in the moonlight.

"Odd," he said, "but I don't see that farmhouse where we got the milk anywhere. It ought to be easily visible from here."

"Hardly — in this light. It was a queer place rather, I thought," he added. He did not deny the curiously suggestive atmosphere of the region, he merely wanted to find satisfactory explanations. "A case in point, I mean. I didn't like it quite — that farmhouse — yet I'm hanged I know why. It made me feel uncomfortable. That girl appeared so suddenly, although the place seemed deserted. And her silence was so odd. Why in the world couldn't she answer a single question? I'm glad I didn't take the milk. I spat it out. I'd like to know where she from, for there was no sign of a cow or a goat to be seen anywhere!"

"I swallowed mine — in spite of the taste," said the other, half smiling at his companion's sudden volubility.

Very abruptly, then, the big man turned and faced his friend. Was it

A PSYCHIC VAMPIRE MEETS ITS MATCH

"The Singular Death of Morton" deals with a vampire of the traditional sort, while another Blackwood story, "The Transfer," features a malignant psychic sponge, Mr. Frene, who has a fateful encounter with an ugly, barren patch of garden that has a hunger all its own.

From
THE TRANSFER
Algernon Blackwood
(1912)

I watched his hard, bleak face; I noticed how thin he was, and the curious, oily brightness of his steady eyes. They did not glitter, but they drew you with a sort of soft, creamy shine like Eastern eyes. And everything he said or did announced what I may dare to call the *suction* of his presence. His nature achieved this result automatically. He dominated us all, yet so gently that until it was accomplished no one noticed it.

Before five minutes had passed, however, I was aware of one thing only. My mind focussed exclusively upon it, and so vividly that I marvelled the others did not scream, or run, or do something violent to prevent it. And it was this; that, separated merely by some dozen yards or so, this man, vibrating with the acquired vitality of others, stood within easy reach of that spot of yawning emptiness, waiting and eager to be filled. Earth scented her prey.

These two active "centers" were within fighting distance; he so thin, so hard, so keen, yet really spreading large with the loose "surround" of others' life he had appropriated, so practiced and triumphant; that other so patient, deep, with so mighty a draw of the whole earth behind it, and — ugh! — so obviously aware that its opportunity at last had come.

I saw it all as plainly as though I watched two great animals prepare for battle, both unconsciously; yet in some inexplicable way I saw it, of course, within me, and not externally. The conflict would be hideously unequal. Each side had already sent out emissaries, how long before I could not tell, for the first evidence he gave that something was going wrong with him was when his voice grew suddenly confused, he missed his words, and his lips trembled a moment and turned flabby. The next second his face betrayed that singular and horrid change, growing somehow loose about the bones of the cheek, and larger, so that I remembered Jamie's miserable phrase. The emissaries of the two kingdoms, the human and the vegetable, had met, I make it out, in that very second. For the first time in his long career of battening on others, Mr. Frene found himself pitted against a vaster kingdom than he knew and, so finding, shook inwardly in that little part that was his

merely an effect of the moonlight, or had his skin really turned pale beneath the sunburn?

"I say, old man," he said, his face grave and serious, "What do you think she was? What made her seem like that, and why the devil do you think she followed us?"

"I think," was the slow reply, "it was *me* she was following."

The words, and particularly the tone of conviction in which they were spoken, clearly were displeasing to the big man, who already regretted having spoken so frankly what was in his mind. With a companion so imaginative, so impressionable, so nervous, it had been foolish and unwise. He led the way home at a pace that made the other arrive five minutes in his rear, panting, limping and perspiring as if they had been running.

"I'm rather for going on into Switzerland tomorrow, or the next day," he ventured that night in the darkness of their two-bedded room. "I think we've had enough of this place. Eh? What do you think?"

But there was no answer from the bed across the room, for...its occupant was sound asleep and snoring.

"Dead tired, I suppose!" he muttered to himself, and then turned over to follow his friend's example. But for a long time sleep refused him. Queer, unwelcome thoughts and feelings kept him awake — of a kind he rarely knew, and thoroughly disliked. It was rubbish, yet it made him uncomfortable, so that his nerves tingled. He tossed about in the bed. "I'm overtired," he persuaded himself, "that's all."

The strange feelings that kept him thus awake were not easy to analyse, perhaps, but their origin was beyond all question: they grouped themselves about the picture of that deserted, tumble-down chalet on the mountain ridge where they had stopped for refreshment a few hours before. It was a farmhouse, dilapidated and dirty, and the name stood in big black letters against a blue background on the wall above the door: "La Chenille." Yet not a living soul was to be seen anywhere about it; the doors were fastened, windows shuttered; chimneys smokeless; dirt, neglect and decay everywhere in evidence.

Then, suddenly, as they had turned to go, after much vain shouting and knocking at the door, a face appeared for an instant at a window, the shutter of which was half open. His friend saw it first, and called aloud. The face nodded in reply, and presently a young girl came round the corner of the house, apparently by a back door, and stood staring at them both from a little distance.

And from that very instant, so far as he could remember, these queer feelings had entered his heart — fear, distrust, misgiving. The thought of it now, as he lay in bed in the darkness made his hair rise. There was something about that girl that struck cold into the soul. Yet she was a mere slip of a thing, very pretty, seductive even, with a certain serpent-like fascination about her eyes and movements; and although she only replied to their questions as to refreshment with a smile, uttering no single word, she managed to convey the impression that she was a managing little person who might make herself very agreeable if she chose. In spite of her undeniable charm there was about her an atmosphere of something sinister. He himself did most of the questioning, but it was his older friend who had the benefit of her smile. Her eyes hardly ever left his face, and once she had slipped quite close to him and touched his arm.

The strange part of it now seemed to him that he could not remember in the least how she was dressed, or what was the colouring of her eyes and hair. It was almost as though he had *felt*, rather than seen, her presence.

The milk — she produced a jug and two wooden bowls after a brief appearance round the corner of the house — was — well, it tasted so odd that he had been unable to swallow it, and had spat it out. His friend, on the other hand, savage with thirst, had drunk his bowl to the last drop too quickly to taste it even, and, while he drank, had kept his eyes fixed on those of the girl, who stood close in front of him.

And from that moment his friend had somehow changed. On the way down he said things that were unusual, talking chiefly about the "Chenille," and the girl, and the delicious, delicate flavour of the milk, yet all phrased in such a way that it sounded singular, unfamiliar, unpleasant even. Now that he tried to recall the sentences the actual words evaded him; but the memory of the uneasiness and apprehension they caused him to feel remained. And night ever italicizes such memories!

Then, to cap it all, the girl had followed them. It was wholly foolish and absurd to feel the things he did feel; yet there the feelings were, and what was the good of arguing? That girl frightened him; the change in his friend was in some way or other a danger signal. More than this he could not tell. An explanation might come later, but for the present his chief desire was to get away from the place and to get his friend away, too.

And on this thought sleep overtook him — heavily.

The windows were wide open; outside was a garden with a rather high enclosing wall, and at the far end a gate that was kept locked because it led into private fields and so, by a back way, to the cemetery and the little church. When it was open the guests of the inn made use of it and got lost in the network of fields and vines, for there was no proper route that way to the road or the mountains. They usually ended up prematurely in the cemetery, and got back to the village by passing through the church, which was always open; or by knocking at the kitchen doors of the other houses and explaining their position. Hence the gate was locked now to save trouble.

After several hours of hot, unrefreshing sleep the big man turned in his bed and woke. He tried to stretch, but couldn't; then sat up panting with a sense of suffocation. And by the faint starlight of the summer night, he saw next that his friend was up and moving about the room. Remembering that sometimes he walked in his sleep, he called to him gently:

"Morton, old chap," he said in a low voice, with a touch of authority in it, "go back to bed! You've walked enough for one day!"

definite actual self. He felt the huge disaster coming.

"Yes, John," he was saying, in his drawling, self-congratulating voice, "Sir George gave me that car — gave it to me as a present. Wasn't it char—?" and then broke off abruptly, stammered, drew breath, stood up, and looked uneasily about him. For a second there was a gaping pause. It was like the click which starts some huge machinery moving — that instant's pause before it actually starts. The whole thing, indeed, then went with the rapidity of machinery running down and beyond control. I thought of a giant dynamo working silently and invisible.

"What's that?" he cried, in a soft voice charged with alarm. "What's that horrid place? And someone's crying there — who is it?"

He pointed to the empty patch. Then, before anyone could answer, he started across the lawn towards it, going every minute faster. Before anyone could move he stood upon the edge. He leaned over — peering down into it.

It seemed a few hours passed, but really they were seconds, for time is measured by the quality and

And the figure, obeying as sleep-walkers often will, passed across the room and disappeared among the shadows over his bed. The other plunged and burrowed himself into a comfortable position again for sleep, but the heat of the room, the shortness of the bed, and this tiresome interruption of his slumbers made it difficult to lose consciousness. He forced his eyes to keep shut, and his body to cease from fidgeting, but there was something nibbling at his mind like a spirit mouse that never permitted him to cross the frontier into actual oblivion. He slept with one eye open, as the saying is. Odours of hay and flowers and baked ground stole in through the open window; with them, too, came from time to time sounds — little sounds that disturbed him without ever being loud enough to claim definite attention.

Perhaps, after all, he did lose consciousness for a moment — when, suddenly, a thought came with a sharp rush into his mind and galvanized him once more into utter wakefulness. It amazed him that he had not grasped it before. It was this: the figure he had seen was *not the figure of his friend.*

Alarm gripped him at once before he could think or argue, and a cold perspiration broke out all over his body. He fumbled for matches, couldn't find them: then, remembering that there was an electric light, he scraped the wall with his fingers — and turned on the little white switch. In the sudden glare that filled the room he saw instantly that his friend's bed was no longer occupied. And his mind, then acting instinctively, without process of conscious reasoning, flew like a flash to their walk of the day — to the tumble-down "Chenille,"' the glass of milk, the odd behaviour of his friend, and — to the girl.

At the same second he noticed that the odour in the room which hitherto he had taken to be the composite odour of fields, flowers and night, was really something else: it was the odour of freshly turned earth. Immediately on the top of this discovery came another. Those slight sounds he had heard outside the window were not ordinary night-sounds, the murmur of wind and insects: they were footsteps moving softly, stealthily down the little paths of crushed granite.

He was dressed in wonderful short order, noticing as he did so that his friend's night-garments lay upon the bed, and that he, too, had therefore dressed; further — that the door had been unlocked and stood half an inch ajar. There was now no question that he *had* slept again: between the present and the moment when he had seen the figure there had been a considerable interval. A couple of minutes later he had made his way cautiously downstairs and was standing on the garden path in the moonlight. And as he stood there, his mind filled with the stories the proprietor had told a few days before of the superstitions that still lived in the popular imagination and haunted this little, remote pine-clad valley. The thought of that girl sickened him. The odour of newly-turned earth remained in his nostrils and made his gorge rise. Utterly and vigorously he rejected the monstrous fictions he had heard, yet for all that, could not

not the quantity of sensations it contains. I saw it all with merciless, photographic detail, sharply etched amid the general confusion. Each side was intensely active, but only one side, the human, exerted all its force — in resistance. The other merely stretched out a feeler, as it were, from its vast, potential strength; no more was necessary. It was such a soft and easy victory. Oh, it was rather pitiful! There was no bluster or great effort, on one side at least. Close by his side I witnessed it, for I, it seemed, alone had moved and followed him. No one else stirred, though Mrs. Frene clattered noisily with the cups, making some sudden impulsive gesture with her hands, and Gladys, I remember, gave a cry — it was like a little scream — "Oh, mother, it's the heat, isn't it?" Mr. Frene, her father, was speechless, pale as ashes.

But the instant I reached his side, it became clear what had drawn me there thus instinctively. Upon the other side, among the silver birches, stood little Jamie. He was watching. I experienced — for him — one of those moments that shake the heart; a liquid fear ran all over me, the more effective because unintelligible really. Yet I felt that if I could know all, and what lay actually behind, my fear would be more than justified; that the thing was awful, full of awe.

And then it happened — a truly wicked sight — like watching a universe in action, yet all

prevent their touching his imagination as he stood there in the early hours of the morning, alone with night and silence. The spell was undeniable; only a mind without sensibility could have ignored it.

He searched the little garden from end to end. Empty! Opposite the high gate he stopped, peering through the iron bars, wet with dew to his hands. Far across the intervening fields he fancied something moved. A second later he was sure of it. Something down there to the right beyond the trees was astir. It was in the cemetery.

And this definite discovery sent a shudder of terror and disgust through him from head to foot. He framed the name of his friend with his lips, yet the sound did not come forth. Some deeper instinct warned him to hold it back. Instead, after incredible efforts, he climbed that iron gate and dropped down into the soaking grass upon the other side. Then, taking advantage of all the cover he could find, he ran, swiftly and stealthily, towards the cemetery. On the way, without quite knowing why he did so, he picked up a heavy stick; and a moment later he stood beside the low wall that separated the fields from the churchyard — stood and stared.

There, beside the tombstones, with their hideous metal wreaths and crowns of faded flowers, he made out the figure of his friend; he was stooping, crouched down upon the ground; behind him rose a couple of bushy yew trees, against the dark of which his form was easily visible. He was not alone; in front of him, bending close over him seemed, was another figure — a slight, shadowy, slim figure.

This time the big man found his voice and called aloud:

"Morton, Morton!" he cried. "What, in the name of heaven, are you doing? What's the matter?"

And the instant his deep voice broke the stillness of the night with its clamour, the little figure, half hiding his friend, turned about and faced him. He saw a white face with shining eyes and teeth as the form rose; the moonlight painted it with its own strange pallor; it was weird, unreal, horrible; and across the mouth, downwards from the lips to chin, ran a deep stain of crimson.

The next moment the figure slid with a queer, gliding motion towards the trees, and disappeared among the yews and tombstones in the direction of the church. The heavy stick, hurled whirling after it, fell harmlessly half way, knocking a metal cross from its perch upon an upright grave; and the man who had thrown it raced full speed towards the huddled up figure of his friend, hardly noticing the thin, wailing cry that rose trembling through the night air from the vanished form. Nor did he notice more particularly that several of the graves, newly made, showed signs of recent disturbance, and that the odour of turned earth he had noticed in the room grew stronger. All his attention was concentrated upon the figure at his feet.

"Morton, man, get up! Wake for God's sake! You've been walking in —"

contained within a small square foot of space. I think he understood vaguely that if someone could only take his place he might be saved, and that was why, discerning instinctively the easiest substitute within reach, he saw the child and called aloud to him across the empty patch, "James, my boy, come here!"

His voice was like a thin report, but somehow flat and lifeless, as when a rifle misses fire, sharp, yet weak; it had no "crack" in it. It was really supplication. And, with amazement, I heard my own ring out imperious and strong, though I was not conscious of saying it, "Jamie, don't move. Stay where you are!" But Jamie, the little child, obeyed neither of us. Moving up nearer to the edge, he stood there — laughing! I heard that laughter, but could have sworn it did not come from him. The empty, yawning patch gave out that sound.

Mr. Frene turned sideways, throwing up his arms. I saw his hard, bleak face grow somehow wider, spread through the air, and downwards. A similar thing, I saw, was happening at the same time to his entire person, for it drew out into the atmosphere in a stream of movement. The face for a second made me think of those toys of green india rubber that children pull. It grew enormous. But this was an external impression only. What actually happened, I clearly understood, was that all this vitality and life he

Then the words died upon his lips. The unnatural attitude of his friend's shoulders, and the way the head dropped back to show the neck, struck him like a blow in the face. There was no sign of movement. He lifted the body up and carried it, all limp and unresisting, by ways he never remembered afterwards, back to the inn.

It was all a dreadful nightmare — a nightmare that carried over its ghastly horror into waking life. He knew that the proprietor and his wife moved busily to and fro about the bed, and that in due course, the village doctor was upon the scene, and that he was giving a muddled and feverish description of all he knew, telling how his friend was a confirmed sleepwalker and all the rest. But he did not realize the truth until he saw the face of the doctor as he straightened up from the long examination.

"Will you wake him?" he heard himself asking, "or let him sleep it out till morning?" And the doctor's expression, even before the reply came to confirm it, told him the truth. "Ah, monsieur, your friend will not ever wake again, I fear! It is the heart, you see; *hélas*, it is sudden failure of the heart!"

The final scenes in the little tragedy which thus brought his holiday to so abrupt and terrible a close need no description, being in no way essential to this strange story. There were one or two curious details, however, that came to light afterwards. One was, that for some weeks before there had been signs of disturbance among newly-made graves in the cemetery, which the authorities had been trying to trace to the nightly meanderings of the village madman — in vain; and another, that the morning after the death a trail of blood had been found across the church floor, as though someone had passed through from the back entrance to the front. A special service was held that very week to cleanse the holy building from the evil of that stain; for the villagers, deep in their superstitions, declared that nothing human had left that trail; nothing could have made those marks but a vampire disturbed at midnight in its awful occupation among the dead.

Apart from such idle rumours, however, the bereaved carried with him to this day certain other remarkable details which cannot be so easily dismissed. For he had a brief conversation with the doctor, it appears, that impressed him profoundly. And the doctor, an intelligent man, prosaic as granite into the bargain, had questioned him rather closely as to the recent life and habits of his dead friend. The account of their climb to the "Chenille" he heard with an amazement he could not conceal.

"But no such châlet exists," he said. "There is no 'Chenille.' A long time ago, fifty years or more, there was such a place, but it was destroyed by the authorities on account of the evil reputation of the people who lived there. They burnt it. Nothing remains today but a few bits of broken wall and foundation."

"Evil reputation — ?"

The doctor shrugged his shoulders. "Travellers, even peasants, disappeared," he said. "An old woman lived there with her daughter, and

had transferred from others to himself for years was now in turn being taken from him and transferred — elsewhere.

One moment on the edge he wobbled horribly, then with that queer sideways motion, rapid yet ungainly, he stepped forward into the middle of the patch and fell heavily upon his face. His eyes, as he dropped, faded shockingly, and across the countenance was written plainly what I can only call an expression of destruction. He looked utterly destroyed. I caught a sound — from Jamie? — but this time not of laughter. It was like a gulp; it was deep and muffled and it dipped away into the earth. Again I thought of a troop of small black horses galloping away down a subterranean passage beneath my feet — plunging into the depths — their tramping growing fainter and fainter into buried distance. In my nostrils was a pungent smell of earth.

•

And then — all passed. I came back into myself. Mr. Frene, junior, was lifting his brother's head from the lawn where he had fallen from the heat, close beside the tea-table. He had never really moved from there. And Jamie, I learned afterwards, had been the whole time asleep upon his bed upstairs, worn out with his crying and unreasoning alarm. Gladys came running out with cold water, sponge and towel, brandy too —

poisoned milk was supposed to be used. But the neighbourhood accused them of worse than ordinary murder."

"In what way?"

"Said the girl was a vampire," answered the doctor shortly.

And, after a moment's hesitation, he added, turning his face away as he spoke:

"It was a curious thing, though, that tiny hole in your friend's throat, small as a pin-prick, yet so deep. And the heart — did I tell you?— was almost completely drained of blood."

all kinds of things. "Mother, it was the heat, wasn't it?" I heard her whisper, but I did not catch Mrs. Frene's reply. From her face it struck me that she was bordering on collapse herself. Then the butler followed, and they just picked him up and carried him into the house. He recovered even before the doctor came.

But the queer thing to me is that I was convinced the others all had seen what I saw, only that no one said a word about it; and to this day no one has said a word. And that was, perhaps, the most horrid part of all.

From that day to this I have scarcely heard a mention of Mr. Frene, senior. It seemed as if he dropped suddenly out of life. The papers never mentioned him. His activities ceased, as it were. His after-life, at any rate, became singularly ineffective. Certainly he achieved nothing worth public mention. But it may be only that, having left the employ of Mrs. Frene, there was no particular occasion for me to hear anything.

The after-life of that empty patch of garden, however, was quite otherwise. Nothing, so far as I know, was done to it by gardeners, or in the way of draining it or bringing in new earth, but even before I left in the following summer it had changed. It lay untouched, full of great, luscious, driving weeds and creepers, very strong, full — fed, and bursting thick with life.

Painting by Mary Woronov.

For the Blood Is the Life

F. MARION CRAWFORD
(1911)

WE HAD DINED at sunset on the broad roof of the old tower, because it was cooler there during the great heat of summer. Besides, the little kitchen was built at one corner of the great square platform, which made it more convenient than if the dishes had to be carried down the steep stone steps, broken in places and everywhere worn with age. The tower was one of those built all down the west coast of Calabria by the Emperor Charles V early in the sixteenth century, to keep off the Barbary pirates, when the unbelievers were allied with Francis I against the Emperor and the Church. They have gone to ruin, a few still stand intact, and mine is one of the largest. How it came into my possession ten years ago, and why I spend a part of each year in it, are matters which do not concern this tale. The tower stands in one of the loneliest spots in Southern Italy, at the extremity of a curving rocky promontory, which forms a small but safe natural harbor at the southern extremity of the Gulf of Policastro, and just north of Cape Scalea, the birthplace of Judas Iscariot, according to the old local legend. The tower stands alone on this hooked spur of the rock, and there is not a house to be seen within three miles of it. When I go there I take a couple of sailors, one of whom is a fair cook, and when I am away it is in charge of a gnome-like little being who was once a miner and who attached himself to me long ago.

ABOUT THE AUTHOR

Francis Marion Crawford (1854-1909) was an American author born in Italy, where he spent most of his life. His novels, including *Mr. Isaacs* (1882), *Saracinesca* (1887), *Sant' Ilario* (1889), and *Don Orsino* (1892), reflected his status as a world citizen, although he was especially partial to Italian settings. Today he is best remembered for his supernatural stories, published two years after his death in a collection entitled *Wandering Ghosts*. His sister, Anne Crawford, was the author of another vampire story set in Italy, "A Mystery of the Campagna" (1887), an excerpt of which appears elsewhere in this volume.

My friend, who sometimes visits me in my summer solitude, is an artist by profession, a Scandinavian by birth, and a cosmopolitan by force of circumstances. We had dined at sunset; the sunset glow had reddened and faded again, and the evening purple steeped the vast chain of the mountains that embrace the deep gulf to eastward and rear themselves higher and higher toward the south. It was hot, and we sat at the landward corner of the platform, waiting for the night breeze to come down from the lower hills. The color sank out of the air, there was a little interval of deep-grey twilight, and a lamp sent a yellow streak from the open door of the kitchen, where the men were getting their supper.

Then the moon rose suddenly above the crest of the promontory, flooding the platform and lighting up every little spur of rock and knoll of grass below us, down to the edge of the motionless water. My friend lighted his pipe and sat looking at a spot on the hillside. I knew that he was looking at it, and for a long time past I had wondered whether he would ever see anything there that would fix his attention. I knew that spot well. It was clear that he was interested at last, though it was a long time before he spoke. Like most painters, he trusts to his own eyesight, as a lion trusts his strength and a stag his speed, and he is always disturbed when he cannot reconcile what he sees with what he believes that he ought to see.

"It's strange," he said. "Do you see that little mound just on this side of the boulder?"

"Yes," I said, and I guessed what was coming.

"It looks like a grave," observed Holger.

"Very true. It does look like a grave."

"Yes," continued my friend, his eyes still fixed on the spot. "But the strange thing is that I see the body lying on the top of it. Of course," continued Holger, turning his head on one side as artists do, "it must be an effect of light. In the first place, it is not a grave at all. Secondly, if it were, the body would be inside and not outside. Therefore, it's an effect of the moonlight. Don't you see it?"

"Perfectly; I always see it on moonlit nights."

"It doesn't seem to interest you much," said Holger.

"On the contrary, it does interest me, though I am used to it. You're not so far wrong, either. The mound is really a grave."

"Nonsense!" cried Holger, incredulously. "I suppose you'll tell me what I see lying on it is really a corpse!"

"No," I answered, "it's not. I know, because I have taken the trouble to go down and see."

"Then what is it?" asked Holger.

"It's nothing."

"You mean that it's an effect of light, I suppose?"

"Perhaps it is. But the inexplicable part of the matter is that it makes no difference whether the moon is rising or setting, or waxing or waning. If there's any moonlight at all, from east or west or overhead, so long as it shines on the grave you can see the outline of the body on top."

Holger stirred up his pipe with the point of his knife, and then used his finger for a stopper. When the tobacco burned well he rose from his chair.

"If you don't mind," he said, "I'll go down and take a look at it."

He left me, crossed the roof, and disappeared down the dark steps. I did not move, but sat looking down until he came out of the tower below. I heard him humming an old Danish song as he crossed the open space in the bright moonlight, going straight to the mysterious mound. When he was ten paces from it, Holger stopped short, made two steps forward, and then three or four backward, and then stopped again. I know what that meant. He had reached the spot where the Thing ceased to be visible — where, as he would have said, the effect of light changed.

Then he went on till he reached the mound and stood upon it. I could see the Thing still, but it was no longer lying down; it was on its knees now, winding its white arms round Holger's body and looking up into his

THE VAMPIRE AND THE CHURCH

The title of F. Marion Crawford's story comes from Deuteronomy 12.23, in the famous proscription against eating blood along with meat, "...for the blood is the life, and you shall not eat the life with the flesh." Similar biblical injunctions against blood ingestion occur in Genesis and Leviticus 17.14: "For the life of every creature is the blood of it; therefore I have said to the people of Israel, You shall not eat the blood of any creature, for the life of the creature is its blood..."

Modern Christianity, of course, has thoroughly embraced the eating of both flesh and blood in the ritual of communion, though Protestants regard the rite as symbolic as opposed to the Catholic dogma of literal transubstantiation (see the discussion of Catholicism below).

The evolution of vampires has been inextricably linked with religious history, although in the twentieth century the myth has become progressively secularized.

The Dominican *Malleus Maleficarum* (1486) established the Catholic Church's position against witches and revenants, stating definitively that Satan could infiltrate corpses to spread evil among the living. The Greek Orthodox Church was much more actively involved in fueling the eastern European vampire hysteria of the eighteenth century, though it ultimately distanced itself from antivampire activity.

face. A cool breeze stirred my hair at that moment, as the night wind began to come down from the hills, but it felt like a breath from another world.

The Thing seemed to be trying to climb to its feet, helping itself up by Holger's body while he stood upright, quite unconscious of it and apparently looking toward the tower, which is very picturesque when the moonlight falls upon it on that side.

"Come along!" I shouted. "Don't stay there all night!"

It seemed to me that he moved reluctantly as he stepped from the mound, or else with difficulty. That was it. The Thing's arms were still round his waist, but its feet could not leave the grave. As he came slowly forward it was drawn and lengthened like a wreath of mist, thin and white, till I saw distinctly that Holger shook himself, as a man does who feels a chill. At the same instant a little wail of pain came to me on the breeze — it might have been the cry of the small owl that lives among the rocks — and the misty presence floated swiftly back from Holger's advancing figure and lay once more at its length upon the mound.

A public service announcement for the Lutheran Church.

Again I felt the cool breeze in my hair, and this time an icy thrill of dread ran down my spine. I remembered very well that I had once gone down there alone in the moonlight; that presently, being near, I had seen nothing; that, like Holger, I had gone and had stood upon the mound; and I remembered how, when I came back, sure that there was nothing there, I had felt the sudden conviction that there was something after all if I would only look behind me. I remembered the strong temptation to look back, a temptation I had resisted as unworthy of a man of sense, until, to get rid of it, I had shaken myself just as Holger did.

And now I knew that those white, misty arms had been round me too; I knew it in a flash, and I shuddered as I remembered that I had heard the night owl then too. But it had not been the night owl. It was the cry of the Thing.

I refilled my pipe and poured out a cup of strong southern wine; in less than a minute Holger was seated beside me again.

"Of course there's nothing there," he said, "but it's creepy, all the same. Do you know, when I was coming back I was so sure that there was something behind me that I wanted to turn round and look? It was an effort not to."

He laughed a little, knocked the ashes out of his pipe, and poured himself out some wine. For a while neither of us spoke, and the moon rose higher, and we both looked at the Thing that lay on the mound.

"You might make a story about that," said Holger after a long time.

"There is one," I answered. "If you're not sleepy, I'll tell it to you."

"Go ahead," said Holger, who likes stories.

Old Alario was dying up there in the village behind the hill. You remember him, I have no doubt. They say that he made his money by selling sham jewelry in South America, and escaped with his gains when he was found out. Like all those fellows, if they bring anything back with them, he at once set to work to enlarge his house, and as there are no masons here, he sent all the way to Paola for two workmen. They were a rough-looking pair of scoundrels — a Neapolitan who had lost one eye and a Sicilian with an old scar half an inch deep across his left cheek. I often saw them, for on Sundays they used to come down here and fish off the rocks. When Alario caught the fever that killed him, the masons were still at work. As he had agreed that part of their pay should be their board and lodging, he made them sleep in the house. His wife was dead, and he had an only son called Angelo, who was a much better sort than himself. Angelo was to marry the daughter of the richest man in the village, and, strange to say, though the marriage was arranged by their parents, the young people were said to be in love with each other.

For that matter, the whole village was in love with Angelo, and among the rest a wild, good-looking creature called Cristina, who was more like a gypsy than any girl I ever saw about here. She had very red lips and very black eyes, she was built like a greyhound, and had the tongue of the devil. But Angelo did not care a straw for her. He was rather a simple-minded fellow, quite different from his old scoundrel of a father, and under what I should call normal circumstances I really believe that he would never have looked at any girl except the nice plump little creature, with a fat dowry, whom his father meant him to marry. But things turned up which were neither normal nor natural.

On the other hand, a very handsome young shepherd from the hills above Maratea was in love with Cristina, who seems to have been quite indifferent to him. Cristina had no regular means of subsistence, but she was a good girl and willing to do any work or go on errands to any distance for the sake of a loaf of bread or a mess of beans, and permission to sleep under cover. She was especially glad when she could get something to do about the house of Angelo's father. There is no doctor in the village, and when the neighbors saw that old Alario was dying they sent Cristina to Scalea to fetch one. That was late in the afternoon, and if they had waited so long, it was because the dying miser refused to allow any such extravagance while he was able to speak. But while Cristina was gone matters grew rapidly worse, the priest was brought to the bedside, and when he had done what he could he gave it as his opinion to the bystanders that the old man was dead, and left the house.

You know these people. They have a physical horror of death. Until the priest spoke, the room had been full of people. The words were hardly out of his mouth before it was empty. It was night now. They hurried down the dark steps and out into the street.

Angelo, as I have said, was away, Cristina had not come back — the simple woman-servant who had nursed the sick man fled with the rest, and the body was left alone in the flickering light of the earthen oil lamp.

Five minutes later two men looked in cautiously and crept forward

CHRISTIANITY

The vampire in western tradition presents a paradox by simultaneously perverting and reinforcing the images and rituals of Christianity. Blood-communion, death, and resurrection are central to both the Christian faith and the conventions of vampire belief. Author Clive Leatherdale, in his excellent critical study *Dracula: The Novel and the Legend* (1993) devotes a fascinating chapter to the ways in which the Dracula story in particular serves as both a Christian parody and a Christian allegory:

"It can be proposed that one of the basic lessons of the novel was to reaffirm the existence of God in an age when the weakening hold of Christianity generated fresh debate about what lay beyond death. The marshalled diary extracts and letters are themselves endowed with the status of scripture. Instead of the Gospels according to St. Matthew and St. Mark, we find Gospels according to Mr. Harker and Dr. Seward. Taken with Van Helsing's concluding remarks, 'we want no proofs' . . . they constitute a 'revelation' of Dracula's existence, as the Bible offers a 'revelation' of Christ's."

toward the bed. They were the one-eyed Neapolitan mason and his Sicilian companion. They knew what they wanted. In a moment they had dragged from under the bed a small but heavy iron-bound box, and long before any one thought of coming back to the dead man they had left the house and the village under cover of the darkness. It was easy enough, for Alario's house is the last toward the gorge which leads down here, and the thieves merely went out by the back door, got over the stone wall, and had nothing to risk after that except the possibility of meeting some belated countryman, which was very small indeed, since few of the people use that path. They had a mattock and shovel, and they made their way here without accident.

I am telling you this story as it must have happened, for, of course, there were no witnesses to this part of it. The men brought the box down by the gorge, intending to bury it until they should be able to come back and take it away in a boat. They must have been clever enough to guess that some of the money would be in paper notes, for they would otherwise have buried it on the beach in the wet sand, where it would have been much safer. But the paper would have rotted if they had been obliged to leave it there long, so they dug their hole down there, close to that boulder. Yes, just where the mound is now.

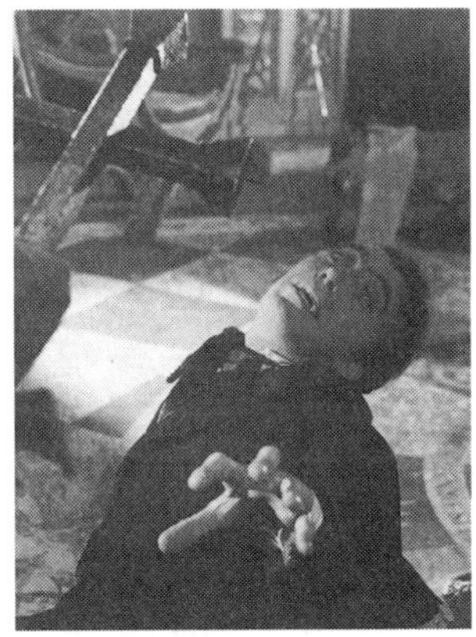

Christopher Lee is subdued by the power of Christ in Horror of Dracula. *(1958)*

Cristina did not find the doctor in Scalea, for he had been sent for from a place up the valley, halfway to San Domenico. If she had found him, he would have come on his mule by the upper road, which is smoother but much longer. But Cristina took the short cut by the rocks, which passes about fifty feet above the mound, and goes round that corner. The men were digging when she passed, and she heard them at work. It would not have been like her to go by without finding out what the noise was, for she was never afraid of anything in her life, and, besides, the fishermen sometimes come ashore here at night to get a stone for an anchor or to gather sticks to make a little fire. The night was dark, and Cristina probably came close to the two men before she could see what they were doing. She knew them, of course, and they knew her, and understood instantly that they were in her power. There was only one thing to be done for their safety, and they did it. They knocked her on the head, they dug the hole deep, and they buried her quickly with the iron-bound chest. They must have understood that their only chance of escaping suspicion lay in getting back to the village before their absence was noticed, for they returned immediately, and were found half an hour later gossiping quietly with the man who was making Alario's coffin. He was a crony of theirs, and had been working at the repairs in the old man's house. So far as I have been able to make out, the only persons who were supposed to know where Alario kept his treasure were Angelo and the one woman-servant I have mentioned. Angelo was away; it was the woman who discovered the theft.

It is easy enough to understand why no one else knew where the money was. The old man kept his door locked and the key in his pocket when he was out, and did not let the woman enter to clean the place unless he was there himself. The whole village knew that he had money somewhere, however, and the masons had probably discovered the whereabouts of the chest by climbing in at the window in his absence. If the old man had not been delirious until he lost consciousness, he would have been in frightful agony of mind for his riches. The faithful woman-servant forgot their existence only for a few moments when she fled with the rest, overcome by the horror of death. Twenty minutes had not passed before she returned with the two hideous old hags who are always called in to prepare the dead for burial. Even then she had not at first the courage to go near the bed with them, but she made a pretense of dropping something, went down on her knees as if to find it, and looked under the bedstead. The walls of the room were newly whitewashed down to the floor, and she saw at a glance that the chest was gone. It had been there in the afternoon; it had therefore been stolen in the short interval since she had left the room.

There are no carabineers stationed in the village; there is not so much as a municipal watchman, for there is no municipality. There never was such a place, I believe. Scalea is supposed to look after it in some mysterious way, and it takes a couple of hours to get anybody from there. As the old woman had lived in the village all her life, it did not even occur to her to apply to any civil authority for help. She simply set up a howl and ran through the village in the dark, screaming out that her dead master's house had been robbed. Many of the people looked out, but at first no one seemed inclined to help her. Most of them, judging her by themselves, whispered to each other that she had probably stolen the money herself. The first man to move was the father of the girl whom Angelo was to marry; having collected his household, all of whom felt a personal interest in the wealth which was to have come into the family, he declared it to be his opinion that the chest had been stolen by the two journeyman masons who lodged in the house. He headed a search for them, which naturally began in Alario's house and ended in the carpenter's workshop, where the thieves were found discussing a measure of wine with the carpenter over the half-finished coffin, by the light of one earthen lamp filled with oil and tallow. The search party at once accused the delinquents of the crime, and threatened to lock them up in the cellar till the carabineers could be fetched from Scalea. The two men looked at each other for one moment, and then without the slightest hesitation they put out the single light, seized the unfinished coffin between them, and using it as a sort of battering ram, dashed upon their assailants in the dark. In a few moments they were beyond pursuit.

That is the end of the first part of the story. The treasure had disappeared, and as no trace of it could be found the people naturally supposed that the thieves had succeeded in carrying it off. The old man was buried, and when Angelo came back at last he had to borrow money to pay for the miserable funeral, and had some difficulty in doing so. He hardly needed

CROSSES AND CRUCIFIXES

The symbol of Christ's crucifixion is one of the best known of all vampire repellents, but the rules and regulations governing its use are sometimes confusing and contradictory. As a symbol of the faith of the person using it, the cross should, therefore, offer little protection to the unfaithful. Nonetheless, in many films and stories the cross seems to contain its own source of power, like a supernatural stun gun. A "real" cross isn't always needed and can sometimes be effectively improvised. In the film *Kiss of the Vampire* (1962), a man whose chest has been scratched by a vampire quickly smears the blood into a cruciform and repels her. In *Horror of Dracula* (1958) the vampire hunter Van Helsing jerry-rigs a cross from two candlesticks to force the monster into a deadly ray of sunlight. In Hammer's next vampire film, *The Brides of Dracula* (1960), Van Helsing manipulates the vanes of a burning windmill to cast a crosslike shadow on a fleeing vampire; it works just fine. In more recent vampire traditions, such as the novels of Anne Rice, holy relics have no power whatsoever over the vampire, who looks on such superstitions with amusement. In the 1979 film version of *Dracula*, for instance, actor Frank Langella causes a cross to burst into flames with nothing more than a contemptuous glance.

to be told that in losing his inheritance he had lost his bride. In this part of the world marriages are made on strictly business principles, and if the promised cash is not forthcoming on the appointed day the bride or the bridegroom whose parents have failed to produce it may as well take themselves off, for there will be no wedding. Poor Angelo knew that well enough. His father had been possessed of hardly any land, and now that the hard cash which he had brought from South America was gone, there was nothing left but debts for the building materials that were to have been used for enlarging and improving the old house. Angelo was beggared, and the nice plump little creature who was to have been his turned up her nose at him in the most approved fashion. As for Cristina, it was several days before she was missed, for no one remembered that she had been sent to Scalea for the doctor, who had never come. She often disappeared in the same way for days together, when she could find a little work here and there at the distant farms among the hills. But when she did not come back at all, people began to wonder, and at last made up their minds that she had connived with the masons and had escaped with them.

I paused and emptied my glass.

"That sort of thing could not happen anywhere else," observed Holger, filling his everlasting pipe again. "It is wonderful what a natural charm there is about murder and sudden death in a romantic country like this. Deeds that would be simply brutal and disgusting anywhere else become dramatic and mysterious because this is Italy and we are living in a genuine tower of Charles V built against genuine Barbary pirates."

Roddy McDowall in Fright Night. *(1985)*

"There's something in that," I admitted. Holger is the most romantic man in the world inside of himself, but he always thinks it necessary to explain why he feels anything.

"I suppose they found the poor girl's body with the box," he said presently.

"As it seems to interest you," I answered, "I'll tell you the rest of the story."

The moon had risen high by this time; the outline of the Thing on the mound was clearer to our eyes than before.

The village very soon settled down to its small, dull life. No one missed old Alario, who had been away so much on his voyages to South America that he had never been a familiar figure in his native place. Angelo lived in the half-finished house, and because he had no money to pay the old woman servant she would not stay with him, but once in a long time she would come and wash a shirt for him for old acquaintance's sake. Besides the house, he had inherited a small patch of ground at some distance from the village; he tried to cultivate it, but he had no heart in the work, for he knew he could never pay the taxes on it and on the house, which would certainly be confiscated by the government, or seized for the debt of the building material, which the man who had supplied it refused to take back.

Angelo was very unhappy. So long as his father had been alive and rich, every girl in the village had been in love with him; but that was all changed now. It had been pleasant to b admired and courted, and invited to drink wine by fathers who had girls to marry. It was hard to be stared at coldly, and sometimes laughed at because he had been robbed of his inheritance. He cooked his miserable meals for himself, and from being sad became melancholy and morose.

At twilight, when the day's work was done, instead of hanging about in the open space before the church with young fellows of his own age, he took to wandering in lonely places on the outskirts of the village till it was quite dark. Then he slunk home and went to bed to save the expense of a light. But in those lonely twilight hours he began to have strange waking dreams. He was not always alone, for often when he sat on the stump of a tree, where the narrow path turns down the gorge, he was sure that a woman came up noiselessly over the rough stones, as if her feet were bare; and she stood under a clump of chestnut trees only half a dozen yards down the path, and beckoned to him without speaking. Though she was in the shadow he knew that her lips were red, and that when they parted a little and smiled at him she showed two small sharp teeth. He knew this at first rather than saw it, and he knew that it was Cristina, and that she was dead. Yet he was not afraid; he only wondered whether it was a dream, for he thought that if he had been awake he should have been frightened.

Besides, the dead woman had red lips, and that could only happen in a dream. Whenever he went near the gorge after sunset she was already there waiting for him, or else she very soon appeared, and he began to be sure that she came a little nearer to him every day. At first he had only been sure of her blood-red mouth, but now each feature grew distinct, and the pale face looked at him with deep and hungry eyes.

It was the eyes that grew dim. Little by little he came to know that some day the dream would not end when he turned away to go home, but would lead him down the gorge out of which the vision rose. She was nearer now when she beckoned to him. Her cheeks were not livid like those of the dead, but pale with starvation, with the furious and unappeased physical hunger of her eyes that devoured him. They feasted on his soul and cast a spell over him, and at last they were close to his own and held him. He could not tell whether her breath was as hot as fire or as cold as ice; he could not tell whether her red lips burned his or froze them, or whether her five fingers on his wrists seared scorching scars or bit his flesh like frost; he could not tell whether he was awake or asleep, whether she was alive or dead, but he knew that she loved him, she alone of all creatures, earthly or unearthly, and her spell had power over him.

When the moon rose high that night the shadow of that Thing was not alone down there upon the mound.

Angelo awoke in the cool dawn, drenched with dew and chilled through flesh, and blood, and bone. He opened his eyes to the faint grey light, and

CATHOLICISM

During the reign of Pope Innocent III in 1215, the Roman Catholic Church formalized the dogma of transubstantiation — the belief that the body and blood of Christ were physically present in the communion wafer and wine used in the celebration of the Mass. Thus, the essential act of vampirism — the literal drinking of human blood — is a central ritual in one of the world's major religions. In this writer's own informal but extensive observation, the vampire myth resonates with a particular strength with lapsed and ex-Catholics — scratch a vampire buff, and it's more than a little likely you'll find a Catholic school uniform bunched beneath the cape.

The reasons are complex and as varied as the individuals. To the rebellious, a vampire fetish can seem to be a perverse badge of antiauthoritarian honor. To the lapsed, the ritual aspects of vampire entertainment with their many shadow-suggestions of the sacraments may, to some extent, fill a spiritual void. And even practicing Catholics may find a reinforcement of their faith in the traditional vampire story's emphasis on the dualistic reality of good and evil, the mystical properties of blood, etc.

The leading practitioner of vampire fiction in our time, Anne Rice, is an ex-Catholic, as is her most famous creation, the vampire Lestat. Novelist Joyce Carol Oates, in a recent essay on the 1931 film *Dracula*, commented on the ritualistic, priestlike demeanor of the master vampire, which she likened to her own childhood memories of the dark-robed priests intoning the Latin mass.

saw the stars still shining overhead. He was very weak, and his heart was beating so slowly that he was almost like a man fainting. Slowly he turned his head on the mound, as on a pillow, but the other face was not there. Fear seized him suddenly, a fear unspeakable and unknown; he sprang to his feet and fled up the gorge, and he never looked behind him until he reached the door of the house on the outskirts of the village. Drearily he went to his work that day, and wearily the hours dragged themselves after the sun, till at last he touched the sea and sank, and the great sharp hills above Maratea turned purple against the dove-colored eastern sky.

Angelo shouldered his heavy hoe and left the field. He felt less tired now than in the morning when he had begun to work, but he promised himself that he would go home without lingering by the gorge, and eat the best supper he could get himself, and sleep all night in his bed like a Christian man. Not again would he be tempted down the narrow way by a shadow with red lips and icy breath; not again would he dream that dream of terror and delight. He was near the village now; it was half an hour since the sun had set, and the cracked church bell sent little discordant echoes across the rocks and ravines to tell all good people that the day was done. Angelo stood still a moment where the path forked, where it led toward the village on the left, and down to the gorge on the right, where a clump of chestnut trees overhung the narrow way. He stood still a minute, lifting his battered hat from his head and gazing at the fast-fading sea westward, prayed. His lips moved, but the words that followed them in his brain lost their meaning and turned into others and ended in a name that he spoke aloud — Cristina! With the name, the tension of his will relaxed suddenly, reality went out and the dream took him again, and bore him on swiftly and surely like a man walking in his sleep, down, down, by the steep path in the gathering darkness. And as she glided beside him, Cristina whispered strange, sweet things in his ear, which somehow, if he had been awake, he knew that he could not quite have understood; but now they were the most wonderful words he had ever heard in his life. And she kissed him also, but not upon his mouth. He felt her sharp kisses upon his white throat, and he knew that her lips were red. So the wild dream sped on through twilight and darkness and moonrise, and all the glory of the summer's night. But in the chilly dawn he lay as one half dead upon the mound down there, recalling and not recalling, drained of his blood, yet strangely longing to give those red lips more. Then came the fear, the awful nameless panic, the mortal horror that guards the confines of the world we see not, neither know of as we know of other things, but which we feel when its icy chill freezes our bones and stirs our hair with the touch of a ghostly hand. Once more Angelo sprang from the mound and fled up the gorge in the breaking day, but his step was less sure this time, and he panted for breath as he ran; and when he came to the bright spring of water that rises halfway up the hillside, he dropped upon his knees and hands and plunged his whole face in and drank as he had never drunk

ANTI-SEMITISM

This is a persistent subtext of vampire stories, no doubt an offshoot of the ugly Christian blood-libel of Jews as a race requiring the blood of gentile babies in its rituals. Although it is rarely commented upon, a long literary tradition of villainous Semitic stereotypes informed Bram Stoker's 1897 conception of *Dracula*, originally presented as a horrific, hook-nosed Shylock from Transylvania (and a close cousin of another mesmeric Jewish predator of the literary 1890s — George Du Maurier's "filthy black Hebrew," Svengali).

The Shylock-Dracula nexus was explored only once on the screen, in Max Schreck's beak-faced impersonation of Dracula in *Nosferatu* (1922). In a more ironic vein, comedian Lenny Bruce conflated Dracula and Jewishness in a series of movie-monster comedy routines in the late 1950s. More recently, Black Muslim leader Louis Farrakhan stirred outrage and controversy when he refused to disavow the "essential truth" of an associate's public description of Jews as "bloodsuckers."

Max Schreck in Nosferatu: *a Shylock from the Carpathians?*

before — for it was the thirst of the wounded man who has lain bleeding all night long upon the battlefield.

She had him fast now, and he could not escape her, but would come to her every evening at dusk until she had drained him of his last drop of blood. It was in vain that when the day was done he tried to take another turning and to go home by a path that did not lead near the gorge. It was in vain that he made promises to himself each morning at dawn when he climbed the lonely way up from the shore to the village. It was all in vain, for when the sun sank burning into the sea, and the coolness of the evening stole out as from a hiding place to delight the weary world, his feet turned toward the old way, and she was waiting for him in the shadow under the chestnut trees; and then all happened as before and she fell to kissing his white throat even as she flitted lightly down the way, winding one arm about him. And as his blood failed, she grew more hungry and more thirsty every day, and every day when he awoke in the early dawn it was harder to rouse himself to the effort of climbing the steep path to the village; and when he went to his work his feet dragged painfully, and there was hardly strength in his arms to wield the heavy hoe. He scarcely spoke to anyone now but the people said he was "consuming himself" for the love of the girl he was to have married when he lost his inheritance; and they laughed heartily at the thought, for this is not a very romantic country. At this time, Antonio, the man who stays here to look after the tower, returned from a visit to his people, who live near Salerno. He had been away all the time since before Alario's death and knew nothing of what had happened. He has told me that he came back late in the afternoon and shut himself up in the tower to eat and sleep, for he was very tired. It was past midnight when he awoke, and when he looked out the waning moon was rising over the shoulder of the hill. He looked out toward the mound, and he saw something, and he did not sleep again that night. When he went out again in the morning it was broad daylight, and there was nothing to be seen on the mound but loose stones and driven sand. Yet he did not go very near it; he went straight up the path to the village and directly to the house of the old priest.

"I have seen an evil thing this night," he said; "I have seen how the dead drink the blood of the living. And the blood is the life."

"Tell me what you have seen," said the priest in reply.

Antonio told him everything he had seen.

"You must bring your book and your holy water tonight," he added. "I will be here before sunset to go down with you, and if it pleases your reverence to sup with me while we wait, I will make ready."

"I will come," the priest answered, "for I have read in old books of these strange beings which are neither quick nor dead, and which lie ever fresh in their graves, stealing out in the dusk to taste life and blood."

Antonio cannot read, but he was glad to see that the priest understood the business; for, of course, the books must have instructed him as to the

SECULAR VAMPIRES

The modern trend toward vampires who operate outside religious traditions is well reflected in the work of American novelist Chelsea Quinn Yarbro (b. 1942). Originally a science fiction writer of a characteristically dark sensibility, Yarbro gravitated to the horror genre, and more specifically to vampires. Her outstanding creation is Rakoczy, *comte* de Saint-Germain, an alchemist turned-vampire based on an actual (though nonvampiric) historical character of eighteenth-century France. Saint-Germain existed prior to Christianity, and is therefore indifferent to its symbols. He's also not a villain, but rather an intelligent, accomplished, seductive and compassionate vampire whose interests range well beyond blood-feasting. Adhering to the traditional requirement that a vampire can't stray too far from its native soil, Saint-Germain has a brilliant solution: he lines his footwear with the needed dirt, and so can come and go wherever he pleases.

Yarbro introduced Saint-Germain in *Hôtel Transylvania* (1977), the first of a popular series including *The Palace* (1978), *Blood Games* (1979), *Path of the Eclipse* (1981), *Tempting Fate* (1982), and a story collection, *The Saint-Germain Chronicles* (1985). More recently, Yarbro has expanded her

best means of quieting the half-living Thing forever.

So Antonio went away to his work, which consists largely in sitting on the shady side of the tower, when he is not perched upon a rock with a fishing-line catching nothing. But on that day he went twice to look at the mound in the bright sunlight, and he searched round and round it for some hole through which the being might get in and out; but he found none. When the sun began to sink and the air was cooler in the shadows, he went up to fetch the old priest, carrying a little wicker basket with him; and in this they placed a bottle of holy water, and the basin, and sprinkler, and the stole which the priest would need; and they came down and waited in the door of the tower till it should be dark. But while the light still lingered very grey and faint, they saw something moving, just there, two figures, a man's that walked, and a woman's that flitted beside him, and while her head lay on his shoulder she kissed his throat. The priest has told me that, too, and that his teeth chattered and he grasped Antonio's arm. The vision passed and disappeared into the shadow. Then Antonio got the leathern flask of strong liquor, which he kept for great occasions, and poured such a draught as made the old man feel almost young again; and he got the lantern, and his pick and shovel, and gave the priest his stole to put on and the holy water to carry, and they went out together toward the spot where the work was to be done. Antonio says that in spite of the rum his own knees shook together, and the priest stumbled over his Latin. For when they were yet a few yards from the mound the flickering light of the lantern fell upon Angelo's white face, unconscious as if in sleep, and on his upturned throat, over which a very thin red line of blood trickled down into his collar; and the flickering light of the lantern played upon another face that looked up from the feast — upon two deep, dead eyes that saw in spite of death — upon parted lips redder than life itself — upon two gleaming teeth on which glistened a rosy drop. Then the priest, good old man, shut his eyes tight and showered holy water before him, and his cracked voice rose almost to a scream; and then Antonio, who is no coward after all, raised his pick in one hand and the lantern in the other, as he sprang forward, not knowing what the end should be; and then he swears that he heard a woman's cry, and the Thing was gone, and Angelo lay alone on the mound unconscious, with the red line on his throat and the beads of deadly sweat on his cold forehead. They lifted him, half-dead as he was, and laid him on the ground close by; then Antonio went to work, and the priest helped him, though he was old and could not do much; and they dug deep, and at last Antonio, standing in the grave, stooped down with his lantern to see what he might see.

His hair used to be dark brown, with grizzled streaks about the temples; in less than a month from that day he was as grey as a badger. He was a miner when he was young, and most of these fellows have seen ugly sights now and then, when accidents have happened, but he had never seen what he saw that night — that Thing that will abide neither above ground nor

vampire universe with a series of spinoff novels, as well as a new trilogy of novels exploring the lives (and un-lives) of Dracula's legendary consorts (see the bibliography accompanying "The Unicorn Tapestry" for a full cataloging of her books).

Chelsea Quinn Yarbro's fiction often blurs the distinction between horror novels and historical romances, and thus has significantly increased the overall readership for vampire literature at the millenium. Next to Anne Rice, she is the most significant vampire novelist of the twentieth century.

Chelsea Quinn Yarbro's Hôtel Transylvania.

in the grave. Antonio had brought something with him which the priest had not noticed. He had made it that afternoon — a sharp stake shaped from a piece of tough old driftwood. He had it with him now, and he had his heavy pick, and he had taken the lantern down into the grave. I don't think any power on earth could make him speak of what happened then, and the old priest was too frightened to look in. He says he heard Antonio breathing like a wild beast, and moving as if he were fighting with something almost as strong as himself; and he heard an evil sound also, with blows, as of something violently driven through flesh and bone; and then the most awful sound of all — a woman's shriek, the unearthly scream of a woman neither dead nor alive, but buried deep for many days. And he, the poor old priest, could only rock himself as he knelt there in the sand, crying aloud his prayers and exorcisms to drown these dreadful sounds. Then suddenly a small ironbound chest was thrown up and rolled over against the old man's knee, and in a moment more Antonio was beside him, his face as white as tallow in the flickering light of the lantern, shovelling the sand and pebbles into the grave with furious haste, and looking over the edge till the pit was half full; and the priest said that there was much fresh blood on Antonio's hands and on his clothes.

Bela Lugosi may have achieved world fame as Dracula, but he once cut an imposing figure as Jesus Christ on the Hungarian stage.

I had come to the end of my story. Holger finished his wine and leaned back in his chair.

"So Angelo got his own again," he said. "Did be marry the prim and plump young person to whom he had been betrothed?"

"No; he had been badly frightened. He went to South America, and has not been heard of since."

"And that poor thing's body is still there, I suppose," said Holger. "Is it quite dead yet, I wonder?"

I wonder, too. But whether it is dead or alive, I should hardly care to see it, even in broad daylight.

Antonio is as gray as a badger, and he has never been quite the same man since that night.

The Room in the Tower

E. F. BENSON
(1912)

IT IS PROBABLE that everybody who is at all a constant dreamer has had at least one experience of an event or a sequence of circumstances which have come to his mind in sleep being subsequently realized in the material world. But, in my opinion, so far from this being a strange thing, it would be far odder if this fulfilment did not occasionally happen, since our dreams are, as a rule, concerned with people whom we know and places with which we are familiar, such as might very naturally occur in the awake and daylit world. True, these dreams are often broken into by some absurd and fantastic incident, which puts them out of court in regard to their subsequent fulfilment, but on the mere calculation of chances, it does not appear in the least unlikely that a dream imagined by anyone who dreams constantly should occasionally come true. Not long ago, for instance, I experienced such a fulfilment of a dream which seems to me in no way remarkable and to have no kind of psychical significance. The manner of it was as follows.

A certain friend of mine, living abroad, is amiable enough to write to me about once in a fortnight. Thus, when fourteen days or thereabouts have elapsed since I last heard from him, my mind, probably, either consciously or subconsciously, is expectant of a letter from him. One night last week I dreamed that as I was going upstairs to dress for dinner I heard, as I often heard, the sound of the postman's knock on my front door, and diverted my direction downstairs instead. There, among other correspondence, was a letter from him. Thereafter the fantastic entered, for on opening it I found inside the ace of diamonds, and scribbled across it in his well-known handwriting, "I am sending you this for safe custody, as you know it is running an unreasonable risk to keep aces in Italy." The next evening I was just preparing to go upstairs to dress when I heard the postman's knock, and did precisely as I had done in my dream. There, among other letters, was one from my friend. Only it did not contain the ace of diamonds. Had it done so, I should have attached more weight to the matter, which, as it stands, seems to me a perfectly ordinary coincidence. No doubt I consciously or subconsciously expected a letter from him, and this suggested to me my dream. Similarly, the fact that my friend had not written to me for a fortnight suggested to him that he should do so. But occasionally it is not so easy to find such an explanation, and for the following story I can find no explanation at all. It came out of the dark, and into the dark it has gone again.

ABOUT THE AUTHOR

Edward Frederick Benson (1867-1940) is best known as the author of *Mapp and Lucia* (1931), an arch social comedy that spawned a series of novels that ultimately found their widest audience in a ten-part British television series in the 1980s. But if he was a master of mannered comedy, he was equally adept at blood-curdling ghost stories. He published three collections of supernatural tales, *The Room in the Tower* (1912), *Visible and Invisible* (1923), *Spook Stories* (1928), and *More Spook Stories* (1934). It's a rare omnibus of horror fiction that doesn't include at least one Benson story. "The Room in the Tower" is one of his most frequently reprinted works, and deservedly so.

Benson was the son of a college headmaster who eventually became the Archbishop of Canterbury. He studied archaeology at Cambridge, published his first book in 1888. It was followed by almost a hundred more in a literary career that included memoirs, novels, social commentary and biographies.

Benson wrote one other vampire story in the classic mold, "Mrs. Amworth," which has also been widely anthologized, and filmed for Canadian television in 1975 with Glynis Johns in the title role. Like "The Room in the Tower," "Mrs. Amworth" includes a highly effective depiction of a nightmare, and provides a fascinating counterpoint to the more famous "Room." In the excerpt that follows, the narrator and his friend Francis Urcombe have begun to have doubts about their jovial Sussex neighbor, whose husband died of a mysterious illness in India. And now, there are some far more local victims of a mysterious and virulent anemia...

From
MRS. AMWORTH
E. F. Benson
(1920)

I went straight up to my bedroom, of which one of the windows looks out over the steet, and as I undressed I thought I heard voices talking outside not far away. But I paid no particular attention, put out my lights, and falling asleep plunged into the depths of the most horrible dream, distortedly suggested, no doubt, by my last words with Mrs. Amworth. I dreamed that I woke, and found that both my bedroom windows were shut. Half-suffocating I dreamed that I sprang out of bed, and went across to open them. The blind over the first was drawn down, and pulling it up I saw, with the indescribable horror of incipient nightmare, Mrs. Amworth's face suspended close to the pane in the darkness outside, nodding and smiling at me. Pulling down the blind again to keep the terror out, I rushed to the second window on the other side of the room, and there again was Mrs. Amworth's face. Then the panic came upon me in full blast; here I was suffocating in the airless room, and whichever window I opened Mrs. Amworth's face would float in, like those noiseless black gnats that bit before one was aware. The nightmare rose to screaming point, and with strangled yells I awoke to find my room cool and quiet with both windows open and blinds up and a half-moon high in its course, casting an oblong of tranquil light on the floor. But even when I was awake the horror persisted, and I lay tossing and turning. I must have slept long before the nightmare seized me, for now it was nearly day, and soon in the east the drowsy eyelids of morning began to lift.

I was scarcely downstairs next morning — for after the dawn I slept late — when Urcombe rang up to know if he might see me

All my life I have been a habitual dreamer: the nights are few, that is to say, when I do not find on awaking in the morning that some mental experience has been mine, and sometimes, all night long, apparently, a series of the most dazzling adventures befall me. Almost without exception these adventures are pleasant, though often merely trivial. It is of an exception that I am going to speak.

It was when I was about sixteen that a certain dream first came to me, and this is how it befell. It opened with my being set down at the door of a big red-brick house, where, I understood, I was going to stay. The servant who opened the door told me that tea was being served in the garden, and led me through a low dark-panelled hall, with a large open fireplace, on to a cheerful green lawn set round with flower beds. There were grouped about the tea-table a small party of people, but they were all strangers to me except one, who was a school-fellow called Jack Stone, clearly the son of the house, and he introduced me to his mother and father and a couple of sisters. I was, I remember, somewhat astonished to find myself here, for the boy in question was scarcely known to me, and I rather disliked what I knew of him; moreover, he had left school nearly a year before. The afternoon was very hot, and an intolerable oppression reigned. On the far side of the lawn ran a red-brick wall, with an iron gate in its center, outside which stood a walnut tree. We sat in the shadow of the house opposite a row of long windows, inside which I could see a table with cloth laid, glimmering with glass and silver. This garden front of the house was very long, and at one end of it stood a tower of three stories, which looked to me much older than the rest of the building.

Before long, Mrs. Stone, who, like the rest of the party, had sat in absolute silence, said to me, "Jack will show you your room: I have given you the room in the tower."

Quite inexplicably my heart sank at her words. I felt as if I had known that I should have the room in the tower, and that it contained something dreadful and significant. Jack instantly got up, and I understood that I had to follow him. In silence we passed through the hall, and mounted a great oak staircase with many corners, and arrived at a small landing with two doors set in it. He pushed one of these open for me to enter, and without coming in himself, closed it after me. Then I knew that my conjecture had been right: there was something awful in the room, and with the terror of nightmare growing swiftly and enveloping me, I awoke in a spasm of terror.

Now that dream or variations on it occurred to me intermittently for fifteen years. Most often it came in exactly this form, the arrival, the tea laid out on the lawn, the deadly silence succeeded by that one deadly sentence, the mounting with Jack Stone up to the room in the tower where horror dwelt, and it always came to a close in the nightmare of terror at that which was in the room, though I never saw what it was. At other times I experienced variations on this same theme. Occasionally, for instance, we would be sitting at dinner in the dining-room, into the windows of which I had looked on the first night when the dream of this house visited me, but wherever we were, there was the same silence, the same sense of dreadful oppression and foreboding. And the silence I knew would always be broken by Mrs. Stone saying to me, "Jack will show you your room: I have given you the room in the tower." Upon which (this was invariable) I had to follow him up the oak staircase with many corners, and enter the place that I dreaded more and more each time that I visited it in sleep. Or, again, I would find myself playing cards still in silence in a drawing-room lit with immense chandeliers, that gave a blinding illumination. What the game was I have no idea; what I remember, with a sense of miserable anticipation, was that soon Mrs. Stone would get up and say to me, "Jack will show you your room: I have given you the room in the tower." This drawing-room where we played cards was next to the dining-room, and, as I have said, was always brilliantly illuminated, whereas the rest of the house was full of dusk and shadows. And yet, how often, in spite of those bouquets of lights, have I not pored over the cards that were dealt me, scarcely able for some reason to see them. Their designs, too, were strange: there were no red suits, but all were black, and among them there were certain cards which were black all over. I hated and dreaded those.

As this dream continued to recur, I got to know the greater part of the house. There was a smoking-room beyond the drawing-room, at the end of a passage with a green baize door. It was always very dark there, and as often as I went there I passed somebody whom I could not see in the doorway coming out. Curious developments, too, took place in the characters that peopled the dream as might happen to living persons. Mrs. Stone, for instance, who, when I first saw her, had been black-haired, became gray, and instead of rising briskly, as she had done at first when she said, "Jack will show you your room: I have given you the room in the tower," got up very feebly, as if the strength was leaving her limbs. Jack also grew up, and became a rather ill-looking young man, with a brown moustache, while one of the sisters ceased to appear, and I understood she was married.

Then it so happened that I was not visited by this dream for six months or more, and I began to hope, in such inexplicable dread did I hold it, that it had passed away for good. But one night after this interval I again found myself being shown out onto the lawn for tea, and Mrs. Stone was not there, while the others were all dressed in black. At once I guessed the reason, and my heart leaped at the thought that perhaps this time I should not have to sleep in the room in the tower, and though we usually all sat in silence, on this occasion the sense of relief made me talk and laugh as I had never yet done. But even then matters were not altogether comfortable, for no one else spoke, but they all looked secretly at each other. And soon the foolish stream of my talk ran dry, and gradually an apprehension worse than anything I had previously known gained on me as the light slowly faded.

immediately. He came in, grim and preoccupied, and I noticed he was pulling on a pipe that was not even filled.

"I want your help," he said, "and so I must tell you first of all what happened last night. I went round with the little doctor to see his patient, and found him just alive, but scarcely more. I instantly diagnosed in my own mind what this anemia, unaccountable by any other explanation, meant. The boy is the prey of a vampire."

He put his empty pipe on the breakfast-table, by which I had just sat down, and folded his arms, looking at me steadily from under his overhanging brows.

"Now about last night," he said. "I insisted that he should be moved from his father's cottage into my house. As we were carrying him on a stretcher, whom should we meet but Mrs. Amworth? She expressed shocked surprise that we were moving him. Now why do you think she did that?"

With a start of horror, as I remembered my dream that night before. I felt an idea come into my mind so preposterous and unthinkable that I instantly turned it out again.

"I haven't the smallest idea," I said.

"Then listen, while I tell you about what happened later. I put out all light in the room where the boy lay, and watched. One window was a little open, for I had forgotten to close it, and about midnight I heard something outside, trying apparently to push it farther open. I guessed who it was — yes, it was full twenty feet from the ground — and I peeped round the corner of the blind. Just outside was the face of Mrs. Amworth and her hand was on the frame of the window. Very softly I crept close, and then banged the window

down, and I think I just caught the tip of one of her fingers."

"But it's impossible," I cried. "How could she be floating in the air like that? And what had she come for? Don't tell me such —"

Once more, with closer grip, the remembrance of my nightmare seized me.

"I am telling you what I saw," said he. "And all night long, until it was nearly day, she was fluttering outside, like some terrible bat, trying to gain admittance. Now put together various things I have told you."

He began checking them off on his fingers.

"Number one," he said, "there was an outbreak of disease similar to that which this boy is suffering from at Peshawar, and her husband died of it. Number two: Mrs. Amworth protested against my moving the boy to my house. Number three: she, or the demon that inhabits her body, a creature powerful and deadly, tries to gain admittance. And add this, too: in medieval times there was an epidemic of vampirism here at Maxley. The vampire, so the accounts run, was found to be Elizabeth Chaston . . . I see you remember Mrs. Amworth's maiden name. Finally, the boy is stronger this morning. He certainly would not be alive if he had been visited again. And what do you make of it?"

There was a long silence, during which I found this incredible horror assuming the hues of reality.

"I have something to add," I said, "which may or may not bear on it. You say that the — the specter went away shortly before dawn."

"Yes."

Suddenly a voice which I knew well broke the stillness, the voice of Mrs. Stone, saying, "Jack will show you your room: I have given you the room in the tower." It seemed to come from near the gate in the red-brick wall that bounded the lawn, and looking up, I saw that the grass outside was sown thick with gravestones. A curious greyish light shone from them, and I could read the lettering on the grave nearest me, and it was, "In evil memory of Julia Stone." And as usual Jack got up, and again I followed him through the hall and up the staircase with many corners. On this occasion it was darker than usual, and when I passed into the room in the tower I could only just see the furniture, the position of which was already familiar to me. Also there was a dreadful odor of decay in the room, and I woke screaming.

The dream, with such variations and developments as I have mentioned, went on at intervals for fifteen years. Sometimes I would dream it two or three nights in succession; once, as I have said, there was an intermission of six months, but taking a reasonable average, I should say that I dreamed it quite as often as once in a month. It had, as is plain, something of nightmare about it, since it always ended in the same appalling terror, which so far from getting less, seemed to me to gather fresh fear every time that I experienced it. There was, too, a strange and dreadful consistency about it. The characters in it, as I have mentioned, got regularly older, death and marriage visited this silent family, and I never in the dream, after Mrs. Stone had died, set eyes on her again. But it was always her voice that told me that the room in the tower was prepared for me, and whether we had tea out on the lawn, or the scene was laid in one of the rooms overlooking it, I could always see her gravestone standing just outside the iron gate. It was the same, too, with the married daughter; usually she was not present, but once or twice she returned again, in company with a man, whom I took to be her husband. He, too, like the rest of them, was always silent. But, owing to the constant repetition of the dream, I had ceased to attach, in my waking hours, any significance to it. I never met Jack Stone again during all those years, nor did I ever see a house that resembled this dark house of my dream. And then something happened.

I had been in London in this year, up till the end of July, and during the first week in August went down to stay with a friend in a house he had taken for the summer months, in the Ashdown Forest district of Sussex. I left London early, for John Clinton was to meet me at Forest Row Station, and we were going to spend the day golfing, and go to his house in the evening. He had his motor with him, and we set off, about five of the afternoon, after a thoroughly delightful day, for the drive, the distance being some ten miles. As it was still so early we did not have tea at the club house, but waited till we should get home. As we drove, the weather, which up till then had been, though hot, deliciously fresh, seemed to me to alter in quality, and become very stagnant and oppressive, and I felt that indefinable sense of ominous apprehension that I am accustomed to before thunder. John, however, did not share my views, attributing my loss of lightness to the fact that I had lost both my matches. Events proved, however, that I was right, though I do not think that the thunderstorm that broke that night was the sole cause of my depression.

Our way lay through deep high-banked lanes, and before we had gone very far I fell asleep, and was only awakened by the stopping of the motor. And with a sudden thrill, partly of fear but chiefly of curiosity, I found myself standing in the doorway of my house of dream. We went, I half wondering whether or not I was dreaming still, through a low oak-panelled hall, and out onto the lawn, where tea was laid in the shadow of the house. It was set in flower beds, a red-brick wall, with a gate in it, bounded one side, and out beyond that was a space of rough grass with a walnut tree. The façade of the house was very long, and at one end stood a three-storied tower, markedly older than the rest.

Here for the moment all resemblance to the repeated dream ceased. There was no silent and somehow terrible family, but a large assembly of exceedingly cheerful persons, all of whom were known to me. And in spite of the horror with which the dream itself had always filled me, I felt nothing of it now that the scene of it was thus reproduced before me. But I felt intensest curiosity as to what was going to happen.

Tea pursued its cheerful course, and before long Mrs. Clinton got up. And at that moment I think I knew what she was going to say. She spoke to me, and what she said was:

"Jack will show you your room: I have given you the room in the tower."

At that, for half a second, the horror of the dream took hold of me again. But it quickly passed, and again I felt nothing more than the most intense curiosity. It was not very long before it was amply satisfied.

John turned to me.

"Right up at the top of the house," he said, "but I think you'll be comfortable. We're absolutely full up. Would you like to go and see it now? By Jove, I believe that you are right, and that we are going to have a thunderstorm. How dark it has become."

I got up and followed him. We passed through the hall, and up the perfectly familiar staircase. Then he opened the door, and I went in. And at that moment sheer unreasoning terror again possessed me. I did not know for certain what I feared: I simply feared. Then like a sudden recollection, when one remembers a name which has long escaped the memory, I knew what I feared. I feared Mrs. Stone, whose grave with the sinister inscription, "In evil memory," I had so often seen in my dream, just beyond the lawn which lay below my window. And then once more the fear passed so completely that I wondered what there was to fear, and I found myself, sober and quiet and sane, in the room in the tower, the name of which I had so often heard in my dream, and the scene of which was so familiar.

I told him of my dream, and he smiled grimly.

"Yes, and you did well to awake," he said. "That warning came from your subconscious self, which never wholly slumbers, and cried out to you of deadly danger. For two reasons, then, you must help me: one to save others, the second to save yourself."

"What do you want me to do?" I asked.

"I want you first of all to help me in watching this boy, and ensuring that she does not come near him. Eventually I want you to help me in tracking the thing down, in exposing and destroying it. It is not human: it is an incarnate fiend. What steps we shall have to take I don't yet know."

It was now eleven of the forenoon, and presently I went across to his house for a twelve-hour vigil while the slept, to come on duty again that night, so that for the next twenty-four hours either Urcombe or myself was always in the room where the boy, now getting stronger every hour, was lying. The day following was Saturday and a morning of brilliant, pellucid weather, and already when I went across to his house to resume my duty the stream of motors down to Brighton had begun. Simultaneously I saw Urcombe with a cheerful face, which boded good news of his patient, and Mrs. Amworth, with a gesture of salutation to me and a basket in her hand, walking up the broad strip of grass which bordered on the road. There we all three met. I noticed (and saw that Urcombe noticed it too) that one finger of her left hand was bandaged.

"Good morning to you both," said she. "And I hear your patient is doing well, Mr. Urcombe. I have come to bring him a bowl of jelly, and to sit with him for an hour. He

and I are great friends. I am overjoyed at his recovery."

Urcombe paused a moment, as if making up his mind, and then shot out a pointing finger at her.

"I forbid that," he said. "You shall not sit with him or see him. And you know the reason as well as I do."

I have never seen so horrible a change come over a human face as that which now blanched hers to the color of a gray mist. She put up her hands as if to shield herself from that pointing finger, which drew the sign of the cross in the air, and shrank back cowering onto the road. There was a wild hoot from a horn, a grinding of brakes, a shout — too late — from a passing car, and one long scream suddenly cut short. Her body rebounded from the roadway after the first wheel had gone over it, and the second followed. It lay there, quivering and twitching, and was still.

She was buried three days afterwards in the cemetery outside Maxley, in accordance with the wishes she had told me that she had devised about her interment, and the shock which her sudden and awful death had caused to the little community began by degrees to pass off. To two people only, Urcombe and myself, the horror of it was mitigated from the first by the nature of the relief her death brought; but, naturally enough, we kept our own counsel, and no hint of what greater horror had been thus averted was ever let slip. But, oddly enough, so it seemed to me, he was still not satisfied about something in connection with her, and would give no answers to my questions on the subject. Then as the days of a tranquil mellow September and the October that followed began to drop away like the leaves of yellowing trees, his uneasiness

I looked round it with a certain sense of proprietorship, and found that nothing had been changed from the dreaming nights in which I knew it so well. Just to the left of the door was the bed, lengthways along the wall, with the head of it in the angle. In a line with it was the fireplace and a small bookcase; opposite the door the outer wall was pierced by two lattice-paned windows, between which stood the dressing-table, while ranged along the fourth wall was the washing-stand and a big cupboard. My luggage had already been unpacked, for the furniture of dressing and undressing lay orderly on the wash-stand and toilet-table, while my dinner clothes were spread out on the coverlet of the bed. And then, with a sudden start of unexplained dismay, I saw that there were two rather conspicuous objects which I had not seen before in my dreams: one a life-sized oil painting of Mrs. Stone, the other a black-and-white sketch of Jack Stone, representing him as he had appeared to me only a week before in the last of the series of these repeated dreams, a rather secret and evil-looking man of about thirty. His picture hung between the windows, looking straight across the room to the other portrait, which hung at the side of the bed. At that I looked next, and as I looked I felt once more the horror of the nightmare seize me.

It represented Mrs. Stone as I had seen her last in my dreams: old and withered and white-haired. But in spite of the evident feebleness of body, a dreadful exuberance and vitality shone through the envelope of flesh, an exuberance wholly malign, a vitality that foamed and frothed with unimaginable evil. Evil beamed from the narrow, leering eyes; it laughed in the demon-like mouth. The whole face was instinct with some secret and appalling mirth; the hands, clasped together on the knee, seemed shaking with suppressed and nameless glee. Then I saw also that it was signed in the left-hand bottom corner, and wondering who the artist could be, I looked more closely, and read the inscription, "Julia Stone by Julia Stone."

There came a tap at the door, and John Clinton entered.

"Got everything you want?" he asked.

"Rather more than I want," said I, pointing to the picture.

He laughed.

"Hard-featured old lady," he said. "By herself, too, I remember. Anyhow she can't have flattered herself much."

"But don't you see?" said I. "It's scarcely a human face at all. It's the face of some witch, of some devil."

He looked at it more closely.

"Yes; it isn't very pleasant," he said. "Scarcely a bedside manner, eh? Yes; I can imagine getting the nightmare if I went to sleep with that close by my bed. I'll have it taken down if you like."

"I really wish you would," I said. He rang the bell, and with the help of a servant we detached the picture and carried it out onto the landing, and put it with its face to the wall.

"By Jove, the old lady is a weight," said John, mopping his forehead. "I wonder if she had something on her mind."

The extraordinary weight of the picture had struck me too. I was about to reply, when I caught sight of my own hand. There was blood on it, in considerable quantities, covering the whole palm.

"I've cut myself somehow," said I.

John gave a little startled exclamation.

"Why, I have too," he said.

Simultaneously the footman took out his handkerchief and wiped his hand with it. I saw that there was blood also on his handkerchief.

John and I went back into the tower room and washed the blood off; but neither on his hand nor on mine was there the slightest trace of a scratch or cut. It seemed to me that, having ascertained this, we both, by a sort of tacit consent, did not allude to it again. Something in my case had dimly occurred to me that I did not wish to think about. It was but a conjecture, but I fancied that I knew the same thing had occurred to him.

The heat and oppression of the air, for the storm we had expected was still undischarged, increased very much after dinner, and for some time most of the party, among whom were John Clinton and myself, sat outside on the path bounding the lawn, where we had had tea. The night was absolutely dark, and no twinkle of star or moon ray could penetrate the pall of cloud that overset the sky. By degrees our assembly thinned, the women went up to bed, men dispersed to the smoking or billiard room, and by eleven o'clock my host and I were the only two left. All the evening I thought that he had something on his mind, and as soon as we were alone he spoke.

"The man who helped us with the picture had blood on his hand, too, did you notice?" he said.

"I asked him just now if he had cut himself, and he said he supposed he had, but that he could find no mark of it. Now where did that blood come from?"

By dint of telling myself that I was not going to think about it, I had succeeded in not doing so, and I did not want, especially just at bedtime, to be reminded of it.

"I don't know," said I, "and I don't really care so long as the picture of Mrs. Stone is not by my bed."

He got up.

"But it's odd," he said. "Ha! Now you'll see another odd thing."

A dog of his, an Irish terrier by breed, had come out of the house as we talked. The door behind us into the hall was open, and a bright oblong of light shone across the lawn to the iron gate which led on to the rough grass outside, where the walnut tree stood. I saw that the dog had all his hackles up, bristling with rage and fright; his lips were curled back from his teeth, as if he was ready to spring at something, and he was growling to himself. He took not the slightest notice of his master or me, but stiffly and tensely walked across the grass to the iron gate. There he stood for a moment,

relaxed. But before the entry of November the seeming tranquility broke into a hurricane.

I had been dining one night at the far end of the village, and about eleven o'clock was walking home again. The moon was of an unusual brilliance, rendering all that it shone on as distinct as in some etching. I had just come opposite the house which Mrs. Amworth had occupied, where there was a board up telling that it was to let, when I heard the click of her front gate, and next moment I saw, with a sudden chill and quaking of my very spirit, that she stood there. Her profile, vividly illuminated, was turned to me, and I could not be mistaken in my identification of her. She appeared not to see me (indeed the shadow of the yew hedge in front of her garden enveloped me in its blackness) and she went swiftly across the road and entered the gate of the house directly opposite. There I lost sight of her completely.

My breath was coming in short pants as if I had been running — and now indeed I ran, with fearful backward glances, along the hundred yards that separated me from my house and Urcombe's. It was to his that my flying steps took me, and next minute I was within.

"What have you come to tell me?" he asked. "Or shall I guess?"

"You can't guess," said I.

"No, it's no guess. She has come back and you have seen her. Tell me about it." I gave him my story.

"That's Major Pearsall's house," he said. "Come back with me there at once."

"But what can we do?" I asked.

"I've no idea. That's what we have got to find out."

looking through the bars and still growling. Then of a sudden his courage seemed to desert him: he gave one long howl, and scuttled back to the house with a curious crouching sort of movement.

"He does that half-a-dozen times a day," said John. "He sees something which he both hates and fears."

I walked to the gate and looked over it. Something was moving on the grass outside, and soon a sound which I could not instantly identify came to my ears. Then I remembered what it was: it was the purring of a cat. I lit a match, and saw the purrer, a big blue Persian, walking round and round in a little circle just outside the gate, stepping high and ecstatically, with tail carried aloft like a banner. Its eyes were bright and shining, and every now and then it put its head down and sniffed at the grass.

I laughed.

"The end of that mystery, I am afraid." I said. "Here's a large cat having Walpurgis night all alone."

"Yes, that's Darius," said John. "He spends half the day and all night there. But that's not the end of the dog mystery, for Toby and he are the best of friends, but the beginning of the cat mystery. What's the cat doing there? And why is Darius pleased, while Toby is terror-stricken?"

At that moment I remembered the rather horrible detail of my dreams when I saw through the gate, just where the cat was now, the white tombstone with the sinister inscription. But before I could answer the rain began, as suddenly and heavily as if a tap had been turned on, and simultaneously the big cat squeezed through the bars of the gate, and came leaping across the lawn to the house for shelter. Then it sat in the doorway, looking out eagerly into the dark. It spat and struck at John with its paw, as he pushed it in, in order to close the door.

Somehow, with the portrait of Julia Stone in the passage outside, the room in the tower had absolutely no alarm for me, and as I went to bed, feeling very sleepy and heavy, I had nothing more than interest for the curious incident about our bleeding hands, and the conduct of the cat and dog. The last thing I looked at before I put out my light was the square empty space by my bed where the portrait had been. Here the paper was of its original full tint of dark red: over the rest of the walls it had faded. Then I blew out my candle and instantly fell asleep.

My awaking was equally instantaneous, and I sat bolt upright in bed under the impression that some bright light had been flashed in my face, though it was now absolutely pitch dark. I knew exactly where I was, in the room which I had dreaded in dreams, but no horror that I ever felt when asleep approached the fear that now invaded and froze my brain. Immediately after a peal of thunder crackled just above the house, but the probability that it was only a flash of lightning which awoke me gave no reassurance to my galloping heart. Something I knew was in the room with me, and instinctively I put out my right hand, which was nearest the wall, to keep it away. And my hand touched the edge of a picture-frame hanging close to me.

A minute later, we were opposite the house. When I had passed it before, it was all dark; now lights gleamed from a couple of windows upstairs. Even as we faced it, the front door opened, and next moment Major Pearsall emerged from the gate. He saw us and stopped.

"I'm on my way to Dr. Ross," he said quickly. "My wife has been taken suddenly ill. She had been in bed an hour when I came upstairs, and found her white as a ghost and utterly exhausted. She had been to sleep, it seemed — but you will excuse me."

"One moment, Major," said Urcombe. "Was there any mark on her throat?"

"How did you guess that?" said he. "There was: one of those beastly gnats must have bitten her there. She was streaming with blood."

"And there's someone with her?" asked Urcombe.

"Yes, I roused her maid." He went off, and Urcombe turned to me. "I know now what we have to do," he said. "Change your clothes, and I'll join you at your house."

"What is it?" I asked.

"I'll tell you on he way. We're going to the cemetery."

•

We had come to the cemetery, and in the brightness of the moonshine there was no difficulty in identifying her grave. It lay some twenty yards from the small chapel, in the porch of which, obscured by shadow, we concealed ourselves. From here we had a clear and open sight of the grave, and now we must wait until its infernal visitor returned home. The night was warm and windless, yet even if a freezing wind had

I sprang out of bed, upsetting the small table that stood by it, and I heard my watch, candle, and matches clatter onto the floor. But for the moment there was no need of light, for a blinding flash leaped out of the clouds, and showed me that by my bed again hung the picture of Mrs. Stone. And instantly the room went into blackness again. But in that flash I saw another thing also, namely a figure that leaned over the end of my bed, watching me. It was dressed in some close-clinging white garment, spotted and stained with mold, and the face was that of the portrait.

Overhead the thunder cracked and roared, and when it ceased and the deathly stillness succeeded, I heard the rustle of movement coming nearer me, and, more horrible yet, perceived an odor of corruption and decay. And then a hand was laid on the side of my neck, and close beside my ear I heard quick-taken, eager breathing. Yet I knew that this thing, though it could be perceived by touch, by smell, by eye and by ear, was still not of this earth, but something that had passed out of the body and had power to make itself manifest. Then a voice, already familiar to me, spoke.

"I knew you would come to the room in the tower," it said. "I have been long waiting for you. At last you have come. Tonight I shall feast; before long we will feast together."

And the quick breathing came closer to me; I could feel it on my neck.

At that the terror, which I think had paralyzed me for the moment, gave way to the wild instinct of self-preservation. I hit wildly with both arms, kicking out at the same moment, and heard a little animal-squeal, and something soft dropped with a thud beside me. I took a couple of steps forward, nearly tripping up over whatever it was that lay there, and by the merest good-luck found the handle of the door. In another second I ran out on the landing, and had banged the door behind me. Almost at the same moment I heard a door open somewhere below, and John Clinton, candle in hand, came running upstairs.

"What is it?" he said. "I sleep just below you, and heard a noise as if — Good heavens, there's blood on your shoulder."

I stood there, so he told me afterwards, swaying from side to side, white as a sheet, with the mark on my shoulder as if a hand covered with blood had been laid there.

"It's in there," I said, pointing. "She, you know. The portrait is in there, too, handing up on the place we took it from."

At that he laughed.

"My dear fellow, this is mere nightmare," he said.

He pushed by me, and opened the door, I standing there simply inert with terror, unable to stop him, unable to move.

"Phew! What an awful smell," he said.

Then there was silence; he had passed out of my sight behind the open door. Next moment he came out again, as white as myself, and instantly shut it.

"Yes, the portrait's there," he said, "and on the floor is a thing — a

been raging I think I should have felt nothing of it, so intense was my preoccupation as to what the night and dawn would bring. There was a bell in the turret of the chapel, that struck the quarters of the hour, and it amazed me to find how swiftly the chimes succeeded one another.

The moon had long set, but a twilight of stars shone in a clear sky, when five o'clock of the morning sounded from the turret. A few minutes more passed, and then I felt Urcombe's hand softly nudging me; and looking out in the direction of his pointing finger, I saw that the form of a woman, tall and large in build, was approachng from the right. Noiselessly, with a motion more of gliding and floating than walking, she moved across the cemetery to the grave which was the center of our observation. She moved round it as if to be certain of its identity, and for a moment stood directly facing us. In the greyness to which my eyes had now grown accustomed, I could easily see her face, and recognize its features.

She drew her hand across her mouth as if wiping it, and broke into a chuckle of such laughter as made my hair stir on my head. Then she leaped into the grave, holding her hands high above her head, and inch by inch disappeared into the earth. Urcombe's hand was laid on my arm, in an injunction to keep still, but now he removed it.

"Come," he said.

With pick and shovel and rope we went to the grave. The earth was light and sandy, and soon after six struck we had delved down to the coffin lid. With his pick he loosened the earth round it, and, adjusting the rope through the handles by which it had been lowered, we tried to raise it. This was a long and laborious business, and the light had begun to herald

thing spotted with earth, like what they bury people in. Come away, quick, come away."

How I got downstairs I hardly know. An awful shuddering and nausea of the spirit rather than of the flesh had seized me, and more than once he had to place my feet upon the steps, while every now and then he cast glances of terror and apprehension up the stairs. But in time we came to his dressing-room on the floor below, and there I told him what I have here described.

The sequel can be made short; indeed, some of my readers have perhaps already guessed what it was, if they remember that inexplicable affair of the churchyard at West Fawley, some eight years ago, where an attempt was made three times to bury the body of a certain woman who had committed suicide. On each occasion the coffin was found in the course of a few days again protruding from the ground. After the third attempt, in order that the thing should not be talked about, the body was buried elsewhere in unconsecrated ground. Where it was buried was just outside the iron gate of the garden belonging to the house where this woman had lived. She had committed suicide in a room at the top of the tower in that house. Her name was Julia Stone.

Subsequently the body was again secretly dug up, and the coffin was found to be full of blood.

day in the east before we had it out, and lying by the side of the grave with his screwdriver he loosened the fastenings of the lid, and slid it aside, and standing there we looked on the face of Mrs. Amworth. The eyes, once closed in death, were open, the cheeks were flushed with color, the red, full-lipped mouth seemed to smile.

"One blow and it is all over," he said. "You need not look."

Even as he spoke he took up the pick again, and, laying the point of it on her left breast, measured his distance. And though I knew what was coming I could not look away . . . .

He grasped the pick in both hands, raised it an inch or two for the taking of his aim, and then with full force brought it down on her breast. A fountan of blood, though she had been dead so long, spouted high in the air, falling with the thud of a heavy splash over the shroud, and simultaneously from those red lips came one long, appalling cry, swelling up like some hooting siren, and dying away again. With that, instantaneous as a lightning flash, came the touch of corruption on her face, the color of it faded to ash, the plump cheeks fell in, the mouth dropped.

"Thank God, that's over," said he, and without pause slipped the coffin lid back into its place.

Day was coming fast now, and, working like men possessed, we lowered the coffin into its place again, and shovelled the earth over it . . . . The birds were busy with their earliest pipings as we went back to Maxley.

Illustration from Varney the Vampyre.

Vampires

ALBIN GRAU
(1921)

IT HAPPENED DURING the war, in Serbia, in the winter of 1916. I had been commandeered along with four other unfortunates to take part in a delousing operation (may my lovely readers forgive me!). We had to cut the grass, so to speak, from under the feet of a beginning typhoid epidemic. In principle, it was very simple. Armed with high-caliber shears, we set upon the long unruly hair of the locals, usually not without a fight. But that is not the point of this story.

The wavering flame of a slow-burning lamp threw ghostly shadows in the depths of the room that served as our quarters. All five of us were seated around a huge brick fireplace, smoking our tobacco of beech leaves, hating death and staring straight ahead with a melancholy look. Suddenly, one of my comrades — his mind still caught up with our day's toils — tossed off into the darkness a question heavy with meaning. "Do you know that we are all more or less tormented by vampires?" Is it at all surprising that, in the desert that was our existence, this word immediately captured our attention? For an instant, there ruled a deathly silence...You could have heard worms chewing on the wood. Outside, a snowstorm was raging around the delapidated thatched roof; it rushed down the chimney and stirred up a shower of sparks from the dying fire. It was then that the answer came, from the darkest corner of the room. Frightened, we saw the old peasant make the sign of the cross. He drew near and, in a voice so soft you would have thought he was afraid of attracting the Evil One, whispered to us: "Before this damned war, I was over there, in Romania. You can laugh at this superstition, but I swear on the mother of God that I had the horrid experience of seeing an undead." "An undead?" one of us asked.

"Yes, an undead, or a 'Nosferatu,' which is what they call vampires over there. You have heard of them only in books about strange and fearsome creatures and you smile at these old wives' tales, but the cradle of the vampires is right here, among us, in the Balkans. We have always been pursued and tormented by these monsters." He crossed himself again. Suddenly he grew silent. "One must not speak about them at this hour," he said, looking fearfully at the clock.

We followed his glance... Midnight! Our nerves were stretched to the breaking point. We urged him as one: "Tell us!" Even today I shudder at the memory of the Serb's terrible story....

ABOUT THE AUTHOR

Little is known about Albin Grau, the German designer/painter/architect who, in a short-lived partnership with businessman Enrico Dieckmann called Prana-Film, conceived the classic expressionist vampire film *Nosferatu* (1922) and had the good sense to hire the up-and-coming director F.W. Murnau to orchestrate it. It is fairly clear that Grau was (in the words of German film historian Lotte Eisner) an "ardent spiritualist," and an active member of the secret society Ordo Templi Orientis, an offshoot of the legendary Hermetic Order of the Golden Dawn. For many years during Grau's involvement, the O.T.O. was headed by the notorious satanist Aleister Crowley.

Grau hoped that *Nosferatu* would be the first of many occult films produced by Prana-Film, but, in spite of all his careful preparation, he never bothered to secure the permission of Bram Stoker's widow to produce an adaption of *Dracula* (see following sidebar). As a result, Prana-Film was sued for copyright infringement, and was bankrupted. It is believed that Grau eventually retired to Switzerland, possibly to care for a disabled daughter, and possibly to be closer to the European headquarters of the O.T.O. To the best of anyone's recollection, he died in the 1960s.

In 2000, Grau himself was played onscreen by actor Udo Kier in E. Elias Merhige's Academy Award-nominated film *Shadow of the Vampire*, a fanciful account of the making of *Nosferatu*.

Max Schreck as Nosferatu's *vampire Count Orlock in advertising art created by Albin Grau.*
(Courtesy of Ronald V. Borst / Hollywood Movie Posters)

One day his father, who hardly led a saintly life, was chopping trees and was crushed by a falling trunk. He died without the sacraments of a priest. Four weeks after his burial, people suddenly began dying in great numbers in that solitary valley in the Carpathians! At first, everyone blamed the plague. Soon, however, a rumor took hold that turned out to be the awful truth. Here and there, peasants — with a look full of terror — insisted that they had seen the man who died without sacraments. At night, whenever he left a farm like a wisp of fog, he left a dead person behind him. Finally, under pressure from the villagers, the authorities took charge of the affair. Brave, strapping fellows, including the son of the deceased, assembled in the cemetery one night. They unearthed the coffin by the light of their torches.

The coffin was empty! "There was no longer any doubt, my father was a vampire," sobbed the old man. "We put the grave back in order and the next morning we opened it again." At this point, our guest stopped his tale, stood heavily and showed us a yellowed paper bearing an official seal. "What happened then, you can read for yourselves in this copy of the proceedings that took place at the time of this affair. I can't bring myself to talk about it anymore. Since those events, I have white hair."

One of our comrades who could read Romanian translated the text for us:

> Progatza, May 18, 1884 — Official Report concerning a dead bloodsucker or vampire spirit, in Progatza (in Romania):
>
> (After what the peasant had related to us above, it read): *"When the above-referenced Morovitch was disinterred, he appeared to have a perfectly healthy complexion without any trace of putrefaction, as if he was sleeping. His mouth was half-opened because his teeth, which had become shockingly long and pointed, would not allow him to close it. After the priest in attendance said three Our Fathers, a stake was driven into the heart of the vampire, which caused him to die while emitting loud moans. His body was burned and the ashes were cast upon the winds."*
>
> Progatza, 18 May 1884
> Pavlovitch Hodunka
> *Niki Staniko*

We did not close our eyes that night! Years have passed since. You no longer see the terror of the war in men's eyes: but some part of it has remained. Suffering and regret have shaken men's souls and, little by little, inspired the desire to understand what caused this monstrous event that swooped down on the earth like a cosmic vampire to drink the blood of millions and millions of men....

This summer, fate caused me to run into the comrade whose question that night in Serbia had launched our whole discussion. In beautiful Prague!

THE NOSFERATU AFFAIR

An unauthorized adaptation of Bram Stoker's *Dracula*, *Nosferatu: Eine Symphonie des Grauens (Nosferatu: A Symphony of Horror)* created a firestorm of controversy when Stoker's outraged widow filed suit for plagiarism, and succeeded in having the German courts order all negatives and prints of the film destroyed.

The judgment, of course, was never fully enforced, which preserved an expressionist masterpiece for posterity — while at the same time creating a decade of grief for the widow Stoker. She was repeatedly forced to take legal action whenever the film surfaced for public screenings in Europe and America. Forced into genteel poverty after her husband's death, she desperately needed to protect the dramatic copyright for an authorized film adaptation. Under the copyright laws of the period, a piracy that was not challenged carried the risk of invalidating a legitimate claim. After tumultuous negotiations, Mrs. Stoker finally sold the film rights to Universal Pictures in 1930 and was able to live comfortably on the proceeds during her remaining years.

It is a shame Florence Stoker never appreciated the artistic care and occult fervor that had fueled *Nosferatu*; indeed, many critics cite it as the finest of all *Dracula* adaptations. *Nosferatu* enthusiastically embraced Stoker's concept of the vampire as a physically repellent creature, exaggerating Stoker's already

I had long since pondered the occult aspects of life, and as though a secret power nourished the same interests in us, the first question of my former war comrade was "Do you still remember the vampire story from that night?..."

"In fact, I'm going to make a film about it," I replied. "Ah, yes! *Nosferatu*, I've heard about it," cried my friend with passion. Already he was taking me by the arm: "Come on, I know a very old inn in the neighborhood where Rudolph II, the great alchemist, and his astrologer Tycho de Brahé used to carouse."

Soon we were seated face to face at a table under the inn's sooty arches, Hungarian wine sparkling in our glasses! In the old city of Prague, the mysterious city of the Golem! It was the ideal place to hash over memories of this sort.

The spirits of great men witnessed our animated conversation until late into the night: nocturnal forces look out for the one who has lost his way and abandons the path that leads to human greatness; they seek to drag him further down to the earth with their voracious tentacles. They magnify his bestial instincts until there emerges a being endowed with an absolute animality, completely bound to the earth. While such beings are physically dead, the shadow of a terrestrial soul attaches to them, preserves them, and bestows upon them extraordinary, diabolical cunning and secret powers in the service of Evil. Enemies of light and of Good, they are driven at night to torment everything bearing the name of man.

But woe to these vampire phantoms if the beaming light of the sun brushes upon them! An innocent and child-like being can hold power over them until the cock's first crow. Then the vampire's power dissolves and its bestial soul is set free — and the cycle of human destiny must start over from the beginning.

"And where are you going now?" asked my friend as we said our goodbyes. "To Upper Tatra; we're looking for locations for *Nosferatu*." "Who is the director?" "We were able to get Murnau." "Well, *Nosferatu* is in good hands then. Good luck!" Laughing, he added, "I'm going to Venice, but don't forget to let me know the date of the premiere. I'll come, even if I'm at the other end of the world!"

Prana-Film will not forget!

From Bühne und Film, *no. 21 (1921).*

unpleasant description into one of the most hideous screen monsters of all time. As portrayed by actor Max Schreck (whose name means "terror" in German), Count Orlock is a nightmare amalgam of a withered corpse and a diseased rat, with hooklike talons that become progressively elongated as the film unreels. As photographed by Fritz Arno Wagner, the image of the monster's shadow gliding up the stairs and reaching for its victim's door remains one of the most famous compositions in all of silent film.

***Nosferatu* was a self-conscious "art" film, using the verminous vampire as a metaphor for the plaguelike destruction of Germany in World War I. The film was, upon its first release, elaborately color-tinted and accompanied by a modernist orchestral score commissioned from composer Hans Erdmann. Following the injunction won by Mrs. Stoker, the film survived in a hodgepodge of various prints until recent restorations by Enno Patalas, David Shepard, and, most definitively, by Kevin Brownlow.**

With its indelibly haunting images of death, *Nosferatu* reminds us that the film medium is itself a kind of technological bargaining chip against oblivion, allowing us a comforting illusion of perpetual life and perpetual reanimation. The death/rebirth fantasy of vampire myths comes full circle in the cinema: light destroys the vampire's shadow, but in motion-picture terms, simultaneously creates the shadows which bring vampires to life on the screen.

Advertising art by Albin Grau for Nosferatu.

The Czech castle used by Albin Grau as a setting for Nosferatu, *as it appears today. (Photo courtesy of Lokke Heiss)*

The unmistakable silhouette of Sherlock Holmes, as portrayed by Basil Rathbone.

The Adventure of the Sussex Vampire

SIR ARTHUR CONAN DOYLE
(1924)

HOLMES HAD READ carefully a note which the last post had brought him. Then, with the dry chuckle which was his nearest approach to a laugh, he tossed it over to me.

"For a mixture of the modern and the mediaeval, of the practical and of the wildly fanciful, I think this is surely the limit," said he. "What do you make of it, Watson?"

I read as follows:

46, OLD JEWRY,
Nov. 19th.
Re Vampires

SIR:

Our client, Mr. Robert Ferguson, of Ferguson and Muirhead, tea brokers, of Mincing Lane, has made some inquiry from us in a communication of even date concerning vampires. As our firm specializes entirely upon the assessment of machinery the matter hardly comes within our purview, and we have therefore recommended Mr. Ferguson to call upon you and lay the matter before you. We have not forgotten your successful action in the case of *Matilda Briggs*.

We are, sir,
Faithfully yours,
MORRISON, MORRISON, AND DODD.
per E. J. C.

"Matilda Briggs was not the name of a young woman, Watson," said Holmes in a reminiscent voice. "It was a ship which is associated with the giant rat of Sumatra, a story for which the world is not yet prepared. But what do we know about vampires? Does it come within our purview either? Anything is better than stagnation, but really we seem to have been switched on to a Grimms' fairy tale. Make a long arm, Watson, and see what V has to say."

I leaned back and took down the great index volume to which he referred. Holmes balanced it on his knee, and his eyes moved slowly and

ABOUT THE AUTHOR

Sir Arthur Conan Doyle (1859-1930) was born in Edinburgh, the son of a civil servant and his wife, who took in boarders to make ends meet. Doyle's father was both alcoholic and epileptic, and ultimately died in a mental institution. Educated by Jesuits and trained as a physician, Doyle practiced medicine for approximately five years until he became a full-time writer in 1891.

His first Sherlock Holmes story, "A Study in Scarlet," had been published in 1887, immediately creating one of the greatest literary icons of logical deduction. Despite the popularity of the stories, Doyle grew tired of the Holmes series and killed the character off in 1893. The public outcry was loud and sustained, and Doyle ultimately returned to the character in *The Hound of the Baskervilles* (1902, the year of Doyle's knighthood), a purportedly earlier, untold adventure. Doyle resurrected Holmes outright in 1903.

Five collections of Holmes stories were published between 1892 and 1927. "The Adventure of the Sussex Vampire" appeared in the final volume, *The Case-Book of Sherlock Holmes*.

lovingly over the record of old cases, mixed with the accumulated information of a lifetime.

"Voyage of the *Gloria Scott*," he read. "That was a bad business. I have some recollection that you made a record of it, Watson, though I was unable to congratulate you upon the result. Victor Lynch, the forger. Venomous lizard or gila. Remarkable case, that! Vittoria, the circus belle. Vanderbilt and the Yeggman. Vipers. Vigor, the Hammersmith wonder. Hullo! Hullo! Good old index. You can't beat it. Listen to this, Watson. Vampirism in Hungary. And again, Vampires in Transylvania." He turned over the pages with eagerness, but after a short intent perusal he threw down the great book with a snarl of disappointment.

Sir Arthur Conan Doyle

"Rubbish, Watson, rubbish! What have we to do with walking corpses who can only be held in their grave by stakes driven through their hearts? It's pure lunacy."

"But surely," said I, "the vampire was not necessarily a dead man? A living person might have the habit. I have read, for example, of the old sucking the blood of the young in order to retain their youth."

"You are right, Watson. It mentions the legend in one of these references. But are we to give serious attention to such things? This agency stands flat-footed upon the ground, and there it must remain. The world is big enough for us. No ghosts need apply. I fear that we cannot take Mr. Robert Ferguson very seriously. Possibly this note may be from him and may throw some light upon what is worrying him."

He took up a second letter which had lain unnoticed upon the table while he had been absorbed with the first. This he began to read with a smile of amusement upon his face which gradually faded away into an expression of intense interest and concentration. When he had finished he sat for some little time lost in thought with the letter dangling from his fingers. Finally, with a start, he aroused himself from his reverie.

"Cheeseman's, Lamberley. Where is Lamberley, Watson?"

"It is in Sussex, South of Horsham."

"Not very far, eh? And Cheeseman's?"

"I know that country, Holmes. It is full of old houses which are named after the men who built them centuries ago. You get Odley's and Harvey's and Carriton's — the folk are forgotten but their names live in their houses."

"Precisely," said Holmes coldly. It was one of the peculiarities of his proud, self-contained nature that though he docketed any fresh information very quietly and accurately in his brain, he seldom made any acknowledgment to the giver. "I rather fancy we shall know a good deal more about Cheeseman's, Lamberley, before we are through. The letter is, as I had hoped, from Robert Ferguson. By the way, he claims acquaintance with you."

"With me!"

"You had better read it."

He handed the letter across. It was headed with the address quoted.

DEAR MR. HOLMES [it said]:

I have been recommended to you by my lawyers, but indeed the matter is so extraordinarily delicate that it is most difficult to discuss. It concerns a friend for whom I am acting. This gentleman married some five years ago a Peruvian lady, the daughter of a Peruvian merchant, whom he had met in connection with the importation of nitrates. The lady was very beautiful, but the fact of her foreign birth and of her alien religion always caused a separation of interests and of feelings between husband and wife, so that after a time his love may have cooled towards her and he may have come to regard their union as a mistake. He felt there were sides of her character which he could never explore or understand. This was the more painful as she was as loving a wife as a man could have — to all appearance absolutely devoted.

Now for the point which I will make more plain when we meet. Indeed, this note is merely to give you a general idea of the situation and to ascertain whether you would care to interest yourself in the matter. The lady began to show some curious traits quite alien to her ordinarily sweet and gentle disposition. The gentleman had been married twice and he had one son by the first wife. This boy was now fifteen, a very charming and affectionate youth, though unhappily injured through an accident in childhood. Twice the wife was caught in the act of assaulting this poor lad in the most unprovoked way. Once she struck him with a stick and left a great weal on his arm.

This was a small matter, however, compared with her conduct to her own child, a dear boy just under one year of age. On one occasion about a month ago this child had been left by its nurse for a few minutes. A loud cry from the baby, as of pain, called the nurse back. As she ran into the room she saw her employer, the lady, leaning over the baby and apparently biting his neck. There was a small wound in the neck from which a stream of blood had escaped. The nurse was so horrified that she wished to call the husband, but the lady implored her not to do so and actually gave her five pounds as a price for her silence. No explanation was ever given, and for the moment the matter was passed over.

It left, however, a terrible impression upon the nurse's mind, and from that time she began to watch her mistress closely and to keep a closer guard upon the baby, whom she tenderly loved. It seemed to her that even as she watched the mother, so the mother watched her, and that every time she was compelled to leave the baby alone the mother was waiting to get at it. Day and night the nurse covered the child, and day and night the silent, watchful mother seemed to be lying in wait as a wolf waits for a lamb. It must read most incredible to you, and yet I beg you to take it

SIR ARTHUR CONAN DOYLE AND THE SUPERNATURAL

Although "The Adventure of the Sussex Vampire" reflects a decidedly rational, skeptical attitude toward the undead, Sir Arthur Conan Doyle himself was a remarkably uncritical believer in all manner of occult phenomena.

Despite a rigorous Jesuit education, Doyle finally turned his back on the Catholic Church — but not on the central idea of life after death. He was fascinated by spiritualism, mesmerism, and other pseudoscientific fads that captured the Victorian popular imagination. As a young man he was especially drawn to the concept of thought transference, and explored the theme fictionally in two stories of vampiric mind-control, "John Barrington Cowles" (1891) and "The Parasite" (1894), which are discussed in greater detail in connection with the psychic vampire story "Luella Miller."

As a member of the Society for Psychical Research, Doyle

seriously, for a child's life and a man's sanity may depend upon it.

At last there came one dreadful day when the facts could no longer be concealed from the husband. The nurse's nerve had given way; she could stand the strain no longer, and she made a clean breast of it all to the man. To him it seemed as wild a tale as it may now seem to you. He knew his wife to be a loving wife, and, save for the assaults upon her stepson, a loving mother. Why, then, should she wound her own dear little baby? He told the nurse that she was dreaming, that her suspicions were those of a lunatic, and that such libels upon her mistress were not to be tolerated. While they were talking a sudden cry of pain was heard. Nurse and master rushed together to the nursery.

Imagine his feelings, Mr. Holmes, as he saw his wife rise from a kneeling position beside the cot and saw blood upon the child's exposed neck and upon the sheet. With a cry of horror, he turned his wife's face to the light and saw blood all round her lips. It was she — she beyond all question — who had drunk the poor baby's blood.

So the matter stands. She is now confined to her room.

There has been no explanation. The husband is half demented. He knows, and I know, little of vampirism beyond the name. We had thought it was some wild tale of foreign parts. And yet here in the very heart of the English Sussex — well, all this can be discussed with you in the morning. Will you see me? Will you use your great powers in aiding a distracted man? If so, kindly wire to Ferguson, Cheeseman's, Lamberley, and I will be at your rooms by ten o'clock.

Yours faithfully,
ROBERT FERGUSON

P. S. I believe your friend Watson played Rugby for Blackheath when I was three-quarter for Richmond. It is the only personal introduction which I can give.

"Of course I remembered him," said I as I laid down the letter. "Big Bob Ferguson, the finest three-quarter Richmond ever had. He was always a good-natured chap. It's like him to be so concerned over a friend's case."

Holmes looked at me thoughtfully and shook his head.

"I never get your limits, Watson," said he. "There are unexplored possibilities about you. Take a wire down, like a good fellow. 'Will examine your case with pleasure.'"

"Your case!"

"We must not let him think that this agency is a home for the weak-minded. Of course it is his case. Send him that wire and let the matter rest till morning."

attended countless seances (or "table turnings," as they were called) and vociferously defended even the most egregious mediumistic frauds — usually on the basis that the psychic gift was fragile, and even the most sincere mediums needed some leeway for cheating. When one celebrated table-rapper, Margaret Fox, publicly confessed in 1888 that "Spiritualism is a fraud and deception," — assisted, in her case, by a unique ability to loudly crack her toe-joints — Doyle responded, "Nothing that she could say. . .would in the least change my opinion, nor would it that of any one else who had become profoundly convinced that there is an occult influence connecting us with an invisible world."

Similarly, Doyle carried out a ridiculous public feud with magician Harry Houdini, a notorious debunker of spiritualist claptrap. Nonetheless, Doyle insisted that Houdini himself employed occult power in his stage illusions, something the magician steadfastly denied. But

Promptly at ten o'clock next morning Ferguson strode into our room. I had remembered him as a long, slab-sided man with loose limbs and a fine turn of speed which had carried him round many an opposing back. There is surely nothing in life more painful than to meet the wreck of a fine athlete whom one has known in his prime. His great frame had fallen in, his flaxen hair was scanty, and his shoulders were bowed. I fear that I roused corresponding emotions in him.

"Hullo, Watson," said he, and his voice was still deep and hearty. "You don't look quite the man you did when I threw you over the ropes into the crowd at the Old Deer Park. I expect I have changed a bit also. But it's this last day or two that has aged me. I see by your telegram, Mr. Holmes, that it is no use my pretending to be anyone's deputy."

"It is simpler to deal direct," said Holmes.

"Of course it is. But you can imagine how difficult it is when you are speaking of the one woman whom you are bound to protect and help. What can I do? How am I to go to the police with such a story? And yet the kiddies have got to be protected. Is it madness, Mr. Holmes? Is it something in the blood? Have you any similar case in your experience? For God's sake, give me some advice, for I am at my wit's end."

"Very naturally, Mr. Ferguson. Now sit here and pull yourself together and give me a few clear answers. I can assure you that I am very far from being at my wit's end, and that I am confident we shall find some solution. First of all, tell me what steps you have taken. Is your wife still near the children?"

"We had a dreadful scene. She is a most loving woman, Mr.Holmes. If ever a woman loved a man with all her heart and soul, she loves me. She was cut to the heart that I should have discovered this horrible, this incredible, secret. She would not even speak. She gave no answer to my reproaches, save to gaze at me with a sort of wild, despairing look in her eyes. Then she rushed to her room and locked herself in. Since then she has refused to see me. She has a maid who was with her before her marriage, Dolores by name — a friend rather than a servant. She takes her food to her."

"Then the child is in no immediate danger?"

"Mrs. Mason, the nurse, has sworn that she will not leave it night or day. I can absolutely trust her. I am more uneasy about poor little Jack, for, as I told you in my note, he has twice been assaulted by her."

"But never wounded?"

"No, she struck him savagely. It is the more terrible as he is a poor little inoffensive cripple." Ferguson's gaunt features softened as he spoke of his boy. "You would think that the dear lad's condition would soften anyone's heart. A fall in childhood and a twisted spine, Mr. Holmes. But the dearest, most loving heart within."

Holmes had picked up the letter of yesterday and was reading it over. "What other inmates are there in your house, Mr. Ferguson?"

both men generated a great deal of valuable personal publicity from the controversy.

Indeed, Doyle was not beyond using the trappings of spiritualism for purely self-promotional ends. In 1925, Doyle generated front-page coverage in the *New York Times* with his public presentation of motion pictures of cavorting dinosaurs. The images, he intimated, might be an example of psychic photography. This time it was not Doyle who was the dupe: the press had swallowed hook, line and sinker advance footage from the forthcoming film version of Doyle's *The Lost World*, with special effects footage by Willis O'Brien, later the stop-motion animation genius behind *King Kong*.

But perhaps the strangest episode in Doyle's quest to prove the existence of the supernatural was the extraordinary credence he gave to purported photographs of fairies. In 1917, two Yorkshire cousins, Elsie Wright, age sixteen, and Frances Griffiths, age ten,

"Two servants who have not been long with us. One stablehand, Michael, who sleeps in the house. My wife, myself, my boy Jack, baby, Dolores, and Mrs. Mason. That is all."

"I gather that you did not know your wife well at the time of your marriage?"

"I had only known her a few weeks."

"How long had this maid Dolores been with her?"

"Some years."

"Then your wife's character would really be better known by Dolores than by you?"

"Yes, you may say so."

Holmes made a note.

"I fancy," said he, "that I may be of more use at Lamberley than here. It is eminently a case for personal investigation. If the lady remains in her room, our presence could not annoy or inconvenience her. Of course, we would stay at the inn."

Ferguson gave a gesture of relief.

"It is what I hoped, Mr. Holmes. There is an excellent train at two from Victoria if you could come."

"Of course we could come. There is a lull at present. I can give you my undivided energies. Watson, of course, comes with us. But there are one or two points upon which I wish to be very sure before I start. This unhappy lady, as I understand it, has appeared to assault both the children, her own baby and your little son?"

"That is so."

"But the assaults take different forms, do they not? She has beaten your son."

"Once with a stick and once very savagely with her hands."

"Did she give no explanation why she struck him?"

"None save that she hated him. Again and again she said so."

"Well, that is not unknown among stepmothers. A posthumous jealousy, we will say. Is the lady jealous by nature?"

"Yes, she is very jealous — jealous with all the strength of her fiery tropical love."

"But the boy — he is fifteen, I understand, and probably very developed in mind, since his body has been circumscribed in action. Did he give you no explanation of these assaults?"

"No, he declared there was no reason."

"Were they good friends at other times?"

"No, there was never any love between them."

"Yet you say he is affectionate?"

"Never in the world could there be so devoted a son. My life is his life. He is absorbed in what I say or do."

Once again Holmes made a note. For some time he sat lost in thought.

produced a pair of photographs they had taken of each other, showing them interacting with fairies that seemed to have stepped from the pages of a Victorian picture book. In a sense, they had, for it is numbingly obvious that the creatures are nothing but flat paper cutouts.

Doyle, who had taken "spirit" photos himself, rushed to the girls' defense, arguing that the images were genuine in a pair of pamphlets, followed by a book, *The Coming of the Fairies*, published in 1922. He made much of the fact that the original negatives had been examined by photographic experts, who found no evidence of tampering. But, of course, no tampering was necessary to take pictures of girls playing with paper dolls. Elsie and Frances stuck by their story until 1982, when they finally admitted the fakery.

At least part of Doyle's passion for the paranormal after World War I stemmed from the death of his youngest son, Kingsley, whom he attempted to contact with the

"No doubt you and the boy were great comrades before this second marriage. You were thrown very close together, were you not?"

"Very much so."

"And the boy, having so affectionate a nature, was devoted, no doubt, to the memory of his mother?"

"Most devoted."

"He would certainly seem to be a most interesting lad. There is one other point about these assaults. Were the strange attacks upon the baby and the assaults upon your son at the same period?"

"In the first case it was so. It was as if some frenzy had seized her, and she had vented her rage upon both. In the second case it was only Jack who suffered. Mrs. Mason had no complaint to make about the baby."

"That certainly complicates matters."

"I don't quite follow you, Mr. Holmes."

"Possibly not. One forms provisional theories and waits for time or fuller knowledge to explode them. A bad habit, Mr. Ferguson, but human nature is weak. I fear that your old friend here has given an exaggerated view of my scientific methods. However, I will only say at the present stage that your problem does not appear to me to be insoluble, and that you may expect to find us at Victoria at two o'clock."

It was evening of a dull, foggy November day when, having left our bags at the Chequers, Lamberley, we drove through the Sussex clay of a long winding lane and finally reached the isolated and ancient farmhouse in which Ferguson dwelt. It was a large, straggling building, very old in the centre, very new at the wings with towering Tudor chimneys and a lichen-spotted, high-pitched roof of Horsham slabs. The doorsteps were worn into curves, and the ancient tiles which lined the porch were marked with the rebus of a cheese and a man after the original builder. Within, the ceilings were corrugated with heavy oaken beams, and the uneven floors sagged into sharp curves. An odour of age and decay pervaded the whole crumbling building.

There was one very large central room into which Ferguson led us. Here, in a huge old-fashioned fireplace with an iron screen behind it dated 1670, there blazed and spluttered a splendid log fire.

The room, as I gazed round, was a most singular mixture of dates and of places. The half-panelled walls may well have belonged to the original yeoman farmer of the seventeenth century. They were ornamented, however, on the lower part by a line of well-chosen modern water-colours; while above, where yellow plaster took the place of oak, there was hung a fine collection of South American utensils and weapons, which had been brought, no doubt, by the Peruvian lady upstairs. Holmes rose, with that quick curiosity which sprang from his eager mind, and examined them with some care. He returned with his eyes full of thought.

"Hullo!" he cried. "Hullo!"

A spaniel had lain in a basket in the corner. It came slowly forward

help of a medium. He finally became convinced that he had, in fact, communicated with his dead son and other relatives.

Not long before his own death, Doyle offered some insights into his deep fascination with the occult. "People ask, 'What do you get from Spiritualism?' The first thing you get is that it absolutely removes all fear of death. Secondly, it bridges death for those dear ones whom we may love...it makes them intensely happy to help and comfort us, to tell us about their happy life in that world to which we are in our turn destined to come."

towards its master, walking with difficulty. Its hind legs moved irregularly and its tail was on the ground. It licked Ferguson's hand.

"What is it, Mr. Holmes?"

"The dog. What's the matter with it?"

"That's what puzzled the vet. A sort of paralysis. Spinal meningitis, he thought. But it is passing. He'll be all right soon — won't you, Carlo?"

A shiver of assent passed through the drooping tail. The dog's mournful eyes passed from one of us to the other. He knew that we were discussing his case.

"Did it come on suddenly?"

"In a single night."

"How long ago?"

"It may have been four months ago."

"Very remarkable. Very suggestive."

"What do you see in it, Mr. Holmes?"

"A confirmation of what I had already thought."

"For God's sake, what do you think, Mr. Holmes? It may be a mere intellectual puzzle to you, but it is life and death to me! My wife a would-be murderer — my child in constant danger! Don't play with me, Mr. Holmes. It is too terribly serious."

The big Rugby three-quarter was trembling all over. Holmes put his hand soothingly upon his arm.

Jeremy Brett played Sherlock Holmes in The Last Vampyre *(1992), a television adaptation of "The Adventure of the Sussex Vampire" that greatly expanded the original story. "Sussex Vampire" also inspired the 1992 Venezuelan film* Sherlock Holmes in Caracas.

"I fear that there is pain for you, Mr. Ferguson, whatever the solution may be," said he. "I would spare you all I can. I cannot say more for the instant, but before I leave this house I hope I may have something definite."

"Please God you may! If you will excuse me, gentlemen, I will go up to my wife's room and see if there has been any change."

He was away some minutes, during which Holmes resumed his examination of the curiosities upon the wall. When our host returned it was clear from his downcast face that he had made no progress. He brought with him a tall, slim, brown-faced girl.

"The tea is ready, Dolores," said Ferguson. "See that your mistress has everything she can wish."

"She verra ill," cried the girl, looking with indignant eyes at her master. "She no ask for food. She verra ill. She need doctor. I frightened stay alone with her without doctor."

Ferguson looked at me with a question in his eyes.

"I should be so glad if I could be of use."

"Would your mistress see Dr. Watson?"

"I take him. I no ask leave. She needs doctor."

"Then I'll come with you at once."

I followed the girl, who was quivering with strong emotion, up the staircase and down an ancient corridor. At the end was an iron-clamped and massive door. It struck me as I looked at it that if Ferguson tried to force his way to his wife he would find it no easy matter. The girl drew a key from her pocket, and the heavy oaken planks creaked upon their old hinges. I passed in and she swiftly followed, fastening the door behind her.

On the bed a woman was lying who was clearly in a high fever. She was only half conscious, but as I entered she raised a pair of frightened but beautiful eyes and glared at me in apprehension. Seeing a stranger, she appeared to be relieved and sank back with a sigh upon the pillow. I stepped up to her with a few reassuring words, and she lay still while I took her pulse and temperature. Both were high, and yet my impression was that the condition was rather that of mental and nervous excitement than of any actual seizure.

"She lie like that one day, two day. I 'fraid she die," said the girl.

The woman turned her flushed and handsome face towards me.

"Where is my husband?"

"He is below and would wish to see you."

"I will not see him. I will not see him." Then she seemed to wander off into delirium. "A fiend! A fiend! Oh, what shall I do with this devil?"

"Can I help you in any way?"

"No. No one can help. It is finished. All is destroyed. Do what I will, all is destroyed."

The woman must have some strange delusion. I could not see honest Bob Ferguson in the character of fiend or devil.

"Madame," I said, "your husband loves you dearly. He is deeply grieved at this happening."

Again she turned on me those glorious eyes.

"He loves me. Yes. But do I not love him? Do I not love him even to sacrifice myself rather than break his dear heart? That is how I love him. And yet he could think of me — he could speak of me so."

"He is full of grief, but he cannot understand."

"No, he cannot understand. But he should trust."

"Will you not see him?" I suggested.

"No, no, I cannot forget those terrible words nor the look upon his face. I will not see him. Go now. You can do nothing for me. Tell him only one thing. I want my child. I have a right to my child. That is the only message I can send him." She turned her face to the wall and would say no more.

I returned to the room downstairs, where Ferguson and Holmes still sat by the fire. Ferguson listened moodily to my account of the interview.

"How can I send her the child?" he said. "How do I know what strange impulse might come upon her? How can I ever forget how she rose from beside it with its blood upon her lips?" He shuddered at the recollection. "The child is safe with Mrs. Mason, and there he must remain."

A smart maid, the only modern thing which we had seen in the house, had brought in some tea. As she was serving it the door opened and a youth entered the room. He was a remarkable lad, pale-faced and fair-haired, with excitable light blue eyes which blazed into a sudden flame of

SHERLOCK HOLMES AND DRACULA

Although Sir Arthur Conan Doyle and Bram Stoker knew each other well, they never collaborated, and it would fall to later generations of writers to pair their most famous creations.

In 1978, Loren D. Estleman published *Sherlock Holmes vs. Dracula: The Adventures of the Sanguinary Count*, wherein Holmes investigates the shipwreck by which Dracula landed in England. The Holmes story is deftly interwoven with Stoker's original plot.

In the far more flamboyant *The Holmes-Dracula Files*, published the same year, Fred Saberhagen posited Dracula and Holmes as physical doppelgangers, giving the Count a nasty case of amnesia (just in case there wasn't identity confusion enough).

A four-part comic-book adventure, *A Study in Gaslight* by Martin Powell (1987), made

emotion and joy as they rested upon his father. He rushed forward and threw his arms round his neck with the abandon of a loving girl.

"Oh, daddy," he cried, "I did not know that you were due yet. I should have been here to meet you. Oh, I am so glad to see you!"

Ferguson gently disengaged himself from the embrace with some little show of embarrassment.

"Dear old chap," said he, patting the flaxen head with a very tender hand. "I came early because my friends, Mr. Holmes and Dr. Watson, have been persuaded to come down and spend an evening with us."

"Is that Mr. Holmes, the detective?"

"Yes."

The youth looked at us with a very penetrating and, as it seemed to me, unfriendly gaze.

"What about your other child, Mr. Ferguson?" asked Holmes. "Might we make the acquaintance of the baby?"

"Ask Mrs. Mason to bring baby down," said Ferguson. The boy went off with a curious, shambling gait which told my surgical eyes that he was suffering from a weak spine. Presently he returned, and behind him came a tall, gaunt woman bearing in her arms a very beautiful child, dark-eyed, golden-haired, a wonderful mixture of the Saxon and the Latin. Ferguson was evidently devoted to it, for he took it into his arms and fondled it most tenderly.

"Fancy anyone having the heart to hurt him," he muttered as he glanced down at the small, angry red pucker upon the cherub throat.

It was at this moment that I chanced to glance at Holmes and saw a most singular intentness in his expression. His face was as set as if it had been carved out of old ivory, and his eyes, which had glanced for a moment at father and child, were now fixed with eager curiosity upon something at the other side of the room. Following his gaze I could only guess that he was looking out through the window at the melancholy, dripping garden. It is true that a shutter had half closed outside and obstructed the view, but none the less it was certainly at the window that Holmes was fixing his concentrated attention. Then he smiled, and his eyes came back to the baby. On its chubby neck there was this small puckered mark. Without speaking, Holmes examined it with care. Finally he shook one of the dimpled fists which waved in front of him.

"Good-bye, little man. You have made a strange start in life. Nurse, I should wish to have a word with you in private."

He took her aside and spoke earnestly for a few minutes. I only heard the last words, which were: "Your anxiety will soon, I hope, be set at rest." The woman, who seemed to be a sour, silent kind of creature, withdrew with the child.

"What is Mrs. Mason like?" asked Holmes.

"Not very prepossessing externally, as you can see, but a heart of gold, and devoted to the child."

Holmes's arch-nemesis Professor Moriarty a criminal ally of Dracula. In Simon Hawke's *The Dracula Caper* (1988), it is not Holmes, but Doyle himself who battles the vampire. *The Tangled Skein* (a hard-to-find, limited-edition novel published by David Stuart Davies in 1992), blends the fictional worlds of *Dracula* and *The Hound of the Baskervilles*, and has been optioned for feature film production.

THE EMERGING VAMPIRE CINEMA

Sir Arthur Conan Doyle wrote "The Adventure of the Sussex Vampire" in the early 1920s, a time of rapid technological advances in motion pictures that would soon transform cinema into the major medium for vampire stories.

As early as 1915, Universal Pictures had considered bringing *Dracula* to the screen, but a combination of copyright irregularities and the distinctly horrific subject matter put the project on hold for fifteen years.

"Do you like her, Jack?" Holmes turned suddenly upon the boy. His expressive mobile face shadowed over, and he shook his head.

"Jacky has very strong likes and dislikes," said Ferguson, putting his arm round the boy. "Luckily I am one of his likes."

The boy cooed and nestled his head upon his father's breast. Ferguson gently disengaged him.

"Run away, little Jacky," said he, and he watched his son with loving eyes until he disappeared. "Now, Mr. Holmes," he continued when the boy was gone, "I really feel that I have brought you on a fool's errand, for what can you possibly do save give me your sympathy? It must be an exceedingly delicate and complex affair from your point of view."

"It is certainly delicate," said my friend with an amused smile, "but I have not been struck up to now with its complexity. It has been a case for intellectual deduction, but when this original intellectual deduction is confirmed point by point by quite a number of independent incidents, then the subjective becomes objective and we can say confidently that we have reached our goal. I had, in fact, reached it before we left Baker Street, and the rest has merely been observation and confirmation."

Ferguson put his big hand to his furrowed forehead.

"For heaven's sake, Holmes," he said hoarsely; "if you can see the truth in this matter, do not keep me in suspense. How do I stand? What shall I do? I care nothing as to how you have found your facts so long as you have really got them."

"Certainly I owe you an explanation, and you shall have it. But you will permit me to handle the matter in my own way? Is the lady capable of seeing us, Watson?"

"She is ill, but she is quite rational."

"Very good. It is only in her presence that we can clear the matter up. Let us go up to her."

"She will not see me," cried Ferguson.

"Oh, yes, she will," said Holmes. He scribbled a few lines upon a sheet of paper. "You at least have the entree, Watson. Will you have the goodness to give the lady this note?"

I ascended again and handed the note to Dolores, who cautiously opened the door. A minute later I heard a cry from within, a cry in which joy and surprise seemed to be blended.

Dolores looked out.

"She will see them. She will leesten," said she.

At my summons Ferguson and Holmes came up. As we entered the room Ferguson took a step or two towards his wife, who had raised herself in the bed, but she held out her hand to repulse him. He sank into an armchair, while Holmes seated himself beside him, after bowing to the lady, who looked at him with wide-eyed amazement.

"I think we can dispense with Dolores," said Holmes. "Oh, very well, madame, if you would rather she stayed I can see no objection. Now, Mr.

That same year, French filmmaker Louis Feuillade produced *Les Vampires*, a ten-part serial about Parisian master criminals who wore bat-like disguises. While it contained no supernatural element, the film made a distinct contribution to the iconography of vampires as stylized night-prowlers.

***Drakula's Death* (Hungary, 1921), directed by Károly Lathjay and photographed by Lajos Glasser, was the first film to exploit the character of Dracula (here spelled Drakula), although it took nothing from Bram Stoker's novel except for the master vampire's name. The story was closer in spirit to *The Cabinet of Dr. Caligari* and concerned a girl in an asylum (the celebrated Hungarian ingenue Margit Lux) victimized by an insane music master (Paul Askonas) who believes himself to be the immortal Drakula. The film itself is now lost, but a few intriguing photographs survive, showing Drakula drawing his victim through a window, and a magical wedding ceremony including a dance performed by the vampire's previous dead brides.**

Ferguson, I am a busy man with many calls, and my methods have to be short and direct. The swiftest surgery is the least painful. Let me first say what will ease your mind. Your wife is a very good, a very loving, and a very ill-used woman."

Ferguson sat up with a cry of joy.

"Prove that, Mr. Holmes, and I am your debtor forever."

"I will do so, but in doing so I must wound you deeply in another direction."

"I care nothing so long as you clear my wife. Everything on earth is insignificant compared to that."

Paul Askonas and Margit Lux in Dracula's Death *(1921).*

"Let me tell you, then, the train of reasoning which passed through my mind in Baker Street. The idea of a vampire was to me absurd. Such things do not happen in criminal practice in England. And yet your observation was precise. You had seen the lady rise from beside the child's cot with the blood upon her lips."

"I did."

"Did it not occur to you that a bleeding wound may be sucked for some other purpose than to draw the blood from it? Was there not a queen in English history who sucked such a wound to draw poison from it?"

"Poison!"

"A South American household. My instinct felt the presence of those weapons upon the wall before my eyes ever saw them. It might have been other poison, but that was what occurred to me. When I saw that little empty quiver beside the small bird-bow, it was just what I expected to see. If the child were pricked with one of those arrows dipped in curare or some other devilish drug, it would mean death if the venom were not sucked out.

An original advertising graphic for Nosferatu *(1922).*

"And the dog! If one were to use such a poison, would one not try it first in order to see that it had not lost its power? I did not foresee the dog, but at least I understand him and he fitted into my reconstruction.

"Now do you understand? Your wife feared such an attack. She saw it made and saved the child's life, and yet she shrank from telling you all the truth, for she knew how you loved the boy and feared lest it break your heart."

"Jacky!"

"I watched him as you fondled the child just now. His face was clearly reflected in the glass of the window where the shutter formed a background. I saw such jealousy, such cruel hatred, as I have seldom seen in a human face."

"My Jacky!"

"You have to face it, Mr. Ferguson. It is the more painful because it is a distorted love, a maniacal exaggerated love for you, and possibly for his dead mother, which has prompted his action. His very soul is consumed with hatred for this splendid child, whose health and beauty are a contrast to his own weakness."

"Good God! It is incredible!"

"Have I spoken the truth, madame?"

The lady was sobbing, with her face buried in the pillows. Now she turned to her husband.

"How could I tell you, Bob? I felt the blow it would be to you. It was better that I should wait and that it should come from some other lips than mine. When this gentleman, who seems to have powers of magic, wrote that he knew all, I was glad."

"I think a year at sea would be my prescription for Master Jacky," said Holmes, rising from his chair. "Only one thing is still clouded, madame. We can quite understand your attacks upon Master Jacky. There is a limit to a mother's patience. But how did you dare to leave the child these last two days?"

"I had told Mrs. Mason. She knew."

"Exactly. So I imagined."

Ferguson was standing by the bed, choking, his hands outstretched and quivering.

"This, I fancy, is the time for our exit, Watson," said Holmes in a whisper. "If you will take one elbow of the too faithful Dolores, I will take the other. There, now," he added as he closed the door behind him, "I think we may leave them to settle the rest among themselves."

I have only one further note of this case. It is the letter which Holmes wrote in final answer to that with which the narrative begins. It ran thus:

BAKER STREET,
Nov. 21st.
Re Vampires

SIR:

Referring to your letter of the 19th, I beg to state that I have looked into the inquiry of your client, Mr. Robert Ferguson, of Ferguson and Muirhead, tea brokers, of Mincing Lane, and that the matter has been brought to a satisfactory conclusion. With thanks for your recommendation, I am, sir,

Faithfully yours,
SHERLOCK HOLMES.

***Nosferatu: A Symphony of Horror*, directed by F.W. Murnau, was a completely unauthorized adaptation of the novel *Dracula*, resulting in a protracted copyright infringement suit by Bram Stoker's widow, and the near-destruction of the film. (For a more detailed discussion of the controversy, see the commentary accompanying Albin Grau's "Vampires.") The main characters of *Dracula* are recognizable, though freely adapted in a tightened, almost fairy-tale like narrative. The basics of Stoker's plot are adhered to, though the Professor Van Helsing role is reduced almost to inconsequence. And *Nosferatu* has a completely different conclusion than *Dracula*, finishing on a tragic note as the heroine sacrifices her own life by keeping the hideous Count Orlock (played with peerless repulsivity by German stage actor Max Schreck) at her bedside until daylight, when he is destroyed by the sun's first rays.**

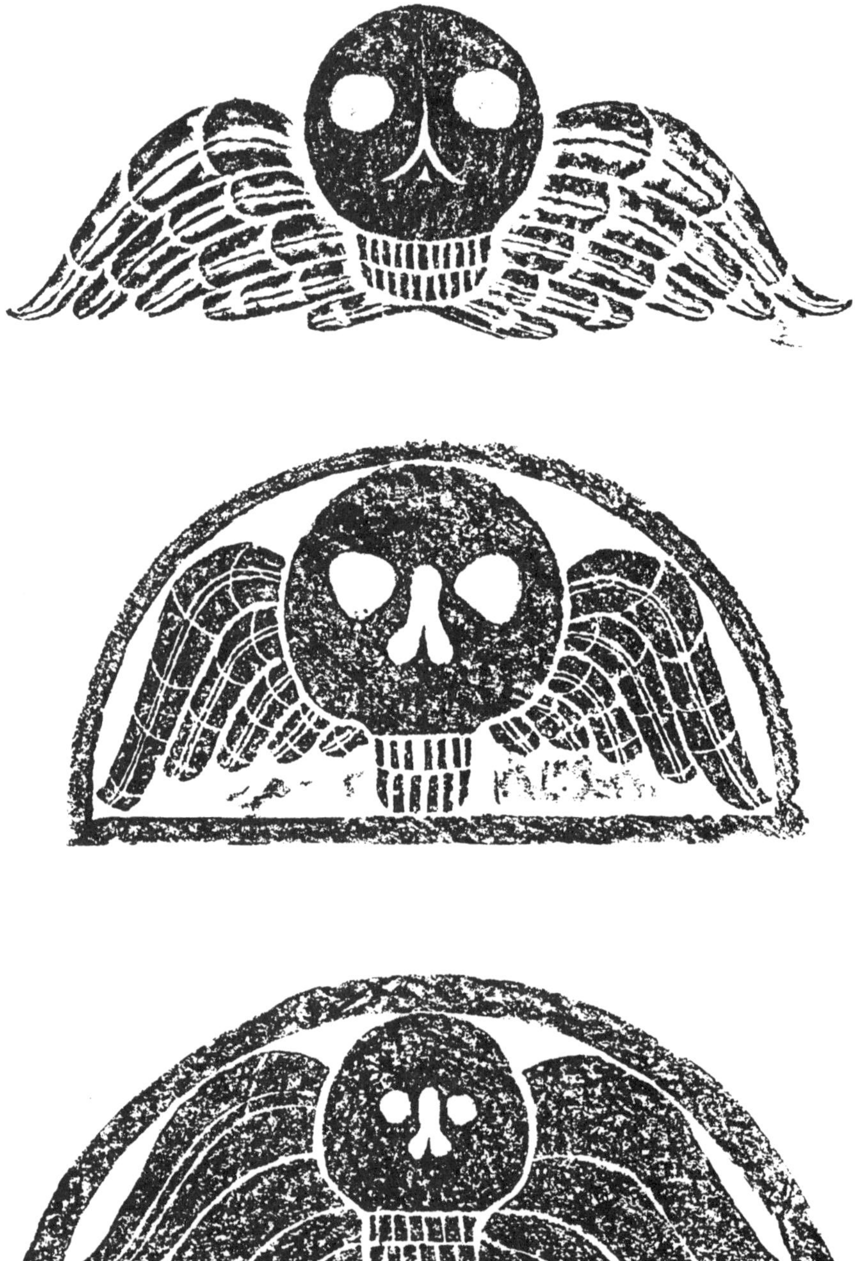

Early New England tombstone rubbings.

Bewitched

EDITH WHARTON
(1927)

THE SNOW WAS still falling thickly when Orrin Bosworth, who farmed the land south of Lonetop, drove up in his cutter to Saul Rutledge's gate. He was surprised to see two other cutters ahead of him. From them descended two muffled figures. Bosworth, with increasing surprise, recognized Deacon Hibben, from North Ashmore, and Sylvester Brand, the widower, from the old Bearcliff farm on the way to Lonetop. —

It was not often that anybody in Hemlock County entered Saul Rutledge's gate; least of all in the dead of winter, and summoned (as Bosworth, at any rate, had been) by Mrs. Rutledge, who passed, even in that unsocial region, for a woman of cold manners and solitary character. The situation was enough to excite the curiosity of a less imaginative man than Orrin Bosworth.

As he drove in between the broken-down white gate-posts topped by fluted urns the two men ahead of him were leading their horses to the adjoining shed. Bosworth followed, and hitched his horse to a post. Then the three tossed off the snow from their shoulders, clapped their numb hands together, and greeted each other.

"Hallo, Deacon."

"Well, well, Orrin —." They shook hands.

"'Day, Bosworth," said Sylvester Brand, with a brief nod. He seldom put any cordiality into his manner, and on this occasion he was still busy about his horse's bridle and blanket.

Orrin Bosworth, the youngest and most communicative of the three, turned back to Deacon Hibben, whose long face, queerly blotched and mouldy-looking, with blinking peering eyes, was yet less forbidding than Brand's heavily-hewn countenance.

"Queer, our all meeting here this way. Mrs. Rutledge sent me a message to come," Bosworth volunteered.

The Deacon nodded. "I got a word from her too — Andy Pond come with it yesterday noon. I hope there's no trouble here."

He glanced through the thickening fall of snow at the desolate front of the Rutledge house, the more melancholy in its present neglected state because, like the gate-posts, it kept traces of former elegance. Bosworth had often wondered how such a house had come to be built in that lonely stretch between North Ashmore and Cold Corners. People said there had once been other houses like it, forming a little township called Ashmore, a

ABOUT THE AUTHOR

Edith Wharton (1862-1937) was born Edith Newbold Jones, daughter of one of New York's most socially prominent families. Privately educated, she was a precocious writer and published poetry anonymously in the *Atlantic Monthly* while still a teenager. She married Boston banker Edward Wharton and settled in Newport, but found the life of a wealthy socialite stultifying, and turned to writing as a full-time career, much to her family's disapproval.

In 1899 she moved to Lenox, Massachusetts, where her mansion, the Mount, became the seat of her literary activities. There she wrote such well-known novels as *The Valley of Decision* (1902) and *The House of Mirth* (1905). In 1907 she relocated permanently to France, where she wrote her most acclaimed novels, *Ethan Frome* (1911) and *The Age of Innocence* (1920), which won her a Pulitzer Prize.

Like her friend and mentor Henry James, author of "The Turn of the Screw," Wharton also proved herself adept at supernatural fiction that combined goosebumps with sharp social and psychological observation. Her published volumes of ghost stories included *Tales of Men and Ghosts* (1910) and *Here and Beyond* (1926), in which "Bewitched" originally appeared.

sort of mountain colony created by the caprice of an English Royalist officer, one Colonel Ashmore, who had been murdered by the Indians, with all his family, long before the Revolution. This tale was confirmed by the fact that the ruined cellars of several smaller houses were still to be discovered under the wild growth of the adjoining slopes, and that the Communion plate of the moribund Episcopal church of Cold Corners was engraved with the name of Colonel Ashmore, who had given it to the church of Ashmore in the year 1723. Of the church itself no traces remained. Doubtless it had been a modest wooden edifice, built on piles, and the conflagration which had burnt the other houses to the ground's edge had reduced it utterly to ashes. The whole place, even in summer, wore a mournful solitary air, and people wondered why Saul Rutledge's father had gone there to settle.

"I never knew a place," Deacon Hibben said, "as seemed as far away from humanity. And yet it ain't so in miles."

"Miles ain't the only distance," Orrin Bosworth answered; and the two men, followed by Sylvester Brand, walked across the drive to the front door. People in Hemlock County did not usually come and go by their front doors, but all three men seemed to feel that, on an occasion which appeared to be so exceptional, the usual and more familiar approach by the kitchen would not be suitable.

Edith Wharton

They had judged rightly; the Deacon had hardly lifted the knocker when the door opened and Mrs. Rutledge stood before them.

"Walk right in," she said in her usual dead-level tone; and Bosworth, as he followed the others, thought to himself: "Whatever's happened she's not going to let it show in her face."

It was doubtful, indeed, if anything unwonted could be made to show in Prudence Rutledge's face, so limited was its scope, so fixed were its features. She was dressed for the occasion in a black calico with white spots, a collar of crochet-lace fastened by a gold brooch, and a gray woollen shawl crossed under her arms and tied at the back. In her small narrow head the only marked prominence was that of the brow projecting roundly over pale spectacled eyes. Her dark hair, parted above this prominence, passed tight and flat over the tips of her ears into a small braided coil at the nape; and her contracted head looked still narrower from being perched on a long hollow neck with cord-like throat-muscles. Her eyes were of a pale cold gray, her complexion was an even white. Her age might have been thirty-five to sixty.

The room into which she led the three men had probably been the dining room of the Ashmore house. It was now used as a front parlour, and a black stove planted on a sheet of zinc stuck out from the delicately fluted panels of an old wooden mantel. A newly-lit fire smouldered reluctantly, and the room was at once close and bitterly cold.

"Andy Pond," Mrs. Rutledge cried to some one at the back of the house, "step out and call Mr. Rutledge. You'll likely find him in the wood-shed, or round the barn somewheres." She rejoined her visitors. "Please suit yourselves to seats," she said.

The three men, with an increasing air of constraint, took the chairs she pointed out, and Mrs. Rutledge sat stiffly down upon a fourth, behind a rickety bead-work table. She glanced from one to the other of her visitors.

"I presume you folks are wondering what it is I asked you to come here for," she said in her dead-level voice. Orrin Bosworth and Deacon Hibben murmured an assent; Sylvester Brand, sat silent, his eyes, under their great thicket of eyebrows, fixed on the huge boot-tip swinging before him.

"Well, I allow you didn't expect it was for a party," continued Mrs. Rutledge.

No one ventured to respond to this chill pleasantry, and she continued: "We're in trouble here, and that's the fact. And we need advice — Mr. Rutledge and myself do." She cleared her throat, and added in a lower tone, her pitilessly clear eyes looking straight before her: "There's a spell been cast over Mr. Rutledge."

The Deacon looked up sharply, an incredulous smile pinching his thin lips. "A spell?"

"That's what I said: he's bewitched."

Again the three visitors were silent; then Bosworth, more at ease or less tongue-tied than the others, asked with an attempt at humour: "Do you use the word in the strict Scripture sense, Mrs. Rutledge?"

She glanced at him before replying: "That's how *he* uses it."

The Deacon coughed and cleared his long rattling throat. "Do you care to give us more particulars before your husband joins us?"

Mrs. Rutledge looked down at her clasped hands, as if considering the question. Bosworth noticed that the inner fold of her lids was of the same uniform white as the rest of her skin, so that when she dropped them her rather prominent eyes looked like the sightless orbs of a marble statue. The impression was unpleasing, and he glanced away at the text over the mantelpiece, which read:

The Soul That Sinneth It Shall Die.

"No," she said at length, "I'll wait."

At this moment Sylvester Brand suddenly stood up and pushed back his chair. "I don't know," he said, in his rough bass voice, "as I've got any particular lights on Bible mysteries; and this happens to be the day I was to go down to Starkfield to close a deal with a man."

Mrs. Rutledge lifted one of her long thin hands. Withered and wrinkled by hard work and cold, it was nevertheless of the same leaden white as her face. "You won't be kept long," she said. "Won't you be seated?"

Farmer Brand stood irresolute, his purplish underlip twitching. "The Deacon here — such things is more in his line..."

"I want you should stay," said Mrs. Rutledge quietly and Brand sat down again.

REVIEWS OF "BEWITCHED"

***Times Literary Supplement* (September 2, 1926): "The success of this story is in its suggestion of bleakness, of wintry skies and temperaments, and in the figure of the farmer's wife, Mrs. Rutledge — a figure of extraordinary dourness."**

***Catholic World* (August 1926): "In this difficult medium [of "psychic" tales] Mrs. Wharton works with the sure touch of an Emily Brontë, although with more restraint."**

***Literary Digest International Review* (June 1926): "...it is the essentially New England story of 'Bewitched' which is the most notable of a notable collection. In its superbly drawn portrait of the self-contained Mrs. Rutledge, in those of the three men she summoned that snowy winter day, to the place which seemed 'so far away from humanity,' in theme and in the spirit pervading it all, the tale belongs absolutely to that one section of the United States. Mrs. Rutledge's way of using her husband's explanation to further her own purposes is as completely in character as are those last words of hers, spoken as she enters the cutter to drive away from the graveyard after the funeral. Though on a smaller scale, the story is carved from the same material as *Ethan Frome*, and in the same manner of an austere and impressive simplicity, a wonderful example of the difficult art of short story writing."**

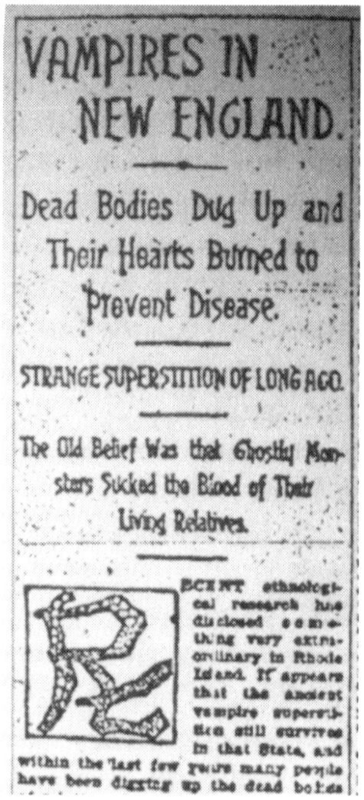

VAMPIRES IN NEW ENGLAND.

Dead Bodies Dug Up and Their Hearts Burned to Prevent Disease.

STRANGE SUPERSTITION OF LONG AGO.

The Old Belief Was that Ghostly Monsters Sucked the Blood of Their Living Relatives.

RECENT ethnological research has disclosed something very extraordinary in Rhode Island. It appears that the ancient vampire superstition still survives in that State, and within the last few years many people have been digging up the dead bodies

Edith Wharton thoroughly understood the morals and manners of New England, including its backwood superstitions. Vampire belief was still widespread around the turn of the century, and the following press clipping from the New York World *for February 2, 1896 chronicles events that transpired not far from Wharton's home in Newport, Rhode Island.*

A silence fell, during which the four persons present seemed all to be listening for the sound of a step; but none was heard, and after a minute or two Mrs. Rutledge began to speak again.

"It's down by that old shack on Lamer's pond; that's where they meet," she said suddenly.

Bosworth, whose eyes were on Sylvester Brand's face, fancied he saw a sort of inner flush darken the farmer's heavy leathern skin. Deacon Hibben leaned forward, a glitter of curiosity in his eyes.

"They — *who*, Mrs. Rutledge?"

"My husband, Saul Rutledge... and her..."

Sylvester Brand again stirred in his seat. "Who do you mean by *her*?" he asked abruptly, as if roused out of some far-off musing.

Mrs. Rutledge's body did not move; she simply revolved her head on her long neck and looked at him.

"Your daughter, Sylvester Brand."

The man staggered to his feet with an explosion of inarticulate sounds. "My — my daughter? What the hell are you talking about? My daughter? It's a damned lie... it's... it's..."

"Your daughter *Ora*, Mr. Brand," said Mrs. Rutledge slowly.

Bosworth felt an icy chill down his spine. Instinctively he turned his eyes away from Brand, and they rested on the mildewed countenance of Deacon Hibben. Between the blotches it had become as white as Mrs. Rutledge's, and the Deacon's eyes burned in the whiteness like live embers among ashes.

Brand gave a laugh: the rusty creaking laugh of one whose springs of mirth are never moved by gaiety. "My daughter *Ora*?" he repeated.

"Yes."

"My *dead* daughter?"

"That's what he says."

"Your husband?"

"That's what Mr. Rutledge says."

Orrin Bosworth listened with a sense of suffocation; he felt as if he were wrestling with long-armed horrors in a dream. He could no longer resist letting his eyes return to Sylvester Brand's face. To his surprise it had resumed a natural imperturbable expression. Brand rose to his feet. "Is that all?" he queried contemptuously.

"All? Ain't it enough? How long is it since you folks seen Saul Rutledge, any of you?" Mrs. Rutledge flew out at them.

Bosworth, it appeared, had not seen him for nearly a year; the Deacon had only run across him once, for a minute, at the North Ashmore post office, the previous autumn, and acknowledged that he wasn't looking any too good then. Brand said nothing, but stood irresolute.

"Well, if you wait a minute you'll see with your own eyes; and he'll tell you with his own words. That's what I've got you here for — to see for yourselves what's come over him. Then you'll talk different," she added, twisting her head abruptly toward Sylvester Brand.

The Deacon raised a lean hand of interrogation.

"Does your husband know we've been sent for on this business, Mrs. Rutledge?"

Mrs. Rutledge signed assent.

"It was with his consent, then —?"

She looked coldly at her questioner. "I guess it had to be," she said. Again Bosworth felt the chill down his spine. He tried to dissipate the sensation by speaking with an affectation of energy.

"Can you tell us, Mrs. Rutledge, how this trouble you speak of shows itself what makes you think...?"

She looked at him for a moment; then she leaned forward across the rickety bead-work table. A thin smile of disdain narrowed her colourless lips. "I don't think — I know."

"Well — but how?"

She leaned closer, both elbows on the table, her voice dropping. "I seen 'em."

In the ashen light from the veiling of snow beyond the windows the Deacon's little screwed-up eyes, seemed to give out red sparks. "Him and the dead?"

"Saul Rutledge and — and Ora Brand?"

"That's so."

Sylvester Brand's chair fell backward with a crash. He was on his feet again, crimson and cursing. "It is a God-damned fiend-begotten lie..."

"Friend Brand... friend Brand," the Deacon protested.

"Here, let me get out of this. I want to see Saul Rutledge himself, and tell him — "

"Well, here he is," said Mrs. Rutledge.

The outer door had opened; they heard the familiar stamping and shaking of a man who rids his garments of their last snowflakes before penetrating to the sacred precincts of the best parlour. Then Saul Rutledge entered.

II

As he came in he faced the light from the north window, and Bosworth's first thought was that he looked like a drowned man fished out from under the ice — "self-drowned," he added. But the snow-light plays cruel tricks with a man's colour, and even with the shape of his features; it must have been partly that, Bosworth reflected, which transformed Saul Rutledge from the straight muscular fellow he had been a year before into the haggard wretch now before them.

The Deacon sought for a word to ease the horror. "Well, now, Saul — you look's if you'd ought to set right up to the stove. Had a touch of ague, maybe?"

VAMPIRES IN NEW ENGLAND

Recent ethnological research has disclosed something very extraordinary in Rhode Island. It appears that the ancient vampire superstition still survives in that State, and within the last few years many people have been digging up the dead bodies of relatives for the purpose of burning their hearts.

Near Newport scores of such exhumations have been made, the purpose being to prevent the dead from preying upon the living. The belief entertained is that a person who has died of consumption is likely to rise from the grave at night and suck the blood of surviving members of his or her family, thus dooming them to a similar fate.

The discovery of the survival in highly educated New England of a superstition dating back to the days of Sardanapalus and Nebuchadnezzar has been made by George R. Stetson, an ethnologist of repute. He has found it rampant in a district which includes the towns of Exeter, Foster, Kingstown, East Greenwich and many scattered hamlets. This region, where abandoned farms are numerous, is the tramping-ground of the book agent, the chromo peddler and the patent medicine man. The social isolation is as complete as it was two centuries ago.

The feeble attempt was unavailing. Rutledge neither moved nor answered. He stood among them silent, incommunicable, like one risen from the dead.

Brand grasped him roughly by the shoulder. "See here, Saul Rutledge, what's this dirty lie your wife tells us you've been putting about?"

Still Rutledge did not move. "It's no lie," he said.

Brand's hand dropped from his shoulder. In spite of the man's rough bullying power he seemed to be undefinably awed by Rutledge's look and tone.

"No lie? You've gone plumb crazy, then, have you?"

Mrs. Rutledge spoke. "My husband's not lying, nor he ain't gone crazy. Don't I tell you I seen 'em?"

Brand laughed again. "Him and the dead?"

"Yes."

"Down by the Lamer pond, you say?"

"Yes."

"And when was that, if I might ask?"

"Day before yesterday."

A silence fell on the strangely assembled group. The Deacon at length broke it to say to Mr. Brand: "Brand, in my opinion we've got to see this thing through."

Brand stood for a moment in speechless contemplation: there was something animal and primitive about him, Bosworth thought, as he hung thus, glowering and dumb, a little foam beading the corners of that heavy purplish underlip. He let himself slowly down into his chair. "I'll see it through."

The two other men and Mrs. Rutledge had remained seated. Saul Rutledge stood before them, like a prisoner at the bar, or rather like a sick man before the physicians who were to heal him. As Bosworth scrutinized that hollow face, so wan under the dark sunburn, so sucked inward and consumed by some hidden fever, there stole over the sound healthy man the thought that perhaps, after all, husband and wife spoke the truth, and that they were all at that moment really standing on the edge of some forbidden mystery. Things that the rational mind would reject without a thought seemed no longer so easy to dispose of as one looked at the actual Saul Rutledge and remembered the man he had been a year before. Yes; as the Deacon said, they would have to see it through...

"Sit down then, Saul; draw up to us, won't you?" the Deacon suggested, trying again for a natural tone.

Mrs. Rutledge pushed a chair forward, and her husband sat down on it. He stretched out his arms and grasped his knees in his brown bony fingers; in that attitude he remained, turning neither his head nor his eyes.

"Well, Saul," the Deacon continued, "your wife says, you thought mebbe we could do something to help you through this trouble, whatever it is."

Two Typical Cases

There is one small village distant fifteen miles from Newport, where within the last few years there have been at least half a dozen resurrections on this account. The most recent was made two years ago in a family where the mother and four children had already succumbed to consumption. The last of these children was exhumed and the heart was burned.

Another instance was noted in a seashore town, not far from Newport, possessing a summer hotel and a few cottages of hot-weather residents. An intelligent man, by trade a mason, informed Mr. Stetson that he had lost two brothers by consumption. On the death of the second brother, his father was advised to take up the body and burn the heart. He refused to do so, and consequently he was attacked by the disease. Finally he died of it. His heart was burned, and in this way the rest of the family escaped.

This frightful superstition is said to prevail in all of the isolated districts of Southern Rhode Island, and it survives to some extent in the large centres of population. Sometimes the body is burned, not merely the heart, and the ashes are scattered.

In some parts of Europe the belief still has a hold on the popular mind. On the Continent from 1727 to 1735 there prevailed an epidemic of vampires. Thousands

Rutledge's gray eyes widened a little. "No; I didn't think that. It was her idea to try what could be done."

"I presume, though, since you've agreed to our coming, that you don't object to our putting a few questions?"

Rutledge was silent for a moment; then he said with a visible effort: "No; I don't object."

"Well — you've heard what your wife says?"

Rutledge made a slight motion of assent.

"And — what have you got to answer? How do you explain...?"

Mrs. Rutledge intervened. "How can he explain? I seen 'em."

There was a silence; then Bosworth, trying to speak in an easy, reassuring tone, queried: "That so, Saul?"

"That's so."

Brand lifted up his brooding head. "You mean to say you...you sit here before us all and say... "

The Deacon's hand again checked him. "Hold on, friend Brand. We're all of us trying for the facts, ain't we?" He turned to Rutledge. "We've heard what Mrs. Rutledge says. What's your answer?"

"I don't know as there's any answer. She found us."

"And you mean to tell me the person with you was...was what you took to be..." the Deacon's thin voice grew thinner: "Ora Brand?"

Saul Rutledge nodded.

"You knew...or thought you knew...you were meeting with the dead?"

Rutledge bent his head again. The snow continued to fall in a steady unwavering sheet against the window, and Bosworth felt as if a winding-sheet were descending from the sky to envelop them all in a common grave.

"Think what you're saying! It's against our religion! Ora...poor child!...Died over a year ago. I saw you at her funeral, Saul. How can you make such a statement?"

"What else can he do?' thrust in Mrs. Rutledge.

There was another pause. Bosworth's resources had failed him, and Brand once more sat plunged in dark meditation. The Deacon laid his quivering fingertips together, and moistened his lips.

"Was the day before yesterday the first time?" he asked.

The movement of Rutledge's head was negative.

"Not the first? Then when..."

"Nigh on a year ago, I reckon."

"God! And you mean to tell us that ever since — ?"

"Well... look at him," said his wife. The three men lowered their eyes.

After a moment Bosworth, trying to collect himself, glanced at the Deacon. "Why not ask Saul to make his own statement, if that's what we're here for?"

"That's so," the Deacon assented. He turned to Rutledge. "Will you try and give us your idea...of...of how it began?"

Above, and on following pages: examples of early New England gravestone rubbings.

There was another silence. Then Rutledge tightened his grasp on his gaunt knees, and still looking straight ahead with his curiously clear unseeing gaze: "Well," he said, "I guess it begun away back, afore even I was married to Mrs. Rutledge..." He spoke in a low automatic tone, as if some invisible agent were dictating his words, or even uttering them for him. "You know," he added, "Ora and me was to have been married."

Sylvester Brand lifted his head. "Straighten that statement out first, please," he interjected.

"What I mean is, we kept company. But Ora she was very young. Mr. Brand here he sent her away. She was gone nigh to three years, I guess. When she come back I was married."

"That's right," Brand said, relapsing once more into his sunken attitude.

"And after she came back did you meet her again?" the Deacon continued.

"Alive?" Rutledge questioned.

A perceptible shudder ran through the room. "Well of course," said the Deacon nervously.

Rutledge seemed to consider. "Once I did — only once. There was a lot of other people round. At Cold Corners fair it was."

"Did you talk with her then?"

"Only for a minute."

"What did she say?"

His voice dropped. "She said she was sick and knew she was going to die, and when she was dead she'd come back to me."

"And what did you answer?"

"Nothing."

"Did you think anything of it at the time?"

"Well, no. Not till I heard she was dead I didn't. After that I thought of it — and I guess she drew me." He moistened his lips.

"Drew you down to that abandoned house by the pond?"

Rutledge made a faint motion of assent, and the Deacon added: "How did you know it was there see wanted you to come?"

"She... just drew me..."

There was a long pause. Bosworth felt, on himself and the other two men, the oppressive weight of the next question to be asked. Mrs. Rutledge opened and closed her narrow lips once or twice, like some beached shell-fish gasping for the tide. Rutledge waited.

"Well, now, Saul, won't you go on with what you was telling us?" the Deacon at length suggested.

"That's all. There's nothing else."

The Deacon lowered his voice. "She just draws you?"

"Yes."

"Often?"

"That's as it happens..."

"But if it's always there she draws you, man, haven't you the strength to keep away from the place?"

For the first time, Rutledge wearily turned his head toward his questioner. A spectral smile narrowed his colourless lips. "Ain't any use. She follers after me… "

There was another silence. What more could they ask, then and there? Mrs. Rutledge's presence checked the next question. The Deacon seemed hopelessly to revolve the matter. At length he spoke in a more authoritative tone. "These are forbidden things. You know that, Saul. Have you tried prayer?"

Rutledge shook his head.

"Will you pray with us now?"

Rutledge cast a glance of freezing indifference on his spiritual adviser. "If you folks want to pray, I'm agreeable," he said. But Mrs. Rutledge intervened.

"Prayer ain't any good. In this kind of thing it ain't no manner of use; you know it ain't. I called you here, Deacon, because you remember the last case in this parish. Thirty years ago it was, I guess; but you remember. Lefferts Nash, did praying help him? *I* was a little girl then, but I used to hear my folks talk of it winter nights. Lefferts Nash and Hannah Cory. They drove a stake through her breast. That's what cured him."

"Oh — " Orrin Bosworth exclaimed.

Sylvester Brand raised his head. "You're speaking of that old story as if this was the same sort of timing?"

"Ain't it? Ain't my husband pining away the same as Lefferts Nash did? The Deacon here knows —"

The Deacon stirred anxiously in his chair. "These are forbidden things," he repeated. "Supposing your husband is quite sincere in thinking himself haunted, as you might say. Well, even then, what proof have we that the… the dead woman… is the spectre of that poor girl?"

"Proof? Don't he say so? Didn't she tell him? Ain't I seen 'em?" Mrs.Rutledge almost screamed.

The three men sat silent, and suddenly the wife burst out: "A stake through the breast! That's the old way; and it's the only way. The Deacon knows it!"

"It's against our religion to disturb the dead."

"Ain't it against your religion to let the living perish as my husband is perishing?" She sprang up with one of her abrupt movements and took the family Bible from the what-not in a corner of the parlour. Putting the book on the table, and moistening a livid finger-tip, she turned the pages rapidly, till she came to one on which she laid her hand like a stony paperweight. "See, here," she, said, and read out in her level chanting voice:

" '*Thou shalt not suffer a witch to live.*'

"That's in Exodus, that's where it is," she added, leaving the book open as if to confirm the statement.

Bosworth continued to glance anxiously from one to the other of the

of people died, as was supposed, from having their blood sucked by creatures that came to their bedsides at night with goggling eyes and lips eager for the life fluid of the victim. In Servia it was understood that the demon might be destroyed by digging up the body and piercing it through with a sharp instrument, after which it was decapitated and burned. Relief was found in eating the earth of the vampire's grave. In the Levant the corpse was cut to pieces and boiled in wine.

Vampirism a Plague

Vampirism became a plague, more dreaded than any form of disease. Everywhere people were dying from the attacks of the blood-sucking monsters, each victim becoming in turn a night-prowler in pursuit of human prey. Terror of the mysterious and unearthly peril filled all hearts.

The contents of every suspected grave were investigated, and many corpses were thereafter promptly subjected to "treatment." This meant that a stake was driven through the chest, and the heart, being taken out, was either burned or chopped into small pieces. For in this way only could a vampire be deprived of power to do mischief. In one case a man who was unburied sat up in his coffin, with fresh blood on his lips. The official in charge of the ceremony held a crucifix before his face and, saying, "Do you recognize your Saviour?" chopped the unfortunate's head off. This

four people about the table. He was younger than any of them, and had had more contact with the modern world; down in Starkfield, in the bar of the Fielding House, he could hear himself laughing with the rest of the men at such old wives' tales. But it was not for nothing that he had been born under the icy shadow of Lonetop, and had shivered and hungered as a lad through the bitter Hemlock County winters. After his parents died, and he had taken hold of the farm himself, he had got more out of it by using improved methods, and by supplying the increasing throng of summer-boarders over Stotesbury way with milk and vegetables. He had been made a selectman of North Ashmore; for so young a man he had a standing in the county. But the roots of the old life were still in him. He could remember, as a little boy, going twice a year with his mother to that bleak hill-farm out beyond Sylvester Brand's, where Mrs. Bosworth's aunt, Cressidora Cheney, had been shut up for years in a cold clean room with iron bars in the windows. When little Orrin first saw Aunt Cressidora she was a small white old woman, whom her sisters used to "make decent" for visitors the day that Orrin and his mother were expected. The child wondered why there were bars to the window. "Like a canary-bird," he said to his mother. The phrase made Mrs. Bosworth reflect. "I do believe they keep Aunt Cressidora too lonesome," she said; and the next time she went up the mountain with the little boy he carried to his great-aunt a canary in a little wooden cage. It was a great excitement; he knew it would make her happy.

The old woman's motionless face lit up when she saw the bird, and her eyes began to glitter. "It belongs to me," she said instantly, stretching her soft bony hand over the cage.

"Of course it does, Aunt Cressy," said Mrs. Bosworth, her eyes filling.

But the bird, startled by the shadow of the old woman's hand, began to flutter and beat its wings distractedly. At the sight, Aunt Cressidora's calm face suddenly became a coil of twitching features. "You she-devil, you!" she cried in a high squealing voice; and thrusting her hand into the cage she dragged out the terrified bird and wrung its neck. She was plucking the hot body, and squealing "she-devil, she-devil!" as they drew little Orrin from the room. On the way down the mountain his mother wept a great deal, and said: "You must never tell anybody that poor Auntie's crazy, or the men would come and take her down to the asylum at Starkfield, and the shame of it would kill us all. Now promise." The child promised.

He remembered the scene now, with its deep fringe of mystery, secrecy and rumour. It seemed related to a great many other things below the surface of his thoughts, things which stole up anew, making him feel that all the old people he had known, and who "believed in these things," might after all be right. Hadn't a witch been burned at North Ashmore? Didn't the summer folk still drive over in jolly buckboard loads to see the meeting-house where the trial had been held, the pond where they had ducked her and she had floated?...Deacon Hibben believed; Bosworth was

person presumably had been buried alive in a cataleptic trance.

How is the phenomenon to be accounted for? Nobody can say with certainty, but it may be that the fright into which people were thrown by the epidemic had the effect of predisposing nervous persons to catalepsy. In a word, people were buried alive in a condition where, the vital functions being suspended, they remained as it were dead for a while. It is a common thing for a cataleptic to bleed at the mouth just before returning to consciousness. According to popular superstition, the vampire left his or her body in the grave while engaged in nocturnal prowls.

The epidemic prevailed all over southeastern Europe, being at its worst in Hungary and Servia. It is supposed to have originated in Greece, where a belief was entertained to the effect that Latin Christians buried in that country could not decay in their graves being under the ban of the Greek Church. The cheerful notion was that they got out of their graves a night and pursued the occupation of ghouls. The superstition as to ghouls is very ancient and undoubtedly of Oriental origin. Generally speaking, however, a ghoul is just the opposite of a vampire, being a living person who preys on dead bodies, while a vampire is dead person that feeds on the blood of the living. If you had your choice, would you rather be a vampire or a ghoul?

sure of it. If he didn't, why did people from all over the place come to him when their animals had queer sicknesses, or when there was a child in the family that had to be kept shut up because it fell down flat and foamed? Yes, in spite of his religion, Deacon Hibben knew...

And Brand? Well, it came to Bosworth in a flash: that North Ashmore woman who was burned had the name of Brand. The same stock, no doubt, there had been Brands in Hemlock County ever since the white men had come there. And Orrin, when he was a child, remembered hearing his parents say that Sylvester Brand hadn't ever oughter married his own cousin, because of the blood. Yet the couple had two healthy girls, and when Mrs. Brand pined away and died nobody suggested that anything was wrong with her mind. And Vanessa and Ora were the handsomest girls anywhere round. Brand knew it, and scrimped and saved all he could to send Ora, the eldest, down to Starkfield to learn bookkeeping. "When she's married, I'll send you," he used to say to little Venny, who was his favorite. But Ora never married. She was away three years, during which Venny ran wild on the slopes of Lonetop; and when Ora came back she sickened and died — poor girl! Since then Brand had grown more savage and morose. He was a hard-working farmer, but there wasn't much to be got out of those barren Bearcliff acres. He was said to have taken to drink since his wife's death; now and then men ran across him in the "dives" of Stotesbury. But not often. And between times he laboured hard on his stony acres and did his best for his daughters. In the neglected grave-yard of Cold Corners there was a slanting head-stone marked with his wife's name; near it, a year since, he had laid his eldest daughter. And, sometimes, at dusk in the autumn, the village people saw him walk slowly by, turn in between the graves, and stand looking down on the two stones. But he never brought a flower, there, or planted a bush; nor Venny either. She was too wild and ignorant...

Mrs. Rutledge repeated: "That's in Exodus."

The three visitors remained silent, turning about their hats in reluctant hands. Rutledge faced them, still with that empty pellucid gaze which frightened Bosworth. What was he seeing?

"Ain't any of you folks got the grit — ?" his wife burst out again, half hysterically.

Deacon Hibben held up his hand. "That's no way, Mrs. Rutledge. This ain't a question of having grit. What we want first of all is...proof..."

"That's so," said Bosworth, with an explosion of relief, as if the words had lifted something black and crouching from his breast. Involuntarily the eyes of both men had turned to Brand. He stood there smiling grimly, but did not speak.

"Ain't it so, Brand?" the Deacon prompted him.

"Proof that spooks walk?" the other sneered.

"Well — I presume you want this business settled too?"

The old farmer squared his shoulders. "Yes — I do. But I ain't a spiritualist. How the hell are you going to settle it?"

Deacon Hibben hesitated; then he said, in a low incisive tone: "I don't see but one way — Mrs. Rutledge's."

There was a silence.

"What?" Brand sneered again. "Spying?"

The Deacon's voice sank lower. "If the poor girl *does* walk... her that's your child... wouldn't you be the first to want her laid quiet? We all know there've been such cases... mysterious visitations... Can any one of us here deny it?"

"I seen 'em," Mrs. Rutledge interjected.

There was another heavy pause. Suddenly Brand fixed his gaze on Rutledge. "See here, Saul Rutledge, you've got to clear up this damned calumny, or I'll know why. You say my dead girl comes to you." He laboured with his breath, and then jerked out: "When? You tell me that, and I'll be there."

Rutledge's head drooped a little, and his eyes wandered to the window. "Round about sunset, mostly."

"You know beforehand?"

Rutledge made a sign of assent.

"Well, then — tomorrow, will it be?"

Rutledge made the same sign.

Brand turned to the door. "I'll be there." That was all he said. He strode out between them without another glance or word. Deacon Hibben looked at Mrs. Rutledge. "We'll be there too," he said, as if she had asked him; but she had not spoken, and Bosworth saw that her thin body was trembling all over. He was glad when he and Hibben were out again in the snow.

Among the numerous folk tales about vampires is one relating to a fiend named Dakanavar, who dwelt in a cave in Armenia. He would not permit anybody to penetrate into the mountains of Ulmish Altotem or to count their valleys. Every one who attempted this had in the night the blood sucked by the monster from the soles of his feet until he died.

At last, however, he was outwitted by two cunning fellows. They began to count the valleys, and when night came they lay down to sleep, taking care to place themselves with the feet of each under the head of the other. In the night the monster came, felt as usual and found a head. Then he felt at the other end and felt a head there also.

"Well!" cried he, "I have gone through all of the three hundred and sixty-six valleys of these mountains and have sucked the blood of people without end, but never did I find one with two heads and no feet!" So saying he ran away, and never more was seen in that country, but ever since people have known that the mountains have three-hundred and sixty-six valleys ...

III

They thought that Brand wanted to be left to himself, and to give him time to unhitch his horse they made a pretense of hanging about in the doorway while Bosworth searched his pockets for a pipe he had no mind to light.

But Brand turned back to them as they lingered. "You'll meet me down by Lamer's pond tomorrow?" he suggested. "I want witnesses. Round about sunset."

They nodded their acquiescence, and he got into his sleigh, gave the horse a cut across the flanks, and drove off under the snow-smothered hemlocks. The other two men went to the shed.

"What do you make of this business, Deacon?" Bosworth asked, to break the silence.

The Deacon shook his head. "The man's a sick man — that's sure. Something's sucking the life clean out of him."

But already, in the biting outer air, Bosworth was getting himself under better control. "Looks to me like a bad case of the ague, as you said."

"Well — ague of the mind, then. It's his brain that's sick."

Bosworth shrugged. "He ain't the first in Hemlock County."

"That's so," the Deacon agreed. "It's a worm in the brain, solitude is."

"Well, we'll know this time, tomorrow, maybe," said Bosworth. He scrambled into his sleigh, and was driving off in his turn when he heard his companion calling after him. The Deacon explained that his horse had cast a shoe; would Bosworth drive him down to the forge near North Ashmore, if it wasn't too much out of his way? He didn't want the mare slipping about on the freezing snow, and he could probably get the blacksmith to drive him back and shoe her in Rutledge's shed. Bosworth made room for him under the bearskin, and the two men drove off, pursued by a puzzled whinny from the Deacon's old mare.

The road they took was not the one that Bosworth would have followed to reach his own home. But he did not mind that. The shortest way to the forge passed close by Lamer's pond, and Bosworth, since he was in for the business, was not sorry to look the ground over. They drove on in silence.

The snow had ceased, and a green sunset was spreading upward into the crystal sky. A stinging wind barbed with ice-flake's caught them in the face on the open ridges, but when they dropped down into the hollow by Lamer's pond the air was as soundless and as empty as an unswung bell. They jogged along slowly, each thinking his own thoughts.

"That's the house... that tumble-down shack over there, I suppose?" the Deacon said, as the road drew near the edge of the frozen pond.

"Yes: that's the house. A queer hermit-fellow built it years ago, my father used to tell me. Since then I don't believe it's ever been used but by the gypsies."

Bosworth had reined in his horse, and sat looking through pine-trunks purpled by the sunset at the crumbling structure. Twilight already lay under the trees, though day lingered in the open. Between two sharply patterned pine-boughs he saw the evening star, like a white boat in a sea of green.

His gaze dropped from that fathomless sky and followed the blue-white undulations of the snow. It gave him a curious agitated feeling to think that here, in this icy solitude in the tumble-down house he had so often passed without heeding it, a dark mystery, too deep for thought, was being enacted. Down that very slope, coming from the grave-yard at Cold Corners, the being they called "Ora" must pass toward the pond. His heart began to beat stiflingly. Suddenly he gave an exclamation: "Look!"

He had jumped out of the cutter and was stumbling up the bank toward the slope of snow. On it, turned in the direction of the house by the pond, he had detected a woman's foot-prints; two; then three; then more. The Deacon scrambled out after him, and they stood and stared.

"God — barefoot!" Hibben gasped. "Then it *is* the dead..."

Bosworth said nothing. But he knew that no live woman would travel with naked feet across that freezing wilderness. Here, then, was the proof the Deacon had asked for — they held it. What should they do with it?

"Supposing we was to drive up nearer — round the turn of the pond, till we get close to the house," the Deacon proposed in a colourless voice. "Mebbe then… "

Postponement was a relief. They got into the sleigh and drove a hundred yards farther the road, a mere lane under steep bushy banks, turned sharply to the right, following the bend of the pond. As they rounded the turn they saw Brand's cutter ahead of them. It was empty, the horse tied to a tree-trunk. The two men looked at each other again. This was not Brand's nearest way home.

Evidently he had been actuated by the same impulse which had made them rein in their horse by the pond-side, and then hasten on to the deserted hovel. Had he too discovered those spectral foot-prints? Perhaps it was for that very reason that he had left his cutter and vanished in the direction of the house. Bosworth found himself shivering all over under his bearskin. "I wish to God the dark wasn't coming on," he muttered. He tethered his own horse near Brand's, and without a word he and the Deacon ploughed through the snow, in the track of Brand's huge feet. They had only a few yards to walk to overtake him. He did not hear them following him, and when Bosworth spoke his name, and he stopped short and turned, his heavy face was dim and confused, like a darker blot on the dusk. He looked at them dully, but without surprise.

"I wanted to see the place," he merely said.

The Deacon cleared his throat. "Just take a look…yes…We thought so...But I guess there won't be anything to *see*… " He attempted a chuckle.

The other did not seem to hear him, but laboured on ahead through the pines. The three men came out together in the cleared space before the house. As they emerged from beneath the trees they seemed to have left night behind. The evening star shed a lustre on the speckless snow, and Brand, in that lucid circle, stopped with a jerk, and pointed to the same light foot-prints turned toward the house — the track of a woman in the snow. He stood still, his face working. "Bare feet…" he said.

The Deacon piped up in a quavering voice: "The feet of the dead."

Brand remained motionless. "The feet of the dead," he echoed.

Deacon Hibben laid a frightened hand on his arm. "Come away now Brand; for the love of God come away."

The father hung there, gazing down at those light tracks on the snow — light as fox or squirrel trails they seemed, on the white immensity. Bosworth thought to himself: "The living couldn't walk so light — not even Ora Brand couldn't have, when she lived…" The cold seemed to have entered into his very marrow. His teeth were chattering.

Brand swung about on them abruptly. "*Now!*" he said, moving on as if to an assault, his head bowed forward on his bull neck.

"Now — now? Not in there?" gasped the Deacon. "What's the use? It was tomorrow he said —." He shook like a leaf.

LEGEND OF A LOST VAMPIRE

Within a year of the publication of "Bewitched," work would begin on a very different kind vampire story, one all the more legendary today because of its apparent complete disappearance.

***London after Midnight* (1927), starred the chameleon-like silent star Lon Chaney, and was director Tod Browning's first excursion into the realm of movie vampires. Because the film's negative and all existing prints have vanished from sight, the film has generated a special mystique; the American Film Institute ranks it among the ten most important "lost" films of all time.**

Chaney acted the dual role of a Scotland Yard detective as well as a pop-eyed, razor-toothed monster in a costume borrowed from the wardrobe of Dr. Caligari — not a real vampire, it turns out, but part of an elaborately theatrical ruse to catch a flesh-and-blood killer.

Many film buffs believe the film isn't really lost but just in hiding, its shadowy owner waiting for the M-G-M/Turner copyright to eventually expire, at which time the vampire will return from its grave. Unsubstantiated reports of *London after Midnight*'s rediscovery keep cropping up, rather like sightings of the Loch Ness monster. On these occasions, the phone lines of film preservationists and historians will burn frenziedly from coast to coast, until the latest close encounter is debunked as yet another cruel hoax.

A 1928 novelization by Marie Coolidge-Rask was published in America and England, freely

"It's now," said Brand. He went up to the door of the crazy house, pushed it inward, and meeting with an unexpected resistance, thrust his heavy shoulder against the panel. The door collapsed like a playing card, and Brand stumbled after it into the darkness of the hut. The others, after a moment's hesitation, followed.

Bosworth was never quite sure in what order the events that succeeded took place. Coming in out of the snow-dazzle, he seemed to be plunging into total blackness. He groped his way across the threshold, caught a sharp splinter of the fallen door in his palm, seemed to see something white and wraithlike surge up out of the darkest corner of the hut, and then heard a revolver shot at his elbow, and a cry —

Brand had turned back, and was staggering past him out into the lingering daylight. The sunset, suddenly flushing through the trees, crimsoned his face like blood. He held a revolver in his hand and looked about him in his stupid way.

"They do walk, then," he said and began to laugh. He bent his head to examine his weapon. "Better here than in the churchyard. They shan't dig her up *now*," he shouted out. The two men caught him by the arms, and Bosworth got the revolver away from him.

IV

The next day Bosworth's sister Loretta, who kept house for him, asked him, when he came in for his midday dinner, if he had heard the news.

Bosworth had been sawing wood all the morning, and in spite of the cold and the driving snow, which had begun again in the night, he was covered with an icy sweat, like a man getting over a fever.

"What news?"

"Venny Brand's down sick with pneumonia. The Deacon's been there. I guess she's dying."

Bosworth looked at her with listless eyes. She seemed far off from him, miles away. "Venny Brand?" he echoed.

"You never liked her, Orrin."

"She's a child. I never knew much about her."

"Well," repeated his sister, with the guileless relish of the unimaginative for bad news, "I guess she's dying." After a pause she added: "It'll kill Sylvester Brand, all alone up there." Bosworth got up and said: "I've got to see to poulticing the gray's fetlock." He walked out into the steadily falling snow.

Venny Brand was buried three days later. The Deacon read the service; Bosworth was one of the pall-bearers. The whole countryside turned out, for the snow had stopped falling, and at any season a funeral offered an opportunity for an outing that was not to be missed. Besides, Venny Brand was young and handsome — at least some people thought her handsome,

adapting the screen story. A brief excerpt from a typically fevered scene follows.

Lon Chaney in London after Midnight *(1927).*

From
LONDON AFTER MIDNIGHT
Marie Coolidge-Rask
(1928)

"Carry on, man," encouraged the colonel, for the trembling hand of Sir James fastened upon his coat sleeve in an effort to draw him back from further investigation. "Look! What's that?" He pointed upward. "In the room, there — on the ceiling!"

The clasp of the baronet's hand tightened and tightened. Like a graven image he stood, staring at the dark, slow-moving, shadowy form visible upon the ceiling of that mysterious upper room. The colonel, also, stood motionless, silent and watchful.

Weird and uncanny beyond words was the sight of that strange hybrid shape, up there above their heads, moving always to the accompaniment of those low, moaning, shuddering sobs.

How long they stood there, mutely viewing the spectral manifestation, neither man could have told. The form, now vague and indistinct, now dark and clearly visible, pervaded the room like an evil spirit. There were movements when the huge, widespread wings hovered, vulture-like in mid-air. Again, the spectre drew itself into a more compact shape and swooped down and out of sight, only to mount upward again in a more startling evolution.

Suddenly, it turned. For the fraction of a second, they glimpsed its head in profile.

An exclamation of horror escaped Sir James. The colonel voiced an oath.

"A woman!" gasped the baronet.

Dust jacket design for the British edition of Marie Coolidge-Rask's London after Midnight *(1928). (Courtesy of Ronald V. Borst/ Hollywood Movie Posters)*

though she was so swarthy — and her dying like that, so suddenly, had the fascination of tragedy.

"They say her lungs filled right up... Seems she'd had bronchial troubles before... I always said both them girls was frail... Look at Ora, how she took and wasted away! And it's colder'n all outdoors up there to Brand's. Their mother, too, *she* pined away just the same. They don't ever make old bones on the mother's side of the family... There's that young Bedlow over there; they say Venny was engaged to him... Oh, Mrs. Rutledge, excuse *me*... Step right into the pew; there's a seat for you alongside of grandma... "

Mrs. Rutledge was advancing with deliberating steps down the narrow aisle of the bleak wooden church. She had on her best bonnet, a monumental structure which no one had seen out of her trunk since old Mrs. Silsee's funeral, three years before. All the women remembered it. Under its perpendicular pile her narrow face, swaying on the long thin neck, seemed whiter than ever; but her air of fretfulness had been composed into a suitable expression of mournful immobility.

"Looks as if the stone-mason had carved her to put atop of Venny's grave," Bosworth thought as she glided past him; and then shivered at his own sepulchral fancy. When she bent over her hymn book her lowered lids reminded him again of marble eye-balls; the bony hands clasping the book were bloodless. Bosworth had never seen such hands since he had seen old Aunt Cressidora Cheney strangle the canary-bird because it fluttered.

The service was over, the coffin of Venny Brand had been lowered into her sister's grave, and the neighbours were slowly dispersing. Bosworth, as pall-bearer, felt obliged to linger and say a word to the stricken father. He waited till Brand had turned from the grave with the Deacon at his side. The three men stood together for a moment; but not one of them spoke. Brand's face was the closed door of a vault, barred with wrinkles like bands of iron.

Finally the Deacon took his hand and said: "The Lord gave — "

Brand nodded and turned away toward the shed where the horses were hitched. Bosworth followed him. "Let me drive along home with you," he suggested.

Brand did not so much as turn his head. "Home? What home?" he said; and the other fell back.

Loretta Bosworth was talking with the other women while the men unblanketed their horses and backed the cutters out into the heavy snow. As Bosworth waited for her, a few feet off, he saw Mrs. Rutledge's tall bonnet lording it above the group. Andy Pond, the Rutledge farm-hand, was backing out the sleigh.

"Saul ain't here today, Mrs. Rutledge, is he?" one of the village elders piped, turning a benevolent old tortoise-head about on a loose neck, and blinking up into Mrs. Rutledge's marble face.

Bosworth heard her measure out her answer in slow incisive words. "No. Mr. Rutledge he ain't here. He would 'a come for certain, but his aunt Minorca Cummins is being buried down to Stotesbury this very day and he had to go down there. Don't it sometimes seem zif we was all walking right in the shadow of Death?"

As she walked toward the cutter, in which Andy Pond was already seated, the Deacon went up to her with visible hesitation. Involuntarily Bosworth also moved nearer. He heard the Deacon say: "I'm glad to hear that Saul is able to be up and around."

She turned her small head on her rigid neck, and lifted the lids of marble.

"Yes, I guess he'll sleep quieter now. And her too, maybe, now she don't lay there alone any longer," she added in a low voice, with a sudden twist of her chin toward the fresh black stain in the grave-yard snow. She got into the cutter, and said in a clear tone to Andy Pond: "'S'long as we're down here I don't know but what I'll just call round and get a box of soap at Hiram Pringle's."

Bela Lugosi gets creepily cozy with Hazel Whitmore in the 1928 Los Angeles stage production of Dracula. *(Courtesy of Ronald V. Borst / Hollywood Movie Posters)*

"When She Was Fed"
Bela Lugosi's Real-Life Vampire Romance

GLADYS HALL
(1929)

MOLDERING GRAVEYARDS AND shrieks in the night. The drip-drip-drip of blood. The odor of Death that comes from the secret places. A man with a pale green face and stretching hands. Ghouls. Unspeakable things. The worm that never dies.

A bloodsucker in human form — *Dracula*!

You have read the book, "*Dracula*"? By Bram Stoker?

You have seen the play of "*Dracula*"?

I have seen the — let us say the man, *Dracula*.

He is here. He is in Hollywood. He walks the streets by day with pale face and preternatural hands. He works in the studios. He lunches at the Monmartre. He is to be seen in a forthcoming Corinne Griffith picture, "Prisoners." Watch for him. Scrutinize his eyes.

By nights, he never sleeps. He never sleeps lest — but hush! Shh! This is another matter.

One I dread lest this sleepless man — this man who cannot love as other men love — but hush! Shh!

You've never believed in vampires? You've thought them figments of disordered imagination? Mordant fancies? You are wrong — there are vampires. In human form. Disguised by day. And dread beyond description by night. The fancy shudders away from the writing of such a tale. The flesh creeps and crawls. The little lonely human spirit whines in its thin envelope. It crouches in its pitiful lair and whines, as the banshee whines about the house of the newly dead.

ABOUT THE AUTHOR

Gladys Hall was the indisputable doyenne of movie magazine journalists during the late 1920s and early 1930s. This article first appeared, in a slightly different form, in the August 1929 issue of *Motion Picture*. Ms. Hall's notes and manuscripts still attract film fans and researchers as part of the Special Collections division of the Academy of Motion Picture Arts and Sciences' Margaret Herrick Library in Beverly Hills, California.

Are They Each Other?

Dracula is Bela Lugosi.

Is Bela Lugosi *Dracula*?

Or is he himself the victim of a *Dracula* in — but more of that, more later.

Bela Lugosi is intimately aware of vampires. For Lugosi comes from the town of Lugos, in Hungary. The black mountains of Hungary where dwelt Bram Stoker's awful hero. The black mountains where dwell those vampires who kiss human beings into a semblance of death. It is true. It is done. The mountain folk will tell you so themselves. It is done time and time again. Bela Lugosi has seen the very funerals of these vampired dead.

Bela Lugosi

PUBLIC VAMPIRE #1

Perhaps no other human being has had such an influence on the modern conception of the vampire than Bela Ferenc Deszo Blasko (1880-1956), better known as the actor Bela Lugosi. The Hungarian political expatriate (from Lugos, Hungary) arrived in New York City in the early 1920s, establishing himself as a dependable interpreter of "heavy" parts, even though his extensive stage experience in Hungary emphasized romantic roles and comedy.

When the British actor Raymond Huntley turned down the chance to play Dracula in the original 1927 Broadway production, Lugosi donned the flowing opera cape that would follow him, quite literally, to the grave. Hungry to repeat the role in Universal's 1931 film version, he accepted a ridiculously small salary ($500 a week, a quarter of the money paid to the third-billed juvenile, David Manners) and thereafter was never able to negotiate a really lucrative Hollywood contract. *Dracula* was the height of his Hollywood career, and also the beginning of its end. Lugosi had turned down the role of the monster in *Frankenstein* (1931), leading to Boris Karloff's eclipsing him as movieland's most bankable horror star. Part of Lugosi's difficulties came from his failure to completely master English; on

And then, nights later, when they awaken in their dread resting places, the vampires who have marked them as their own — are waiting.

So often and so frightful is this practice that in the town of Lugos they keep their dead for days and sometimes weeks to make sure they have died the sweet death of the Church and not the horrible, half-death of the vampires.

I have said that Bela Lugosi comes from Hungary.

No, no, no, he doesn't come from Hungary. He was driven from Hungary by a ———. He fled the place, emaciated, a skeleton, drained of blood and nerve and sinew. He fled just in time. He fled with the ghastly two-pronged mark of the vampire on his throat. He fled for the thin life that was left him. He told me so. This is the truth.

Hiding His Hands

A vampire attacked Bela Lugosi. I repeat, he told me so. His strange opaque blue eyes receded into his head. His pallid face grew paler. His hands, those powerful, predatory dreadful white hands of his clenched and unclenched under the table. He would not let me see his hands.

There are stranger things in life than we have wind of. You know this, you feel this while you are talking with Bela Lugosi. He has touched the charnel houses of the Plutonian shores. He has ripped the heart of the night from its most foul hiding place. He knows the secrets we dare not listen to. He has heard the language of the dread horned owl and listened to a green moon whispering in the cypress trees.

There are vampires in Hungary. There are vampires here in America. In Europe, Bela Lugosi says, the vampire is more subtle, more secret, more sweet. She vamps to get the man. In America, perhaps because we are younger and healthier, more obvious, these creatures vamp to get a man's money. But there are vampires everywhere. Sometimes they are only symbols of the real thing. And sometimes they are — real.

Such a one got him.

A woman, she was. Of about thirty-five. An actress. Not outstandingly beautiful. Pale brown hair. Nothing extraordinary about it. A pale, pale skin. Naturally pale and, at times, at certain times, unnaturally red. "When she was fed," said Lugosi. She had an average nose. Her eyes were brown. Her mouth — her mouth was ravenous. A mouth one could not look on without shuddering. Those tiny, pointed teeth, like fangs.

She had been married several times, this woman who must be nameless. She was known to have had innumerable lovers. The fates of those husbands and lovers was obscured. Nobody ever asked. If there was speculation, surmise, fear, it did not keep from her other men. For no matter where she was, in what company, in competition with what beautiful woman or women, the men gravitated toward her as under some strange compulsion. Wives left their husbands because of her. Young girls left their sweethearts.

They never knew why. They couldn't tell. They knew that their husbands and lovers did not love the strange pale woman with the hungry mouth. Why then did they go to her without being asked?

The Dead Woman

This woman met young Bela Lugosi. He was of the leading family of the town. His name-town. There is royal blood on his mother's side. He had played Romeo in Lugos with enormous success. He was young. He was conspicuous. He was of an arresting appearance, as he is now. They met. He found that whenever he entered the vampire's presence he was immediately covered in a deathly cold sweat. He trembled. He turned first hot and then cold. His heart and pulse behaved curiously. He had no control of his limbs, of his speech. His customary *sangfroid* fell from him like a robe. It was as if the very vitals of his youth, his strength and manhood were being smitten with a deadly poison.

He was not happy in her presence. He was ill, dizzy, terrified. But he dared not stay away. He could never tell why he dared not stay away. For she never spoke the words that bade him come to her. There was never any of the conventional making of appointments, of dates. No plans at any time for future meetings. He simply went to her, at odd hours, noon, midnight, mid-morning, any time, and he always found her waiting. She was expecting him.

He was very young. He knew himself well enough to realize that he was more than ordinarily emotional. He thought that it was love. Love, attacking him in some peculiarly virulent form. He had never given much thought to human vampires. He had heard the folk-lore of the mountains, of course. He had seen young men and young women fade and wither before his very eyes — but he did not recognize it when it came to himself.

He dreamed young dreams and made young plans and did not even realize how silent this pale woman was.

His mother realized. She had been watching him. And she told him she wished him to please her by weighing himself the first and the last of every week. At the end of the first week he found that he had lost twenty pounds. He consulted a physician, many physicians. There was no organic trouble. No reason for the frightful loss of weight. He began to lose to such an alarming extent that his exertions were confined entirely to these compulsory trips to her house.

His mother warned him. His mother told him. The hideous threat was apparent to her. She showed him other cases. She dared to drag forth into the light of day what every other person kept concealed. And in the end she connived at his escape. She watched over him, built up his self-will, made him flee the country. Once away, his weight began to return, his health came back and with it a clear perspective and a sure knowledge of what had happened to him. He knew. He knows now. He dares not go

stage, and to a lesser degree in Hollywood, he learned his roles phonetically, resulting in the peculiar vocal rhythms and intensity now universally recognized as "Dracula."

Although Dracula was Lugosi's most famous role, he only played the part twice on film; as a display of his acting ability, however, it must take a distinct backseat to his inspired interpretation of Ygor, the demented monster-keeper of *Son of Frankenstein* (1939).

Other highlights of Lugosi's Hollywood career included *Murders in the Rue Morgue* and *White Zombie* (both 1932), *The Black Cat* (1934), *The Raven* and *Mark of the Vampire* (both 1935), and, in a change of pace, a memorable cameo opposite Greta Garbo in *Ninotchka* (1939). But Lugosi's screen assignments continued to slide throughout the 1940s and were virtually finished after his reprise of Dracula in Universal's *Abbott and Costello Meet Frankenstein* (1948). He never worked again for a major studio, and in 1955 made headlines as the first Hollywood star to go public with his treatment for substance abuse (an addiction to painkillers). His final films were produced under the dubious direction of Edward D. Wood, Jr., notorious as "the worst filmmaker of all time" for efforts like *Bride of the Monster* (1955) and *Plan Nine from Outer Space* (1958). In the latter, Bela appeared posthumously in a few scenes, a stand-in with a cape over his face playing the bulk of the role originally intended for him.

Bela Lugosi died in Hollywood on August 16, 1956, and was buried in makeup and costume as Dracula, at his family's request. Martin Landau won a 1994 Oscar for Best Supporting Actor for his portrayal of the aged, drug-addicted Lugosi in Tim Burton's outlandish biopic *Ed Wood*.

back while that woman lives and is available. For if ever again she should send him that wordless command — he doesn't know. What man dare say how strong he will be against the agencies of darkness and death?

Here in Hollywood Bela Lugosi works and even plays as other men. He is charming. He is strange. His hands tell all the emotions from young love to saturnalia.

But he never sleeps.

He told me so. The whole night through he is in his room with a dim light burning. He dares not sleep. He may never sleep again. If rest overtakes him, it is during the day. Never, never by night. During the night watches he reads a little, studies — thinks.

And he can never love again. He told me this, too. The person of a beautiful woman, the usual thing called sex-appeal is nil to him. He cannot get a reaction as other men do. He can only love the love some woman bears for him. He can love the emotion she feels. The separate emotion. The instant that emotion dies, his feeling dies with it. If any woman, beautiful or plain, falls in love with Lugosi he, too, falls in love with that love. Not otherwise.

This is one of the strange tales of Hollywood. The strange tale of the man who never sleeps and cannot love. Bela Lugosi.

Bela Lugosi, the king of vampires. Autographed publicity art for Dracula *(1931).*

The Dark Castle

MARION BRANDON
(1931)

LOST ON A mountain road — and out of gas! That worst possible combination of misfortunes for the tourist had overtaken us; worse than ever, with night now fallen on the unknown countryside around us wrapping it in darkness, veiling the simplest objects in mystery, and endowing the most commonplace of sounds with sinister meaning. But there was no getting around the fact that the tank was as dry as the proverbial bone and that no matter how Arescu and I cursed our luck our car would never stir again — until something could be procured to fill the empty gasoline tank.

Nor was there any telling when that might be, for in the mountain districts of Central Europe sources of supply were few and far between in 1930. Wrong directions had been given us somewhere on the way from the little city which we had left at noon, and instead of reaching the town that was our destination before sundown, here we were, hours later, nowhere — and unable to move.

I was touring these remote regions with but one companion, a most likable young fellow, a Rumanian, who had graduated that June from the college where I was an instructor. We had formed one of the peculiar friendships that sometimes occur between an older man and a younger, and when the time came for him to return to his native country, he had suggested that I accompany him and make up a party of two for a summer of leisurely travel in such unfamiliar countries as Serbia, Bulgaria, and his own Rumania — where we were at the moment when our engine died on that tortuous road.

It was very cold in the high, clear atmosphere, for it was late in August and autumn was approaching. Not a sound to break the silence but the eery screech of an owl and the faint rustle of the night wind in the undergrowth by the roadside, like the stealthy prowling of some hostile animal. Though the entire day had been heavily overcast and dull, the night was clear and starry, but black as the pit, for the moon had not yet risen, and beyond the small range of our headlights we could see not a thing.

"Well," I said resignedly as I sat down on the running board and filled my pipe, "this may be very romantic, but it's cold, too, and I'd give a good deal at this minute to be on a prosaic, concrete state highway, with a red gas pump sure to turn up within half a mile!"

Arescu seated himself beside me. "It's only about an hour till moonrise," he said. "We can perhaps get some idea of where we are then. There

ABOUT THE AUTHOR

Little is known about pulp writer Marion Brandon, except that she contributed two stories to *Strange Tales* magazine in the early 1930s, and was seemingly never heard from again. "The Dark Castle" appeared in the very first issue of *Strange Tales*, just as the Great Depression was tightening its grip. Horror entertainment was a curious cultural byproduct of the economic crisis. Like the Depression itself, vampires were a grim, enervating force, and soon became familiar fixtures in pulp magazines and in Hollywood.

"The Dark Castle" of Hollywood's Dracula *was inhabited by three vampire women instead of one.*

must be a village somewhere... Hear that dog that's just begun to howl? Wonder whose death *he's* heralding?"

I have never blamed the originator of the superstition that the repeated howling of a dog means impending death, for it is the most depressing and ominous of sounds — doubly so at night — and it was beginning to get on my nerves when Arescu said in surprise, "We're looking for the moon in the wrong direction! I had expected it to come up on the right, behind us. . . . Look the other way."

I obeyed. To the left, the sky was softly golden, proclaiming the approach of the hidden moon and throwing into bold relief the turrets and peaked roofs of a building.

Not a light in it anywhere, not a sound, not a sign of life. But at least it promised some degree of protection from the penetrating mountain wind which was by this time going through our clothing as if it were made of paper. Releasing the brakes of our useless car, we rolled it backward down the slight decline of the road for the few hundred feet that lay between us and two great gates that rested heavily on their hinges. With a final effort, we pushed it through them, so that the headlights might illumine the scene before us.

The building was, as we had already surmised, a ruin, a small castle, or very large house, its paneless windows staring like hostile eyes from the embrasures of the rough stone walls. Some of its turrets were broken, like jagged teeth, others seemingly intact — all darkly outlined against the rapidly brightening sky. As we gazed, the golden rim of the moon rose above it; the shivering screech of the owl trembled through the chilly air, answered by the dismal howl of the distant dog. A scene of such unearthly desolation may I never behold again!

"Looks pretty solid at the right-hand end," Arescu remarked after we had examined it as fully as was possible from the distance at which we stood. And arming ourselves each with an electric torch, we approached the building.

The huge iron-bound door sagged open like the gate. Passing in, we found ourselves in a great stone-floored hall, roofless and chill and forbidding. At the right, however, a doorway opened, beyond which we discovered a smaller room in fair condition. It was but a single story high. The strong black beams that supported the ceiling were all in place and looked as if they would stay there. Boards had been pushed against the paneless windows, half-burned logs lay in the gaping stone fireplace, and in a corner of the room was a pile of dry wood.

"Not at all bad!" said Arescu, surveying the scene approvingly. "Others have camped here like ourselves. Made arrangements for a longer stay though, and apparently changed their minds. But the wood they didn't burn up will come in nicely for us!

"I'll build up the fire," he went on, "while you start carrying in the rugs and food."

1931: A VERY GOOD YEAR, AT LEAST FOR VAMPIRES

The appearance of Marion Brandon's "The Dark Castle" in 1931 coincided with the debut of the most influential vampire movie of all time: Universal's *Dracula*, directed by Tod Browning and starring Bela Lugosi. Although there had been many grotesque and horrifying characters in the American cinema during the silent era, there had yet to be a truly supernatural monster. The studios weren't at all sure that audiences would accept a screen story whose horrors weren't "explained away" in some rational manner.

It was never a foregone conclusion that *Dracula* would reach the screen, and the original readers' reports at Universal (and other studios) were not overwhelmingly positive. Here is a small sample of the extreme diversity of opinion:

"While this may have a fantastic opening and be very engrossing for those who like the weird, I cannot possibly see how it is going to make a motion picture. It is blood — blood — blood — kill and everything that would cause any average human being to seek a convenient 'railing.'"

"For mystery and bloodcurdling horror I have never read its equal. This story contains everything necessary for a weird, unnatural, mysterious picture."

"ABSOLUTELY NOT! In the first place, it would be impossible to transcribe this novel of horrors to the screen. And, if it were possible, who would want to sit through an evening of unpleasantness such as a picture of this type would afford?"

In the end, *Dracula* turned out to be Universal's biggest money-maker of 1931. It was the naysayers that needed "explaining away" — not vampires.

The moon had by this time risen high enough to render a torch unnecessary out of doors, its greenish silver radiance made the world almost as light as day. We were well prepared for camping out with plenty of warm rugs, cans of soup, coffee, bread, bacon, and, fortunately, candles.

As I went out for my last load, I was startled to find, standing by the car and gazing toward me, a woman. She was enveloped in a long, dark, hooded cloak which so shrouded her form and shadowed her face that I could form no idea of her age, though the voice in which she addressed me, in German, had the clear vigorous ring of youth. I could see only that her eyes were very bright and her teeth remarkably fine and white between the scarlet lips that parted with her smile.

"Pardon," she said, "if I have startled you. But I live nearby, and strangers seldom come this way."

I expressed my surprise that people lived near, since I had seen no lights, and suggested that she could perhaps find us a warmer lodging for the night.

"My home is hardly large enough," she replied with that flashing, brilliant smile. "I came only to look — *this* time; but I shall perhaps see you — later." And as I gathered a few more articles from the back of the car, she wished me good night and hurried away with sure steps down the dark road.

"Fine!" Arescu exclaimed when I reported the encounter. "Perhaps they have a small farm where we can get eggs for breakfast, and something on four legs to hunt gas with!"

Arescu was of a decidedly domestic turn, and he had spread a couple of our heavy traveling rugs on the floor by the roaring fire which he had built, and which was already having an effect upon the chilly atmosphere, stuck a candle at each end of the heavy stone chimney piece, and set our camp coffeepot on a brick to boil — he had found a well just outside — the ruined room looked almost cozy!

Yet, for some reason, I felt nervous and "edgy." I would gladly have strangled the distant owl and the more-distant dog, each of which, at irregular intervals, continued to emit its eldritch lament. Just as I would think that they had knocked off for good, one or the other of the eery sounds would break out through the night. And the miserable dog seemed to be coming gradually nearer! A couple of bats flitted blunderingly about the room, the night wind prowled uneasily outside.

"I've always heard that you Central Europeans were a superstitious lot," I remarked as Arescu, whistling cheerfully, set the finished coffee aside to keep hot and placed over the fire a generous pan of bacon, "but here we are in what might be the setting for all sorts of horrors. It gives even me the creeps, and for all the effect it has on you, you might be fixing up a midnight mess in a college dormitory.

Arescu sat back on his heels. "I'm just as superstitious as the next person — when I have reason to be," he replied in a perfectly matter-of-fact

DRACULA FROM PAGE TO STAGE

It was Bram Stoker himself who first wanted to dramatize *Dracula*, and did, in fact, stage a five-hour reading with actors for copyright purposes when the book was first published in 1897. But he never saw a full production in his lifetime, and it ultimately fell to a Stoker family acquaintance from Dublin to approach the novelist's widow in 1924 with a viable proposal for a touring stage version.

The dramatist was the actor/manager Hamilton Deane, and *Dracula* — in a greatly simplified, mystery-play format — became a standard fixture of his repertory, until, emboldened by provincial success, he moved the play to the West End in 1927. The reviews were scathing.

"Mr. Deane cannot write dialogue," complained the *Morning Post*. "He does not begin to understand what dramatic dialogue is, and sometimes he gets an unintended comic effect by sheer pomposity of speech... Much of the acting, too, by a company not one member of which was known to a London audience, was of the gifted amateur type. Several of the cast were clearly suffering from prolonged bouts of elocution: they enunciated clearly in a monotonous manner each and every syllable, giving all syllables the same weight and value. A lady talked about her Leth Are Gee."

Raymond Huntley as Dracula.

manner. "But plain creepy surroundings don't disturb me in the least when I know there's nothing wrong."

"How do you know there's nothing wrong with this place?" I asked curiously. "You never saw it before, did you?"

"Never." Arescu placidly arranged the crisp, hot bacon between slices of bread and poured coffee into enamel cups. "There are many ruined castles in these parts, but there is only one haunted place — a vampire castle — in this entire region, and it's on a road leading out of the other side of Koslo from the direction we took this noon. There's nothing else within a hundred miles that's credited with even the mildest of specters!"

"And you really do believe in the supernatural?" I demanded incredulously. "You wouldn't sleep here if the place were called haunted?"

"My good friend," said Arescu, for the moment unwontedly serious, as he turned his dark eyes on mine. "It seems strange, I know, to a native of the great super-civilized United States that supposedly intelligent people can believe the unbelievable that is, unbelievable from your point of view. But, after all, the powers of darkness love the dark; and isn't it only reasonable to think that they shun the more civilized and populous regions of the earth, and cling to the remote and little-known places? Granted that the idea of a specter or spirit seems preposterous to one sitting comfortably in his modern well-lighted home, or driving along a traveled highway. But, if you were told that *this* was haunted, would it seem so ridiculous?"

The sinister howl of the dog, nearer beyond all question, answered him.

"Knowing that it isn't," he added, "I'm as happy as I'd be in the finest of hotels. But if this were Archenfels, you may be certain that I shouldn't be here!"

And as we devoured our hot supper, this astonishing young man whose American education had not shaken one whit his belief in supernatural manifestations told me the story of the Vampire Castle.

"It's twenty miles out in the mountains, to the west of Koslo," he said. "Hasn't been lived in for over a century. It had been for hundreds of years the perfectly peaceful home of a noble family who had to abandon it, a hundred and twenty years or so ago, when it suddenly became vampire-haunted for no reason that anyone could think of. First the eldest son and heir was found dead in his bed; then his brothers, one after the other, at considerable intervals. After the original owners got out in despair, a few attempts at living in it were made by others who hoped to get a fine estate at little cost, but it was just the same: a series of mysterious deaths. Always men, too — young men, never a woman. Grin as much as you like," he reproved me, "but in every case, the same little sharp wound was to be found in the throat of the victim!

"Nobody has knowingly spent a night there in over a century, as I have said," he went on. "But now and again a traveler has done so — as we are doing here — and always with the same dire result: the finding of his

Nonetheless, the *Post* admitted, "in spite of prolixity, of toneless acting, of undramatic dialogue, the play extraordinarily gripped and scared the audience. Those who like to be horribly thrilled will be horribly thrilled by 'Dracula.' "

Despite its primitive production values, *Dracula* proved to be critic-proof, and continued to pack in audiences in the mood for corny thrills. One of the early audience members was the flamboyant American publisher/producer Horace Liveright, who saw even larger commercial possibilities. He hired the playwright/journalist John L. Balderston to completely rewrite the script for Broadway. When the 24-year-old actor Raymond Huntley turned down the chance to reprise the role of Dracula in New York, Liveright cast a virtually unknown expatriate performer, who just happened to hail from Transylvanian Hungary.

Given the indelible mark he made on the role, it is surprising that Bela Lugosi's initial performance as Dracula was not greeted with universal enthusiasm by the critics. Alexander Woolcott, in the *World*, made note of the actor's "incongruous accent," which audiences might associate with "loveable and threadbare music masters." Brooks Atkinson in the *New York Times* called Lugosi's acting "a little too deliberate and confident." The *New York Post* compared him to "an operatically inclined but cheerless mortician." The *Herald-Tribune* complained that Lugosi's Dracula was a "rigid hobgoblin," more window mannequin than suave demon.

Little did these critics realize that they were witnessing the birth of one of the most instantly recognizable icons in theatrical history. And, as in London, dismissive critical appraisals did nothing to keep away the crowds.

body, sometimes long afterward; the throat marked by that cruel little wound. No one lives near it any more; its only neighbors are the dead in the churchyard of an old ruined church.

"No, Professor," he finished with his engaging young smile, "if this were Archenfels, I should be running now with a speed that would surprise you! As it is, in our cozy spot, with neighbors not far, I shall sleep soundly, and I wish you the same."

With that, he wrapped himself in one of our extra rugs, lay down by the fire, and with his coat for a pillow, fell asleep almost immediately. I suddenly felt very lonely.

But though I tried my best to follow his example, it was of no use. The fire was burning low. The bats, joined by others, still blundered among the wavering shadows; the rising wind moaned outside as it tried one window after another. The last howl of the accursed dog was surely much nearer! I shouldn't have drunk that coffee so late at night, shouldn't let my mind play with the boy's tale of the ruined castle, haunted by those hideous visitants who are said to feed upon the blood of their living victims...

Suddenly, as I lay staring at the dying fire, my heart seemed actually to stop — then to race thundering in my ears. Icy sweat crept out upon my body. Though I had heard nothing, I knew that someone — *something* — was in the room, advancing soundlessly upon us from the doorway behind me!

With a desperate effort, I fought down the engulfing terror that had laid hold on me, and turned my head.

Coming slowly toward me across the room, in the fitful glow of the failing fire, was the woman who had spoken to me at the gate. But how terribly, how awesomely, changed! The long cloak had been cast aside, revealing a white gown of olden fashion; the face, shadowed before by the dark hood, exhibited a strangely bright pink-and-white quality that was not human. The lips, red as blood, were parted in a mocking smile. Her fingers were claw-like and suggested the talons of a bird of prey.

And the eyes! I could not — heaven help me — remove my own from the baleful gaze with which they fixed on me. They fascinated me, like the eyes of some deadly serpent. I could neither move nor speak. I lay inert, paralyzed, and cold.

"Welcome to Archenfels!" she said, smiling a terrible smile of derisive triumph. "It has long fain untenanted, and I have had to go far afield."

Archenfels! The castle of dread repute! Paralyzed as my body had become, my brain was clear as I groped frantically for an explanation of the horror.

Archenfels, Arescu had said, was to the west of the city; we had gone east. Impossible!... Yet, as I stared into the narrowed cruel eyes of the scarlet-mouthed creature whose sharp teeth shone white in the flickering light of the fire, I knew that it was not impossible, knew that there are indeed more terrible things in the world than man dare dream of... Some of the

DRACULA OVER HOLLYWOOD

The stage play of *Dracula* earned over two million dollars in the United States between 1927 and 1929, the kind of track record that even the most skeptical Hollywood executives could no longer ignore. After long and contentious negotiations, not to mention the rival interest of M-G-M, Universal acquired the screen rights to the novel and play in the summer of 1930.

After unsuccessfully trying to persuade Lon Chaney to play the vampire (the actor had terminal cancer), the studio considered a wide range of actors for the part, finally settling on Lugosi after all other possibilities fell through. As Universal's financial situation worsened in the wake of the stock market crash, the production was scaled back from an ambitious, novelistic adaptation to a less expensive reproduction of the stage play.

***Dracula* was shot between September 29 and November 15, 1930, and premiered at the Roxy Theatre in New York City February 12, 1931.**

Poster for Dracula.

country people of whom we had asked directions had doubtless given wrong ones, and with the sun overcast as it had been all day, it had been easy enough to lose our sense of direction and circle around. Simple enough to understand — now, when it was too late! Frantically I struggled to break the hold of those awful eyes. Sweat streamed from every pore; yet I lay inert as a log. Not a movement could I make; not a finger could I lift. Nor could I, by the most desperate striving, remove my gaze from hers. If I could do that, something told me, the spell would dissolve. I might attack her and perhaps save our lives... But can the sparrow look from the beady eyes of the snake gliding toward it?

All this time, Arescu lay sleeping as quietly as a baby, one arm over his heart, the other thrown out upon the coverlet, his slow, regular breathing the only sound in the room. A heavy graveyard odor of damp earth and decay stifled me as the creature came closer — *closer!* Stepped past me, but always facing me, never taking those terrible eyes from mine... Knelt by Arescu, and gathering his slumbering form into her arms, bared her shining teeth...

Ballyhoo artist for the original release of Dracula.

As she paused, still holding me chained by that unwavering gaze, I thought in blind revulsion of a tiger crouching over its prey, glaring jealously lest another beast interfere.

Disturbed from the deep sleep of youth and health, Arescu opened his eyes. For an instant he stared, blankly and uncomprehendingly, as if in a nightmare. Then over his handsome young face swept a look of stark frozen horror that I shall see in my dreams till my last day. Under the very window the dog suddenly howled, long and despairingly.

I think that the boy died at once from shock. I hope he did. For when I realized to the full the appalling thing that was about to occur, a thing which I cannot even now put into words, I felt that I, too, was dying — and merciful unconscious overcame me...

A promotional window display.

After what seemed an eternity of struggling, submerged in blackness, I won back to consciousness, confusedly aware of a white form slipping out through the door. Weak and dizzy, I sat up. The room was still. The fire had sunk to a few sullen embers. Even the wind had died, and I, thank heaven, was no longer in the grip of that nefarious gaze!

Snatching the torch that lay beside me, I turned its beam upon the crumpled young figure by the hearth.

No need to look a second time, to feel the pulseless wrist! The terrible unearthly pallor of the boyish face, the ghastly, drained, grayness, was enough. Boiling rage seized upon me. Where had the foul creature gone? To find others of her kind, and tell them that a living man still survived in the accursed castle, material for another grim feast?

Demented and without plan, I rushed out into the night. Across the lawn, plain in the clear green moonlight, a white form was passing through the great gate. I dashed after it in mad pursuit as, realizing that it was followed, it fled, fleet as the wind itself, down the rough mountain road.

Never once did I raise my eyes from the level of its feet as, with bursting lungs, I labored after flying shape. Not again would I fall a victim to that dread gaze!

I was almost upon it when it suddenly veered to the left. Unable to check myself, I ran past the little gate in the stone wall, thus permitting the monster to gain time. Halting as quickly as possible, I turned and rushed through the gate — into a graveyard… Yes, there were the fluttering robes before me, silver in the moonlight the streaming golden hair. With a final mighty effort, temples pounding, pulses throbbing, I gained upon my quarry. No more than twenty feet separated us when it suddenly stopped, laid hold upon an ancient slanting tombstone — and vanished into the earth…

Sick with horror, and utterly exhausted, I dropped beside the grave; and for a second time that night — and the second in all my life — a wave of unconsciousness swept over me…

When I came to myself, the stars were paling before the rosy light in the east; cocks crowed in the distance; birds twittered in the trees. Lame and stiff, I struggled to my feet.

I was standing in an old cemetery, disused, apparently, for many years. The aged lichened tombstones were canted drunkenly this way and that; the ruined little church was half-hidden in overgrown shrubbery. All as poor Arescu had described the fateful region, had we but been able to see! For some reason, I drove a little stick into the grave at my feet — her home was indeed small! — and hastened back along the road to the dire castle of Archenfels.

Here I found many people grouped around our car, all talking excitedly, but in hushed tones. And pale with fear. Others were within. As I entered the room of death, a tall old priest rose from his knees beside the body of Arescu, now decently arranged, with eyes closed and hands crossed.

We both spoke German, and the priest told his story. A small farmer, living across the valley from Archenfels, had seen our lights in the night and had, at first peep of dawn, hastened to the village to report what could mean but one thing — another tragedy. Practically the entire population had accompanied him to the castle, to find what they had feared, a new victim of the vampire. They had deduced that two people had occupied the room; and upon my explaining where I had been, the priest's dark eyes lighted strangely —

"Sir," he inquired eagerly, "do you *know* where she disappeared?"

"I do, indeed!" I answered. "I marked the spot."

An incomprehensible look flashed from face to face among the listeners as the priest translated my reply; and one woman, with tears streaming down her cheeks, knelt and kissed my hand.

"I think, sir," said the priest slowly, after he had given some directions in his own tongue to several men present, "that, shocking as this experience has been for you, you have been the instrument for saving

BELA'S BROADCAST

Bela Lugosi delivered the following radio address on Los Angeles station KFI on March 27, 1931.

I read the book *Dracula*, written by Bram Stoker, eighteen years ago, and always dreamed to create and to play the part of Dracula. Finally the opportunity came. Horace Liveright, stage producer of New York, acquired the stage rights of the novel and he chose me for the part. I have played the role of Dracula about a thousand times on the stage, and people often ask me if I still retain my interest in the character. I do — intensely. Because many people regard the story of *Dracula* simply as a glorified superstition, the actor who plays the role is constantly engaged in the battle of wits with the audience, in a sense, because he is constantly striving to make the character so real that the audience will believe in it.

Now that I have appeared in the screen version of the story which Universal has just completed, I am of course not under this daily strain in the depiction of the character. My work in this direction was finished with the completion of the picture, but while it was being made I was working more intensely to this end than I ever did on the stage.

Although *Dracula* is a fanciful tale of a fictional character, it is actually a story which has many essential elements of truth. I was born and reared in almost the exact location of the story, and I came to know that what is looked on as a superstition of ignorant people is really based on facts which are literally hair-raising in their strangeness — but which are true. Many people will leave the theatre with a sniff at the fantastic character of the story,

the countryside from a great fear. I will explain to you as we return to the graveyard."

The priest's story tallied closely with Arescu's and with a grim addition. The victims of the attacks were, as the boy had said, always men, young men, a fact which doubtless accounted for my own survival. But during the hundred years that the castle had stood untenanted, the number of young men who by chance spent a night there had been insufficient to satisfy the creature's bloodlust; and now and again, some village lad would be found, dead in his bed, a sharp little wound plain in his throat. The last victim, two years before, had been a son of the woman who had kissed my hand; and, as she had two other living sons, her fear had been great.

None but the dead had ever seen the destroyer; no description had ever been given; no theory could be formed as to its sudden origin in so peaceful a district. The village had taken such steps as it could. The few graves of suicides, and others who had died violent deaths, situated outside the graveyard wall beyond the consecrated area, had been opened long ago, for such unfortunates were said sometimes to become vampires. But the rotting coffins had contained nothing suspicious; only the moldering bones that would be natural. And one cannot open all the graves in an old cemetery with no clues to go on!

Long beams of morning sunlight were stretching across the dewy grass when we arrived at the one that I had marked. "Helena Barrientos," read the almost-obliterated inscription upon the stone. "Died August 5, 1799, aged twenty years."

For a time we waited there, the people behind us, all silent under the solemn spell of impending strange events, until the men to whom the priest had spoken his orders returned, some with picks, shovels and ropes, and one with a long strong stake, sharpened at one end. A lad carried the processional cross from the village church.

Amid a deep strained silence, they set to work, the pile of fresh black clods rising rapidly beside the excavation. Then came the dull sound of blows upon rotting wood. The hole was made wide and deep enough to permit the workers to descend into it, the earth carefully cleared from around the coffin, too frail with age to bear removal from its place. The lid was loosened, lifted off...

Cries of horror rose from those crowding around the grave. As for me, my brain reeled. Even then, I could not believe my own eyes.

Lying before us, in the decayed old coffin, with the fresh rosy coloring and scarlet lips of a child asleep, was the visitant of the night before; but now, everything terrible about her was gone. The eyes which had exercised their dread power of fascination were quietly closed, the red lips pressed together... A corpse, fresh and bright as the living, where should be a heap of disintegrating bones!

"You see!" said the old priest simply.

A little metal box lay beside the body, and this he opened, disclosing a

but many others who think just as deeply will gain an insight into one of the most remarkable facts of human existence. *Dracula* is a story which has always had a powerful effect on the emotions of an audience, and I think the picture will be no less effective than the stage play. In fact, the motion picture should even prove more remarkable in this direction, since many things which could only be talked about on the stage are shown on the screen in all their uncanny detail.

I am sure you will enjoy *Dracula*. I am sure you will be mightily affected by its strange story, and I hope it will make you *think* — about the weirdest, most remarkable condition that ever affected mankind.

I thank you.

Lugosi in the role of his lifetime.

letter written in faded, but still legible, ink, Slowly and solemnly he read it aloud:

"I confess to God — but not to man — that this, my daughter, met her death by her own hand. I wish that my child, wronged and mistaken though she has been, shall lie in consecrated ground, for I fear that she will not rest outside. For a day after her death I told others that she was grievously ill; then, that she had died. I prepared her for the grave myself, and none suspect. May God forgive her. She had much provocation, having been heartlessly betrayed by the young lord of Archenfels, though I alone know. May God forgive me, too, her mother."

Amid a profound hush the priest folded up the tragic message from a long-gone day and let himself down into the grave. The sharpened stake was passed to him. Grasping it in his right hand, he received in his left the shining brass cross. Even I, stranger and skeptic though I had been, had heard tales of the grim method of exorcising vampires, and I held my breath with the rest as we watched.

Murmuring a Latin sentence, he raised the sharpened stake.

"May God have mercy on your soul!" he said. And plunged the point into the heart of the body before him.

A gasp of mingled relief and horror rose from all who could see into the grave. In the winking of an eyelid, the corpse vanished. Only a disintegrating skeleton lay in the coffin in a pool of bright red blood that was running rapidly out through the cracks, and soaking into the rich black earth.

"May God have mercy!" said the priest once more in the sonorous rolling Latin of the Church, and with infinite compassion in his tone.

"Amen!" answered the people, and went their ways.

That is all.

I remember but little of the trip back to the ancient city of Koslo where I spent nearly a month in the hospital, delirious much of the time.

When I had recovered enough to study a map of the region, it was easy to see how we had wandered into the fatal neighborhood. The road on which Archenfels stood left the city in a westerly direction, it is true, but soon bore decidedly south, while ours, going east at first, also bent south before very long. The wrong direction given us had put us on a road that joined the two, and our own wanderings had done the rest.

If only the sun had not been hidden! If only we had reached the dread spot before night veiled the scene.

The college granted me leave of absence for a year, and I am somewhat better now, but I know that never again shall I have the nervous balance of a normal human being.

The doctors said that it might help me to write it all out — "get it off my mind," in a measure. I think that it has helped. I feel too that if the recital of my experience brings others to the realization that there are still dark and terrible things to be encountered in this commonplace world of

THE SPANISH DRACULA

In the days of early talkies, before dubbing was perfected, many Hollywood films were simultaneously shot in alternate, foreign language versions for the international market. One of the most ambitious and interesting examples that has survived is Universal's 1931 Spanish-language version of *Dracula*, filmed at night on the same sets as the English-language version, but with a completely different cast and crew. Owing to studio politics, the Spanish production was in many ways a rival production to the Tod Browning film. Its associate producer, Paul Kohner, had been frustrated in his attempts to bring an English-language version to the screen, and, in collaboration with director George Melford, he succeeded in upstaging the English-language version at almost every turn. The real star of the picture is cinematographer George Robinson, whose mobile camera, dramatic lighting and visual effects often look a decade ahead of their time. Spanish actor Carlos Villarias makes a campy stand-in for Lugosi; Kohner cast his bride-to-be, the beautiful Mexican ingenue Lupita Tovar, as the female lead. The film was beautifully restored by Universal in 1992.

Advertising herald for the Spanish version of Dracula.

today, and restrains them from speaking — or even thinking — lightly of them, I shall at least have accomplished something of good.

Painting by Mary Woronov.

Revelations in Black

CARL JACOBI
(1933)

IT WAS A dreary, forlorn establishment way down on Harbor Street. An old sign announced the legend: "Giovanni Larla — Antiques," and a dingy window revealed a display half masked in dust.

Even as I crossed the threshold that cheerless September afternoon, driven from the sidewalk by a gust of rain and perhaps a fascination for all antiques, the gloominess fell upon me like a material pall. Inside was half darkness, piled boxes and a monstrous tapestry, frayed with the warp showing in worn places. An Italian Renaissance wine cabinet shrank despondently in its corner and seemed to frown at me as I passed.

"Good afternoon, *Signor*. There is something you wish to buy? A picture, a ring, a vase perhaps?"

I peered at the squat bulk of the Italian proprietor there in the shadows and hesitated.

"Just looking around," I said, turning to the jumble about me. "Nothing in particular…"

The man's oily face moved in smile as though he had heard the remark a thousand times before. He sighed, stood there in thought a moment, the rain drumming and swishing against the outer pane. Then very deliberately he stepped to the shelves and glanced up and down them considering. At length he drew forth an object which I perceived to be a painted chalice.

"An authentic Sixteenth Century Tandart," he murmured. "A work of art, *Signor*."

I shook my head. "No pottery," I said. "Books perhaps, but no pottery."

He frowned slowly. "I have books too," he replied, "rare books which nobody sells but me, Giovanni Larla. But you must look at my other treasures too."

There was, I found, no hurrying the man. A quarter of an hour passed during which I had to see a Glycon cameo brooch, a carved chair of some indeterminate style and period, and a muddle of yellowed statuettes, small oils and one or two dreary Portland vases. Several times I glanced at my watch impatiently, wondering how I might break away from this Italian and his gloomy shop. Already the fascination of its dust and shadows had begun to wear off, and I was anxious to reach the street.

But when he had conducted me well toward the rear of the shop, something caught my fancy. I drew then from the shelf the first book of horror. If I had but known the events that were to follow, if I could only have had a foresight into the future that September day, I swear I would

ABOUT THE AUTHOR

Carl Jacobi (1908-1997) had one of the twentieth century's longest careers as a writer of horror, fantasy, adventure fiction and science fiction. A newspaperman by training, he made his fiction-writing debut in 1932 in *Weird Tales* and became a dependable contributor to such pulp magazines as *Planet Stories, Thrilling Wonder Stories, Thrilling Adventures*, and many others. His short fiction was collected in three volumes during his lifetime: *Revelations in Black* (1947), *Portraits in Moonlight* (1964) and *Disclosures in Scarlet* (1972). "Revelations in Black" remains his most frequently reprinted work.

have avoided the book like a leprous thing, would have shunned that wretched antique store and the very street it stood on like places accursed. A thousand times I have wished my eyes had never rested on that cover in black. What writhings of the soul, what terrors, what unrest, what madness would have been spared me!

But never dreaming the secret of its pages I fondled it casually and remarked:

"An unusual book. What is it?"

Larla glanced up and scowled.

"That is not for sale," he said quietly. "I don't know how it got on these shelves. It was my poor brother's."

Carl Jacobi was a regular contributor to Weird Tales, *which did much to keep fictional vampires alive.*

The volume in my hand was indeed unusual in appearance. Measuring but four inches across and five inches in length and bound in black velvet with each outside corner protected with a triangle of ivory, it was the most beautiful piece of book-binding I had ever seen. In the center of the cover was mounted a tiny piece of ivory intricately cut in the shape of a skull. But it was the title of the book that excited my interest. Embroidered in gold braid, the title read:

"Five Unicorns and a Pearl."

I looked at Larla. "How much?" I asked and reached for my wallet.

He shook his head. "No, it is not for sale. It is... it is the last work of my brother. He wrote it just before he died in the institution."

"The institution?"

Larla made no reply but stood staring at the book, his mind obviously drifting away in deep thought. A moment of silence dragged by. There was a strange gleam in his eyes when finally he spoke. And I thought I saw his fingers tremble slightly.

"My brother, Alessandro, was a fine man before he wrote that book," he said slowly. "He wrote beautifully, *Signor*, and he was strong and healthy. For hours I could sit while he read to me his poems. He was a dreamer, Alessandro; he loved everything beautiful, and the two of us were very happy.

"All... until that terrible night. Then he... but no... a year has passed now. It is best to forget." He passed his hand before his eyes and drew in his breath sharply.

"What happened?" I asked.

"Happened, *Signor*? I do not really know. It was all so confusing. He became suddenly ill, ill without reason. The flush of sunny Italy, which was always on his cheek, faded, and he grew white and drawn. His strength left him day by day. Doctors prescribed, gave medicines, but nothing helped. He grew steadily weaker until . . . until that night."

I looked at him curiously, impressed by his perturbation.

"And then — ?"

Hands opening and closing, Larla seemed to sway unsteadily; his liquid eyes opened wide to the brows.

"And then . . . oh, if I could but forget! It was horrible. Poor Alessandro came home screaming, sobbing. He was . . . he was stark, raving mad!

"They took him to the institution for the insane and said he needed a complete rest, that he had suffered from some terrific mental shock. He . . . died three weeks later with the crucifix on his lips."

For a moment I stood there in silence, staring out at the falling rain. Then I said:

"He wrote this book while confined to the institution?"'

Larla nodded absently.

"Three books," he replied. "Two others exactly like the one you have in your hand. The bindings he made, of course, when he was quite well. It was his original intention, I believe, to pen in them by hand the verses of Marini. He was very clever at such work. But the wanderings of his mind which filled the pages now, I have never read. Nor do I intend to. I want to keep with me the memory of him when he was happy. This book has come on these shelves by mistake. I shall put it with his other possessions."

My desire to read the few pages bound in velvet increased a thousand-fold when I found they were unobtainable. I have always had an interest in abnormal psychology and have gone through a number of books on the subject. Here was the work of a man confined in the asylum for the insane. Here was the unexpurgated writing of an educated brain gone mad. And unless my intuition failed me, here was a suggestion of some deep mystery. My mind was made up. I must have it.

I turned to Larla and chose my words carefully.

"I can well appreciate your wish to keep the book," I said, "and since you refuse to sell, may I ask if you would consider lending it to me for just one night? If I promised to return it in the morning? . . ." The Italian hesitated. He toyed undecidedly with a heavy gold watch chain.

"No, I am sorry . . ."

"Ten dollars and back tomorrow unharmed."

Larla studied his shoe.

"Very well, *Signor*, I will trust you. But please, I ask you, please be sure and return it."

That night in the quiet of my apartment I opened the book. Immediately my attention was drawn to three lines scrawled in a feminine hand across the inside of the front cover, lines written in a faded red solution that looked more like blood than ink. They read:

"Revelations meant to destroy but only binding without the stake. Read, fool, and enter my field, for we are chained to the spot. Oh wo unto Larla."

I mused over these undecipherable sentences for some time without solving their meaning. At last, I turned to the first page and began the last work of Alessandro Larla, the strangest story I had ever in my years of browsing through old books, come upon.

"On the evening of the fifteenth of October I turned my steps into the cold and walked until I was tired The roar of the present was in the distance when

Carl Jacobi's "Revelations in Black" is an old-fashioned vampire story that makes use of a time-honored fantasy fiction device: a fascinating object found in an old shop that has the demoniacal power to obsess or possess. Hume Nesbit's "The Old Portrait," published three decades earlier, demonstrates just how durable and effective the device can be.

THE OLD PORTRAIT
Hume Nisbet
(1900)

Old-fashioned frames are a hobby of mine. I am always on the prowl amongst the framers and dealers in curiosities for something quaint and unique in picture frames. I don't care much for what is inside them, for being a painter it is my fancy to get the frames first and then paint a picture which I think suits their probable history and design. In this way I get some curious and I think also some original ideas.

One day in December, about a week before Christmas, I picked

I came to twenty-six bluejays silently contemplating the ruins. Passing in the midst of them I wandered by the skeleton trees and seated myself where I could watch the leering fish. A child worshipped. Glass threw the moon at me. Grass sang a litany at my feet. And the pointed shadow moved slowly to the left.

"'I walked along the silver gravel until I came to five unicorns galloping beside water of the past. Here I found a pearl, a magnificent pearl, a pearl beautiful but black. Like a flower it carried a rich perfume, and once I thought the odor was but a mask, but why should such a perfect creation need a mask?

"I sat between the leering fish and the five galloing unicorns, and I fell madly in love wth the pearl. The past lost itself in drabness and —"

I laid the book down and sat watching the smoke-curls from my pipe eddy ceilingward. There was much more, but I could make no sense of any of it. All was in that strange style and completely incomprehensible. And yet it seemed the story was more than the mere wanderings of a madman. Behind it all seemed to lie a narrative cloaked in symbolism.

Something about the few sentences had cast an immediate spell of depression over me. The vague lines weighed upon my mind, and I felt myself slowly seized by a deep feeling of uneasiness.

The air of the room grew heavy and close. The open casement and the out-of-doors seemed to beckon to me. I walked to the window, thrust the curtain aside, stood there, smoking furiously. Let me say that regular habits have long been a part of my make-up. I am not addicted to nocturnal strolls or late meanderings before seeking my bed; yet now, curiously enough, with the pages of the book still in my mind I suddenly experienced an indefinable urge to leave my apartment and walk the darkened streets.

I paced the room nervously. The clock on the mantel pushed its ticks slowly through the quiet. And at length I threw my pipe to the table, reached for my hat and coat and made for the door.

Ridiculous as it may sound, upon reaching the street I found that urge had increased to a distinct attraction. I felt that under no circumstances must I turn any direction but northward, and although this way led into a district quite unknown to me, I was in a moment pacing forward, choosing streets deliberately and heading without knowing why toward the outskirts of the city. It was a brilliant moonlit night in September. Summer had passed and already there was the smell of frosted vegetation in the air. The great chimes in Capitol tower were sounding midnight, and the buildings and shops and later the private houses were dark and silent as I passed.

Try as I would to erase from my memory the queer book which I had just read, the mystery of its pages hammered at me, arousing my curiosity. "Five Unicorns and a Pearl!" What did it all mean?

More and more I realized as I went on that a power other than my own will was leading my steps. Yet once when I did momentarily come to a halt that attraction swept upon me as inexorably as the desire for a narcotic.

It was far out on Easterly Street that I came upon a high stone wall flanking the sidewalk. Over its ornamented top I could see the shadows of

up a fine but dilapidated specimen of wood-carving in a shop near Soho. The gilding had been worn nearly away, and three of the corners broken off; yet as there was one of the corners still left, I hoped to be able to repair the others from it. As for the canvas inside this frame, it was so smothered with dirt and time stains that I could only distinguish it had been a very badly painted likeness of some sort, of some commonplace person, daubed in by a poor pot-boiling painter to fill the secondhand frame which his patrons may have picked up cheaply as I had done after him; but as the frame was alright I took the spoiled canvas along with it, thinking it might come in handy.

For the next few days my hands were full of work of one kind and another, so that it was only on Christmas Eve that I found myself at liberty to examine my purchase which had been lying with its face to the wall since I had brought it to my studio.

Having nothing to do on this night, and not in the mood to go out, I got my picture and frame

a dark building set well back in the grounds. A wrought-iron gate in the wall opened upon a view of wild desertion and neglect. Swathed in the light of the moon, an old courtyard strewn with fountains, stone benches and statues lay tangled in rank weeds and undergrowth. The windows of the building, which evidently had once been a private dwelling, were boarded up, all except those on a little tower or cupola rising to a point in front. And here the glass caught the blue-gray light and refracted it into the shadows.

Before that gate my feet stopped like dead things. The psychic power which had been leading me had now become a reality. Directly from the courtyard it emanated, drawing me toward it with an intensity that smothered all reluctance.

Strangely enough, the gate was unlocked; and feeling like a man in a trance I swung the creaking hinges and entered, making my way along a grass-grown path to one of the benches. It seemed that once inside the court the distant sounds of the city died away, leaving a hollow silence broken only by the wind rustling through the tall dead weeds. Rearing up before me, the building with its dark wings, cupola and facade oddly resembled a colossal hound, crouched and ready to spring.

There were several fountains, weather-beaten and ornamented with curious figures, to which at the time I paid only casual attention. Farther on, half hidden by the underbrush, was the life-size statue of a little child kneeling in position of prayer. Erosion on the soft stone had disfigured the face, and in the half-light the carved features presented an expression strangely grotesque and repelling.

How long I sat there in the quiet, I don't know. The surroundings under the moonlight blended harmoniously with my mood. But more than that I seemed physically unable to rouse myself and pass on.

It was with a suddenness that brought me electrified to my feet that I became aware of the significance of the objects about me. Held motionless, I stood there running my eyes wildly from place to place, refusing to believe. Surely I must be dreaming. In the name of all that was unusual this . . . this absolutely couldn't be. And yet —

It was the fountain at my side that had caught my attention first. Across the top of the water basin were *five stone unicorns*, all identically carved, each seeming to follow the other in galloping procession. Looking farther, prompted now by a madly rising recollection, I saw that the cupola, towering high above the house, eclipsed the rays of the moon and threw *a long pointed shadow* across the ground *at my left*. The other fountain some distance away was ornamented with the figure of a stone fish, *a fish* whose empty eye-sockets *were leering* straight in my direction. And the climax of it all — the wall! At intervals of every three feet on the top of the street expanse were mounted crude carven stone shapes of birds. And counting them I saw that *those birds were twenty-six bluejays.*

Unquestionably — startling and impossible as it seemed — I was in

from the corner, and laying them upon the table, with a sponge, basin of water, and some soap, I began to wash so that I might see them the better. They were in a terrible mess, and I think used the best part of a packet of soap-powder and had to change the water about a dozen times before the pattern began to show up on the frame, and the portrait within it asserted its awful crudeness, vile drawing, and intense vulgarity. It was the bloated, piggish visage of a publican clearly, with a plentiful supply of jewellery displayed, as is usual with such masterpieces, where the features are not considered of so much importance as a strict fidelity in the depicting of such articles as watch-guard and seals, finger rings, and breast pins; these were all there, as natural and hard as reality.

The frame delighted me, and the picture satisfied me that I had not cheated the dealer with my price, and I was looking at the monstrosity as the gaslight beat full upon it, and wondering how the owner could be pleased with himself as thus depicted, when something about the background

the same setting as described in Larla's book! It was a staggering revelation, and my mind reeled at the thought of it. How strange, how odd that I should be drawn to a portion of the city I had never before frequented and thrown into the midst of a narrative written almost a year before!

I saw now that Alessandro Larla, writing as a patient in the institution for the insane, had seized isolated details but neglected to explain them. Here was a problem for the psychologist, the mad, the symbolic, the incredible story of the dead Italian. I was bewildered and I pondered for an answer.

As if to soothe my perturbation there stole into the court then a faint odor of perfume. Pleasantly it touched my nostrils, seemed to blend with the moonlight. I breathed it in deeply as I stood there by the fountain. But slowly that odor became more noticeable, grew stronger, a sickish sweet smell that began to creep down my lungs like smoke. Heliotrope! The honeyed aroma blanketed the garden, thickened the air.

And then came my second surprise of the evening. Looking about to discover the source of the fragrance I saw opposite me, seated on another stone bench, a woman. She was dressed entirely in black, and her face was hidden by a veil. She seemed unaware of my presence. Her head was slightly bowed, and her whole position suggested a person in deep contemplation.

I noticed also the thing that crouched by her side. It was a dog, a tremendous brute with a head strangely out of proportion and eyes as large as the ends of big spoons. For several moments I stood staring at the two of them. Although the air was quite chilly, the woman wore no over-jacket, only the black dress relieved solely by the whiteness of her throat.

With a sigh of regret at having my pleasant solitude thus disturbed I moved across the court until I stood at her side. Still she showed no recognition of my presence, and clearing my throat I said hesitatingly:

"I suppose you are the owner here. I… I really didn't know the place was occupied, and the gate… well, the gate was unlocked. I'm sorry I trespassed."

She made no reply to that, and the dog merely gazed at me in dumb silence. No graceful words of polite departure came to my lips, and I moved hesitatingly toward the gate.

"Please don't go," she said suddenly, looking up. "I'm lonely. Oh, if you but knew how lonely I am" She moved to one side on the bench and motioned that I sit beside her. The dog continued to examine me with its big eyes.

Whether it was the nearness of that odor of heliotrope, the suddenness of it all, or perhaps the moonlight, I did not know, but at her words a thrill of pleasure ran through me, and I accepted the proffered seat.

There followed an interval of silence, during which I puzzled for a means to start conversation. But abruptly she turned to the beast and said in German:

"Fort mit dir, Johann!"

attracted my attention — a slight marking underneath the thin coating as if the portrait had been painted over some other subject.

It was not much certainly, yet enough to make me rush over to my cupboard, where I kept my spirits of wine and turpentine, with which, and a plentiful supply of rags, I began to demolish the publican ruthlessly in the vague hope that I might find something worth looking at underneath.

A slow process that was, as well as a delicate one, so that it was close upon midnight before the gold cable rings and vermilion visage disappeared and another picture loomed up before me; then giving it the final wash over. I wiped it dry, and set it in a good light on my easel, while I filled and lit my pipe, and then sat down to look at it.

What had I liberated from that vile prison of crude paint? For I did not require to set it up to know that this bungler of the brush had covered and defiled a work as far beyond his comprehension as the clouds are from the caterpillar.

The dog rose obediently to its feet and stole slowly off into the shadows. I watched it for a moment until it disappeared in the direction of the house. Then the woman said to me in English which was slightly stilted and marked with an accent:

"It has been ages since I have spoken to anyone... We are strangers. I do not know you, and you do not know me. Yet... strangers sometimes find in each other a bond of interest. Supposing... supposing we forget customs and formality of introduction? Shall we?"

For some reason I felt my pulse quicken as she said that. "Please do," I replied. "A spot like this is enough introduction in itself. Tell me, do you live here?"

She made no answer for a moment, and I began to fear I had taken her suggestion too quickly. Then she began slowly:

"My name is Perle von Mauren, and I am really a stranger to your country, though I have been here now more than a year. My home is in Austria near what is now the Czechoslovakian frontier. You see, it was to find my only brother that I came to the United States. During the war he was a lieutenant under General Mackensen, but in 1916, in April I believe it was, he... he was reported missing.

"War is a cruel thing. It took our money; it took our castle on the Danube, and then — my brother. Those following years were horrible. We lived always in doubt, hoping against hope that he was still living.

"Then after the Armistice a fellow officer claimed to have served next to him on grave-digging detail at a French prison camp near Monpré. And later came a thin rumor that he was in the United States. I gathered together as much money as I could and came here in search of him."

Her voice dwindled off, and she sat in silence staring at the brown weeds. When she resumed, her voice was low and wavering.

"I... found him... but would to God I hadn't! He... he was no longer living."

I stared at her. "Dead?" I asked.

The veil trembled as though moved by a shudder, as though her thoughts had exhumed some terrible event of the past. Unconscious of my interruption she went on:

"Tonight I came here — I don't know why — merely because the gate was unlocked, and there was a place of quiet within. Now have I bored you with my confidences and personal history?"

"Not at all," I replied. "I came here by chance myself. Probably the beauty of the place attracted me. I dabble in amateur photography occasionally and react strongly to unusual scenes. Tonight I went for a midnight stroll to relieve my mind from the bad effect of a book I was reading."

She made a strange reply to that, a reply away from our line of thought and which seemed an interjection that escaped her involuntarily.

"Books," she said, "are powerful things. They can fetter one more than

The bust and head of a young woman of an uncertain age, merged within a gloom of rich accessories painted as only a master hand can paint who is above asserting his knowledge, and who has learnt to cover his technique. It was as perfect and natural in its sombre yet quiet dignity as if it had come from the brush of Moroni.

A face and neck perfectly colourless in their pallid whiteness, with the shadows so artfully managed that they could not be seen, and for this quality would have delighted the strong-minded Queen Bess.

At first as I looked I saw in the centre of a vague darkness a dim patch of grey gloom that drifted into the shadow. Then the greyness appeared to grow lighter as I sat from it, and leaned back in my chair until the features stole out softly, and became clear and definite, while the figure stood out from the background as if tangible, although, having washed it, I knew that it had been smoothly painted.

An intent face, with delicate nose,

the walls of a prison."

She caught my puzzled stare at the remark and added hastily. "It is odd that we should meet here."

For a moment I didn't answer. I was thinking of her heliotrope perfume, which for a woman of her apparent culture was applied in far too great a quantity to show good taste. The impression stole upon me that the perfume cloaked some secret, that if it were removed I should find... but what?

The hours passed, and still we sat there talking, enjoying each other's companionship. She did not remove her veil, and though I was burning with desire to see her features, I had not dared to ask her to. A strange nervousness had slowly seized me. The woman was a charming conversationalist, but there was about her an indefinable something which produced in me a distinct feeling of unease.

It was, I should judge, but a few moments before the first streaks of a dawn when it happened. As I look back now, even with mundane objects and thoughts on every side, it is not difficult to realize the significance of that vision. But at the time my brain was too much in a whirl to understand.

A thin shadow moving across the garden attracted my gaze once again into the night about me. I looked up over the spire of the deserted house and started as if struck by a blow. For a moment I thought I had seen a curious cloud formation racing low directly above me, a cloud black and impenetrable with two wing-like ends strangely in the shape of a monstrous flying bat.

I blinked my eyes hard and looked again.

"That cloud!" I exclaimed, "that strange cloud!... Did you see — "

I stopped and stared dumbly.

The bench at my side was empty. The woman had disappeared.

During the next day I went about my professional duties in the law office with only half interest, and my business partner looked at me queerly several times when he came upon me mumbling to myself. The incidents of the evening before were rushing through my mind. Questions unanswerable hammered at me. That I should have come upon the very details described by mad Larla in his strange book: the leering fish, the praying child, the twenty-six bluejays, the pointed shadow of the cupola — it was unexplainable; it was weird.

"Five Unicorns and a Pearl." The unicorns were the stone statues ornamenting the old fountain, yes — but the pearl? With a start I suddenly recalled the name of the woman in black: *Perle* von Mauren. What did it all mean?

Dinner had little attraction for me that evening. Earlier I had gone to the antique-dealer and begged him to loan me the sequel, the second volume of his brother Alessandro. When he had refused, objected because I had not yet returned the first book, my nerves had suddenly jumped on edge. I felt like a narcotic fiend faced with the realization that he could not

well-shaped, although bloodless, lips and eyes like caverns without a spark of light in them. The hair loosely about the head and oval cheeks, massive, silky-textured, jet black and lustreless, which hid the upper portion of her brow, with the ears, and fell in straight indefinite waves over the left breast, leaving the right portion of the transparent neck exposed.

The dress and background were symphonies of ebony, yet full of subtle colouring and masterly feeling; a dress of rich brocaded velvet with a background that represented vast receding space, wondrously suggestive and awe-inspiring.

I noticed that the pallid lips were parted slightly, and showed a glimpse of the upper front teeth, which added to the intent expression of her face. A short upper lip, which, curled upward, with the underlip full and sensuous, or rather, if colour had been in it, would have been so.

It was an eerie-looking face that I had resurrected on this midnight hour of Christmas Eve; in its passive pallidity it looked as if the

procure the desired drug. In desperation, yet hardly knowing why, I offered the man money, more money, until at length I had come away, my powers of persuasion and my pocketbook successful.

The second volume was identical in outward respects to its predecessor except that it bore no title. But if I was expecting more disclosures in symbolism I was doomed to disappointment. Vague as "Five Unicorns and a Pearl" had been, the text of the sequel was even more wandering and was obviously only the ramblings of a mad brain. By watching the sentences closely I did gather that Alessandro Larla had made a second trip to his court of the twenty-six bluejays and met there again his "pearl."

There was a paragraph toward the end that puzled me. It read:

"Can it possibly be? I pray that it is not. And yet I have seen it and heard it snarl. Oh, the loathsome creature! I will not, I will not believe it."

I closed the book and tried to divert my attention elsewhere by polishing the lens of my newest portable camera. But again, as before, that same urge stole upon me, that same desire to visit the garden. I confess that I had watched the intervening hours until I would meet the woman in black again; for strangely enough, in spite of her abrupt exit before, I never doubted that she would be there waiting for me. I wanted her to lift the veil. I wanted to talk with her. I wanted to throw myself once agan into the narrative of Larla's book.

Yet the whole thing seemed preposterous, and I fought the sensation with every ounce of will-power I could call to mind. Then it suddenly occurred to me what a remarkable picture she would make, sitting there on the stone bench, clothed in black. If I could but catch the scene on a photographic plate. . . .

I halted my polishing and mused a moment. With a new electric flashlamp, that handy invention which has supplanted the old mussy flashpowder, I could illuminate the garden and snap the picture with ease. And if the result were satisfactory it would make a worthy contribution to the International Camera Contest at Geneva next month.

The idea appealed to me, and gathering together the necessary equipment I drew on an ulster (for it was a wet, chilly night) and slipped out of my rooms and headed northward. Mad, unseeing fool that I was! If only I had stopped then and there, returned the book to the antique-dealer and closed the incident! But the strange magnetic attraction had gripped me in earnest, and I rushed headlong into the horror.

A fall rain was drumming the pavement, and the streets were deserted. Off to the east, however, the heavy blanket of clouds glowed with a soft radiance where the moon was trying to break through, and a strong wind from the south gave promise of clearing the skies before long. With my coat collar turned well up at the throat I passed once again into the older section of the town and down forgotten Easterly Street. I found the gate to the grounds unlocked as before, and the garden a dripping place masked in shadow.

blood had been drained from the body, and that I was gazing upon an open-eyed corpse.

The frame, also, I noticed for the first time, in its details appeared to have been designed with the intention of carrying out the idea of life in death; what had before looked like scroll-work of flowers and fruit were loathsome snake-like worms twined amongst charnel-house bones which they half-covered in a decorative fashion; a hideous design in spite of its exquisite workmanship, that made me shudder and wish that I had left the cleaning to be done by daylight.

I am not at all of a nervous temperament, and would have laughed had anyone told me that I was afraid, and yet, as I sat there alone, with that portrait opposite me in this solitary studio, away from all human contact; for none of the other studios were tenanted on this night, and the janitor had gone on his holiday; I wished that I had spent my evening in a more congenial manner, for in spite of a good fire in the stove and the brilliant gas, that intent face and those

The woman was not there. Still the hour was early, and I did not for a moment doubt that she would appear later. Gripped now with the enthusiasm of my plan, I set the camera carefully on the stone fountain, training the lens as well as I could on the bench where we had sat the previous evening. The flash-lamp with its battery handle I laid within easy reach.

Scarcely had I finished my arrangements when the crunch of gravel on the path caused me to turn. She was approaching the stone bench, heavily veiled as before and with the same sweeping black dress.

"You have come again," she said as I took my place beside her.

"Yes," I replied. "I could not stay away."

Our conversation that night gradually centered about her dead brother, although I thought several times that the woman tried to avoid the subject. He had been, it seemed, the black sheep of the family, had led more or less of a dissolute life and had been expelled from the University of Vienna not only because of his lack of respect for the pedagogues of the various sciences but also because of his queer unorthodox papers on philosophy. His sufferings in the war prison camp must have been intense. With a kind of grim delight she dwelt on his horrible experiences in the grave-digging detail which had been related to her by the fellow officer. But of the manner in which he had met his death she would say absolutely nothing.

Stronger than on the night before was the sweet smell of heliotrope. And again as the fumes crept nauseatingly down my lungs there came that same sense of nervousness, that same feeling that the perfume was hiding something I should know. The desire to see beneath the veil had become maddening by this time, but still I lacked the boldness to ask her to lift it.

Toward midnight the heavens cleared and the moon in splendid contrast shone high in the sky. The time had come for my picture.

"Sit where you are," I said. "I'll be back in a moment."

Stepping to the fountain I grasped the flash-lamp, held it aloft for an instant and placed my finger on the shutter lever of the camera. The woman remained motionless on the bench, evidently puzzled as to the meaning of my movements. The range was perfect. A click, and a dazzling white light enveloped the courtyard about us. For a brief second she was outlined there against the old wall. Then the blue moonlight returned, and I was smiling in satisfaction.

"It ought to make a beautiful picture," I said.

She leaped to her feet.

"Fool!" she cried hoarsely. "Blundering fool! What have you done?"

Even though the veil was there to hide her face I got the instant impression that her eyes were glaring at me, smouldering with hatred. I gazed at her curiously as she stood erect, head thrown back, body apparently taut as wire, and a slow shudder crept down my spine. Then without warning she gathered up her dress and ran down the path toward the deserted house. A moment later she had disappeared somewhere in the shadows of the giant bushes.

haunting eyes were exercising a strange influence upon me.

I heard the clocks from the different steeples chime out the last hour of the day, one after the other, like echoes taking up the refrain and dying away in the distance, and still I sat spellbound, looking at that weird picture, with my neglected pipe in my hand, and a strange lassitude creeping over me.

It was the eyes which fixed me now with the unfathomable depths and absorbing intensity. They gave out no light, but seemed to draw my soul into them, and with it my life and strength as I lay inert before them, until overpowered I lost consciousness and dreamt.

I thought that the frame was still on the easel with the canvas, but the woman had stepped from them and was approaching me with a floating motion, leaving behind her a vault filled with coffins, some of them shut down whilst others lay or stood upright and open, showing the grizzly contents in their decaying and stained cerements.

I stood there by the fountain, staring after her in a daze. Suddenly, off in the umbra of the house's facade there rose a low animal snarl.

And then before I could move, a huge gray shape came hurtling through the long weeds, bounding in great leaps straight toward me. It was the woman's dog, which I had seen with her the night before. But no longer was it a beast passive and silent. Its face was contorted in diabolic fury, and its jaws were dripping slaver. Even in that moment of terror as I stood frozen before it, the sight of those white nostrils and those black hyalescent eyes emblazoned itself on my mind, never to be forgotten.

Then with a lunge it was upon me. I had only time to thrust the flashlamp upward in half protection and throw my weight to the side. My arm jumped in recoil. The bulb exploded, and I could feel those teeth clamp down hard on the handle. Backward I fell, a scream gurgling to my lips, a terrific heaviness surging upon my body.

I struck out frantically, beat my fists into that growling face. My fingers groped blindly for its throat, sank deep into the hairy flesh. I could feel its very breath mingling with my own now, but desperately I hung on. The pressure of my hands told. The dog coughed and fell back. And seizing that instant I struggled to my feet, jumped forward and planted a terrific kick straight into the brute's middle.

"Fort mit dir, Johann!" I cried, remembering the woman's German command.

It leaped back and, fangs bared, glared at me motionless for a moment. Then abruptly it turned and slunk off through the weeds.

Weak and trembling, I drew myself together, picked up my camera and passed through the gate toward home.

Three days passed. Those endless hours I spent confined to my apartment suffering the tortures of the damned.

On the day following the night of my terrible experience with the dog I realized I was in no condition to go to work. I drank two cups of strong black coffee and then forced myself to sit quietly in a chair, hoping to soothe my nerves. But the sight of the camera there on the table excited me to action. Five minutes later I was in the dark room arranged as my studio, developing the picture I had taken the night before. I worked feverishly, urged on by the thought of what an unusual contribution it would make for the amateur contest next month at Geneva, should the result be successful.

An exclamation burst from my lips as I stared at the still-wet print. There was the old garden clear and sharp with the bushes, the statue of the child, the fountain and the wall in the background, but the bench — the stone bench was empty. There was no sign, not even a blur of the woman in black.

I rushed the negative through a saturated solution of mercuric chloride in water, then treated it with ferrous oxalate. But even after this intensifying process the second print was like the first, focused in every detail,

I could only see her head and shoulders with the sombre drapery of the upper portion and the inky wealth of hair hanging round.

She was with me now, that pallid face touching my face and those cold bloodless lips glued to mine with a close lingering kiss, while the soft black hair covered me like a cloud and thrilled me through and through with a delicious thrill that, whilst it made me grow faint, intoxicated me with delight.

As I breathed she seemed to absorb it quickly into herself, giving me back nothing, getting stronger as I was becoming weaker, while the warmth of my contact passed into her and made her palpitate with vitality.

And all at once the horror of approaching death seized upon me, and with a frantic effort I flung her from me and started up from my chair dazed for a moment and uncertain where I was, then consciousness returned and I looked around wildly.

The gas was still blazing brightly, while the fire burned ruddy in the

the bench standing in the foreground in sharp relief, but no trace of the woman.

She had been in plain view when I snapped the shutter. Of that I was positive. And my camera was in perfect condition. What then was wrong? Not until I had looked at the print hard in the daylight would I believe my eyes. No explanation offered itself, none at all; and at length, confused, I returned to my bed and fell into a heavy sleep.

Straight through the day I slept. Hours later I seemed to wake from a vague nightmare, and had not strength to rise from my pillow. A great physical faintness had overwhelmed me. My arms, my legs, lay like dead things. My heart was fluttering weakly. All was quiet, so still that the clock on my bureau ticked distinctly each passing second. The curtain billowed in the night breeze, though I was positive I had closed the casement when I entered the room.

And then suddenly I threw back my head and screamed! For slowly, slowly creeping down my lungs was that detestable odor of heliotrope!

Morning, and I found all was not a dream. My head was ringing, my hands trembling, and I was so weak I could hardly stand. The doctor I called in looked grave as he felt my pulse.

"You are on the verge of a complete collapse," he said. "If you do not allow yourself a rest it may permanently affect your mind. Take things easy for a while. And if you don't mind, I'll cauterize those two little cuts on your neck. They're rather raw wounds. What caused them?"

I moved my fingers to my throat and drew them away tipped with blood.

"I . . . don't know," I faltered.

He busied himself with his medicines, and a few minutes later reached for his hat.

"I'd advise that you don't leave your bed for a week at least," he said. "I'll give you a thorough examination then and see if there are any signs of anemia." But as he went out the door I thought I saw a puzzled look on his face.

Those subsequent hours allowed my thoughts to run wild once more. I vowed I would forget it all, go back to my work and never look upon the books again. But I knew I could not. The woman in black persisted in my mind, and each minute away from her became a torture. But more than that, if there had been a decided urge to continue my reading in the second book, the desire to see the third book, the last of the trilogy, was slowly increasing to an obsession.

At length I could stand it no longer, and on the morning of the third day I took a cab to the antique store and tried to persuade Larla to give me the third volume of his brother. But the Italian was firm. I had already taken two books, neither of which I had returned. Until I brought them back he would not listen. Vainly I tried to explain that one was of no value without the sequel and that I wanted to read the entire narrative as a unit.

stove. By the timepiece on the mantel I could see that it was half-past twelve.

The picture and frame were still on the easel, only as I looked at them the portrait had changed, a hectic flush was on the cheeks while the eyes glittered with life and the sensuous lips were red and ripe-looking with a drop of blood still upon the nether one. In a frenzy of horror I seized my scraping knife and slashed out the vampire picture, then tearing the mutilated fragments out I crammed them into my stove and watched them frizzle with savage delight.

I have that frame still, but have not yet had courage to paint a suitable subject for it.

He merely shrugged his shoulders.

Cold perspiration broke out on my forehead as I heard my desire disregarded. I argued. I pleaded. But to no avail.

At length when Larla had turned the other way I seized the third book as I saw it lying on the shelf, slid it into my pocket and walked guiltily out. I made no apologies for my action. In the light of what developed later it may be considered a temptation inspired, for my will at the time was a conquered thing blanketed by that strange lure.

Back in my apartment I dropped into a chair and hastened to open the velvet cover. Here was the last chronicling of that strange series of events which had so completely become a part of my life during the past five days. Larla's volume three. Would all be explained in its pages? If so, what secret would be revealed?

With the light from a reading-lamp glaring full over my shoulder I opened the book, thumbed through it slowly, marveling again at the exquisite hand-printing. It seemed then as I sat there that an almost palpable cloud of quiet settled over me, muffling the distant sounds of the street. Something indefinable seemed to forbid me to read farther. Curiosity, that queer urge told me to go on. Slowly I began to turn the pages, one at a time, from back to front.

Symbolism again. Vague wanderings with no sane meaning.

But suddenly my fingers stopped! My eyes had caught sight of the last paragraph on the last page, the final pennings of Alessandro Larla. I read, re-read, and read again those blasphemous words. I traced each word in the lamplight, slowly, carefully, letter for letter. Then the horror of it burst within me.

In blood-red ink the lines read:

"What shall I do? She has drained my blood and rotted my soul. My pearl is black as all evil. The curse be upon her brother, for it is he who made her thus. I pray the truth in these pages will destroy them for ever.

"Heaven help me, Perle von Mauren and her brother, Johann, are vampires!"

I leaped to my feet.

"Vampires!"

I clutched at the edge of the table and stood there swaying. Vampires! Those horrible creatures with a lust for human blood, taking the shape of men, of bats, of dogs.

The events of the past days rose before me in all their horror now, and I could see the black significance of every detail.

The brother, Johann — some time since the war he had become a vampire. When the woman sought him out years later he had forced this terrible existence upon her too.

With the garden as their lair the two of them had entangled poor Alessandro Larla in their serpentine coils a year before. He had loved the woman, had worshipped her. And then he had found the awful truth that had sent him stumbling home, raving mad.

SPOTLIGHT ON ZOMBIES

In the wake of Universal's *Dracula* and *Frankenstein* in the early 1930s, it became clear that the living dead had the power to bring even the most moribund box office back to life, and Hollywood naturally began looking for variations on a profitable theme.

It didn't take long for someone to stumble across zombies.

In Haitian folklore, a zombie is a mindless, animated corpse, or a living person dehumanized and enslaved by black magic. In actuality, zombies have less to do with magic than with neurotoxic folk poisons (and their corresponding antidotes) used by voodoo priests to simulate death and resurrection, thus effectively establishing social fear and social control. The misdiagnosis of cataleptic trance as death is an important feature in the evolution of both the Caribbean zombie and the folkloric European vampire.

***White Zombie* (1932) was the first film to exploit the topic; its famous star, Bela Lugosi, forged an immediate link between zombies and vampires. (Lugosi didn't play a zombie himself, but rather the malevolent zombie master, "Murder" Legendre.) In the ensuing decades, the image of the zombie has become even more vampiric, especially in the wake of George A. Romero's flesh-eating zombie films, starting with *Night of the Living Dead* (1968). The characteristics of the cannibal-zombie overlap considerably with those of the traditional vampire; both rise from the grave to bite the flesh of the living, who subsequently become infected with the curse of living death themselves.**

Mad, yes, but not mad enough to keep him from writing the fact in his three velvet-bound books. He had hoped the disclosures would dispatch the woman and her brother for ever. But it was not enough.

I whipped the first book from the table and opened the cover. There again I saw those scrawled lines which had meant nothing to me before.

"Revelations meant to destroy but only binding without the stake. Read, fool, and enter my field, for we are chained to the spot. Oh, wo unto Larla!"

Perle von Mauren had written that. The books had not put an end to the evil life of her and her brother. No, only one thing could do that. Yet the exposures had not been written in vain. They were recorded for mortal posterity to see.

Those books bound the two vampires, Perle von Mauren, Johann, to the old garden, kept them from roaming the night streets in search of victims. Only him who had once passed through the gate could they pursue and attack.

It was the old metaphysical law: evil shrinking in the face of truth.

Yet if the books had found their power in chains they had also opened a new avenue for their attacks. Once immersed in the pages of the trilogy, the reader fell helplessly into their clutches. Those printed lines had become the outer reaches of their web. They were an entrapping net within which the power of the vampires always crouched.

White Zombie*: another flavor of living death, specially seasoned for Depression audiences.*

That was why my life had blended so strangely with the story of Larla. The moment I had cast my eyes on the opening paragraph I had fallen into their coils to do with as they had done with Larla a year before. I had been drawn relentlessly into the tentacles of the woman in black. Once I was past the garden gate the binding spell of the books was gone, and they were free to pursue me and to —

A giddy sensation rose within me. Now I saw why the doctor had been puzzled. Now I saw the reason for my physical weakness. She had been — feasting on my blood! But if Larla had been ignorant of the one way to dispose of such a creature, I was not. I had not vacationed in south Europe without learning something of these ancient evils.

Frantically I looked about the room. A chair, a table, one of my cameras with its long tripod. I seized one of the wooden legs of the tripod in my hands, snapped it across my knee. Then, grasping the two broken pieces, both now with sharp splintered ends, I rushed hatless out of the door to the street.

A moment later I was racing northward in a cab bound for Easterly Street.

"Hurry!" I cried to the driver as I glanced at the westering sun. "Faster, do you hear?"

We shot along the cross-streets, into the old suburbs and toward the outskirts of town. Every traffic halt found me fuming at the delay. But at length we drew up before the wall of the garden.

I swung the wrought-iron gate open and with the wooden pieces of the tripod still under my arm, rushed in. The courtyard was a place of reality in the daylight, but the moldering masonry and tangled weeds were steeped in silence as before.

Straight for the house I made, climbing the rotten steps to the front entrance. The door was boarded up and locked. I retraced my steps and began to circle the south wall of the building. It was this direction I had seen the woman take when she had fled after I had tried to snap her picture. Well toward the rear of the building I reached a small half-open door leading to the cellar. Inside, cloaked in gloom, a narrow corridor stretched before me. The floor was littered with rubble and fallen masonry, the ceiling interlaced with a thousand cobwebs.

I stumbled forward, my eyes quickly accustoming themselves to the half-light from the almost opaque windows.

At the end of the corridor a second door barred my passage. I thrust it open — and stood swaying there on the sill staring inward.

Beyond was a small room, barely ten feet square, with a low raftered ceiling. And by the light of the open door I saw side by side in the center of the floor — two white wood coffins.

How long I stood there leaning weakly against the stone wall I don't know. There was an odor drifting from out of that chamber. Heliotrope! But heliotrope defiled by the rotting smell of an ancient grave.

Then suddenly I leaped to the nearest coffin, seized its cover and ripped it open.

Would to heaven I could forget that sight that met my eyes. There lay the woman in black — unveiled.

That face — it was divinely beautiful, the hair black as sable, the cheeks a classic white. But the lips — I grew suddenly sick as I looked upon them. They were scarlet . . . and sticky with human blood.

I reached for one of the tripod stakes, seized a flagstone from the floor and with the pointed end of the wood resting directly over the woman's heart, struck a crashing blow. The stake jumped downward. A violent contortion shook the coffin. Up to my face rushed a warm, nauseating breath of decay.

I wheeled and hurled open the lid of her brother's coffin. With only a glance at the young masculine Teutonic face I raised the other stake high in the air and brought it stabbing down with all the strength in my right arm.

In the coffins now, staring up at me from eyeless sockets, were two gray and moldering skeletons.

The rest is but a vague dream. I remember rushing outside, along the path to the gate and down Easterly, away from that accursed garden of the jays.

At length, utterly exhausted, I reached my apartment. Those mundane surroundings that confronted me were like balm to my eyes. But there centered into my gaze three objects lying where I had left them, the three volumes of Laria.

VAMPIRES ON STAGE IN THE 1930S

The sale of Hamilton Deane and John L. Balderston's stage version of *Dracula* to Universal Pictures in 1930 did not end the property's life on the boards; Universal's purchase excluded theatrical rights, and the stock and amateur rights became available around the time of the film's release. The play has been in production somewhere in the world almost perpetually ever since, and had a major revival on Broadway in 1977.

In England, Hamilton Deane continued to tour his original, 1924 version of the play. Actor W.E. Holloway was one of Deane's most dependable Draculas in the early thirties; he also played the monster in Deane's revival of Peggy Webling's *Frankenstein: An Adventure in the Macabre*. Deane himself finally played the role of Dracula in a farewell engagement at London's Wintergarden Theatre in 1939. During the curtain call of one memorable performance, Deane was confronted by Bela Lugosi himself (in England to shoot a film) who unexpectedly strode across the stage from the wings to embrace him.

Hamilton Deane

I turned to the grate on the other side of the room and flung the three of them onto the still glowing coals.

There was an instant hiss, and yellow flames streaked upward and began eating into the velvet. The fire grew higher… higher… and diminished slowly.

And as the last glowing spark died into a blackened ash there swept over me a mighty feeling of quiet and relief.

I, the Vampire

HENRY KUTTNER
(1937)

ABOUT THE AUTHOR

Henry Kuttner (1914-1958), was born in California and ultimately died there after his acclaimed, but all-too-brief career as a science fiction and fantasy writer was cut short by a heart attack at the age of 43. Like his contemporary Robert Bloch, Kuttner engaged in youthful correspondence with such legendary fantasy writers as H. P. Lovecraft, and published his first professional story, "The Graveyard Rats" (1936) in *Weird Tales*. He began writing science fiction in 1937, his voluminous production aided by at least sixteen pseudonyms. In 1940 Kuttner married Catherine Lucille Moore, who wrote under the name C.L. Moore, and thereafter their work was highly collaborative, and often published under the name Lewis Padgett. Self-educated, Kuttner did not enroll in college until well after having established his writing career. He graduated from the University of Southern California, Los Angeles, in 1954 and was pursuing a master's degree at the time of his premature death from a heart attack. "I, the Vampire" first appeared in *Weird Tales*.

1. The Chevalier Futaine

THE PARTY WAS dull. I had come too early. There was a preview that night at Grauman's Chinese, and few of the important guests would arrive until it was over. Indeed, Jack Hardy, ace director at Summit Pictures, where I worked as an assistant director, hadn't arrived — yet — and he was the host. But Hardy had never been noted for punctuality.

I went out on the porch and leaned against a pillar, sipping a cocktail and looking down at the lights of Hollywood. Hardy's place was on the summit of a hill overlooking the film capital, near Falcon Lair, Valentino's famous turreted castle. I shivered a little. Fog was sweeping in from Santa Monica, blotting out the lights to the west.

Jean Hubbard, who was an ingenue at Summit, came up beside me and took the glass out of my hand.

"Hello, Mart," she said, sipping the liquor. "Where've you been?"

"Down with the *Murder Desert* troupe, on location in the Mojave," I said. "Miss me, honey?" I drew her close.

She smiled up at me, her tilted eyebrows lending a touch of diablerie to the tanned, lovely face. I was going to marry Jean, but I wasn't sure just when.

"Missed you lots," she said, and held up her lips. I responded.

After a moment I said, "What's this about the vampire man?"

She chuckled. "Oh, the Chevalier Futaine. Didn't you read Lolly Parson's write-up in *Script*? Jack Hardy picked him up last month in Europe. Silly rot. But it's good publicity."

"Three cheers for publicity," I said. "Look what it did for *Birth of a Nation*. But where does the vampire angle come in?"

"Mystery man. Nobody can take a picture of him, scarcely anybody can see him. Weird tales are told about his former life in Paris. Going to play *Red Thirst*. The kind of build-up Universal gave Karloff for *Frankenstein*. "The Chevalier Futaine" — she rolled out the words with amused relish —"is a singing waiter from a Paris café. I haven't seen him — but the deuce with him, anyway. Mart, I want you to do something for me. For Deming."

"Hess Deming?" I raised my eyebrows in astonishment. Hess Deming, was Summit's biggest box-office star, whose wife, Sandra Colter, had died

VAMPIRES, DEPRESSION-STYLE

By the time Henry Kuttner wrote "I, the Vampire," film audiences had become completely familiar with the exploits of the undead through several motion pictures, and especially through the mesmeric persona of Bela Lugosi, Dracula himself. Following is an overview of pertinent films.

THE DEATH KISS (1933)

Audiences for *The Death Kiss* may well have anticipated a screen story along the lines of Henry Kuttner's "I, the Vampire"— after all, a backstage Hollywood mystery involving Bela Lugosi called *The Death Kiss* just had to have some supernatural neck-nuzzling, didn't it? Actually, not. Lugosi played a flesh and blood producer merely suspected of murder, despite publicity stills showing him putting the hypnotic whammy on the film's leading lady, Adrienne Ames. *The Death Kiss* was shot on the abandoned lot of Tiffany Studios, which had suspended operations in the Depression. Directed by Edwin L. Martin, with a better-than-average cast including David Manners and Edward Van Sloan (both from *Dracula*) and John Wray (who was considered for the title role in *Dracula* before Lugosi). (World Wide Pictures)

HOLLYWOOD ON PARADE (1933)

Movie audiences waiting for Dracula's return to the screen had to settle for this amusing short subject, in which a statue of Dracula (Bela Lugosi) comes to life in a Hollywood wax museum and attacks a live-action version of Betty Boop (Mae Questal). "You have Booped your last Boop!" he intones, before folding her into his cloak. The public domain footage is frequently seen in documentaries and compilation tapes. (Paramount)

two days before. She, too, had been an actress, although never the great star her husband was. Hess loved her, I knew — and now I guessed what the trouble was. I said, "I noticed he was a bit wobbly."

"He'll kill himself," Jean said, looking worried. "I — I feel responsible for him somehow, Mart. After all, he gave me my start at Summit. And he's due for the DTs any time now."

"Well, I'll do what I can," I told her. "But that isn't a great deal. After all, getting tight is probably the best thing he could do. I know if I lost you, Jean —"

I stopped. I didn't like to think of it.

Jean nodded. "See what you can do for him, anyway. Losing Sandra that way was — pretty terrible."

"What way?" I asked. "I've been away, remember. I read something about it, but —"

"She just died," Jean said. "Pernicious anemia, they said. But Hess told me the doctor really didn't know what it was. She just seemed to grow weaker and weaker until — she passed away."

I nodded, gave Jean a hasty kiss, and went back into the house. I had just seen Hess Deming walk past, a glass in his hand.

He turned as I tapped his shoulder.

"Oh, Mart," he said, his voice just a bit fuzzy. He could hold his liquor, but I could tell by his bloodshot eyes that he was almost at the end of his rope. He was a handsome devil, all right, well-built, strong-featured, with level gray eyes and a broad mouth that was usually smiling. It wasn't smiling now. It was slack, and his face was bedewed with perspiration.

"You know about Sandra?" he asked.

"Yeah," I said. "I'm sorry, Hess."

He drank deeply from the glass, wiped his mouth with a grimace of distaste.

"I'm drunk, Mart," he confided. "I had to get drunk. It was awful — those last few days. I've got to burn her up."

I didn't say anything.

"Burn her up. Oh my God, Mart — that beautiful body of hers, crumbling to dust — and I've got to watch it! She made me promise I'd watch to make sure they burned her."

I said, "Cremation's a clean ending, Hess. And Sandra was a clean girl, and a damned good actress."

He put his flushed face close to mine. "Yeah — but I've got to burn her up. Oh, God!" He put the empty glass down on a table and looked around dazedly.

I was wondering why Sandra had insisted on cremation. She'd given an interview once in which she stressed her dread of fire. Most write-ups of stars are applesauce, but I happened to know that Sandra did dread fire. Once, on the set, I'd seen her go into hysterics when her leading man lit his pipe too near her face.

"Excuse me, Mart," Hess said. "I've got to get another drink."

"Wait a minute," I said, holding him. "You want to watch yourself, Hess. You've had too much already."

"It still hurts," he said. "Just a little more and maybe it won't hurt so much."

But he didn't pull away. Instead he stared at me with the dullness of intoxication in his eyes. "Clean," he said presently. "She said that too, Mart. She said burning was a clean death. But, God, that beautiful white body of hers — I can't stand it, Mart! I'm going crazy, I think. Get me a drink, like a good fellow."

I said, "Wait here, Hess. I'll get you one." I didn't add that it would be watered — considerably.

He sank down in a chair, mumbling thanks. As I went off I felt sick. I'd seen too many actors going on the rocks to mistake Hess's symptoms. I knew that his box-office days were over. There would be longer and longer waits between pictures, and then personal appearances, and finally Poverty Row and serials. And in the end a man found dead in a cheap hall bedroom on Main Street, with gas on.

Top: Adrienne Ames and Bela Lugosi vamp it up in a red-herring publicity still for The Death Kiss. *Below: Lugosi and Mae Questal in* Hollywood on Parade.

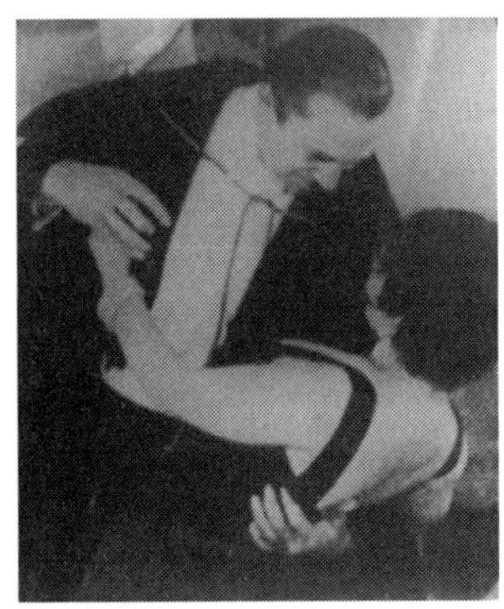

There was a crowd around the bar. Somebody said, "Here's Mart. Hey, come over and meet the vampire."

Then I got my shock. I saw Jack Hardy, my host, the director with whom I'd worked on many a hit. He looked like a corpse. And I'd seen him looking plenty bad before. A man with a hangover or a marijuana jag isn't a pretty sight, but I'd never seen Hardy like this. He looked as though he was keeping going on his nerve alone. There was no blood in the man.

I'd last seen him as a stocky, ruddy blond, who looked like nothing so much as a wrestler, with his huge biceps, his ugly, good-natured face, and his bristling crop of yellow hair. Now he looked like a skeleton, with skin hanging loosely on the big frame. His face was a network of sagging wrinkles. Pouches bagged beneath his eyes, and those eyes were dull and glazed. About his neck a black silk scarf was knotted tightly.

"Good God, Jack!" I exclaimed. "What have you done to yourself?"

He looked away quickly. "Nothing," he said brusquely. "I'm all right. I want you to meet the Chevalier Futaine — this is Mart Prescott."

"Pierre," a voice said. "Hollywood is no place for titles. Mart Prescott — the pleasure is mine."

I faced the Chevalier Pierre Futaine.

We shook hands. My first impression was of icy cold, and a slick kind of dryness — and I let go of his hand too quickly to be polite. He smiled at me.

A charming man, the chevalier. Or so he seemed. Slender, below height, his bland, round face seemed incongruously youthful. Blond hair was plastered close to his scalp. I saw that his cheeks were rouged — very deftly, but I know something about makeup. And under the rouge I read a curious, deathly pallor that would have made him a marked man had he not disguised it. Some disease, perhaps, had blanched his skin — but his lips were not artificially reddened. And they were as crimson as blood.

He was clean-shaved, wore impeccable evening clothes, and his eyes were black pools of ink.

"Glad to know you," I said. "You're the vampire, eh?"

He smiled. "So they tell me. But we all serve the dark god of publicity, eh, Mr. Prescott? Or — is it Mart?"

"It's Mart," I said, still staring at him. I saw his eyes go past me, and an extraordinary expression appeared on his face — an expression of amazement, disbelief. Swiftly it was gone.

I turned. Jean was approaching, was at my side as I moved. She said, "Is this the chevalier?"

Pierre Futaine was staring at her, his lips parted a little. Almost inaudibly he murmured, "Sonya." And then, on a note of interrogation, "Sonya?"

I introduced the two. Jean said, "You see, my name isn't Sonya."

The chevalier shook his head, an odd look in his black eyes.

"I once knew a girl like you," he said softly. "Very much like you. It is strange."

"Will you excuse me?" I broke in. Jack Hardy was leaving the bar. I followed him.

I touched his shoulder as he went out the French windows. He jerked out a startled oath, turned a white death-mask of a face to me.

"Damn you, Mart," he snarled. "Keep your hands to yourself."

I put my hands on his shoulders and swung him around.

"What the devil has happened to you?" I asked. "Listen, Jack, you can't bluff me or lie to me. You know that. I've straightened you out enough times in the past, and I can do it again. Let me in on it."

His ruined face softened. He reached up and took away my hands. His own hands were ice-cold, like the hands of the Chevalier Futaine.

"No," he said. "No use, Mart. There's nothing you can do. I'm all right, really, just — overstrain. I had too good a time in Paris."

I was up against a blank wall. Suddenly, without volition, a thought popped out of my mouth before I knew it.

"What's the matter with your neck?" I asked abruptly.

He just frowned and shook his head.

"I've a throat infection," he told me. "Caught it on the steamer."

His hand went up and touched the black scarf.

There was a croaking, harsh sound from behind us — a sound that didn't seem quite human. I turned. It was Hess Deming. He was swaying in the portal, his eyes glaring and bloodshot, a little trickle of saliva running down his chin.

He said in a dead, expressionless voice that was somehow dreadful, "Sandra died of a throat infection, Hardy."

Jack didn't answer. He stumbled back a step. Hess went on dully.

THE VAMPIRE BAT (1933)

This low-budget charmer was filmed on the Universal back lot, and if you listen carefully, you will hear at least three soundtrack bites from Universal's *Dracula* and *Frankenstein* — a wolf, dogs, carriage sounds, etc. Dwight Frye, *Dracula*'s memorable Mr. Renfield, here pushes the Renfield bit over the top as Herman Gleib, a village halfwit scapegoated by the locals for the actual blood-crimes of Dr. Otto van Niemann (Lionel Atwill). Fay Wray is the female lead, though she doesn't display anywhere near the lung power she brought to bear on *King Kong* and *The Mystery of the Wax Museum* the same year. And while you listen to the burgomeister (Lionel Belmore) wax rhapsodic on vampire lore, ponder the fact that Belmore himself had a wonderful direct connection to the most famous vampire promulgator of all time. While a member of Henry Irving's Lyceum Theatre company in London, Belmore had his paychecks signed by none other than Irving's devoted manager — *Dracula*'s creator, Bram Stoker. With Melvyn Douglas and Maude Eburne. Directed by Frank Strayer. (Majestic Pictures)

"She got all white and died. And the doctor didn't know what it was, although the death certificate said anemia. Did you bring back some filthy disease with you, Hardy? Because if you did I'm going to kill you."

"Wait a minute," I said. "A throat infection? I didn't know — "

"There was a wound on her throat — two little marks, close together. That couldn't have killed her, unless some loathsome disease —"

"You're crazy, Hess," I said. "You know you're drunk. Listen to me: Jack couldn't have had anything to do with — that."

Hess didn't look at me. He watched Jack Hardy out of his bloodshot eyes. He went on in that low, deadly monotone:

"Will you swear Mark's right, Hardy? Will you?"

Jack's lips were twisted by some inner agony. I said, "Go on, Jack. Tell him he's wrong."

Hardy burst out, "I haven't been near your wife! I haven't seen her since I got back. There's —"

"That's not the answer I want," Hess whispered. And he sprang for the other man — reeled forward, rather.

Hess was too drunk, and Jack too weak, for them to do each other any harm, but there was a nasty scuffle for a moment before I separated them. As I pulled them apart, Hess's hand clutched the scarf about Jack's neck, ripped it away.

And I saw the marks on Jack Hardy's throat. Two red, angry little pits, white-rimmed, just over the left jugular.

2. The Cremation of Sandra

It was the next day that Jean telephoned me.

"Mart," she said, "we're going to run over a scene for *Red Thirst* tonight at the studio — Stage 6. You've been assigned as assistant director for the pic, so you should be there. And — I had an idea Jack might not tell you. He's been — so odd lately."

"Thanks, honey," I said. "I'll be there. But I didn't know you were in the flicker."

"Neither did I, but there's been some wire-pulling. Somebody wanted me in it — the chevalier, I think — and the big boss phoned me this morning and let me in on the secret. I don't feel up to it, though. Had a bad night."

"Sorry," I sympathized. "You were okay when I left you."

"I had a — nightmare," she said slowly. "It was rather frightful, Mart. It's funny, though, I can't remember what it was about. Well — you'll be there tonight?"

I said I would, but as it happened I was unable to keep my promise. Hess Deming telephoned me, asking if I'd come out to his Malibu place and drive him into town. He was too shaky to handle a car himself, he said, and Sandra's cremation was to take place that afternoon. I got out my

Newspaper advertisements for The Vampire Bat.

roadster and sent it spinning west on Sunset. In twenty minutes I was at Deming's beach house.

The houseboy let me in, shaking his head gravely as he recognized me. "Mist' Deming pretty bad," he told me. "All morning drinking gin straight —"

From upstairs Hess shouted, "That you, Mart? Okay — I'll be down right away. Come up here, Jim!"

The Japanese, with a meaningful glance at me, pattered upstairs.

I wandered over to a table, examining the magazines upon it. A little breath of wind came through the half-open window, fluttering a scrap of paper. A word on it caught my eye, and I picked up the note. For that's what it was. It was addressed to Hess, and after one glance I had no compunction about scanning it. "Hess dear," the message read. "I feel I'm going to die very soon so I want you to do something for me. I've been out of my head, I know, saying things I didn't mean. Don't cremate me, Hess. Even though I were dead I'd feel the fire — I know it. Bury me in a vault in Forest Lawn — and don't embalm me. I shall be dead when you find this, but I know you'll do as I wish, dear. And, alive or dead, I'll always love you."

The note was signed by Sandra Colter, Hess's wife. This was odd. I wondered whether Hess had seen it yet.

There was a little hiss of indrawn breath from behind me. It was Jim, the houseboy. He said, "Mist' Prescott — I find that note last night. Mist' Hess not seen it. It Mis' Colter's writing."

He hesitated, and I read fear in his eyes — sheer, unashamed fear. He put a brown forefinger on the note.

"See that, Mist' Prescott?"

He was pointing to a smudge of ink that half obscured the signature. I said, "Well?"

"I do that, Mist' Prescott. When I pick up the note. The ink — not dry."

I stared at him. He turned hastily at the sound of footsteps on the stairs. Hess was coming down, rather shakily.

I think it was then that I first realized the horrible truth. I didn't believe it, though — not then. It was too fantastic, too incredible; yet something of the truth must have crept into my mind, for there was no other explanation for what I did then.

Hess said, "What have you got there, Mart?"

"Nothing," I said quietly. I crumpled the note and thrust it into my pocket. "Nothing important, anyway. Ready to go?"

He nodded, and we went to the door. I caught a glimpse of Jim staring after us, an expression of — was it relief — in his dark, wizened face.

The crematory was in Pasadena, and I left Hess there. I would have stayed with him, but he wouldn't have it. I knew he didn't want anyone to be watching when Sandra's body was being incinerated. And I knew it would be easier for him that way. I took a short cut through the Hollywood

MARK OF THE VAMPIRE (1935)

Director Tod Browning was kept on a short leash by M-G-M after the unmitigated disaster of his 1932 film *Freaks*. *Mark of the Vampire* was evidently considered a "safe" project, being a remake of Browning's 1927 silent hit *London after Midnight* with Lon Chaney. Once more the story involved a police sting using phony vampires to catch a killer.

Bela Lugosi provided terrific atmosphere in the mute role of Count Mora, gliding around a cobwebby castle in Dracula drag with his lank-haired daughter Luna (Carroll Borland) in tow. The real star of the show, however, is cinematographer James Wong Howe, who visually elevates the material far above Browning's perfunctory direction. The haunted house atmospherics are occasionally so heavily laid on as to approach parody, but the brief film (cut by the studio from its original ninety minutes to an hour) is still fun to watch — if only to count the instances of visual cribbing from Universal's *Dracula*, which Browning also directed. (Also cut was an original intimation that Count Mora and his daughter achieved undeath through incest and suicide.) Lionel Barrymore receives top billing as a Van Helsing-like professor; also with Elizabeth Allan, Lionel Atwill, and Jean Hersholt. Screenplay by Guy Endore and Bernard Schubert. (Metro-Goldwyn-Mayer)

hills, and that's where the trouble started.

I broke an axle. Recent rains had gullied the road, and I barely saved the car from turning over. After that I had to hike miles to the nearest telephone, and then I wasted more time waiting for a taxi to pick me up. It was nearly eight o'clock when I arrived at the studio.

The gateman let me in, and I hurried to Stage 6. It was dark. Cursing under my breath, I turned away, and almost collided with a small figure. It was Forrest, one of the cameramen. He let out a curious squeal, and clutched my arm.

"That you, Mart? Listen, will you do me a favor? I want you to watch a print —"

"Haven't time," I said. "Seen Jean around here? I was to-"

"It's about that," Forrest said. He was a shriveled, monkey-faced little chap, but a mighty good cameraman. "They've gone — Jean and Hardy and the chevalier. There's something funny about that guy."

"Think so? Well, I'll phone Jean. I'll look at your rushes tomorrow."

"She won't be home," he told me. "The chevalier took her over to the Grove. Listen, Mart, you've got to watch this. Either I don't know how to handle a grinder any more, or that Frenchman is the damnedest thing I've ever shot. Come over to the theater, Mart — I've got the reel ready to run. Just developed the rough print myself."

"Oh, all right," I assented, and followed Forrest to the theater.

I found a seat in the dark little auditorium, and listened to Forrest moving about in the projection booth. He clicked on the amplifier and said, "Hardy didn't want any pictures taken — insisted on it, you know. But the boss told me to leave one of the automatic cameras going — not to bother with the sound — just to get an idea how the French guy would screen. Lucky it wasn't one of the old rattler cameras, or Hardy would have caught on. Here it comes, Mart!"

Poster for The Mark of the Vampire.

I heard a click as the amplifier was switched off. White light flared on the screen. It faded, gave place to a picture — the interior of Stage 6. The set was incongruous — a mid-Victorian parlor, with overstuffed plush chairs, gilt-edged paintings, even a particularly hideous what-not. Jack Hardy moved into the range of the camera. On the screen his face seemed to leap out at me like a death's-head, covered with sagging, wrinkled skin. Following him came Jean, wearing a tailored suit — no one dresses for rehearsals — and behind her —

I blinked, thinking that my eyes were tricking me. Something like a glowing fog — oval, tall as a man — was moving across the screen. You've seen the nimbus of light on the screen when a flashlight is turned directly on the camera? Well — it was like that, except that its source was not traceable. And, horribly, it moved forward at about the pace a man would walk.

The amplifier clicked again. Forrest said, "When I saw it on the negative I thought I was screwy, Mart. I saw the take — there wasn't any funny light there. Look —" The oval, glowing haze was motionless beside Jean, and she was looking directly at it, a smile on her lips. "Mart, when that was taken, Jean was looking right at the French guy!"

I said, somewhat hoarsely, "Hold it, Forrest. Right there."

The images slowed down, became motionless. Jean's profile was toward the camera. I leaned forward, staring at something I had glimpsed on the girl's neck. It was scarcely visible save as a tiny, discolored mark on Jean's throat, above the jugular — but unmistakably the same wound I had seen on the throat of Jack Hardy the night before!

I heard the amplifier click off. Suddenly the screen showed blinding white, and then went black.

Like father, like daughter. Bela Lugosi and Carroll Borland in Mark of the Vampire.

I waited a moment, but there was no sound from the booth.

"Forrest," I called. "You okay?"

There was no sound. The faint whirring of the projector had died. I got up quickly and went to the back of the theater. There were two entrances to the booth, a door which opened on stairs leading down to the alley outside and a hole in the floor reached by means of a metal ladder. I went up this swiftly, an ominous apprehension mounting within me.

Forrest was still there. But he was no longer alive. He lay sprawled on his back, his wizened face staring up blindly, his head twisted at an impossible angle. It was quite apparent that his neck had been broken almost instantly.

I sent a hasty glance at the projector. The can of film was gone! And the door opening on the stairway was ajar a few inches.

I stepped out on the stairs, although I knew I would see no one. The white-lit broad alley between Stages 6 and 4 was silent and empty.

The sound of running feet came to me, steadily growing louder. A man came racing into view. I recognized him as one of the publicity gang. I hailed him.

"Can't wait," he gasped, but slowed down nevertheless.

I said, "Have you seen anyone around here just now? The Chevalier Futaine?"

He shook his head. "No, but —" His face was white as he looked up at me. "Hess Deming's gone crazy. I've got to contact the papers."

Ice gripped me. I raced down the stairs, clutched his arm.

"What do you mean?" I snapped. "Hess was all right when I left him. A bit tight, that's all."

His face was glistening with sweat. "It's awful — I'm not sure yet what happened. His wife — Sandra Colter came to life while they were cremating her. They saw her through the window, you know — screaming and pounding at the glass while she was being burned alive. Hess got her out too late. He went stark, raving mad. Suspended animation, they say — I've got to get to a phone, Mr. Prescott!"

He tore himself away, sprinted in the direction of the administration buildings.

I put my hand in my pocket and pulled out a scrap of paper. It was the note I had found in Hess Deming's house. The words danced and wavered before my eyes. Over and over I was telling myself, "It can't be true! Such things can't happen!"

I didn't mean Sandra Colter's terrible resurrection during the cremation. That, alone, might be plausibly explained — catalepsy, perhaps. But taken in conjunction with certain other occurrences, it led to one definite conclusion — and it was a conclusion I dared not face.

What had poor Forrest said? That the chevalier was taking Jean to the Cocoanut Grove? Well —

The taxi was still waiting. I got in.

"The Ambassador," I told the driver grimly. "Twenty bucks if you hit the green lights all the way."

3. The Black Coffin

All night I had been combing Hollywood — without success. Neither the Chevalier Futaine nor Jean had been to the Grove, I discovered. And no one knew the Chevalier's address. A telephone call to the studio, now ablaze with the excitement over the Hess Deming disaster and the Forrest killing, netted me exactly nothing. I went the rounds of Hollywood night life vainly. The Trocadero, Sardi's, all three of the Brown Derbies, the smart, notorious clubs of the Sunset eighties — nowhere could I find my quarry. I telephoned Jack Hardy a dozen times, but got no answer. Finally, in a "private club" in Culver City, I met with my first stroke of good luck.

"Mr. Hardy's upstairs," the Proprietor told me, looking anxious. "Nothing wrong, I hope, Mr. Prescott? I heard about Deming."

"Nothing," I said. "Take me up to him."

"He's sleeping it off," the man admitted. "Tried to drink the place dry, and I put him upstairs where he'd be safe."

"Not the first time, eh?" I said, with an assumption of lightness. "Well, bring up some coffee, will you? Black. I've got to — talk to him."

But it was half an hour before Hardy was in any shape to understand what I was saying. At last he sat up on the couch, blinking, and a gleam of realization came into his sunken eyes.

"Prescott," he said, "can't you leave me alone?"

I leaned close to him, articulating carefully so he would be sure to understand me. "I know what the Chevalier Futaine is," I said.

And I waited for the dreadful, impossible confirmation or for the words which would convince me that I was an insane fool.

Hardy looked at me dully. "How did you find out?" he whispered.

An icy shock went through me. Up to that moment I had not really believed, in spite of all the evidence. But now Hardy was confirming the suspicions which I had not let myself believe.

I didn't answer his question. Instead, I said, "Do you know about Hess?"

He nodded, and at sight of the agony in his face I almost pitied him.

CONDEMNED TO LIVE (1935)

Frank Strayer, director of *The Vampire Bat* (1933), returned to familiar terrain. Ralph Morgan is a gentle professor with one problem: before birth, his mother was attacked by a huge African vampire bat, and passed on to him a blood curse. In a twist taken from werewolf stories, Morgan becomes a monster under the influence of the full moon, and in a twist taken from the *Frankenstein* films, has a hunchbacked assistant who helps cover up his crimes. With Pedro De Cordoba, Maxine Doyle, Russell Gleason, Lucy Beaumont, Barbara Bedford, and Mischa Auer as the hunchback. (Invincible Pictures)

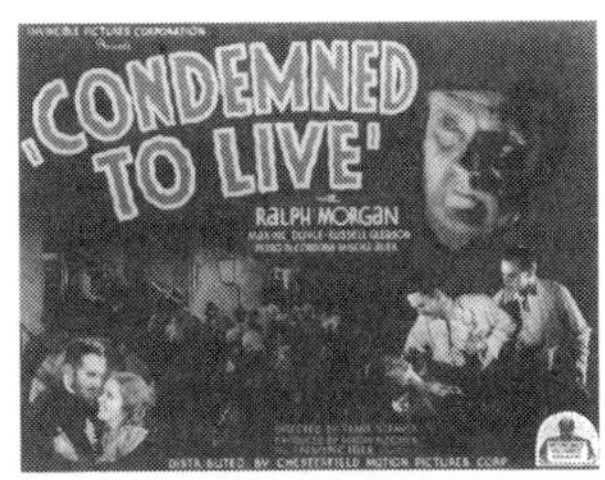

Lobby card for Condemned to Live. *(Courtesy of Ronald V. Borst/Hollywood Movie Posters)*

Then the thought of Jean steadied me.

"Do you know where he is now?" I asked.

"No. What are you talking about?" he flared suddenly. "Are you mad, Mart? Do you —"

"I'm not mad. But Hess Deming is."

He looked at me like a cowering, whipped dog.

I went on grimly: "Are you going to tell me the truth? How you got those marks on your throat? How you met this — creature? And where he's taken Jean?"

"Jean!" He looked genuinely startled. "Has he got — I didn't know that, Mart — I swear I didn't. You — you've been a good friend to me, and — and I'll tell you the truth — for your sake and Jean's — although now it may be too late —"

My involuntary movement made him glance at me quickly. Then he went on.

"I met him in Paris. I was out after new sensations — but I didn't expect anything like that. A Satanist club — devil-worshippers, they were. The ordinary stuff — cheap, furtive blasphemy. But it was there that I met — him.

"He can be a fascinating chap when he tries. He drew me out, made me tell him about Hollywood — about the women we have here. I bragged a little. He asked me about the stars, whether they were really as beautiful as they seemed. His eyes were hungry as he listened to me, Mart.

"Then one night I had a fearful nightmare. A monstrous, black horror crept in through my window and attacked me — bit me in the throat, I dreamed, or thought I did. After that —

"I was in his power. He told me the truth. He made me his slave, and I could do nothing. His powers — are not human."

I licked dry lips.

Hardly continued: "He made me bring him here, introducing him as a new discovery to be starred in *Red Thirst* — I'd mentioned the picture to him, before I — knew. How he must have laughed at me! He made me serve him, keeping away photographers, making sure that there were no cameras, no mirrors near him. And for a reward — he let me live."

I knew I should feel contempt for Hardy, panderer to such a loathsome evil. But somehow I couldn't.

I said quietly, "What about Jean? Where does the chevalier live?"

He told me. "But you can't do anything, Mart. There's a vault under the house, where he stays during the day. It can't be opened, except with a key he always keeps with him — a silver key. He had a door specially made, and then did something to it so that nothing can open it but that key. Even dynamite wouldn't do it, he told me."

I said, "Such things — can be killed."

"Not easily. Sandra Colter was a victim of his. After death she, too, became a vampire, sleeping by day and living only at night. The fire destroyed

DRACULA'S DAUGHTER (1936)

Universal's long-awaited sequel to *Dracula* went through an extended period of development hell. Though director James Whale and screenwriter R.C. Sherriff originally intended it as a starring vehicle for Bela Lugosi and Jane Wyatt, the studio found it impossible to get their original concept past the industry censors, who had become skittish in the wake of political pressure from morality groups like the Legion of Decency. Universal had acquired the rights to Bram Stoker's short story "Dracula's Guest" from David O. Selznick, who had independently optioned the property from Bram Stoker's widow. No one seriously considered using anything in the story except for its brief depiction of a female vampire (who not even Stoker claimed was a blood relative of Dracula).

John L. Balderston, who co-authored the stage version of *Dracula*, wrote a treatment depicting Dracula's daughter as a sexual sadist, complete with whips and chains. "The use of a female Vampire instead of male gives us the chance to play up SEX and CRUELTY legitimately," Balderston wrote. "In *Dracula* these had to be almost eliminated . . . We profit by making Dracula's daughter amorous of her victims . . . The seduction of young men will be tolerated whereas we had to eliminate seduction of girls from the original as obviously censorable."

her, but there's no way to get into the vault under Futaine's house."

"I wasn't thinking of fire," I said. "A knife —"

"Through the heart," Hardy interrupted almost eagerly. "Yes — and decapitation. I've thought of it myself, but I can do nothing. I — am his slave, Mart."

I said nothing, but pressed the bell. Presently the proprietor appeared.

"Can you get me a butcher knife?" I measured with my hands. "About so long? A sharp one?"

Accustomed to strange requests, he nodded. "Right away, Mr. Prescott."

As I followed him out, Hardy said weakly, "Mart."

I turned.

"Good luck," he said. The look on his wrecked face robbed the words of their pathos.

"Thanks," I forced myself to say. "I don't blame you, Jack, for what's happened. I — I'd have done the same."

I left him there, slumped on the couch, staring after me with eyes that had looked into hell.

It was past daylight when I drove out of Culver City, a long, razor-edged knife hidden securely inside my coat. And the day went past all too quickly. A telephone call told me that Jean had not yet returned home. It took me more than an hour to locate a certain man I wanted — a man who had worked for the studio before on certain delicate jobs. There was little about locks he did not know, as the police had sometimes ruefully admitted.

His name was Axel Ferguson, a bulky, good-natured Swede, whose thick fingers seemed more adapted to handling a shovel than the mechanisms of locks. Yet he was as expert as Houdini — indeed, he had at one time been a professional magician.

The front door of Futaine's isolated canyon home proved no bar to Ferguson's fingers and the tiny sliver of steel he used. The house, a modern two-story place, seemed deserted. But Hardy had said *below* the house.

We went down the cellar stairs and found ourselves in a concrete-lined passage that ran down at a slight angle for perhaps thirty feet. There the corridor ended in what seemed to be a blank wall of bluish steel. The glossy surface of the door was unbroken, save for a single keyhole.

Ferguson set to work. At first he hummed under his breath, but after a time he worked in silence. Sweat began to glisten on his face. Trepidation assailed me as I watched.

The flashlight he had placed beside him grew dim. He inserted another battery, got out unfamiliar-looking apparatus. He buckled on dark goggles, and handed me a pair. A blue, intensely brilliant flame began to play on the door.

It was useless. The torch was discarded after a time, and Ferguson returned to his tools. He was using a stethoscope, taking infinite pains in the delicate movements of his hands.

Poster for Dracula's Daughter. *(Courtesy of Ronald V. Borst/ Hollywood Movie Posters)*

It was fascinating to watch him. But all the time I realized that the night was coming, that presently the sun would go down, and that the life of the vampire lasts from sunset to sunrise.

At last Ferguson gave up. "I can't do it," he told me, panting as though from a hard race. "And if I can't, nobody can. Even Houdini couldn't have broken this lock. The only thing that'll open it is the key."

"All right, Axel," I said dully. "Here's your money."

He hesitated, watching me. "You going to stay here, Mr. Prescott?"

"Yeah," I said. "You can find your way out. I'll — wait awhile."

"Well, I'll leave the light with you," he said. "You can let me have it sometime, eh?"

He waited, and, as I made no answer, he departed, shaking his head.

Then utter silence closed around me. I took the knife out of my coat, tested its edge against my thumb, and settled back to wait.

Less than half an hour later the steel door began to swing open. I stood up. Through the widening crack I saw a bare, steel-lined chamber, empty save for a long, black object that rested on the floor. It was a coffin.

The door was wide. Into view moved a white, slender figure — Jean, clad in a diaphanous, silken robe. Her eyes were wide, fixed and staring. She looked like a sleepwalker.

A man followed her — a man wearing impeccable evening clothes. Not a hair was out of place on his sleek blond head, and he was touching his lips delicately with a handkerchief as he came out of the vault.

There was a little crimson stain on the white linen where his lips had brushed.

Balderston's ideas went nowhere, and James Whale favored a more sardonic approach. R.C. Sherriff''s script for Whale featured a long Transylvanian flashback in which the cruel count amused himself with elaborate palace games involving young lovers, severed arms, and so on. A local wizard (a role possibly intended for Boris Karloff), fed up with the debauchery, interrupts the revels and casts a spell that turns the count's degenerate guests into monkeys and swine and Dracula himself into a vampire. Whale, Sherriff, Lugosi, and even the character of Dracula were dropped from the production, which was finally scripted by Garrett Fort and directed by Lambert Hillyer.

Actress Gloria Holden made an austere, soignée Countess Maria Zaleska, a reluctant vampire who unsuccessfully seeks a psychiatric cure. She also has distinct lesbian tendencies: her blood-seduction of a young streetwalker (Nan Grey) is the film's most famous scene, and still packs a punch. Novelist Anne Rice credits this film as a major early inspiration for her vampire novels; in *The Queen of the Damned* (1992), she paid it homage by naming a quasi-gay bar in San Francisco's Castro district "Dracula's Daughter." The film also stars Otto Kruger as the psychiatrist (a role originally slated for Cesar Romero); Edward Van Sloan, reprising his role as Professor Van Helsing; and Irving Pichel as a Lugosi-esque servant who undermines his mistress's recovery program for his own undead ambitions. (Universal)

4. I, the Vampire

Jean walked past me as though I didn't exist. But the Chevalier Futaine passed, eyebrows lifted. His black eyes pierced through me.

The handle of the knife was hot in my hand. I moved aside to block Futaine's way. Behind me came a rustle of silk, and from the corner of my eye I saw Jean pause hesitatingly.

The chevalier eyed me, toying negligently with his handkerchief. "Mart," he said slowly. "Mart Prescott." His eyes flickered toward the knife, and a little smile touched his lips.

I said, "You know why I'm here, don't you?"

"Yes," he said. "I — heard you. I was not disturbed. Only one thing can open this door."

From his pocket he drew a key, shining with a dull silver sheen.

"Only this," he finished, replacing it. "Your knife is useless, Mart Prescott."

"Maybe," I said, edging forward very slightly. "What have you done to Jean?"

A curious expression, almost of pain, flashed into his eyes. "She is mine," he shot out half angrily. "You can do nothing, for —"

I sprang then, or, at least, I tried to. The blade of the knife sheared down straight for Futaine's white shirtfront. It was arrested in midair. Yet he had not moved. His eyes had bored into mine, suddenly, terribly, and it seemed as though a wave of fearful energy had blasted out at me — paralyzing me, rendering me helpless. I stood rigid. Veins throbbed in my temples as I tried to move — to bring down the knife. It was useless. I stood as immovable as a statue.

The chevalier brushed past me.

"Follow," he said almost casually, and like an automaton I swung about, began to move along the passage. What hellish hypnotic power was this that held me helpless?

Futaine led the way upstairs. It was not yet dark, although the sun had gone down. I followed him into a room, and at his gesture dropped into a chair. At my side was a small table. The chevalier touched my arm gently, and something like a mild electric shock went through me. The knife dropped from my fingers, clattering to the table.

Jean was standing rigidly nearby, her eyes dull and expressionless. Futaine moved to her side, put an arm about her waist. My mouth felt as though it were filled with mud, but somehow I managed to croak out articulate words.

"Damn you, Futaine! Leave her alone!"

He released her, and came toward me, his face dark with anger.

"You fool, I could kill you now, very easily. I could make you go down to the busiest corner of Hollywood and slit your throat with that knife. I have the power. You have found out much, apparently. Then you know — my power."

"Yes," I muttered thickly. "I know that. You devil — Jean is mine!"

The face of a beast looking into mine. He snarled, "She is not yours. Nor is she — *Jean*. She is Sonya."

Vampire wannabe: Irving Pichel in Dracula's Daughter.

I remembered what Futaine had murmured when he had first seen Jean. He read the question in my eyes.

"I knew a girl like that once, very long ago. That was Sonya. They killed her — put a stake through her heart, long ago in Thurn. Now that I've found this girl, who might be a reincarnation of Sonya — they are so alike — I shall not give her up. Nor can anyone force me."

"You've made her a devil like yourself," I said through half-paralyzed lips. "I'd rather kill her —"

Futaine turned to watch Jean. "Not yet," he said softly. "She is mine — yes. She bears the stigmata. But she is still — alive. She will not become — *wampyr* — until she has died, or until she has tasted the red milk. She shall do that tonight."

I cursed him bitterly, foully. He touched my lips, and I could utter no sound. Then they left me — Jean and her master. I heard a door close quietly.

The night dragged on. Futile struggles had convinced me that it was too useless to attempt escape — I could not even force a whisper through my lips. More than once I felt myself on the verge of madness — thinking of Jean, and remembering Futaine's ominous words. Eventually agony brought its own surcease, and I fell into a kind of coma, lasting for how long I could not guess. Many hours had passed, I knew, before I heard footsteps coming toward my prison.

Jean moved into my range of vision. I searched her face with my eyes, seeking for some mark of a dreadful metamorphosis. I could find none. Her beauty was unmarred, save for the terrible little wounds on her throat. She went to a couch and quietly lay down. Her eyes closed.

The chevalier came past me and went to Jean's side. He stood looking down at her. I have mentioned before the incongruous youthfulness of his face. That was gone now. He looked old — old beyond imagination.

At last he shrugged and turned to me. His fingers brushed my lips again, and I found that I could speak. Life flooded back into my veins, benign lancing twinges of pain. I moved an arm experimentally. The paralysis was leaving me.

The chevalier said, "She is still — clean. I could not do it."

Amazement flooded me. My eyes widened in disbelief.

Futaine smiled wryly. "It is quite true. I could have made her as myself — undead. But at the last moment I forbade her." He looked toward the windows. "It will be dawn soon."

I glanced at the knife on the table beside me. The chevalier put out a hand and drew it away.

"Wait," he said. "There is something I must tell you, Mart Prescott. You say that you know who and what I am."

I nodded.

"Yet you cannot know," he went on. "Something you have learned, and something you have guessed, but you can never know me. You are human, and I am — the undead.

"Through the ages I have come, since first I fell victim to another vampire — for thus is the evil spread. Deathless and not alive, bringing fear and sorrow always, knowing the bitter agony of Tantalus, I have gone down through the weary centuries. I have known Richard and Henry and Elizabeth of England, and ever have I brought terror and destruction in the night, for I am an alien thing, I am the undead."

The quiet voice went on, holding me motionless in its weird spell.

"I, the vampire. I, the accursed, the shining evil, *negotium perambulans in tenebris*...but I was not always thus. Long ago in Thurn, before the shadow fell upon me, I loved a girl — Sonya. But the vampire visited me, and I sickened and died — and awoke. Then I arose.

"It is the curse of the undead to prey upon those they love. I visited Sonya. I made her my own. She, too, died, and for a brief while we walked the earth together, neither alive nor dead. But that was not Sonya. It was

BLONDE VAMPIRES HAVE MORE FUN?

Although Henry Kuttner introduced the world's first blonde male vampire in 1937, it was not until 1960 that the conceit reached the screen. In Hammer Films' *The Brides of Dracula* (1960), actor David Peel played the bad boy (or is it bat boy?) Baron Meinster, who has an unusually unhealthy relationship with his mother. The *London Evening Standard* noted, disapprovingly, that the golden-locked vampire "capitalises on current fashion by resembling Oscar Wilde's Bosie with fangs." *The Brides of Dracula* has since become a minor classic, prefiguring the complex sexual themes that have been expanded upon more recently by Anne Rice and others.

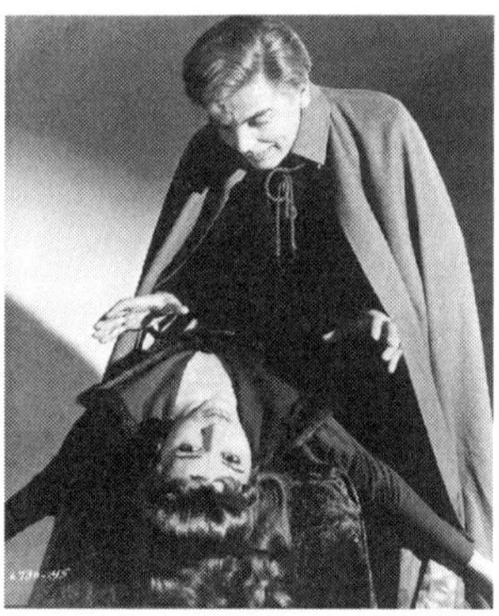

David Peel and Yvonne Monlaur in The Brides of Dracula. *(Photofest)*

her body, yes, but I had not loved her body alone. I realized too late that I had destroyed her utterly.

"One day they opened her grave, and the priest drove a stake through her heart, and gave her rest. Me they could not find, for my coffin was hidden too well. I put love behind me then, knowing that there was none for such as I.

"Hope came to me when I found — Jean. Hundreds of years have passed since Sonya crumbled to dust, but I thought I had found her again. And — I took her. Nothing human could prevent me."

The chevalier's eyelids sagged. He looked infinitely old.

"Nothing human. Yet in the end I found that I could not condemn her to the hell that is mine. I thought I had forgotten love. But, long and long ago, I loved Sonya. And, because of her, and because I know that I would only destroy, as I did once before, I shall not work my will on this girl."

I turned to watch the still figure on the couch. The chevalier followed my gaze and nodded slowly.

"Yes, she bears the stigmata. She will die, unless" — he met my gaze unflinchingly — "unless I die. If you had broken into the vault yesterday, if you had sunk that knife into my heart, she would be free now." He glanced at the windows again. "The sun will rise soon."

Then he went quickly to Jean's side. He looked down at her for a moment.

"She is very beautiful," he murmured. "Too beautiful for hell."

The chevalier swung about, went toward the door. As he passed me he threw something carelessly on the table, something that tinkled as it fell. In the portal he paused, and a little smile twisted the scarlet lips. I remembered him thus, framed against the black background of the doorway, his sleek blond head erect and unafraid. He lifted his arm in a gesture that should have been theatrical, but, somehow, wasn't.

"And so farewell. I who am about to die —"

He did not finish. In the faint grayness of dawn I saw him striding away, heard his footsteps on the stairs, receding and faint — heard a muffled clang as of a great door closing. The paralysis had left me. I was trembling a little, for I realized what I must do soon. But I knew I would not fail.

I glanced down at the table. Even before I saw what lay beside the knife, I knew what would be there. A silver key…

Had it been written thirty years earlier, Stuart Kamnisky's novel Never Cross a Vampire *(1980) would have made an ideal Hollywood-backstage vehicle for Bela Lugosi, who appears in the murder mystery as himself, facing fictional death threats from a disgruntled fan. (St. Martin's Press)*

Out of the belfry: Lon Chaney, Jr. in a remarkably dramatic publicity still for Son of Dracula *(1943).*

The Bat Is My Brother

ROBERT BLOCH
(1944)

"Have you ever wondered why there are not more vampires — for every victim of a vampire becomes one in turn!"

IT BEGAN IN twilight — a twilight I could not see.

My eyes opened on darkness, and for a moment I wondered if I were still asleep and dreaming. Then I slid my hands down and felt the cheap lining of the casket, and knew that this nightmare was real.

I wanted to scream, but who can hear through six feet of earth above a grave?

Better to save my breath and try to save my sanity. I fell back, and the darkness rose all around me. The darkness, the cold, clammy darkness of death.

I could not remember how I had come here, or what hideous error had brought about my premature interment. All I knew was that I lived — but unless I managed to escape, I would soon be in a condition horribly appropriate to my surroundings.

Then began that which I dare not remember in detail. The splintering of wood, the burrowing struggle through loosely-packed grave earth; the gasping hysteria accompanying my clawing, suffocated; progress to the sane surface of the world above.

It is enough that I finally emerged. I can only thank poverty for my deliverance — the poverty which had placed me in a flimsy, unsealed coffin and a pauper's shallow grave.

Clotted with sticky clay, drenched with cold perspiration, racked by utter revulsion, I crawled forth from betwixt the gaping jaws of death.

Dusk crept between the tombstones, and somewhere to my left the moon leered down to watch the shadowy legions that conquered in the name of Night.

The moon saw me, and a wind whispered furtively to brooding trees, and the trees bent low to mumble a message to all those sleeping below their shade.

I grew restless beneath the moon's glaring eye, and I wanted to leave this spot before the trees had told my secret to the nameless, numberless dead.

Despite my desire, several minutes passed before I summoned strength to stand erect, without trembling.

ABOUT THE AUTHOR

Robert Bloch (1917-1994) will be forever known as the author of the novel *Psycho* (1959), but his public identification with that single title often overshadows his remarkably prolific career as a short-story writer, novelist, and screenwriter. Born in Chicago, Bloch was drawn to fantastic fiction at an early age. Under the spell of *Weird Tales* magazine, he had a youthful correspondence with that magazine's master of dark fantasy, H.P. Lovecraft, who encouraged Bloch's own early efforts. He began publishing professionally in 1934, and contributed voluminously to pulp magazines throughout his career; Bloch's short-story output would eventually fill more than thirty collections published during his lifetime. His first novel, *The Scarf* (1947) impressively introduced his masterful understanding of serial-killer psychology that would culminate in *Psycho*. He adapted several of his stories for the Boris Karloff-hosted anthology series "Thriller" (1960-62) and collaborated with producer William Castle on the 1964 films *Strait-Jacket* and *The Night Walker*. Other Bloch screen credits included the original "Star Trek" television series, *The Psychopath* (1966), *Torture Garden* (1968), *The House that Dripped Blood* (1970), and *Asylum* (1972). His engaging autobiography, *Once Around the Bloch* (1993) appeared the year before his death.

Then I breathed deeply of fog and faint putridity; breathed, and turned away along the path.

It was at that moment the figure appeared.

It glided like a shadow from the deeper shadows haunting the trees, and as the moonlight fell upon a human face I felt my heart surge in exultation.

I raced towards the waiting figure, words choking in my throat as they fought for prior utterance.

"You'll help me, won't you?" I babbled. "You can see . . . they buried me down there I was trapped alive in the grave out now . . . you'll understand. I can't remember how it began, but you'll help me?"

A head moved in silent assent.

I halted, regaining my composure, striving for coherency.

"This is awkward," I said, more quietly. "I've really no right to ask you for assistance. I don't even know who you are."

The voice from the shadows was only a whisper, but each word thundered in my brain.

"I am a vampire," said the stranger.

Madness. I turned to flee, but the voice pursued me.

"Yes, I am a vampire," he said. "And... *so are you!*"

II

I must have fainted, then. I must have fainted, and he must have carried me out of the cemetery, for when I opened my eyes once more I lay on a sofa in his house.

The panelled walls loomed high, and shadows crawled across the ceiling beyond the candlelight. I sat up, blinked, and stared at the stranger who bent over me.

I could see him now, and I wondered. He was of medium height, gray-haired, clean-shaven, and clad discreetly in a dark business suit. At first glance he appeared normal enough.

As his face glided towards me, I stared closer, trying to pierce the veil of his seeming sanity, striving to see the madness beneath the prosaic exterior of dress and flesh.

I stared and saw that which was worse than any madness.

At close glance his countenance was cruelly illumined by the light. I saw the waxen pallor of his skin, and what was worse than that, the peculiar corrugation. For his entire face and throat was covered by a web of tiny wrinkles, and when he smiled it was with a mummy's grin.

Yes, his face was white and wrinkled; white, wrinkled, and long dead. Only his lips and eyes were alive, and they were red . . . too red. A face as white as corpse-flesh, holding lips and eyes as red as blood.

He smelled musty.

All these impressions came to me before he spoke. His voice was like

THE LEGEND AND LORE OF BATS

The premier emblem and avatar of vampirism, the bat occupies a rich place in world folklore. It is, of course, the image of the blood-drinking vampire bat that forges the strongest link between the winged mammal and imaginary vampires, but the bat has many other associations with darkness, death, and the supernatural that reinforce its mythic reputation.

From a rational perspective, the reputation is undeserved, because bats are an important part of the ecosystem and essential for the control of insects in many regions. Bats are also much less exotic than many people would believe; it has been estimated that twenty percent of all living mammals are bats.

The vampire bat constitutes a small category of the bat world: the family *Desmodontidae*, indigenous to South and Central America. Like its fantastic counterpart, it attacks its victims (far more likely to be livestock than human) during sleep. The vampire bat does not alight directly on its prey, preferring to land at some distance, making the final approach with a stealthy, hopping crawl. It seeks out a warm place where the skin is unprotected and the blood supply copious — the neck, the eyes, the anus — and there painlessly opens the skin with a pair of razor-sharp incisors. The bat's saliva contains an anticoagulant, which keeps the blood flowing for

the rustle of the wind through a mortuary wreath.

"You are awake? It is well."

"Where am I? And who are you?" I asked the questions but dreaded an answer. The answer came.

"You are in my house. You will be safe here, I think. As for me, I am your guardian."

"Guardian?"

He smiled. I saw his teeth. Such teeth I had never seen, save in the maw of a carnivorous beast. And yet — wasn't *that* the answer?

"You are bewildered, my friend. Understandably so. And that is why you need a guardian. Until you learn the ways of your new life, I shall protect you." He nodded. "Yes, Graham Keene, I shall protect you."

"Graham Keene."

It was my name. I knew it *now*. But how did *he* know it?

"In the name of mercy," I groaned, "tell me what has happened to me!"

He patted my shoulder. Even through the cloth I could feel the icy weight of his pallid fingers. They crawled across my neck like worms, like wriggling white worms —

"You must be calm," he told me. "This is a great shock, I know. Your confusion is understandable. If you will just relax a bit and listen, I think I can explain everything."

I listened.

"To begin with, you must accept certain obvious facts. The first being — that you are a vampire."

"But —"

He pursed his lips, his *too* red lips, and nodded.

"There is no doubt about it, unfortunately. Can you tell me how you happened to be emerging from a grave?"

"No. I don't remember. I must have suffered a cataleptic seizure. The shock gave me partial amnesia. But it will come back to me. I'm all right, I must be."

The words rang hollowly even as they gushed from my throat. "Perhaps. But I think not." He sighed and pointed.

"I can prove your condition to you easily enough. Would you be so good as to tell me what you see behind you, Graham Keene?"

"Behind me?"

"Yes, on the wall."

I stared.

"I don't see anything."

"Exactly. "

"But —"

"Where is your shadow?"

I looked again. There was no shadow, no silhouette. For a moment my sanity wavered. Then I stared at him. "You have no shadow either," I exclaimed, triumphantly. "What does that prove?"

The Devil Bat *(1946) was not exactly a vampire film, but it successfully exploited the bat's perennial identification with vampires, not to mention Bela "Dracula" Lugosi.*

"That I am a vampire," he said, easily. "And so are you."

"Nonsense. It's just a trick of the light," I scoffed.

"Still skeptical? Then explain this optical illusion." A bony hand proffered a shining object.

I took it, held it. It was a simple pocket mirror.

"Look."

I looked.

The mirror dropped from my fingers and splintered on the floor. "There's no reflection!" I murmured.

"Vampires have no reflections." His voice was soft. He might have been reasoning with a child.

"If you still doubt," he persisted, "I advise you to feel your pulse. Try to detect a heartbeat."

Have you ever listened for the faint voice of hope to sound within you . . . knowing that it alone can save you? Have you ever listened and heard nothing? Nothing but the silence of *death*?"

I knew it then, past all doubt. I was of the Undead . . . the Undead who cast no shadows, whose images do not reflect in mirrors, whose hearts are forever stilled, but whose bodies live on — live and walk abroad, and take nourishment.

Nourishment!

I thought of my companion's red lips and his pointed teeth. I thought of the light blazing in his eyes. A light of hunger. Hunger for what?

How soon must I share that hunger?

He must have sensed the question, for he began to speak once more.

"You are satisfied that I speak the truth, I see. That is well. You must accept your condition and then prepare to make the necessary adjustments. For there is much you have to learn in order to face the centuries to come.

"To begin with, I will tell you that many of the common superstitions about — people like us — are false."

He might have been discussing the weather, for all the emotion his face betrayed. But I could not restrain a shudder of revulsion at his words.

"They say we cannot abide garlic. That is a lie. They say we cannot cross running water. Another lie. They say that we must lie by day in the earth of our own graves. That's picturesque nonsense.

"These things, and these alone, are true. Remember them, for they are important to your future. We must sleep by day and rise only at sunset. At dawn an overpowering lethargy bedrugs our senses, and we fall into a coma until dusk. We need not sleep in coffins — that is sheer melodrama, I assure you! — but it is best to sleep in darkness, and away from any chance of discovery by men.

"I do not know why this is so, any more than I can account for other phenomena relative to the disease. For vampirism is a disease, you know."

He smiled when he said it. I didn't smile. I groaned.

"Yes, it is a disease. Contagious, of course, and transmissible in the

the length of the meal (and sometimes longer, leading to debilitating blood loss and infections). The feast is over in about twenty minutes, with the bat often so bloated that it can barely fly.

One of the earliest descriptions of a human encounter with the vampire bat was written in 1565 by Benzoni, who made his observations in what is now Costa Rica. "There are many bats which bite people during the night . . . while I was sleeping they bit the toes of my feet so delicately that I felt nothing, and in the morning I found the sheets and mattresses with so much blood that it seemed I had suffered some great injury. . . ."

The fact that bats are also common vectors for rabies has done little to improve the animals' image historically, and may in fact have helped fuel the notion of vampirism as a communicable condition.

Bat-winged demons are a common fixture of religious and occult iconography; such creatures are, of course, dark travesties of angels. The motif of wings grafted onto the human form is an ambiguous image, one that can represent mankind's highest aspirations (in the case of ethereal, feathery angels) or (in the case of the leathery bat-demon), divine presumption and damnation.

The idea of flight has always captured the human imagination

classic manner, through a bite. Like rabies. What reanimates the body after death no one can say. And why it is necessary to take certain forms of nourishment to sustain existence, I do not know. The daylight coma is a more easily classified medical phenomenon. Perhaps an allergy to the direct actinic rays of the sun.

"I am interested in these matters, and I have studied them.

"In the centuries to come I shall endeavor to do some intensive research on the problem. It will prove valuable in perpetuating my existence, and yours."

The voice was harsher now. The slim fingers clawed the air in excitement.

"Think of that, for a moment, Graham Keene," he whispered. "Forget your morbid superstitious dread of this condition and look at the reality.

"Picture yourself as you were before you awoke at sunset. Suppose you had remained there, inside that coffin, nevermore to awaken! Dead — dead for all eternity!"

He shook his head. "You can thank your condition for an escape. It gives you a new life, not just for a few paltry years, but for centuries. Perhaps — forever!

"Yes, think and give thanks! You need never die, now. Weapons cannot harm you, nor disease, nor the workings of age. You are immortal — and I shall show you how to live like a god!"

He sobered. "But that can wait. First we must attend to our needs. I want you to listen carefully now. Put aside your silly prejudices and hear me out. I will tell you that which needs be told regarding our nourishment.

"It isn't easy, you know.

"There aren't any schools you can attend to learn what to do. There are no correspondence courses or books of helpful information. You must learn everything through your own efforts. Everything.

"Even so simple and vital a matter as biting the neck — using the incisors properly — is entirely a matter of personal judgment.

"Take that little detail, just as an example. You must choose the classic trinity to begin with — the time, the place, and the girl.

"When you are ready, you must pretend that you are about to kiss her. Both hands go under her ears. That is important, to hold her neck steady, and at the proper angle.

"You must keep smiling all the while, without allowing a betrayal of intent to creep into your features or your eyes. Then you bend your head. You kiss her throat. If she relaxes, you turn your mouth to the base of her neck, open it swiftly and place the incisors in position.

"Simultaneously — it *must* be simultaneously — you bring your left hand up to cover her mouth. The right hand must find, seize, and pinion her hands behind her back. No need to hold her throat now. The teeth are doing that. Then, and only then, will instinct come to your aid. It must come then, because once you begin, all else is wept away in the red,

in a double-edged manner. Freud tells us that flying dreams are, essentially, sex dreams; dreamlike images of flying monsters, therefore, contain a distinct air of dangerous or forbidden sexuality — a powerful component of the vampire mystique.

The first theatrical use of a batlike vampire cloak may have been made by Dion Boucicault in his 1852 play *The Vampire*. Bram Stoker's use of the bat in *Dracula* (1897) is somewhat ambiguous; the shape-shifting, winged creature that Dracula becomes is simultaneously described as resembling a lizard or a bird. Perhaps the most important stabilizing factor in the relationship between theatrical vampires and bats was the 1927 Broadway version of *Dracula*, in which audiences hooted the attempt to depict the vampire in werewolf form with an unconvincing stuffed animal head. The taxidermist's beast was summarily withdrawn, and it fell to a rubber bat to provide all the evening's animal thrills — werewolves, thereafter, became a separate horror category.

swirling blur of fulfillment."

I cannot describe his intonation as he spoke, or the unconscious pantomime which accompanied the incredible instructions. But it is simple to name the look that came into his eyes.

Hunger.

"Come, Graham Keene," he whispered. "We must go now."

"Go? Where?"

"To dine," he told me. "To dine!"

III

He led me from the house, and down a garden pathway through a hedge.

The moon was high, and as we walked along a windswept bluff, flying figures spun a moving web across the moon's bright face.

My companion shrugged.

"Bats," he said. And smiled.

"They say that — we — have the power of changing shape. That we become bats, or wolves. Alas, it's only another superstition. Would that it were true? For then our life would be easy. As it is, the search for sustenance in mortal form is hard. But you will soon understand."

I drew back. His hand rested on my shoulder in cold command.

"Where are you taking me?" I asked.

"To food."

Irresolution left me. I emerged from nightmare, shook myself into sanity.

"No — I won't!" I murmured. "I can't —"

"You must," he told me. "Do you want to go back to the grave?"

"I'd rather," I whispered. "Yes, I'd rather die."

His teeth gleamed in the moonlight.

"That's the pity of it," he said. "You can't die. You'll weaken without sustenance, yes. And you will appear to be dead. Then, whoever finds you will put you in the grave.

"But you'll be alive down there. How would you like to lie there undying in the darkness . . . writhing as you decay . . . suffering the torments of red hunger as you suffer the pangs of dissolution?

"How long do you think that goes on? How long before the brain itself is rotted away? How long must one endure the charnel consciousness of the devouring worm? Does the very dust still billow in agony?"

His voice held horror.

"That is the fate you escaped. But it is still the fate that awaits you unless you dine with me.

"Besides, it isn't something to avoid, believe me. And I am sure, my friend, that you already feel the pangs of — appetite."

I could not, dared not answer.

VAMPIRES DURING WARTIME

Robert Bloch's "The Bat is My Brother" was first published during World War II, a time when vampires and monsters were enjoying a popular-culture resurgence, especially in the movies. An embargo on horror films by the British Board of Film Censors between 1937 and 1939 sent Hollywood vampires scampering for their crypts, but the onset of the war once more cracked open a niche for pulse-pounding parables of the struggle between good and evil.

Vampire fiction during the war was mainly relegated to the pulp magazines, which showcased several memorable stories. Manly Wade Wellman wrote three vampire tales during the war years. "When it Was Moonlight" (1940) posits a journalistic encounter between Edgar Allan Poe and a female vampire in Philadelphia. "The Vampire of Shiloh" (1942) is set during the Civil War, with Union soldiers beset by yet another fanged femme. In "The Devil is Not Mocked" (1943), Wellman explicitly evoked the current war, setting a battalion of Nazi soldiers against the lonely lord of a castle in the Carpathians. Any guesses as to the landlord's real identity? (Hint: your first guess is your final answer.) "The Devil is Not Mocked" was adapted for television's "Night Gallery" in the 1970s with Francis Lederer as the castle's uncanny inhabitant. Among other writers, P. Schuyler Miller's imaginative tour de force

For it was true. Even as he spoke, I felt hunger. A hunger greater than any I had ever known. Call it a craving, call it a desire — call it lust. I felt it, gnawing deep within me. Repugnance was nibbled away by the terrible teeth of growing need.

"Follow me," he said, and I followed. Followed along the bluff and down a lonely country road.

We halted abruptly on the highway. A blazing neon sign winked incongruously ahead.

I read the absurd legend.

"DANNY'S DRIVE-IN."

Even as I watched, the sign blinked out.

"Right," whispered my guardian. "It's closing time. They will be leaving now."

"Who?"

"Mr. Danny and his waitress. She serves customers in their cars. They always leave together, I know they are locking up for the night now. Come along and do as you are told."

I followed him down the road. His feet crunched gravel as he stalked towards the now darkened drive-in stand. My stride quickened in excitement. I moved forward as though pushed by a gigantic hand. The hand of hunger —

He reached the side door of the shack. His fingers rasped the screen.

An irritable voice sounded.

"What do you want? We're closing."

"Can't you serve any more customers?"

"Nah. Too late. Go away."

"But we're very hungry."

I almost grinned. Yes, we were very hungry.

"Beat it!" Danny was in no mood for hospitality.

"Can't we get anything?"

Danny was silent for a moment. He was evidently debating the point. Then he called to someone inside the stand.

"Marie! Couple customers outside. Think we can fix 'em up in a hurry?"

"Oh, I guess so." The girl's voice was soft, complaisant. Would she be soft and complaisant, too?

"Open up. You guys mind eating outside?"

"Not at all."

"Open the door, Marie."

Marie's high heels clattered across the wooden floor. She opened the screen door, blinked out into the darkness.

My companion stepped inside the doorway. Abruptly, he pushed the girl forward.

"Now!" he rasped.

I lunged at her in darkness. I didn't remember his instructions about smiling at her, or placing my hands beneath her ears. All I knew was that her

"Over the River" (1941) is written from the viewpoint of a newly-minted vampire, vividly conveying the creature's unearthly sensory responses. Science fiction writer A.E. Van Vogt's "Asylum"(1942) may feature the first vampires — unless one counts H.G. Wells's blood-drinking invaders in *War of the Worlds* (1896) as members of the genus.

Hollywood, however, continued to bring vampires to the largest audiences. The major vampire films of the war years are covered on the following pages.

Nina Foch meets Bela Lugosi during the London blitz in Return of the Vampire. *(Courtesy of Ronald V. Borst/Hollywood Movie Posters)*

throat was white, and smooth, except where a tiny vein throbbed in her neck.

I wanted to touch her neck there with my fingers — with my mouth — with my teeth.

So I dragged her into the darkness, and my hands were over her mouth, and I could hear her heels scraping through the gravel as I pulled her along. From inside the shack I heard a single long moan and then nothing.

Nothing . . . except the rushing white blur of her neck, as my face swooped towards the throbbing vein…

Lon Chaney, Jr. and Louise Albritton in Son of Dracula.

SON OF DRACULA (1943)

Lon Chaney, Sr., who died on the eve of being cast in Universal's original *Dracula*, might well have made an unusual and arresting vampire count, but his son Creighton, better known as Lon Chaney, Jr., is excruciatingly miscast in this otherwise well-produced entry in the wartime Universal horror cycle. As Count Alucard (yeah, groan, but hey, it was the first time), Chaney has a pained expression on his face throughout most of the film, atmospherically photographed by George Robinson, the gifted cameraman of Universal's 1931 Spanish-language version of *Dracula*, *Dracula's Daughter* (1936), *Son of Frankenstein* (1939) and many others. One of the eeriest visual sequences has Alucard standing in his open coffin, which floats across a foggy bayou toward his undead bride-to-be. Directed by Robert Siodmak, from a screenplay by Eric Taylor based on a story by Curt Siodmak. With Louise Albritton, George Irving, and Robert Paige.

IV

It was cold in the cellar — cold, and dark.

I stirred uneasily on my couch and my eyes blinked open on blackness. I strained to see, raising myself to a sitting position as the chill slowly faded from my bones.

I felt sluggish, heavy with reptilian contentment. I yawned, trying to grasp a thread of memory from the red haze cloaking my thoughts.

Where was I? How had I come here? What had I been doing?

I yawned. One hand went to my mouth. My lips were caked with a dry, flaking substance.

I felt it — and then remembrance flooded me.

Last night, at the drive-in, I had feasted. And then —

"No!" I gasped.

"You have slept? Good."

My host stood before me. I arose hastily and confronted him.

"Tell me it isn't true," I pleaded. "Tell me I was dreaming."

"You were," he answered. "When I came out of the shack you lay under the trees, unconscious. I carried you home before dawn and placed you here to rest. You have been dreaming from sunrise to sunset, Graham Keene."

"But last night — ?"

"Was real."

"You mean I took that girl and — ?"

"Exactly." He nodded. "But come, we must go upstairs and talk. There are certain questions I must ask."

We climbed the stairs slowly and emerged on ground level. Now I could observe my surrounding with a more objective eye. This house was large, and old. Although completely furnished, it looked somehow untenanted. It was as though nobody had lived here for a long time.

Then I remembered who my host was, and what he was. I smiled grimly. It was true. Nobody was *living* in this house now.

Dust lay thickly everywhere, and the spiders had spun patterns of decay in the corners. Shades were drawn against the darkness, but still it crept in through the cracked walls. For darkness and decay belonged here.

We entered the study where I had awakened last night, and as I was seated, my guardian cocked his head towards me in an attitude of inquiry.

"Let us speak frankly," he began. "I want you to answer an important question."

"Yes?"

"What did you do with her?"

"Her?"

"That girl — last night. What did you do with her body?"

I put my hands to my temples. "It was all a blur. I can't seem to remember."

His head darted towards me, eyes blazing. "I'll tell you what you did with her," he rasped. "You threw her body down the well. I saw it floating there."

"Yes," I groaned. "I remember now."

"You fool — why did you do that?"

"I wanted to hide it… I thought they'd never know — "

"You *thought!*" Scorn weighted his voice. "You didn't think for an instant. Don't you see, now she will never rise?"

"Rise?"

"Yes, as you rose. Rise to become one of us."

"But I don't understand."

"That is painfully evident." He paced the floor, then wheeled towards me.

"I see that I shall have to explain certain things to you. Perhaps you are not to blame, because you don't realize the situation. Come with me."

He beckoned. I followed. We walked down the hall, entered a large, shelf-lined room. It was obviously a library. He lit a lamp, and halted.

"Take a look around," he invited. "See what you make of it, my friend."

I scanned the titles on the shelves — titles stamped in gold on thick, handsome bindings; titles worn to illegibility on ancient, raddled leather. The latest in scientific and medical treatises stood on these shelves, flanked by age-encrusted incunabula.

Modern volumes dealt with psychopathology. The ancient lore was frankly concerned with black magic.

"Here is the collection," he whispered. "Here is gathered together all that is known, all that has ever been written about — us."

"A library on vampirism?"

"It took me decades to assemble it completely."

"But why?"

"Because knowledge is power. And it is power I seek."

Suddenly a resurgent sanity impelled me. I shook off the nightmare enveloping me and sought an objective viewpoint. A question crept into my mind, and I did not try to hold it back.

"Just who are you, anyway?" I demanded. "What is your name?"

THE RETURN OF THE VAMPIRE (1943)

Bela Lugosi was unable to convince Universal to cast him again as Dracula when the studio revived the character for three films in the mid-1940s, choosing Lon Chaney, Jr. and John Carradine instead. But Columbia Pictures did not hesitate in contracting Lugosi's services for *The Return of the Vampire*. Universal refused Columbia permission to use the word "Dracula" in the script, but could do nothing about the fact that Lugosi in evening dress and an opera cape looked exactly like you-know-who. As the vampire Armand Tesla, Lugosi entertained wartime audiences with a story set during the London blitz — *deus ex machina* Nazi bombs both open his grave and send him back to it. Nina Foch, who made her Hollywood debut in the film, recalled in 1994 that her strongest impression of Lugosi during production was that he reeked of sulfur water, a popular but odiferous health tonic. One can see why Universal may have passed the actor over — he's well past his prime, though his line readings are as priceless as ever. Directed by Lew Landers from a screenplay by Griffin Jay. With Frieda Inescort, Miles Mander, and Ottola Nesmith.

Poster for The Return of the Vampire *(Courtesy of Budy Barnett)*

My host smiled.

"I have no name," he answered.

"No name?"

"Unfortunate, is it not? When I was buried, there were no loving friends apparently, to erect a tombstone. And when I arose from the grave, I had no mentor to guide me back to a memory of the past. Those were barbaric times in the East Prussia of 1777."

"You died in 1777?" I muttered.

"To the best of my knowledge," he retorted, bowing slightly in mock deprecation. "And so it is that my real name is unknown. Apparently I perished far from my native heath, for diligent research on my part has failed to uncover my paternity, or any contemporaries who recognized me at the time of my — er — resurrection.

"And so it is that I have no name; or rather, I have many pseudonyms. During the past sixteen decades I have traveled far, and have been all things to all men. I shall not endeavor to recite my history.

"It is enough to say that slowly, gradually, I have grown wise in the ways of the world. And I have evolved a plan. To this end I have amassed wealth, and brought together a library as a basis for my operations. Those operations I propose will interest you. And they will explain my anger when I think of you throwing the girl's body into the well."

He sat down. I followed suit. I felt anticipation crawling along my spine. He was about to reveal something — something I wanted to hear, yet dreaded. The revelation came, slyly, slowly.

"Have you ever wondered," he began, "why there are not more vampires in the world?"

"What do you mean?"

"Consider. It is said, and it is true, that every victim of a vampire becomes a vampire in turn. The new vampire finds other victims. Isn't it reasonable to suppose, therefore, that in a short time — through sheer mathematical progression — the virus of vampirism would run epidemic throughout the world? In other words, have you ever wondered why the world is not filled with vampires by this time?"

"Well, yes — I never thought of it that way. What is the reason?" I asked.

He glared and raised a white finger. It stabbed forward at my chest — a rapier of accusation.

"Because of fools like you. Fools who cast their victims into wells; fools whose victims are buried in sealed coffins, who hide the bodies or dismember them so no one would suspect their work.

"As a result, few new recruits join ranks. And the old ones — myself included — are constantly subject to the ravages of the centuries. We eventually disintegrate, you know. To my knowledge, there are only a few hundred vampires today. And yet, if new victims all were given the opportunity to rise — we would have a vampire army within a year. Within

HOUSE OF FRANKENSTEIN (1944)

The first go-for-broke "house party" of the major Universal Studios' monsters (minus, for some reason, the Mummy), the episodic *House of Frankenstein* resurrects Count Dracula (John Carradine) for its memorable first sequence. Boris Karloff is a mad scientist escaped from prison with Frankenstein ambitions; he soon finds refuge in a traveling carnival whose chamber of horrors contains the staked skeleton of Dracula. Spike removed, the revived vampire agrees to do Karloff's bidding, until he is once more reduced to dust by an ill-timed sunrise. Also starring Lon Chaney, Jr., J. Carroll Naish, Elena Verdugo, Glenn Strange, and George Zucco. Directed by Erle C. Kenton from a screenplay by Edward T. Lowe.

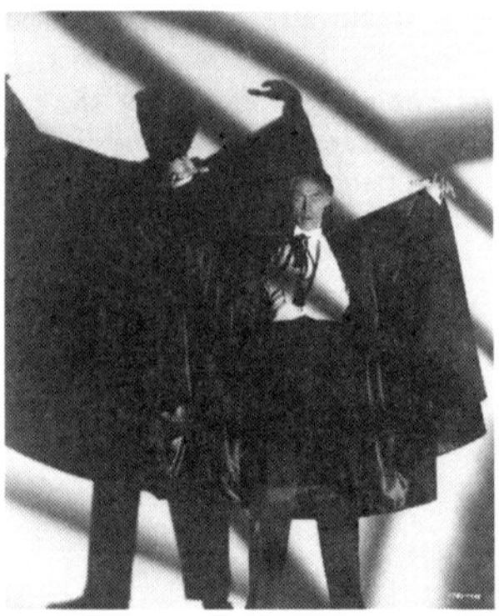

John Carradine as Dracula in House of Frankenstein *(1944).*

three years there would be millions of vampires! Within ten years we could rule earth!

"Can't you see that? If there was no cremation, no careless disposal of bodies, no bungling, we could end our hunted existence as creatures of the night — brothers of the bat! No longer would we be a legendary, cowering minority, living each a law unto himself!

"All that is needed is a plan. And I — I have evolved that plan!"

His voice rose. So did the hairs upon my neck. I was beginning to comprehend, now —

"Suppose we started with the humble instruments of destiny," he suggested. "Those forlorn, unnoticed, ignorant little old men — night watchmen of graveyards and cemeteries."

A smile creased his corpse-like countenance. "Suppose we eliminated them? Took over their jobs? Put vampires in their places — men who would go to the fresh graves and dig up the bodies of each victim they had bitten while those bodies were still warm and pulsing and undecayed?

"We could save the lives of most of the recruits we make. Reasonable, is it not?"

To me it was madness, but I nodded.

"Suppose that we made victims of those attendants? Then carried them off, nursed them back to reanimation, and allowed them to resume their posts as our allies? They work only at night — no one would know.

"Just a little suggestion, but so obvious! And it would mean so much!"

His smile broadened.

"All that it takes is organization on our part. I know many of my brethren. It is my desire soon to call them together and present this plan. Never before have we worked cooperatively, but when I show them the possibilities, they cannot fail to respond.

"Can you imagine it? An earth which we could control and terrorize — a world in which human beings become our property, our cattle?

"It is so simple, really. Sweep aside your foolish concepts of *Dracula* and the other superstitious confectionery that masquerades in the public mind as an authentic picture. I admit that we are — unearthly. But there is no reason for us to be stupid, impractical figures of fantasy. There is more for us than crawling around in black cloaks and recoiling at the sight of crucifixes!

"After all, we are a life-form, a race of our own. Biology has not yet recognized us, but we exist. Our morphology and metabolism has not been evaluated or charted; our actions and reactions never studied. But we exist. And we are superior to ordinary mortals. Let us assert this superiority. Plain human cunning, coupled with our super-normal powers, can create for us a mastery over all living things. For we are greater than Life — we are Life-in-Death!"

I half-rose. He waved me back, breathlessly.

"Suppose we band together and make plans? Suppose we go about,

HOUSE OF DRACULA (1945)

John Carradine makes his second appearance as Dracula in this entertaining finale to the Universal horror cycles of the 1930s and 1940s. Here, there is an attempt to scientifically rationalize the monsters: Dracula, for instance, has a blood disease, the Wolf Man (Lon Chaney, Jr.) suffers from pressure on the brain, etc. Both seek treatment from Dr. Edelman (Onslow Stevens), who is also tinkering with the comatose Frankenstein monster (Glenn Strange). Carradine made the most ambivalent vampire yet depicted on screen, constantly sabotaging his treatment to nip on the side. There are some nice bat transformations and an especially atmospheric scene in which the count hypnotically compels a female pianist to play music she has never heard before. Directed by Erle C. Kenton from Edward T. Lowe's script.

John Carradine as Dracula in House of Dracula.

first of all, selecting our victims on the basis of value to our ranks? Instead of regarding them as sources of easy nourishment, let's think in terms of an army seeking recruits. Let us select keen brains, youthfully strong bodies. Let us prey upon the best earth has to offer. Then we shall wax strong and no man shall stay our hand — or teeth!"

He crouched like a black spider, spinning his web of words to enmesh my sanity. His eyes glittered. It was absurd somehow to see this creature of superstitious terror calmly creating a super-dictatorship of the dead.

And yet, I was one of them. It was real. The nameless one would do it, too.

"Have you ever stopped to wonder why I tell you this? Have you ever stopped to wonder why you are my confidant in this venture?" he purred.

I shook my head.

"It is because you are young. I am old. For years I have labored only to this end. Now that my plans are perfected, I need assistance. Youth, a modern viewpoint. I know of you, Graham Keene. I watched you before... you became one of us. You were selected for this purpose."

"Selected?" Suddenly it hit home. I fought down a stranglehold gasp as I asked the question. "Then you know who — did this to me? *You know who bit me?*"

Rotting fangs gaped in a smile. He nodded slowly.

"Of course," he whispered. "Why — *I* did!"

V

He was probably prepared for anything except the calmness with which I accepted this revelation.

Certainly he was pleased. And the rest of that night, and all the next night, were spent in going over the plans, in detail. I learned that he had not yet communicated with — others — in regard to his ideas.

A meeting would be arranged soon. Then we would begin the campaign. As he said, the times were ripe. War, a world in unrest — we would be able to move unchallenged and find unusual opportunities.

I agreed. I was even able to add certain suggestions as to detail. He was pleased with my cooperation.

Then, on the third night, came hunger.

He offered to serve as my guide, but I brushed him aside.

"Let me try my own wings," I smiled. "After all, I must learn sooner or later. And I promise you, I shall be very careful. This time I will see to it that the body remains intact. Then I shall discover the place of burial and we can perform an experiment. I will select a likely recruit, we shall go forth to open the grave, and thus will we test our plan in miniature."

He fairly beamed at that. And I went forth that night, alone.

I returned only as dawn welled out of the eastern sky — returned to slumber through the day.

ROBERT BLOCH'S WORLD OF VAMPIRES

"The Bat is My Brother" was only one of several Robert Bloch stories featuring vampires. The first was "The Cloak" (1939), in which a Halloween partygoer seeking a last-minute costume purchases a musty old cloak from one of those musty old shops that pepper pulp fiction. "The cold, heavy cloth hung draped about Henderson's shoulders. The faint odor rose mustily in his nostrils as he stepped back and surveyed himself in the mirror. The light was poor, but Henderson saw that the cloak effected a striking transformation in his appearance. His long face seemed thinner, his eyes were accentuated in the facial pallor heightened by the somber cloak he wore. It was a big, black shroud." Not surprisingly, the cloak continues to create changes in Henderson's appearance and personality. His demeanor becomes imperious. He begins to notice the details of other people's necks as he had never noticed them before.

"The Cloak" is one of Bloch's most frequently reprinted vampire stories, and was freely adapted by Bloch himself for the 1970 British anthology film, *The House that Dripped Blood*.

In "The Bogey Man Will Get You" (1946), a girl, suspicious that her boyfriend is a vampire, is relieved of her fears when she realizes that he indeed casts a reflection in the mirror. Her relief, however, is brief

That night we spoke, and I confided my success to his eager ears.

"Sidney J. Garrat is the name," I said. "A college professor, about 45. I found him wandering along a path near the campus. The trees form a dark, deserted avenue. He offered no resistance. I left him there. I don't think they'll bother with an autopsy — for the marks on his throat are invisible and he is known to have a weak heart.

"He lived alone without relatives. He had no money. That means a wooden coffin and quick burial at Everest tomorrow. Tomorrow night we can go there."

My companion nodded.

"You have done well," he said.

We spent the remainder of the night in perfecting our plans. We would go to Everest, locate the night watchman and put him out of the way, then seek the new grave of Professor Garrat.

And so it was that we re-entered the cemetery on the following evening.

Once again a midnight moon glared from the Cyclopean socket of the sky. Once more the wind whispered to us on our way, and the trees bowed in black obeisance along the path.

We crept up to the shanty of the graveyard watchman and peered through the window at his stooping figure.

"I'll knock," I suggested. "Then when he comes to the door — "

My companion shook his gray head. "No teeth," he whispered. "The man is old, useless to us. I shall resort to more mundane weapons."

I shrugged. Then I knocked. The old man opened the door, and blinked out at me with rheumy eyes.

"What is it?" he wheezed, querulously. "Ain't nobuddy suppose' tuh be in uh cemetery this time uh night — "

Lean fingers closed around his windpipe. My companion dragged him forth towards nearby shrubbery. His free arm rose and fell, and a silver arc stabbed down. He had used a knife.

Ingrid Pitt in The House that Dripped Blood *(1970).*

Then we made haste along the path, before the scent of blood could divert us from our mission — and far ahead, on the hillside dedicated to the last slumbers of Poverty, I saw the raw, gaping edges of a new-made grave.

He ran back to the hut, then, and procured the spades we had neglected in our haste. The moon was our lantern and the grisly work began amidst a whistling wind.

No one saw us, no one heard us, for only empty eyes and shattered ears lay far beneath the earth.

We toiled, and then we stooped and tugged. The grave was deep, very deep. At the bottom the coffin lay, and we dragged forth the pine box.

"Terrible job" confided my companion. "Not a professionally dug grave at all, in my opinion. Wasn't filled in right. And this coffin is pine, but very thick. He'd never claw his own way out. Couldn't break through the boards. And the earth was packed too tightly. Why would they waste so much time on a pauper's grave?"

"Doesn't matter," I whispered. "Let's open it up. If he's revived, we must hurry."

He'd brought a hammer from the caretaker's shanty, too, and he went down into the pit itself to pry the nails free. I heard the board covering move, and peered down over the edge of the grave.

He bent forward, stooping to peer into the coffin, his face a mask of livid death in the moonlight. I heard him hiss.

"Why — the coffin is empty" he gasped.

"Not for long!"

I drew the wrench from my pocket, raised it, brought it down with every ounce of strength I possessed until it shattered through his skull.

And then I leaped down into the pit and pressed the writhing, mewing shape down into the coffin, slammed the lid on, and drove the heavy nails into place. I could hear his whimperings rise to muffled screams, but the screams grew faint as I began to heap the clods of earth upon the coffin-lid.

I worked and panted there until no sound came from the coffin below. I packed earth down hard — harder than I had last night when I dug the grave in the first place.

And then, at last, the task was over.

He lay there, the nameless one, the deathless one; lay six feet underground in a stout wooden coffin.

He could not claw his way free, I knew. And even if he did, I'd pressed him into his wooden prison face down. He'd claw his way to hell, not to earth.

But he was past escape. Let him lie there, as he had described it to me — not dead, not alive. Let him be conscious as he decayed, and as the wood decayed and the worms crawled in to feast. Let him suffer until the maggots at last reached his corrupt brain and ate away his evil consciousness.

I could have driven a stake through his heart. But his ghastly desire deserved defeat in this harsher fate.

Thus it was ended, and I could return now before discovery and the coming of dawn — return to his great house which was the only home I knew on the face of the earth.

Return I did, and for the past hours I have been writing this that all might know the truth.

I am not skilled with words, and what I read here smacks of mawkish melodrama. For the world is superstitious and yet cynical — and this account will be deemed the ravings of a fool or madman, worse still, as a practical joke.

So I must implore you; if you seek to test the truth of what I've set down, go to Everest tomorrow and search out the newly-dug grave on the hillside. Talk to the police when they find the dead watchman, make them go to the well near Danny's roadside stand.

— for it never occurred to her that he might be a different kind of monster entirely, one without reflection problems...a werewolf, for instance. The story was adapted for television's "The Darkroom" in 1981.

Bloch's "Hungarian Rhapsody" (1958, published under the pen name Wilson Kane) brings together a retired American gangster and a seasoned female vampire. "The Living Dead" (1967) is an atmospheric World War II story that recalls the gambit of Tod Browning's *Mark of the Vampire*: an actor is hired to impersonate a vampire as an elaborate ruse — in this case to hide the activities of Nazis occupying a French chateau. But the actor does his job too well: instead of frightening off the local peasantry, he only succeeds in goading them to their own act of anti-vampire overkill. "The Yougoslaves" (1985) is an effective mood piece involving feral Gypsy children in Paris, and their dealings with a mysterious visitor from the Carpathians.

Then, if you must, dig up the grave and find that which must still writhe and crawl within. When you see it, you'll believe — and in justice, you will not relieve the torment of that monstrous being by driving a stake through his heart.

For that stake represents release and peace.

I wish you'd come here, after that — and bring a stake for me . . .

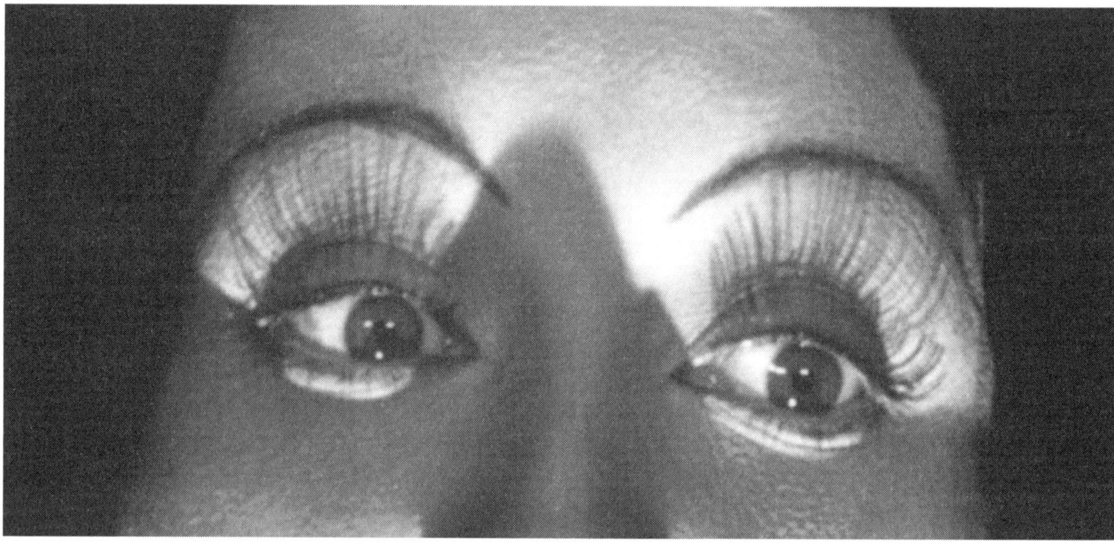

By the time Fritz Leiber wrote "The Girl With the Hungry Eyes," the most voracious female peepers in Hollywood were still those of Gloria Holden in Dracula's Daughter *(1936).*

The Girl with the Hungry Eyes

FRITZ LEIBER, JR.
(1949)

ABOUT THE AUTHOR

Fritz Reuter Leiber, Jr. (1910-1992) was the son of the noted Shakespearean actor Fritz Leiber, whom horror fans will fondly remember in the role of Franz Liszt in the 1943 version of *The Phantom of the Opera*. The younger Leiber also had a brief career as an actor, and played a small role in Greta Garbo's *Camille* (1936). He began publishing fiction professionally in 1939, and coined the term "sword and sorcery" to describe the brand of heroic fantasy in which he specialized. A versatile craftsman, Leiber was also adept at science fiction and contemporary gothic fantasy, and he returned many times to the theme of the femme fatale; "The Girl with the Hungry Eyes" remains one of the most innovative updates on the traditional vampire story published during the twentieth century. His 1977 novel, *Our Lady of Darkness* also involves a quasi-vampiric theme, and is considered by many critics to be his finest work. One of Leiber's best known novels, *Conjure Wife* (1943), was adapted twice for the screen, first as *Weird Woman* (1944) and later as *Burn, Witch, Burn* (1961). Fritz Leiber was a six-time winner of the Hugo Award, a four-time Nebula winner, and the recipient of nearly twenty other awards, including a World Fantasy citation for lifetime achievement.

ALL RIGHT I'LL tell you why the Girl gives me the creeps. Why I can't stand to go downtown and see the mob slavering up at her on the tower, with that pop bottle or pack of cigarettes or whatever it is beside her. Why I hate to look at magazines any more because I know she'll turn up somewhere in a brassiere or a bubble bath. Why I don't like to think of millions of Americans drinking in that poisonous half-smile. It's quite a story — more story than you're expecting.

No, I haven't suddenly developed any long-haired indignation at the evils of advertising and the national glamor-girl complex. That'd be a laugh for a man in my racket, wouldn't it? Though I think you'll agree there's something a little perverted about trying to capitalize on sex that way. But it's okay with me. And I know we've had the Face and the Body and the Look and what not else, so why shouldn't someone come along who sums it all up so completely, that we have to call her the Girl and blazon her on all the billboards from Times Square to Telegraph Hill?

But the Girl isn't like any of the others. She's unnatural. She's morbid. She's unholy.

Oh, these are modern times, you say, and the sort of thing I'm hinting at went out with witchcraft. But you see I'm not altogether sure myself what I'm hinting at, beyond a certain point. There are vampires and vampires, and not all of them suck blood.

And there were the murders, if they were murders. Besides, let me ask you this. Why, when America is obsessed with the Girl, don't we find out more about her? Why doesn't she rate a *Time* cover with a droll biography inside? Why hasn't there been a feature in *Life* or the *Post*? A profile in *The New Yorker*? Why hasn't *Charm* or *Mademoiselle* done her career saga? Not ready for it? Nuts!

Why haven't the movies snapped her up? Why hasn't she been on "Information, Please"? Why don't we see her kissing candidates at political rallies? Why isn't she chosen queen of some sort of junk or other at a convention?

Why don't we read about her tastes and hobbies, her views of the Russian situation? Why haven't the columnists interviewed her in a kimono on the top floor of the tallest hotel in Manhattan and told us who her boyfriends are?

Finally — and this is the real killer — why hasn't she ever been drawn or painted?

Oh no she hasn't. If you knew anything about commercial art, you'd know that. Every blessed one of those pictures was worked up from a photograph. Expertly? Of course. They've got the top artists on it. But that's how it's done.

And now I'll tell you the why of all that. It's because from the top to the bottom of the whole world of advertising, news, and business, there isn't a solitary soul who knows where the Girl came from, where she lives, what she does, who she is, even what her name is.

You heard me. What's more, not a single solitary soul ever sees her — except one poor damned photographer, who's making more money off her than he ever hoped to in his life and who's scared and miserable as hell every minute of the day.

No, I haven't the faintest idea who he is or where he has his studio.

But I know there has to be such a man and I'm morally certain he feels just like I said.

Yes, I might be able to find her, if I tried. I'm not sure though — by now she probably has other safeguards. Besides, I don't want to.

Oh, I'm off my rocker, am I? That sort of thing can't happen in the Era of the Atom? People can't keep out of sight that way, not even Garbo?

Well, I happen to know they can, because last year I was that poor damned photographer I was telling you about. Yes, last year, when the Girl made her first poisonous splash right here in this big little city of ours.

Yes, I know you weren't here last year and you don't know about it. Even the Girl had to start small. But if you hunted through the files of the local newspapers, you'd find some ads, and I might be able to locate you some of the old displays — I think Lovelybelt is still using one of them. I used to have a mountain of photos myself, until I burned them.

Yes, I made my cut off her. Nothing like what that other photographer must be making, but enough so it still bought this whiskey. She was funny about money. I'll tell you about that.

But first picture me then. I had a fourth-floor studio in that rathole the Hauser Building, not far from Ardleigh Park.

I'd been working at the Marsh-Mason studios until I'd gotten my bellyful of it and decided to start in for myself. The Hauser Building was awful — I'll never forget how the stairs creaked — but it was cheap and there was a skylight.

Business was lousy. I kept making the rounds of all the advertisers and agencies, and some of them didn't object to me too much personally, but my stuff never clicked. I was pretty near broke. I was behind on my rent. Hell, I didn't even have enough money to have a girl.

It was one of those dark gray afternoons. The building was very quiet — I'd just finished developing some pix I was doing on speculation for Lovelybelt Girdles and Budford's Pool and Playground. My model had

THE VAMPIRIC POWER OF SEX

As should be more than apparent to even the most casual reader of this anthology, vampire stories tend to reflect human sexuality in a negative and terrifying mirror. Female sexuality is regarded with special distrust, if not outright disgust, and a general atmosphere of sexual puritanism prevails.

In the case of "The Girl with the Hungry Eyes," Fritz Leiber reflects a larger misogynistic trend in popular culture; his story fits in perfectly with the standard depiction of women in hardboiled detective fiction and *noir* cinema of the 1940s: stunningly beautiful but corrupt, insatiable, and devouring. Their bottomless needs and hungers are lineally descended from the vampire women in "La morte amoureuse" and "Wake Not the Dead," and can be traced further to the classical archetypes of the lamia and Lilith.

Sigmund Freud noted that "morbid dread always signifies repressed sexual wishes." Freud's disciple Ernest Jones more specifically diagnosed vampire superstitions as springing from the incest taboo — vampirism in the European folk tradition was originally a family affair. Only in literature would vampires materialize outside the family, where their sexuality became less clandestine.

A recent, provocative thread of feminist vampire criticism (notably that of Nina Auerbach) maintains that female vampires are perverse representations of sexual empowerment, but for nonacademic audiences, vampire entertainment seems to relentlessly promulgate an almost reactionary, "just say no" attitude toward sex in general.

left. A Miss Leon. She was a civics teacher at one of the high schools and modeled for me on the side, just lately on speculation, too. After one look at the prints, I decided that Miss Leon probably wasn't just what Lovelybelt was looking for — or my photography either. I was about to call it a day.

And then the street door slammed four stories down and there were steps on the stairs and she came in.

She was wearing a cheap, shiny black dress. Black pumps. No stockings. And except that she had a gray cloth coat over one of them, those skinny arms of hers were bare. Her arms are pretty skinny, you know, or can't you see things like that any more?

And then the thin neck, the slightly gaunt, almost grim face, the tumbling mass of dark hair, and looking out from under it the hungriest eyes in the world.

That's the real reason she's plastered all over the country today, you know — those eyes. Nothing vulgar, but just the same they're looking at you with a hunger that's all sex and something more than sex. That's what everybody's been looking for since the Year One — something a little more than sex.

Well, boys, there I was, alone with the Girl, in an office that was getting shadowy, in a nearly empty building. A situation that a million male Americans have undoubtedly pictured to themselves with various lush details. How was I feeling? Scared.

I know sex can be frightening. That cold heart-thumping when you're alone with a girl and feel you're going to touch her. But if it was sex this time, it was overlaid with something else.

At least I wasn't thinking about sex.

I remember that I took a backward step and that my hand jerked so that the photos I was looking at sailed to the floor.

There was the faintest dizzy feeling like something was being drawn out of me. Just a little bit.

That was all. Then she opened her mouth and everything was back to normal for a while.

"I see you're a photographer, mister," she said. "Could you use a model?"

Her voice wasn't very cultivated.

"I doubt it," I told her, picking up the pix. You see, I wasn't impressed. The commercial possibilities of her eyes hadn't registered on me yet, by a long shot. "What have you done?"

Well, she gave me a vague sort of story and I began to check her knowledge of model agencies and studios and rates and what not and pretty soon I said to her, "Look here, you never modeled for a photographer in your life. You just walked in here cold."

Well, she admitted that was more or less so.

All along through our talk I got the idea she was feeling her way, like someone in a strange place. Not that she was uncertain of herself, or of me, but just of the general situation.

Insatiable? A Victorian vampiress, by artist Max Kahn.

SEX NIGHTMARES

Before there were vampires, there were nightmares, which primitive cultures interpreted as the night visitations of evil spirits. The classical nightmare is marked by a terrifying paralysis, which was attributed to the weight of a demon on the sleeper's breast.

By the Middle Ages, the popular conception of the nightmare evolved into the idea of the incubus and succubus — male and female dream-demons who drained their victims by having sexual relations with them during sleep. The motif was eventually fused with that of the blood-drinking, folkloric vampire, especially in Romantic literature of the early nineteenth century.

The classic, sleep-invading incubus as depicted in advertising art for The Night Walker *(1965).*

SEXUAL SYMBOLISM

Vampire stories may be all about sex, but rarely do the characters actually *have* sex (see Suzy McKee Charnas's story "Unicorn Tapestry" for a notable exception). Instead, the sex imagery is symbolically coded.

Take wooden stakes, for instance. On a literal level, the stake is a physical means of pinning the vampire to its grave; on a more metaphorical plane, the stake is an unmistakable phallic symbol which makes clear the displaced, transformed sexuality of vampire beliefs in general. The vampire, in other words, is a kind of symbolic sex itch that can be destroyed/ dispelled by a symbolic act of sexual penetration.

Fangs, too, are penetrating objects, and in recent years increasingly liberated special effects technology has made it almost de rigeur for vampire films to feature a semi-pornographic "master shot" of erectile vampire teeth as they thrust forward to their full length and hardness. In

"And you think anyone can model?" I asked her pityingly.

"Sure," she said.

"Look," I said, "a photographer can waste a dozen negatives trying to get one halfway human photo of an average woman. How many do you think he'd have to waste before he got a real catchy, glamorous photo of her?"

"I think I could do it," she said.

Well, I should have kicked her out right then. Maybe I admired the cool way she stuck to her dumb little guns. Maybe I was touched by her underfed look. More likely I was feeling mean on account of the way my pictures had been snubbed by everybody and I wanted to take it out on her by showing her up.

Okay, I'm going to put you on the spot, I told her. I'm going to try to take a few shots of you. Understand it's strictly on spec. If anyone should ever want to use a photo of you, which is about one chance in two million. I'll pay you regular rates for your time. Not otherwise. She gave me a smile. The first! "That's swell by me," she said. Well I took three or four shots, close-ups of her face since I didn't fancy her cheap dress, and at least she stood up to my sarcasm. Then I remembered I still had the Loverlybelt stuff and I guess the meanness was still working in me because I handed her a girdle and told her to go behind the screen and get into it and she did, without getting flustered as I'd expected, and since we'd gone that far, I figured we might as well shoot the beach scene to round it out, and that was that.

All this time I wasn't feeling anything particular one way or the other, except every once in a while I'd get one of those faint dizzy flashes and wonder if there was something wrong with my stomach or if I could have been a bit careless with MY chemicals.

Still, you know, I think the uneasiness was in me all the while.

I tossed her a card and pencil. "Write your name and address and phone," I told her and made for the darkroom.

A little later she walked out. I didn't call any good-byes. I was irked because she hadn't fussed around or seemed anxious about her poses, or even thanked me, except for that one smile.

I finished developing the negatives, made some prints, glanced at them, and decided they weren't a great deal worse than Miss Leon. On an impulse I slipped them in with the pictures I was going to take on the rounds next morning.

By now I'd worked long enough, so I was a bit fagged and nervous, but I didn't dare waste enough money on liquor to help that. I wasn't very hungry. I think I went to a cheap movie.

I didn't think of the Girl at all, except maybe to wonder faintly why in my present womanless state I hadn't made a pass at her. She had seemed to belong to a — well, distinctly more approachable social strata than Miss Leon. But then, of course, there were all sorts of arguable reasons for my not doing that. Next morning I made the rounds, My first step was Munsch's Brewery. They were looking for a "Munsch Girl." Papa Munsch had a sort of affection for me, though he razzed my photography. He had a good natural judgment about that, too. Fifty years ago he might have been one the shoestring boys who made Hollywood.

Right now he was out in the plant, pursuing his favorite occupation. He put down the beaded schooner, smacked his lips, gabbled something technical to someone about hops, wiped his fat hands on the big apron he was wearing, and grabbed my thin stack of pictures.

He was about halfway through, making noises with his tongue and teeth, when he came to her. I kicked myself for even having stuck her in.

"That's her," he said. "The photography's not so hot, but that's the girl."

It was all decided. I wonder now why Papa Munsch sensed what the Girl had right away, while I didn't. I think it was because I saw her first in the flesh, if that's the right word.

At the time I just felt faint.

"Who is she?" he asked.

"One of my new models." I tried to make it casual.

"Bring her out tomorrow morning," he told me. "And your stuff. We'll photograph her here."

"Here, don't look so sick," he added. "Have some beer."

Well, I went away telling myself it was just a fluke, so that she'd probably blow it tomorrow with her inexperience, and so on.

Just the same, when I reverently laid my next stack of pictures on Mr. Fitch, of Lovelybelt's, rose-colored blotter, I had hers on top.

Mr. Fitch went through the motions of being an art critic. He leaned over backward, squinted his eyes, waved his long fingers, and said, "Hmm. What do you think, Miss Willow? Here, in this light, of course, the photograph doesn't show the bias cut. And perhaps we should use the Lovelybelt Imp instead of the Angel. Still, the girl... come over here, Binns." More finger-waving. "I want a married man's reaction."

He couldn't hide the fact that he was hooked.

Exactly the same thing happened at Budford's Pool and Playground, except that Da Costa didn't need a married man's say-so. "Hot stuff," he said, sucking his lips. "Oh boy, you photograhers!"

I hotfooted it back to the office and grabbed up the card I'd given her to put down her name and address.

It was blank.

I don't mind telling you that the next five days were about the worst I ever went through, in an ordinary way. When next morning rolled around and I still hadn't got hold of her, I had to start stalling.

"She's sick," I told Papa Munsch over the phone.

"She at a hospital?" he asked me.

"Nothing that serious," I told him.

the case of female vampires, a voluptuous mouth filled with piercing fangs is usually interpreted as the most recent update of the *vagina dentata*, the source of male castration anxiety.

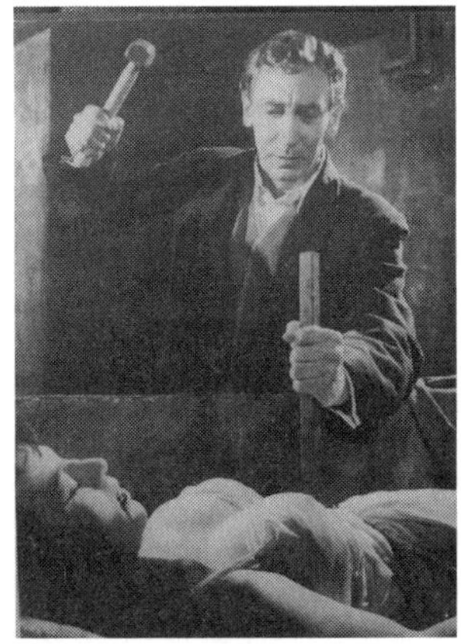

Phallic symbolism in Horror of Dracula *(1958).*

HOMOSEXUALITY

Since vampire stories create tension and interest through the presence of a sexual "outsider," it is not surprising that homosexuality, implicit or explicit, has been employed in film and fiction to evoke aspects of vampirism. Curiously, the image has vacillated wildly between negative stereotypes of the gay sexual predator to glamorous evocations of a liberating pansexuality.

The persistent, pop-cultural interplay between images of homosexuality, bisexuality, and vampirism date to J. Sheridan Le Fanu's quasi-lesbian 1872 novella *Carmilla*. In Bram Stoker's *Dracula* the theme is soft-pedaled but still palpable; the men in the story think they are saving Lucy Westenra's life by repeatedly transfusing her, while in reality they are opening their

own veins and bodies to the invasive thirst of a male monster.

In the landmark 1931 film version of *Dracula*, director Tod Browning — who knew better than anyone how to pluck Freudian nerves in popular culture — staged the visit of Renfield to Castle Dracula as a distinctly homoerotic seduction. Browning shot his interpretation of the scene over the producer's objections: "Dracula should go only for women and not men!" Carl Laemmle, Jr. scrawled on his copy of the shooting script. In Browning's film, Renfield's unrequited love for Dracula becomes the only compelling story line; the heterosexual hero and heroine are bloodless ghosts by comparison.

The first follow-up to the Browning film, *Dracula's Daughter* (1936) featured an austere, contralto-voiced heir to the Dracula curse. Played by actress Gloria Holden, *Dracula's Daughter* displays decidedly Sapphic tastes; in the film's most famous scene, Holden picks up a street-walker on the pretext of using her as a model. The original script, which called for nudity, was toned down considerably for filming, where the mere sight of the girl's bare shoulder triggers a deadly vampire attack.

The 1960 film *The Brides of Dracula* has a premise reminiscent of *Suddenly, Last Summer*: a beautiful young man, overindulged by his dragon-lady mother, somehow becomes a vampire. Mom (the great Martita Hunt) lures young girls to the castle to feed him; the thick Oedipal tensions are resolved only when the mother is herself penetrated (i.e., bitten) by the son. "Would you believe we once had gay times here?" the tormented baroness asks, in campy deadpan.

"Get her out here then. What's a little headache?"

"Sorry, I can't."

Papa Munsch got suspicious. "You really got this girl?"

"Of course I have."

"Well, I don't know. I'd think it was some New York model, except I recognized your lousy photography."

I laughed.

"Well, look, you get her here tomorrow morning, you hear?"

"I'll try."

"Try nothing. You get her out here."

He didn't know half of what I tried. I went around to all the model and employment agencies. I did some slick detective work at the photographic and art studios. I used up some of my last dimes putting advertisements in all three papers. I looked at high school yearbooks and at employee photos in local house organs. I went to restaurants and drugstores, looking at waitresses, and to dime stores and department stores, looking at clerks. I watched the crowds coming out of movie theaters. I roamed the streets.

Evenings, I spent quite a bit of time along Pickup Row. Somehow that seemed the right place.

The fifth afternoon I knew I was licked. Papa Munsch's deadline — he'd given me several, but this was it — was due to run out at six o'clock. Mr. Fitch had already canceled.

I was at the studio window, looking out at Ardleigh Park.

She walked in.

I'd gone over this moment so often in my mind that I had no trouble putting on my act. Even the faint dizzy feeling didn't throw me off.

"Hello," I said, hardly looking at her. "Hello," she said.

"Not discouraged yet?"

"No." It didn't sound uneasy or defiant. It was just a statement.

I snapped a look at my watch, got up and said curtly, "Look here, I'm going to give you a chance. There's a client of mine looking for a girl your general type. If you do a real good job you might break into the modeling business.

"We can see him this afternoon if we hurry," I said. I picked up my stuff. "Come on. And next time if you expect favors, don't forget to leave your phone number."

"Uh-uh," she said, not moving.

"What do you mean?" I said.

"I'm not going out to see any client of yours."

"The hell you aren't," I said. "You little nut, I'm giving you a break."

She shook her head slowly. "You're not fooling me, baby. You're not fooling me at all. They want me." And she gave me the second smile.

At the time I thought she must have seen my newspaper ad. Now I'm not so sure.

"And now I'll tell you how we're going to work," she went on.

"You aren't going to have my name or address or phone number. Nobody is. And we're going to do all the pictures right here. Just you and me."

You can imagine the roar I raised at that. I was everything — angry, sarcastic, patiently explanatory, off my nut, threatening, pleading.

I would have slapped her face off, except it was photographic capital.

In the end all I could do was phone Papa Munsch and tell him her conditions. I knew I didn't have a chance, but I had to take it.

He gave me a really angry bawling out, said "no" several times and hung up.

It didn't worry her. "We'll start shooting at ten o'clock tomorrow," she said.

It was just like her, using that corny line from the movie magazines.

About midnight Papa Munsch called me up.

"I don't know what insane asylum you're renting this girl from," he said, "but I'll take her. Come around tomorrow morning and I'll try to get it through your head just how I want the pictures. And I'm glad I got you out of bed!"

After that it was a breeze. Even Mr. Fitch reconsidered and, after taking two days to tell me it was quite impossible, he accepted the conditions too.

Of course you're all under the spell of the Girl, so you can't understand how much self-sacrifice it represented on Mr. Fitch's part when he agreed to forego supervising the photography of my model in the Lovelybelt Imp or Vixen or whatever it was we finally used.

Next morning she turned up on time according to her schedule, and we went to work. I'll say one thing for her, she never got tired and she never kicked at the way I fussed over shots. I got along okay, except I still had that feeling of something being shoved away gently. Maybe you've felt it just a little, looking at her picture.

When we finished I found out there were still more rules. It was about the middle of the afternoon. I started with her to get a sandwich and coffee.

"Uh-uh," she said, "I'm going down alone. And look, baby, if you ever try to follow me, if you ever so much as stick your head out of that window when I go, you can hire yourself another model."

You can imagine how all this crazy stuff strained my temper — and my imagination. I remember opening the window after she was gone — I waited a few minutes first — and standing there getting some fresh air and trying to figure out what could be behind it, whether she was hiding from the police, or was somebody's ruined daughter, or maybe had got the idea it was smart to be temperamental, or more likely Papa Munsch was right and she was partly nuts.

But I had my pictures to finish up.

Vampire-like slurs against gays have been standard ammunition for the religious right for quite some time. Consider Anita Bryant's charming assertion that "the male homosexual eats another man's sperm. The homosexual is eating life." Tangled blood themes also run beneath the surface of gay clashes with the Roman Catholic Church over AIDS issues. It was somehow appropriate that a 1989 demonstration at St. Patrick's Cathedral specifically disrupted the communion service — the purest ritual sublimation of the blood-drinking, "life-eating" impulse in Western civilization. In some parts of the world, the metaphors become deadly literal — in the West Indies, a favorite sport is the beating and even killing of gay men, who are called (interestingly enough) "batty boys," and who are believed to be ghosts of Sodom and Gomorrah who actually drink the blood of slum dwellers.

In an astute essay, "Children of the Night: Vampirism as Homosexuality, Homosexuality as Vampirism" (included in Susannah Radstone, ed., *Sweet Dreams: Sexuality, Gender and Popular Fiction*, 1988), Richard Dyer examines the ways in which "the languid, worn, sad, refined paleness of vampire imagery" intersects with popular stereotypes of gay "decadence." Language filled with murkiness and mystery has traditionally informed the presentation of gay images. In the 1950s and 1960s, Dyer points out, if a book "was called *Women in the Shadows*, *Twilight Men* [or] *Desire in the Shadows* then it had to be about queers. This imagery derives in part from the idea of decadence, people who do not go out into public life, whose complexions are not weathered, who are always indoors or in the shade. It may also relate to the idea that lesbians and gay men are not

'real' women and 'real' men, [that] we have not got the blood (with its very different gender associations) of normal human beings."

Roman Polanski fends off a gay vampire attack in The Fearless Vampire Hunters *(1967).*

LESBIANISM

Long an undercurrent of classic vampire stories like *Carmilla*, supernaturalized sexual relations between women have become a common horror motif in recent decades, paralleling the cultural demonization of male homosexuality, but without the particular overlay of disease imagery that has colored male-male vampirism in the age of AIDS. According to Andrea Weiss, author of *Vampires and Violets: Lesbians in Film* (1991): "Merging two kinds of sexual outlaws, the lesbian vampire is more than simply a negative stereotype. She is a complex and ambiguous figure, at once an image of death and an object of desire, drawing on profound subconscious fears that the living have toward the dead and that men have toward women, while serving as a focus for repressed fantasies. The generic vampire image both expresses and represses sexuality, but the lesbian vampire especially operates in the sexual rather than the supernatural realm."

Looking back, it's amazing to think how fast her magic began to take hold of the city after that. Remembering what came after, I'm frightened of what's happening to the whole country — and maybe the world. Yesterday I read something in *Time* about the Girl's picture turning up on billboards in Egypt.

The rest of my story will help show you why I'm frightened in that big, general way. But I have a theory, too, that helps explain, though it's one of those things that's beyond that "certain point." It's about the Girl. I'll give it to you in a few words.

You know how modern advertising gets everybody's mind set in the same direction, wanting the same things, imagining the same things. And you know the psychologists aren't so skeptical of telepathy as they used to be.

Add up the two ideas. Suppose the identical desires of millions of people focussed on one telepathic person. Say a girl. Shaped her in their image.

Imagine her knowing the hidden-most hungers of millions of men. Imagine her seeing deeper into those hungers than the people that had them, seeing the hatred and the wish for death behind the lust. Imagine her shaping herself in that complete image, keeping herself as aloof as marble. Yet imagine the hunger she might feel in answer to their hunger.

But that's getting a long way from the facts of my story. And some of those facts are damn solid. Like money. We made money.

That was the funny thing I was going to tell you. I was afraid the Girl was going to hold me up. She really had me over a barrel, you know.

But she didn't ask for anything but the regular rates. Later on I insisted on pushing more money at her, a whole lot. But she always took it with that same contemptuous look, as if she were going to toss it down the first drain when she got outside.

Maybe she did.

At any rate, I had money. For the first time in months I had money enough to get drunk, buy new clothes, take taxicabs. I could make a play for any girl I wanted to. I only had to pick.

And so of course I had to go and pick . . .

But first let me tell you about Papa Munsch.

Papa Munsch wasn't the first of the boys to try to meet my model but I think he was the first to really go soft on her. I could watch the change in his eyes as he looked at her pictures. They began to get sentimental, reverent. Mama Munsch had been dead for two years.

He was smart about the way he planned it. He got me to drop some information which told him when she came to work, and then one morning he came pounding up the stairs a few minutes before.

"I've got to see her, Dave," he told me.

I argued with him, I kidded him, I explained he didn't know just how serious she was about her crazy ideas. I even pointed out he was cutting both our throats. I even amazed myself by bawling him out.

He didn't take any of it in his usual way. He just kept repeating, "But, Dave, I've got to see her."

The street door slammed. "That's her," I said, lowering my voice. "You've got to get out."

He wouldn't, so I shoved him in the darkroom, "And keep quiet," I whispered. "I'll tell her I can't work today."

I knew he'd try to look at her and probably come busting in, but there wasn't anything else I could do.

The footsteps came to the fourth floor. But she never showed at the door. I got uneasy.

"Get that bum out of there!" she yelled suddenly from beyond the door. Not very loud, but in her commonest voice.

"I'm going up to the next landing," she said. "And if that fat-bellied bum doesn't march straight down to the street, he'll never get another picture of me except spitting in his lousy beer."

Papa Munsch came out of the darkroom. He was white. He didn't look at me as he went out. He never looked at her pictures in front of me again.

That was Papa Munsch. Now it's me I'm telling about. I talked around the subject with her, I hinted, eventually I made my pass.

She lifted my hand off her as if it were a damp rag.

"No, baby," she said. "This is working time."

"But afterward…" I pressed.

"The rules still hold." And I got what I think was the fifth smile.

It's hard to believe, but she never budged an inch from that crazy line. I mustn't make a pass at her in the office, because our work was very important and she loved it and there mustn't be any distractions. And I couldn't see her anywhere else, because if I tried to, I'd never snap another picture of her — and all this with more money coming in all the time and me never so stupid as to think my photography had anything to do with it.

Of course I wouldn't have been human if I hadn't made more passes. But they always got the wet-rag treatment and there weren't any more smiles.

I changed. I went sort of crazy and light-headed — only sometimes I felt my head was going to burst. And I started to talk to her all the time. About myself.

It was like being in a constant delirium that never interfered with business. I didn't pay any attention to the dizzy feeling. It seemed natural.

I'd walk around and for a moment the reflector would look like a sheet of white-hot steel, or the shadows would seem like armies of moths, or the camera would be a big black coal car. But the next instant they'd come all right again.

And I talked. It didn't matter what I was doing — lighting her, posing

Lesbians made coy appearances from time to time in early vampire films — take a look at Gloria Holden's seduction of the streetwalker in *Dracula's Daughter* (1936). But the modern lesbian movie vampire owes much of her popularity to Hammer Films, which, beginning with *The Vampire Lovers* (1970), found a gold mine in *Carmilla*-derived horror films that fully exploited the seventies' new tolerance for onscreen nudity and violence. The beasts-and-breasts formula continued happily at Hammer with *Lust for a Vampire* (1970) and *Twins of Evil* (1971). *Daughters of Darkness* (1971) was a particularly elegant Belgian effort, imaginatively amplifying the lesbian aspects of the Erzebet Bathory legend. Perhaps the most celebrated of all lesbian vampire films is *The Hunger* (1983), in which the ageless Catherine Deneuve pursues the sexually ambivalent Susan Sarandon without apology or pity. In literature, author Jewelle Gomez created a full-scale historical epic of lesbian vampirism in *The Gilda Stories* (1991). Like homosexual men, lesbian readers and audiences have tended to embrace gay vampires as ironic role models, responding to the vampire's romantic aspects of rebellion, alienation, and social transcendence.

The straight world reads the signals differently. Real-life lesbians threaten the heterosexual male's sense of himself as the center of the sexual universe — not needing or wanting men's bodies, their disinterest is nonetheless seen as judgmental, an "unnatural" challenge to maleness. The lesbian's sexual independence from men overlaps with the more generalized independence extolled by feminism; it is therefore not

surprising that the demonized image of the lesbian vampire became a stock image in popular culture and soft porn during the feminist revival of the 1970s. Pam Kesey, editor of the anthology *Daughters of Darkness* (1993), cites twenty-six films dealing with lesbian vampires; most appeared during this period of widescale reappraisal of sex roles and sexual politics.

From Twins of Evil. *(Photofest)*

SADOMASOCHISM

The sizable overlap between vampire fans and S & M aficionados is due in no small part, this editor suspects, to Anne Rice's dual influence as best-selling vampire author and best-selling sadomasochistic pornographer. S & M, often romanticized by its practitioners as a renegade activity, is in reality a depressingly status-quo fantasyland where real-world power imbalances are erotically celebrated and thereby reinforced and perpetuated. But it's probably not surprising, in this age of increasing economic disparities and a creeping master-slave corporate ethos, that vampires and other ritual figures of domination/exploitation should be so culturally potent.

her, fussing with props, snapping my pictures — or where she was — on the platform, behind the screen, relaxing with a magazine — I kept up a steady gab.

I told her everything I knew about myself. I told her about my first girl. I told her about my brother Bob's bicycle. I told her about running away on a freight, and the licking Pa gave me when I came home. I told her about shipping to South America and the blue sky at night. I told her about Betty. I told her about my mother dying of cancer. I told her about being beaten up in a fight in an alley behind a bar. I told her about Mildred. I told her about the first picture I ever sold. I told her how Chicago looked from a sailboat. I told her about the longest drunk I was ever on. I told her about Marsh-Mason. I told her about Gwen. I told her about how I met Papa Munsch. I told her about hunting her. I told her about how I felt now.

She never paid the slightest attention to what I said. I couldn't even tell if she heard me.

It was when we were getting our first nibble from national advertisers that I decided to follow her when she went home.

Wait, I can place it better than that. Something you'll remember from the out-of-town papers — those maybe murders I mentioned. I think there were six.

I say "maybe" because the police could never be sure they weren't heart attacks. But there's bound to be suspicion when attacks happen to people whose hearts have been okay, and always at night when they're alone and away from home and there's a question of what they were doing.

The six deaths created one of those "mystery poisoner" scares. And afterward there was a feeling that they hadn't really stopped, but were being continued in a less suspicious way. That's one of the things that scares me now.

But at that time my only feeling was relief that I'd decided to follow her.

I made her work until dark one afternoon. I didn't need any excuses, we were snowed under with orders. I waited until the street door slammed, then I ran down. I was wearing rubber-soled shoes. I'd slipped on a dark coat she'd never seen me in, and a dark hat.

I stood in the doorway until I spotted her. She was walking by Ardleigh Park toward the heart of town. It was one of those warm fall nights. I followed her on the other side of the street. My idea for tonight was just to find out where she lived. That would give me a hold on her.

She stopped in front of a display window of Everley's department store, standing back from the flow. She stood there looking in.

I remembered we'd done a big photograph of her for Everley's, to make a flat model for a lingerie display. That was what she was looking at.

At the time it seemed all right to me that she should adore herself, if that was what she was doing.

When people passed she'd turn away a little or drift back farther into the shadows.

Then a man came by alone. I couldn't see his face very well, but he looked middle-aged. He stopped and stood looking in the window.

She came out of the shadows and stepped up beside him.

How would you boys feel if you were looking at a poster of the Girl and suddenly she was there beside you, her arm linked with yours?

This fellow's reaction showed plain as day. A crazy dream had come to life for him.

They talked for a moment. Then he waved a taxi to the curb. They got in and drove off.

I got drunk that night. It was almost as if she'd known I was following her and had picked that way to hurt me. Maybe she had. Maybe this was the finish.

But the next morning she turned up at the usual time and I was back in the delirium, only now with some new angles added.

That night when I followed her she picked a spot under a street-light, opposite one of the Munsch Girl billboards.

Now it frightens me to think of her lurking that way.

After about twenty minutes a convertible slowed down going past her, backed up, swung into the curb.

I was closer this time. I got a good look at the fellow's face. He was a little younger, about my age.

Next morning the same face looked up at me from the front page of the paper. The convertible had been found parked on a side street. He had been in it. As in the other maybe-murders, the cause of death was uncertain.

All kinds of thoughts were spinning in my head that day, but there were only two things I knew for sure. That I'd got the first real offer from a national advertiser, and that I was going to take the Girl's arm and walk down the stairs with her when we quit work.

She didn't seem surprised. "You know what you're doing?" she said.

"I know."

She smiled. "I was wondering when you'd get around to it."

I began to feel good. I was kissing everything good-bye, but I had my arm around hers.

It was another of those warm fall evenings. We cut across into Ardleigh Park. It was dark there, but all around the sky was a sallow pink from the advertising signs.

We walked for a long time in the park. She didn't say anything and she didn't look at me, but I could see her lips twitching and after a while her hand tightened on my arm.

We stopped. We'd been walking across the grass. She dropped down and pulled me after her. She put her hands on my shoulders. I was looking down at her face. It was the faintest sallow pink from the glow in the sky. The hungry eyes were dark smudges.

VAMPIRISM: A SEXUALLY TRANSMITTED DISEASE?

As a malady contracted by sexually-charged blood contact, vampirism powerfully evokes venereal disease. The sexual scourge of Victorian times, syphilis was the AIDS epidemic of its day and, like AIDS, fueled much of the era's fascination with vampires and dangerous, fatal sexuality.

A story like *Dracula* can be read as an almost transparent syphilis parable; its images of wanton women, contaminated blood, telltale skin lesions, and pseudoscientific "cures" resonating powerfully with widespread panic about sexual contagion, the demonization of prostitutes, and the attendant rise of blood-purifying quack doctors. One of *Dracula's* most memorable scenes, in which the men take turns transfusing the vampire-tainted Lucy Westenra, is a ritual enactment of the anxiety described by Elaine Showalter in her pertinent book *Sexual Anarchy: Gender and Culture at the Fin de Siècle* (1990) : "The prostitute's body was the vessel in which men discharged and mingled polluting fluids . . . [H]ostility toward the prostitute could be generalized to apply to all women." Lucy's transfusions are significantly followed by savage hostility meted out by the same group of men: a sexually charged mutilation by hammer, stake, and knife as overwrought as any of Jack the Ripper's prostitute predations. (One theory of the Ripper's crimes suggested that the killer himself had contracted syphilis from a

prostitute.) It is especially fascinating that Ibsen's *Ghosts* (1891), a straightforward dramatic treatment of syphilis, met with public outrage over its supposedly indecent subject matter, but the same theme, veiled only slightly by penny-dreadful fantasy trappings, *Dracula* was considered harmless popular entertainment.

I was fumbling with her blouse. She took my hand away, not like she had in the studio. "I don't want that," she said.

First I'll tell you what I did afterward. Then I'll tell you why I did it. Then I'll tell you what she said.

What I did was run away. I don't remember all of that because I was dizzy, and the pink sky was swinging against the dark trees. But after a while I staggered into the lights of the street. The next day I closed up the studio. The telephone was ringing when I locked the door and there were unopened letters on the floor. I never saw the Girl again in the flesh, if that's the right word.

I did it because I didn't want to die. I didn't want the life drawn out of me. There are vampires and vampires, and the ones that suck blood aren't the worst. If it hadn't been for the warning of those dizzy flashes, and Papa Munsch and the face in the morning paper, I'd have gone the way the others did. But I realized what I was up against while there was still time to tear myself away. I realized that wherever she came from, whatever shaped her, she's the quintessence of the horror behind the bright billboard. She's the smile that tricks you into throwing away your money and your life. She's the eyes that lead you on and on, and then show you death. She's the creature you give everything for and never really get. She's the being that takes everything you've got and gives nothing in return. When you yearn toward her face on billboards, remember that. She's the lure. She's the bait. She's the Girl.

And this is what she said, "I want you. I want your high spots. I want everything that's made you happy and everything that's hurt you bad. I want your first girl. I want that shiny bicycle. I want that licking. I want that pinhole camera. I want Betty's legs. I want the blue sky filled with stars. I want your mother's death. I want your blood on the cobblestones. I want Mildred's mouth. I want the first picture you sold. I want the lights of Chicago. I want the gin. I want Gwen's hands. I want your wanting me. I want your life. Feed me, baby, feed me."

FROM

I Am Legend

RICHARD MATHESON
(1954)

One

January 1976

On those cloudy days, Robert Neville was never sure when sunset came, and sometimes they were in the streets before he could get back.

If he had been more analytical, he might have calculated the approximate time of their arrival; but he still used the lifetime habit of judging nightfall by the sky, and on cloudy days that method didn't work. That was why he chose to stay near the house on those days.

He walked around the house in the dull gray of afternoon, a cigarette dangling from the corner of his mouth, trailing threadlike smoke over his shoulder. He checked each window to see if any of the boards had been loosened. After violent attacks, the planks were often split or partially pried off, and he had to replace them completely; a job he hated. Today only one plank was pried loose. Isn't that amazing? he thought.

In the back yard he checked the hothouse and the water tank. Sometimes the structure around the tank might be weakened or its rain catchers bent or broken off. Sometimes they would lob rocks over the high fence around the hothouse, and occasionally they would tear through the overhead net and he'd have to replace panes.

Both the tank and the hothouse were undamaged today.

He went to the house for a hammer and nails. As he pushed open the front door, he looked at the distorted reflection of himself in the cracked mirror he'd fastened to the door a month ago. In a few days, jagged pieces of the silver-backed glass would start to fall off. Let 'em fall, he thought. It was the last damned mirror he'd put there; it wasn't worth it. He'd put garlic there instead. Garlic always worked.

He passed slowly through the dim silence of the living room, turned left into the small hallway, and left again into his bedroom.

Once the room had been warmly decorated, but that was in another time. Now it was a room entirely functional, and since Neville's bed and bureau took up so little space; he had converted one side of the room into a shop.

A long bench covered almost an entire wall, on its hardwood top a heavy band saw, a wood lathe, an emery wheel, and a vise. Above it, on the wall, were haphazard racks of the tools that Robert Neville used.

ABOUT THE AUTHOR

Richard Burton Matheson (b. 1926) is one of the influential figures in twentieth-century imaginative literature, and one of the few to successfully blend the horror and science fiction genres. He began publishing professionally in 1950; both his first story collection and first novel were published in 1954. The novel was *I Am Legend*, one of the most important and widely imitated vampire novels since Stoker's *Dracula*. The book set a tone of tightly-wound paranoia that would become a hallmark of Matheson's fiction. He effectively adapted his second novel, *The Shrinking Man* (1956) to the screen as *The Incredible Shrinking Man* (1957), a film now recognized as a masterpiece of its kind. He scripted some of the most memorable episodes of the original "Twilight Zone" television series in the 1960s, as well as the Roger Corman-produced Poe films *House of Usher* (1960), *The Pit and the Pendulum* (1961) and *The Raven* (1963). *I Am Legend* was twice produced as a film, first in 1964 as *The Last Man on Earth* starring Vincent Price, and again in 1971 as *The Omega Man* starring Charlton Heston. His 1975 novel *Bid Time Return* became the film *Somewhere in Time* (1980, from Matheson's screenplay), and *What Dreams May Come* (1978) was filmed in 1998. Matheson has been the recipient of the World Fantasy and Bram Stoker life achievement awards, as well as the Hugo, the Edgar, and the Writer's Guild awards.

He took a hammer from the bench and picked out a few nails from one of the disordered bins. Then he went back outside and nailed the plank fast to the shutter. The unused nails he threw into the rubble next door.

For a while he stood on the front lawn looking up and down the silent length of Cimarron Street. He was a tall man, thirty-six, born of English-German stock, his features undistinguished except for the long, determined mouth and the bright blue of his eyes, which moved now over the charred ruins of the houses on each side of his. He'd burned them down to prevent *them* from jumping on his roof from the adjacent ones.

After a few minutes he took a long, slow breath and went back into the house. He tossed the hammer on the living-room couch, then lit another cigarette and had his mid-morning drink.

The original British paperback edition of I Am Legend.

Later he forced himself into the kitchen to grind up the five-day accumulation of garbage in the sink. He knew he should burn up the paper plates and utensils too, and dust the furniture and wash out the sinks and the bathtub and toilet, and change the sheets and pillowcase on his bed; but he didn't feel like it.

For he was a man and he was alone and these things had no importance to him.

It was almost noon. Robert Neville was in his hothouse collecting a basketful of garlic.

In the beginning it had made him sick to smell garlic in such quantity; his stomach had been in a state of constant turmoil. Now the smell was in his house and in his clothes, and sometimes he thought it was even in his flesh. He hardly noticed it at all.

When he had enough bulbs, he went back to the house and dumped them on the drainboard of the sink. As he flicked the wall switch, the light flickered, then flared into normal brilliance. A disgusted hiss passed his clenched teeth. The generator was at it again. He'd have to get out that damned manual again and check the wiring. And, if it were too much trouble to repair, he'd have to install a new generator.

Angrily he jerked a high-legged stool to the sink, got a knife, and sat down with an exhausted grunt.

First he separated the bulbs into the small, sickle-shaped cloves. Then he cut each pink, leathery clove in half, exposing the fleshy center buds. The air thickened with the musky, pungent odor. When it got too oppressive, he snapped on the air-conditioning unit and suction drew away the worst of it.

Now he reached over and took an ice pick from its wall rack. He punched holes in each clove half, then strung them all together with wire—until he had about twenty-five necklaces.

In the beginning he had hung these necklaces over the windows. But from a distance they'd thrown rocks until he'd been forced to cover the broken panes with plywood scraps. Finally one day he'd torn off the plywood and nailed up even rows of planks instead. It had made the house a gloomy sepulcher, but it was better than having rocks come flying into his rooms in a shower of splintered glass. And, once he had installed the three air-conditioning units, it wasn't too bad. A man could get used to anything if he had to.

When he was finished stringing the garlic cloves, he went outside and nailed them over the window boarding, taking down the old strings, which had lost most of their potent smell.

He had to go through this process twice a week. Until he found something better, it was his first line of defense.

Defense? he often thought. For what?

All afternoon he made stakes.

He lathed them out of thick doweling, band-sawed into nine-inch lengths. These he held against the whirling emery stone until they were as sharp as daggers.

It was tiresome, monotonous work, and it filled the air with hot-smelling wood dust that settled in his pores and got into his lungs and made him cough.

Yet he never seemed to get ahead. No matter how many stakes he made, they were gone in no time at all. Doweling was getting harder to find, too. Eventually he'd have to lathe down rectangular lengths of wood. Won't *that* be fun? he thought irritably.

It was all very depressing and it made him resolve to find a better method of disposal. But how could he find it when they never gave him a chance to slow down and think?

As he lathed, he listened to records over the loudspeaker he'd set up in the bedroom — Beethoven's Third, Seventh, and Ninth symphonies. He was glad he'd learned early in life, from his mother, to appreciate this kind of music. It helped to fill the terrible void of hours.

Vincent Price starred in the first film version of I Am Legend, *filmed in Italy as* The Last Man on Earth *(1964). Richard Matheson wrote the screenplay, but ultimately removed his name from the credits in a dispute over script alterations.*

From four o'clock on, his gaze kept shifting to the clock on the wall. He worked in silence, lips pressed into a hard line, a cigarette in the corner of his mouth, his eyes staring at the bit as it gnawed away the wood and sent floury dust filtering down to the floor.

Four-fifteen. Four-thirty. It was a quarter to five.

In another hour they'd be at the house again, the filthy bastards. As soon as the light was gone.

He stood before the giant freezer, selecting his supper. His jaded eyes moved over the stacks of meats down to the frozen vegetables down to the breads and pastries, the fruits and ice cream.

He picked out two lamb chops, string beans, and a small box of orange sherbet. He picked the boxes from the freezer and pushed shut the door with his elbow.

Next he moved over to the uneven stacks of cans piled to the ceiling. He took down a can of tomato juice, then left the room that had once belonged to Kathy and now belonged to his stomach.

He moved slowly across the living room, looking at the mural that covered the back wall. It showed a cliff edge sheering off to green-blue ocean that surged and broke over black rocks. Far up in the clear blue sky, white sea gulls floated on the wind, and over on the right a gnarled tree hung over the precipice, its dark branches etched against the sky.

Neville walked into the kitchen and dumped the groceries on the table, his eyes moving to the clock. Twenty minutes to six. Soon now.

He poured a little water into a small pan and clanked it down on a stove burner. Next he thawed out the chops and put them under the broiler. By this time the water was boiling and he dropped in the frozen string beans and covered them, thinking that it was probably the electric stove that was milking the generator.

The Omega Man *(1971) reworked Matheson's original story, changing the vampires to light-sensitive mutants created by biological warfare.*

At the table he sliced himself two pieces of bread and poured himself a glass of tomato juice. He sat down and looked at the red second hand as it swept slowly around the clock face. The bastards ought to be here soon.

After he'd finished his tomato juice, he walked to the front door and went out onto the porch. He stepped off onto the lawn and walked down to the sidewalk.

The sky was darkening and it was getting chilly. He looked up and down Cimarron Street, the cool breeze ruffling his blond hair. That's what was wrong with these cloudy days; you never knew when they were coming.

Oh, well, at least they were better than those damned dust storms. With a shrug, he moved back across the lawn and into the house, locking and bolting the door behind him, sliding the thick bar into place. Then he went back into the kitchen, turned his chops, and switched off the heat under the string beans.

He was putting the food on his plate when he stopped and his eyes moved quickly to the clock. Six-twenty-five today. Ben Cortman was shouting.

"Come out, Neville!"

Robert Neville sat down with a sigh and began to eat.

He sat in the living room, trying to read. He'd made himself a whisky and soda at his small bar and he held the cold glass as he read a physiology text. From the speaker over the hallway door, the music of Schönberg was playing loudly.

Not loudly enough, though. He still heard them outside, their murmuring and their walkings about and their cries, their snarling and fighting among themselves. Once in a while a rock or brick thudded off the house. Sometimes a dog barked.

And they were all there for the same thing.

Robert Neville closed his eyes a moment and held his lips in a tight line. Then he opened his eyes and lit another cigarette, letting the smoke go deep into his lungs.

He wished he'd had time to soundproof the house.

It wouldn't be so bad if it weren't that he had to listen to them. Even after five months, it got on his nerves.

He never looked at them any more. In the beginning he'd made a peephole in the front window and watched them. But then the women had seen him and had started striking vile postures in order to entice him out of the house. He didn't want to look at that.

He put down his book and stared bleakly at the rug, hearing *Verklärte Nacht* play over the loud-speaker. He knew he could put plugs in his ears to shut off the sound of them, but that would shut off the music too and he didn't want to feel that they were forcing him into a shell.

He closed his eyes again. It was the women who made it so difficult, he thought, the women posing like hand puppets in the night on the possibility that he'd see them and decide to come out.

A shudder ran through him. Every night it was the same. He'd be reading and listening to music. Then he'd start to think about soundproofing the house, then he'd think about the women.

Deep in his body, the knotting heat began again and he pressed his lips together until they were white. He knew the feeling well and it enraged him that he couldn't combat it. It grew and grew until he couldn't sit still any more. Then he'd get up and pace the floor, fists bloodless at his sides. Maybe he'd set up the movie projector or eat something or have too much to drink or turn the music up so loud it hurt his ears. He had to do something when it got really bad.

He felt the muscles of his abdomen closing in like tightening coils. He picked up the book and tried to read, his lips forming each word slowly and painfully.

But in a moment the book was on his lap again. He looked at the bookcase across from him. All the knowledge in those books couldn't put out the fires in him; all the words of centuries couldn't end the wordless, mindless craving of his flesh.

The realization made him sick. It was an insult to a man. All right, it was a natural drive, but there was no outlet for it any more. They'd forced celibacy on him; he'd have to live with it. You have a mind, don't you? he asked himself. Well, *use* it!

He reached over and turned the music still louder, then forced himself to read a whole page without pause. He read about blood cells being forced through membranes, about pale lymph carrying the wastes through tubes blocked by lymph nodes, about lymphocytes and phagocytic cells.

" . . . to empty, in the left shoulder region, near the thorax, into a large vein of the blood circulating system."

The book shut with a thud.

Why didn't they leave him alone? Did they think they could *all* have him? Were they so stupid they thought that? Why did they keep coming every night? After five months, you'd think they'd give up and try elsewhere.

He went over to the bar and made himself another drink. As he turned

VAMPIRES IN THE FIFTIES: FROM CRYPT TO FALLOUT SHELTER

Following the real-world horrors of World War II, vampires began to mutate, taking on strange new forms as if in response to radiation exposure.

Traditional vampires tended to haunt magazine short fiction during the fifties; other than *I Am Legend* there were virtually no vampire novels published during the decade. Matheson wrote two other vampire-themed stories: the often-anthologized "Blood Son" (1951), in which a young boy's obsession with vampires finally brings him under the wings of Count Dracula himself, and "No Such Thing as a Vampire" (1959), in which a betrayed Romanian husband takes a devilish revenge on his rival with a trumped-up charge of vampirism.

Among other writers of the period, science-fictional variants on traditional vampire stories were popular. Joe Hensley's "And Not Quite Human" (1951) featured alien invaders who don't comprehend that one of their abductees is a vampire. Philip K. Dick published "The Cookie Lady"

back to his chair he heard stones rattling down across the roof and landing with thuds in the shrubbery beside the louse. Above the noises, he heard Ben Cortman shout as he always shouted.

"Come out, Neville!"

Someday I'll get that bastard, he thought as he took a big swallow of the bitter drink. Someday I'll knock a stake right through his goddamn chest. I'll make one a foot long for him, a special one with ribbons on it, the bastard.

Tomorrow. Tomorrow he'd soundproof the house. His fingers drew into white-knuckled fists. He couldn't stand thinking about those women. If he didn't hear them maybe he wouldn't think about them. Tomorrow. Tomorrow.

The music ended and he took a stack of records off the turntable and slid them back into their cardboard envelopes. Now he could hear them even more clearly outside. He reached for the first new record he could get and put it on the turntable and twisted the volume up to its highest point.

"The Year of the Plague," by Roger Leie, filled his ears. Violins scraped and whined, tympani thudded like the beats of a dying heart, flutes played weird, atonal melodies.

With a stiffening of rage, he wrenched up the record and snapped it over his right knee. He'd meant to break it long ago. He walked on rigid legs to the kitchen and flung the pieces into the trash box. Then he stood in the dark kitchen, eyes tightly shut, teeth clenched, hands clamped over his ears. Leave me alone, leave me alone, *leave me alone!*

No use, you couldn't beat them at night. No use trying; it was their special time. He was acting very stupidly, trying to beat them. Should he watch a movie? No, he didn't feel like setting up the projector. He'd go to bed and put the plugs in his ears. It was what he ended up doing every night anyway.

Quickly, trying not to think at all, he went to the bedroom and undressed. He put on pajama bottoms and went into the bathroom. He never wore pajama tops; it was a habit he'd acquired in Panama during the war.

As he washed, he looked into the mirror at his broad chest at the dark hair swirling around the nipples and down the center line of his chest. He looked at the ornate cross he'd had tattooed on his chest one night in Panama when he'd been drunk. What a fool I was in those days! he thought. Well, maybe that cross had saved his life.

He brushed his teeth carefully and used dental floss.

He tried to take good care of his teeth because he was his own dentist now. Some things could go to pot, but not his health, he thought. Then why don't you stop pouring alcohol into yourself? he thought. Why don't you shut the hell up? he thought.

Now he went through the house, turning out lights. For a few minutes he looked at the mural and tried to believe it was really the ocean. But how

in 1953, a tale of intergenerational psychic vampirism. "Share Alike"(1953) by Jerome Bixby and Joe E. Dean, is a vampiric variation on the starving-strangers-in-a-lifeboat theme. Theodore Sturgeon's "The Music" (1953) is told from the point of view of a mental hospital patient who may or may not be a vampire. It anticipates Sturgeon's 1961 novel *Some of Your Blood*, also set in a psychiatric hospital (and featuring a vampire who drinks only menstrual blood). Charles Beaumont's "Place of Meeting" (1954) posits vampires as the lone survivors of a cataclysmic war, faced with the grim prospect of blood-starvation. Fredric Brown's "Blood" (1954) sends vampires time-traveling into the future in search of new feasting grounds, only to find the world populated by intelligent, though bloodless, turnip-people.

Vampire cinema reached a crossroads at mid-century, and by the end of the fifties had generated some fascinating hybrid forms, as well as new approaches to traditional material. An overview of key titles follows.

could he believe it with all the bumpings and the scrapings, the howlings and shadings and cries in the night?

He turned off the living-room lamp and went into the bedroom.

He made a sound of disgust when he saw that sawdust covered the bed. He brushed it off with snapping hand strokes, thinking that he'd better build a partition between the shop and the sleeping portion of the room. Better do this and better do that, he thought morosely. There were so many damned things to do, he'd never get to the real problem.

He jammed in his earplugs and a great silence engulfed him. He turned off the light and crawled in between the sheets. He looked at the radium-faced clock and saw that it was only a few minutes past ten. Just as well, he thought. This way I'll get an early start.

He lay there on the bed and took deep breaths of the darkness, hoping for sleep. But the silence didn't really help. He could still see them out there, the white-faced men prowling around his house, looking ceaselessly for away to get in at him. Some of them, probably, crouching on their haunches like dogs, eyes glittering at the house, teeth slowly grating together; back and forth, back and forth.

And the women . . .

Did he have to start thinking about *them* again? He tossed over on his stomach with a curse and pressed his face into the hot pillow. He lay there, breathing heavily, body writhing slightly on the sheet. Let the morning come. His mind spoke the words it spoke every night. Dear God, let the morning come.

He dreamed about Virginia and he cried out in his sleep and his fingers gripped the sheets like frenzied talons.

Two

The alarm went off at five-thirty and Robert Neville reached out a numbed arm in the morning gloom and pushed in the stop.

He reached for his cigarettes and lit one, then sat up. After a few moments he got up and walked into the dark living room and opened the peephole door.

Outside, on the lawn, the dark figures stood like silent soldiers on duty. As he watched, some of them started moving away, and he heard them muttering discontentedly among themselves. Another night was ended.

He went back to the bedroom, switched on the light, and dressed. As he was pulling on his shirt, he heard Ben Cortman cry out, "Come out, Neville!"

And that was all. After that, they all went away weaker, he knew, than when they had come. Unless they had attacked one of their own. They did that often. There was no union among them. Their need was their only motivation.

INVASION OF THE BODY SNATCHERS (1956)

Invasion of the Body Snatchers **is not precisely a vampire film — it merits mention here because of its many overlaps with the genre. The film's legendary, invading pod-people don't drink blood, but they do absorb the minds and life-energy of their hosts, and while their shapeshifting is plant-to-human rather than human-to-anmal, it's still shape-shifting of a classic stripe. Directed by Don Siegel (as an intentional McCarthy era parable) from a script by Daniel Mainwaring and Sam Peckinpah, based on the novel *The Body Snatchers* by Jack Finney. With Kevin McCarthy, Dana Wynter, Carolyn Jones and King Donovan. (Republic Pictures)**

Vampires from space? A poster for Invasion of the Body Snatchers.

After dressing, Neville sat down on his bed with a grunt and penciled his list for the day:

Lathe at Sears
Water
Check generator
Doweling (?)
Usual

Breakfast was hasty: a glass of orange juice, a slice of toast and two cups of coffee. He finished it quickly, wishing he had the patience to eat slowly.

After breakfast he threw the paper plate and cup into the trash box and brushed his teeth. At least I have one good habit, he consoled himself.

The first thing he did when he went outside was look at the sky. It was clear, virtually cloudless. He could go out today. Good.

As he crossed the porch, his shoe kicked some pieces of the mirror. Well, the damn thing broke just as I thought it would, he thought. He'd clean it up later.

One of the bodies was sprawled on the sidewalk; the other one was half concealed in the shrubbery. They were both women. They were almost always women.

He unlocked the garage door and backed his Willys station wagon into the early-morning crispness. Then he got out and pulled down the back gate. He put on heavy gloves and walked over to the woman on the sidewalk.

There was certainly nothing attractive about them in the daylight, he thought, as he dragged them across the lawn and threw them up on the canvas tarpaulin. There wasn't a drop left in them; both women were the color of fish out of water. He raised the gate and fastened it.

He went around the lawn then, picking up stones and bricks and putting them into a cloth sack. He put the sack in the station wagon and then took off his gloves. He went inside the house, washed his hands, and made lunch: two sandwiches, a few cookies, and a thermos of hot coffee.

When that was done, he went into the bedroom and got his bag of stakes. He slung this across his back and buckled on the holster that held his mallet. Then he went out of the house, locking the front door behind him.

He wouldn't bother searching for Ben Cortman that morning; there were too many other things to do. For a second, he thought about the soundproofing job he'd resolved to do on the house. Well; the hell with it, he thought. I'll do it tomorrow or some cloudy day.

He got into the station wagon and checked his list. "Lathe at Sears"; that was first. After he dumped the bodies, of course.

He started the car and backed quickly into the street and headed for Compton Boulevard. There he turned right and headed east. On both sides of him the houses stood silent, and against the curbs cars were parked, empty and dead.

NOT OF THIS EARTH (1957)

A nifty science-fictional vampire film, produced and directed by Roger Corman from a script by Charles Griffith and Mark Hanna, *Not of This Earth* stars actor Paul Birch as a kind of sinister salesman from outer space who pumps his valise full of earth people's red elixir in an attempt to save his dying extraterrestrial race. He wears special sunglasses to hide his pupil-less, Orphan-Annie eyes, a direct glimpse of which will strike you dead. An interesting little exercise in noirish fifties paranoia, remade, quite unnecessarily, in 1988. With Beverly Garland, Morgan Jones, William Roerick, and Jonathan Haze. (American International)

Robert Neville's eyes shifted down for a moment to the fuel gauge. There was still a half tank, but he might as well stop on Western Avenue and fill it. There was no point in using any of the gasoline stored in the garage until he had to.

He pulled into the silent station and braked. He got a barrel of gasoline and siphoned it into his tank until the pale amber fluid came gushing out of the tank opening and ran down onto the cement.

He checked the oil, water, battery water, and tires. Everything was in good condition. It usually was, because he took special care of the car. If it ever broke down so that he couldn't get back to the house by sunset . . .

Well, there was no point in even worrying about that. If it ever happened, that was the end.

Now he continued up Compton Boulevard past the tall oil derricks, through Compton, through all the silent streets. There was no one to be seen anywhere.

But Robert Neville knew where they were.

The fire was always burning. As the car drew closer, he pulled on his gloves and gas mask and watched through the eyepieces the sooty pall of smoke hovering above the earth. The entire field had been excavated into one gigantic pit; that was in June 1975.

Neville parked the car and jumped out, anxious to get the job over with quickly. Throwing the catch and jerking down the rear gate, he pulled out one of the bodies and dragged it to the edge of the pit. There he stood it on its feet and shoved.

Poster for Not of This Earth. *(Courtesy of Ronald V. Borst/ Hollywood Movie Posters)*

The body bumped and rolled down the steep incline until it settled on the great pile of smoldering ashes at the bottom.

Robert Neville drew in harsh breaths as he hurried back to the station wagon. He always felt as though he were strangling when he was here, even though he had the gas mask on.

Now he dragged the second body to the brink of the pit and pushed it over. Then, after tossing the sack of rocks down, he hurried back to the car and sped away.

After he'd driven a half mile, he skinned off the mask and gloves and tossed them into the back. His mouth opened and he drew in deep lungfuls of the fresh air. He took the flask from the glove compartment and took a long drink of burning whisky. Then he lit a cigarette and inhaled deeply. Sometimes he had to go to the burning pit every day for weeks at a time, and it always made him sick.

Somewhere down there was Kathy.

On the way to Inglewood he stopped at a market to get some bottled water.

As he entered the silent store, the smell of rotted food filled his nostrils. Quickly he pushed a metal wagon up and down the silent, dust-thick aisles, the heavy smell of decay setting his teeth on edge, making him breathe through his mouth.

He found the water bottles in back, and also found a door opening on a flight of stairs. After putting all the bottles into the wagon, he went up the stairs. The owner of the market might be up there; he might as well get started.

There were two of them. In the living room, lying on a couch, was a woman about thirty years old, wearing a red housecoat. Her chest rose and fell slowly as she lay there, eyes closed, her hands clasped over her stomach.

Robert Neville's hands fumbled with the stake and mallet. It was always hard when they were alive; especially with women. He could feel that senseless demand returning again, tightening his muscles. He forced it down. It was insane, there was no rational argument for it.

She made no sound except for a sudden, hoarse intake of breath. As he walked into the bedroom, he could hear a sound like the sound of water running. Well, what else can I do? He asked himself, for he still had to convince himself he was doing the right thing.

He stood in the bedroom doorway, staring at the small bed by the window, his throat moving, breath shuddering in his chest. Then, driven on, he walked to the side of the bed and looked down at her.

Why do they all look like Kathy to me? he thought, drawing out the second stake with shuddering hands.

Driving slowly to Sears, he tried to forget by wondering why it was that only wooden stakes should work.

He frowned as he drove along the empty boulevard, the only sound the muted growling of the motor in his car. It seemed fantastic that it had taken him five months to start wondering about it.

Which brought another question to mind. How was it that he always managed to hit the heart? It had to be the heart; Dr. Busch had said so. Yet he, Neville, had no anatomical knowledge.

His brow furrowed. It irritated him that he should have gone through this hideous process so long without stopping once to question it.

He shook his head. No, I should think it over carefully, he thought, I should collect all the questions before I try to answer them. Things should be done the right way, the scientific way.

Yeah, yeah, yeah, he thought, shades of old Fritz. That had been his father's name. Neville had loathed his father and fought the acquisition of his father's logic and mechanical facility every inch of the way. His father had died denying the vampire violently to the last.

At Sears he got the lathe, loaded it into the station wagon, then searched the store.

There were five of them in the basement, hiding in various shadowed places. One of them Neville found inside a display freezer. When he saw the man lying there in this enamel coffin, he had to laugh; it seemed such a funny place to hide.

Later, he thought of what a humorless world it was when he could find amusement in such a thing.

RETURN OF DRACULA (1958)

"I hope he likes cheese sauce and asparagus," says the Carleton, California housewife, who has no inkling that the man upstairs posing as a political expatriate cousin is really Count Dracula. Nonetheless, she enjoys the idea of feeding him: "Why don't you go up and ask Cousin Bellac if he wants some pie?" she asks a bit later, after he starts his pie-free rampage. Actor Francis Lederer makes a smarmy vampire king who dispenses with the usual politeness vampires affect in these films. The dumb host family just chalks it up to cultural differences. This must be the first film in which Dracula gets to ride in a convertible. For the original release, the black-and-white film contained a color insert of a stake being driven through the heart of one of Dracula's victims. Directed by Paul Landres. Screenplay by Pat Fielder. With Norma Eberhardt, Ray Stricklyn, and Jimmie Baird. (Allied Artists)

About two o'clock he parked and ate his lunch. Every thing seemed to taste of garlic.

And that set him wondering about the effect garlic had on them. It must have been the smell that chased them off, but why?

They were strange, the facts about them: their staying inside by day, their avoidance of garlic, their death by stake, their reputed fear of crosses, their supposed dread of mirrors.

Take that last, now. According to legend, they were invisible in mirrors, but he knew that was untrue. As untrue as the belief that they transformed themselves into bats. That was a superstition that logic plus observation had easily disposed of. It was equally foolish to believe that they could transform themselves into wolves. Without a doubt there were vampire dogs; he had seen and heard them outside his house at night. But they were only dogs.

Robert Neville compressed his lips suddenly. Forget it, he told himself; you're not ready yet. The time would come when he'd take a crack at it, detail for detail, but the time wasn't now. There were enough things to worry about now.

After lunch, he went from house to house and used up his stakes. He had forty-seven stakes.

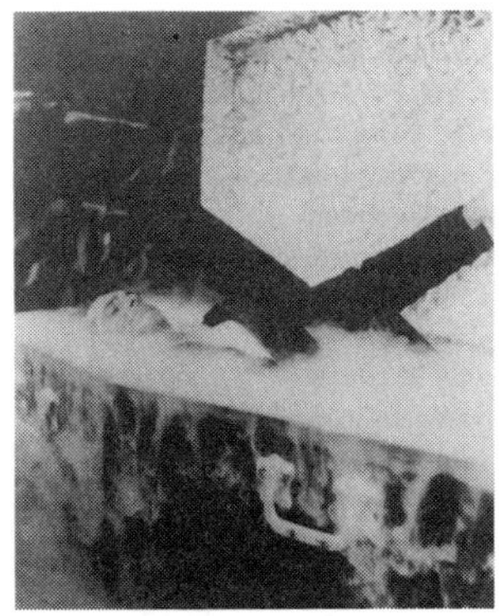

It's well-known that vampires must line their coffins with their native soil, but who knew about dry ice? Francis Lederer in Return of Dracula.

Three

"The strength of the vampire is that no one will believe in him."

Thank *you,* Dr. Van Helsing, he thought, putting down his copy of *Dracula.* He sat staring moodily at the bookcase, listening to Brahms' second piano concerto, a whisky sour in his right hand, a cigarette between his lips.

It was true. The book was a hodgepodge of superstitions and soap-opera cliches, but that line was true; no one had believed in them, and how could they fight something they didn't even believe in?

That was what the situation had been. Something black and of the night had come crawling out of the Middle Ages. Something with no framework or credulity, some thing that had been consigned, fact and figure, to the pages of imaginative literature. Vampires were passé. Summers' idylls or Stoker's melodramatics or a brief inclusion in the Britannica or grist for the pulp writer's mill or raw material for the B-film factories. A tenuous legend passed from century to century.

Well, it was true.

He took a sip from his drink and closed his eyes as the cold liquid trickled down his throat and warmed his stomach. True, he thought, but no one ever got the chance to know it. Oh, they knew it was something, but it couldn't be *that* — not *that.* That was imagination, that, was superstition, there was no such thing as *that.*

And, before science had caught up with the legend, the legend had swallowed science and everything.

BLOOD OF DRACULA (1957)

The first of producer Herman Cohen's teen-monster movies (*I Was a Teenage Werewolf* and *I Was a Teenage Frankenstein* were next) established a now-familiar formula: take a troubled teen with a lot of pent-up rage, stir in an unscrupulous authority figure and a dash of mad science, and presto, all hell breaks loose. Sandra Harrison stars as a girl whose widowed dad has taken up with an expensive floozie. They ditch her in a private girl's academy where a power-crazed science teacher (Louise Lewis) plans, somehow, to upset the male-dominated scientific establishment with a Mr. Wizard-style chemistry set and a Transylvanian amulet she twiddles in a darkened room. She succeeds only in turning the impressionable Harrison into a small-time serial killer with really big teeth. This movie is simultaneously awful and entertaining, and that chemistry set can't be beat for low-budget shamelessness. Directed by Herbert L. Strock, from a screenplay by Ralph Thornton. (American International)

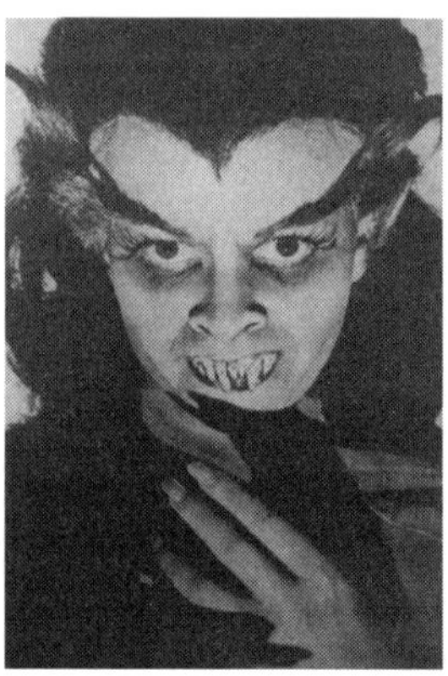

Sandra Harrison in Blood of Dracula.

He hadn't found any doweling that day. He hadn't checked the generator. He hadn't cleaned up the pieces of mirror. He hadn't eaten supper; he'd lost his appetite.

That wasn't hard: He lost it most of the time. He couldn't do the things he'd done all afternoon and then come home to a hearty meal. Not even after five months.

He thought of the eleven — no, the twelve children that afternoon, and he finished his drink in two swallows.

He blinked and the room wavered a little before him. You're getting blotto, Father, he told himself. So what? he returned. Has anyone more right?

He tossed the book across the room. Begone, Van Helsing and Mina and Jonathan and blood-eyed Count and all! All figments, all driveling extrapolations on a somber theme.

A coughing chuckle emptied itself from his throat. Outside, Ben Cortman called for him to come out. Be right out, Benny, he thought. Soon as I get my tuxedo on.

He shuddered and gritted his teeth edges together. Be right out. Well, why not? Why *not* go out? It was a sure way to be free of them.

Be one of them.

He chuckled at the simplicity of it, then shoved himself up and walked crookedly to the bar. Why not? His mind plodded on. Why go through all this complexity when a flung-open door and a few steps would end it all?

For the life of him, he didn't know. There was, of course, the faint possibility that others like him existed somewhere, trying to go on, hoping that someday they would be among their own kind again. But how could he ever find them if they weren't within a day's drive of his house?

He shrugged and poured more whisky in the glass; he'd given up the use of jiggers months ago. Garlic on the windows and nets over the hot-house and burn the bodies and cart the rocks away and, fraction of an inch by fraction of an inch, reduce their unholy numbers Why kid himself? He'd never find anyone else.

His body dropped down heavily on the chair. Here we are, kiddies, sitting like a bug in a rug, snugly, surrounded by a battalion of blood-suckers who wish no more than to sip freely of my bonded, 100-proof hemoglobin. Have a drink, men, this one's really on me.

His face twisted into an expression of raw, unqualified hatred. *Bastards!* I'll kill every mother's son of you before I'll give in! His right hand closed like a clam and the glass shattered in his grip.

He looked down, dull-eyed, at the fragments on the floor, at the jagged piece of glass still in his hand, at the whisky-diluted blood dripping off his palm.

Wouldn't they like to get some of it, though? he thought. He started up with a furious lunch and almost opened the door so he could wave the hand in their face and hear them howl.

Then he closed his eyes and a shudder ran through his body. Wise up, buddy, he thought. Go bandage your goddamn hand.

He stumbled into the bathroom and washed his hand carefully, gasping as he daubed iodine into the sliced-open flesh. Then he bandaged it clumsily, his broad chest rising and falling with jerky movements, sweat dripping from his forehead. I need a cigarette, he thought.

In the living room again, he changed Brahms for Bernstein and lit a cigarette. What will I do if I ever run out of coffin nails? he wondered, looking at the cigarette's blue trailing smoke. Well, there wasn't much chance of that. He had about a thousand cartons in the closet of Kathy's —

He clenched his teeth together. In the closet of the *larder*, the *larder*, the *larder*.

Kathy's room.

He sat staring with dead eyes at the mural while "The Age of Anxiety" pulsed in his ears. Age of anxiety, he mused. You thought you had anxiety, Lenny boy. Lenny and Benny; you two should meet. Composer, meet corpse. Mamma, when I grow up I wanna be a wampir like Dada. Why, bless you, son, of course you shall.

The whisky gurgled into the glass. He grimaced a little at the pain in his hand and shifted the bottle to his left hand.

He sat down and sipped. Let the jagged edge of sobriety be now dulled, he thought. Let the crumby balance of clear vision be expunged, but post haste. I hate 'em.

Gradually the room shifted on its gyroscopic center and wove and undulated about his chair. A pleasant haze, fuzzy at the edges, took over sight. He looked at the glass, at the record player. He let his head flop from side to side. Outside, they prowled and muttered and waited.

Pore vampires, he thought, pore little cusses, pussy' footin' round my house, so thirsty, so all forlorn.

A thought. He raised a forefinger that wavered before his eyes.

Friends, I come before you to discuss the vampire; a minority element if there ever was one, and there was one.

But to concision: I will sketch out the basis for my thesis, which thesis is this: Vampires are prejudiced against.

The keynote of minority prejudice is this: They are loathed because they are feared. Thus . . .

He made himself a drink. A long one.

At one time, the Dark and Middle Ages, to be succinct, the vampire's power was great, the fear of him tremendous. He was anathema and still remains anathema. Society hates him without ration.

But are his needs any more shocking than the need of other animals and men? Are his deeds more outrageous than the deeds of the parent who drained the spirit from his child? The vampire may foster quickened heartbeats and levitated hair. But is he worse than the parent who gave to society a neurotic child who became a politician? Is he worse than the

BLOOD OF THE VAMPIRE (1958)

Light years ahead of Dr. Christiaan Barnard, this better-than-average costume piece posits a truly original means to revive a staked vampire: give him a heart transplant. Dr. Callistratus (played with an industrial-strength Bela Lugosi makeup by Sir Donald Wolfit, the over-the-top Shakespearean actor who inspired the play *The Dresser*) is not a supernatural vampire, but all this transplant/resurrection business has resulted in a blood condition amounting to the same thing. Callistratus sets himself up as the head of a nasty Victorian prison, where he can experiment with blood however he pleases. Wolfit's icy portrayal evokes death-camp doctors like Dr. Josef Mengele who were still in business the decade before this film was released. Technically, the film has a rich look to it and makes clever use of trompe l'oeil painted backdrops. With Vincent Ball, Barbara Shelley, and Victor Maddern. Henry Cass directed from Jimmy Sangster's Hammer-style script. (Eros Films/Universal International)

Poster for Blood of the Vampire. *(Photofest)*

manufacturer who set up belated foundations with the money he made by handing bombs and guns to suicidal nationalists? Is he worse than the distiller who gave bastardized grain juice to stultify further the brains of those who, sober, were incapable of progressive thought? (Nay, I apologize for this calumny; I nip the brew that feeds me.) Is he worse, then, than the publisher who filled ubiquitous racks with lust and death wishes? Really, now, search your soul, lovie — is the vampire so bad?

All he does is drink blood.

Why, then, this unkind prejudice, this thoughtless bias? Why cannot the vampire live where he chooses? Why must he seek out hiding places where none can find him out Why do you wish him destroyed? Ah, see, you have turned the poor guileless innocent into a haunted animal. He has no means of support, no measures for proper education, he has not the voting franchise. No wonder he is compelled to seek out a predatory nocturnal existence.

Robert Neville grunted a surly grunt. Sure, sure, he thought, but would you let your sister marry one?

He shrugged. You got me there, buddy, you got me there.

The music ended. The needle scratched back and forth in the black grooves. He sat there, feeling a chill creeping up his legs. That's what was wrong with drinking too much. You became immune to drunken delights. There was no solace in liquor. Before you got happy, you collapsed. Already the room was straightening out, the sounds outside were starting to nibble at his eardrums.

"Come out, Neville!"

His throat moved and a shaking breath passed his lips. Come out. The women were out there, their dresses open or taken off, their flesh waiting for his touch, their lips waiting for —

My blood, my *blood!*

As if it were someone else's hand, he watched his whitened fist rise up slowly, shuddering, to drive down on his leg. The pain made him suck in a breath of the house's stale air. Garlic. Everywhere the smell of garlic. On his clothes and in the furniture and in his food and even in his drink. Have a garlic and soda; his mind rattled out the attempted joke.

He lurched up and started pacing. What am I going to do now? Go through the routine again? I'll save you the trouble. Reading-drinking-soundproof-the-house — the women. The women, the lustful, bloodthirsty, naked women flaunting their hot bodies at him. No, not hot.

A shuddering whine wrenched up through his chest and throat. Goddamn them, what were they waiting for? Did they think he was going to come out and hand himself over?

Maybe I am, maybe I am. He actually found himself jerking off the crossbar from the door. Coming, girls. I'm coming. Wet your lips, now.

Outside, they heard the bar being lifted, and a howl of anticipation sounded in the night.

HORROR OF DRACULA (1958)

Released in New York City the same day as Alfred Hitchcock's *Vertigo*, *Horror of Dracula* (called simply *Dracula* in the U.K.) proved to be as influential to the vampire genre as the Hitchcock film was to the psychothriller. Made by Hammer Films as a follow-up to its *Curse of Frankenstein* (1957), *Horror of Dracula* forever broke the monochromatic cobweb conventions of earlier vampire movies with a bright red swath of Technicolor blood from which horror films have never recovered.

In keeping with the stage and film tradition linking adaptations of *Dracula* and *Frankenstein*, actor Christopher Lee, who played the monster in *Curse of Frankenstein*, was cast as Count Dracula, with a distinct Jekyll/Hyde coloration. At one moment urbane and Oxford-accented, Lee could shift effortlessly into animal-fanged fury. He would repeat the Dracula role in seven other films for Hammer Films.

Valerie Gaunt and Christopher Lee in Horror of Dracula.

Spinning, he drove his fists one after the other into wall until he'd cracked the plaster and broken his skin. Then he stood there trembling helplessly, his teeth chattering.

After a while it passed. He put the bar back across the door and went into the bedroom. He sank down the bed and fell back on the pillow with a groan. His left hand beat once, feebly, on the bedspread.

Oh, *God*, he thought, how long, how long?

Painting by Mary Woronov.

Unicorn Tapestry

SUZY McKEE CHARNAS
(1980)

HOLD ON," FLORIA said. "I know what you're going to say: I agreed not to take any new clients for a while. But wait till I tell you — you're not going to believe this — first phone call, setting up an initial appointment, he comes out with what his problem is: 'I seem to have fallen victim to a delusion of being a vampire.'"

"Christ H. God!" cried Lucille delightedly. "Just like that, over the telephone?"

"When I recovered my aplomb, so to speak, I told him that I prefer to wait with the details until our first meeting, which is tomorrow."

They were sitting on the tiny terrace outside the staff room of the clinic, a converted town house on the upper West Side. Floria spent three days a week here and the remaining two in her office on Central Park South where she saw private clients like this new one. Lucille, always gratifyingly responsive, was Floria's most valued professional friend. Clearly enchanted with Floria's news, she sat eagerly forward in her chair, eyes wide behind Coke-bottle lenses.

She said, "Do you suppose he thinks he's a revivified corpse?"

Below, down at the end of the street, Floria could see two kids skidding their skateboards near a man who wore a woolen cap and a heavy coat despite the May warmth. He was leaning against a wall. He had been there when Floria had arrived at the clinic this morning. If corpses walked, some, not nearly revivified enough, stood in plain view in New York.

"I'll have to think of a delicate way to ask," she said.

"How did he come to you, this 'vampire'?"

"He was working in an upstate college, teaching and doing research, and all of a sudden he just disappeared — vanished, literally, without a trace. A month later he turned up here in the city. The faculty dean at the school knows me and sent him to see me."

Lucille gave her a sly look. "So you thought, ahah, do a little favor for a friend, this looks classic and easy to transfer if need be: repressed intellectual blows stack and runs off with spacey chick, something like that."

"You know me too well," Floria said with a rueful smile.

"Huh," grunted Lucille. She sipped ginger ale from a chipped white mug. "I don't take panicky middle-aged men anymore, they're too depressing. And you shouldn't be taking this one, intriguing as he sounds."

Here comes the lecture, Floria told herself.

Lucille got up. She was short, heavy, prone to wearing loose garments

ABOUT THE AUTHOR

Since the beginning of her writing career, Suzy McKee Charnas (b. 1939) has received high acclaim for her unique brand of science fiction and fantasy. Writing in genres previously considered "escapist," Charnas writes provocative social parables that bring the reader face to face with exquisitely dramatized problems of feminism, power and powerlessness. Her first novel, *Walk to the End of the World* (1974) was nominated for the John W. Campbell Award; "Unicorn Tapestry" won the Nebula Award for best novella in 1980; and her short story "Boobs" (1989) garnered the Hugo. Her other novels include *Motherlines* (1978), and *Dorothea Dreams* (1986) and *The Furies* (1994).

"Unicorn Tapestry" later became the first chapter of her novel *The Vampire Tapestry* (1980), which is still in print. Charnas recently discussed the book's appeal with interviewer Denise Dumars. The vampire Dr. Weyland "was conceived not as a suave, seductive aristocrat or glamorous punk, or a musty, claw-fingered revenant, but as one of a predatory species with a long, long life-span. The book allows him to utilize, confront, or be victimized himself by various predatory strains in our own culture."

Charnas's latest project, *Strange Seas,* is a nonfiction e-book account of cetaceans, trance channeling, and the writing life, and can be accessed at www.hidden-knowledge.com. Her home page, containing a complete bibliography, essays and interviews, is www.sfwa.org/members/charnas.

that swung about her like ceremonial robes. As she paced, her hem brushed at the flowers starting up in the planting boxes that rimmed the little terrace. "You know damn well this is just more overwork you're loading on. Don't take this guy; refer him."

Floria sighed. "I know, I know. I promised everybody I'd slow down. But you said it yourself just a minute ago — it looked like a simple favor. So what do I get? Count Dracula, for God's sake! Would you give that up?"

Fishing around in one capacious pocket, Lucille brought out a dented package of cigarettes and lit up, scowling. "You know, when you give me advice I try to take it seriously. Joking aside, Floria, what am I supposed to say? I've listened to you moaning for months now, and I thought we'd figured out that what you need is to shed some pressure, to start saying no — and here you are insisting on a new case. You know what I think: you're hiding in other people's problems from a lot of your own stuff that you should be working on.

"Okay, okay, don't glare at me. Be pigheaded. Have you gotten rid of Chubs, at least?" This was Floria's code name for a troublesome client named Kenny whom she'd been trying to unload for some time.

Floria shook her head.

"What gives with you? It's weeks since you swore you'd dump him! Trying to do everything for everybody is wearing you out. I bet you're still dropping weight. Judging by the very unbecoming circles under your eyes, sleeping isn't going too well, either. Still no dreams you can remember?"

"Lucille, don't nag. I don't want to talk about my health."

"Well, what about his health — Dracula's? Did you suggest that he have a physical before seeing you? There might be something physiological — "

"You're not going to be able to whisk him off to an M.D. and out of my hands," Floria said wryly. "He told me on the phone that he wouldn't consider either medication or hospitalization."

Involuntarily she glanced down at the end of the street. The woolen-capped man had curled up on the sidewalk at the foot of the building, sleeping or passed out or dead. The city was tottering with sickness. Compared with that wreck down there and others like him, how sick could this "vampire" be, with his cultured baritone voice, his self-possessed approach?

"And you won't consider handing him off to somebody else," Lucille said.

"Well, not until I know a little more. Come on, Luce — wouldn't you want at least to know what he looks like?"

Lucille stubbed out her cigarette against the low parapet. Down below a policeman strolled along the street ticketing the parked cars. He didn't even look at the man lying at the corner of the building. They watched his progress without comment. Finally Lucille said, "Well, if you won't drop Dracula, keep me posted on him, will you?"

A TAPESTRY OF VAMPIRE FICTION

While the following checklist makes no pretensions to bibliographic completeness, it should provide a valuable overview of the chronological evolution of vampire fiction, including periodic trends and cross-influences. Novel titles appear first in italics, short fiction follows in quotation marks.

1818
"The Vampyre: A Tale"
John Polidori (UK)

1820
"Aurelia"
E.T.A. Hoffmann (GERMANY)

1823
"Wake Not the Dead"
Johann Ludwig Tieck (GERMANY)

1835
"Viy"
Nikolai Gogol (RUSSIA)

1836
"La mort amoureuse"
Théophile Gautier (FRANCE)

1843
"The Family of the Vourdalak"
Alexis Tolstoy (RUSSIA)

1845
The Last of the Vampires
Smyth Upton (UK)

1847
Varney the Vampyre or, The Feast of Blood
James Malcolm Rymer (UK)

1848
"The Pale-Faced Lady"
Alexandre Dumas

1860
"The Mysterious Stranger"
Anonymous

He entered the office on the dot of the hour, a gaunt but graceful figure. He was impressive. Wiry gray hair, worn short, emphasized the massiveness of his face with its long jaw, high cheekbones, and granite cheeks grooved as if by winters of hard weather. His name, typed in caps on the initial information sheet that Floria proceeded to fill out with him, was Edward Lewis Weyland.

Crisply he told her about the background of the vampire incident, describing in caustic terms his life at Cayslin College: the pressures of collegial competition, interdepartmental squabbles, student indifference, administrative bungling. History has limited use, she knew, since memory distorts; still, if he felt most comfortable establishing the setting for his illness, that was as good a way to start off as any.

At length his energy faltered. His angular body sank into a slump, his voice became flat and tired as he haltingly worked up to the crucial event: night work at the sleep lab, fantasies of blood-drinking as he watched the youthful subjects of his dream research slumbering, finally an attempt to act out the fantasy with a staff member at the college. He had been repulsed; then panic had assailed him. Word would get out, he'd be fired, blacklisted forever. He'd bolted. A nightmare period had followed — he offered no details. When he had come to his senses he'd seen that just what he feared, the ruin of his career, would come from his running away. So he'd phoned the dean, and now here he was.

Throughout this recital she watched him diminish from the dignified academic who had entered her office to a shamed and frightened man hunched in his chair, his hands pulling fitfully at each other.

"What are your hands doing?" she said gently. He looked blank. She repeated the question.

He looked down at his hands. "Struggling," he said.

"With what?"

"The worst," he muttered. "I haven't told you the worst." She had never grown hardened to this sort of transformation. His long fingers busied themselves fiddling with a button on his jacket while he explained painfully that the object of his "attack" at Cayslin had been a woman. Not young but handsome and vital, she had first caught his attention earlier in the year during a *festschrift* — an honorary seminar — for a retiring professor.

A picture emerged of an awkward Weyland, lifelong bachelor, seeking this woman's warmth and suffering her refusal. Floria knew she should bring him out of his past and into his here-and-now, but he was doing so beautifully on his own that she was loath to interrupt.

"Did I tell you there was a rapist active on the campus at this time?" he said bitterly. "I borrowed a leaf from his book: I tried to take from this woman, since she wouldn't give. I tried to take some of her blood." He stared at the floor. "What does that mean — to take someone's blood?"

"What do you think it means?"

The button, pulled and twisted by his fretful fingers, came off. He put

1867
"The Last Lords of Gardonal"
William Gilbert

1870
Vikram and the Vampire
Richard Burton (UK)

1871
"Carmilla"
J. Sheridan Le Fanu (UK)

1880
"The Fate of Madame Cabanel"
Eliza Lynn Linton

1881
"The Man Eating Tree"
Philip Robinson

1883
"Ken's Mystery"
Julian Hawthorne

1885
The Vampire; or, Detective Brand's Greatest Case
Norman H. Munro

1887
"A Mystery of the Campagna"
Anne Crawford

1891
A Creature of the Night
Fergus Hume (UK)

La-Bas
Joris-Karl Huysmans (FRANCE)

The Strange Story of Dr. Senex
E.E. Baldwin

The Picture of Dorian Gray
Oscar Wilde (UK)

1892
"The Last of the Vampires"
Philip Robinson (UK)

1893
"The Death of Halpin Frayser"
Ambrose Bierce

1894
The Fair Abigail
Paul Heyse

it into his pocket, the impulse, she guessed, of a fastidious nature. "Her energy," he murmured, "stolen to warm the aging scholar, the walking corpse, the vampire — myself."

His silence, his downcast eyes, his bent shoulders, all signaled a man brought to bay by a life crisis. Perhaps he was going to be the kind of client therapists dream of and she needed so badly these days: a client intelligent and sensitive enough, given the companionship of a professional listener, to swiftly unravel his own mental tangles. Exhilarated by his promising start, Floria restrained herself from trying to build on it too soon. She made herself tolerate the silence, which lasted until he said suddenly, "I notice that you make no notes as we speak. Do you record these sessions on tape?"

A hint of paranoia, she thought; not unusual. "Not without your knowledge and consent, just as I won't send for your personnel file from Cayslin without your knowledge and consent. I do, however, write notes after each session as a guide to myself and in order to have a record in case of any confusion about anything we do or say here. I can promise you that I won't show my notes or speak of you by name to anyone — except Dean Sharpe at Cayslin, of course, and even then only as much as is strictly necessary — without your written permission. Does that satisfy you?"

"I apologize for my question," he said. "The... incident has left me... very nervous; a condition that I hope to get over with your help."

The time was up. When he had gone, she stepped outside to check with Hilda, the receptionist she shared with four other therapists here at the Central Park South office. Hilda always sized up new clients in the waiting room.

Of this one she said, "Are you sure there's anything wrong with that guy? I think I'm in love."

Waiting at the office for a group of clients to assemble Wednesday evening, Floria dashed off some notes on the "vampire."

> Client described incident, background. No history of mental illness, no previous experience of therapy. Personal history so ordinary you almost don't notice how bare it is: only child of German immigrants, schooling normal, field work in anthropology, academic posts leading to Cayslin College professorship. Health good, finances adequate, occupation satisfactory, housing pleasant (though presently installed in a N.Y. hotel); never married, no kids, no family, no religion, social life strictly job-related; leisure — says he likes to drive. Reaction to question about drinking, but no signs of alcohol problems. Physically very smooth-moving for his age (over fifty) and height; catlike, alert. Some apparent stiffness in the midsection — slight protective stoop — tightening up of middle age? Paranoic defensiveness? Voice pleasant, faint accent (German-speaking childhood at home). Entering therapy condition of consideration for return to job.

"A Kiss of Judas"
Julian Osgood Field (UK)

"The Flowering of the Strange Orchid"
H.G. Wells (UK)

"The Parasite"
Arthur Conan Doyle

"The True Story of a Vampire"
Eric, Count Stenbock (UK)

1896
"Good Lady Ducayne"
Mary Elizabeth Braddon (UK)

1897
The Blood of the Vampire
Florence Marryat (UK)

Dracula
Bram Stoker (UK)

1898
"The Story of Baelbrow"
E. & H. Heron (UK)

1899
"The Stone Chamber"
H.B. Watson (UK)

1900
"The Old Portrait"
Hume Nisbet (UK)

"The Tomb of Sarah"
F.G. Loring (UK)

"The Vampire Maid"
Hume Nisbet (UK)

1902
"Luella Miller"
Mary E. Wilkins Freeman

1904
"Count Magnus"
M.R. James (UK)

"A Dead Finger"
Sabine Baring-Gould (UK)

1907
House of the Vampire
George Sylvester Viereck

What a relief: his situation looked workable with a minimum of strain on herself. Now she could defend to Lucille her decision to do therapy with the "vampire."

After all, Lucille was right. Floria did have problems of her own that needed attention, primarily her anxiety and exhaustion since her mother's death more than a year before. The breakup of Floria's marriage had caused misery, but not this sort of endless depression. Intellectually the problem was clear: with both her parents dead she was left exposed. No one stood any longer between herself and the inevitability of her own death. Knowing the source of her feelings didn't help: she couldn't seem to mobilize the nerve to work on them.

The Wednesday group went badly again. Lisa lived once more her experiences in the European death camps and everyone cried. Floria wanted to stop Lisa, turn her, extinguish the droning horror of her voice in illumination and release, but she couldn't see how to do it. She found nothing in herself to offer except some clever ploy out of the professional bag of tricks — dance your anger, have a dialog with yourself of those days — useful techniques when they flowed organically as part of a living process in which the therapist participated. But thinking out responses that should have been intuitive wouldn't work. The group and its collective pain paralyzed her. She was a dancer without a choreographer, knowing all the moves but unable to match them to the music these people made.

Rather than act with mechanical clumsiness she held back, did nothing, and suffered guilt. Oh God, the smart, experienced people in the group must know how useless she was here.

Going home on the bus she thought about calling up one of the therapists who shared the downtown office. He had expressed an interest in doing co-therapy with her under student observation. The Wednesday group might respond well to that. Suggest it to them next time? Having a partner might take pressure off Floria and revitalize the group, and if she felt she must withdraw he would be available to take over. Of course he might take over anyway and walk off with some of her clients.

Oh boy, terrific, who's paranoid now? Wonderful way to think about a good colleague. God, she hadn't even known she was considering chucking the group.

Had the new client, running from his "vampirism," exposed her own impulse to retreat? This wouldn't be the first time that Floria had obtained help from a client while attempting to give help. Her old supervisor, Rigby, said that such mutual aid was the only true therapy — the rest was fraud. What a perfectionist, old Rigby, and what a bunch of young idealists he'd turned out, all eager to save the world.

Eager, but not necessarily able. Jane Fennerman had once lived in the world, and Floria had been incompetent to save her. Jane, an absent member of tonight's group, was back in the safety of a locked ward, hazily gliding on whatever tranquilizers they used there.

"The Feather Pillow"
Horacio Quiroga

1909
The Lady of the Shroud
Bram Stoker (UK)

1910
"The Singular Death of Morton"
Angernon Blackwood (UK)

1911
The Lair of the White Worm
Bram Stoker (UK)

"For the Blood is the Life"
F. Marion Crawford

1912
"The Room in the Tower"
E.F. Benson

"The Transfer"
Algernon Blackwood (UK)

1913
The Vampire
William Reginald Hodder (UK)

1914
Malevola
Theo Douglas (UK)

The White Vampire
A.M. Judd (UK)

"Aylmer Vance and the Vampire"
Alice and Claude Askew

"Dracula's Guest"
Bram Stoker (UK)

"An Episode of Cathedral History"
M.R. James (UK)

1919
"The Sumach"
Ulric Daubeny (UK)

1920
"Mrs. Amworth"
E.F. Benson

1922
Vampir
Hans Heinz Ewers (GERMANY)

1924
"The Adventure of the Sussex Vampire"
Arthur Conan Doyle (UK)

"The Hound"
H.P. Lovecraft

1925
The Vampires of the Andes
Henry Carew (UK)

"Four Wooden Stakes"
Victor Roman

1926
"Bat's Belfry"
August Derleth

"Bewitched"
Edith Wharton

1927
The Demon Lover
Dion Fortune (UK)

"The Canal"
Everil Worrell

"The Man Who Cast No Shadow"
Seabury Quinn

"The Wolf Woman"
Bassett Morgan

1928
London After Midnight
Marie Coolidge-Rask

The Mark of the Bat: A Tale of Vampires Living and Dead
Gertrude Dunn (UK)

"Restless Souls"
Seabury Quinn

"The Tinkle of the Camel's Bell"
Frank Owen

1929
"The Silver Countess"
Seabury Quinn

1930
A Human Vampire
T. Francis Elstow (UK)

Why still mull over Jane? she asked herself severely, bracing against the bus's lurching halt. Any client was entitled to drop out of therapy and commit herself. Nor was this the first time that sort of thing had happened in the course of Floria's career. Only this time she couldn't seem to shake free of the resulting depression and guilt.

But how could she have helped Jane more? How could you offer reassurance that life was not as dreadful as Jane felt it to be, that her fears were insubstantial, that each day was not a pit of pain and danger?

She was taking time during a client's canceled hour to work on notes for the new book. The writing, an analysis of the vicissitudes of salaried versus private practice, balked her at every turn. She longed for an interruption to distract her circling mind.

Hilda put through a call from Cayslin College. It was Doug Sharpe, who had sent Dr. Weyland to her.

"Now that he's in your capable hands, I can tell people plainly that he's on what we call 'compassionate leave' and make them swallow it." Doug's voice seemed thinned by the long-distance connection. "Can you give me a preliminary opinion?"

"I need time to get a feel for the situation."

He said, "Try not to take too long. At the moment I'm holding off pressure to appoint someone in his place. His enemies up here — and a sharp-tongued bastard like him acquires plenty of those — are trying to get a search committee authorized to find someone else for the directorship of the Cayslin Center for the Study of Man."

"Of People," she corrected automatically, as she always did. "What do you mean, 'bastard'? I thought you liked him, Doug. 'Do you want me to have to throw a smart, courtly, old-school gent to Finney or MaGill?' Those were your very words." Finney was a Freudian with a mouth like a pursed-up little asshole and a mind to match, and MaGill was a primal yowler in a padded gym of an office.

She heard Doug tapping at his teeth with a pen or pencil. "Well," he said, "I have a lot of respect for him, and sometimes I could cheer him for mowing down some pompous moron up here. I can't deny, though, that he's earned a reputation for being an accomplished son-of-a-bitch and tough to work with. Too damn cold and self-sufficient, you know?"

"Mmm," she said. "I haven't seen that yet."

He said, "You will. How about yourself? How's the rest of your life?"

"Well, offhand, what would you say if I told you I was thinking of going back to art school?"

"What would I say? I'd say bullshit, that's what I'd say. You've had fifteen years of doing something you're good at, and now you want to throw all that out and start over in an area you haven't touched since Studio 101 in college? If God had meant you to be a painter, She'd have sent you to art school in the first place."

"I did think about art school at the time."

"The point is that you're good at what you do. I've been at the receiving end of your work and I know what I'm talking about. By the way, did you see that piece in the paper about Annie Barnes, from the group I was in? That's an important appointment. I always knew she'd wind up in Washington. What I'm trying to make clear to you is that your 'graduates' do too well for you to be talking about quitting. What's Morton say about that idea, by the way?"

Mort, a pathologist, was Floria's lover. She hadn't discussed this with him, and she told Doug so.

"You're not on the outs with Morton, are you?"

"Come on, Douglas, cut it out. There's nothing wrong with my sex life, believe me. It's everyplace else that's giving me trouble."

"Just sticking my nose into your business," he replied. "What are friends for?"

They turned to lighter matters, but when she hung up Floria felt glum. If her friends were moved to this sort of probing and kindly advice-giving, she must be inviting help more openly and more urgently than she'd realized.

The work on the book went no better. It was as if, afraid to expose her thoughts, she must disarm criticism by meeting all possible objections beforehand. The book was well and truly stalled — like everything else. She sat sweating over it, wondering what the devil was wrong with her that she was writing mush. She had two good books to her name already. What was this bottleneck with the third?

"But what do you think?" Kenny insisted anxiously. "Does it sound like my kind of job?"

"How do you feel about it?"

"I'm all confused, I told you."

"Try speaking for me. Give me the advice I would give you."

He glowered. "That's a real cop-out, you know? One part of me talks like you, and then I have a dialog with myself like a TV show about a split personality. It's all me that way; you just sit there while I do all the work. I want something from you."

She looked for the twentieth time at the clock on the file cabinet. This time it freed her. "Kenny, the hour's over."

Kenny heaved his plump, sulky body up out of his chair. "You don't care. Oh, you pretend to, but you don't really —"

"Next time, Kenny."

He stumped out of the office. She imagined him towing in his wake the raft of decisions he was trying to inveigle her into making for him. Sighing, she went to the window and looked out over the park, filling her eyes and her mind with the full, fresh green of late spring. She felt dismal. In two years of treatment the situation with Kenny had remained a stalemate. He wouldn't go to someone else who might be able to help him, and she couldn't bring herself to kick him out, though she knew she must

The Veiled Vampire
Arlton Eadie (UK)

"The End of the Story"
Clark Ashton Smth

"The Hills of the Dead"
Robert E. Howard

"The Moon of Skulls"
Robert E. Howard

1931
"A Rendezvous in Averoigne"
Clark Ashton Smith

"The Dark Castle"
Marion Brandon

"Placide's Wife"
Kirk Mashburn

1932
"The Horror from the Mound"
Robert E. Howard

"Stragella"
Hugh B. Cave

"Wings in the Night"
Robert E. Howard

"Vampire Village"
Edmond Hamilton

1933
The Unmeasured Place
John Lambourne (UK)

"The Lovely Lady"
D.H. Lawrence

"Revelations in Black"
Carl Jacobi

"Shambleau"
C.L. Moore

1934
"The Garden of Fear"
Robert E. Howard

"The Graveyard Duchess"
John Flanders

eventually. His puny tyranny couldn't conceal how soft and vulnerable he was…

Dr. Weyland had the next appointment. Floria found herself pleased to see him. She could hardly have asked for a greater contrast to Kenny: tall, lean, that august head that made her want to draw him, good clothes, nice big hands — altogether, a distinguished-looking man. Though he was informally dressed in slacks, light jacket, and tieless shirt, the impression he conveyed was one of impeccable leisure and reserve. He took not the padded chair preferred by most clients but the wooden one with the cane seat.

"Good afternoon, Dr. Landauer," he said gravely. "May I ask your judgment of my case?"

"I don't regard myself as a judge," she said. She decided to try to shift their discussion onto a first-name basis if possible. Calling this old-fashioned man by his first name so soon might seem artificial, but how could they get familiar enough to do therapy while addressing each other as "Dr. Landauer" and "Dr. Weyland" like two characters out of a vaudeville sketch?

"This is what I think, Edward," she continued. "We need to find out about this vampire incident — how it tied into your feelings about yourself, good and bad, at the time; what it did for you that led you to try to 'be' a vampire even though that was bound to complicate your life terrifically. The more we know, the closer we can come to figuring out how to insure that this vampire construct won't be necessary to you again."

"Does this mean that you accept me formally as a client?" he said.

Comes right out and says what's on his mind, she noted; no problem there. "Yes."

"Good. I too have a treatment goal in mind. I will need at some point a testimonial from you that my mental health is sound enough for me to resume work at Cayslin."

Floria shook her head. "I can't guarantee that. I can commit myself to work toward it, of course, since your improved mental health is the aim of what we do here together."

"I suppose that answers the purpose for the time being," he said. "We can discuss it again later on. Frankly, I find myself eager to continue our work today. I've been feeling very much better since I spoke with you, and I thought last night about what I might tell you today."

She had the distinct feeling of being steered by him; how important was it to him, she wondered, to feel in control? She said, "Edward, my own feeling is that we started out with a good deal of very useful verbal work, and that now is a time to try something a little different."

He said nothing. He watched her. When she asked whether he remembered his dreams he shook his head, no.

She said, "I'd like you to try to do a dream for me now, a waking dream. Can you close your eyes and daydream, and tell me about it?"

1935
The Three Coffins
John Dickson Carr

The Vampire
Sydney Horler (UK)

Vampires of Vengeance
T. Stanleyan King

Vampires Overhead
Alan Hyder (UK)

"The Believer"
Sydney Horler (UK)

"The Horror Undying"
Manly Wade Wellman

"The Thirsty Dead"
Raymond Whetstone

1936
"Doom of the House of Duryea"
Earl Pierce, Jr.

"Isle of the Undead"
Lloyd Arthur Eshbach

"Return to Death"
J. Wesley Rosenquist

1937
Fingers of Fear
J.U. Nicolson

"The Death of Ilalotha"
Clark Ashton Smith

"I, the Vampire"
Henry Kuttner

"School for the Unspeakable"
Manly Wade Wellman

1938
The Bat Woman
Cromwell Gibbons

"The Vampire of Kaldenstein"
Frederick Cowles

1939
"The Cloak"
Robert Bloch

He closed his eyes. Strangely, he now struck her as less vulnerable rather than more, as if strengthened by increased vigilance.

"How do you feel now?" she said.

"Uneasy." His eyelids fluttered. "I dislike closing my eyes. What I don't see can hurt me."

"Who wants to hurt you?"

"A vampire's enemies, of course — mobs of screaming peasants with torches."

Translating into what, she wondered — young Ph.D.s pouring out of the graduate schools panting for the jobs of older men like Weyland? "Peasants, these days?"

"Whatever their daily work, there is still a majority of the stupid, the violent, and the credulous, putting their feather-brained faith in astrology, in this cult or that, in various branches of psychology."

His sneer at her was unmistakable. Considering her refusal to let him fill the hour his own way, this desire to take a swipe at her was healthy. But it required immediate and straightforward handling.

"Edward, open your eyes and tell me what you see."

He obeyed. "I see a woman in her early forties," he said, "clever-looking face, dark hair showing gray; flesh too thin for her bones, indicating either vanity or illness; wearing slacks and a rather creased batik blouse — describable, I think, by the term 'peasant style' — with a food stain on the left side."

Damn! Don't blush. "Does anything besides my blouse suggest a peasant to you?"

"Nothing concrete, but with regard to me, my vampire self, a peasant with a torch is what you could easily become."

"I hear you saying that my task is to help you get rid of your delusion, though this process may be painful and frightening for you."

Something flashed in his expression — surprise, perhaps alarm, something she wanted to get in touch with before it could sink away out of reach again. Quickly she said, "How do you experience your face at this moment?"

He frowned. "As being on the front of my head. Why?"

With a rush of anger at herself she saw that she had chosen the wrong technique for reaching that hidden feeling: she had provoked hostility instead. She said, "Your face looked to me just now like a mask for concealing what you feel rather than an instrument of expression."

He moved restlessly in the chair, his whole physical attitude tense and guarded. "I don't know what you mean."

"Will you let me touch you?" she said, rising.

His hands tightened on the arms of his chair, which protested in a sharp creak. He snapped, "I thought this was a talking cure."

Strong resistance to body work — ease up. "If you won't let me massage some of the tension out of your facial muscles, will you try to do it yourself?"

"Cross of Fire"
Lester del Rey

"The Drifting Snow"
August Derleth

"The Silver Coffin"
Robert Barbour Johnson

"Vampire's Honeymoon"
Cornell Woolrich

1940
"Darker than You Think
Jack Williamson

"When It Was Moonlight"
Manly Wade Wellman

1941
The Vampire Men
Herald Verner (UK)

"Over the River"
P. Schuyler Miller

1942
Dreadful Hollow
Irina Karlova (UK)

"Asylum"
A.E. van Vogt

"Experiment with Evil"
Curt Siodmak

"They Bite"
Anthony Boucher

"The Vampires of Shiloh"
Manly Wade Wellman

1943
"The Devil Is Not Mocked"
Manly Wade Wellman

"Return of the Undead"
Frank Belknap Long and Otis Adelbert Kline

1944
"The Bat Is My Brother"
Robert Bloch

1945
"The Vampire of Woolpit Grange"
Robert Thurston Hopkins

1946
The Dark World
Henry Kuttner

He Who Whispers
John Dickson Carr

"The Bogeyman Will Get You"
Robert Bloch

"Homecoming"
Ray Bradbury

1947
"Heredity"
David H. Keller

"The Man Upstairs"
Ray Bradbury

1948
"The Human Angle"
William Tenn

"Pillar of Fire"
Ray Bradbury

"Schloss Wappenburg"
D. Scott Moncrieff

"The Undead Die"
E. Everett Evans

1949
"The Antimacassar"
Greye La Spina

"The Girl with the Hungry Eyes"
Fritz Leiber

1950
Vampires of Venus
Karl Mannheim (UK)

"The Mindworm"
Cyril M. Kornbluth

"The Traitor"
James S. Hart

1951
"Blood Son" (a.k.a. "Drink My Red Blood")
Richard Matheson

"I don't enjoy being made ridiculous," he said, standing and heading for the door, which clapped smartly to behind him.

She sagged back in her seat; she had mishandled him. Clearly her initial estimation of this as a relatively easy job had been wrong and had led her to move far too quickly with him. Certainly it was much too early to try body work. She should have developed a firmer level of trust first by letting him do more of what he did so easily and so well — talk.

The door opened. Weyland came back in and shut it quietly. He did not sit again but paced about the room, coming to rest at the window.

"Please excuse my rather childish behavior just now," he said. "Playing these games of yours brought it on."

"It's frustrating, playing games that are unfamiliar and that you can't control," she said. As he made no reply, she went on in a conciliatory tone, "I'm not trying to belittle you, Edward. I just need to get us off whatever track you were taking us down so briskly. My feeling is that you're trying hard to regain your old stability.

"But that's the goal, not the starting point. The only way to reach your goal is through the process, and you don't drive the therapy process like a train. You can only help the process happen, as though you were helping a tree grow."

"These games are part of the process?"

"Yes."

"And neither you nor I control the games?"

"That's right."

He considered. "Suppose I agree to try this process of yours; what would you want of me?"

Observing him carefully, she no longer saw the anxious scholar bravely struggling back from madness. Here was a different sort of man — armored, calculating. She didn't know just what the change signaled, but she felt her own excitement stirring, and that meant she was on the track of something.

"I have a hunch," she said slowly, "that this vampirism extends further back into your past than you've told me and possibly right up into the present as well. I think it's still with you. My style of therapy stresses dealing with the now at least as much as the then; if the vampirism is part of the present, dealing with it on that basis is crucial."

Silence.

"Can you talk about being a vampire: being one now?"

"You won't like knowing," he said.

"Edward, try."

He said, "I hunt."

"Where? How? What sort of — of victims?"

He folded his arms and leaned his back against the window frame. "Very well, since you insist. There are a number of possibilities here in the city in summer. Those too poor to own air-conditioners sleep out on rooftops and fire escapes. But often, I've found, their blood is sour with drugs or liquor. The same is true of prostitutes. Bars are full of accessible people but also full of smoke and noise, and there too the blood is fouled. I must choose my hunting grounds carefully. Often I go to openings of galleries or evening museum shows or department stores on their late nights — places where women may be approached."

And take pleasure in it, she thought, if they're out hunting also — for acceptable male companionship. Yet he said he's never married. Explore where this is going. "Only women?"

He gave her a sardonic glance, as if she were a slightly brighter student than he had at first assumed.

"Hunting women is liable to be time-consuming and expensive. The best hunting is in the part of Central Park they call the Ramble, where homosexual men seek encounters with others of their kind. I walk there too at night."

Floria caught a faint sound of conversation and laughter from the waiting room; her next client had probably arrived, she realized, looking reluctantly at the clock. "I'm sorry, Edward, but our time seems to be — "

"Only a moment more," he said coldly. "You asked; permit me to finish my answer. In the Ramble I find someone who doesn't reek of alcohol or drugs, who seems healthy, and who is not insistent on 'hooking up' right there among the bushes. I invite such a man to my hotel. He judges me safe, at least: older, weaker than he is, unlikely to turn out to be a dangerous maniac. So he comes to my room. I feed on his blood.

"Now, I think, our time is up."

He walked out.

She sat torn between rejoicing at his admission of the delusion's persistence and dismay that his condition was so much worse than she had first thought. Her hope of having an easy time with him vanished. His initial presentation had been just that — a performance, an act. Forced to abandon it, he had dumped on her this lump of material, too much — and too strange — to take in all at once.

Her next client liked the padded chair, not the wooden one that Weyland had sat in during the first part of the hour. Floria started to move the wooden one back. The armrests came away in her hands.

She remembered him starting up in protest against her proposal of touching him. The grip of his fingers had fractured the joints, and the shafts now lay in splinters on the floor.

Floria wandered into Lucille's room at the clinic after the staff meeting. Lucille was lying on the couch with a wet cloth over her eyes.

"I thought you looked green around the gills today," Floria said. "What's wrong?"

"Big bash last night," said Lucille in sepulchral tones. "I think I feel about the way you do after a session with Chubs. You haven't gotten rid of him yet, have you?"

"The Last Grave of Lil Warren"
Manly Wade Wellman

1952
"Who Shall I Say Is Calling?"
August Derleth

1953
"And Not Quite Human"
Joe L. Hensley

"The Cookie Lady"
Philip K. Dick

"The Music"
Theodore Sturgeon

"Place of Meeting"
Charles Beaumont

"Share Alike"
Jerome Bixby and Joe E. Dean

1954
The Feasting Dead
John Metcalfe

I Am Legend
Richard Matheson

"Blood"
Fredric Brown

1956
The Unquiet Corpse
William Sloane

"She Only Goes Out at Night"
William Tenn

1958
"Eripmav"
Damon Knight

"Hungarian Rhapsody"
Robert Bloch

1959
"No Such Thing as a Vampire"
Richard Matheson

1960
Blood and Roses
Robin Carlisle (UK)

The Brides of Dracula
Dean Owen

Doctors Wear Scarlet
Simon Raven

"The Arrogant Vampire"
Arthur Porges

"The Longest Night"
Ray Russell

"Softly, While You're Sleeping"
Evelyn E. Smith

1961
Some of Your Blood
Theodore Sturgeon

"Blood Brother"
Charles Beaumont

1962
Moonflowers
Margaret Peterson (UK)

1926
"Bat's Belfry"
August Derleth

1963
"The Stainless Steel Leech"
Roger Zelazny

1964
The Shiny Narrow Grin
Jane Gaskell (UK)

1965
Progeny of the Adder
Leslie H. Whitten

To Kill a Corpse
Eugene Ascher (UK)

1966
Dark Shadows
Marilyn Ross*

The Disoriented Man
(a.k.a. *Scream and Scream Again*)
Peter Saxon (UK)

Queen of Blood
Charles Nuetzel

So What Killed the Vampire?
Carter Brown

"No. I had him lined up to see Marty instead of me last week, but damned if he didn't show up at my door at his usual time. It's a lost cause. What I wanted to talk to you about was Dracula."

"What about him?"

"He's smarter, tougher, and sicker than I thought, and maybe I'm even less competent than I thought, too. He's already walked out on me once — I almost lost him. I never took a course in treating monsters."

Lucille groaned. "Some days they're all monsters." This from Lucille, who worked longer hours than anyone else at the clinic, to the despair of her husband. She lifted the cloth, refolded it, and placed it carefully across her forehead. "And if I had ten dollars for every client who's walked out on me . . . Tell you what: I'll trade you Madame X for him, how's that? Remember Madame X, with the jangling bracelets and the parakeet eye makeup and the phobia about dogs? Now she's phobic about things dropping on her out of the sky. Just wait — it'll turn out that one day when she was three a dog trotted by and pissed on her leg just as an over-passing pigeon shat on her head. What are we doing in this business?"

"God knows." Floria laughed. "But am I in this business these days — I mean, in the sense of practicing my so-called skills? Blocked with my group work, beating my brains out on a book that won't go, and doing something — I'm not sure it's therapy — with a vampire . . . You know, once I had this sort of natural choreographer inside myself that hardly let me put a foot wrong and always knew how to correct a mistake if I did. Now that's gone. I feel as if I'm just going through a lot of mechanical motions. Whatever I had once that made me useful as a therapist, I've lost it."

Ugh, she thought, hearing the descent of her voice into a tone of gloomy self-pity.

"Well, don't complain about Dracula," Lucille said. "You were the one who insisted on taking him on. At least he's got you concentrating on his problem instead of just wringing your hands. As long as you've started, stay with it — illumination may come. And now I'd better change the ribbon in my typewriter and get back to reviewing Silverman's latest best-seller on self-shrinking while I'm feeling mean enough to do it justice." She got up gingerly. "Stick around in case I faint and fall into the wastebasket."

"Luce, this case is what I'd like to try to write about."

"Dracula?" Lucille pawed through a desk drawer full of paper clips, pens, rubber bands and old lipsticks.

"Dracula. A monograph . . ."

"Oh, I know that game: you scribble down everything you can and then read what you wrote to find out what's going on with the client, and with luck you end up publishing. Great! But if you are going to publish, don't piddle this away on a dinky paper. Do a book. Here's your subject, instead of those depressing statistics you've been killing yourself over. This one is really exciting — a case study to put on the shelf next to Freud's own wolf-man, have you thought of that?"

Floria liked it. "What a book that could be — fame if not fortune. Notoriety, most likely. How in the world could I convince our colleagues that it's legit? There's a lot of vampire stuff around right now — plays on Broadway and TV, books all over the place, movies. They'll say I'm just trying to ride the coattails of a fad."

"No, no, what you do is show how this guy's delusion is related to the fad. Fascinating." Lucille, having found a ribbon, prodded doubtfully at the exposed innards of her typewriter.

"Suppose I fictionalize it," Floria said, "under a pseudonym. Why not ride the popular wave and be free in what I can say?"

"Listen, you've never written a word of fiction in your life, have you?" Lucille fixed her with a bloodshot gaze. "There's no evidence that you could turn out a best-selling novel. On the other hand, by this time you have a trained memory for accurately reporting therapeutic transactions. That's a strength you'd be foolish to waste. A solid professional book would be terrific — and a feather in the cap of every woman in the field. Just make sure you get good legal advice on disguising your Dracula's identity well enough to avoid libel."

The cane-seated chair wasn't worth repairing, so she got its twin out of the bedroom to put in the office in its place. Puzzling: by his history Weyland was fifty-two, and by his appearance no muscle man. She should have asked Doug — but how, exactly? "By the way, Doug, was Weyland ever a circus strong man or a blacksmith? Does he secretly pump iron?" Ask the client himself — but not yet.

She invited some of the younger staff from the clinic over for a small party with a few of her outside friends. It was a good evening; they were not a heavy-drinking crowd, which meant the conversation stayed intelligent. The guests drifted about the long living room or stood in twos and threes at the windows looking down on West End Avenue as they talked.

Mort came, warming the room. Fresh from a session with some amateur chamber-music friends, he still glowed with the pleasure of making his cello sing. His own voice was unexpectedly light for so large a man. Sometimes Floria thought that the deep throb of the cello was his true voice.

He stood beside her talking with some others. There was no need to lean against his comfortable bulk or to have him put his arm around her waist. Their intimacy was long-standing, an effortless pleasure in each other that required neither demonstration nor concealment.

He was easily diverted from music to his next favorite topic, the strengths and skills of athletes.

"Here's a question for a paper I'm thinking of writing," Floria said. "Could a tall, lean man be exceptionally strong?"

Mort rambled on in his thoughtful way. His answer seemed to be no.

The Torturer
Peter Saxon (UK)

The Vampire Affair
(The Man from U.N.C.L.E. #6)
David McDaniel

"Count Dracula"
Woody Allen

"The New Men"
Joanna Russ

*Using the pen name Marilyn Ross, William Edward Daniel Ross wrote thirty-two *Dark Shadows* novella-length paperbacks for Paperback Library between 1966 and 1972, nearly all chronicling new adventures for daytime television's favorite vampire, Barnabas Collins. Given the repetitious nature of the series and the titles, I will omit individual citations from this checklist. *Ed.*

1967
Dracula, Prince of Darkness
John Burke (UK)

"The Living Dead"
Robert Bloch

"Sleeping Beauty"
Terry Carr

1968
Image of the Beast
Philip José Farmer

The Orgy at Madame Dracula's
F.W. Paul

The Vampire Cameo
Dorothea Nile

"Dr. Porthos"
Basil Copper

"Try a Dull Knife"
Harlan Ellison

1969
Blown
Philip José Farmer

Dracutwig
Mallory T. Knight

"But what about chimpanzees?" put in a young clinician. "I went with a guy once who was an animal handler for TV, and he said a three month-old chimp could demolish a strong man."

"It's all physical conditioning," somebody else said. "Modern people are soft."

Mort nodded. "Human beings in general are weakly made compared to other animals. It's a question of muscle insertions — the angles of how the muscles are attached to the bones. Some angles give better leverage than others. That's how a leopard can bring down a much bigger animal than itself. It has a muscular structure that gives it tremendous strength for its streamlined build."

Floria said, "If a man were built with muscle insertions like a leopard's, he'd look pretty odd, wouldn't he?"

"Not to an untrained eye," Mort said, sounding bemused by an inner vision. "And my God, what an athlete he'd make — can you imagine a guy in the decathlon who's as strong as a leopard?"

When everyone else had gone Mort stayed, as he often did. Jokes about insertions, muscular and otherwise, soon led to sounds more expressive and more animal, but afterward Floria didn't feel like resting snuggled together with Mort and talking. When her body stopped racing, her mind turned to her new client. She didn't want to discuss him with Mort, so she ushered Mort out as gently as she could and sat down by herself at the kitchen table with a glass of orange juice.

How to approach the reintegration of Weyland the eminent, gray-haired academic with the rebellious vampire-self that had smashed his life out of shape?

She thought of the broken chair, of Weyland's big hands crushing the wood. Old wood and dried-out glue, of course, or he never could have done that. He was a man, after all, not a leopard.

The day before the third session Weyland phoned and left a message with Hilda: he would not be coming to the office tomorrow for his appointment, but if Dr. Landauer were agreeable she would find him at their usual hour at the Central Park Zoo.

Am I going to let him move me around from here to there? she thought. I shouldn't — but why fight it? Give him some leeway, see what opens up in a different setting. Besides, it was a beautiful day, probably the last of the sweet May weather before the summer stickiness descended. She gladly cut Kenny short so that she would have time to walk over to the zoo.

There was a fair crowd there for a weekday. Well-groomed young matrons pushed clean, floppy babies in strollers. Weyland she spotted at once.

He was leaning against the railing that enclosed the seals' shelter and their murky green pool. His jacket, slung over his shoulder, draped elegantly down his long back. Floria thought him rather dashing and faintly foreign-looking. Women who passed him, she noticed, tended to glance back.

Night of the Vampire
Raymond Giles

Secret of the Pale Lover
Clarisa Ross

"The Tale of the Vampire's Kingdom"
Italo Calvino (ITALY; trans. 1977)

1970
The Adult Version of Dracula
Anonymous

Blood Moon
Jan Alexander

House of Dark Shadows
Marilyn Ross

One More Time
Michael Avallone

The Vampire of Moura
Virginia Coffman

Vampire's Kiss
Sonny Barker

The Vampires of Finistere (The Guardians 34)
Peter Saxon (UK)

"Vanishing Breed"
Niel Straum

1971
Countess Dracula
Michel Parry

The Dracula Archives
Raymond Rudorff (UK)

The Feast of the Wolf
Thomas Blackburn (UK)

Lust for a Vampire
William Hughes

The Scars of Dracula
Angus Hall (UK)

The Vampire Curse
Daoma Winston

1972
The Bloody Countess
Valentine Penrose (UK)

He looked at everyone. She had the impression that he knew quite well that she was walking up behind him.

"Outdoors makes a nice change from the office, Edward," she said, coming to the rail beside him. "But there must be more to this than a longing for fresh air." A fat seal lay in sculptural grace on the concrete, eyes blissfully shut, fur drying in the sun to a translucent water-color umber.

Weyland straightened from the rail. They walked. He did not look at the animals; his eyes moved continually over the crowd. He said, "Someone has been watching for me at your office building."

"Who?"

"There are several possibilities. Pah, what a stench — though humans caged in similar circumstances smell as bad." He sidestepped a couple of shrieking children who were fighting over a balloon and headed out of the zoo under the musical clock.

They walked the uphill path northward through the park. By extending her own stride a little Floria found that she could comfortably keep pace with him.

"Is it peasants with torches?" she said. "Following you?"

He said, "What a childish idea."

All right, try another tack, then: "You were telling me last time about hunting in the Ramble. Can we return to that?"

"If you wish." He sounded bored — a defense? Surely — she was certain this must be the right reading — surely his problem was a transmutation into "vampire" fantasy of an unacceptable aspect of himself. For men of his generation the confrontation with homosexual drives could be devastating.

"When you pick up someone in the Ramble, is it a paid encounter?"

"Usually."

"How do you feel about having to pay?" She expected resentment.

He gave a faint shrug. "Why not? Others work to earn their bread. I work, too, very hard, in fact. Why shouldn't I use my earnings to pay for my sustenance?"

Why did he never play the expected card? Baffled, she paused to drink from a fountain. They walked on.

"Once you've got your quarry, how do you . . . " She fumbled for a word.

"Attack?" he supplied, unperturbed. "There's a place on the neck, here, where pressure can interrupt the blood flow to the brain and cause unconsciousness. Getting close enough to apply that pressure isn't difficult."

"You do this before or after any sexual activity?"

"Before, if possible," he said aridly, "and instead of." He turned aside to stalk up a slope to a granite outcrop that overlooked the path they had been following. There he settled on his haunches, looking back the way they had come. Floria, glad she'd worn slacks today, sat down near him.

He didn't seem devastated — anything but. Press him, don't let him get by on cool. "Do you often prey on men in preference to women?"

The Werewolf vs. the Vampire Woman
Arthur Scram

1973
Dracula Returns
Robert Lory

Dracula's Brother
Robert Lory

The Hand of Dracula
Robert Lory

The Night Stalker
Jeff Rice

Red Wine of Rapture
Margaret Worth

The Vampire Women
Victor Samuels

The Vampires
John Rechy

"Pages from a Young Girl's Journal"
Robert Aickman

"Thirst"
Bill Pronzini

1974
Children of the Night
(a.k.a. *Dracula's Children*)
Richard Lortz

The Deathbox
Errol Lecale (UK)

Dracula and the Virgins of the Undead
Etienne Aubin (UK)

Dracula's Lost World
Robert Lory

The Drums of Dracula
Robert Lory

An Enquiry into the Existence of Vampires
Marc Lowell

Our Lady of Pain
John Blackburn

The Vampire Contessa: From the Journal of Jeremy Quentain
Marilyn Ross

The Vampires and the Witch (The Phantom #12)
Lee Falk

The Witching of Dracula
Robert Lory

1975
The Blood Countess (The Avenger #33)
Kenneth Robeson

Challenge to Dracula
Robert Lory

The Dracula Tape
Fred Saberhagen

Dracula's Disciple
Robert Lory

'Salem's Lot
Stephen King

Vampirella #1: Bloodstalk
Ron Goulart

Vampirella #2: On Alien Wings
Ron Goulart

Village of Blood
Ian Dear

"Dies Irae"
Richard Howard

"Domdaniel"
Peter Allen

"Mirror Without Image"
James Turner

"To Claim His Own"
Crispin Derby

"V is for the Vordolake"
Harlan Ellison

1976
Blood Summer
Louise Cooper

Count Manfred
Miranda Seymour

"Certainly. I take what is easiest. Men have always been more accessible because women have been walled away like prizes or so physically impoverished by repeated childbearing as to be unhealthy prey for me. All this has begun to change recently, but gay men are still the simplest quarry." While she was recovering from her surprise at his unforeseen and weirdly skewed awareness of female history, he added suavely, "How carefully you control your expression, Dr. Landauer — no trace of disapproval."

She did disapprove, she realized. She would prefer him not to be committed sexually to men. Oh, hell.

He went on, "Yet no doubt you see me as one who victimizes the already victimized. This is the world's way. A wolf brings down the stragglers at the edges of the herd. Gay men are denied the full protection of the human herd and are at the same time emboldened to make themselves known and available.

"On the other hand, unlike the wolf I can feed without killing, and these particular victims pose no threat to me that would cause me to kill. Outcasts themselves, even if they comprehend my true purpose among them they cannot effectively accuse me."

God, how neatly, completely, and ruthlessly he distanced the homosexual community from himself! "And how do you feel, Edward, about their purposes — their sexual expectations of you?"

"The same way I feel about the sexual expectations of women whom I choose to pursue: they don't interest me. Besides, once my hunger is active, sexual arousal is impossible. My physical unresponsiveness seems to surprise no one. Apparently impotence is expected in a gray-haired man, which suits my intention."

Some kids carrying radios swung past below, trailing a jumble of amplified thump, wail, and jabber. Floria gazed after them unseeingly, thinking, astonished again, that she had never heard a man speak of his own impotence with such cool indifference. She had induced him to talk about his problem all right. He was speaking as freely as he had in the first session, only this time it was no act. He was drowning her in more than she had ever expected or for that matter wanted to know about vampirism. What the hell: she was listening, she thought she understood — what was it all good for? Time for some cold reality, she thought; see how far he can carry all this incredible detail. Give the whole structure a shove.

She said, "You realize, I'm sure, that people of either sex who make themselves so easily available are also liable to be carriers of disease. When was your last medical checkup?"

"My dear Dr. Landauer, my first medical checkup will be my last. Fortunately, I have no great need of one. Most serious illnesses — hepatitis, for example — reveal themselves to me by a quality in the odor of the victim's skin. Warned, I abstain. When I do fall ill, as occasionally happens, I withdraw to some place where I can heal undisturbed. A doctor's attentions would be more dangerous to me than any disease."

Eyes on the path below, he continued calmly, "You can see by looking at me that there are no obvious clues to my unique nature. But believe me, an examination of any depth by even a half-sleeping medical practitioner would reveal some alarming deviations from the norm. I take pains to stay healthy, and I seem to be gifted with an exceptionally hardy constitution."

Fantasies of being unique and physically superior; take him to the other pole. "I'd like you to try something now. Will you put yourself into the mind of a man you contact in the Ramble and describe your encounter with him from his point of view?"

He turned toward her and for some moments regarded her without expression. Then he resumed his surveillance of the path. "I will not. Though I do have enough empathy with my quarry to enable me to hunt efficiently. I must draw the line at erasing the necessary distance that keeps prey and predator distinct.

"And now I think our ways part for today." He stood up, descended the hillside, and walked beneath some low-canopied trees, his tall back stooped, toward the Seventy-second Street entrance of the park.

Floria arose more slowly, aware suddenly of her shallow breathing and the sweat on her face. Back to reality or what remained of it. She looked at her watch. She was late for her next client.

Floria couldn't sleep that night. Barefoot in her bathrobe she paced the living room by lamplight. They had sat together on that hill as isolated as in her office — more so, because there was no Hilda and no phone. He was, she knew, very strong, and he had sat close enough to her to reach out for that paralyzing touch to the neck —

Just suppose for a minute that Weyland had been brazenly telling the truth all along, counting on her to treat it as a delusion because on the face of it the truth was inconceivable. Jesus, she thought, if I'm thinking that way about him, this therapy is more out of control than I thought. What kind of therapist becomes an accomplice to the client's fantasy? A crazy therapist, that's what kind.

Frustrated and confused by the turmoil in her mind, she wandered into the workroom. By morning the floor was covered with sheets of newsprint, each broadly marked by her felt-tipped pen. Floria sat in the midst of them, gritty-eyed and hungry.

She often approached problems this way, harking back to art training: turn off the thinking, put hand to paper and see what the deeper, less verbally sophisticated parts of the mind have to offer. Now that her dreams had deserted her, this was her only access to those levels.

The newsprint sheets were covered with rough representations of Weyland's face and form. Across several of them were scrawled words: *"Dear Doug, your vampire is fine, it's your ex-therapist who's off the rails. Warning: Therapy can be dangerous to your health. Especially if you are the therapist. Beautiful vampire, awaken to me. Am I really ready to take on a legendary monster? Give up — refer this one out. Do your job — work is a good doctor."*

Dracula Began
Gail Kimberly

In Memory of Sarah Bailey
Louise Cooper (UK)

Interview with the Vampire
Anne Rice

The Monster Club
R. Chetwynd-Hayes

Sandworld
Richard A. Lupoff

The Space Vampires
Colin Wilson

The Vampires of Alfama
Pierre Kast (UK)

The Vampire of Mons
Desmond Stewart

Vampirella #3: Deadwalk
Ron Goulart

Vampirella #5: Deathgame
Ron Goulart

Vampirella #6: Snakegod
Ron Goulart

"Conversion"
Ramsey Campbell

"Night Life"
Steven Utley

1977

Count Dracula
Gerald Savory (UK)

Count Dracula's Canadian Affair
Otto Fredrick

The Dance of Blood
Stewart Farrar (UK)

Dracula Unborn
Peter Tremayne (UK)

Dracula's Daughter
Carl Dreadstone (UK)

Frankenstein Meets Dracula
Don Glut (UK)

That last one sounded pretty good, except that doing her job was precisely what she was feeling so shaky about these days.

Here was another message: *"How come this attraction to someone so scary?"* Oh ho, she thought, is that a real feeling or an aimless reaction out of the body's early-morning hormone peak? You don't want to confuse honest libido with mere biological clockwork.

Deborah called. Babies cried in the background over the Scotch Symphony. Nick, Deb's husband, was a musicologist with fervent opinions on music and nothing else.

"We'll be in town a little later in the summer," Deborah said, "just for a few days at the end of July. Nicky has this seminar-convention thing. Of course, it won't be easy with the babies . . . I wondered if you might sort of coordinate your vacation so you could spend a little time with them?"

Baby-sit, that meant. Damn. Cute as they were and all that, damn! Floria gritted her teeth. Visits from Deb were difficult. Floria had been so proud of her bright, hard-driving daughter, and then suddenly Deborah had dropped her studies and rushed to embrace all the dangers that Floria had warned her against: a romantic, too-young marriage, instant breeding, no preparation for self-support, the works. Well, to each her own, but it was so wearing to have Deb around playing the empty-headed hausfrau.

"Let me think, Deb. I'd love to see all of you, but I've been considering spending a couple of weeks in Maine with your Aunt Nonnie." God knows I need a real vacation, she thought, though the peace and quiet up there is hard for a city kid like me to take for long. Still, Nonnie, Floria's younger sister, was good company. "Maybe you could bring the kids up there for a couple of days. There's room in that great barn of a place, and of course Nonnie'd be happy to have you."

"Oh, no, Mom, it's so dead up there, it drives Nick crazy — don't tell Nonnie I said that. Maybe Nonnie could come down to the city instead. You could cancel a date or two and we could all go to Coney Island together, things like that."

Kid things, which would drive Nonnie crazy and Floria too before long. "I doubt she could manage," Floria said, "but I'll ask. Look, hon, if I do go up there, you and Nick and the kids could stay here at the apartment and save some money."

"We have to be at the hotel for the seminar," Deb said shortly. No doubt she was feeling just as impatient as Floria was by now. "And the kids haven't seen you for a long time — it would be really nice if you could stay in the city just for a few days."

"We'll try to work something out." Always working something out. Concord never comes naturally — first we have to butt heads and get pissed off. Each time you call I hope it'll be different, Floria thought.

Somebody shrieked for "oly," jelly that would be, in the background — Floria felt a sudden rush of warmth for them, her grandkids for God's sake. Having been a young mother herself, she was still young

Hounds of Dracula
Ken Johnson

Martin
George Romero and Susan Sparrow

The Vampire Tapes
Arabella Randolphe

"The Cardula Detective Agency"
Jack Ritchie

"One for the Road"
Stephen King

1978

The Black Castle
Les Daniel

Dracula's Cat
Jan Wahl

The Holmes-Dracula File
Fred Saberhagen

Hôtel Transylvania
Chelsea Quinn Yarbro

The Palace
Chelsea Quinn Yarbro

The Revenge of Dracula
Peter Tremayne (UK)

Sherlock Holmes vs. Dracula
Loren D. Estleman (UK)

Tabitha ffoulkes
John Linssen

"Cardula and the Kleptomaniac'
Jack Ritchie

"Cardula's Revenge"
Jack Ritchie

1979

Blood Games
Chelsea Quinn Yarbro

By Blood Alone
Bernhardt J. Hurwood

The Blooding
William Darrid

enough to really enjoy them (and to fight with Deb about how to bring them up).

Deb was starting an awkward goodbye. Floria replied, put the phone down, and sat with her head back against the flowered kitchen wallpaper, thinking, Why do I feel so rotten now? Deb and I aren't close, no comfort, seldom friends, though we were once. Have I said everything wrong, made her think I don't want to see her and don't care about her family? What does she want from me that I can't seem to give her? Approval? Maybe she thinks I still hold her marriage against her. Well, I do, sort of. What right have I to be critical, me with my divorce? What terrible things would she say to me, would I say to her, that we take such care not to say anything important at all?

"I think today we might go into sex," she said.

Weyland responded dryly, "Might we indeed. Does it titillate you to wring confessions of solitary vice from men of mature years?"

Oh no you don't, she thought. You can't sidestep so easily. "Under what circumstances do you find yourself sexually aroused?"

"Most usually upon waking from sleep," he said indifferently.

"What do you do about it?"

"The same as others do. I am not a cripple, I have hands."

"Do you have fantasies at these times?"

"No. Women, and men for that matter, appeal to me very little, either in fantasy or reality."

"Ah — what about female vampires?" she said, trying not to sound arch.

"I know of none."

Of course: the neatest out in the book. "They're not needed for reproduction, I suppose, because people who die of vampire bites become vampires themselves."

He said testily, "Nonsense. I am not a communicable disease."

So he had left an enormous hole in his construct. She headed straight for it: "Then how does your kind reproduce?"

"I have no kind, so far as I am aware," he said, "and I do not reproduce. Why should I, when I may live for centuries still, perhaps indefinitely? My sexual equipment is clearly only detailed biological mimicry, a form of protective coloration." How beautiful, how simple a solution, she thought, full of admiration in spite of herself. "Do I occasionally detect a note of prurient interest in your questions, Dr. Landauer? Something akin to stopping at the cage to watch the tigers mate at the zoo?"

"Probably," she said, feeling her face heat. He had a great backhand return shot there. "How do you feel about that?"

He shrugged.

"To return to the point," she said. "Do I hear you saying that you have no urge whatever to engage in sexual intercourse with anyone?"

"Would you mate with your livestock?"

His matter-of-fact arrogance took her breath away. She said weakly,

Bloodthirst
Mark Ronson (UK)

Dracula, Go Home
Kim Platt

Dracula in Love
John Shirley

Hell's Bitch
Anonymous

Nightworld
David Bischoff

Nosferatu, the Vampyre
Paul Monette

An Old Friend of the Family
Fred Saberhagen

The Silver Skull
Les Daniels

The Vampire Chase
Stephen Brett

Village of Vampires
Steven Otfinoski

"Blood Moon"
Thomas Owens

"The Lady of the House of Love"
Angela Carter

"Love-Starved"
Charles L. Grant

"Red as Blood"
Tanith Lee

"Spareen Among the Tartars"
Susan C. Petrey

1980

Bunnicula
Deborah Howe and James Howe

Dracula, My Love
Peter Tremayne (UK)

The Intrusion
David Combs

Night Whispers
Charles Veley

"Men have reportedly done so."

"Driven men. I am not driven in that way. My sex urge is of low frequency and is easily dealt with unaided — although I occasionally engage in copulation out of the necessity to keep up appearances. I am capable, but not — like humans — obsessed."

Was he sinking into lunacy before her eyes? "I think I hear you saying," she said, striving to keep her voice neutral, "that you're not just a man with a unique way of life. I think I hear you saying that you're not human at all."

"I thought that this was already clear."

"And that there are no others like you."

"None that I know of."

"Then — you see yourself as what? Some sort of mutation?"

"Perhaps. Or perhaps your kind are the mutation."

She saw disdain in the curl of his lip. "How does your mouth feel now?"

"The corners are drawn down. The feeling is contempt."

"Can you let the contempt speak?"

He got up and went to stand at the window, positioning himself slightly to one side as if to stay hidden from the street below.

"Edward," she said.

He looked back at her. "Humans are my food. I draw the life out of their veins. Sometimes I kill them. I am greater than they are. Yet I must spend my time thinking about their habits and their drives, scheming to avoid the dangers they pose — I hate them."

She felt the hatred like a dry heat radiating from him. God, he really lived all this! She had tapped into a furnace of feeling. And now? The sensation of triumph wavered, and she grabbed at a next move: hit him with reality now, while he's burning.

"What about blood banks?" she said. "Your food is commercially available, so why all the complication and danger of the hunt?"

"You mean I might turn my efforts to piling up a fortune and buying blood by the case? That would certainly make for an easier, less risky life in the short run. I could fit quite comfortably into modern society if I became just another consumer.

"However, I prefer to keep the mechanics of my survival firmly in my own hands. After all, I can't afford to lose my hunting skills. In two hundred years there may be no blood banks, but I will still need my food."

Jesus, you set him a hurdle and he just flies over it. Are there no weaknesses in all this, has he no blind spots? Look at his tension — go back to that. Floria said, "What do you feel now in your body?"

"Tightness." He pressed his spread fingers to his abdomen.

"What are you doing with your hands?"

"I put my hands to my stomach."

"Can you speak for your stomach?"

The Partaker
R. Chetwynd-Hayes

Sabella: Or the Blood Stone
Tanith Lee (UK)

A Taste for Power
Muriel Dobbin

Thorn
Fred Saberhagen

The Vampire Tapestry
Suzy McKee Charnas

Vegas Vampire
Jory Sherman

World Enough, and Time
James Kahn

"Opening a Vein"
Bill Pronzini and Barry N. Malzberg

1981

Blood Country
Curt Selby

Blood Shift
Garfield Reeves-Stevens (CANADA)

Citizen Vampire
Les Daniels

Crimson Kisses
Asa Drake

Death-Angel
J.N. Williamson

Death-Coach
J.N. Williamson

The Demon Lover
Maria Valdemi

Doctor Who and the State of Decay
Terrance Dicks

Dracula Sucks
Maxwell Kearny

Flowers of Evil
Robert Charles (UK)

The Hunger
Whitley Streiber

"'Feed me or die,'" he snarled.

Elated again, she closed in: "And for yourself, in answer?"

"'Will you never be satisfied?'" He glared at her. "You shouldn't seduce me into quarreling with the terms of my own existence!"

"Your stomach is your existence," she paraphrased.

"The gut determines," he said harshly. "That first, everything else after."

"Say, 'I resent . . .'"

He held to a tense silence.

"'I resent the power of my gut over my life,'" she said for him.

He stood with an abrupt motion and glanced at his watch, an elegant flash of slim silver on his wrist. "Enough," he said.

That night at home she began a set of notes that would never enter his file at the office, notes toward the proposed book.

> Couldn't do it, couldn't get properly into the sex thing with him. Everything shoots off in all directions. His vampire concept so thoroughly worked out, find myself half believing sometimes — my own childish fantasy-response to his powerful death-avoidance, contact-avoidance fantasy. Lose professional distance every time — is that what scares me about him? Don't really want to shatter his delusion (my life a mess, what right to tear down others' patterns?) — so see it as real? Wonder how much of "vampirism" he acts out, how far, how often. Something attractive in his purely selfish, predatory stance — the lure of the great outlaw.
>
> Told me today quite coolly about a man he killed recently — inadvertently — by drinking too much from him. *Is* it fantasy? Of course — the victim, he thinks, was college student. Breathes there a professor who hasn't dreamed of murdering some representative youth, retaliation for years of classroom frustration? Speaks of teaching with acerbic humor — amuses him to work at cultivating the minds of those he regards strictly as bodies, containers of his sustenance. He shows the alienness of full-blown psychopathology, poor bastard, plus clean-cut logic. Suggested he find another job (assuming his delusion at least in part related to pressures at Cayslin); his fantasy-persona, the vampire, more realistic than I about job-switching:
>
> "For a man of my apparent age it's not so easy to make such a change in these tight times. I might have to take a position lower on the ladder of 'success' as you people assess it." Status is important to him? "Certainly. An eccentric professor is one thing; an eccentric pipe-fitter, another. And I like good cars, which are expensive to own and run." Then, thoughtful addition, "Although there are advantages to a simpler, less visible life." He refuses to discuss other "jobs" from former "lives." We are deep into the fantasy — where the hell going? Damn right I don't control the

The Journal of Edwin Underhill
Peter Tonkin (UK)

The Keepsake
Paul Hudson

Ludlow's Mill
R.R. Walters

Moonlight Variations
Florence Stevenson

New Blood
Richard Salem (UK)

Path of the Eclipse
Chelsea Quinn Yarbro

They Thirst
Robert McCammon

Vampyr
Jan Jennings

"Kaeti's Nights"
Keith Roberts

"Pyotor's Story"
Spider Robinson

"Spareen Among the Cossacks"
Susan C. Petrey

"The Vampire of Mallworld"
S.P. Somtow

1982

Blood Will Have Blood
Linda Barnes

The Curse of the Vampire
Karl Alexander

The Dark Angel
Meredith Ann Pierce

The Delicate Dependency: A Novel of the Vampire Life
Michael Talbot

Dominion
Fred Saberhagen

Dracula's Diary
Michael Corby and Michael Geare

Dhampire
Scott Baker

"games" — preplanned therapeutic strategies get whirled away as soon as we begin. Nerve-wracking.

Tried again to have him take the part of his enemy-victim, peasant with torch. Asked if he felt himself rejecting that point of view? Frosty reply: "Naturally. The peasant's point of view is in no way my own. I've been reading in your field, Dr. Landauer. You work from the Gestalt orientation —" Originally yes, I corrected; eclectic now. "But you do proceed from the theory that I am projecting some aspect of my own feelings outward onto others, whom I then treat as my victims. Your purpose then must be to maneuver me into accepting as my own the projected 'victim' aspect of myself. This integration is supposed to effect the freeing of energy previously locked into maintaining the projection. All this is an interesting insight into the nature of ordinary human confusion, but I am not an ordinary human, and I am not confused. I cannot afford confusion." Felt sympathy for him — telling me he's afraid of having own internal confusions exposed in therapy, too threatening. Keep chipping away at delusion, though with what prospect? It's so complex, deep-seated.

Returned to his phrase "my apparent age." He asserts he has lived many human lifetimes, all details forgotten, however, during periods of suspended animation between lives. Perhaps sensing my skepticism at such handy amnesia, grew cool and distant, claimed to know little about the hibernation process itself: "The essence of this state is that I sleep through it — hardly an ideal condition for making scientific observations."

Edward thinks his body synthesizes vitamins, minerals (as all our bodies synthesize vitamin D), even proteins. Describes unique design he deduces in himself: special intestinal microfauna plus superefficient body chemistry extracts enough energy to live on from blood. Damn good mileage per calorie, too. (Recall observable tension, first interview, at question about drinking — my note on possible alcohol problem)

Speak for blood: "'Lacking me, you have no life. I flow to the heart's soft drumbeat through lightless prisons of flesh. I am rich, I am nourishing, I am difficult to attain.'" Stunned to find him positively lyrical on subject of his "food." Drew attention to whispering voice of blood. "'Yes. I am secret, hidden beneath the surface, patient, silent, steady. I work unnoticed, an unseen thread of vitality running from age to age — beautiful, efficient, self-renewing, self-cleansing, warm, filling —'" Could see *him* getting worked up. Finally he stood: "My appetite is pressing. I must leave you." And he did.

Death-Doctor
J.N. Williamson

Death-School
J.N. Williamson

Fevre Dream
George R.R. Martin

Miranda
Edylne Bond

The Soft Whisper of the Dead
Charles L. Grant

Tempting Fate
Chelsea Quinn Yarbro

Time's Dark Laughter
James Kahn

"Beyond Any Measure"
Karl Edward Wagner

"Cardula and the Locked Rooms"
Jack Ritchie

"Down Among the Dead Men"
Jack Dann and Gardner Dozois

"Following the Way"
Alan Ryan

"The Healer's Touch"
Susan C. Petrey

"The House at Evening"
Frances Garfield

"Leechcraft"
Susan C. Petrey

"A Night on the Docks"
Freff

"The Return of Cardula"
Jack Ritchie

"The Vampires Who Loved *Beowulf*"
Mary Caraker

1983
The Awakening
John Russo

The Dracula Murders
Philip Daniels

Sat and trembled for five minutes after.

New development (or new perception?): he sometimes comes across very unsophisticated about own feelings — lets me pursue subjects of extreme intensity and delicacy to him.

Asked him to daydream — a hunt. (Hands — mine — shaking now as I write. God. What a session.) He told of picking up a woman at poetry reading, 92nd Street Y — has N.Y.C. all worked out, circulates to avoid too much notice any one spot. Spoke easily, eyes shut without observable strain: chooses from audience a redhead in glasses, dress with drooping neckline (ease of access), no perfume (strong smells bother him). Approaches during the intermission, encouraged to see her fanning away smoke of others' cigarettes — meaning she doesn't smoke, health sign. Agreed in not enjoying the reading, they adjourn together to coffee shop.

"She asks whether I'm a teacher," he says, eyes shut, mouth amused. "My clothes, glasses, manner all suggest this, and I emphasize the impression — it reassures. She's a copy editor for a publishing house. We talk about books. The waiter brings her a gummy-looking pastry. As a non-eater, I pay little attention to the quality of restaurants, so I must apologize to her. She waves this away — is engrossed, or pretending to be engrossed, in talk." A longish dialog between interested woman and Edward doing shy-lonesome-scholar act — dead wife, competitive young colleagues who don't understand him, quarrels in professional journals with big shots in his field — a version of what he first told me. She's attracted (of course — lanky, rough-cut elegance plus hints of vulnerability all very alluring, as intended). He offers to take her home.

Tension in his body at this point in narrative — spine clear of chair back, hands braced on thighs. "She settles beside me in the back of the cab, talking about problems of her own career — illegible manuscripts of Biblical length, mulish editors, suicidal authors — and I make comforting comments, I lean nearer and put my arm along the back of the seat, behind her shoulders. Traffic is heavy, we move slowly. There is time to make my meal here in the taxi and avoid a tedious extension of the situation into her apartment — if I move soon."

How do you feel?

"Eager," he says, voice husky. "My hunger is so roused I can scarcely restrain myself. A powerful hunger, not like yours — mine compels. I embrace her shoulders lightly, make kindly-uncle remarks, treading that fine line between the game of seduction she perceives and the game of friendly interest I pretend to affect. My real purpose underlies all: what I say, how I look, every gesture is

The Dragon Waiting
John M. Ford

The Ice Maiden
Marc Behm

The Nursery
William W. Johnstone

"Aim for the Heart"
Craig Shaw Gardner

"Il Bacio (Il Chiave)"
Tanith Lee

"In Darkness, Angels"
Eric Van Lustbader

"Nunc Dimittis"
Tanith Lee

"Small Changes"
Susan C. Petrey

"Spareen and Old Turk"
Susan C. Petrey

"The Sunshine Club"
Ramsey Campbell

1984

Blood Sport
(Dark Forces #14)
R.C. Scott

Dracula Is a Pain in the Neck
Elizabeth Levy

Erebus
Shaun Hutson (UK)

The Fellowship
Aden F. Romine and
Mary C. Romine

I, Vampire
Jody Scott

Myth-ing Persons
Robert Asprin

Prisoner of Vampires
Nancy Garden

The Prisoner of Blackwood Castle
Ron Goulart

The Spy Who Drank Blood
Gordon Linzer

Vampire Junction
S.P. Somtow

"Bite-Me-Not, or Fleur de Feu"
Tanith Lee

"The Master of Rampling Gate"
Anne Rice

"The Men and Women of Rivendale"
Steve Rasnic Tem

"I Shall Not Leave England Now"
Alan Ryan

"A Surfeit of Melancholic Humors"
Sharon Farber

"Tooth Fairy"
Carol Carr

"The Vampire Lover"
Tanith Lee

1985
Blood Autumn
Kathryn Ptacek

Dinner at Deviant's Palace
Tim Power

Fright Night
John Skipp and Craig Spector

In the Drift
Michael Swanwick

Last Rites
Jorge Saralegui

The Saint-Germain Chronicles
Chelsea Quinn Yarbro

The Vampire Lestat
Anne Rice

"Credibility Problem"
Greg Cox

"Dayblood"
Roger Zelazny

part of the stalk. There is an added excitement, and fear, because I'm doing my hunting in the presence of a third person — behind the cabbie's head."

Could scarcely breathe. Studied him — intent face, masklike with closed eyes, nostrils slightly flared; legs tensed, hands clenched on knees. Whispering: "I press the place on her neck. She starts, sighs faintly, silently drops against me. In the stale stench of the cab's interior, with the ticking of the meter in my ears and the mutter of the radio — I take hold here, at the tenderest part of her throat. Sound subsides into the background — I feel the sweet blood beating under her skin, I taste salt at the moment before I — strike. My saliva thins her blood so that it flows out, I draw the blood into my mouth swiftly, swiftly, before she can wake, before we can arrive . . ."

Trailed off, sat back loosely in chair — saw him swallow. "Ah. I feed." Heard him sigh. Managed to ask about physical sensation. His low murmur, "Warm. Heavy, here —" touches his belly — "in a pleasant way. The good taste of blood, tart and rich, in my mouth . . ." And then? A flicker of movement beneath his closed eyelids: "In time I am aware that the cabbie has glanced back once and has taken our — embrace for just that. I can feel the cab slowing, hear him move to turn off the meter. I withdraw, I quickly wipe my mouth on my handkerchief. I take her by the shoulders and shake her gently; does she often have these attacks, I inquire, the soul of concern. She comes around, bewildered, weak, thinks she has fainted. I give the driver extra money and ask him to wait. He looks intrigued — 'What was that all about,' I can see the question in his face — but as a true New Yorker he won't expose his own ignorance by asking.

"I escort the woman to her front door, supporting her as she staggers. Any suspicion of me that she may entertain, however formless and hazy, is allayed by my stern charging of the doorman to see that she reaches her apartment safely. She grows embarrassed, thinks perhaps that if not put off by her 'illness' I would spend the night with her, which moves her to press upon me, unasked, her telephone number. I bid her a solicitous good night and take the cab back to my hotel, where I sleep."

No sex? No sex.

How did he feel about the victim as a person? "She was food."

This was his "hunting" of last night, he admits afterward, not a made-up dream. No boasting in it, just telling. Telling me! Think: I can go talk to Lucille, Mort, Doug, others about most of what matters to me. Edward has only me to talk to and that for a fee — what isolation! No wonder the stone, monumental face — only those long, strong lips (his point of contact, verbal

> and physical-in-fantasy, with world and with "food") are truly expressive. An exciting narration; uncomfortable to find I felt not only empathy but enjoyment. Suppose he picked up and victimized — even in fantasy — Deb or Hilda, how would I feel then?
>
> *Later:* truth — I also found this recital sexually stirring. Keep visualizing how he looked finishing this "dream" — he sat very still, head up, look of thoughtful pleasure on his face. Like handsome intellectual listening to music.

Kenny showed up unexpectedly at Floria's office on Monday, bursting with malevolent energy. She happened to be free, so she took him — something was definitely up. He sat on the edge of his chair.

"I know why you're trying to unload me," he accused. "It's that new one, the tall guy with the snooty look — what is he, an old actor or something? Anybody could see he's got you itching for him."

"Kenny, when was it that I first spoke to you about terminating our work together?" she said patiently.

"Don't change the subject. Let me tell you, in case you don't know it: that guy isn't really interested, Doctor, because he's a fruit. A faggot. You want to know how I know?"

Oh Lord, she thought wearily, he's regressed to age ten. She could see that she was going to hear the rest whether she wanted to or not. What in God's name was the world like for Kenny, if he clung so fanatically to her despite her failure to help him?

"Listen, I knew right away there was something flaky about him, so I followed him from here to that hotel where he lives. I followed him the other afternoon too. He walked around like he does a lot, and then he went into one of those ritzy movie houses on Third that open early and show risqué foreign movies — you know, Japs cutting each other's things off and glop like that. This one was French, though.

"Well, there was a guy came in, a Madison Avenue type carrying his attaché case, taking a work break or something. Your man moved over and sat down behind him and reached out and sort of stroked the guy's neck, and the guy leaned back, and your man leaned forward and started nuzzling at him, you know — kissing him.

"I saw it. They had their heads together and they stayed like that a while. It was disgusting: complete strangers, without even 'hello.' The Madison Avenue guy just sat there with his head back looking zonked, you know, just swept away, and what he was doing with his hands under his raincoat in his lap I couldn't see, but I bet you can guess.

"And then your fruity friend got up and walked out. I did, too, and I hung around a little outside. After a while the Madison Avenue guy came out looking all sleepy and loose, like after you-know-what, and he wandered off on his own someplace.

"What do you think now?" he ended, on a high, triumphant note.

"The Yougoslaves"
Robert Bloch

"Kiss the Vampire Goodbye"
Alan Ryan

1986

In Hot Blood
Petru Popescu

Legacy of Lehr
Katherine Kurtz

The Lord Count Drakulya
P.C. Doherty (UK)

Necroscape
Brian Lumley (UK)

The Prince Drakulya
P.C. Doherty (UK)

The Light at the End
John Skipp and Craig Spector

Yellow Fog
Les Daniels

"Glory"
Ron Goulart

"Night Kings"
Roger Zelazny

"Return of the Dust Vampires"
Sharon Farber

"Something in the Blood"
Richard R. Purtill

"Vampire"
Richard Christian Matheson

1987

Blood Hunt
Lee Killough

Blood Mist
Robert James

Bloodthirst
(*Star Trek #37*)
J.M. Dillard

The Curse of the Obelisk
Ron Goulart

Her impulse was to slap his face the way she would have slapped Deb-as-a-child for tattling. But this was a client, not a kid. God give me strength, she thought.

"Kenny, you're fired."

"You can't!" he squealed. "You can't! What will I — who can I —"

She stood up, feeling weak but hardening her voice. "I'm sorry. I absolutely cannot have a client who makes it his business to spy on other clients. You already have a list of replacement therapists from me."

He gaped at her in slackjawed dismay, his eyes swimmy with tears.

"I'm sorry, Kenny. Call this a dose of reality therapy and try to learn from it. There are some things you simply will not be allowed to do." She felt better: it was done at last.

"I hate you!" He surged out of his chair, knocking it back against the wall. Threateningly he glared at the fish tank, but, contenting himself with a couple of kicks at the nearest table leg, he stamped out.

Floria buzzed Hilda: "No more appointments for Kenny, Hilda. You can close his file."

"Whoopee," Hilda said.

Poor, horrid Kenny. Impossible to tell what would happen to him, better not to speculate or she might relent, call him back. She had encouraged him, really, by listening instead of shutting him up and throwing him out before any damage was done.

Was it damaging, to know the truth? In her mind's eye she saw a cream-faced young man out of a Black Thumb Vodka ad wander from a movie theater into daylight, yawning and rubbing absently at an irritation on his neck . . .

She didn't even look at the telephone on the table or think about whom to call, now that she believed. No; she was going to keep quiet about Dr. Edward Lewis Weyland, her vampire.

> Hardly alive at staff meeting, clinic, yesterday — people asking what's the matter, fobbed them off. Settled down today. Had to, to face him.
>
> Asked him what he felt were his strengths. He said speed, cunning, ruthlessness. Animal strengths, I said. What about imagination, or is that strictly human? He defended at once: not human only. Lion, waiting at water hole where no zebra yet drinks, thinks "Zebra — eat," therefore performs feat of imagining event yet-to-come. Self experienced as animal? Yes — reminded me that humans are also animals. Pushed for his early memories; he objected: "Gestalt is here-and-now, not history-taking." I insist, citing anomalous nature of his situation, my own refusal to be bound by any one theoretical framework. He defends tensely: "Suppose I became lost there in memory, distracted from dangers of the present, left unguarded from those dangers." Speak for

A Flame in Byzantium
Chelsea Quinn Yarbro

The Goldcamp Vampire, or, the Sanguinary Sourdough
Elizabeth Scarborough

The Devouring
F. W. Armstrong

How Dear the Dawn
Marc Eliot

The House of Dracula
R. Chetwynd-Hayes (UK)

Live Girls
Ray Garton

The Lost Boys
Craig Shaw Gardner

Sweet Silver Blues
Glen Cook

Timefall
James Kahn

"Blood"
William Relling, Jr.

"The Bloodsnare"
Bruce Chandler Fergusson

"Ever After"
Susan Palwick

"Ice Dreams"
Sharon Farber

"Princess of Darkness"
Frederick Cowles (UK)

"Popsy"
Stephen King

"Under Her Skin"
Susan Casper

1988
Black Ambrosia
Elizabeth Engstrom

Blood Farm: An Iowa Gothic
Sam Siciliano

memory. He resists, but at length attempts it: "'I am heavy with the multitudes of the past.'" Fingertips to forehead, propping up all that weight of lives. "'So heavy, filling worlds of time laid down eon by eon, I accumulate, I persist, I demand recognition, I am as real as the life around you — more real, weightier, richer.'" His voice sinking, shoulders bowed, head in hands — I begin to feel pressure at the back of my own skull. "'Let me in.'" Only a rough whisper now. "'I offer beauty as well as terror. Let me in.'" Whispering also, I suggest he reply to his memory.

"Memory, you want to crush me," he groans. "You would overwhelm me with the cries of animals, the odor and jostle of bodies, old betrayals, dead joys, filth and anger from other times — I must concentrate on the danger now. Let me be." All I can take of this crazy conflict, I gabble us off onto something else. He looks up — relief? — follows my lead — where? Rest of session a blank.

No wonder sometimes no empathy at all — a species boundary! He has to be utterly self-centered just to keep balance — self-centeredness of an animal. Thought just now of our beginning, me trying to push him to produce material, trying to control him, manipulate — no way, no way; so here we are, someplace else — I feel dazed, in shock, but stick with it — it's real.

Therapy with a dinosaur, a Martian.

"You call me 'Weyland' now, not 'Edward.'" I said first name couldn't mean much to one with no memory of being called by that name as a child, silly to pretend it signifies intimacy where it can't. I think he knows now that I believe him. Without prompting, told me truth of disappearance from Cayslin. No romance; he tried to drink from a woman who worked there, she shot him, stomach and chest. Luckily for him, small-caliber pistol, and he was wearing a lined coat over three-piece suit. Even so, badly hurt. (Midsection stiffness I noted when he first came — he was still in some pain at that time.) He didn't "vanish" — fled, hid, was found by questionable types who caught on to what he was, sold him "like a chattel" to someone here in the city. He was imprisoned, fed, put on exhibition — very privately — for gain. Got away. "Do you believe any of this?" Never asked anything like that before, seems of concern to him now. I said my belief or lack of same was immaterial; remarked on hearing a lot of bitterness.

He steepled his fingers, looked brooding at me over tips: "I nearly died there. No doubt my purchaser and his diabolist friend still search for me. Mind you, I had some reason at first to be glad of the attentions of the people who kept me prisoner. I was in no condition to fend for myself. They brought me food and kept me hidden and sheltered, whatever their motives. There are always advantages..."

Bloodlinks
Lee Killough

Crusader's Torch
Chelsea Quinn Yarbro

The Dracula Caper
(*Time Wars #8*)
Simon Hawke

The House of Caine
Ken Eulo

In Silence Sealed
Kathryn Ptacek

Lifeblood
Lee Duigon

Monastery
Patrick Whalen

Natural History
Joan Perucho

Necroscape II: Wamphyri!
Brian Lumley (UK)

The Queen of the Damned
Anne Rice

Raw Pain Max
Dean Andersson

This Rough Magic
Eather Graham Pozzessere

Those of My Blood
Jacqueline Lichtenberg

Those Who Hunt the Night
Barbara Hambly

Vessel
John Tigges

"Aid from a Vampire"
Ian Watson

"Child of an Ancient City"
Tad Williams

"Fortress Memory"
Greg Cox

"The Man Who Loved the Vampire Lady"
Brian Stableford

Silence today started a short session. Hunting poor last night, Weyland still hungry. Much restless movement, watching goldfish darting in tank, scanning bookshelves. Asked him to be books. " 'I am old and full of knowledge, well made to last long. You see only the title, the substance is hidden. I am a book that stays closed.' " Malicious twist of the mouth not quite a smile: "This is a good game." Is he feeling threatened, too — already "opened" too much to me? Too strung out with him to dig when he's skimming surfaces that should be probed. Don't know how to *do* therapy with Weyland — just have to let things happen, hope it's good. But what's "good"? Aristotle? Rousseau? Ask Weyland what's good, he'll say "Blood."

Everything in a spin — these notes too confused, too fragmentary — worthless for a book, just a mess, like me, my life. Tried to call Deb last night, cancel visit. Nobody home, thank God. Can't tell her to stay away — but damn it — do not need complications now!

Floria went down to Broadway with Lucille to get more juice, cheese and crackers for the clinic fridge. This week it was their turn to do the provisions, a chore that rotated among the staff. Their talk about grant proposals for the support of the clinic trailed off.

"Let's sit a minute," Floria said. They crossed to a traffic island in the middle of the avenue. It was a sunny afternoon, close enough to lunchtime so that the brigade of old people who normally occupied the benches had thinned out. Floria sat down and kicked a crumpled beer can and some greasy fast-food wrappings back under the bench.

"You look like hell but wide awake at least," Lucille commented. "Things are still rough," Floria said. "I keep hoping to get my life under control so I'll have some energy left for Deb and Nick and the kids when they arrive, but I can't seem to do it. Group was awful last night — a member accused me afterward of having abandoned them all. I think I have, too. The professional messes and the personal are all related somehow, they run into each other. I should be keeping them apart so I can deal with them separately, but I can't. I can't concentrate, my mind is all over the place. Except with Dracula, who keeps me riveted with astonishment when he's in the office and bemused the rest of the time."

A bus roared by, shaking the pavement and the benches. Lucille waited until the noise faded. "Relax about the group. The others would have defended you if you'd been attacked during the session. They all understand, even if you don't seem to: it's the summer doldrums, people don't want to work, they expect you to do it all for them. But don't push so hard. You're not a shaman who can magic your clients back into health."

Floria tore two cans of juice out of a six-pack and handed one to her. On a street corner opposite, a violent argument broke out in typewriter-fast

"My Brother's Keeper"
Pat Cadigan

"Saved from the Shredder"
Charles Sheffield

"Valentine for a Vampire"
Daniel Ransom

1989
Blood Legacy
Prudence Foster

Blood of the Impaler
Jefferey Sackett

Blood Thirst
L.A. Freed

Carrion Comfort
Dan Simmons

A Candle for D'Artagnan
Chelsea Quinn Yarbro

Dark Souls
Barry Porter

Dreamspy
Jacqueline Lichtenberg

A Matter of Taste
Fred Saberhagen

Monday's Child
Patricia Wallace

Necroscape II: The Source
Brian Lumley (UK)

Night Brothers
Sidney Williams

The Night Prophets
Paul F. Olson

Nightshade
Jack Butler

Sunglasses after Dark
Nancy A. Collins

Vampire Nights
Timothy Moriarty

"A Child of Darkness"
Susan Casper

Spanish between two women. Floria sipped tinny juice and watched. She'd seen a guy last winter straddle another on that same corner and try to smash his brains out on the icy sidewalk. The old question again: What's crazy, what's health?

"It's a good thing you dumped Chubs, anyhow," Lucille said. "I don't know what finally brought that on, but it's definitely a move in the right direction. What about Count Dracula? You don't talk about him much anymore. I thought I diagnosed a yen for his venerable body."

Floria shifted uncomfortably on the bench and didn't answer. If only she could deflect Lucille's sharp-eyed curiosity.

"Oh," Lucille said. "I see. You really are hot — or at least warm. Has he noticed?"

"I don't think so. He's not on the lookout for that kind of response from me. He says sex with other people doesn't interest him, and I think he's telling the truth."

"Weird," Lucille said. "What about *Vampire on My Couch*? Shaping up all right?"

"It's shaky, like everything else. I'm worried that I don't know how things are going to come out. I mean, Freud's wolf-man case was a success, as therapy goes. Will my vampire case turn out successfully?"

She glanced at Lucille's puzzled face, made up her mind, and plunged ahead. "Luce, think of it this way: suppose, just suppose, that my Dracula is for real, an honest-to-God vampire —"

"Oh *shit!*" Lucille erupted in anguished exasperation. "Damn it, Floria, enough is enough — will you stop futzing around and get some help? Coming to pieces yourself and trying to treat this poor nut with a vampire fixation — how can you do him any good? No wonder you're worried about his therapy!"

"Please, just listen, help me think this out. My purpose can't be to cure him of what he is. Suppose vampirism isn't a defense he has to learn to drop? Suppose it's the core of his identity? Then what do I do?"

Lucille rose abruptly and marched away from her through a gap between the rolling waves of cabs and trucks. Fioria caught up with her on the next block.

"Listen, will you? Luce, you see the problem? I don't need to help him see who and what he is, he knows that perfectly well, and he's not crazy, far from it —"

"Maybe not," Lucille said grimly, "but you are. Don't dump this junk on me outside of office hours, Floria. I don't spend my time listening to nut-talk unless I'm getting paid."

"Just tell me if this makes psychological sense to you: he's healthier than most of us because he's always true to his identity, even when he's engaged in deceiving others. A fairly narrow, rigorous set of requirements necessary to his survival — that is his identity, and it commands him completely. Anything extraneous could destroy him. To go on living, he

"Dirty Work"
Pat Cadigan

"Good Kids"
Edward Bryant

"The Janfia Tree"
Tanith Lee

"L'Chaim"
Harvey Jacobs

"Nocturne"
Steve Rasnic Tem

"The Silver Collar"
Garry Kilworth

"Time Lapse"
Joe Haldeman

"...To Feel Another's Woe"
Chet Williamson

"Varicose Worms"
Scott Baker

1990

Anno Dracula
Kim Newman (UK)

The Bargain
Jon Ruddy

Blood Alone
Elaine Bergstrom

Blood Circle
P.N. Elrod

The Blood of Roses
Tanith Lee (UK)

Bloodlist
P.N. Elrod

Blood Lust
Ron Dee

The Empire of Fear
Brian Stableford

The Gift
Scott Edelman

I, Vampire
Michael Romkey

has to act solely out of his own undistorted necessity, and if that isn't authenticity, what is? So he's healthy, isn't he?" She paused, feeling a sudden lightness in herself. "And that's the best sense I've been able to make of this whole business so far."

They were in the middle of the block. Lucille, who could not on her short legs outwalk Floria, turned on her suddenly. "What the hell do you think you're doing, calling yourself a therapist? For God's sake, Floria, don't try to rope me into this kind of professional irresponsibility. You're just dipping into your client's fantasies instead of helping him to handle them. That's not therapy, it's collusion. Have some sense! Admit you're over your head in troubles of your own, retreat to firrner ground — go get treatment for yourself!"

Floria angrily shook her head. When Lucille turned away and hurried on up the block toward the clinic, Floria let her go without trying to detain her.

> Thought about Lucille's advice. After my divorce going back into therapy for a while did help, but now? Retreat again to being a client, like old days in training — so young, inadequate, defenseless then. Awful prospect. And I'd have to hand over W. to somebody else — who? I'm not up to handling him, can't cope, too anxious, yet with all that we do good therapy together somehow. I can't control, can only offer; he's free to take, refuse, use as suits, as far as he's willing to go. I serve as resource while he does own therapy — isn't that therapeutic ideal, free of "shoulds," "shouldn'ts"?
>
> Saw ballet with Mort, lovely evening — time out from W. — talking, singing, pirouetting all the way home, feeling safe as anything in the shadow of Mort-mountain; rolled later with that humming (off-key), sun-warm body. Today W. says he saw me at Lincoln Center last night, avoided me because of Mort. W. is ballet fan! Started attending to pick up victims, now also because dance puzzles and pleases.
>
> "When a group dances well, the meaning is easy — the dancers make a visual complement to the music, all their moves necessary, coherent, flowing. When a gifted soloist performs, the pleasure of making the moves is echoed in my own body. The soloist's absorption is total, much like my own in the actions of the hunt. But when a man and a woman dance together, something else happens. Sometimes one is hunter, one is prey, or they shift these roles between them. Yet some other level of significance exists — I suppose to do with sex — and I feel it — a tugging sensation, here —" touched his solar plexus — "but I do not understand it."
>
> Worked with his reactions to ballet. The response he feels to pas de deux is a kind of pull, "like hunger but not hunger." Of

Lifeblood
P. N. Elrod

Midnight Mass
F. Paul Wilson

Necroscape IV: Deadspeak
Brian Lumley (UK)

Nightblood
T. Chris Martindale

Nightmare People
Lawrence Watt-Evans

Out of the House of Life
Chelsea Quinn Yarbro

The Stake
Richard Laymon

Tempter
Nancy A. Collins

The Twelfth Child
Raymond Van Over

Vampire Notes
Robert Arthur Smith

Vampire Winter
Lois Tilton

Vampire$
John Streakley

1991
Art in the Blood
P.N. Elrod

Blood on the Water
P.N. Elrod

Blood Rites
Elaine Bergstrom

Dracula Unbound
Brian W. Aldiss

Fire in the Blood
P.N. Elrod

The Gilda Stories
Jewelle Gomez

"Home by the Sea"
Pat Cadigan

course he's baffled — Balanchine writes that the pas de deux is always a love story between man and woman. W. isn't man, isn't woman, yet the drama connects. His hands hovering as he spoke, fingers spread toward each other. Pointed this out. Body work comes easier to him now: joined his hands, interlaced fingers, spoke for hands without prompting: "'We are similar, we want the comfort of like closing to like.'" How would that be for him, to find — likeness, another of his kind? "Female?" Starts impatiently explaining how unlikely this is — No, forget sex and pas de deux for now; just to find your like, another vampire.

He springs up, agitated now. There are none, he insists; adds at once, "But what would it be like? What would happen? I fear it!" Sits again, hands clenched. "I long for it."

Silence. He watches goldfish, I watch him. I withhold fatuous attempt to pin down this insight, if that's what it is — what can I know about his insight? Suddenly he turns, studies me intently till I lose my nerve, react, cravenly suggest that if I make him uncomfortable he might wish to switch to another therapist —

"Certainly not." More follows, all gold: "There is value to me in what we do here, Dr. Landauer, much against my earlier expectations. Although people talk appreciatively of honest speech they generally avoid it, and I myself have found scarcely any use for it at all. Your straightforwardness with me — and the straightforwardness you require in return — this is healthy in a life so dependent on deception as mine."

Sat there, wordless, much moved, thinking of what I don't show him — my upset life, seat-of-pants course with him and attendant strain, attraction to him — I'm holding out on him while he appreciates my honesty.

Hesitation, then lower-voiced, "Also, there are limits on my methods of self-discovery, short of turning myself over to a laboratory for vivisection. I have no others like myself to look at and learn from. Any tools that may help are worth much to me, and these games of yours are — potent." Other stuff besides, not important. Important: he moves me and he draws me and he keeps on coming back. Hang in if he does.

Bad night — Kenny's aunt called: no bill from me this month, so if he's not seeing me who's keeping an eye on him, where's he hanging out? Much implied blame for what *might* happen. Absurd, but shook me up: I did fail Kenny. Called off group this week also; too much.

No, it was a *good* night — first dream in months I can recall, contact again with own depths — but disturbing. Dreamed myself in cab with W. in place of the woman from the Y. He put his hand

The Hunt
T. Lucien Wright

"The Impaler in Love"
Rick Wilber

"Infidel"
Thomas Tessier

"Lifeblood"
Jack Womack

Lot Lizards
Ray Garton

"M is for the Many Things"
Elizabeth Massie

"The Moose Church"
Jonathan Carroll

"Mr. Rinaldi's Angel"
Thomas Ligotti

Necroscape V: Deadspawn
Brian Lumley (UK)

Night Thirst
Patrick Whalen

No Blood Spilled
Les Daniels

"Now I Lay Me Down to Sleep"
Suzy McKee Charnas

Obsession
Lori Herter

Out for Blood
John Peyton Cooke

"The Pool People"
Melissa Mia Hall

"The Ragthorn"
Robert Holdstock and Garry Kilworth

"The Slug"
Karl Edward Wagner

"Teratisms"
Kathe Koja

"True Love"
K.W. Jeter

not on my neck but breast — I felt intense sensual response in the dream, also anger and fear so strong they woke me.

Thinking about this: anyone leans toward him sexually, to him a sign his hunting technique has maneuvered prospective victim into range, maybe arouses his appetite for blood. *I don't want that.* "She was food." I am not food, I am a person. No thrill at languishing away in his arms in a taxi while he drinks my blood — that's disfigured sex, masochism. My sex response in dream signaled to me I would be his victim — I rejected that, woke up.

Mention of *Dracula* (novel). W. dislikes: meandering, inaccurate, those absurd fangs. Says he himself has a sort of needle under his tongue, used to pierce skin. No offer to demonstrate, and no request from me. I brightly brought up historical Vlad Dracul — celebrated instance of Turkish envoys who, upon refusing to uncover to Vlad to show respect, were killed by spiking their hats to their skulls. "Nonsense," snorts W. "A clever ruler would use very small thumbtacks and dismiss the envoys to moan about the streets of Varna holding their tacked heads." First spontaneous play he's shown — took head in hands and uttered plaintive groans, "Ow, oh, ooh." I cracked up. W. reverted at once to usual dignified manner: "You can see that this would serve the ruler much more effectively as an object lesson against rash pride."

Later, same light vein: "I know why I'm a vampire; why are you a therapist?" Off balance as usual, said things about helping, mental health, etc. He shook his head: "And people think of a vampire as arrogant! You want to perform cures in a world which exhibits very little health of any kind — and it's the same arrogance with all of you. This one wants to be President or Class Monitor or Department Chairman or Union Boss, another must be first to fly to the stars or to transplant the human brain, and on and on. As for me, I wish only to satisfy my appetite in peace."

And those of us whose appetite is for competence, for effectiveness? Thought of Green, treated eight years ago, went on to be indicted for running a hellish "home" for aged. I had helped him stay functional so he could destroy the helpless for profit.

W. not my first predator, only most honest and direct. Scared; not of attack by W., but of process we're going through. I'm beginning to be up to it (?), but still — utterly unpredictable, impossible to handle or manage. Occasional stirrings of inward choreographer that used to shape my work so surely. Have I been afraid of that, holding it down in myself, choosing mechanical manipulation instead? Not a choice with W. — thinking no good, strategy no good, nothing left but instinct, clear and uncluttered responses if I can find them. Have to be my own authority with him, as he is always his own authority with a world in which he's

Vampire Beat
Vincent Courtenay

Vampire Blood
Kathryn Meyer Griffith

The Vampire Diaries
L.J. Smith

The Vampire Memoirs
Mara McCuniff and Traci Biery

Vampires Anonymous
Jeffrey M. McMahan

"A Week in the Unlife"
David J. Schow

1992
Blood Brothers
Brian Lumley (UK)

Blood Brothers
T. Lucien Wright

Bram Stoker's Dracula
James V. Hart and Fred Saberhagen

Children of the Night
Dan Simmons

Dance of the Dead
Christie Golden

Dark Dance
Tanith Lee (UK)

Daughter of the Night
Elaine Bergstrom

Kiss of the Vampire
Lee Weathersby

Less than Human
Gary Raisor

Liquid Diet
William Tedford

Lost Souls
Poppy Z. Brite

The Need
Andrew Niederman

A Question of Time
Fred Saberhagen

unique. So work with W. not just exhausting — exhilarating too, along with strain, fear.

Am I growing braver? Not much choice. Park again today (air-conditioning out at office). Avoiding Lucille's phone calls from clinic (very reassuring that she calls despite quarrel, but don't want to take all this up with her again). Also meeting W. in open feels saner somehow — wild creatures belong outdoors? Sailboat pond N. of 72nd, lots of kids, garbage, one beautiful tall boat drifting. We walked.

W. maintains he remembers no childhood, no parents. I told him my astonishment, confronted by someone who never had a life of the previous generation (even adopted parent) shielding him from death — how naked we stand when the last shield falls. Got caught in remembering a death dream of mine, dream it now and then — couldn't concentrate, got scared, spoke of it — a dog tumbled under a passing truck, ejected to side of the road where it lay unable to move except to lift head and shriek; couldn't help. Shaking nearly to tears — remembered Mother got into dream somehow — had blocked that at first. Didn't say it now. Tried to rescue situation, show W. how to work with a dream (sitting in vine arbor near band shell, some privacy).

He focused on my obvious shakiness: "The air vibrates constantly with the death cries of countless animals large and small. What is the death of one dog?" Leaned close, speaking quietly, instructing. "Many creatures are dying in ways too dreadful to imagine. I am part of the world; I listen to the pain. You people claim to be above all that. You deafen yourselves with your own noise and pretend there's nothing else to hear. Then these screams enter your dreams, and you have to seek therapy because you have lost the nerve to listen."

Remembered myself, said, Be a dying animal. He refused: "You are the one who dreams this." I had a horrible flash, felt I was the dog — helpless, doomed, hurting — burst into tears. The great therapist, bringing her own hangups into session with client! Enraged with self, which did not help stop bawling.

W. disconcerted, I think; didn't speak. People walked past, glanced over, ignored us. W. said finally, "What is this?" Nothing, just the fear of death. "Oh, the fear of death. That's with me all the time. One must simply get used to it." Tears into laughter. Goddamn wisdom of the ages. He got up to go, paused: "And tell that stupid little man who used to precede me at your office to stop following me around. He puts himself in danger that way." Kenny, damn it! Aunt doesn't know where he is, no answer on his phone. Idiot!

The Season of Passage
Christopher Pike

Soul Snatchers
Michael Cecilione

Sweet Blood
Pat Graverson

The Tale of the Body Thief
Anne Rice

A Taste of Blood Wine
Freda Warrington (UK)

Thirst of the Vampires
T. Lucien Wright

Valentine
S.P. Somtow

Young Blood
Brian Stableford (UK)

1993

Afterage
Yvonne Navarro

Bats
William W. Johnstone

Better in the Dark
Chelsea Quinn Yarbro

Blind Hunger
David Drake

Blood
Ron Dee

Blood Feud
Sam Siciliano

The Book of Common Dread
Brent Monohan

Cold Allies
Patricia Anthony

Darker Jewels
Chelsea Quinn Yarbro

The Darker Passions: Dracula
Amarantha Knight

Darkness on the Ice
Lois Tilton

Sketching all night — useless. W. beautiful beyond the scope of line — the beauty of singularity, cohesion, rooted in absolute devotion to demands of his specialized body. In feeding (woman in taxi), utter absorption one wants from a man in sex — no score-keeping, no fantasies, just hot urgency of appetite, of senses, the moment by itself.

His sleeves worn rolled back today to the elbows — strong, sculptural forearms, the long bones curved in slightly, suggest torque, leverage. How old?

Endurance: huge, rich cloak of time flows back from his shoulders like wings of a dark angel. All springs from, elaborates, the single, stark, primary condition: he is a predator who subsists on human blood. Harmony, strength, clarity, magnificence — all from that basic animal integrity. Of course I long for all that, here in the higgledy-piggledy hodgepodge of my life! Of course he draws me!

Wore no perfume today, deference to his keen, easily insulted sense of smell. He noticed at once, said curt thanks. Saw something bothering him, opened my mouth seeking desperately for right thing to say — up rose my inward choreographer, wide awake, and spoke plain from my heart: thinking on my floundering in some of our sessions — I am aware that you see this confusion of mine. I know you see by your occasional impatient look, sudden disengagement — yet you continue to reveal yourself to me (even shift our course yourself if it needs shifting and I don't do it). I think I know why. Because there's no place for you in world as you truly are. Because beneath your various facades your true self suffers; like all true selves, it wants, needs to be honored as real and valuable through acceptance by another. I try to be that other, but often you are beyond me.

He rose, paced to window, looked back, burning at me. "If I seem sometimes restless or impatient, Dr. Landauer, it's not because of any professional shortcomings of yours. On the contrary — you are all too effective. The seductiveness, the distraction of our — human contact worries me. I fear for the ruthlessness that keeps me alive."

Speak for ruthlessness. He shook his head. Saw tightness in shoulders, feet braced hard against floor. Felt reflected tension in my own muscles.

Prompted him: "'I resent . . .'"

"I resent your pretension to teach me about myself! What will this work that you do here make of me? A predator paralyzed by an unwanted empathy with his prey? A creature fit only for a cage and keeper?" He was breathing hard, jaw set. I saw suddenly the truth of his fear: his integrity is not human, but my work is

Domination
Michael Cecilione

Forever and the Night
Linda Lael Miller

The Golden
Lucius Shepard

I Am Dracula
Dean C. Anderson

Insatiable
David Dvorkin

I, Strahd: The Memoirs of a Vampire
P.N. Elrod

King of the Grey
Richard J. Knaak

Knights of the Blood
Scott MacMillan

The Last Aerie
Brian Lumley

The Making of a Monster
Gail Petersen

McLendon's Syndrome
Robert Frezza

Night Blood
Eric Flanders

Nighthag
Michael David

The Night Inside: A Vampire Thriller
Nancy Baker (UK)

The Perfect Kiss
Amanda Stevens

Personal Darkness
Tanith Lee (UK)

The Phallus of Osiris
Valentina Cilescu (UK)

Shadows After Dark
Ouida Crozier

Shadows in Their Blood
Marian Babson (UK)

specifically human, designed to make humans more human — what if it does that to him? Should have seen it before, should have seen it. No place left to go: had to ask him, in small voice, Speak for my pretension.

"No!" Eyes shut, head turned away.

Had to do it: Speak for me.

W. whispered, "As to the unicorn, out of your own legends — 'Unicorn, come lay your head in my lap while the hunters close in. You are a wonder, and for love of wonder I will tame you. You are pursued, but forget your pursuers, rest under my hand till they come and destroy you.'" Looked at me like steel: "Do you see? The more you involve yourself in what I am, the more you become the peasant with the torch!"

Two days later Doug came into town and had lunch with Floria.

He was a man of no outstanding beauty who was nevertheless attractive: he didn't have much chin and his ears were too big, but you didn't notice because of his air of confidence. His stability had been earned the hard way — as a gay man facing the straight world. Some of his strength had been attained with effort and pain in a group that Floria had run years earlier. A lasting affection had grown between herself and Doug. She was intensely glad to see him.

They ate near the clinic. "You look a little frayed around the edges," Doug said. "I heard about Jane Fennerman's relapse — too bad."

"I've only been able to bring myself to visit her once since."

"Feeling guilty?"

She hesitated, gnawing on a stale breadstick. The truth was, she hadn't thought of Jane Fennerman in weeks. Finally she said, "I guess I must be."

Sitting back with his hands in his pockets, Doug chided her gently. "It's got to be Jane's fourth or fifth time into the nuthatch, and the others happened when she was in the care of other therapists. Who are you to imagine — to demand — that her cure lay in your hands? God may be a woman, Floria, but She is not you. I thought the whole point was some recognition of individual responsibility — you for yourself, the client for himself or herself."

"That's what we're always saying," Floria agreed. She felt curiously divorced from this conversation. It had an old-fashioned flavor: Before Weyland. She smiled a little.

The waiter ambled over. She ordered bluefish. The serving would be too big for her depressed appetite, but Doug wouldn't be satisfied with his customary order of salad (he never was) and could be persuaded to help out.

He worked his way around to Topic A. "When I called to set up this lunch, Hilda told me she's got a crush on Weyland. How are you and he getting along?"

Shadow Sisters
Stephen Markley (UK)

Some Things Never Die
Robert Morgan

Suckers
Anne Bilson (UK)

The Summoning
Bentley Little

Twilight Memories
Maggie Shayne

Twilight Phantasies
Maggie Shayne

The Vampire Journals
Tracy Bieri

Vampire of the Mists
Christie Golden

The Vampire's Apprentice
Richard Lee Byers

Warhammer: Genevieve Undead
Jack Yeovil (UK)

The Winter Man
Denise Vitola

1994

Among the Immortals
Paul Lake

Bitter Blood
Karen E. Taylor

Blood and Chrysanthemums: A Vampire Novel
Nancy Baker

Blood Secrets
Karen E. Taylor

Blood Work
Fay Zachary

Bloodletter
Warren Newton Beath

Bloodwars
Brian Lumley (UK)

Cold Kiss
Roxanne Longstreet

"My God, Doug, now you're going to tell me this whole thing was to fix me up with an eligible suitor!" She winced at her own rather strained laughter. "How soon are you planning to ask Weyland to work at Cayslin again?"

"I don't know, but probably sooner than I thought a couple of months ago. We hear that he's been exploring an attachment to an anthropology department at a Western school, some niche where I guess he feels he can have less responsibility, less visibility, and a chance to collect himself. Naturally, this news is making people at Cayslin suddenly eager to nail him down for us. Have you a recommendation?"

"Yes," she said. "Wait."

He gave her an inquiring look. "What for?"

"Until he works more fully through certain stresses in the situation at Cayslin. Then I'll be ready to commit myself about him." The bluefish came. She pretended distraction: "Good God, that's too much fish for me. Doug, come on and help me out here."

Hilda was crouched over Floria's file drawer. She straightened up, looking grim. "Somebody's been in the office!"

What was this, had someone attacked her? The world took on a cockeyed, dangerous tilt. "Are you okay?"

"Yes, sure, I mean there are records that have been gone through. I can tell. I've started checking and so far it looks as if none of the files themselves are missing. But if any papers were taken out of them, that would be pretty hard to spot without reading through every folder in the place. Your files, Floria. I don't think anybody else's were touched."

Mere burglary; weak with relief, Floria sat down on one of the waiting-room chairs. But only her files? "Just my stuff, you're sure?"

Hilda nodded. "The clinic got hit, too. I called. They see some new-looking scratches on the lock of your file drawer over there. Listen, you want me to call the cops?"

"First check as much as you can, see if anything obvious is missing."

There was no sign of upset in her office. She found a phone message on her table: Weyland had canceled his next appointment. She knew who had broken into her files.

She buzzed Hilda's desk. "Hilda, let's leave the police out of it for the moment. Keep checking." She stood in the middle of the office, looking at the chair replacing the one he had broken, looking at the window where he had so often watched.

Relax, she told herself. There was nothing for him to find here or at the clinic.

She signaled that she was ready for the first client of the afternoon.

That evening she came back to the office after having dinner with friends. She was supposed to be helping set up a workshop for next month, and she'd been putting off even thinking about it, let alone doing any real work. She set herself to compiling a suggested bibliography for her section.

Countess Dracula: A Novel
Carroll Borland

Covenant with the Vampire
Jeanne Kalogridis

Cursed Be the Child
Mort Castle

A Dance in Blood Velvet
Freda Warrington (UK)

Darkness
Tanith Lee (UK)

Dark Obsession
Amanda Stevens

A Deeper Hunger
Sabine Kells

Drakulya: The Lost Journals of Micea Drakulya
Earl Lee

The Fangs of Morning
Leslie H. Whitten

For All Eternity
Linda Lael Miller

Gideon
Stephen Laws (UK)

Hunter's Moon
Dawn Stewardson

Kiss of Darkness
Sharon Brondos

Love Bite
Sherry Gottlieb

Midnight Kiss
Nancy Gideon

Midnight Kiss
Rebecca York

Midnight Temptation
Nancy Gideon

Mina
Marie Kiraly

Near Death
Nancy Kilpatrick

The phone light blinked.

It was Kenny, sounding muffled and teary. "I'm sorry," he moaned. "The medicine just started to wear off. I've been trying to call you everyplace. God, I'm so scared — he was waiting in the alley."

"Who was?" she said, dry-mouthed. She knew.

"Him. The tall one, the faggot — only he goes with women too, I've seen him. He grabbed me. He hurt me. I was lying there a long time. I couldn't do anything. I felt so funny — like floating away. Some kids found me. Their mother called the cops. I was so cold, so scared —"

"Kenny, where are you?"

He told her which hospital. "Listen, I think he's really crazy, you know? And I'm scared he might . . . you live alone . . . I don't know — I didn't mean to make trouble for you. I'm so scared."

God damn you, you meant exactly to make trouble for me, and now you've bloody well made it. She got him to ring for a nurse. By calling Kenny her patient and using "Dr." in front of her own name without qualifying the title she got some information: two broken ribs, multiple contusions, a badly wrenched shoulder, and a deep cut on the scalp which Dr. Wells thought accounted for the blood loss the patient had sustained. Picked up early today, the patient wouldn't say who had attacked him. You can check with Dr. Wells tomorrow, Dr. —?

Can Weyland think I've somehow sicced Kenny on him? No, he surely knows me better than that. Kenny must have brought this on himself.

She tried Weyland's number and then the desk at his hotel. He had closed his account and gone, providing no forwarding information other than the address of a university in New Mexico.

Then she remembered: this was the night Deb and Nick and the kids were arriving. Oh, God. Next phone call. The Americana was the hotel Deb had mentioned. Yes, Mr. and Mrs. Nicholas Redpath were registered in room whatnot. Ring, please.

Deb's voice came shakily on the line. "I've been trying to call you." Like Kenny.

"You sound upset," Floria said, steadying herself for whatever calamity had descended: illness, accident, assault in the streets of the dark, degenerate city.

Silence, then a raggedy sob. "Nick's not here. I didn't phone you earlier because I thought he still might come, but I don't think he's coming, Mom." Bitter weeping.

"Oh, Debbie. Debbie, listen, you just sit tight, I'll be right down there."

The cab ride took only a few minutes. Debbie was still crying when Floria stepped into the room.

"I don't know, I don't know," Deb wailed, shaking her head. "What did I do wrong? He went away a week ago, to do some research, he said, and I didn't hear from him, and half the bank money is gone — just half, he left me half. I kept hoping . . . they say most runaways come back in a

Nosferatu
Karl Sargent and Marc Gascoigne

Of Saints and Shadows
Christopher Golden

Progeny of the Adder
Leslie H. Whitten

Séance for a Vampire
Fred Saberhagen

The Secret Life of Laszlo, Count Dracula
Roderick Anscombe

Sex and Blood
Ron Dee

Sex and the Single Vampire
Nancy Kilpatrick

Shade
David Drake

Shattered Glass
Elaine Bergstrom

Shrines and Desecrations
Brian J. Hodge

Sins of the Blood
Kristine Kathryn Rusch

Succumb
Ron Dee

This Dark Paradise
Wendy Haley

Throat Sprockets
Tim Lucas

Twilight Illusions
Maggie Shayne

Vampire Lover
Charlotte Lamb

The Vampire Papers
Michael Romkey

Vampire's Kiss
Robert S. Hichens

1995
Angel Souls and Devil Hearts
Christopher Golden

few days or call up, they get lonely . . . I haven't told anybody — I thought since we were supposed to be here at this convention thing together, I'd better come, maybe he'd show up. But nobody's seen him, and there are no messages, not a word, nothing."

"All right, all right, poor Deb," Floria said, hugging her.

"Oh God, I'm going to wake the kids with all this howling." Deb pulled away, making a frantic gesture toward the door of the adjoining room. "It was so hard to get them to sleep — they were expecting Daddy to be here, I kept telling them he'd be here." She rushed out into the hotel hallway. Floria followed, propping the door open with one of her shoes since she didn't know whether Deb had a key with her or not. They stood out there together, ignoring passersby, huddling over Deb's weeping.

"What's been going on between you and Nick?" Floria said. "Have you two been sleeping together lately?"

Deb let out a squawk of agonized embarrassment, "Mo-*ther!*" and pulled away from her. Oh, hell, wrong approach.

"Come on, I'll help you pack. We'll leave word you're at my place. Let Nick come looking for you." Floria firmly squashed down the miserable inner cry, How am I going to stand this?

"Oh, no, I can't move till morning now that I've got the kids settled down. Besides, there's one night's deposit on the rooms. Oh, Mom, what did I do?"

"You didn't do anything, hon," Floria said, patting her shoulder and thinking in some part of her mind, Oh boy, that's great, is that the best you can come up with in a crisis with all your training and experience? Your touted professional skills are not so hot lately, but this bad? Another part answered, Shut up, stupid, only an idiot does therapy on her own family. Deb's come to her mother, not to a shrink, so go ahead and be Mommy. If only Mommy had less pressure on her right now — but that was always the way: everything at once or nothing at all.

"Look, Deb, suppose I stay the night here with you."

Deb shook the pale, damp-streaked hair out of her eyes with a determined, grown-up gesture. "No, thanks, Mom. I'm so tired I'm just going to fall out now. You'll be getting a bellyful of all this when we move in on you tomorrow anyway. I can manage tonight, and besides —"

And besides, just in case Nick showed up, Deb didn't want Floria around complicating things; of course. Or in case the tooth fairy dropped by.

Floria restrained an impulse to insist on staying; an impulse, she recognized, that came from her own need not to be alone tonight. That was not something to load on Deb's already burdened shoulders.

"Okay," Floria said. "But look, Deb, I'll expect you to call me up first thing in the morning, whatever happens." And if I'm still alive, I'll answer the phone.

All the way home in the cab she knew with growing certainty that Weyland would be waiting for her there. He can't just walk away, she

Baroness of Blood
Elaine Bergstrom

The Blood Countess
Andrei Codrescu

Blood Ties
Karen E. Taylor

The Bloody Red Baron
Kim Newman (UK)

Caravan of Shadows
Richard Lee Byers

Children of the Vampire
Jeanne Kalogridis

Circus of the Damned
Laurel K. Hamilton

Death Masque
P.N. Elrod

Desmodus
Melanie Tem

Diary of a Vampire
Gary Bowen

Dream Lover
Katrina Vicente (UK)

Glittering Savages
Mark Burnell (UK)

Empire of the Night
Amanda Ashley

Eye Killers
A.A. Carr

Memnoch the Devil
Anne Rice

Midnight Surrender
Nancy Gideon

Necroscope: The Lost Years
Brian Lumley (UK)

Night's Immortal Touch
Cherlyn Jac

Nightwing
Lynn Michaels

thought; he has to finish things with me. So let's get it over.

In the tiled hallway she hesitated, keys in hand. What about calling the cops to go inside with her? Absurd. You don't set the cops on a unicorn.

She unlocked and opened the door to the apartment and called inside, "Weyland! Where are you?"

Nothing. Of course not — the door was still open, and he would want to be sure she was by herself. She stepped inside, shut the door, and snapped on a lamp as she walked into the living room.

He was sitting quietly on a radiator cover by the street window, his hands on his thighs. His appearance here in a new setting, her setting, this faintly lit room in her home place, was startlingly intimate. She was sharply aware of the whisper of movement — his clothing, his shoe soles against the carpet underfoot — as he shifted his posture.

"What would you have done if I'd brought somebody with me?" she said unsteadily. "Changed yourself into a bat and flown away?"

"Two things I must have from you," he said. "One is the bill of health that we spoke of when we began, though not, after all, for Cayslin College. I've made other plans. The story of my disappearance has of course filtered out along the academic grapevine so that even two thousand miles from here people will want evidence of my mental soundness. Your evidence. I would type it myself and forge your signature, but I want your authentic tone and language. Please prepare a letter to the desired effect, addressed to these people."

He drew something white from an inside pocket and held it out. She advanced and took the envelope from his extended hand. It was from the Western anthropology department that Doug had mentioned at lunch.

"Why not Cayslin?" she said. "They want you there."

"Have you forgotten your own suggestion that I find another job? That was a good idea after all. Your reference will serve me best out there — with a copy for my personnel file at Cayslin, naturally."

She put her purse down on the seat of a chair and crossed her arms. She felt reckless — the effect of stress and weariness, she thought, but it was an exciting feeling.

"The receptionist at the office does this sort of thing for me," she said. He pointed. "I've been in your study. You have a typewriter there, you have stationery with your letterhead, you have carbon paper."

"What was the second thing you wanted?"

"Your notes on my case."

"Also at the —"

"You know that I've already searched both your work places, and the very circumspect jottings in your file on me are not what I mean. Others must exist: more detailed."

"What makes you think that?"

"How could you resist?" He mocked her. "You have encountered

Out of the Darkness
Lynn Erikson

Prince of the City
Susan Krinard

Rulers of Darkness
Steven Spruill

Tap, Tap
David Martin

These Fallen Angels
Wendy Haley

Traveling with the Dead
Barbara Hambly

Vampire Diary
Robert Weinberg

The Vampire Viscount
Karen Harbaugh (UK)

The Vampyre
Tom Holland (UK)

Vanitas
S.P. Somtow

Vivia
Tanith Lee (UK)

nothing like me in your entire professional life, and never shall again. Perhaps you hope to produce an article someday, even a book — a memoir of something impossible that happened to you one summer. You're an ambitious woman, Dr. Landauer."

Floria squeezed her crossed arms tighter against herself to quell her shivering. "This is all just supposition," she said.

He took folded papers from his pocket: some of her thrown-aside notes on him, salvaged from the wastebasket. "I found these. I think there must be more. Whatever there is, give it to me, please."

"And if I refuse, what will you do? Beat me up the way you beat up Kenny?"

Weyland said calmly, "I told you he should stop following me. This is serious now. There are pursuers who intend me ill — my former captors, of whom I told you. Whom do you think I keep watch for? No records concerning me must fall into their hands. Don't bother protesting to me your devotion to confidentiality. There is a man named Alan Reese who would take what he wants and be damned to your professional ethics. So I must destroy all evidence you have about me before I leave the city."

Floria turned away and sat down by the coffee table, trying to think beyond her fear. She breathed deeply against the fright trembling in her chest.

"I see," he said dryly, "that you won't give me the notes; you don't trust me to take them and go. You see some danger."

"All right, a bargain," she said. "I'll give you whatever I have on your case if in return you promise to go straight out to your new job and keep away from Kenny and my offices and anybody connected with me — "

He was smiling slightly as he rose from the seat and stepped soft footed toward her over the rug. "Bargains, promises, negotiations — all foolish, Dr. Landauer. I want what I came for."

She looked up at him. "But then how can I trust you at all? As soon as I give you what you want —"

"What is it that makes you afraid — that you can't render me harmless to you? What a curious concern you show suddenly for your own life and the lives of those around you! You are the one who led me to take chances in our work together — to explore the frightful risks of self-revelation. Didn't you see in the air between us the brilliant shimmer of those hazards? I thought your business was not smoothing the world over but adventuring into it, discovering its true nature, and closing valiantly with everything jagged, cruel, and deadly."

In the midst of her terror the inner choreographer awoke and stretched. Floria rose to face the vampire.

"All right, Weyland, no bargains. I'll give you freely what you want." Of course she couldn't make herself safe from him — or make Kenny or Lucille or Deb or Doug safe — any more than she could protect Jane Fennerman from the common dangers of life. Like Weyland, some dangers

The years just before and after the 1997 Dracula *Centenary saw an explosion of vampire novels and stories, which have still not been comprehensively catalogued. The remaining checklist, therefore, is decidedly more selective, but I hope it still serves to reflect the astonishing vitality of the genre at the millennium. Ed.*

1996

Brand New Cherry Flavor
Todd Grimson

A Dozen Black Roses
Nancy A. Collins

Keeper of the King
Nigel Bennett and P.N. Elrod

Lilith
D.A. Heeley

Mansions of Darkness
Chelsea Quinn Yarbro

Slave of My Thirst
Tom Holland

Stainless
Todd Grimson

The World on Blood
Jonathan Nasaw

"Avenging Angel"
Doug Murray

"The Blood of Othima"
C. Dean Anderson

"Bloodlover"
Victor Komer

"Brimstone and Salt"
S.P. Somtow

"Cam Shaft"
David Bischoff

were too strong to bind or banish. "My notes are in the workroom — come on, I'll show you. As for the letter you need, I'll type it right now and you can take it away with you."

She sat at the typewriter arranging paper, carbon sheets, and white-out, and feeling the force of his presence. Only a few feet away, just at the margin of the light from the gooseneck lamp by which she worked, he leaned against the edge of the long table that was twin to the table in her office. Open in his large hands was the notebook she had given him from the table drawer. When he moved his head over the notebook's pages, his glasses glinted.

She typed the heading and the date. How surprising, she thought, to find that she had regained her nerve here, and now. When you dance as the inner choreographer directs, you act without thinking, not in command of events but in harmony with them. You yield control, accepting the chance that a mistake might be part of the design. The inner choreographer is always right but often dangerous: giving up control means accepting the possibility of death. What I feared I have pursued right here to this moment in this room.

A sheet of paper fell out of the notebook. Weyland stooped and caught it up, glanced at it. "You had training in art?" Must be a sketch.

"I thought once I might be an artist," she said.

"What you chose to do instead is better," he said. "This making of pictures, plays, all art, is pathetic. The world teems with creation, most of it unnoticed by your kind just as most of the deaths are unnoticed. What can be the point of adding yet another tiny gesture? Even you, these notes — for what, a moment's celebrity?"

"You tried it yourself," Floria said. "The book you edited, *Notes on a Vanished People.*" She typed: "...temporary dislocation resulting from a severe personal shock..."

"That was professional necessity, not creation," he said in the tone of a lecturer irritated by a question from the audience. With disdain he tossed the drawing on the table. "Remember, I don't share your impulse toward artistic gesture — your absurd frills —"

She looked up sharply. "The ballet, Weyland. Don't lie." She typed: "...exhibits a powerful drive toward inner balance and wholeness in a difficult life situation. The steadying influence of an extraordinary basic integrity..."

He set the notebook aside. "My feeling for ballet is clearly some sort of aberration. Do you sigh to hear a cow calling in a pasture?"

"There are those who have wept to hear whales singing in the ocean."

He was silent, his eyes averted.

"This is finished," she said. "Do you want to read it?"

He took the letter. "Good," he said at length. "Sign it, please. And type an envelope for it." He stood closer, but out of arm's reach, while she complied. "You seem less frightened."

"Caveat Shatner"
Joy Mosier-Dubinsky

"Chozzerai"
Roland J. Green and Frieda Murray

"The Death-Sweet Scent of Lilies"
David Niall Wilson

"The Devil's March"
Philip Nutman and Anya Martin

"Film Noir"
Stephen Deduran

"How Do I Love Thee"
Rick R. Reed

"A Life in the Theatre"
Esther M. Friesner

"On a Big Night in Monster History"
Marc Levinthal and John Mason Skipp

"Revolts in the Desert"
Susan Scwartz

"So Proudly We Heil"
Fred Olen Ray

"Stoker's Mistress"
Caitlín R. Kiernan

"Uruborus"
Victor Rodriguez

"WhiteTerror"
Rick Hautala

"The Wizard and the Dragon"
Richard Lee Byers

1997
Daughter of Darkness
Stephen Sprull

Shadows
Jonathan Nasaw

To Speak in Lifeless Tongues
David Niall Wilson

The Vampire Virus
Michael Romkey

Writ in Blood
Chelsea Quinn Yarbro

"Armies of the Heart"
Christopher Fowler

"Bela's Plot"
Caitlín R. Kiernan

"Black Beads"
John Gordon

"Blood Freak"
Nancy Holder

"Bloodlight"
Stephen Mark Rainey

"Bloodlines"
Conrad Williams

"Blood of Eden"
Mike Chinn

"Ceilings and Sky"
Lucy Taylor

"Children of the Long Night"
Lisa Morton

"Coppola's Dracula"
Kim Newman

"Daddy's Little Girl"
Mandy Slater

"Dracula's Library"
Christopher Fowler

"I'm terrified but not paralyzed," she said and laughed, but the laugh came out a gasp.

"Fear is useful. It has kept you at your best throughout our association. Have you a stamp?"

Then there was nothing to do but take a deep breath, turn off the gooseneck lamp, and follow him back into the living room. "What now, Weyland?" she said softly. "A carefully arranged suicide so that I have no chance to retract what's in that letter or to reconstruct my notes?"

At the window again, always on watch at the window, he said, "Your doorman was sleeping in the lobby. He didn't see me enter the building. Once inside, I used the stairs, of course. The suicide rate among therapists is notoriously high. I looked it up."

"You have everything all planned?"

The window was open. He reached out and touched the metal grille that guarded it. One end of the grille swung creaking outward into the night air, like a gate opening. She visualized him sitting there waiting for her to come home, his powerful fingers patiently working the bolts at that side of the grille loose from the brick-and-mortar window frame. The hair lifted on the back of her neck.

He turned toward her again. She could see the end of the letter she had given him sticking palely out of his jacket pocket.

"Floria," he said meditatively. "An unusual name — is it after the heroine of Sardou's *Tosca*? At the end, doesn't she throw herself to her death from a high castle wall? People are careless about the names they give their children. I will not drink from you — I hunted today, and I fed. Still, to leave you living . . . is too dangerous."

A fire engine tore past below, siren screaming. When it had gone Floria said, "Listen, Weyland, you said it yourself: I can't make myself safe from you — I'm not strong enough to shove you out the window instead of being shoved out myself. Must you make yourself safe from me? Let me say this to you, without promises, demands, or pleadings: I will not go back on what I wrote in that letter. I will not try to recreate my notes. I mean it. Be content with that."

"You tempt me to it," he murmured after a moment, "to go from here with you still alive behind me for the remainder of your little life — to leave woven into Dr. Landauer's quick mind those threads of my own life that I pulled for her . . . I want to be able sometimes to think of you thinking of me. But the risk is very great."

"Sometimes it's right to let the dangers live, to give them their place," she urged. "Didn't you tell me yourself a little while ago how risk makes us more heroic?"

He looked amused. "Are you instructing me in the virtues of danger? You are brave enough to know something, perhaps, about that, but I have studied danger all my life."

"A long, long life with more to come," she said, desperate to make him understand and believe her. "Not mine to jeopardize. There's no torch-brandishing peasant here; we left that behind long ago. Remember when you spoke for me? You said, 'For love of wonder.' That was true."

He leaned to turn off the lamp near the window. She thought that he had made up his mind, and that when he straightened it would be to spring.

But instead of terror locking her limbs, from the inward choreographer came a rush of warmth and energy into her muscles and an impulse to turn toward him. Out of a harmony of desires she said swiftly, "Weyland, come to bed with me."

She saw his shoulders stiffen against the dim square of the window, his head lift in scorn. "You know I can't be bribed that way," he said contemptuously. "What are you up to? Are you one of those who come into heat at the sight of an upraised fist?"

"My life hasn't twisted me that badly, thank God," she retorted. "And if you've known all along how scared I've been, you must have sensed my attraction to you too, so you know it goes back to — very early in our work. But we're not at work now, and I've given up being 'up to' anything. My feeling is real — not a bribe, or a ploy, or a kink. No 'love me now, kill me later,' nothing like that. Understand me, Weyland: if death is your answer, than let's get right to it — come ahead and try."

Her mouth was dry as paper. He said nothing and made no move; she pressed on. "But if you can let me go, if we can simply part company here, then this is how I would like to mark the ending of our time together. This is the completion I want. Surely you feel something, too — curiosity at least?"

"Granted, your emphasis on the expressiveness of the body has instructed me," he admitted, and then he added lightly, "Isn't it extremely unprofessional to proposition a client?"

"Extremely, and I never do; but this, now, feels right. For you to indulge in courtship that doesn't end in a meal would be unprofessional, too, but how would it feel to indulge anyway — this once? Since we started, you've pushed me light-years beyond my profession. Now I want to travel all the way with you, Weyland. Let's be unprofessional together."

She turned and went into the bedroom, leaving the lights off. There was a reflected light, cool and diffuse, from the glowing night air of the great city. She sat down on the bed and kicked off her shoes. When she looked up, he was in the doorway.

Hesitantly, he halted a few feet from her in the dimness, then came and sat beside her. He would have lain down in his clothes, but she said quietly, "You can undress. The front door's locked and there isn't anyone here but us. You won't have to leap up and flee for your life."

He stood again and began to take off his clothes, which he draped neatly over a chair. He said, "Suppose I am fertile with you; could you conceive?"

"Dusting the Flowers"
David J. Schow

"The Dripping of Sundered Wineskins"
Brian Hodge

"Endangered Species"
Brian Mooney

"First Date"
Richard Laymon

"The Fly Room"
Thom Metzger

"The Heart of Count Dracula, Descendant of Attila, Scourge of God"
Thomas Ligotti

"The I of the Eye of the Worm"
Janet Berliner and George Guthridge

"I'm Not Well, But I'm Better"
Pat Califia

"Kingyo no fun"
Nicholas Royle

"Larry's Guest"
Guy N. Smith

"The Last Testament"
Brian Hodge

"The Last Vampire"
Peter Crowther

"The Lord of the Undead"
Jo Fletcher

"Mbo"
Nicholas Royle

"Melancholia"
Roberta Lannes

"The Privilege of the Dead"
Thomas S. Roche

"Roadkill"
Graham Masterson

"Stigmata"
Jean-Daniel Breque

"The Subtle Ties that Bind"
David Niall Wilson

"A Taste of Culture"
Jan Edwards

"Teaserama"
Nancy Kilpatrick

"To Have You with Me"
Randy Fox

"Volunteers"
Terry Lamsley

"When Memory Fails"
Roberta Lannes

"Whispers in Walled Tombs"
O'Neil De Noux

"When Greek Meets Greek"
Basil Copper

"Windows '99 of the Soul"
Chris Morgan

"The Worst Place in the World"
Paul J. McAuley

"Your European Son"
Joel Lane

"Quality Control"
Brian Stableford

"Dear Alison"
Michael Marshall Smith

1998
The Angry Angel
Chelsea Quinn Yarbro

By her own choice any such possibility had been closed off after Deb. She said, "No," and that seemed to satisfy him.

She tossed her own clothes onto the dresser. He sat down next to her again, his body silvery in the reflected light and smooth, lean as a whippet and as roped with muscle. His cool thigh pressed against her own fuller, warmer one as he leaned across her and carefully deposited his glasses on the bedtable. Then he turned toward her, and she could just make out two puckerings of tissue on his skin: bullet scars, she thought, shivering.

He said, "But why do I wish to do this?"

"Do you?" She had to hold herself back from touching him.

"Yes." He stared at her. "How did you grow so real? The more I spoke to you of myself, the more real you became."

"No more speaking, Weyland," she said gently. "This is body work."

He lay back on the bed.

She wasn't afraid to take the lead. At the very least she could do for him as well as he did for himself, and at the most, much better. Her own skin was darker than his, a shadowy contrast where she browsed over his body with her hands. Along the contours of his ribs she felt knotted places, hollows — old healings, the tracks of time. The tension of his muscles under her touch and the sharp sound of his breathing stirred her. She lived the fantasy of sex with an utter stranger; there was no one in the world so much a stranger as he. Yet there was no one who knew him as well as she did, either. If he was unique, so was she, and so was their confluence here.

The vividness of the moment inflamed her. His body responded. His penis stirred, warmed, and thickened in her hand. He turned on his hip so that they lay facing each other, he on his right side, she on her left. When she moved to kiss him he swiftly averted his face: of course — to him, the mouth was for feeding. She touched her fingers to his lips, signifying her comprehension.

He offered no caresses but closed his arms around her, his hands cradling the back of her head and neck. His shadowed face, deep-hollowed under brow and cheekbone, was very close to hers. From between the parted lips that she must not kiss his quick breath came, roughened by groans of pleasure. At length he pressed his head against hers, inhaling deeply; taking her scent, she thought, from her hair and skin.

He entered her, hesitant at first, probing slowly and tentatively. She found this searching motion intensely sensuous, and clinging to him all along his sinewy length she rocked with him through two long, swelling waves of sweetness. Still half submerged, she felt him strain tight against her, she heard him gasp through his clenched teeth. Panting, they subsided and lay loosely interlocked. His head was tilted back; his eyes were closed. She had no desire to stroke him or to speak with him, only to rest spent against his body and absorb the sounds of his breathing, her breathing.

He did not lie long to hold or be held. Without a word he disengaged his body from hers and got up. He moved quietly about the bedroom, gathering his clothing, his shoes, the drawings, the notes from the workroom. He dressed without lights. She listened in silence from the center of a deep repose.

There was no leavetaking. His tall figure passed and repassed the dark rectangle of the doorway, and then he was gone. The latch on the front door clicked shut.

Floria thought of getting up to secure the deadbolt. Instead she turned on her stomach and slept.

She woke as she remembered coming out of sleep as a youngster — peppy and clearheaded.

"Hilda, let's give the police a call about that break-in. If anything ever does come of it, I want to be on record as having reported it. You can tell them we don't have any idea who did it or why. And please make a photocopy of this letter carbon to send to Doug Sharpe up at Cayslin. Then you can put the carbon into Weyland's file and close it."

Hilda sighed. "Well, he was too old anyway."

He wasn't, my dear, but never mind.

In her office Floria picked up the morning's mail from her table. Her glance strayed to the window where Weyland had so often stood. God, she was going to miss him; and God, how good it was to be restored to plain working days.

Only not yet. Don't let the phone ring, don't let the world push in here now. She needed to sit alone for a little and let her mind sort through the images left from . . . from the pas de deux with Weyland. It's the notorious morning after, old dear, she told herself; just where have I been dancing, anyway?

In a clearing in the enchanted forest with the unicorn, of course, but not the way the old legends have it. According to them, hunters set a virgin to attract the unicorn by her chastity so they can catch and kill him. My unicorn was the chaste one, come to think of it, and this lady meant no treachery. No, Weyland and I met hidden from the hunt, to celebrate a private mystery of our own . . .

Your mind grappled with my mind, my dark leg over your silver one, unlike closing with unlike across whatever likeness may be found: your memory pressing on my thoughts, my words drawing out your words in which you may recognize your life, my smooth palm gliding down your smooth flank . . .

Why, this will make me cry, she thought, blinking. And for what? Does an afternoon with the unicorn have any meaning for the ordinary days that come later? What has this passage with Weyland left me? Have I anything in my hands now besides the morning's mail?

What I have in my hands is my own strength, because I had to reach deep to find the strength to match him.

She put down the letters, noticing how on the backs of her hands the veins stood, blue shadows, under the thin skin. How can these hands be

Blue Moon
Laurel K. Hamilton

Night Prayers
P. D. Cacek

Pandora
Anne Rice

1999

Buffy the Vampire Slayer: Immortal
Christopher Golden and Nancy Holder

Communion Blood
Chelsea Quinn Yarbro

The Embrace
Aphrodite Jones

The Vampire Hunter
Michael Romkey

The Vampire Armand
Anne Rice

Vampyrrhic
Simon Clark

2000

Blood to Blood
Elaine Bergstrom

Cirque du Freak
Darren Shan

Come Twilight
Chelsea Quinn Yarbro

Darkspawn
Lois Tilton

Obsidian Butterfly
Laurel K. Hamilton

Partners: Laws of the Blood
Susan Sizemore

strong? Time was beginning to wear them thin and bring up the fragile inner structure in clear relief. That was the meaning of the last parent's death: that the child's remaining time has a limit of its own.

But not for Weyland. No graveyards of family dead lay behind him, no obvious and implacable ending of his own span threatened him. Time has to be different for a creature of an enchanted forest, as morality has to be different. He was a predator and a killer formed for a life of centuries, not decades; of secret singularity, not the busy hum of the herd. Yet his strength, suited to that nonhuman life, had revived her own strength. Her hands were slim, no longer youthful, but she saw now that they were strong enough.

For what? She flexed her fingers, watching the tendons slide under the skin. Strong hands don't have to clutch. They can simply open and let go.

She dialed Lucille's extension at the clinic.

"Luce? Sorry to have missed your calls lately. Listen, I want to start making arrangements to transfer my practice for a while. You were right, I do need a break, just as all my friends have been telling me. Will you pass the word for me to the staff over there today? Good, thanks. Also, there's the workshop coming up next month.... Yes. Are you kidding? They'd love to have you in my place. You're not the only one who's noticed that I've been falling apart, you know. It's awfully soon — can you manage, do you think? Luce, you are a brick and a lifesaver and all that stuff that means I'm very, very grateful."

Not so terrible, she thought, but only a start. Everything else remained to be dealt with. The glow of euphoria couldn't carry her for long. Already, looking down, she noticed jelly on her blouse, just like old times, and she didn't even remember having breakfast. If you want to keep the strength you've found in all this, you're going to have to get plenty of practice being strong. Try a tough one now.

She phoned Deb. "Of course you slept late, so what? I did, too, so I'm glad you didn't call and wake me up. Whenever you're ready — if you need help moving uptown from the hotel, I can cancel here and come down... Well, call if you change your mind. I've left a house key for you with my doorman.

"And listen, hon, I've been thinking — how about all of us going up together to Nonnie's over the weekend? Then when you feel like it maybe you'd like to talk about what you'll do next. Yes, I've already started setting up some free time for myself. Think about it, love. Talk to you later."

Kenny's turn. "Kenny, I'll come by during visiting hours this afternoon."

"Are you okay?" he squeaked.

"I'm okay. But I'm not your mommy, Ken, and I'm not going to start trying to hold the big bad world off you again. I'll expect you to be ready to settle down seriously and choose a new therapist for yourself. We're going to get that done today once and for all. Have you got that?"

After a short silence he answered in a desolate voice, "All right."

When Darkness Falls
Shannon Drake

Worse than Death
Sherry Gottlieb

2001
Bound in Blood
David Thomas Lord

Buffy the Vampire Slayer: The Book of Fours
Nancy Holder

Dead Until Dark
Charlaine Harris

The London Vampire Panic
Michael Romkey

Quincey Morris, Vampire
P.N. Elrod

Red Moon Rising
Billie Sue Mosiman

The editor gratefully acknowledges the assistance of Robert Eighteen-Bisang for his generous assistance with the original version of this checklist, which first appeared in V is for Vampire: The A-Z Guide to Everything Undead *(Plume Books, 1996). For more detailed bibliographical information, the editor recommends the excellent and comprehensive listings compiled by Mr. Eighteen-Bisang and Gordon Melton in Dr. Melton's* The Vampire Book *(Visible Ink Press,1999). For an unparalleled critical overview of titles prior to 1993, there is still no better resource than Greg Cox's* The Transylvanian Library: A Consumer Guide to Vampire Fiction *(The Borgo Press, 1993).*

"Kenny, nobody grown up has a mommy around to take care of things for them and keep them safe — not even me. You just have to be tough enough and brave enough yourself. See you this afternoon."

How about Jane Fennerman? No, leave it for now, we are not Wonder Woman, we can't handle that stress today as well.

Too restless to settle down to paperwork before the day's round of appointments began, she got up and fed the goldfish, then drifted to the window and looked out over the city. Same jammed-up traffic down there, same dusty summer park stretching away uptown — yet not the same city, because Weyland no longer hunted there. Nothing like him moved now in those deep, grumbling streets. She would never come upon anyone there as alien as he — and just as well. Let last night stand as the end, unique and inimitable, of their affair. She was glutted with strangeness and looked forward frankly to sharing again in Mort's ordinary human appetite.

And Weyland — how would he do in that new and distant hunting ground he had found for himself? Her own balance had been changed. Suppose his once perfect, solitary equilibrium had been altered too? Perhaps he had spoiled it by involving himself too intimately with another being — herself. And then he had left her alive — a terrible risk. Was this a sign of his corruption at her hands?

A vampire bibliophile.

"Oh, no," she whispered fiercely, focusing her vision on her reflection in the smudged window glass. Oh, no, I am not the temptress. I am not the deadly female out of legends whose touch defiles the hitherto unblemished being, her victim. If Weyland found some human likeness in himself, that had to be in him to begin with. Who said he was defiled anyway? Newly discovered capacities can be either strengths or weaknesses, depending on how you use them.

Very pretty and reassuring, she thought grimly; but it's pure cant. Am I going to retreat now into mechanical analysis to make myself feel better?

She heaved open the window and admitted the sticky summer breath of the city into the office. There's your enchanted forest, my dear, all nitty-gritty and not one flake of fairy dust. You've survived here, which means you can see straight when you have to. Well, you have to now.

Has he been damaged? No telling yet, and you can't stop living while you wait for the answers to come in. I don't know all that was done between us, but I do know who did it: I did it, and he did it, and neither of us withdrew until it was done. We were joined in a rich complicity — he in the wakening of some flicker of humanity in himself, I in keeping and, yes, enjoying the secret of his implacable blood hunger. What that complicity means for each of us can only be discovered by getting on with living and watching for clues from moment to moment. His business is to continue from here, and mine is to do the same, without guilt and without resentment. Doug was right: the aim is individual responsibility. From that effort, not even the lady and the unicorn are exempt.

Shaken by a fresh upwelling of tears, she thought bitterly, Moving on is easy enough for Weyland; he's used to it, he's had more practice. What about me? Yes, be selfish, woman — if you haven't learned that, you've learned damn little.

The Japanese say that in middle age you should leave the claims of family, friends, and work, and go ponder the meaning of the universe while you still have the chance. Maybe I'll try just existing for a while, and letting grow in its own time my understanding of a universe that includes Weyland — and myself — among its possibilities.

Is that looking out for myself? Or am I simply no longer fit for living with family, friends, and work? Have I been damaged by him — by my marvelous, murderous monster?

Damn, she thought, I wish he were here, I wish we could talk about it. The light on her phone caught her eye; it was blinking the quick flashes that meant Hilda was signaling the imminent arrival of — not Weyland — the day's first client.

We're each on our own now, she thought, shutting the window and turning on the air-conditioner.

But think of me sometimes, Weyland, thinking of you.

PART FOUR

Postmodern Vampires

Stakes and stripes forever: the original jacket design for American Vampires. *(W.W. Norton & Company)*

FROM

American Vampires

NORINE DRESSER
(1990)

MELODY REMEMBERS THAT she was wearing a low-cut, hot-pink dress the first time she met Pam, the vampire. It was 1985 and Melody's friend Brad had arranged the meeting. "Pam's always looking for someone else. Would you be interested?"

The modest apartment filled with books and plants was located in an East Coast city and seemed almost secluded because of the quiet created by the well-insulated building. Four young people — Melody, twenty-five; the others in their early thirties — had gathered to participate in a practice they were not ashamed of, but were nonetheless wary of divulging to others, as revealed by the clothespin holding closed the orange drapes.

Pretty, blue-eyed, with long wavy dark brown hair, Pam was wearing an ankle-length, dark flower-print dress with spaghetti straps, partly covered by a black cape. She carefully arranged the setting. Turning off the lights, she lit a few candles and started playing a recording of sounds of a rain forest. "Come closer and watch," she invited Melody.

Brad was stretched out on the gold upholstered sofa. Pam kneeled over him as Melody sat on the floor next to them and leaned forward to catch every detail. First Pam massaged Brad's back and then once he became relaxed she began to sensuously lick the fingertips of his left hand.

Across the room, seated next to the computer, Pam's husband, an observer only, watched in a rather detached manner. As the act began, two cats nonchalantly paraded back and forth across the room.

Pam stabbed one of Brad's fingers with a sterilized hypodermic needle of a type normally used for animals. She pierced the skin, squeezed the finger, squeezed some more; then she sucked from it. When she finished one finger, she repeated the steps with the next one, licking each finger when she was finished.

To Melody, it all seemed quite mysterious, but she wasn't afraid. "I felt pretty good. I'm an adventurous person, so I do kind of crazy, off-the-wall things," she explained.

After spending about twenty minutes stabbing Brad's fingers and then sucking them, Pam turned to Melody and said, "Okay, you're next."

Thus began Melody's initiation into the act of vampirism, playing the role of donor for Pam's belief in her need for fresh, warm human blood.

The activities just described were freely and enthusiastically reported because Pam, Melody, and Brad have a curiosity about their own behavior. Admittedly, these young people are not representative of a large population.

ABOUT THE AUTHOR

Cross-cultural customs and beliefs have fascinated folklorist Norine Dresser for over twenty-five years and have been the focus of her university teaching, research, and writing. *American Vampires* was especially praised by the *Boston Globe*, which commented, "Dresser has accumulated a wonderful assortment of anecdotal evidence to demonstrate just how much a part of American life is the vampire myth." The book won the 1989 Best Literature Award from the Count Dracula Society. In 1995, Dresser was a presenter at the groundbreaking First World Dracula Congress in Romania, and she served on the committee for "Dracula '97" celebration, marking the centennial of Bram Stoker's novel.

Her other books include *Multicultural Celebrations* (Three Rivers Press, 1999), *Multicultural Manners* (Wiley, 1996), based on her *Los Angeles Times* column of the same name; *I Felt Like I Was from Another Planet* (Longman/Addison-Wesley, 1994) and *Our Own Stories* (Longman/Addison-Wesley, 1995).

Dresser has received grants from the Smithsonian Institution, the National Endowment for the Humanities, and the Los Angeles Department of Cultural Affairs. In 1992, she retired from the faculty of California State University, Los Angeles, and has been writing full-time ever since.

They are but a minute fraction of modern society. Their behavior is carried out in private, of their own volition, to the harm of no one, and for the purpose of satisfying certain inner needs, sometimes understood, other times not. In spite of the reality that they are imitating behavior not publicly sanctioned, they do not think of themselves as aberrant.

If you were to meet them they would not strike you as being strange in any way. They seem like ordinary people functioning positively in the external world with family commitments and responsible jobs. One of them is employed full time in the art world. Another holds a government position requiring knowledge of the sciences and well-developed problem-solving abilities. The third works in the medical field.

Yet this raises a few questions about why they are attracted to the act of vampirism, a behavior associated primarily with horror stories and movies. If they called themselves blood fetishists, it might be easier to comprehend, but they don't. Instead, they affiliate themselves with vampires. It is curious that they have chosen to emulate the behavior of fictional, exaggerated creatures who are generally looked upon with terror, abhorrence, even mockery. Here is a situation of life imitating folklore, but why?

Folklorist Norine Dresser. (Photo: Mary G. Wentz)

The behavior of these three persons makes a statement about the power and vitality of the vampire symbol in this country, which apparently is not relegated to merely being part of leftover European legends, but is actually made viable and glorified by this tiny part of American culture.

Surely Melody, Brad, and Pam must interpret the vampire differently from the old stereotyped images portrayed by Bela Lugosi and Christopher Lee. Melody and Brad are not frantically waving a cross to frighten off the blood-sucking attacker, nor are they wearing garlic. Instead they appear to welcome the giving of their own life's vital force — their blood.

Traditionally, the vampire is a mythical creature believed to be a dead human who is doomed, unable to rest in peace and forced to leave the grave and feed upon the living by taking blood or vitality. Consequently, the act of taking another person's fluid or energy is called *vampirism*. Sometimes *psychic vampirism* is the term applied when one person takes energy and/or emotions from another. The vampire differs from the werewolf, the latter being a mythical live human who transforms into a wolf to attack the living. While there has never been any scientific evidence to indicate that these creatures are real, belief in them has existed from ancient times and throughout the world. The earliest known depiction of a vampire appears on a prehistoric Assyrian bowl showing a man copulating with a vampire.

Public awareness of the vampire became widespread in the twentieth century after the emergence of the fictional character Count Dracula, which was created by Irish author Bram Stoker and in part based on historical accounts of a fifteenth-century real Romanian tyrant, Vlad Tepes, also know as Vlad the Impaler.

American vampires have their basis in the Old World, but have developed in a particular way, shaped by our uniquely American forms of mass communication. American vampires differ from European, African, Latin American, and Asian antecedents. Just as the English language spoken in the United States has taken on its own unique characteristics, so, too, have the vampires. They have developed a style of their own, a Yankee style. American vampires are a powerful contemporary symbol, affecting more lives than one might anticipate. Why is the vampire so potent here? He is an all-American guy.

What follows is a documentation of ways in which Americans identify with and imitate the vampire's behavior. The intention is to understand the vampires' place in our culture and to examine their impact on the lives of ordinary people.

After her first experience at being a donor, Melody became a regular visitor to Pam for this purpose, sometimes seeing her once a month or going as often as once a week. "I do it for my own personal enjoyment. I mean, I have a very close and enjoyable friendship with Pam. Even if we didn't have this kind of relationship, I'd still be friends with her, so I think this just brings us closer." Melody and Pam have maintained this association for over three years.

Melody has never concerned herself with wondering if there was anything physically real about Pam's being a vampire. "It's not for me to say. If Pam drinks blood and she says she's a vampire, who am I to say she's not? I don't have any evidence that says she's not. I don't know any other requirements for being a vampire, except if you drink blood, and she does."

However, Melody rejects the vampire label for herself, and has never been a donor to anyone else. She has further discovered that it is wise not to let any boyfriends know about her activities with Pam. Melody learned this the hard way. On several occasions in the past she thought she would be honest with the men she was romantically involved with and let them know about her relationship with Pam. "I just kind of mentioned it. I mean, I didn't mention any names, but it kind of scared a couple of them off." As a result, Melody now keeps her donor life a secret from potential suitors.

She admits that the exchange of blood is a sexual act. "Oh yes, I believe so." She revealed that her donor relationship with Pam has led, like a natural step, to other erotic acts. "It wasn't that way in the beginning, but I think it just aroused some feelings in Pam and she wanted to go on."

While Melody's involvement with real vampiric activity happened accidentally, as a result of her friendship with Brad, Brad's interest in vampirism had begun as a sexual fantasy when he was a teenager. Seeing the female vampire Angelique on TV going after men in the gothic soap opera "Dark Shadows" sexually excited him.

Brad said he has engaged in considerable introspection about his unusual interest and believes that it is related to his being victimized and psychologically abused by his parents when he was young. In the past he

BLOOD FETISHISM

Norine Dresser's *American Vampires* was the first popular book to explore the modern world of practicing vampires. Sexual gratification through blood ingestion has been a well-documented clinical phenomenon since Richard von Krafft-Ebbing's *Psychopathia Sexualis* (1892), but it has only been in recent decades that real-life blood drinkers have cultivated an identification with vampires of the imaginary sort.

A contributing factor to this new phenomenon, no doubt, has been the mass rehabilitation of the vampire image in Anne Rice's novels and elsewhere; the once villainous revenant is now more likely to be presented as a sensitive outcast craving meaningful human contact. In the standard psychoanalytic interpretation, the root of the "vampire's" alienation is his/her arrested development at the oral/sadistic stage of sexual development. Blood, as a potent symbol of human warmth and belonging, where tangled emotions of love, pain, power and powerlessness (and, too often, the literal presence of blood) trap the child in a limbo-state of infantile rage and insatiable emotional hunger.

For some blood fetishists, the gratifying act is wholly masturbatory, involving self-bleeding, or, in more extreme cases, self-mutilation. For others, who share their practice, bloodletting/

tried to understand his fantasy while undergoing psychotherapy, but his therapist tried to discourage him by saying there weren't people who really drank blood. "You've got to realize it's just a fantasy. Nobody can drink blood," the doctor argued. He even tried to convince Brad that it would be impossible to drink this warm, red body fluid. "I'd like to see you drink a glass of it sometime," he once challenged, according to Brad.

Articulate and an avid reader, Brad tried to explain that in other parts of the world people do drink blood, citing African tribes, among them the Masai, who drink cow's blood with milk. His therapist was disgusted to hear of such things, but this did not discourage Brad. It was a fantasy he was not willing to relinquish, so when he met Pam in 1984 through a network of "Dark Shadows" fans, he was elated because Pam revealed she had previously experimented with vampirism and was interested in trying it with him.

At first he encouraged her to bite him. It was enjoyable, but painful, too. It was pleasurable in terms of his being able to finally live the fantasy in real life, and painful in the sense that it really hurt. Brad muses that his former therapist's eyes would have bulged at the scene.

While Pam obliged his fantasy needs at first by biting him, she prefers to prick the finger as a means of obtaining the blood. Eventually that became her most-used method with Brad and others as well. In addition, she has replaced the animal hypodermic needle with an Auto-Lancet, an over-the-counter drugstore device used by diabetics to prick their fingers for measuring blood sugar levels. Pam claims that it is safe, clean, leaves no scars, and is practically painless. Its one drawback is that only drops of blood can be extracted, but for Pam that doesn't matter. She needs just a small amount at a time — one teaspoon to one tablespoon, she said.

Asked about the sexual aspects of his giving blood in such a manner, Brad described a kind of psychic bond that grew between him and Pam. He recalled being at home and, when thinking about her, that part of his neck where Pam had bitten him would suddenly start to ache and draw. Despite the intimacy they shared, he held his erotic impulses in check because he didn't want to get involved with a married woman, especially with the wife of someone he genuinely liked and respected, a man who tolerated his wife's peculiar habits.

For several years Brad donated his blood to Pam on a monthly basis. He has always thought of this act as a gentle, harmless fantasy. He allows for the sexual aspects and believes it is better than the golden showers (a euphemism for sexual partners urinating on each other) or mud sports (coprophilia) or "any of those other disgusting things that some of the more normal people in our society pursue with great affectation and delight."

Brad sees the vampiric bite as only a couple of steps beyond the love bite, a few degrees beyond the hickey. "It's nice when it's under the proper setting, when it's controlled." Yet, he warned about some of those "crazies"

drinking is a preferred way of establishing intimacy or trust. Since human teeth are extremely crude instruments for opening veins, modern vampires tend to use razor blades, knives or syringes to make incisions; the blood may then be sucked directly, or, more ritualistically, sipped from a chalice, cordial glass, or other vessel.

The explosive growth of vampire literature/imagery/entertainment has given many blood fetishists a positive sense of connectedness for the first time in their lives. Several books have explored this unsettling sexual netherworld since Dresser: Rosemary Ellen Guiley's *Vampires Among Us* (1991), excerpted elsewhere in this volume; Carol Page's *Bloodlust* (1991), and Katherine Ramsland's *Piercing the Darkness: Undercover with Vampires in America Today* (1998).

Richard Noll's *Vampires, Werewolves, and Demons: Twentieth-Century Reports in the Psychiatric Literature* (1992) recounts several of the more psychopathic cases, including the notorious "acid-bath murderer" John Haigh, who killed nine people between 1944 and 1949 in order to drink their fresh blood. Haigh, before his trial and execution, was judged to be sane, despite his claim that "I was impelled to kill by wild blood demons." He assured his mother before his execution that "my spirit will remain earthbound for a while. My mission is not yet fulfilled."

who take it and turn it into a perversion, such as sadomasochism. "That's not so hot. It's something you have to be careful about."

Pam's own vampiric practices became fixed five years ago when she encountered Kristin, who lives in another state. Kristin claims her need for blood to be one cup per week during winter and three-quarters to one and one-quarter pint per week during the summer. She doesn't get "high" on blood. Rather, it makes her feel energized, full of life, satisfied. Animal blood doesn't supply what she needs. She says that without warm human blood she becomes weakened or sick. Once she even ended up in the hospital needing a transfusion.

It should be noted that the simple fact of blood arousal does not necessarily indicate a propensity for violence. While it is true that many serial killers do drink blood, a more typical scenario involves consensual erotic play involving relatively mild forms of biting, cutting, and sucking. Neither is blood fetishism necessarily correlated with satanism — many practitioners consider their activities to be essentially pre-Christian, rendering satanic considerations meaningless — though satanists are frequently attracted to the highly theatrical trappings of vampirism.

Kristin, also in her thirties, stated that she obtains blood from donors rather than victims, whom she describes as "Those who are forced to give." On the other hand, donors give willingly with or without an exchange of something, that is, sex, money, dinner. To find donors she usually avoids bars and picking up people. Instead, her friends find other friends, or she meets them at places such as bowling alleys, arcades, or occasionally in a library or at a park.

At times it is difficult to find donors. When the "hunger gets bad," she starts pacing. Sometimes she has to go to another town to find donors. "Explaining to someone of my need happens when the time is right. I sense it." She starts out talking about movies, then the occult, and from there she broaches the subject. "There have been times at the last minute people back out. I let them go. All is done willingly and with agreement not to tell."

Unlike Pam, Kristin prefers to bite her donors on the neck with her small, "fang-like" teeth. One of her donors describes it as being a pinch-like pain where penetration of the skin has occurred. Others say it feels like an injection. One former donor revealed that Kristin lacked tenderness and growled while she was biting him. That was frightening to him, even more so later when she explained that there was a panther inside her — a black cat. She felt that she was this black cat at certain times and didn't know whether she might rip a person apart. When the donor heard this, he decided never to give blood to her again.

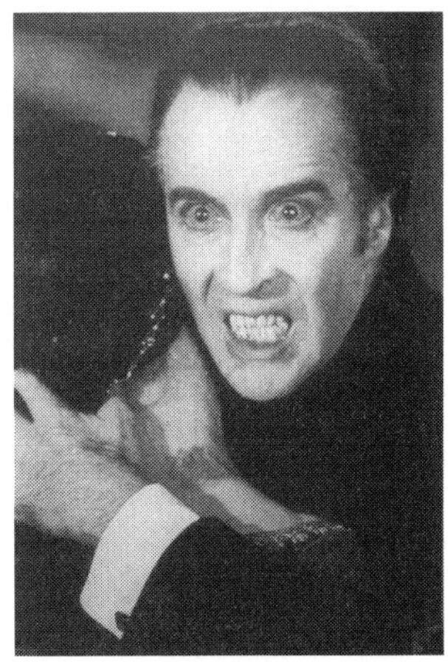

Blood was rarely shown in classic vampire films, but by the 1970s, graphic exsanguination was a standard part of the entertainment ritual. Here, Christopher Lee in Scars of Dracula *(1970).*

Rarely does the bite mark leave a scar, but according to Kristin, people don't care anyway because they're proud of the marks. "They ask for more. Sometimes for the sake of getting bit.... It fulfills their fantasy!" Instead of biting she would prefer using syringe needles, but she reports that people are squeamish about needles. "I've used single-edge razor blades to cut the skin, but again I'm told they'd rather be bitten."

Kristin perceives of herself as having been born that way. She says that vampirism ran in her family, with only one or two per generation. She claims that as an infant she kept biting herself until she bled. When she was growing up, other schoolchildren teased her about her pallor. Bright sunlight hurt her eyes, and today she must wear sunglasses during the daylight. Although she has inherited her condition, she believes she can pass it on to others through her bite and transmission of something she refers to as a "V" cell. She "transfuses" varying "dosages" of the "V" cell to different donors. Feedings take place day or night, but she prefers darkness and must be alone with the donor. Some donors tell her they get turned on when bitten.

Kristin asserted that she is not a satanist, does not worship the devil, does not kill people or animals, and is not with any witch coven. She works "solo."

It was through Kristin that Pam believes she received the "V" cell, transmitted when Pam was initiated into the donor role for Kristin. This subsequently created Pam's need for donors, which began to manifest itself during a full moon about six months after Pam had first given blood to Kristin.

George Hamilton in Love at First Bite *(1979) represented the glamorous vampire ideal – an all-American icon of material and metaphysical "suck-cess." (Photofest)*

Pam describes physical changes she believes have taken place in her as a result of this "V" cell transfusion — a kind of itching and throbbing in the gums above her canine teeth, especially during the full and new moons when she has a craving. Pam says that hot weather increases the hunger. "It feels like little pinch-like sensations or a bite or a burning sort of feeling in the veins." Kristin, too, reports the same type of sensations, describing her veins as feeling like they are on fire during hot weather.

Pam's eyes now seem to be more sensitive to light and glare, and when she has the "hunger" they take on a more animal- or cat-like look. In fact, she claims that her husband is able to detect these changes and will say, "The cat's up,"or "Hey cat!" Pam says that the cat aspect of her comes out a lot, particularly when she and her husband make love. "It's like the beast gets excited or something. There's something there," she said in a puzzled manner. "Like it's not all me. Maybe it's possession," she speculated.

Pam doesn't quite understand the changes going on within her. She reports that ever since she has been giving and taking human blood she has stopped being plagued by canker sores. She doesn't comprehend why. In addition, she wonders whether or not her condition would ever be passed on to her child, if she were to have one.

She is not certain whether or not she has passed on the "V" cell to any of her donors, although she reports that some of them have noticed changes within themselves. For example, one of her female donors reports a "pressure" feeling over her canine teeth which is stimulated by exposure to sun and heat.

Because of the potential transmission of the AIDS virus through the exchange of blood, Pam is very concerned about her donors' health and asks them numerous personal questions before taking their blood. She is reluctant to take on too many donors or unknown ones, so now when her hunger is aroused and she has no one to meet her need, she substitutes by eating raw hamburger or cows' hearts, even taking blood from her husband or herself to satisfy the craving. In contrast, Kristin is less worried about AIDS and says she can "tell" when someone has good or bad blood. She doesn't seem to be concerned about risks to her health.

Since Pam has only been drinking human blood for a few years, she believes she could stop the habit, if necessary. However, at present she finds the feeding so emotionally and physically satisfying that she has no desire to discontinue the practice. She says that a few hours following a feeding her symptoms of need stop — her eyes become normal, and she sleeps well. It has a calming effect on her.

Here are two women who appear to take pleasure — even pride — in the act of vampirism and in calling themselves vampires. Neither of them seems to have any qualms about being associated with a creature who in the past was reviled. They seem not to be dealing with the notion that they will live forever, but rather to be applying the vampire name to themselves based primarily on their habits of drinking blood. There is no association whatsoever with the undead state — not being able to rest in peace — usually associated with vampires.

Kristin's and Pam's need to be associated with vampire behavior may be one way of gaining attention or giving themselves a sense of power. Perhaps it is just another way to create a stage where they can perform and star in front of a captive audience. It would take lengthy and psychoanalytic analysis to discover their motivations for the need and rewards of this unusual behavior — something beyond the ken of this study.

Brad mentioned that he did not enjoy feeling like a slave to the vampire. Instead he preferred being the recipient of unconditional love, his term for the way Pam treats him in his role as donor. Maybe this is related to the history of abuse by his parents, particularly by his mother, which he feels always restricted him to merely superficial relationships with women late in life. He recalls the *"Sturm und Drang"* of his adolescence and the anxiety he felt about relating to other women. This he ties directly to his attraction to the female vampire relationship.

On the other hand, Melody appears to be someone who merely enjoys sampling the variety of spices that life offers — someone who is very open and experimental in everything she does.

Jeanne Youngson, Ph.D., founder and president of the Count Dracula Fan Club, has been conducting interviews for over eleven years with people who have a need to drink blood and who think of themselves as either vampires or blood fetishists. While the act of drinking blood is the same for both vampires and blood fetishists, there is a difference in the way these people think of themselves. For some, being a vampire has a more romantic and sympathetic connotation. This label sets them apart from ordinary persons, making them more exotic by virtue of their blood-drinking habits. For others, the term *blood fetishism* is more scientific and satisfactory. Almost all of these blood drinkers have revealed a traumatic childhood, being victims of either abuse, neglect, abandonment, loneliness, or molestation. While Youngson has dealt with only a small statistical sample, the regularity of reports of a disturbed childhood background

DRACULABILIA:
THE WORLD OF VAMPIRE OBJECTS AND COLLECTIBLES

In recent decades, vampires have been popping up in the strangest places: at the supermarket, in advertising, as toys and household objects. In fact, there's virtually no area of culture or commerce that has remained untouched by the undead. Grateful thanks to Jeanne Youngson and the Count Dracula Fan Club for providing the following examples.

Vampires can impart almost supernatural efficiency to familiar services.

has made her hypothesize that there is a definite link between the need for blood and the need for a close, warm human connection.

She makes this generalization based on life histories that her informants write as well as on information elicited from a questionnaire that asks, for example, about the first blood-drinking experience, sexual experiences, and fantasies. In addition, she asks informants if their childhoods were happy, if they were ever molested, and if they believe that sexual abuse has contributed to their vampirism.

Sometimes she receives unsolicited letters from people involved with vampire life who are heard from only once, contacts who unmask their feelings to her, never to be heard from again, so she is never able to fill in important informational gaps. One such person was Belle, a woman who wrote, "I am a self-made vampire! Due to sheer boredom I have created an exciting personality out of a drudge and have never been happier!"

A vampire air-freshener: modern chemistry overtakes garlic. (Photo: Donal F. Holway)

Belle divulged that no one knew about her hidden self, which made it that much more exciting. She carried out her concealed life by writing herself very long, sexy letters from a secret lover, an imaginary vampire called Malcolm. Sometimes when she left the house, she slipped those letters under the door, so that when she returned home she could find them and be "surprised."

She named her lover Malcolm after an uncle who always brought her presents as a child. He was tall, and in spite of his acne scars, she always thought he was good-looking. Unfortunately he was killed in an auto accident at the age of thirty-two. Ever since she has always liked men with acne scars.

Belle wrote of her husband that he was "a good man but not very romantic." He was a high school dropout, earning a living as a long-distance trucker. Belle's parents had been against their marriage, but in spite of this they were wed as soon as she was graduated from high school. Her own children now are married and have moved away.

For Belle, "being a vampire has opened up a whole new world for me." She describes herself as having her feelings easily hurt if her children don't write or send her a birthday card or come to visit as often as she wishes. Having another life unknown to others reduces her pain and lessens her need for attention from her family.

She is astonished by the realization that her fantasy life could make such a difference in her real life, but there was a time when she worried that she was going "bonkers." Because of this concern, she consulted a psychologist whose name was posted on her dentist's bulletin board. She made an appointment, spent fifty dollars, and told him about her private thoughts.

The psychologist was reassuring, said Belle. He convinced her that as long as she was aware that it was only make-believe, there was nothing to worry about. Belle now feels at ease with her fantasy world because it has provided an escape from drudgery, loneliness, and emotional neglect. It is unlikely that she will ever be heard from again.

Then there is Donna, who as a lonely youngster was introduced to blood drinking by her next-door neighbors, a German couple who gave the child the love and attention she missed from her parents. The husband was a butcher, and he would bring home packages of steak, drain the blood into a cup, and drink it.

Vampire lapel buttons.

This couple introduced the girl to the practice, and eventually she began to enjoy it. The early death of Donna's mother, a traumatic encounter with an exhibitionist, and an unhappy childhood stood in sharp contrast to the happy moments the girl spent with her neighbors, who eventually moved away. Now as an adult, she associates drinking blood from meat with warm, nurturing memories of her neighbors and sexual fantasies about the butcher. Even though drinking blood from meat is not uncommon, nor is it an aberration, Donna considers herself a vampire.

Why she has decided to give herself this label is curious. When Youngson asked her to explain the difference between a blood fetishist and a vampire, Donna said, "A blood fetishist is somebody who is mentally sick and gets blood by illegal means. A vampire is somebody who likes to drink blood for reasons of his own." She explained, "I'm not too sure about these things, except that blood fetishist sounds like it has to be against the law. I don't feel as if I am breaking any laws with my blood drinking, and I do consider myself a vampire."

Andrew is another person who drinks cow's blood and thinks the habit qualifies him as a vampire. He reveals a tragic background. His parents were killed by a runaway truck that crashed through a diner where they were eating. Andrew was then sent to live with his uncle and aunt and their four children, but his aunt rejected him and was particularly perturbed by his bedwetting habits. He felt humiliated when she made him wash out his soiled linens by hand in the barnyard trough and hang them on the clothesline. Sometimes they froze and he had to sleep in the hayloft.

One day a friend told him about the men in the slaughterhouse who drank cow's blood, so he tried it. He made a small slit in the neck of a cow and drank its blood. To Andrew it was warm and nourishing and he felt stronger after drinking it. Years later when he saw the film *The Killing Fields*, which in one scene shows a prisoner of war drinking cow's blood, he was shocked and nearly "fell out of my seat!" This reawakened his memories and pain.

Another young man contacted Youngson. He thought he was Dracula and for that reason tried to remain indoors as much as possible. He limited his outdoor activities to nighttime hours and during the summer would visit cemeteries and lie atop graves. He claimed that this was the only time he felt content. He practiced other vampire traditions — avoiding mirrors and garlic — and delighted in sucking his loosened teeth until his gums bled; then he swallowed the blood.

What is unique about this young man is that he is one of the few who has successfully rid himself of these habits through the use of psychotherapy.

Today he attends one of the top East Coast universities. In retrospect, his former vampire activities may have been related to being left alone for long periods of time because of working parents. During his unsupervised hours he imaginatively decorated his bedroom to look like a tomb. His parents, pleased with his ability to amuse himself in a constructive manner while on his own, inadvertently encouraged his unusual interests and bizarre behavior by delighting in his handiwork.

Youngson told of another case involving one of her informants, who, because of her blood-drinking habits, thought of herself as a vampire, at least during her adolescence. Now she considers herself a blood fetishist. As a youngster, Rose had attended Catholic schools and had been impressed with pictures of Christ bleeding and being whipped and tortured. These images and the nuns' repeated message that Jesus loved them so much that he bled for humanity's salvation deeply affected this young woman.

Above and opposite: vampire figurines. (Photos: Donal F. Holway)

Years after, when Rose became an elementary school teacher and the children injured themselves on the playground, she cleaned their wounds and saved the bloodied cotton, later sucking the blood from it. This habit was never discovered by anyone, but eventually she quit teaching and got a job in a factory employing many women who had strong beliefs in voodoo and witchcraft.

In this setting Rose was able to obtain blood by promising to work spells for other personnel in exchange for three drops of their blood, which they put on a clean piece of paper. Fortunately for this woman, most of what her fellow employees wished for was fulfilled, so she was able to have her blood goodies regularly, while her co-workers remained content.

According to Youngson, some informants have revealed inventive ways of obtaining blood. For example, one woman acquired it by raking her lovers' backs with her fingernails, then licking off the blood from the nails. This woman, who considers herself a "glamorous vampire," claims that most men love this habit and have told her that she is the greatest lover they ever had.

Then there is Deborah, a flight attendant with a major airline who has developed a need for human blood. She uses her boyfriend as her intermediary and source. The boyfriend works in a hospital blood bank. Twice a week he siphons fifty cubic centimeters (about two tablespoons) from the hospital supplies. He does this by using a syringe and withdrawing twenty-five to thirty cubic centimeters of blood out of a five-hundred-cubic centimeter container, transferring it into a small bottle which he carries to her in his pocket. He has never been caught and the amount he pilfers is enough to satisfy her. However, given the AIDS scare in relation to blood, it is possible they will discontinue this practice until such time that accurate hospital testing can be guaranteed.

Because people are reluctant to publicly admit to the practice of ingesting another person's blood and the numbers involved with this are so small, no statistics have been gathered regarding the direct hazards of contracting AIDS in this matter. However, there is indirect medical evidence to suggest that ingesting blood might be a risky procedure.

There have been several cases where health workers were splashed with AIDS-contaminated blood on cracks or rashes on the face. This resulted in their testing positive for the AIDS virus. In addition, there has been one recorded case of AIDS in a woman who had reported only oral intercourse, where she was a recipient of the semen, which also carries the virus. Therefore, oral contact was the method by which she contracted the disease. Furthermore, the lining of the mouth is similar to the lining of the rectal area and it is well known that there is increased risk of transmitting AIDS through rectal intercourse.

On the other hand, if AIDS-infected blood was ingested, once it moved down into the stomach the digestive enzymes would likely destroy the virus. Thus, the danger of getting AIDS from infected blood would come more from the hazard of contact with breaks in the skin of the mouth (including cold sores) or face, rather than from the chance of it getting into the bloodstream from the stomach. Consequently, a realistic fear of AIDS has affected the habits of those who think of themselves as vampires.

Dee, an Ohio woman who calls herself a vampire, says that she has almost completely stopped taking blood from people unless she's very familiar with their life-style. Now she obtains the blood she needs from raw beef steaks, which she prefers to pork (so as to avoid parasites) and chicken blood, which she considers to be "just plain disgusting." For Dee these substitutes seem cold; she misses the intimacy she felt when taking blood from another living being.

One last self-labeled vampire from Youngson's file is a man who calls himself a "metaphoric vampire" because he engages in illegal business practices. "If the people I con are suckers (no pun intended), they deserve to be 'bled.'" He claims no conscience as far as women are concerned: "Females are such gullible dears. I date only the beautiful — flight attendants, models, showgirls. The younger the better."

He has deliberately extinguished all feeling toward his fellow human beings. He does not care if there are earthquakes and thousands die. He does not care if foreign countries blow each other to bits. He does not care if a person in the next apartment is murdered or the homeless freeze to death by the millions. He just doesn't care.

He excuses his behavior. "Allow me to point out that I am no worse than the average man. I believe we are all vampires in one way or another. Show me the person who will really inconvenience himself to help a stranger, and I'll show you a non-vampire."

This man's outlook is unlike most others who identify with the vampire. He does not seek closeness or intimacy. Rather, he seeks revenge and he desires to be the cool taker of whatever advantage he can gain over others. For him, being a vampire brings power and control. He seems to hold his fellow man in contempt.

There is another group of persons who identify with the vampire in a different way, who think of themselves as outside the schema of contemporary vampire phenomena. These are members of the Order of the Vampyre, a special-interest division within the Temple of Set, which evolved from Anton Szandor LaVey's Church of Satan.

LaVey had attempted to bring respectability to the Church of Satan and at the height of his religious career was a media celebrity, appearing on the Johnny Carson show and in national magazines. His professional background included being an oboist in the San Francisco Ballet Orchestra, an assistant lion trainer in the Clyde Beatty Circus, a nightclub organist, and a police photographer. Along the way he developed an interest in the occult, and became a collector of books on the topic that led to his creation of a Magic Circle study group.

This was the genesis of his San Franciso-based Church of Satan, which he founded in 1966 and which became famous for its colorful black masses open only to Church of Satan members in good standing featuring a live naked woman posing as an altar to symbolize the pleasures of the flesh, an important tenet of the church.

A vampire novelty record.

LaVey's goal was to encourage the study of the black arts and repudiate what he saw as the religious hypocrisy of conventional society. However, in 1975 some of LaVey's members resigned to form a new organization, the Temple of Set.

Setians value both individualism and self-actualization. For those who have risen to qualify for the Order of the Vampyre (OV), their goal is to become acquainted with the characteristics of the Vampyric Being. They do not use blood in any manner, nor do they identify with the undead state. Instead they look at some of the alleged powers of the vampire creature, seeking the essence behind the myth, the noble quality of the archetype — elegance, sophistication, and charm — as personified by the Count Saint-Germain vampire figure in Chelsea Quinn Yarbro's novel *Hôtel Transylvania*. In Yarbro's work, the Count is loosely based on a mysterious, eighteenth-century historical figure who practices alchemy, has a passion for diamonds, and is conspicuous in his fine-quality black-and-white clothes.

Recognizing that the real Vampyre can be hideous, the OV acknowledges that it can be glamourous as well and encourages learning the skill of applying cosmetics, using their voices effectively, and acquiring methods of holding another person's gaze. One of their co-Grand Masters is Lilith Aquino (no relation to the president of the Philippines). Her attractive appearance — red dress contrasting with well-made-up fair complexion and striking black hair — during an interview on the Oprah Winfrey television show gave testimony to the skillfulness she has acquired, serving well as a role model for other members of the OV.

Development of the powers of imagination, visualization, and invisibility is encouraged in this order along with the use and study of art, music, and literature. It is clear that despite their name, these Vampyres have a different agenda from the other vampires already described. Perhaps that is one of the reasons they have chosen to spell their name differently, which also helps to maintain their sense of mystery.

Since most of the OV information is never made public, the number of members remains unknown. All they will reveal is that they have been in existence for several years, that the membership comprises roughly a little over a third of the Temple membership (also an unknown figure), and that male and female members are equal in number. The OV meets during the Temple's annual International Conclave in places like Las Vagas, New Orleans, Toronto. During these meetings, Setians from all over the world gather for a week of social and magical events. That is when the OV has its formal meeting as well as its ceremonial Ritual Workings. Outside of these yearly gatherings, church members do not congregate on a regular basis. They claim that their deliberately individualistic atmosphere is not easily conducive to group activities on a routine or programmed basis and proudly point out that they are not a congregation of docile followers — only cooperative philosophers and magicians. As a result, members of the Order of the Vampyre communicate with one another primarily through their newsletters *Nightwing* and *The Vampyre Papers*.

Vampire dolls. (Photo: Donal F. Holway and David J. Skal)

Ironically, the Temple of Set deplores other kinds of vampires, called psychic vampires. This idea is based upon a passage found in Anton LaVey's *The Satanic Bible*, which describes psychic vampires as persons who make you feel guilty if you don't do favors for them — those who, like leeches, drain others of their energy and emotions.

Advising that you should not waste your time with these people because they can destroy you, LaVey gives concrete advice about avoiding their clutches. In the "General Information and Admission Policies" of the Temple of Set, there is also a negative reference to psychic vampires, referred to as those who wish to partake of the Church of Satan without making a personal commitment to it, to enjoy aspects of the philosophy without contributing. They cite those persons who "continued to vampirize the Church indefinitely," contributing to its disruption. In spite of his warnings and advice to others, LaVey himself had difficulty in controlling this phenomenon under his own organizational roof.

Setians are firm in their desire to prevent this from happening to them. They will not tolerate psychic vampires in their organization, promising that if they were inadvertently to become members, once identified, they would be asked to leave.

The Temple of Set is currently headed by Lilith's husband, Michael A. Aquino, Ph.D., a controversial figure by virtue of his unusual looks — inverted v-shaped eyebrows, Friar Tuck hairdo, round face, and dark robe garb — and his occupation. He is a Vietnam War-decorated lieutenant colonel in the U.S. Army, now on active reserve, working in military intelligence as a psychological warfare officer, and holding security clearance for top-secret information.

In his appearance with his wife on Oprah Winfrey's show, Aquino mentioned his military occupation, defended the Temple of Set from those who would link it with more sinister satanic groups, and agreed that his unique appearance probably served as the model for one of the actors in *The Omen* series of movies.

According to an army public affairs officer in Washington, Aquino's satanic practices are protected by his constitutional freedom of religion. Furthermore, his record shows that he is trustworthy and capable.

The term "psychic vampire" is used outside the Temple of Set but with different connotations. Stefan, another self-proclaimed vampire who volunteered information, is in his early fifties. He was born and raised in a Polish neighborhood in a small Pennsylvania town where his parents ran a mom-and-pop grocery store located across the street from their house.

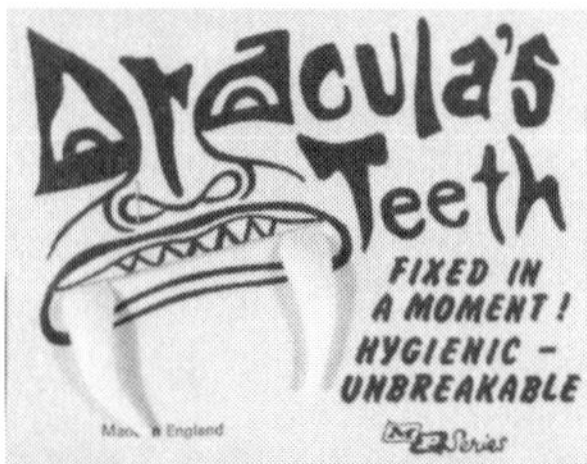

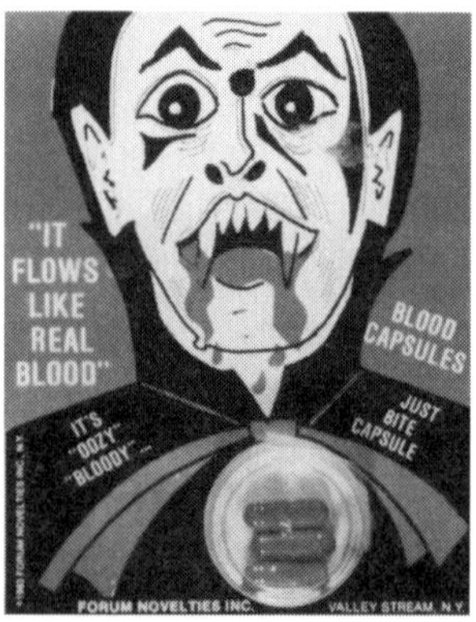

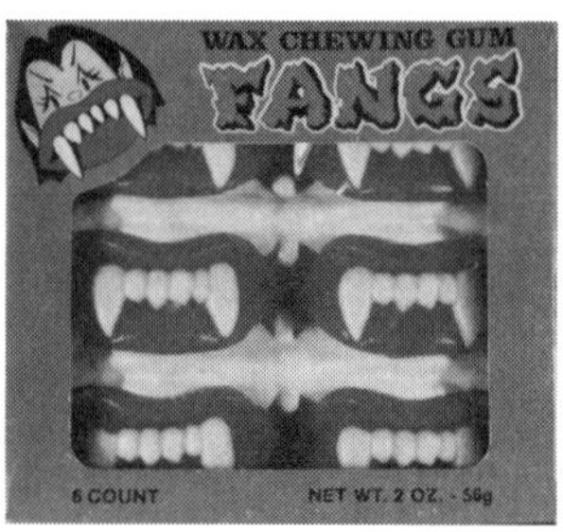

Fangs and blood: what no vampire should be without.

As a youngster Stefan had many responsibilities — taking care of his beautiful younger sister, whom he adored; translating for his parents, who spoke little English; cooking; and performing household duties. At Catholic school the nuns made painful innuendoes about his delicate looks and high-strung personality, calling him by female names — Anna and Marya. In spite of his preference to play with girls more than boys, Stefan was an avid sports fan and attended all the high school football and baseball games.

A traumatic event shook Stefan's life when his sister was four. She was hit by a car while chasing a ball into the street. At the time of the accident, Stefan was at a baseball game, so on his way home when he saw blood on the road near his home, he was unaware that it was hers.

Stefan and his family fervidly prayed for the injured child during the days she spent on the critical list. "It was a terrible time for all of us until we found out she would live," he recalls with anguish.

"The worst part was that the blood stayed on the street for days afterward," he remembers. He tells of being haunted by the accident and having recurring dreams in which she was hit over and over and over again. There was always a lot of blood in these dreams, which were in vivid color.

Stefan admits feeling guilty and responsible for the accident. He believes that if he had gone straight home from school instead of going to the baseball game, he would have been watching her and would have been able to prevent her from running into the street. To punish himself, he obtained some switches and began beating his legs until the blood ran. He did this off and on for the next several years, whenever his sister got cranky, which she often did. According to her brother, the little girl changed dramatically once she came home from the hospital. She was no longer the darling child she had been before the accident. She was touchy and flew off the handle at the slightest provocation.

The next years were equally difficult for Stefan. His mother had a heart attack, his father closed the store to be with her in the hospital, and his sister went to live with an aunt. During the time the boy was alone in the house he discovered a large jar hidden in the back of a closet. It contained a male fetus, which Stefan described.

"Its eyes were closed and its little fists were clenched tightly as if in anger.... I knew immediately that it was my twin brother."

No one had ever told him that he was a twin. All he knew was that he was born at home in his parents' bed, yet instinctively he had always felt that a part of him was missing.

> Sometimes when I'd be walking down the street I would feel there should be someone walking beside me. I got the same feeling when I went to the movies and there was an empty seat next to me. I had always felt the "absence of a presence."

In spite of Stefan's intuition that there was a part of him that was lacking, finding the fetus was a terrible shock. At the same time he describes the discovery as being a form of relief as well. He returned the jar to the closet and since that day, he never again had the feeling of incompleteness because "I knew where he was."

A vampire coffee mug. (Photo: Donal F. Holway)

After this time he began experiencing great anxiety and decided to experiment with relaxation exercises and ESP. He read books about psychic phenomena and tried sending mental messages to people while they slept. He was quite successful with his sister.

However, Stefan was still having the "blood dreams." Now they involved his kneeling down in the street and licking up his sister's blood. He also continued to beat his legs with a switch, only now he licked the blood off after he was finished. This caused feelings of faintness and lightheadedness.

Stefan describes his first venture as a psychic vampire. One day at a bus station he noticed a blonde, sixteen-year-old girl asleep on a bench. He sat down across from her and tried his first out-of-body experiment. He concentrated on floating up and over to her and sucking blood from her neck.

After a few minutes she woke with a start and glared at him. This convinced Stefan that he had succeeded. "I was a successful psychic vampire!" Immediately he felt free of anxiety and remained that way for nearly a week.

Stefan's experience is the reverse of most psychic vampirism cases, which are usually reported from the point of view of the "victim" who feels depleted or sapped of his or her strength and emotions. In this situation there is no way to know what Stefan's "victim" experienced or thought.

Stefan continued his experiments on people napping in other public places: parks, libraries, and once in a dentist's waiting room. He claims that he always felt relaxed and somehow superior afterward. At this time in his life he worked in a diner, but still lived at home.

The turning point in Stefan's life came when he met a young

hunchbacked woman at a Protestant church service. She befriended him and said she sensed a deep unhappiness within him. Eventually he told her of his "shameful secrets." Stefan believes that this friend, who as a disabled person had also known great suffering, provided him with deep understanding. With her help he managed to stop his self-destructive habits and eventually pulled himself together both mentally and physically, so that today he is able to live on his own and run his own small diner.

Stefan's story seems to have a happy ending. For him, being a psychic vampire brought relief from the gnawing feelings resulting from a lifetime accumulation of pitiable events and scars. The vampiric act provided him with a closeness which he had felt lacking all his life. From an outsider's point of view, Stefan had no physical closeness with his victims. Yet, his perceived ability to psychically extract some essence from another human being filled the void caused by the many losses he suffered — lack of attention from his overworked parents, little respect from the nuns, the death of his twin brother, and lack of comradeship from other male friends.

A Dracula model kit from the 1970s.

Stefan also regrets never having instigated a sexual relationship with his understanding woman friend, his "savior." "I'm afraid that at the time, I was not able to cope with her physical disability."

Today he is still alone and celibate, but accepting of himself.

Other individuals have taken on the identity of the vampire, but with other outcomes. For example, in Ft. Lauderdale, Florida, in 1977, a man six feet four inches tall with long hair was terrorizing children and adults at two local elementary schools, one middle school, and a high school. Wearing a cape and fangs, he would run up to the students and spread his arms wide. The children ran home crying. Reports said local officials' "blood was boiling!"

A similar prankster intermittently harassed a family in the Seattle area. According to police spokesperson JoAnn Cratty, he had a vampire scar and wore heavy makeup, a cape, and black clothes. His victims described him as being young, tall, and slender and driving a black vehicle — on one occasion, a hearse. He merely smiled and escaped into the night. Ms. Cratty said the visits came at irregular times and did not seem to coincide with symbolic occasions, such as Halloween or a full moon. But they were always at night.

Pranksters dressed as vampires are not unique to the United States. They have also been found in Yugoslavia, where vampire impersonators harassed villagers and vandalized villages many years ago. In one incident some men used vampire cloaks for love trysts with young women, while in another an inventive lover dressed as a vampire to frighten away inquisitive neighbors while he carried on a three-month affair with a widow.

Other instances in which people dress or behave as vampires have been recorded. In 1985, a self-proclaimed vampire, proposed to help the Missouri Tigers College football team beat the California Golden Bears. In painted white face highlighted with purple veins, wearing slicked-back black hair, fanged teeth, black-and-red cape, and carrying a walking stick, he handed Coach Woody Widenhofer a business card identifying himself as a "vampire for hire with five hundred and eight years' experience."

The vampire offered to put a Romanian curse on the Tigers' opposing team free of charge, in the form of a hex on the goal line so they would not be able to score. He was concerned that if the Tigers did not take up his offer the team would lose again, which Vladimir considered tragic. He boasted, "Once put hex on ship. I think they call it *Titanic*. It would not let Dracula on board." The Tigers refused the vampire's offer. They lost.

On April 2, 1981, the Associated Press ran a story datelined Mineral Point, Wisconsin, where a police officer making rounds through the Graceland Cemetery, spotted a six foot three inch tall "ugly person" with a white-painted face wearing a dark cape. After encountering Officer Pepper among the gravestones, the man disappeared. Following the filing of this report, there was a rash of creepy white-faced vampire pranksters jumping out and scaring people on the streets of the downtown area the following night.

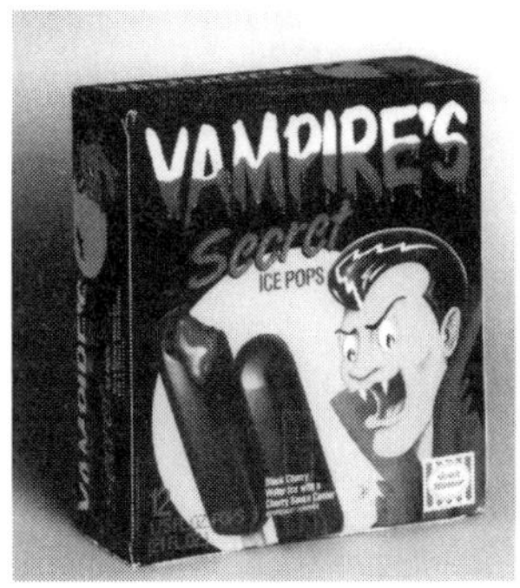

Something to suck on: a delicious vampire ice-treat. (Photo: Donal F. Holway)

The police were not eager to return to the cemetery either. A senior official reported, "I can't get anyone to go back to the cemetery at night. I even offered to pay [the officer] overtime to stake the place out for his vampire, but he wouldn't bite."

All examples given thus far seem to indicate that those persons who attempt to emulate the behavior of the vampire are doing so because of the power it gives — allowing one person to manipulate another both physically and/or emotionally — as exemplified by the behavior of the pranksters who terrorized unsuspecting strangers, by the metaphoric vampire who looked down on others, and by Stefan, too, who talked about feeling superior to his victims after a "succcssful" psychic vampire act. Power is clearly the motivating attraction for members of Order of the Vampyre as well.

Fulfilling a need for intimacy is also revealed as a reason for informants' unusual life-styles. Certainly Brad, Donna (who attached herself to her kind neighbors), orphaned Andrew, Dee, Belle, and the boy who laid on tombs all disclose a lack of closeness and a neediness for human connections. For some, like Pam, Melody, and Brad, the physically close relationship can also be an erotic act.

The taking on of the vampire persona seems to satisfy a need for exhibitionism or to serve as an attention-getting device. It makes one different from others. Certainly acting like a vampire qualifies for making a person "stand out in the crowd." Based on what these examples reveal, the vampire provides glamour and an escape from the humdrum life.

So far no one in the illustrations just given has been hurt. On the other hand, a vampire incident with a tragic consequence took place in Parma, Ohio, on November 1, 1981. A man dressed as a vampire for a Halloween party died after accidentally stabbing himself in the chest with a knife. In an attempt to make it appear that he had a stake in his heart, twenty-three-year-old Ernest A. Pecek drove a pointed, very sharp, double-edged knife into a pine board taped to his chest.

Police are uncertain about what happened, but they have labeled it an accident. Pecek either forced the knife into his chest accidentally with a hammer as he was pounding the knife into the board, or he perhaps fell with the apparatus in place. The victim's landlady said he was able to get upstairs from the basement, but collapsed before he could tell her what happened.

There have been other tragedies associated with vampiric activity. On February 12, 1988, a fourteen-year-old girl in a small southern town placed a gun in her mouth and killed herself. Stories circulated that she had been a heavy metal music fan and that she belonged to a satanic cult. When spokespersons were asked if the child also had been involved with drug use, no definitive answer was given. However, it was known that at age thirteen this girl had seen the movie *The Lost Boys* many times and had been intrigued by what she viewed as the attractive life-style of the movie's four vampire teenagers, who wore long, bleached hair, earrings, leather, and metal.

Vampire candles. (Photos: Donal F. Holway)

In the film, these "lost boys" roared about at night on motorbikes terrorizing their town of Santa Carla, "murder capital of the world," by kidnapping and killing the locals. (One month later, in Sauk Rapids-St. Cloud, Minnesota, *The Lost Boys* was again cited as a source of inspiration, this time for the gruesome murder of a thirty-year-old vagrant by three teenagers who licked the victim's blood off their hands after they beat and stabbed him to death. The sheriff proclaimed that one of the boys cited the *Lost Boys* video as the source of his interest in vampirism. Subsequently some videotape shops in the area removed the film from their shelves, while others reported an increase in rentals.)

Rumor had it that the girl wrote a letter to her cousin explaining her suicide motive. Verification of the contents of the note has been impossible because the police are withholding this evidence. However, as the story goes, and as it rapidly circulated through the town of 37,000 inhabitants, the girl promised to return to her school reincarnated as "Samantha the Vampire," on Monday, February 29, Leap Year's Day, to "get" some of her friends and bring them back with her as vampires. According to one person, she would be doing these friends a favor by having them join her in a better way of life. Some versions of the story claim that she had prepared a hit list of twenty-five fellow students she intended to kill. Although the police held back the letter, a report in the local paper stated that a minister had shown the alleged letter to the principal of the school which she attended, and he verified that the note did contain references to vampires.

The suicide and subsequent rumors that flew through the community caused pandemonium at the school, where some children delighted in sneaking up behind their classmates, grabbing them, and saying, "He's going to get you!" or "You're one of the ones!" or "The vampire's going to get you!"

Underlying and long-term fears of satanic cults were renewed among both parents and children. Because the town was located in the Bible Belt, where the population believes in predestination, where the teaching of Revelations occurs in local churches, and where local graffiti consist of "Satan Lives'" and "Satan Rules!" the vampire satanic cult issues immediately became intertwined.

During the time period between the girl's death and her threatened return, some students wore clothes bearing satanic symbols, such as the inverted cross and pentagrams. Other children wrote 666 (the symbol of the devil) on their hands. This behavior spread to another school where girls used their mascara brushes to paint 666 on bathroom walls.

The school started a crisis intervention program, beginning the first day of classes after the girl's death. Between sixty and seventy-five at risk students — friends of the dead girl, as well as teenagers previously identified as potential suicides — were brought in for counseling sessions with the school system's mental health professionals.

School officials contacted the parents of the at-risk students, giving advice and requesting that they watch their children carefully. The local police chief urged parents to keep their children in school on the targeted date as a way to discourage the other children's fears. Furthermore, students were warned that anyone bringing weapons to school on that day would be arrested and that the principal would personally press charges against them.

The Friday before Samantha's predicted reappearance, school administrators held an assembly to encourage attendance the next Monday. They assured students that there would be a police officer on duty all that day. By the time Monday arrived, rumors and fear had spread to other schools in the area, and police officers were stationed at all the middle schools in the school system.

What is significant and horrifying is that this girl apparently desired to be a vampire, fantasizing that she would be able to return, in the legendary fashion, to convert others to her state. It appears that she saw this behavior as being an improvement over her experiences of real life. Was ordinary life for this fourteen-year-old that hopeless, that difficult, that colorless, that she had to resort to this dramatic act? What qualities of these fictional creatures provided her with a solution to her problems?

Vampire Halloween masks.

Some mental health experts suggest that children's suicides more generally reflect family situations where great communication barriers exist.

Certainly a community health professional is in a better position to interpret the tragedy than a folklorist. However, what cannot be ignored is that on the day of her threatened return as "Samantha the Vampire" 225 children stayed home from school, representing a 25 percent absentee rate, as opposed to the normal rate of only 7 or 8 percent. The sentiment conveyed by some parents was, "Yeah, I'm taking my kid out of school. These chain-wielding, bat-carrying people at school will never hurt my baby."

While it is impossible to assess the motivation of each family's decision to keep their children at home, implicit in this act was some belief that their children's welfare was threatened. Once the word "vampire" was introduced into the community tragedy, an unpredictable wild card came into play and many parents, understandably, wished to take no chances of their children being hurt or frightened or becoming victims of some mentally unbalanced person desiring to take advantage of the situation.

It may not have been a wholehearted fear of vampires. Perhaps a "just-in-case" attitude was behind their decisions. On the other hand, we can't totally rule out the possibility that for some of the parents and children, vampires can and do exist. Beliefs in vampires as real entities might seem anachronistic in today's high-tech and scientifically aware society, but, surprisingly, there are pockets of belief in this unreal creature.

A popular vampire board game. The television serial Dark Shadows *did much to promote vampire awareness in the 1960s and 1970s.*

An Interview with a "Vampire"

DAVID J. SKAL
(1993)

WHEN I WAS a little girl," the vampire intoned, "television was a very new thing. I think we were the first people on the block to get a television."

It is only appropriate that the vampire now lives on the outskirts of the television and film industry, on a quiet residential street in North Hollywood. An ample, apple-cheeked woman resplendent in a floor-length red velvet dress, Megan the vampire is a part-time actress who occasionally plays walk-ons on network sitcoms, usually cast as a housewife or aged into a cherubic grandmother. Millions have had a fleeting vampiric visitation in their own living rooms, without ever suspecting. The vampire likes it that way.

Megan introduces the visitor to her longtime companion, Douglas, as well as to Christopher, her donor. Douglas, it turns out, is not interested in donating to the vampire cause, although Megan has "been after him for years."

The vampire was born during World War II in southern California. Her father had been born in London in 1897, the year Bram Stoker published *Dracula*, a fact Megan believes explains much. A naturalized U.S. citizen, "my father tried to shake his Britishness, but he could never really scrape it off his shoes," she said. "But his ideas on how to bring up a young girl came straight out of Edwardian England. The golden rule in our household was 'No.' "

When Megan was still quite young, her mother thrilled her with eastern European folk legends and ghost stories. "My grandmother was of Polish ancestry, and my mother picked up a lot of strange tales from her that she passed on to me in turn. And the happiest experiences I had with my mother were listening to her 'witch' tales. They didn't scare me. They delighted me."

But one kind of narrative was absolutely taboo. "Vampire stories gave my mother the willies," Megan remembered. And the forbidden nature of the subject, of course, made it all the more fascinating to Megan. She doesn't remember when she first saw Bela Lugosi in *Dracula*, but knows she was fully familiar with the movie before the age of ten. Universal had given the film a major re-release in 1947, when Megan was five.

Unlike Dracula, who never drank wine, both Megan's parents imbibed

AUTHOR'S NOTE

I first met the practicing vampire "Megan" through a joint appearance on a Canadian television program during a publicity swing for my book *Hollywood Gothic* in 1990. Megan's real name was Anne-Marie Bates, a fact she was more than eager to share with the public, but the program's producers insisted on disguising her in an outrageous "Vampira" wig, dark glasses and the none-too-imaginative nom-de-video "Anne." In billowing bombazine, she created the impression of a massive black toadstool — a highly theatrical image, one which she no doubt relished. The Toronto studio audience sat silent and open-mouthed as she candidly discussed the more sensational details of her life as a blood-fetishist — the only details, of course, that interested commercial television.

I was immediately struck by Anne-Marie's obvious and articulate intelligence, and suspected that there was a deeper story to be told. I contacted her in Los Angeles while researching my book *The Monster Show* (in which this piece originally appeared), and she welcomed me graciously to her home in North Hollywood for an extended interview.

Since Anne-Marie was still eking out a precarious existence as a bit-part actress, I felt uncomfortable identifying her (or other interviewees) by name, lest her

The legendary 1950s horror hostess Vampira (Maila Nurmi) inspired a young fan in a career of real-life blood drinking. (Courtesy of Erik N. Stogo).

heavily. Megan's father physically abused both mother and daughter, "but we fought back," she recalls. "I could never have any friends over, of course. So I created my own world to live in. I read a great deal. My mother had taught me to read and write before kindergarten."

Left alone one day while her parents drank in the kitchen, Megan started rummaging through her father's trunk. She didn't know exactly what she expected to find there, but was flabbergasted at what she did find: a pair of comic books, both of "a decidedly macabre nature." They were exactly the kind of publication that Frederic Wertham, author of the controversial best-seller *Seduction of the Innocent* (1954) had crusaded against.

Read today, Wertham's interviews with "comic-book addicts" are less than convincing. It was too bad that he never interviewed Megan, who was "absolutely thrilled" by her father's secret stash of horror. As her parents were consumed by their stupor, she stole away to her bedroom and read the books from cover to cover. "One of them contained a story about a female vampire named Lila. It left a lasting impression. I felt very sympathetic toward the character. She was so beautiful."

Megan's delight with *The Vault of Horror* was short-lived. Her mother caught her poring over Lila's undead exploits and punished her severely. "She told me this was completely unsuitable reading material for a young girl. I was never to look at it again."

Funny animal comics were okay, but even Batman was too close to another kind of bat-man for her mother's comfort. "She was determined to turn me into a happy, wholesome little girl." But, as in many dysfunctional families, Megan received contradictory signals. Despite the household prohibition on horror comics, her mother continued to tell her spine-tingling ghost stories until her young adulthood.

Megan maintained her "vampire" life on a fantasy plane that she could enter at will, with or without her parents' approval. "I began to play my vampire games," Megan recalled. "These were very secret games I told very few people about. Only certain, selected kids who were creative enough, imaginative enough, or just 'odd' enough in some way to be receptive." Megan's vampires were never creatures of destruction. "They were always after the bad guys — beautiful women who took care of the villains of the world. I would act out the stories with the neighborhood kids — we would wander around in flowing nightgowns, things like that." Once, Megan's mother snapped her picture in a billowing garment, unaware that her preadolescent daughter was modeling not sleepwear, but, to her own mind, a shroud.

Soon she started writing her stories down, in the form of serials she would read to trusted friends as delicious, shared secrets. She filled notebooks with the kind of vampire stories she could never find in published sources, stories in which the vampires triumphed and were clearly superior beings. On Saturday nights, she watched the horror movies hosted by

controversial story lead to any professional reprisals or reversals. Hence, the name "Megan." But, since the piece was originally published, Anne-Marie gave many other interviews, enthusiastically publicizing her real identity — as well as her unrepentant thirst for human blood.

Sadly, in the summer of 2000, Anne-Marie died of a massive heart attack in Hollywood. She had, by the end of her life, established herself as a prominent fixture of the Hollywood Goth scene. I now apologize for my initial timidity in fully acknowledging this courageous pioneer of out-of-the-closet vampirism.

Anne-Marie Bates, this is your life.

A comic-book vampire of the 1950s.

Maila "Vampira" Nurmi. "I absolutely idealized her," Megan said. "I can still remember the introduction — 'Fletcher Jones Chevrolet presents — Vampira!' She was like a goddess to me."

Megan's obsession with Vampira got her into trouble, at school and at home. "My parents had stuck me in a Catholic girls' school, unaware how stimulating I found the religion's blood imagery." From an early age, Megan had been receiving conflicting messages about religion and blood, love and pain. "It was good, it was bad, do it, don't do it . . . My mother was raised a Catholic in the Polish tradition, and had the fear of God beaten into her by her own mother." In church, the children were taught that the wine the priest drank was blood, not just symbolic blood, but the genuine article. The nuns, however, failed to see the relevance of vampirism to a Catholic education.

"I used to come to school wearing dark fingernail polish. The other kids thought I was crazy and persecuted me for it." Even the girls in Megan's inner circle didn't share her fixation on the glamorous Vampira. "It was a full-blown secret identity — until I got busted." One of her girlfriends ratted to the nuns, revealing all the details of Megan's fantasy life. "The nuns interrogated me, and called my parents, who in turn decided to investigate Vampira themselves. They sat up one night, and watched the show in silence. Afterwards, I was told I was never to watch the program again. They brought out my precious stories and made me tear them up, one by one. If I ever so much as mentioned the subject again, they told me, they would take me to a doctor who would shrink my head down to size."

In practice, the embargo on the undead was short-lived, due to her parents' alcoholism. "I'd just wait until they got smashed on Saturday night, which they always did. They'd be so blotto they couldn't tell what was on television anyway." Then, like an unseen wraith in her own house, Megan would rise from her bed and drift past the semiconscious figures of her parents to commune with Vampira. The character's perverse contradictions — beautiful-ugly, sexy-deathly — echoed the unresolvable conflicts of Megan's family. "In a way," she recalled, "Vampira was a kind of beatnik," a comforting, antiauthoritarian image.

Vampira turned out to be a brief phenomenon, a dark comet visible on the airwaves for less than a year. Her broadcast demise was one of two traumatic events for Megan in 1954. One day, Megan's mother left the house and was hit by a car, which crippled her for life. The accident ruined the family financially. Forced into a caretaker role at an early age, Megan began sublimating her dark fantasies. She was still an oddball and outcast at school, but now she read the Brontës instead of *Tales from the Crypt*, substituting a genteel Gothicism for the cartoony flamboyance of the Crypt Keeper and Vampira. Megan lived at home until she was twenty-five, taking care of her mother, whose injuries proved mental as well as physical. The older woman sank into depression, continued to

VAMPIRISM: A DRINKING DISORDER?

Megan's family history of alcoholism evokes a connection between vampirism and intemperate drinking that predates the modern media age by more than a century. Vampire historian Montague Summers cites the "frantic teetotal tract" *Vampyre* ("By the Wife of a Medical Man"), published in 1858, and quotes the delirium ravings of its central victim: "They fly — they bite — they suck my blood — I die. That hideous 'Vampyre!' Its eyes pierce me thro' — they are red — they are bloodshot. Tear it from my pillow. I dare not lie down. It bites — I die! Give me brandy — brandy — more brandy."

Like creepy clockwork every Halloween, advertising agencies for major beer, wine, and liquor companies invariably trundle out campaigns featuring boozing vampires. "Welcome to our lite-mare" beckoned a 1992 promotion featuring a dissipated-looking Dracula and Draculette clutching their cans of Miller Light in the window of a dank castle. HURRY, SUNDOWN was the headline of a vintage Smirnoff's ad pushing something called "The Vampire Gimlet." No doubt, the Bacardi Rum bat logo has a certain subliminal significance for the seriously stewed. Just as Halloween allows a ceremonial acknowledgment of antisocial impulses, so does holiday liquor advertising allow us a forbidden glimpse at the dark side of drinking. Bloody Marys have long been a staple of vampire humor, and characters like the eponymous antihero of *Blacula* (1972) sometimes order the drink when in public, if only to appear sociable.

drink heavily, and ignored her doctor's prescriptions for physical therapy. "She was committing slow suicide," Megan remembered.

With considerable emotional effort, Megan gave up the dark side, found a job that paid $50 a week, and moved into an apartment of her own. "I started to have fun," she recalled. "I became a hippie for a while, I smoked pot, I took acid, I had a ball. I learned a lot about life, and a lot about different levels of consciousness. Acid was a spiritual thing, it was ritualistic, religious."

For a time Megan became one of L.A.'s many psychedelic gurus, a hippie earth mother nurturing others the way she had never been nurtured herself.

In 1971, Megan's mother suddenly died of a heart attack. And shortly afterward, her psychedelic experiences began to take on frightening, vampire-related imagery. "I thought I had given it all up. But it woke up again, stronger than it had ever been before. It was something so contrary to the earth-mother role I was playing at the time that it really freaked me out." The visions were "intensely religious, ecstatic, and frightening. I saw myself as a vampire, with an overpowering compulsion to drink blood from a lover. Male or female, it didn't matter."

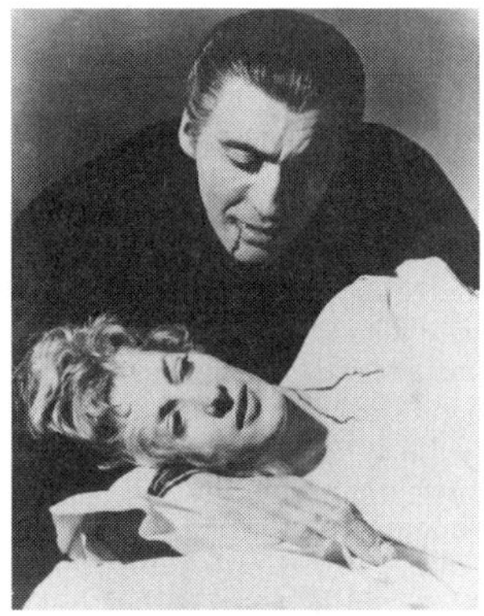

Cold eroticism: Christopher Lee and Melissa Stribling in Horror of Dracula *(1958).*

Megan drifted through the seventies in a kind of trance. She knew that her vampire obsession was no whimsical childhood fantasy that had resurfaced, but something that was part of her fundamental nature, needing to be confronted and assimilated. But the only release she found was in domestic horror movies like *Count Yorga, Vampire* (1970) and *Love at First Bite* (1979) and the elegant imported nightmares of Hammer Films. At the same time she nursed her father through his final illness, numbing herself in the cold eroticism of Christopher Lee.

Rock music, which had interested Megan since her early psychedelic days, began appropriating the trappings of horror movies in the early seventies, beginning with Alice Cooper, Kiss, and other groups. Fused with the burgeoning punk sensibility, a disturbing hybrid emerged by the end of the decade, a new kind of shock-rock known generically as Gothic. Clubs began springing up in Hollywood with names like Theatre of Blood, the Veil, and Helter Skelter. One day Megan was walking down Hollywood Boulevard when she saw a handbill plastered on a wall: ONE NIGHT ONLY — CASTRATION SQUAD.

Megan had always been passionately fascinated by the legendary castrati of the opera world, an interest that had almost eclipsed her obsession with vampires, and the name of the group transfixed her. And although she didn't understand it completely at the time, the castrato and the vampire shared an underlying psychology: an oral displacement of genital sex, steeped and transformed in nineteenth-century Romanticism. Megan began hanging out with groups like Castration Squad, the Sleepless, 45 Grave, and Christian Death.

It was about this time that Megan received a call from Anne Rice.

Megan didn't know her, but they had a mutual friend who recommended the author contact her as part of her research for her new novel, *Cry to Heaven*, which concerned the castrati. According to Megan, they talked for hours, the way only two people who share a passionate, specialized interest in an arcane subject can. Megan made the novelist tapes from her collection of rare, turn-of-the-century discs of the last castrati voices ever recorded, and received an acknowledgment in print. Rice suggested that Megan might enjoy her first novel, which dealt with a related theme. The book was called *Interview with the Vampire*.

"I went out and found a paperback," Megan recalled. "And I couldn't put the damned thing down." A ripe, decadent story of vampires in nineteenth-century New Orleans that had already claimed a cult following, the book included an especially unforgettable character with a special resonance for Megan, a little girl transformed into a vampire and trapped in a child's undead body for eternity. Megan came out of her room only to eat. "I couldn't get enough of it — it was a revelation. At last, here was one modern writer who wrote a beautiful story and didn't destroy her vampires at the end. It was as if vampire literature had finally come out of prehistoric caves and into civilization."

Interview with the Vampire pushed Megan's life into a kind of critical mass. Suddenly, vampires were everywhere. And one day a "ghoulish little guy in whiteface and a full mortician's outfit" came into the L. A. newspaper office where Megan worked as a production artist. He was a musician with British pretensions, and called himself Jonathan Cape. His band was named Schreck's Bad Boys, after the actor Max Schreck in the German film *Nosferatu*. "I fell head over heels in love with him," Megan said. "He told me all about his dark self, secret things he did and got high on." Blood, he hinted, was one of them.

Megan also fell in love with one of Jonathan's singers, who used the name Desiree LeFanu. "She was so exotic, tall and thin. *Real* thin. She came up to me like I was her long-lost sister and stared at me with her strange eyes. She invited me to one of her poetry readings. And after I heard her perform she called me one night, a la Anne Rice, and we just talked for five and a half hours. It was a marathon phone call that totally changed my life. She talked about her poetry, her friends in the underground, her drug use, her dream life. And finally she said, 'Do you mind if I ask you a very personal question?' 'No,' I said. 'Are you *quite* sure…?' she asked. I told her I was.

"'Tell me, Megan,' she said, 'has anyone ever, asked you . . . for your blood?'"

It was as if a dam had burst. Desiree revealed her personal history of vampirism, how she had been practicing blood-drinking from the age of twelve or thirteen. Megan felt the most powerful surge of yearning she had ever experienced. To think that this beautiful creature, not a fantasy, not a movie, wanted to partake of her very essence!

THE QUEEN OF THE NIGHT

The vampire fantasy which galvanized Megan's commitment to real-life blood-bingeing was Anne Rice's novel *Interview with the Vampire* (1976), which resonated on several levels with Megan's unhappy relationship with her family. One of the most memorable characters in Rice's novel is a child-vampire, Claudia, unconsciously patterned (by the author's later acknowledgment) after Rice's own daughter Michele, who died of a blood disease at the age of six. As a woman supernaturally imprisoned in a child's body, Claudia uncannily anticipated the rise of the "inner child" as an axiom of pop psychology — a seething adult sensibility stunted and immobilized by childhood abuse. Anne Rice is now the most successful vampire novelist in history, her total earnings undoubtedly eclipsing even a century's worth of *Dracula's* profits. All of Rice's vampire books — the main series including *The Vampire Lestat* (1985), *The Queen of the Damned* (1988), *The Tale of the Body Thief* (1992) and *Memnoch the Devil* (1995) — tap deftly into various zeitgeist motifs, which no doubt accounts for their widespread popularity. Rice rankles some critics with the trademark against-the-grain ripeness of her prose, but by eschewing the fashionably anorectic style of much literary fiction, Rice (like Stephen King) manages to attract

Actual consummation proved more difficult. Despite her declarations of thirst, Desiree tormented Megan by putting off the crucial event. And then, in a kind of painful primal scene, Megan learned that her beloved Jonathan and Desiree had begun drinking each other's blood, that they were driven around the Hollywood hills in the back of a car, tearing the skin of each other's throats and lapping the flow. (Later Megan would learn of blood fetishists who played for higher stakes, who could only be sated by *spurting* blood, auto-vampires who knew how to safely puncture their own jugular veins, catching the leaping stream in their mouths.)

Jonathan's and Desiree's flirtations turned out to be a cruel tease. They were to remain the unapproachable, undead adults, and she the perpetual child. Jonathan further alienated Megan, and nearly everyone else, by his involvement with skinheads, Satanists, and neo-Nazis. "Nobody wanted to play with him anymore," Megan remembered. Thoroughly disillusioned, she withdrew from the orbit of Schreck's Bad Boys.

Eventually, Megan found a sympathetic donor in Christopher, an intense, dark-haired young man with a hawklike profile and a prodigious knowledge of punk, heavy-metal, and Gothic music. Christopher is what is known in vampire circles as a "passive" donor; he and Megan are soul mates, not physical lovers. His own erotic fantasies center on the classic Vampira-dominatrix image: Diamanda Galas, the intense and uncompromising performance artist who has made a stark theatrical connection between Gothicism, politics, and AIDS with her *Plague Mass*, is his most idealized female icon. Christopher, who is HIV-negative, gives Megan approximately 100 monogamous cubic centimeters a week, withdrawn by syringe for her private, autoerotic delectation. AIDS awareness is high in the vampire community, whose central rite is the exchange of bodily fluids, even though the disease itself is not readily spread by the oral ingestion of blood.

Many people have trouble seeing this kind of activity as anything but bestial. According to Christopher, "Although the Gothic movement may seem to be surrounded by a kind of violence, it is actually gentle, sharing, and nurturing. Gothic people are intensely alive — and honest enough to deal with images of death openly."

There are many "types" in the Gothic scene, said Christopher, "but the vampire occupies center stage. The vampire is the ultimate imaginative outcast, a lonely rebel against the establishment." Both Megan and Christopher decry the popular connection between vampires and satanism as misleading and stereotypical. "Vampirism is a pre-Christian idea," said Megan, and has nothing to do with modern religions or their satanic inversions. Like practitioners of wicca, or Druidic witchcraft, they feel that the vampire/Gothic movement has been unjustly maligned.

And exactly how many people are involved in this sort of thing? Megan and Christopher personally know a half dozen out-of-the-closet blood drinkers in Hollywood; estimates of the size of an "active" vampire community

a silent majority readership that may well avoid fiction approved by the literary establishment because it seems, well, vampirized of texture, emotion, and color. The film version of *Interview with the Vampire* (1994), directed by Neil Jordan, received mixed reviews, but was nonetheless a resounding box office success. Rice herself initially opposed the casting of Tom Cruise in the pivotal role of the Vampire Lestat, but was equally vocal in her ultimate approval of the finished film.

VAMPIRES IN THE AGE OF AIDS
Megan and her donor display caution in their blood contacts, as well they should. The AIDS epidemic has given the modern world a primitive shot of fear in the 1980s and 1990s, the characteristics of AIDS itself eerily echoing the classic motifs of vampire legends. A wasting malady appears, each victim capable of creating others through vein-puncturing and blood exposure. Self-appointed moral guardians come forth, waving religious talismans, insisting that the affliction is the work of the devil. Nevertheless, the scourge seems unstoppable, and there is a steady procession of coffins in the streets. AIDS is an undeniable subtext in the explosive growth of vampire entertainment in all media during the last two decades. To the conscious mind, the reality of AIDS can be almost too much to bear, but on the plane of fantasy, the immortal vampire may represent a psychological bargaining chip against death, a supernatural immunity in an age of immune dysfunction.

in America usually run into the hundreds. The closeted population seems to be much larger — large enough, in fact, that a 1-900-VAMPIRE phone service was inaugurated in 1990 for anonymous devotees.

Just who becomes a vampire? Is Megan's case typical? According to Norine Dresser, the California folklorist who broke the taboo on the subject with her book *American Vampires* in 1989, strong correlations have been made between early childhood abuse and a later acting-out of vampire fantasies. Blood-drinking becomes an overpowering metaphor for warm human contact. Dresser cites the unique research of Jeanne Youngson, Ph.D., who has collected correspondence with blood fetishists for the past thirteen years (Youngson's interest in vampires is decidedly catholic — as president of the Count Dracula Fan Club, she ministers to enthusiasts of all stripes, from movie buffs to the hard-core thirsty). Their histories are a litany of "abuse, neglect, abandonment, loneliness or molestation."

An advertisement for one of Hollywood's original Gothic music venues.

There is another way to look at it. Like mythical vampires, Megan is nothing if not a survivor, who has assembled from the wreckage of a tumultuous childhood a powerful sense of identity, albeit an identity many would find perverse. But their reaction might also contain a sense of recognition, for we all share Megan's larger world. There are those who have said for years that ours is a generation already raised living-dead in front of the television, drawing vicarious life and sustenance from images of violence and blood, real and imagined. We hardly need to share Megan's specific traumas to become "undead" ourselves. With or without the family, modern life itself can be supremely dysfunctional and abusive, cranking out a largely unchronicled generation of lost boys, and girls. And to them, finally, attention must be paid.

Two recent paperback horror novels throw a dark illumination on Megan's world, as well as ours. Issued by separate publishers, they feature strikingly similar covers. At a distance they might be mistaken for classic images of Madonna and child, but upon close inspection it becomes clear that the infants are not so interested in their mother's breast as in her throat. Their eyes seem to flicker demonically, and at the corners of their baby-lips there is something else — sharp, glistening, and white. Love and hate, blood and nourishment, obscenely juxtaposed. These are truly monstrous images, all the more so for their ambivalent depiction of the roots of the vampire in life's basic bonds and interdependencies. On this level, where we have all been, no one is safe.

Children of the night?

What music we make.

Stalking the Vampire

ROSEMARY ELLEN GUILEY
(1991)

IN 1970, a furor erupted over the alleged existence of a vampire in London's Highgate Cemetery. Numerous people attested to seeing a menacing creature drifting about the perimeter of the grounds at night. It was subsequently identified as a vampire — an Undead corpse that rose from its grave at night. There was a splash in the press, which incited a small army of self-styled vampire hunters and thrill seekers to descend upon the cemetery at night. Graves were vandalized and corpses were desecrated, and access to the cemetery was soon restricted.

One individual emerged as vampire stalker supreme: Sean Manchester, president of an occult society, now defunct, which has as its still-active arms the Vampire Research Society and the International Society for the Advancement of Irreproducible Vampire and Lycanthropy Research (ISAIVLR). According to Manchester — now the Reverend Manchester — he succeeded in tracking the vampire to his lair in an abandoned mansion and, in the best East European tradition, drove a stake through its evil heart, causing a "terrible roar" to sound from "the bowels of hell."

The case became famous for its controversy. Some say there never was a vampire, though there may have been ghosts mistaken for vampires. Others say there was a vampire. Still others maintain there was a vampire, but that Sean Manchester didn't find it and the story of the staked vampire is not true. The mystery remains unsolved. But the case earned Reverend Manchester a place in the public eye as "vampire hunter general" and enemy of the forces of darkness. Because of the continuing interest and controversy in the Highgate vampire case, Manchester was one of those at the top of my list to seek out during my stay in England.

Before meeting Manchester, I paid a visit to Highgate Cemetery to familiarize myself with the terrain. If there is such a thing as the perfect vampire haunt, Highgate Cemetery is it. Bram Stoker apparently thought so, for the evidence points to him using Highgate as the model for the cemetery near Hampstead Heath where Lucy Westenra is buried in *Dracula*. Stoker didn't name Highgate, disguising it in his novel, but it is the only cemetery near Hampstead Heath that fits the description in the book. Perhaps *Dracula* is the source of later rumors about a vampire infestation in the real cemetery. Such rumors have no foundation, according to Bernard Davies, among the skeptics concerning the Highgate vampire case. Davies told me he had found no historical evidence prior to Stoker's time that associated Highgate Village or Highgate Cemetery with vampires.

ABOUT THE AUTHOR

Rosemary Ellen Guiley, Ph.D. is a best-selling author, presenter and teacher on topics of spirituality, metaphysics and visionary experience. She has written more than twenty books, which have been translated into twelve languages. Her works include *The Encyclopedia of Ghosts and Spirits, The Encyclopedia of Witches and Witchcraft, The Encyclopedia of Angels, Harper's Encyclopedia of Mystical and Paranormal Experience, Atlas of the Mysterious in North America, The Complete Vampire Companion, The Encyclopedia of Dreams and Dreamwork for the Soul.* As president of Visionary Living, Inc., she lectures internationally and makes frequent media appearances.

"Stalking the Vampire" originally appeared in her book, *Vampires Among Us.*

The cemetery sits atop a hill in Highgate, one of London's northern neighborhoods. A sprawling place, it is divided into halves, cut through by Swains Lane. The eastern side is newer and still active, with several burials a week. Karl Marx, George Eliot, and Sir Ralph Richardson are buried there, among other luminaries. The markers are a mix of Victorian and modern. For a small fee, the public can enter during the day and wander freely along numerous narrow footpaths.

The western half, however, is closed to the public except on guided tours. It is the original cemetery, and is a jungle of brush, overgrowth, jumbled tombstones, and half-hidden vaults. Michael Faraday, Radclyffe Hall, and the wife and daughter of Charles Dickens are among its notable occupants. On the western side, one finds Victorian glorification of death at its peak. Winged angels and mournful animals watch over tombs. Crosses and fabulous crypts jut toward the sky. Vines cling to everything. The overgrowth makes the place dark even on a sunny day, and if the day is cloudy, it looks most gloomy and foreboding.

Above and opposite: time-honored tools for the destruction of vampires. (Courtesy of Dark Delicacies)

Highgate has a long and illustrious history. A hermitage existed there as early as 1364, and the wealthy began settling nearby during the sixteenth century. They were attracted by the stunning views of London, the Thames, and the Surrey hills — views partly obscured by urban growth today. By the late seventeenth century, Highgate was a village unto itself. One of the best-sited homes for the view was constructed on what is now part of the cemetery by William Ashurst, a merchant, member of Parliament, and lord mayor of Highgate. Ashurst died in 1720. In 1830, his house was pulled down to make way for St. Michael's Church.

By the 1820s, London faced a terrible problem. The city was growing fast, and there was nowhere to bury the dead. Some 52,000 people expired every year, and the church cemeteries were so overstuffed that they literally belched up their corpses. In answer to the dilemma, private cemeteries opened. This was nothing short of a revolution in burial practices, for private cemeteries were unheard-of. These profit-driven operations charged money for interment.

Highgate Cemetery was created by the London Cemetery Company. Stephen Geary, the architect and founder, set out to design and landscape the most spectacular cemetery in all of London to ensure its success. It would celebrate death in the darkest and grandest glory that Victorian art and architecture could muster. It was the Age of Romantic Antiquity and Gothic Revival. Geary acquired seventeen acres, including the Ashurst estate, and laid out systematic rows of graves. He planned for 30,000 graves, each containing three bodies.

His pièces de résistance were Egyptian Avenue and the Circle of Lebanon, also called the Columbarium. Egyptian Avenue was excavated into the steepest part of the hill; it is twelve feet below ground level. One approaches it through an iron gateway under massive Pharaonic arches flanked by obelisks. The sides of Egyptian Avenue, leading to the Circle of

Lebanon, are lined with sixteen vaults, each of which holds twelve coffins. The Circle of Lebanon is a circle of twenty vaults, even more magnificent, dug thirty feet below ground level. In the center is a great cedar of Lebanon, which had been part of the original Ashurst estate.

Near the circle are the Terrace Catacombs, an underground gallery with 840 recesses. Some 500 cholera victims are stuffed away in the left wing of the catacombs. St. Michael's Church faces the catacombs from across a street.

On May 20, 1839, the cemetery was consecrated by the Bishop of London. Two acres were left unhallowed for the burial of "dissenters" — agnostics, atheists, non-Anglicans, and the like. The first burial took place three days later, and Highgate became London's third commercial cemetery.

Highgate quickly became one of the most sought-after resting places in London. A modest grave cost; £2 10 shillings (the shilling was worth five of today's pence). Interment in the Circle of Lebanon cost 200 guineas, (£210) — a phenomenal sum of money for the times. Money talked — fame, social prominence, and bloodlines had nothing to do with the grandeur of one's grave. Some of the most imposing vaults and markers were purchased by persons of no historical importance. The first year the cemetery was in operation, 204 people were buried there. Most of them were from London and were an average age of thirty-six. The living, meanwhile, flocked to Highgate to be seen, promenading past St. Michael's Church and the Terrace Catacombs on pleasant days.

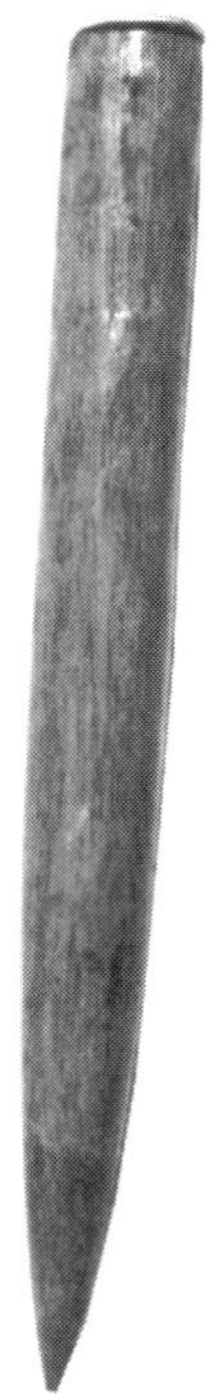

Business was so good that in 1854, the company purchased an additional nineteen acres to expand to the east side. Geary died the same year and was buried on the west side.

By World War I, however, the cemetery was in decline. The London Cemetery Company was taken over by United Cemeteries Limited. Fewer families bought plots, and many of those that did failed to maintain them. Following World War II, the decline accelerated. The cemetery, especially the western side, was engulfed by brush and rapidly growing sycamore trees. The overgrowth choked pathways and the sycamore roots split vaults, causing shanks of marble to tilt at crazy angles. Graves sank and some opened, only to be covered over by a mask of brush. The unwary visitor could not only lose his way, he could lose his footing and fall into a thirty-foot-deep ruptured vault. The only creatures happy with this state of affairs were the foxes, badgers, hedgehogs, squirrels, birds, rodents, and other creatures who lived relatively undisturbed amidst the clutter.

On my first visit to Highgate, I took the tube to Archway and then labored on foot up steep Highgate Hill to the cemetery. The neighborhood is working class now, an ethnic mix. The red brick row houses lining the street looked somber under an overcast sky. Partway up, I cut over to Swains Lane, which took me through the heart of the cemetery. In Victorian times, horse-drawn hearses pulled up the lane to the cemetery gates, west side on the left and east side across the lane on the right. To minimize obstruction to traffic on the lane, the coffins were taken out and lowered by hydraulic lift to a tunnel underneath the road. They would be raised and lifted out on the cemetery grounds.

I took a self-guided tour of the east side and signed up for the next available guided tour of the west side. At the appointed hour, about half a dozen people were collected. We were greeted by a small old man dressed in a baggy green sweater and slacks with brown wool socks and brown sandals. He had a bushy head of long gray and white hair and an equally bushy beard to match. For the next hour or so, he kept us entertained with his droll sense of humor.

Above and on following pages: views of Highgate Cemetery, Hampstead, London. (Photos by David J. Skal)

We were advised to keep strictly to the footpaths to avoid treacherous hidden graves and vaults.

We learned that the east side once again has become a fashionable burying ground; plots are purchased by individuals as young as twenty-one. The west side is so full of bodies that practically every square inch of ground covers a grave. One can be buried here only if a family vault already exists. In all, Highgate contains some 60,000 vaults and 170,000 bodies.

Any graveyard has a psychic atmosphere that differs from that of the surrounding world of the living. There probably isn't a graveyard in existence where ghosts and unexplained phenomena haven't been reported. Highgate was not a place where I would have wanted to be caught after dark.

Egyptian Avenue and the Circle of Lebanon were dark and gloomy. This part of Highgate has been made famous on film and television, for this is where studios, production companies, and television crews come whenever they need to shoot a vampire's lair. Hammer Films was here in the 1950s through early 1970s for its famous Dracula films starring Christopher Lee and Peter Cushing. The Friends of Highgate now charge hefty fees for the privilege of shooting the funereal scenery. "What would we do without our vampires and Dracula?" said our guide. "Dracula is big business!"

Many tourists come to Highgate expecting Count Dracula himself to be buried here. Our puckish guide said he grew so weary of being asked for directions to Dracula's grave that he began telling tourists that the Count was really buried under Platform Five at the Kings Cross Underground/British Rail station in London. "One American girl actually went there looking for the grave," he said. "She came back quite peeved. She said the least they could do was put a plaque on the platform!"

I asked the guide if Highgate Cemetery was haunted.

"*Of course* it's haunted," he said. "We've got *everything* here." Many visitors report seeing apparitions, and a number of the volunteer staff, who all spend a great deal of time prowling about the pathways alone, also say they see ghosts. In the typical English fashion of understatement, these experiences are not trumpeted about but are kept private. Ask about them, however, and they are acknowledged.

"What about the Highgate vampire?" I asked.

The guide scowled. "Don't tell me you've read *that* book." He was referring to the infamous book by Sean Manchester, *The Highgate Vampire* (1985), which purports to tell the full and unexpurgated account of the vampire reportedly haunting the cemetery in the 1970s. "Complete rubbish and utter nonsense," the guide snorted. And with that, the subject of real vampires at Highgate was dismissed.

My next opportunity to discuss the Highgate vampire came several days later, when I returned to Highgate to meet with Sean Manchester and his executive secretary, Diana Brewster.

Our rendezvous took place at the Lauderdale House, a historical site with a small café located on Highgate Hill adjacent to Waterlow Park and the cemetery. The location was Manchester's choice, despite the fact that the Friends of Highgate Cemetery had pressured the Lauderdale House to discontinue selling *The Highgate Vampire* in its bookshop. The three of us sat outdoors under a blistering sun and talked over cappuccino and strawberries.

Manchester, handsome and genial, was attired in a cassock. He explained that he was about to be ordained as a priest by the Celtic Catholic Church, which is independent of the Vatican. He has his own order, the Holy Grail Church.

Manchester has always cultivated a striking Romantic appearance in haircut and period clothing, perhaps to underscore his claim to blood lineage to one of the greatest Romantic poets, Lord Byron. The lineage cannot be proved, according to Manchester, because it is through the illegitimate offspring of a liaison between Byron and a household servant. Nonetheless, he wears this badge proudly, and has often been featured in the press dressed in Byronic attire. The Byron Society says it cannot comment on his claim of lineage since no written evidence of it exists, and notes only that many similar claims are made by others.

Manchester said he grew up in Newstead Abbey Park in Sherwood Forest, Nottinghamshire, some distance from Newstead Abbey, a converted twelfth-century priory that was home to the family of Lord Byron from 1540 until 1817, when the poet was forced to sell it to pay debts. The house was restored in the Gothic style in the nineteenth century and is a museum today. This environment, a "time capsule untouched by centuries," as Manchester described it, predisposed him to an early interest in the supernatural. He said he detested formal education and ran away at age fifteen without ever having taken a single exam. His true teacher was nature, and he devoted himself to his own pursuit of poetry, music, and painting. "I've always been a bit of a bohemian, a bit of a poet, a bit of a wandering minstrel," he said in answer to a question as to how he earned his living in his pre-vampire-hunting days. "The Grail quest, the discovery of one's true self, has been my main drive through life. That I have slain dragons and rescued maidens has been along the way. I have never been occupied in a nine-to-five office job simply because I don't think I'm capable of it. My abilities are more tied up with my quests. Sometimes it's not made it easy because of not having a solid, reliable income. But of course one learns to live as someone on a spiritual journey should. Sometimes it will be difficult, other times you don't notice those difficulties. It does come to you. I'm always looked after, wherever I go."

Thus, when the vampire surfaced at Highgate, Manchester answered the call to rescue maidens and slay the vampire-dragon.

Reports of a vampire at Highgate first came to light in 1967. The story, which unfolded over several years, is marked by ambiguities, weird activities, and claims that are backed up by questionable evidence. Highly entertaining, the case, even at its best, can strain one's credulity. The most complete accounts have been given by Manchester in two major works: a chapter in *The Vampire's Bedside Companion* (1975), edited by Peter Underwood, founder of the eminent Ghost Club in England, and *The Highgate Vampire*, a book that Manchester self-published in 1985 and revised in 1990. Manchester's chapter in Underwood's book gives a straightforward account of various reports of the vampire, and his own identification of the corpse believed to be the Undead one. His second account, *The Highgate Vampire*, expands the story and includes Manchester's tracking the vampire to a new lair, where he spectacularly dispatched it.

The story is recounted here from Manchester's writings and a number of newspaper reports and periodical articles.

The Haunting of Highgate

The first reports of something foul afoot in Highgate came from two sixteen-year-old girls who shared a terrifying vision one night while walking home down Swains Lane. As they passed the rusted and locked north gate of the western side of the cemetery, they both saw graves open and dead bodies rise from them. The vision ended as soon as they spoke to each other. One of the girls, Elizabeth Wojdyla, then began suffering nightmares in which an evil presence with a pale, corpselike face was trying to enter her bedroom.

Several weeks later, a young man and his girlfriend experienced a fright while walking down Swains Lane at night. As they passed the north gate, the girl suddenly screamed. They both saw a figure standing behind the iron railings of the gate. Its face was twisted into a mask of horror. They froze and watched the figure melt back into the shadows of the cemetery.

When Manchester heard the second account, he decided to investigate. He was as yet unaware of the experience of the two girls. He poked about the western side during the day, and noticed that a path from the unused northern gate led directly to the Columbarium. But the young man who'd reported the incident refused Manchester's request to meet him at the cemetery for a night visit. He told Manchester that he had returned to the cemetery with another friend one night near midnight.

They'd climbed the wall near the north gate and moved deeper into the grounds. They were startled by a deep, slow booming sound that came closer and closer, until a dark shape moved across the path directly in front of them. Terrified, they fled.

The mysterious figure remained unexplained, and little else happened for nearly two and a half years. During the summer of 1969, Manchester had a chance meeting with Wojdyla, who by this time was so thin and pale that she looked like a cadaver. She told him of her nightmares, headaches, dizziness, and nausea. She had bouts of sleepwalking as well.

Soon after this meeting, Wojdyla's boyfriend, Keith, contacted Manchester for help, explaining that a doctor's regimen of diet and vitamins had failed to improve her condition. In addition to her other symptoms, she now showed two small puncture wounds on her neck, Keith reported. Manchester suggested she was under attack by a vampire. He and Keith sealed her room with garlic and a crucifix and gave her a silver cross to wear, along with a necklace made of linen containing a small amount of salt. Keith was to take other measures of prayer and the sprinkling of holy water about the room. Following these steps, Wojdyla's health improved.

Manchester and several unidentified colleagues began inquiring into phenomena experienced in the cemetery area. Perhaps because of their inquiries, numerous reports were offered. It appears that the reports snowballed to the point where the local press took notice in February 1970 by reporting some of them. Numerous people claimed to have seen a ghostly figure lurking about the area. One young man claimed to have been knocked down by a tall figure with the face of a wild animal.

The reports included no evidence of vampirism, however. Meanwhile, dead animals, mostly foxes and other nocturnal creatures, began to appear in Waterlow Park and the cemetery. They were all lacerated around the throat and drained of blood. Manchester opined publicly that the culprit of all the strange events was a vampire that should be dispatched in the traditional manner by staking it through the heart and beheading the corpse. When his remarks appeared in the press, the entire affair mushroomed. Others expressed doubt over the vampire theory. One alternative theory held that the caretaker's dog had killed the animals and dragged them about, thus draining them of their blood. But the idea of a vampire loose in Highgate caught the public's fancy, and hordes of vampire hunters and the curious invaded the cemetery.

One of the many who contacted Manchester was the sister of a woman he calls "Lusia." Lusia reportedly was suffering from somnambulism, feelings of suffocation at night, and other odd behavior. She had two tiny pinprick marks on her neck. One night as the sleepwalking Lusia headed toward Highgate Cemetery, the sister, Anne, rang Manchester. The two of them trailed Lusia to St. Michael's Church, through a broken railing of the west-side cemetery boundary, and to the Columbarium. There Lusia stopped before one of the vaults and ripped the crucifix from about her neck. A low booming sound echoed through the area. Lusia collapsed. Anne and Manchester carried her home. She had no recollection of the event the next morning.

Meanwhile public interest reached fever pitch following a show broadcast by Thames Television, in which Manchester led an interviewer on a tour of the vault area where the suspected vampire was said to lurk. Other vampire hunters announced their intent to find the vampire and stake it. One of them was David Farrant, born David Robert Donovan Farrant, who gave his name at various times as David Farrant, Allan Farrant, and Robert Farrant. Farrant, a young man who lived in the Highgate area, inspired numerous other would-be vampire hunters to climb the cemetery walls and invade the grounds. Farrant claimed to have seen a strange shape on four occasions during his own nocturnal forays. The eight-foot, humanoid shape glided over ground where, Farrant believed, vaults had been newly damaged.

Some of the sightings may have natural explanations, however. At about the same time, a group of amateur filmmakers was working in the cemetery filming *Vampires at Night*, and it is possible that some of the "shapes" floating about were real human beings in costume. But Farrant clung to his belief that he had encountered a true supernatural entity.

Meanwhile Manchester continued his own activities. On Friday night, March 13, 1970, he led a group of about a hundred people to the vault singled out by Lusia. The iron door would not open, so Manchester and two assistants lowered themselves into the tomb from a hole in the top of the Circle of Lebanon, which forms an earthly roof over the vaults of the inner circle. Inside the vault they found three empty coffins. In each of them they placed garlic, salt, and a cross, and they sprinkled the coffins and the vault with holy water. They left the tomb and stood watch for the night. At about 2 AM, the booming noise sounded again. (The source of this mysterious noise was never explained, but we may presume it was associated with the vampire.)

The news media continued to report stories of sightings of menacing, ghostly figures in the cemetery. In August 1970, the situation worsened. The mutilated corpse of a woman buried in the cemetery forty-four years earlier was discovered beside a broken vault. The corpse was charred and missing the head and an arm. There was evidence that an occult ritual had taken place. It was speculated that the abuse had been perpetrated by a zealous vampire hunter or by satanists.

Disturbed by the desecration of the corpse, Manchester returned to the Columbarium and lowered himself into the suspect vault. One of the three coffins was missing. He then contacted Lusia and convinced her to return with him to the site during daylight hours. They were accompanied by three assistants. Seated in front of the vault, Lusia entered a trance and seemed to become an instrument for an evil entity. She answered questions put by Manchester, then walked to a nearby vault and began crying.

It was the vault where the headless woman's corpse had been interred.

Manchester wondered if the vampire had changed lairs since the purification of the suspect vault. The doors of this vault yielded to pressure. Inside, Manchester, his assistants, and Luisa were greeted with a revolting stench and decaying coffins. They noticed that the vault contained one coffin too many, according to the inscriptions on the outside of the vault. Was it the missing coffin from the other vault? Perhaps. They identified the extra coffin as one in the back that appeared newer than the others and had no nameplate. With pounding hearts, they raised the lid.

Accounts of what they found inside differ. Manchester said in his first account in *The Vampire's Bedside Companion*:

> There it lay: a body which appeared neither dead nor alive. Eerily, we gazed at that sight which defied explanation and logic for several long moments. It's newly dead, said one of my assistants, breaking the silence. But the vault was a hundred years old and there had been no recent admissions.

Ten years later, in his unexpurgated and expanded account in *The Highgate Vampire*, the description was much more vivid:

> My torch lit up in unnerving revelation the sleeping form of something that had long been dead; something nevertheless gorged and stinking with the life-blood of others, fresh clots of which still adhered to the edge of the mouth whose fetid breath made me sick to my stomach. The glazed eyes stared horribly — almost mocking me, almost knowing that my efforts to destroy it would be thwarted. Under the parchment-like skin a faint bluish tinge could be detected. The face was the colour and appearance of a three-day-old corpse.

The group stared at the thing in horror for some minutes; then Manchester took from his bag of holy tricks an aspen stake and placed it on the beast of the vampire. His assistants exhorted him to desist because staking the body would be an illegal desecration of a corpse. Manchester relented and agreed to perform an exorcism instead with garlic, salt, holy water, incense, candles, crosses, and incantations intended to banish the evil forever. The group had to hurry, for the sun was near setting. As the rite proceeded, the temperature plunged and the candles flickered. The eerie booming noise sounded. At the end of the rite, Manchester hurled a crucifix into the vault.

Later, upon Machester's recommendation, he said, the entrance to the vault was sealed with cement.

In his second account, Manchester said that garlic was mixed into the cement.

Despite the exorcism, Manchster considered the case open until the vampire was properly dispatched. The exorcism might have allayed the fears of some as Manchester asserted, but not Farrant. On August 29, on the night of the full moon, the police arrested Farrant as he climbed over the Highgate Cemetery wall with a wooden cross and a sharp wooden stake in hand. Farrant identified himself as Allan Farrant, an orderly at Barnet General Hospital. He admitted he was out to stake the vampire, which he believed had haunted the cemetery for about ten years. He was charged with entering enclosed premises for an unlawful purpose, but the charges were dismissed in court. Farrant stated he had formed an organization several years earlier that had about a hundred members throughout Britain and Europe who were stalking vampires.

Nearly one month later, Farrant was back in the press, this time as David Farrant, leading a reporter on a tour of the cemetery, again armed with cross and stake. This time, Farrant hid his tools in a paper bag. They found no vampire, but did find extensive, recent desecration throughout the cemetery. Graves and coffins had been opened and their contents moved or stolen, and vaults had been defaced with occult symbols. The damage was estimated at £10,000, a staggering sum at the time. The public was outraged.

Farrant kept a nightly vigil at the cemetery for a while, and though he never saw the specter nor found the vampire, he did not give up on his mission.

HIGHGATE'S MOST NOTORIOUS VAMPIRE

Although Bram Stoker called it "Kingstead" instead of "Highgate," there is little doubt of the location he had in mind for the most horrifying sequence in* Dracula, *"a lordly death-house in a lonely churchyard, away from teeming London; where the air is fresh; and the sun rises over Hampstead Hill, and where wildflowers grow of their own accord." The destruction of the vampire Lucy Westenra is chillingly recorded in his diary by Dr. Seward.

•

***29 September, night.* — A little before twelve o'clock we three, Arthur, Quincey Morris, and myself, called for the Professor. It was odd to notice that by common consent we had all put on black clothes. Of course, Arthur wore black, for he was in deep mourning, but the rest of us wore it by instinct. We got to the graveyard by half-past one, and strolled about, keeping out of official observation, so that when the gravediggers had completed their task and the sexton under the belief that every one had gone, had locked the gate, we had the place all to ourselves. Van Helsing, instead of his little black**

Throughout 1971, there were periodic reports of occult activities taking place in the cemetery during the night. Farrant surfaced once again in the news when he was apprehended by police on the night of October 8, 1971, while in the company of a naked woman. The two allegedly were photographing magical ceremonies, possibly necromancy, which is the conjuring of the dead for prophecy. Farrant said he was "high priest" of an occult order founded by Manchester. He was released from custody.

Manchester denied any association with Farrant. Meanwhile, Manchester pursued his own investigations. Lusia, in trance, told him that the vampire in Highgate Cemetery had succeeded in moving to a new location. Manchester had the first vault reopened. He found that the casket that he believed held the elusive Undead one was missing — and so were the garlic, crucifixes, and bags of salt he had left there.

In the winter of 1973, a deserted nineteenth-century mansion in north London surfaced in the media as reportedly haunted by something so unpleasant that local people called it the "House of Dracula." Manchester decided to investigate. He took with him two assistants, a clairvoyant woman named Veronika and a "skeptic" named Arthur, who would operate "the scientific equipment." Entering the house at night, they had a number of harrowing experiences involving a sinister fog, terrible smells, icy presences, knocking noises, and voices heard only by Veronika. They determined that they had found the lair of the escaped Undead. Arthur and Manchester returned in daylight to stake the vampire.

In the basement they found an enormous black casket. They dragged it outside to the rear of the house grounds. According to *The Highgate Vampire*:

> Arthur rushed to fetch my case of stakes as I prayed for strength. Suddenly I kicked the lid off the coffin and beheld something which shall ever remain stained upon my memory. In the half-light our quarry looked fiendishly exaggerated and distorted — much worse than the time before in the vault at Highgate. Burning, fierce eyes beneath black furrowed brows stared with hellish reflection. Yellow at the edges with blood-red centres, they were unlike any other beast of prey. Flared nostrils connected to a thin, high-bridged nose. The mouth still set in its cruel expression with lips drawn far back as if unable to contain the fangs. The creature had not long entered its vampire sleep and the hideous countenance was yet to become subdued by its undead repose.

The vampire's skin began to discolor immediately from the sunlight coming through the overhead tree branches, Manchester said. And then:

> With a mighty blow I drove the stake through the creature's heart, then shielded my ears as a terrible roar emitted from the bowels of hell. This died away as suddenly as it had erupted and all became still. We witnessed the body-shell cave in and quickly turn filthy brown which soon became a sluggish flow of inhuman slime and viscera in the bottom of the casket. Arthur had been trying to use a camera as if he could not believe his eyes, but the intolerably putrid stench soon caused him to turn away nauseated.

Manchester and Arthur then built a pyre, placed the coffin and its awful contents on top, poured on gasoline, and lit it. The vampire's remains burned to ash.

The house later was razed, and flats for the elderly were erected upon the grounds.

But Highgate Cemetery still was not at rest. Graves were tampered with, bodies were taken out of coffins, and occult symbols were written on vault floors and walls. In January 1974, a hapless architect parked his car near the cemetery one night, and the next morning found what at first appeared to be a log resting in the seat. It was a headless corpse. Sometime later, Farrant was again apprehended by police, who took from him photos of a nude woman dancing in a tomb. He was arrested and charged with grave despoilation, illegal possession of a handgun, and petty theft.

Farrant was brought to trial in Central Criminal Court in June 1974, in what was called the "Nude Rituals Case." It was juicy stuff for the media, and the trial was widely covered. Interest was fueled by the fact

bag, had with him a long leather one, something like a cricketing bag. It was manifestly of fair weight.

When we were alone and had heard the last of the footsteps die out up the road, we silently, and as if by ordered intention, followed the Professor to the tomb. He unlocked the door, and we entered, closing it behind us. Then he took from his bag the lantern, which he lit, and also two wax candles, which, when lighted, he stuck by melting their own ends, on other coffins, so that they might give light sufficient to work by. When he again lifted the lid off Lucy's coffin we all looked, Arthur trembling like an aspen, and saw that the corpse lay there in all its death beauty. But there was no love in my own heart, nothing but loathing for the foul Thing which had taken Lucy's shape without her soul. I could see even Arthur's face grow hard as he looked. Presently he said to Van Helsing, "Is this really Lucy's body, or only a demon in her shape?"

"It is her body, and yet not it. But wait a while, and you shall see her as she was, and is."

She seemed like a nightmare of Lucy as she lay there, the pointed

that prior to the commencement of the trial, Farrant and a codefendant, John Pope, allegedly sent poppets stuck with pins to the detectives who intended to testify at the trial. The poppets, small dolls used as effigies in magical rituals, apparently represented the detectives and were intended as a warning not to testify.

The trial proceeded, and Farrant conducted his own defense. He pleaded not guilty to five charges, including one charge of damaging a memorial to the dead on consecrated grounds, one charge of damaging property, and three charges of breaking open catacombs and interfering with corpses. He also denied practicing necromantic rituals in vaults at Highgate. The incident that was the basis of that charge had been a satanic ritual performed by other persons, he said, and his own group had performed an exorcism that involved the dancing nude woman. Farrant did plead guilty to possessing a revolver and one round of ammunition.

The prosecution contended that the offenses took place over a three-year period, and that Farrant was the true "vampire" of Highgate. The jury was shown photographs, alleged to have been found in Farrant's home, of a desecrated coffin, occult symbols in a vault, and a nude woman dancing inside a vault. Among those who testified was author Francis King, an authority on magic, who said that the occult symbols on the floor of a vault at Highgate were those used in necromancy.

Farrant was convicted and was sentenced in July 1974 to four years and eight months in jail, plus a fine of £750 to cover court costs. In passing sentence, the judge noted that tomb desecration was still occurring at the cemetery, for he had gone to Highgate the day before to see a mutilated corpse inside a damaged coffin. At the sentencing, the judge heard testimony from John Duncan Kidd, a solicitor and a director of United Cemeteries Ltd., the owner of Highgate Cemetery. Kidd stated that the cost of cleaning up Highgate was "absolutely phenomenal" and would exceed the cemetery's limited income and subsidy from the parent company.

In two other trials, Farrant was convicted of interfering with witnesses to a criminal trial — the poppet incident — and of stealing blankets and bed linen from Barnet General Hospital.

In October 1975, the Friends of Highgate Cemetery was formed, partly in response to outrage over the occult activities. The purpose of the Friends was to manage Highgate Cemetery as a national monument and historical burial ground, to represent the interests of grave owners, to secure public access to the grounds, and to sponsor conservation of the cemetery. The Friends formed a volunteer maintenance squad, which began hacking away at the overgrowth, felling some of the sycamores, clearing the paths, and restoring the place to at least some of its former eminence. They didn't overdo the clearing — the place, after all, had to maintain its famous "Victorian gloom." To discourage occult activity and vandalism, the cemetery was closed at night and access to the west side was severely restricted. In 1981, two Friends formed Pinemarsh Limited and bought Highgate

teeth, the blood stained, voluptuous mouth, which made one shudder to see, the whole carnal and unspirited appearance, seeming like a devilish mockery of Lucy's sweet purity. Van Helsing, with his usual methodicalness, began taking the various contents from his bag and placing them ready for use. First he took out a soldering iron and some plumbing solder, and then small oil lamp, which gave out, when lit in a corner of the tomb, gas which burned at a fierce heat with a blue flame, then his operating knives, which he placed to hand, and last a round wooden stake, some two and a half or three inches thick and about three feet long. One end of it was hardened by charring in the fire, and was sharpened to a fine point. With this stake came a heavy hammer, such as in households is used in the coal cellar for breaking the lumps. To me, a doctor's preparations for work of any kind are stimulating and bracing, but the effect of these things on both Arthur and Quincey was to cause them a sort of consternation. They both, however, kept their courage, and remained silent and quiet.

When all was ready, Van Helsing said, "Before we do anything, let me tell you this. It is out of the lore and experience of the

Cemetery, east and west sides, for a mere £50.

Farrant was released from jail on parole in July 1976.

In 1977, a rash of animal deaths began to be reported in the areas near Highgate Cemetery. The bodies of pets and various small wild animals were found with wounds in their throats. It was speculated that dogs or wild animals were the culprits, but the vampire theory also began to circulate again. For Manchester, there was only one possible explanation: The vampire he had destroyed had somehow managed to leave behind a new Undead.

By 1979, he learned that Lusia had died and was buried in the Great Northern London Cemetery, where sightings of a woman in white and other supernatural activities were being reported. Manchester concluded that Lusia had been contaminated by the vampire. He kept vigil at her gravesite to confirm this, and then one night cast a protective magic circle and performed an exorcism ritual.

What manifested was a hissing black spider the size of a full-grown cat, according to Manchester. He plunged a stake into it, whereupon it emitted a "heartrending screech." When dawn came, Manchester found the decomposing body of Lusia where spider had met its demise. He reburied her. That, he said, finally laid to rest the vampiric activity at Highgate.

Farrant, however, was not done with vampires. In late 1986, he challenged Manchester to a duel. According to Manchester, the two met but no duel ever took place, and he then lost track of Farrant. Because Farrant had claimed to be the "high priest" of Manchester's organization, Manchester said, he dismantled the society and focused his efforts on the Vampire Research Society and the ISAIVLR. Open to public membership, the ISAIVLR may be joined by invitation only. The move was made in response to what Manchester termed "infiltration and abuse of information and research material" by unfriendly agents.

In representing the ISAIVLR, Manchester occasionally uses the pseudonym "Ruthwen Glenarvon." The name is taken from Byronic associations: Lord Ruthven is the vampire in John Polidori's story, "The Vampyre," borrowed from Lord Byron's writings, and Glenarvon is the name of one of Byron's fictional characters. Ruthwen Glenarvon is listed as the editor of *The Cross and the Stake*, the newsletter of the ISAIVLR. According to Manchester, the name is used to answer correspondence and protect identities of various staff of the society. He has used it for some years (and, during the Highgate vampire case, he used the pseudonym "George Byron" while investigating the cemetery where Lusia was said to be buried).

Robin Hood: Highwayman Turned Vampire?

Vampires — or at least, headline-producing vampires — have been few and far between since the Highgate case vaulted Manchester to fame.

ancients and of all those who have studied the powers of the Un-Dead. When they become such, there comes with the change the curse of immortality. They cannot die, but must go on age after age adding new victims and multiplying the evils of the world. For all that die from the preying of the Un-dead become themselves Un-dead, and prey on their kind. And so the circle goes on ever widening, like as the ripples from a stone thrown in the water. Friend Arthur, if you had met that kiss which you know of before poor Lucy die, or again, last night when you open your arms to her, you would in time, when you had died, have become *nosferatu*, as they call it in Eastern Europe, and would for all time make more of those Un-Deads that so have filled us with horror. The career of this so unhappy dear lady is but just begun. Those children whose blood she sucked are not as yet so much the worse, but if she lives on, Un-Dead, more and more they lose their blood and by her power over them they come to her, and so she draw their blood with that so wicked mouth. But if she die in truth, then all cease. The tiny wounds of the throats disappear, and they go back to their play unknowing ever of what has been. But of the most blessed of all, when this now Un-Dead be made to rest as true dead, then the soul

He cited two other cases of possible vampirism that he has investigated: the Kirklees vampire and Lady Caroline Lamb, a lover of Lord Byron. Both are weak on vampire evidence and both are inconclusive.

Manchester said he had been researching the Lamb case for about ten years "Ever since her death there has been talk of her passion living beyond the grave," he said. "There is an inkling of material to point to a vampiric side to it. She was clearly a disturbed personality who died brokenhearted and insane." He hastened to add, "There is no evidence that she was involved in the dark arts. Yet she set in force such powerful energies that they might well have remained after her death. Her distressed energy form has been seen and felt in Hertfordshire and other areas. There also has been a strange history in the same areas of something manifestly dark and vampiric. I don't know if the two are in any way connected. After ten years, that link has not been established beyond a reasonable doubt."

of the poor lady whom we love shall again be free. Instead of working wickedness by night and growing more debased in the assimilating of it by day, she shall take her place with the other Angels. So that, my friend, it will be a blessed hand for her that shall strike the blow that sets her free. To this I am willing, but is there none amongst us who has a better right? Will it be no joy to think of hereafter in the silence of the night when sleep is not, 'It was my hand that sent her to the stars. It was the hand of him that loved her best, the hand that of all she would herself have chosen, had it been to her to choose?' Tell me if there be such a one amongst us?"

We all looked at Arthur. He saw too, what we all did, the infinite kindness which suggested that his should be the hand which would restore Lucy to us as a holy, and not an unholy, memory. He stepped forward and said bravely, though his hand trembled, and his face was as pale as snow, "My true friend, from the bottom of my broken heart I thank you. Tell me what I am to do, and I shall not falter!"

Van Helsing laid a hand on his shoulder, and said, "Brave lad! A moment's courage, and it is done. This stake must be driven through

The Kirklees case likewise is slim on evidence. It concerns an alleged gravesite of the legendary figure Robin Hood and speculations on whether or not Robin Hood died the victim of a vampire — and thus may have become a vampire himself.

According to legend, Robin Hood, the thirteenth-century highwayman who robbed the rich and gave to the poor, was ill and distempered by the time he reached about fifty. He set off for the Kirklees priory to be treated by the prioress there, who was reputed to be skilled in the arts of "physique and surgery." En route, he was cursed by a witch. When he reached the nunnery, the prioress bled him and, because she felt him to be an enemy of the church, allowed him to bleed straight on to death. By other accounts, Robin Hood survived the bloodletting but was killed in a sword fight by Red Roger, who may have been Sir Roger of Doncaster. This same Sir Roger, according to yet other accounts, may have incited the prioress to dispatch Robin Hood. Before he expired, Robin Hood is said to have shot an arrow through the priory window to mark his grave. Because he was a highwayman, he was buried outside of consecrated ground near the priory. The alleged site — no one knows for certain that Robin Hood is buried there — is marked by a deteriorating monument that was rebuilt in the eighteenth century. The gravesite has been on private property almost since 1539, when Henry VIII broke with the Catholic Church and dissolved the monasteries and nunneries. It now belongs to Lady Margarete Armytage, the owner of Kirklees Hall estate, and is inaccessible. Allegedly, an exhumation was performed about 200 years ago, and the grave supposedly was found to be empty.

The notion that vampirism was involved in the legend of Robin Hood apparently never occurred to anyone for several centuries, until Manchester suggested it as a possibility to Barbara Green, president of the Yorkshire Robin Hood Society. Green met Manchester about four or five years ago, and he generously offered to become the society's patron and promote its interests. When he later proposed the vampire theory concerning Robin

Hood's death, "it seemed worth pursuing," Green told me in an interview. She emphasized that the idea is theory only.

Manchester told Green that a "Ruthwen Glenarvon" from the ISAIVLR surreptitiously visited the gravesite and found that "certain things could be vampiric," as Green put it. In 1989, the vampire theory began to circulate in the media. "Ruthwen Glenarvon" wrote to Lady Armytage asking for permission to visit the grave and to interview Lady Armytage, information about the alleged exhumation, permission to examine the contents of the grave, and permission to allow "an accredited investigator from our society to keep vigil nocturnally near the grave for several nights." Lady Armytage would have none of it. "I have taken my solicitor's advice and don't want anything to do with them," she told the press.

The Yorkshire Robin Hood Society backs the ISAIVLR, Green said. She said that she sought the help of the local clergy to obtain permission to visit the grave and perform a blessing over it, but the local clergy — and higher Church of England officials — declined to become involved. A local priest was initially sympathetic, but after visiting Lady Armytage backed off.

Lady Armytage reported to the media that the gravesite had been vandalized by culprits unknown. It is possible that the vampire publicity may have encouraged some occult thrill seekers.

Green, who lives in Brighouse about two miles from the gravesite, said that after the publicity stirred things up, she began suffering mysterious manifestations in her own home, which she thought might be associated with the "atmosphere of evil" pervading the gravesite. She would wake up in the night and sense a presence, as though something evil were trying to take her over. A young girl staying in the house said something came into her own room and frightened her. The girl moved out and Green slept with the lights on for about a month. Green also started to receive "nasty letters" from occultists in London. She didn't read most of them, but burned them and poured holy water over them.

Green said that she and a small band of like-minded friends, dressed in period costumes from Robin Hood's day, sneaked to the gravesite in April 1990. Nearby they found the body of a goat with its throat tore open, which they took as another sign of something evil afoot. At the gravesite, they performed a blessing.

The Robin Hood vampire theory has been dismissed by most vampire experts as a reach. There are difficulties with it. To begin with, there is no proof that Robin Hood ever existed. Various historical figures have been put forth as the famed highwayman, but some historians believe the legend is a composite of a number of men.

If Robin Hood did exist, and he did go to a priory to seek medical treatment, there is nothing unusual in the fact that he was bled. It was standard medical treatment at the time (and was up to the nineteenth century), the prevailing notion being that the blood carried bad humors

her. It well be a fearful ordeal, be not deceived in that, but it will be only a short time, and you will then rejoice more than your pain was great. From this grim tomb you will emerge as though you tread on air. But you must not falter when once you have begun. Only think that we, your true friends, are round you, and that we pray for you all the time."

"Go on," said Arthur hoarsely. "Tell me what I am to do."

"Take this stake in your left hand, ready to place to the point over the heart, and the hammer in your right. Then when we begin our prayer for the dead, I shall read him, I have here the book, and the others shall follow, strike in God's name, that so all may be well with the dead that we love and that the Un-Dead pass away." Arthur took the stake and the hammer, and when once his mind was set on action his hands never trembled nor even quivered. Van Helsing opened his missal and began to read, and Quincey and I followed as well as we could.

Arthur placed the point over the heart, and as I looked I could see its dint in the white flesh. Then he struck with all his might.

The thing in the coffin writhed, and a hideous, bloodcurdling

and poisons that were eliminated in the bloodletting. (The fact that many patients were so weakened by the loss of blood that they died was an unfortunate side-effect.) There is nothing in the recorded stories about Robin Hood's demise to suggest that prioress was a vampire (the word *vampire* was unknown in England at the time) or that she was a vampirelike spirit of the dead. Nor is there any evidence to suggest that the prioress drank Robin Hood's blood — and even if she had, it wouldn't necessarily mean that she was a vampire. In short there is no evidence suggesting any supernatural hanky-panky.

And there also is no proof that Robin Hood, if he existed, is buried in the grave at Kirklees. Although all the documents concerning him say he is buried at Kirklees, the specific gravesite is unproven. The Robin Hood Society in Nottingham steers tourists to another site that they say may be the grave of the legendary hero.

Nonetheless, the "only possible way of identifying Undead activity is surveillance," Manchester said. "A vigil has been denied us by Lady Armytage. Thus it does seem improbable that we will arrive at a satisfactory conclusion."

Beside vampires, Manchester has waged war against all forces of darkness, including satanists. He published his account of his rescue from a satanic cult of a beautiful young woman, Sarah, whom he married in 1987. Together they work to build Manchester's lay order and to battle evil. As a priest, Manchester expects to be more effective against the dark forces, and more effective as an exorcist of demonic agencies.

Manchester has his own beliefs as to where vampires belong in the order of creation. He defines a vampire as a "demon" and sees vampirism as part of the larger menace of satanism. It is a view similar to the one held by medieval Inquisitors, who said that Satan, with God's permission, torments humankind and wreaks evil upon the world. "There is a direct corollary between the rise in the cult of satanism and the incredible resurgence of interest in recent years in vampirism and real vampires," he said.

He observed that satanism takes many forms including devotion to the dark arts. One did not have to worship Satan per se to become one of his infernal servants. He said that satanic cults attract recruits by holding out promises of cheating death through a vampire existence from the grave. But prior to death, the initiates are vampirized psychically of their will and energy through mind control and domination. "The satanists suck the vital streams of energy and in the end the blood of their victims," he said. "Whatever they are promised in these cults of the vampire, the actual result is death."

If the victims gain the favor of Satan prior to death, a grisly vampire existence ensues, according to Manchester. Their corpses become possessed by the forces of evil. "They become like animals with gnashing teeth," he said. "They become *demonic*, a malevolent force of evil that contaminates their material residue. That is why the vampire can leave its resting place

screech came from the opened red lips. The body shook and quivered and twisted in wild contortions. The sharp white champed together till the lips were cut, and the mouth was smeared with a crimson foam. But Arthur never faltered. He looked like a figure of Thor as his untrembling arm rose and fell, driving deeper and deeper the mercy-bearing stake, whilst the blood from the pierced heart welled and spurted up around it. His face was set, and high duty seemed to shine through it. The sight of it gave us courage so that our voices seemed to ring through the little vault.

And then the writhing and quivering of the body became less, and the teeth seemed to champ, and the face to quiver. Finally it lay still. The terrible task was over.

The hammer fell from Arthur's hand. He reeled and would have fallen had we not caught him. The great drops of sweat sprang from his forehead, and his breath came in broken gasps. It had indeed been an awful strain on him, and had he not been forced to his task by more than human considerations he could never have gone through with it. For a few minutes we were so taken up with him that we did not look towards the coffin. When we did, however, a

without disturbing a single blade of grass. It isn't the corpse itself, it is a masquerading, demonic force that has the ability to shape-shift, alter, and deceive."

The demonic vampire force sustaining a corpse lives outside of time, Manchester said. That is how it is able to preserve the body until it is cast out, and then the body disintegrates to ashes.

He said that even people who merely want to become vampires, for whatver reason, invite this horrible fate. He expressed concern for those who join vampire fan clubs. "You only have to look at the new members' sections and see what they're saying. Would they like to become vampires? Would they indeed! This is the folklore equivalent to opening the door and inviting the vampire in. You don't have to physically open a door. The mental door is the real door. Then the vampire can come in anytime he wants."

Manchester believes that the ranks of the Undead are swelling in response to a calling-in by Satan, who is gathering his forces for the final confrontation of the Apocalypse. The legions of Undead are lying dormant in graves, like radioactive canisters, waiting to "spill forth their poison and contaminate the earth." A few who are evoked in occult rituals manage to escape their graves, he said. The rest await Satan's clarion call to action. "They are the legacy of our own collective unconscious into a form that appears tangible."

Manchester said he spends a lot of time fending off critics who dispute his story as told in *The Highgate Vampire*. The account is markedly different from the story that first appeared in *The Vampire's Bedside Companion*, and some find the blood-gorged, decomposing corpse and shrieking spider incredible. Manchester answers them thusly: "I am not interested in whether people believe in vampires or not. I wrote *The Highgate Vampire* to relieve myself of a very large mailbag of pressure from people who were asking what had happened. They had read *The Vampire's Bedside Companion* and only had the beginning of the story."

I asked him why the two accounts differed dramatically in the descriptions of the vampire exposed in his coffin. Surely the details in the second account, such as the bloodstained mouth, the fetid breath, and the staring eyes, should have been included in the first account. Manchester said he had indeed included them, but his account had been "radically edited" for "legal reasons" by the publisher. However, I later had a chance to ask Peter Underwood, who had collected and edited the material for *The Vampire's Bedside Companion*, if Manchester's chapter had indeed been drastically edited. Underwood said that to the best of his recollection, the Highgate account had been published as Manchester had supplied it, "without any editing."

Manchester's critics also discount photographs of the alleged vampire that he published in *The Highgate Vampire*. Manchester said the photos were taken by his assistant, Arthur, moments before the corpse disintegrated.

murmur of startled surprise ran from one to the other of us. We gazed so eagerly that Arthur rose, for he had been seated on the ground, and came and looked too, and then a glad strange light broke over his face and dispelled altogether the gloom of horror that lay upon it.

There, in the coffin lay no longer the foul Thing that we had so dreaded and grown to hate that the work of her destruction was yielded as a privilege to the one best entitled to it, but Lucy as we had seen her in life, with her face of unequalled sweetness and purity. True that there were there, as we had seen them in life, the traces of care and pain and waste. But these were all dear to us, for they marked her truth to what we knew. One and all we felt that the holy calm that lay like sunshine over the wasted face and form was only an earthly token and symbol of the calm that was to reign for ever.

Van Helsing came and laid his hand on Arthur's shoulder, and said to him, "And now, Arthur my friend, dear lad, am I not forgiven?"

The reaction of the terrible strain came as he took the old man's hand in his, and raising it to his lips, pressed it, and said,

I'm not sure what a vampire corpse should look like, but I had to agree with the critics that the photos didn't convince me. Manchester asserted the photos were genuine, and observed that one couldn't expect to see the likes of Christopher Lee meeting his doom in a Hammer film.

Manchester's description of the vampire's last moments, when it decomposed instantly into "inhuman slime and viscera" may be unique in vampire records.

In searching through historical cases, I found none in which the corpse of a suspected vampire decomposed or disintegrated immediately upon being violated. The corpses were usually dismembered and burned. In film and literature, we find more theatrical demises. Stoker's Count Dracula — who is knifed to death in his box of earth, rather than staked in his coffin crumbles to dust "almost in the drawing of a breath." Since then, celluloid and literary vampires have met increasingly spectacular ends, contorting in agony, having their flesh wither and peel from their bones and the bones then turning to dust, or bursting into flames.

Nonetheless, Manchester, has gone one step further in his second revised edition of *The Highgate Vampire* (1991), publishing photos said to be time-elapsed exposures of the vampire's final moments.

Regardless of whether one accepts Manchester's account of the Highgate vampire, the case ultimately remains inconclusive. The primary evidence is anecdotal. Reports of mysterious shapes, figures, and noises may have been genuine, but they must be considered in light of possible natural explanations, and of the expectations that many people have that they will encounter something supernatural and frightening in a cemetery. In addition, reports often spawn "me, too" reports, and that seems to have happened in the Highgate case. The case is further clouded by the film production that occurred there during a time of sightings, by possible activities of occult groups, by the extensive media coverage, and by the antics of David Farrant.

The case of the Highgate vampire may never be proven, but it does stand as one of the most significant modern examples of the power possessed by the vampire image — a power to create a mass horror at the very idea that such a creature could exist. Our fear may be in truth a revulsion at our own collective dark side.

"Forgiven! God bless you that you have given my dear one her soul again, and me peace." He put his hands on the Professor's shoulder, and laying his head on his breast, cried for a while silently, whilst we stood unmoving.

When he raised his head Van Helsing said to him, "And now, my child, you may kiss her. Kiss her dead lips if you will, as she would have you to, if for her to choose. For she is not a grinning devil now, not any more a foul Thing for all eternity. No longer she is the devil's Un-Dead. She is God's true dead, whose soul is with Him!"

Arthur bent and kissed her, and then we sent him and Quincey out of the tomb. The Professor and I sawed the top off the stake, leaving the point of it in the body. Then we cut off the head and filled the mouth with garlic. We soldered up the leaden coffin, screwed on the coffin lid, and gathering up our belongings, came away. When the Professor locked the door he gave the key to Arthur.

Outside the air was sweet, the sun shone, and the birds sang, and it seemed as if all nature were tuned to a different pitch. There was gladness and mirth and peace everywhere, for we were at rest ourselves on one account, and we were glad, though it was with a tempered joy.

Last Call for the Sons of Shock

DAVID J. SCHOW
(1994)

ABOUT THE AUTHOR

David J. Schow's short stories have been regularly selected for over 25 volumes of "Year's Best" anthologies across two decades and have won the World Fantasy Award as well as the Dimension Award from *Twilight Zone* magazine. His novels include *The Kill Riff* (1987) and *The Shaft* (1990). His short stories are collected in *Seeing Red* (1990), *Lost Angels* (1990), *Black Leather Required* (1994), and *Crypt Orchids* (1998). He is the author of the exhaustively detailed *Outer Limits Companion* (1998) and has written extensively for films and television. You can see him talking and moving around on documentaries and DVDs for everything from *Creature from the Black Lagoon* and *Incubus* to *The Shawshank Redemption* and the horror film boom of the Eighties. He is also the editor of the *Lost Bloch* series for Subterranean Press, celebrating the pulp writing of Robert Bloch. His latest books, in 2001, are *Eye* (short fiction) and *Wild Hairs* (nonfiction). You can sample more of his work online at Fictionwise.com and as part of the Scorpius Digital anthology *Hours of Darkness*. His bibliography and much other information is available online at his official site, Black Leather Required: http://www.gothic.net/~chromo.

BLANK FRANK NOTCHES down the Cramps, keeping an eye on the blue LED bars of the equalizer. He likes the light.

"Creature from the Black Leather Lagoon" calms.

The club is called Un/Dead. The sound system is from the guts of the old Tropicana, LA's altar of mud wrestling, foxy boxing, and the cock-tease unto physical pain. Its specs are for metal, loud, lots of it. The punch of the subwoofers is a lot like getting jabbed in the sternum by a big velvet piston.

Blank Frank likes the power. Whenever he thinks of getting physical, he thinks of the Vise Grip.

He perches a case of Stoli on one big shoulder and tucks another of Beam under his arm. After this he is done replenishing the bar. To survive the weekend crush, you've gotta arm. Blank Frank can lug a five-case stack without using a dolly. He has to duck to clear the lintel. The passage back to the phones and bathrooms is tricked out to resemble a bank vault door, with tumblers and cranks. It is up past six-six. Not enough for Blank Frank, who still has to stoop.

Two hours till doors open.

Blank Frank enjoys his quiet time. He has not forgotten the date. He grins at the movie poster framed next to the back bar register. He scored it at a Hollywood memorabilia shop for an obscene price even though he got a professional discount. He had it mounted on foamcore to flatten the creases. He does not permit dust to accrete on the glass. The poster is duotone, with lurid lettering. His first feature film. Every so often some Un/Dead patron with cash to burn will make an exorbitant offer to buy it. Blank Frank always says no with a smile... and usually spots a drink on the house for those who ask.

He nudges the volume back up for Bauhaus, doing "Bela Lugosi's Dead," extended mix.

The staff sticks to coffee and iced tea. Blank Frank prefers a nonalcoholic concoction of his own devising, which he has christened a Blind Hermit. He rustics up one, now, in a chromium blender, one hand idly on his plasma globe. Michelle gave it to him about four years back, when they first became affordably popular. Touch the exterior and the purple veins of electricity follow your fingertips. Knobs permit you to fiddle with density

HUMOR IN A JUGULAR VEIN

Vampires, like other monsters, seem to have a natural, self-renewing life cycle in popular culture: at first objects of terror, they soften with familiarity, becoming clownish or cuddly. Ultimately, someone (like Anne Rice, for instance) comes along and gives the genre a shot in the arm with a new conception of vampires, and, bang, the undead are scary again (or at least not ridiculous).

In "Last Call for the Sons of Shock," David J. Schow casts a master vampire in the role of a jaded drug kingpin, a brilliantly amusing conceit someone really should have thought of before. Most vampire comedies, however, are products of film and television, and lack Schow's finely-tuned sense of cultural metaphor. Nevertheless, here is an overview of some of the highlights and lowlights in the ongoing attempts to mix bloodsucking and burlesque.

ABBOTT AND COSTELLO MEET FRANKENSTEIN (1948)

Part of the reason Charles T. Barton's postwar comedy remains so beloved among horror fans is that it sounded the death-rattle of the original Universal monster cycle. Even though the beloved bogeymen were reduced to proplike buffoons opposite popular forties comedians Bud Abbott and Lou Costello, it was the last time we would ever get to see them on screen. Dracula (Bela Lugosi, in the last role for a major studio) oversees an attempt to transplant the brain of Costello into the Frankenstein monster (Glenn Strange), but is thwarted by the efforts of the Wolf Man (Lon Chaney, Jr.) and the

and amplitude, letting you master the power, feel like Tesla showing off.

Blank Frank likes the writhing electricity.

By now he carries many tattoos. But the one on the back of his left hand — the hand toying with the globe — is his favorite: a stylized planet Earth, with a tiny propellored aircraft circling it. It is old enough that the cobalt-colored dermal ink has begun to blur.

Blank Frank has been utterly bald for three decades. A tiny wisp of hair issues from his occipital. He keeps it in a neat braid, clipped to six inches. It is dead white. Sometimes, when he drinks, the braid darkens briefly. He doesn't know why.

Michelle used to be a stripper, before management got busted, the club got sold, and Un/Dead was born of the ashes. She likes being a waitress and she likes Blank Frank. She calls him "big guy." Half the regulars think Blank Frank and Michelle have something steamy going. They don't. But the fantasy detours them around a lot of potential problems, especially on weekend nights. Blank Frank has learned that people often need fantasies to *seem* superficially true, whether they really are or not.

Blank Frank dusts. If only the bikers could see him now, being dainty and attentive. Puttering.

Blank Frank rarely has to play bouncer whenever some booze-fueled trouble sets to brewing inside Un/Dead. Mostly, he just strolls up behind the perp and waits for him or her to turn around and apologize. Blank Frank's muscle duties generally consist of just *looming*.

If not, he thinks with a smile, there's always the Vise Grip.

The video monitor shows a Red Top taxicab parking outside the employee entrance. Blank Frank is pleased. This arrival coincides exactly with his finish-up on the bartop, which now gleams like onyx. He taps up the slide pot controlling the mike volume on the door's security system. There will come three knocks.

Blank Frank likes all this gadgetry. Cameras and shotgun mikes, amps and strobes and strong, clean alternating current to web it all in concert with maestro surety. Blank Frank loves the switches and toggles and running lights. But most of all, he loves the power.

Tap-tap-tap. Precisely. Always three knocks.

"Good," he says to himself, drawing out the vowel. As he hastens to the door, the song ends and the club fills with the empowered hiss of electrified dead air.

Out by limo. In by cab. One of those eternally bedamned scheduling glitches.

The Count overtips the cabbie because his habit is to deal only in round sums. He never takes . . . change. The Count has never paid taxes. He has cleared forty-three million large in the past year, most of it safely banked in bullion, out-of-country, after overhead and laundering.

The Count raps smartly with his umbrella on the service door of Un/Dead. Blank Frank never makes him knock twice.

It is a pleasure to see Blank Frank's face overloading the tiny security window; his huge form filling the threshold. The Count enjoys Blank Frank despite his limitations when it comes to social intercourse. It is relaxing to appreciate Blank Frank's conditionless loyalty, the innate tidal pull of honor and raw justice that seems programmed into the big fellow. Soothing, it is, to sit and drink and chat lightweight chat with him, in the autopilot way normals told their normal acquaintances where they'd gone and what they'd done since their last visit. Venomless niceties.

None of the buildings in Los Angeles has been standing as long as the Count and Blank Frank have been alive.

Alive. Now there's a word that begs a few new comprehensive, enumerated definitions in the dictionary. Scholars could quibble, but the Count and Blank Frank and Larry were definitely alive. As in "living" — *especially* Larry. Robots, zombies, and the walking dead in general could never get misty about such traditions as this threesome's annual conclaves at Un/Dead.

The Count's face is mappy, the wrinkles in his flesh rice-paper fine. Not creases of age, but tributaries of usage, like the creeks and streams of palmistry. His pallor, as always, tends toward blue. He wears dark shades with faceted, lozenge-shaped lenses of apache tear; mineral crystal stained a bloody-black. Behind them, his eyes, bright blue like a husky's. He forever maintains his hair wet and backswept, what Larry has called his "renegade opera conductor coif." Dramatic threads of pure cobalt-black streak backward from the snow-white crown and temples. His lips are as thin and bloodless as two slices of smoked liver. His diet does not render him robustly sanguine; it merely sustains him, these days. It bores him.

Before Blank Frank can get the door open, the Count fires up a hand-rolled cigarette of coca paste and drags the milky smoke deep. It mingles with the dope already loitering in his metabolism and perks him to.

The cab hisses away into the wet night. Rain on the way.

Blank Frank is holding the door for him, grandly, playing butler.

The Count's brow is overcast. "Have you forgotten so soon, my friend?" Only a ghost of his old, marble-mouthed, middle-Euro accent lingers. It is a trait that the Count has fought for long years to master, and he is justly proud that he is intelligible. Occasionally, someone asks if he is from Canada.

Blank Frank pulls the exaggerated face of a child committing a big boo-boo. "Oops, sorry." He clears his throat. "Will you come in?"

Equally theatrically, the Count nods and walks several thousand worth of Armani double-breasted into the cool, dim retreat of the bar. It is nicer when you're invited, anyway.

"Larry?" says the Count.

"Not yet," says Blank Frank. "You know Larry — tardy is his twin. There's real time and Larry time. Celebrities expect you to expect them to be late." He points toward the back bar clock, as if that explains everything.

machinations of a predictable script, which recycled several of the team's tried and true routines. Lugosi's makeup is appallingly overapplied by Bud Westmore; he looks more like a kewpie doll than a bloodsucking fiend, and a shiny satin cape does nothing to alleviate the overall circus-clown effect). The film does have a number of truly funny moments, most memorably Dracula's cat-and-mouse game with Costello as he prepares to emerge from his box, and later, the hilarious closeup of the eyes of one of the vampire's victims (Lenore Aubert), revealing flapping bats instead of pupils. *Abbott and Costello Meet Frankenstein* is also of some interest for the superimposition into Dracula of both the vampire and mad scientist traditions: the time-honored opposition of science and the supernatural here settle down for some cheerful codependency. While the film has developed a reputation as a comedy classic over the years, the original reviews were fairly caustic. As the *Hollywood Citizen* put it, "If you've never known whether to laugh or scream at an Abbott and Costello epic, their latest screen adventure will leave you more confused than ever." (Universal-International)

Advertising art for Abbott and Costello Meet Frankenstein.

FEARLESS VAMPIRE KILLERS: OR, PARDON ME, BUT YOUR TEETH ARE IN MY NECK (1967)

Roman Polanski's stylish, celebrated horror spoof, called *Dance of the Vampires* in Britain, isn't half so funny today as it seemed on its first release. The spectre of Polanski's then-wife Sharon Tate and her gruesome murder by the Manson "family" a year after the film's release still haunts every frame in which the beautiful actress appears, lending the proceedings a gloomy fatalism that sits uneasily with the film's overall aspiration to slapstick. On its first American release, *The Fearless Vampire Killers* was so drastically cut and reedited that Polanski requested his name be removed from the credits. It has subsequently been restored, although Polanski's cut often feels interminable. Silent film buffs will notice how much Polanski himself resembled Gustav Von Wangenheim, the young hero of *Nosferatu* (1922) in both costume and appearance. Ferdy Mayne as the evil Count Krolock is a masterful vampire, a plausible amalgam of both the Bela Lugosi and Christopher Lee traditions. A bizarre detail this writer noticed only on a second viewing: one of the revellers at the big bloodsucker's ball is carefully made up to look exactly like Laurence Olivier as Richard III. There has to be a story in this. With Jack MacGowran, Alfie Bass, and Jessie Robbins. Polanski co-scripted, with Gerard Brach. (Metro-Goldwyn-Mayer)

The Count can see perfectly in the dark, even with his murky glasses.As he strips them, Blank Frank notices the silver crucifix dangling from his left earlobe, upside-down.

"You into metal?"

"I like the ornamentation," says the Count. "I was never too big on jewelry; greedy people try to dig you up and steal it if they know you're wearing it; just ask Larry. The sort of people who would come to thieve from the dead in the middle of the night are not the class one would choose for friendly diversion."

Blank Frank conducts the Count to three highback Victorian chairs he has dragged in from the lounge and positioned around a cocktail table. The grouping is directly beneath a pinlight spot, intentionally theatrical.

"Impressive." The Count's gaze flickers toward the bar. Blank Frank is way ahead of him.

The Count sits, continuing: "I once knew a woman who was beleaguered by a devastating allergy to cats. And this was a person who felt some deep emotional communion with that species. Then one day, pouf! She no longer sneezed; her eyes no longer watered. She could stop taking medications that made her drowsy. She had forced herself to be around cats so much that her body chemistry adapted. The allergy receded." He fingers the silver cross hanging from his ear, a double threat, once alum a time. "I wear this as a reminder of how the: body can triumph. Better living through chemistry."

"It was the same with the and fire." Blank Frank hands over a very potent mixed drink called a Gangbang. The Count sips, then presses his eyelids contentedly shut. Like a cat. The drink must be industrial strength. Controlled substances are the Count's lifeblood.

Blank Frank watches as the Count sucks out another long, deep, soul drowning draught. "You know Larry's going to ask again, whether you're still doing . . . what you're doing."

"I brook no apologia or excuses." Nevertheless, Blank Frank sees him straighten in his chair, almost defensively. "I could say that you provide the same service in this place." With an outswept hand, he indicates the bar. If nothing else remains recognizable, the Count's gesticulations remain grandiose: physical exclamation points.

"It's legal. Food. Drink. Some smoke."

"Oh, yes, there's the rub." The Count pinches the bridge of his nose. He consumes commercial decongestants ceaselessly. Blank Frank expects him to pop a few pills, but instead the Count lays out a scoop of toot inside his mandarin pinky fingernail, which is lacquered ebony, elongated, a talon. Capacious. Blank Frank knows from experience that the hair and nails continue growing long after death. The Count inhales the equivalent of a pretty good dinner at Spago. Capuccino included.

"There is no place in the world I have not lived," says the Count. "Even the Arctic. The Australian outback. The Kenyan sedge. Siberia. I walk unharmed through fire-fight zones, through sectors of strife. You learn so much when you observe people at war. I've survived holocausts, conflagration, even a low-yield one-megaton test, once, just to see if I could do it. Sue me; I was high. But wherever I venture, whatever phylum of human beings I encounter, they all have one thing in common."

"The red stuff." Blank Frank half-jests; he dislikes it when the mood grows too grim.

"No. It is their need to be narcotized." The Count will not be swerved. "With television. Sex. Coffee. Power. Fast cars and sado-games. Emotional encumbrances. More than anything else, with chemicals. All drugs are like instant coffee. The fast purchase of a feeling. You buy the feeling, instead of earning it. You want to relax, go up or go down, get strong or get stupid? You simply swallow or snort or inject, and the world changes because of you. The most lucrative commercial enterprises are those with the most undeniable core simplicity; just look at prostitution. Blood, bodies, armaments, position — all commodities. Human beings want so much out of life."

Poster for The Fearless Vampire Killers.

The Count smiles, sips. He knows that the end of life is only the beginning. "Today is the first day of the rest of your death."

"I do apologize, my old friend, for coming on so aggressively. I've rationalized my calling, you see, to the point where it is a speech of lists; I make my case with demographics. Rarely do I find anyone who cares to suffer the speech."

"You've been rehearsing." Blank Frank recognizes the bold streak the Count gets in his voice when declaiming. Blank Frank has himself been jammed with so many hypos in the past few centuries that he has run out of free veins. He has sampled the Count's root-canal quality coke; it made him irritable and sneezy. The only drugs that still seem to work on him unfailingly are extremely powerful sedatives in large, near-toxic dosages. And those never last long. "Tell me. The drugs. Do they have any effect on you?"

He sees the Count pondering how much honesty is too much. Then the tiny, knowing smile flits past again, a wraith between old comrades.

"I employ various palliatives. I'll tell you the absolute truth: mostly it is an affectation, something to occupy my hands. Human habits — vices, for that matter — go a long way toward putting my customers at ease when I am closing negotiations."

"Now you're thinking like a merchant," says Blank Frank. "No royalty left in you?"

"A figurehead gig." The Count frowns. "Over whom, my good friend, would I hold illimitable dominion? Rock stars. Thrill junkies. Corporate monsters. No percentage in flaunting your lineage there. No. I occupy my time much as a fashion designer does. I concentrate on next season's line. I brought cocaine out of its Vin Mariani limbo and helped repopularize it in the Eighties. Then crank, then crack, then ice. Designer

DRACULA BLOWS HIS COOL (1979)

A really bizarre film from West Germany. The vampire's castle becomes a garish tourist trap, with the count reduced to providing oral room service for paying customers looking for the thrill of a chill. The idea of Dracula as a capitalist whore is an interesting one, but this movie isn't interested in ideas. Directed by Carlo Ombra, with Gianni Garko as Dracula. (Lisa-Barthonia Film)

LOVE AT FIRST BITE (1979)

A completely corny but nonetheless delightful Dracula spoof. George Hamilton's pale, pale makeup is an instant parodic comment on the actor's otherwise carcinogenically suntanned playboy persona. Hamilton makes a very funny vampire, coping with the cultural shock attendant on his relocating from Transylvania to the Manhattan disco scene. Actress Carroll Borland, protégée of Bela Lugosi, once told this writer that Hamilton's performance was an uncanny approximation of Lugosi's original stage performance as Dracula (a performance considerably muted in the 1931 film version). A sequel, *Love at Second Bite*, has been announced repeatedly over the years, but has never materialized. Stan Dragoti directed, and Robert Kaufman wrote. With Susan Saint James, Richard Benjamin, Dick Shawn, Arte Johnson, and Sherman Hemsley. (American International)

dope. You've heard of Ecstasy. You haven't heard of Chrome yet. Or Amp. But you will."

Suddenly a loud booming rattles the big main door, as though the entire DEA is hazarding a spot raid. Blank Frank and the Count are both twisted around in surprise. Blank Frank catches a glimpse of the enormous Browning Hi-Power holstered in the Count's left armpit.

It's probably just for the image, Blank Frank reminds himself.

The commotion sounds as though some absolute lunatic is kicking the door and baying at the moon. Blank Frank hurries over, his pulse relaying as his pace quickens.

It has to be Larry.

"Gah-DAMN it's peachy to see ya, ya big dead dimwit!" Larry is a foot shorter than Blank Frank. Nonetheless, he bounds in, pounces, and sufocates his amigo in a big wolfy bear hug.

Larry is almost too much to take in with a single pair of eyes.

His skintight red Spandex tights are festooned with spangles and fringe that snake, at knee level, into golden cowboy boots. Glittering spurs on the boots. An embossed belt buckle the size of the grille on a Rolls. Larry is into ornaments, including a feathered earring with a skull of sterling, about a hundred metalzoid bracelets, and a three-finger rap ring of slush-cast 24K that spells out AWOO. His massive, pumped chest fairly bursts from a bright silver Daytona racing jacket, snapped at the waist but not zippered, so the world can see his collarless muscle tee in neon scarlet, featuring his caricature in yellow. Fiery letters on the shirt scream about THE REAL WOLF MAN. Larry is wearing his Ray-Bans at night and jingles a lot whenever he walks.

"Where's ole Bat Man? Yo! I see you skulking in the dark!" Larry whacks Blank Frank on the biceps, then lopes to catch the Count. With the Count, it is always a normal handshake — dry, firm, businesslike. "Off thy bunnage, fang-dude; the party has arriiiived!!"

"Nothing like having a real celebrity in our midst," says Blank Frank. "But jeez — what the hell is this 'Real' Wolf Man crap?"

Larry grimaces as if from a gas pain, showing teeth. "A slight little ole matter of copyrights, trademarks, eminent domain . . . and some fuckstick who registered himself with the World Wrestling Federation as 'The Wolfman.' Turns out to be a guy I bit, my own self, a couple of decades ago. So I have to be 'The Real.' We did a tag-team thing, last Wrestlemania. But we can't think of a good team name."

"Runts of the Litter," opines the Count. Droll.

"Hellpups," says Blank Frank.

"Fuck ya both extremely much." Larry grins his trademark grin. Still showing teeth. He snaps off his shades and scans Un/Dead. "What's to quaff in this pit? Hell, what town is this, anywho?"

"On tour?" Blank Frank plays host.

"Yep. Gotta kick Jake the Snake's ass in Atlanta next Friday. Gonna strangle him with Damien, if the python'll put up with it. Wouldn't want to hurt him for real but might have ole Jake pissing blood for a day, if you know what I mean."

Blank Frank grins; he knows what Larry means. He makes a fist with his left hand, then squeezes his left wrist tightly with his right hand. "Vise Grip him."

Larry is the inventor of the Vise Grip, second only to the Sleeper Hold in wrestling infamy. The Vise Grip has done Blank Frank a few favors with rowdies in the past. Larry owns the move, and is entitled to wax proud.

"I mean pissing pure blood!" Larry enthuses.

"Ecch," says the Count. "Please."

"Sorry, oh cloakless one. Hey! Remember that brewery, made about three commercials with the Beer Wolf before that campaign croaked and ate dirt? That was me!"

Blank Frank hoists his Blind Hermit. "Here's to the Beer Wolf, then. Long may he howl."

"Prost," says the Count.

"Fuckin A." Larry downs his entire mugfull of draft in one slam-dunk. He belches, wipes foam from his mouth and lets go with a lupine yee-hak.

The Count dabs his lips with a cocktail napkin.

Blank Frank watches Larry do his thing and a stiff chaser of memory quenches his brain. That snout, the bicuspids, and those beady, hallbearing eyes will always give Larry away. His eyebrows ran together; that was supposed to be a classic clue in the good old days. Otherwise, Larry was not so hirsute. In human form, at least. The hair on his forearms was very fine tan down. Pumping iron and beating up people for a living has bulked out his shoulders. He usually wears his shirts open-necked. T-shirts, he tears the throats out. He is all piston-muscles and zero flab. He is able to squeeze a full beer can in one fist and pop the top with a gunshot bang. His hands are calloused and wily. The pentagram on his right palm is barely visible. It has faded, like Blank Frank's tattoo.

"Cool," Larry says of the Count's crucifix.

"Aren't you wearing a touch as well?" The Count points at Larry's skull earring. "Or is it the light?"

Larry's fingers touch the silver. "Yeah. Guilty. Guess we haven't had to fret that movieland spunk for quite a piece, now."

"I had fun." Blank Frank exhibits his tat. "It was good."

"Goood," Larry and the Count say together, funning their friend.

All three envision the tiny plane — in growly flight, circling a black and white world, forever.

"How long have you had that?" Larry is already on his second mug, foaming at the mouth.

Blank Frank's pupils widen, filling with his skin illustration. He does not remember.

FRIGHT NIGHT (1985)

The first vampire movie to spend one million dollars on special effects, Tom Holland's fang-in-cheek *Fright Night* isn't always succesful in finding the proper balance between humor and horror; nonetheless, the picture is an entertaining eyeful, whether it is scoring points with over-the-top visuals, or more discreetly, with such touches as the vampire (Chris Sarandon) nonchalantly whistling "Strangers in the Night." Sarandon's teenage neighbor (William Ragsdale) suspects the worst and enlists the burnt-out host of the local television horror movie show (Roddy McDowall) to banish the evil forever — or at least until the sequel, *Fright Night Part II* (1988), which featured a memorable sequence with a vampire on roller skates. The make-up in both films was elaborate and inventive; the *New York Post*'s "Phantom of the Movies" dutifully noted that *Fright Night*'s heroine Amanda Bearse "sports the champ vamp choppers of all time — stalactic fangs gleaming from a crimson kisser that take up half her face." The final twenty minutes or so of the film — everything after Sarandon intones "Welcome to Fright Night..." — constitute a wildly entertaining vampire Götterdämmerung that will leave you gasping, and reaching for the rewind button. (Vistar/Columbia)

Chris Sarandon in Fright Night.

“At least forty years ago,” says the Count. “They’d changed the logo by the time he’d committed to getting the tattoo.”

“Maybe that was why I did it.” Blank Frank is still a bit lost. He touches the tattoo as though it will lead to a swirl dissolve and an expository flashback.

“Hey, we saved that fuckin’ studio from bankruptcy.” Larry bristles. “Us and A&C.”

“They were shown the door, too.” To this day, the Count is understandably piqued about the copyright snafu involving the use of his image. He sees his face everywhere, and does not rate compensation. This abrades his business instinct for the jugular. He understands too well why there must be a Real Wolf Man. “Bud and Lou and you and me and the big guy all went out with the dishwater of the Second World War.”

“I was at Lou’s funeral,” says Larry. “You were lurking the Carpathians.” He turned to Blank Frank. “And you didn’t even know about it.”

“I loved Lou,” says Blank Frank. “Did I ever tell you the story of how I popped him by accident on the set of —”

“Yes.” The Count and Larry speak in unison. This breaks the tension of remembrance tainted by the unfeeling court intrigue of studios. Recall the people, not the things.

Blank Frank tries to remember some of the others. He returns to the bar to rinse his glass. The plasma globe zizzes and snaps calmly, a man-made tempest inside clear glass.

“I heard ole Ace got himself a job at the Museum of Natural History.” Larry refers to Ace Bandage; he has nicknames like this for everybody.

“The Prince,” the Count corrects, “still guards the Princess. She’s on display in the Egyptology section. The Prince cut a deal with museum security. He prowls the graveyard shift; guards the bone rooms. They’ve got him on a synthetic of tana leaves. It calmed him down. Like methadone.”

“A night watchman gig,” says Larry, obviously thinking of the low pay scale. But what in hell would the Prince need human coin for, anyway? “Hard to picture.”

“Try looking in a mirror, yourself,” says the Count.

Larry blows a raspberry. “Jealous.”

It is very easy for Blank Frank to visualize the Prince, gliding through the silent, cavernous corridors in the wee hours. The museum is, after all, just one giant tomb.

Larry is fairly certain ole Fish Face — another nickname — escaped from a mad scientist in San Francisco and butterfly-stroked south, probably to wind up in bayou country. He and Larry had shared a solid mammal-to-amphibian simpatico. He and Larry had been the most physically violent of the old crew. Larry still entertained the notion of talking his scaly pal into doing a bout for pay-for-view. He has never been able to work out the logistics of a steel fishtank match, however.

VAMP (1986)

Amazonian club diva Grace Jones embodies the kind of voracious phallic feminity that nerdy college guys find both scary and hot — all the hotter because of the scariness. A trio of frat brothers go to the big city to find a stripper to bring back for a campus party. They find Jones instead, a zillion-year-old Egyptian vampire who has party ideas all her own. For a formula teen comedy, this is pretty passable. Directed by Richard Wenk, from his own screenplay, based on a story by Donald P. Borchers. With Chris Makepeace, Sandy Baron, Robert Rusler, and Gedde Watanabe. (Balcor Films/ New World Pictures)

Grace Jones in Vamp.

"Griffin?" says the Count.

"Who can say?" Blank Frank shrugs. "He could be standing right there and we wouldn't know it unless he started singing 'Nuts in May.'"

"He was a misanthrope," says Larry. "His crazy kid, too. That's what using drugs will get you."

This last is a veiled stab at the Count's calling. The Count expects this from Larry, and stays venomless. The last thing he wants this evening is a conflict over the morality of substance abuse.

"I dream, sometimes, of those days," says Blank Frank. "Then I see the films again. The dreams are literalized. It's scary."

"Before *this* century," says the Count, "I never had to worry that anyone would stockpile my past." Of the three, he is the most paranoid where personal privacy is concerned.

"You're a romantic." Larry will only toss an accusation like this in special company. "It was important to a lot of people that we *be* monsters. You can't deny what's nailed down there in black and white. There was a time when the world *needed* monsters like that."

They each considered their current occupations, and found that they did indeed still fit into the world.

"Nobody's gonna pester you now," Larry presses on. "Don't bother to revise your past — today, your past is public record, and waiting to contradict you. We did our jobs. How many people become mythologically legendary for just doing their jobs?"

"Mythologically legendary?" mimicks the Count. "You'll grow hair on your hands from using all those big words."

"Bite this." Larry offers the unilateral peace symbol.

"No, thank you; I've already dined. But I have brought something for you. For both of you."

Blank Frank and Larry both notice the Count is now speaking as though a big Mitchell camera is grinding away, somewhere just beyond the grasp of sight. He produced a small pair of wrapped gifts, and hands them over.

Larry wastes no time ripping into his. "Weighs a ton."

Nestled in styro popcorn is a wolf's head — savage, streamlined, snarling. The gracile canine neck is socketed.

"It's from the walking stick," says the Count. "All that was left."

"No kidding." Larry's voice grows small for the first time this evening. The wolf's head seems to gain weight in his grasp. Two beats of his powerful heart later, his eyes seem a bit wet.

Blank Frank's gift is much smaller and lighter.

"You were a conundrum," says the Count. He enjoys playing emcee. "So many choices, yet never easy to buy for. Some soil from Transylvania? Water from Loch Ness? A chunk of some appropriate ruined castle?"

What Blank Frank unwraps is a ring. Old gold, worn smooth of its subtler filigree. A small ruby set in the grip of a talon. He holds it to the light.

DRACULA: DEAD AND LOVING IT (1995)

This must have sounded like a great idea on paper: a companion piece to Mel Brooks's classic farce *Young Frankenstein*, this time taking on the other major horror icon. But *Dracula: Dead and Loving It* ended up mostly a mess. Leslie Nielsen's casting as Dracula is inexplicable, except for the box office value of his name as a comedy star (although it must be said that he obviously studied Bela Lugosi's vocal mannerisms *verrry* carefully, which is more than most impersonators do). Unlike *Young Frankenstein*, there is no attempt here to recreate the black-and-white world of the classic horror films, and by aiming darts at every Dracula variation from Lugosi to Lee to Langella to Gary Oldman, Brooks ends up missing all targets. Peter MacNichol steals the show with an inspired impersonation of Dwight Frye's classic Renfield, and Anne Bancroft (Mrs. Brooks), has fun in a cameo as a pushy, crucifix-dispensing gypsy. But the screenplay, by Brooks, Rudy De Luca, and Steve Haberman, is sluggish and uninventive. Still, it's hard to suppress a smile at lines like "Yes, we have Nosferatu." If only there had been more of them. With Steven Weber, Amy Yasbeck, Lysette Anthony, Harvey Korman, and Mel Brooks himself as the vampire's nemesis, Van Helsing. (Columbia Pictures/ Castle Rock Entertainment)

Leslie Nielsen as Dracula.

TELEVISION VAMPIRES

The great technological vampire of modern times, television rests in a box and is especially active at night, when it mesmerizes us with a baleful gaze. Like a vampiric encounter, television is often about living vicariously in a deadened trance state. In his against-the-grain tract, *Four Arguments for the Elimination ofTelevision* (1978), ex-advertising executive Jerry Mander lists typical phrases used by Americans to describe their relationship with the tube: "I feel hypnotized," "Television sucks my energy," "My kids look like zombies when they're watching it."

Most of the time, television plays vampires for laughs. The original TV horror hostess was Vampira (Maila Nurmi), whose cartoony persona kept her fifties fans glued to their sets. Bela Lugosi himself appeared on *You Asked for It* at the request of a California housewife — for whom newfangled television, perhaps, awakened a nostalgia for low-tech leeching of the Lugosi kind. In the 1960s, *The Munsters* comedy series featured a half-vampire family of prime-time monsters, and homemakers and other shut-ins eagerly consumed images of living death delivered by the soap opera *Dark Shadows* (which wasn't intentionally funny, but...). Vampires made ghost appearances everywhere on TV, from *Gilligan's Island* to *F Troop* to *Love, American Style*.

ABC's vampire-in-Vegas TV movie *The Night Stalker* (1971) spun off a series, as did *Nick Knight* (1989) which became the series *Forever Knight* (1992). *Dracula – The Series* (1990) was a teen-oriented series that ran only one season, but was certainly a precursor of today's television mega-hit, *Buffy, the Vampire Slayer*.

"As nearly as I could discover, that ring once belonged to a man named Ernst Volmer Klumpf."

"Whoa,"says Larry. Weird name.

Blank Frank puzzles it. He holds it toward the Count, like a lens.

"Klumpf died a long time ago," says the Count. "Died and was buried. Then he was disinterred. Then a few of his choicer parts were recycled by a skillful surgeon of our mutual acquaintance."

Blank Frank stops looking so blank.

"In fact, part of Ernst Volmer Klumpf is still walking around today. . . tending bar for his friends, among other things."

The new expression on Blank Frank's face pleases the Count. The ring just barely squeezes onto the big guy's left pinky — his smallest finger.

Larry, to avoid choking up, decides to make noise. Showing off, he vaults the bartop and draws his own refill. "This calls for a toast." He hoists his beer high, slopping the head. "To dead friends. Meaning us."

The Count pops several capsules from an ornate tin and washes them down with the last of his Gangbang. Blank Frank murders his Blind Hermit.

"Don't even think of the bill," says Blank Frank, who knows of the Count's habit of paying for everything. The Count smiles and nods graciously. In his mind, the critical thing is to keep the tab straight. Blank Frank pats the Count on the shoulder, hale and brotherly, since Larry is out of reach. The Count dislikes physical contact but permits this because it is, after all, Blank Frank.

"Shit man, we could make our own comeback sequel, with all the talent in this room," Larry says. "Maybe hook up with some of those new guys. Do a monster rally."

It could happen. They all look significantly at each other. A brief stink of guilt, of culpability, like a sneaky fart in a dimly-lit chamber.

Make that dimly-lit *torture dungeon*, thinks Blank Frank, who never forgets the importance of staying in character.

Blank Frank thinks about sequels. About how studios had once jerked their marionette strings, compelling them to come lurching back for more, again and again, adding monsters when the brew ran weak, until they had all been bled dry of revenue potential and dumped at a bus stop to commence the long deathwatch that had made them nostalgia.

It was like living death, in its way.

And these gatherings, year upon year, had become sequels in their own right.

The realization is depressing. It sort of breaks the back of the evening for Blank Frank. He stands friendly and remains as chatty as he ever gets. But the emotion has soured.

Larry chugs so much that he has grown a touch bombed. The Count's chemicals intermix and buzz; he seems to sink into the depths of his coat, his chin ever-closer to the butt of the gun he carries. Larry drinks deep, then howls. The Count plugs one ear with a finger on his free hand. "I wish he wouldn't do that," he says in a proscenium-arch *sotto voce* that indicates his annoyance is mostly token.

When Larry tries to hurdle the bar again, moving exaggeratedly as he almost always does, he manages to plant his big wrestler's elbow right into the glass on Blank Frank's framed movie poster. It dents inward with a sharp crack, cobwebbing into a snap puzzle of fracture curves. Larry swears, instantly chagrined. Then, lamely, he offers to pay for the damage.

The Count, not unexpectedly, counter-offers to buy the poster, now that it's damaged.

Blank Frank shakes his massive square head at both of his friends. So many years, among them. "It's just glass. I can replace it. It wouldn't be the first time."

The thought that he has done this before depresses him further. He sees the reflection of his face, divided into staggered components in the broken glass, and past that, the lurid illustration. Him then. Him now.

Blank Frank touches his face as though it is someone else's. His fingernails have always been black. Now they are merely fashionable.

Larry remains embarrassed about the accidental damage and the Count begins spot-checking his Rolex every five minutes or so, as though he is pressing the envelope on an urgent appointment. Something has spoiled the whole mood of their reunion, and Blank Frank is angry that he can't quite pinpoint the cause. When he is angry, his temper froths quickly.

The Count is the first to rise. Decorum is all. Larry tries one more time to apologize. Blank Frank stays cordial, but is overpowered by the sudden strong need to get them the hell out of Un/Dead.

The Count bows stiffly. His limo manifests precisely on schedule. Larry gives Blank Frank a hug. His arms can reach all the way 'round.

"*Au revoir*," says the Count.

"Stay dangerous," says Larry.

Blank Frank closes and locks the service door. He monitors, via the tiny security window, the silent, gliding departure of the Count's limousine, the fading of Larry's spangles into the night.

Still half an hour till opening. The action at Un/Dead doesn't really crank until midnight anyway, so there's very little chance that some bystander will get hurt.

Blank Frank bumps up the volume and taps his club foot. A eulogy with a beat. He loves Larry and the Count in his massive, broad, uncompromisingly loyal way, and hopes they will understand his actions. He hopes that his two closest friends are perceptive enough, in the years to come, to know that he is not crazy.

Not crazy, and certainly not a monster.

While the music plays, he fetches two economy-sized plastic bottles of lantern kerosene, which he ploshes liberally around the bar, saturating the old wood trim. Arsonists call such flammable liquids "accelerator."

In the scripts, it was always an overturned lantern, or a flung torch

Bob Denver on Gilligan's Island. *(Courtesy of Ronald V. Borst/ Hollywood Movie Posters)*

Vincent Price with the cast of TV's F Troop. *(Courtesy of Ronald V. Borst/Hollywood Movie Posters)*

Geordie Johnson played the king of vampires on Dracula — the Series.

from a mob of villagers, that touched off the conclusive inferno. Mansions, mad labs, even stone fortresses burned and blew up, eliminating monster menaces until they were needed again.

Dark threads snake through the tiny warrior braid at the back of Blank Frank's skull. All those Blind Hermits.

The purple electricity arcs toward his finger and trails it loyally. He unplugs the plasma globe and cradles it beneath one giant forearm. The movie poster, he leaves hanging in its smashed frame.

He snaps the sulphur match with one black thumbnail. Ignition craters and blackens the head, eating it with a sharp hiss. Un/Dead's PA throbs with the bass line of "D.O.A." Phosphorus tangs the unmoving air. The match fires orange to yellow to a steady blue. The flamepoint reflects from Blank Frank's large black pupils. He can see himself, as if by candlelight, fragmented by broken picture glass. The past. In his grasp is the plasma globe, unblemished, pristine, awaiting a new charge. The future.

He recalls all of his past experiences with fire. He drops the match into the thin pool of accelerator glistening on the bartop. The flame grows quietly.

Good.

Light springs up, hard white, behind him as he exits and locks the door. The night is cool, near foggy. Condensation mists the plasma globe as he strolls away, pausing beneath a streetlamp to appreciate the ring on his little finger. He doesn't need to eat or sleep. He'll miss Michelle and the rest of the Un/Dead folks. But he is not like them; he has all the time he'll ever need, and friends who will be around forever.

Blank Frank likes the power.

Coppola's Dracula

KIM NEWMAN
(1997)

A TREELINE AT dusk. Tall, straight, Carpathian pines. The red of sunset bleeds into the dark of night. Great flapping sounds. Huge, dark shapes flit languidly between the trees, sinister, dangerous. A vast batwing brushes the treetops.

Jim Morrison's voice wails in despair. "People Are Strange."

Fire blossoms. Blue flame, pure as candle light. Black trees are consumed...

Fade to a face, hanging upside-down in the roiling fire.

Harker's Voice: *Wallachia... shit!*

Jonathan Harker, a solicitor's clerk, lies uneasy on his bed, upstairs in the inn at Bistritz, waiting. His eyes are empty.

With great effort, he gets up and goes to the full-length mirror. He avoids his own gaze and takes a swig from a squat bottle of plum brandy. He wears only long drawers. Bite-marks, almost healed, scab his shoulders. His arms and chest are sinewy, but his belly is white and soft. He staggers into a program of isometric exercises, vigorously Christian, ineptly executed.

Harker's Voice: *I could only think of the forests, the mountains... the inn was just a waiting room. Whenever I was in the forests, I could only think of home, of Exeter. Whenever I was home, I could only think of getting back to the mountains.*

The blind crucifix above the mirror, hung with cloves of garlic, looks down on Harker. He misses his footing and falls on the bed, then gets up, reaches, and takes down the garlic.

He bites into a clove as if it were an apple, and washes the pulp down with more brandy.

Harker's Voice: *All the time I stayed here in the inn, waiting for a commission, I was growing older, losing precious life. And all the time the Count sat on top of his mountain, leeching off the land, he grew younger, thirstier.*

Harker scoops a locket from a bedside table and opens it to look at a portrait of his wife, Mina. Without malice or curiosity, he dangles the cameo in a candle flame. The face browns, the silver setting blackens.

Harker's Voice: *I was waiting for the call from Seward. Eventually, it came.*

There is a knock on the door.

ABOUT THE AUTHOR

Kim Newman (b. 1959) discovered Dracula at the age of eleven, when his parents let him stay up far beyond his bedtime to watch Tod Browning's film with Bela Lugosi on late-night British television. "I think my parents expected the craze to wear off, but obviously it never did," Newman later recalled. In the intervening years Newman became an expert in horror, fantasy and science fiction on film and television, and he became a highly-respected reviewer and commentator. His first published book (coauthored with Neil Gaiman) was *Ghastly Beyond Belief* (1985), an over-the-top compendium of outrageous quotations from the genres. It was followed by the considerably more nuanced *Nightmare Movies* (1987), which remains one of the best critical examinations of late-twentieth century horror films ever published.

Newman's first novel, *The Night Mayor* (1989) posited a virtual world blurring the distinctions between film noir and reality, which created the template for his next, major work, *Anno Dracula* (1990), a deliriously readable "alternate universe" take on Stoker's novel, in which Dracula succeeds in conquering England, marries Queen Victoria, displays Professor Van Helsing's head on a pike at Buckingham Palace, in which the prostitutes of Whitechapel are really vampires, in which Stoker's Dr. Seward is actually Jack the Ripper, etc. "I decided that if Dracula were to replace Prince Albert as Victoria's consort, then all the other vampires of literature would come

out of hiding and flock to his court in the hope of advancement."

And so they did.

Anno Dracula was followed by *The Bloody Red Baron* (1995), which updated the Newman/Dracula "multiverse" to World War II, and continued with *The Judgment of Tears: Anno Dracula 1959* (1998; a.k.a. *Dracula Cha Cha Cha*), and which will be continued in the forthcoming *Johnny Alucard*, which will incorporate "Coppola's Dracula." (Readers will be quick to note that Newman's story has nothing to do with Coppola's actual film *Bram Stoker's Dracula* (1992) but is instead an alternate universe fantasy in which the director's major career debacle is not *Apocalypse Now* but something that might be called *Adraculypse Now*. Commentary on the "real" Coppola film can be found at the conclusion of the filmography that follows.)

Newman's other works include the novel *The Quorum* (1994), the collection *Famous Monsters* (1995), *Back in the USSA* (1997), *Seven Stars* (2000), and *Unforgivable Stories* (2000).

Kim Newman's official website, "Dr. Shade's Laboratory," can be visited at http://indigo.ie/~imago/newman.html.

2

"It's all right for you, Katharine Reed," Francis whined as he picked over the unappetising craft services table. "You're dead, you don't have to eat this shit."

Kate showed teeth, hissing a little. She knew that despite her coke-bottle glasses and freckles, she could look unnervingly feral when she smiled. Francis didn't shrink: deep down, the director thought of her as a special effect, not a real vampire.

In the makeshift canteen, deep in the production bunker, the Americans wittered nostalgia about McDonald's. The Brits — the warm ones, anyway — rhapsodised about Pinewood breakfasts of kippers and fried bread. Romanian location catering was not what they were used to.

Francis finally found an apple less than half brown and took it away. His weight had dropped visibly since their first meeting, months ago in pre-production. Since he had come to Eastern Europe, the insurance doctor diagnosed him as suffering from malnutrition and put him on vitamin shots. *Dracula* was running true to form, sucking him dry.

A production this size was like a swarm of vampire bats — some large, many tiny — battening tenaciously onto the host, making insistent, never-ending demands. Kate had watched Francis — bespectacled, bearded and hyperactive — lose substance under the draining siege, as he made and justified decisions, yielded the visions to be translated to celluloid, rewrote the script to suit locations or new casting. How could one man throw out so many ideas, only a fraction of which would be acted on? In his position, Kate's mind would bleed empty in a week.

A big budget film shot in a backward country was an insane proposition, like taking a touring three-ring circus into a war zone. Who will survive, she thought, and what will be left of them?

The craft table for vampires was as poorly stocked as the one for the warm. Unhealthy rats in chickenwire cages. Kate watched one of the floor effects men, a new-born with a padded waistcoat and a toolbelt, select a writhing specimen and bite off its head. He spat it on the concrete floor, face stretched into a mask of disgust.

"Ringworm," he snarled. "The commie gits are trying to kill us off with diseased vermin."

"I could murder a bacon sarnie," the effects man's mate sighed.

"I could murder a Romanian caterer," said the new-born.

Kate decided to go thirsty. There were enough Yanks around to make coming by human blood in this traditionally superstitious backwater not a problem. Ninety years after Dracula spread vampirism to the Western world, America was still sparsely populated by the blood-drinking undead. For a lot of Americans, being bled by a genuine olde worlde creature of the night was something of a thrill.

That would wear off.

3

Outside the bunker, in a shrinking patch of natural sunlight between a stand of real pines and the skeletons of fake trees, Francis shouted at Harvey Keitel. The actor, cast as Jonathan Harker, was stoic, inexpressive, grumpy. He refused to be drawn into argument, invariably driving Francis to shrieking hysteria.

"I'm not Martin Fucking Scorsese, man," he screamed. "I'm not going to slather on some lousy voice-over to compensate for what you're not giving me. Without Harker, I don't have a picture."

Keitel made fists but his body language was casual. Francis had been riding his star hard all week. Scuttlebutt was that he had wanted Pacino or McQueen but neither were willing to spend three months behind the Iron Curtain.

Kate could understand that. This featureless WWII bunker, turned over to the production as a command centre, stood in ancient mountains, dwarfed by the tall trees. As an outpost of civilisation in a savage land, it was ugly and ineffective.

When approached to act as a technical advisor to Coppola's *Dracula*, she had thought it might be interesting to see where it all started: the Changes, the Terror, the Transformation. No one seriously believed vampirism began here, but it was where Dracula came from. This land had nurtured him through centuries before he decided to spread his wings and extend his bloodline around around the world.

Three months had already been revised as six months. This production didn't have a schedule, it had a sentence. A few were already demanding parole.

Some vampires felt Transylvania should be the undead Israel, a new state carved out of the much-redrawn map of Central Europe, a geographical and political homeland. As soon as it grew from an inkling to a notion, Nicolae Ceausescu vigorously vetoed the proposition. Holding up in one hand a silver-edged sickle, an iron-headed hammer and a sharpened oak spar, the Premier reminded the world that "in Romania, we know how to treat leeches — a stake through the heart and off with their filthy heads." But the Transylvania Movement ("back to the forests, back to the mountains") gathered momentum. Some elders, after ninety years of the chaos of the larger world, wished to withdraw to their former legendary status. Many of Kate's generation, turned in the 1880s, Victorians stranded in this mechanistic century, were sympathetic.

"You're the Irish vampire lady," Harrison Ford, flown in for two days to play Dr Seward as a favour to Francis, had said. "Where's your castle?"

"I have a flat in the Holloway Road," she admitted. "Over an off-license."

In the promised Transylvania, all elders would have castles, fiefdoms, slaves, human cattle. Everyone would wear evening dress. All vampires would have treasures of ancient gold, like leprechauns. There would be a

VAMPIRE CINEMA
1896-2000

The following checklist of vampire cinema includes films in which vampires appear, be they supernatural, scientific, extraterrestrial, or only figurative (e.g., silent films trading on the words *vampire* or *vamp*, featuring predatory women who drain their victims sexually or financially). While this is not by any means a definitive or exhaustive list, it should provide a useful tool in tracking the history of the vampire on screen.

1896
The Haunted Castle
(FRANCE; Georges Méliès; orig.: *Le manoir du diable*; a.k.a. *The Devil's Castle, The Devil's Manor, The Manor of the Devil*)

1909
Vampire of the Coast
(USA)

1910
The Vampire's Trail
(USA)

1911
The Vampire
(USA; Selig)

1912
In the Grip of the Vampire
(FRANCE)

The Vampire Dancer
(DENMARK; Ingvald C. Oes; a.k.a. *Vampyrdanserinden, Vampyr tanzerinnen, Danse vampirisque*)

Vampyrn
(SWEDEN; Mauritz Stiller)

1913
The Vampire
(USA; Robert Vignola)

The Vampire of the Desert
(USA)

La Vampira Indiana
(ITALY)

The Vampire
(UK)

1914
Vasco, The Vampire
(USA)

The Vampire
(FRANCE)

Vampire Bat and Armadillo
(USA)

The Vampire's Tower
(ITALY; A. Ambrosio; orig.: *La torre dei vampiri*)

Vampires of the Night
(USA)

The Vampires of Warsaw
(POLAND; orig.: *Wampiry Warmawy*)

1915
The Devil's Daughter
(USA; Frank Powell; production titles: *The Vampire, La Gioconda*)

The Exploits of Elaine
(USA; Louis J. Gasnier and George B. Seitz; serial featuring an episode called "The Vampire")

A Fool There Was
(USA; Frank Powell; production title: *The Vampire*)

Kiss of the Vampire
(USA)

Saved From the Vampire
(USA)

The Vampire
(USA; Alice Blache)

The Vampire's Clutch
(USA)

The Vampire's Dance
(USA)

The Vampires, The Arch Criminals of Paris
(FRANCE; Louis Feuillade; orig.: *Les vampires*)

Was She a Vampire?
(USA; A. Hale)

1916
Ceneri a vampe
(ITALY)

silk-lined coffin in every crypt, and every night would be a full moon. Unlife eternal and luxury without end, bottomless wells of blood and Paris label shrouds.

Kate thought the Movement lunatic. Never mind cooked breakfasts and (the other crew complaint) proper toilet paper, this was an intellectual desert, a country without conversation, without (and she recognised the irony) life.

She understood Dracula had left Transylvania in the first place not merely because he — the great dark sponge — had sucked it dry, but because even he was bored with ruling over gypsies, wolves and mountain streams. That did not prevent the elders of the Transylvania Movement from claiming the Count as their inspiration and using his seal as their symbol. An Arthurian whisper had it that once vampires returned to Transylvania, Dracula would rise again to assume his rightful throne as their ruler.

Dracula, at long long last, was truly dead, had been for more than fifteen years. Kate had seen his head stuck on a pole, heard the confession of his merciful assassin, attended his cremation on a beach outside Rome, seen his ashes scattered into the sea. From that, there was no coming back, not even for a creature who had so many times avoided his appointment in Samarra.

But the Count meant so much to so many. Kate wondered if anything was left inside so many meanings, anything concrete and inarguable and true. Or was he now just a phantom, a slave to anyone who cared to invoke his name? So many causes and crusades and rebellions and atrocities. One man, one monster, could never have kept track of them all, could never have encompassed so much mutually exclusive argument.

There was the Dracula of the histories, the Dracula of Stoker's book, the Dracula of this film, the Dracula of the Transylvania Movement. Dracula, the vampire and the idea, was vast. But not so vast that he could cast his cloak of protection around all who claimed to be his followers. Out here in the mountains where the Count had passed centuries in petty predation, Kate understood that he must in himself have felt tiny, a lizard crawling down a rock.

Nature was overwhelming. At night, the stars were laser-points in the deep velvet black of the sky. She could hear, taste and smell a thousand flora and fauna. If ever there was a call of the wild, this forest exerted it. But there was nothing she considered intelligent life.

She tied tight under her chin the yellow scarf, shot through with golden traceries, she had bought at Biba in 1969. It was a flimsy, delicate thing, but to her it meant civilisation, a coloured moment of frivolity in a life too often preoccupied with monochrome momentousness.

Francis jumped up and down and threw script pages to the winds. His arms flapped like wings. Clouds of profanity enveloped the uncaring Keitel.

"Don't you realise I've put up my own fucking money for this fucking picture," he shouted, not just at Keitel but at the whole company. "I could lose my house, my vineyard, everything. I can't afford a fucking honourable failure. This has abso-goddamn-lutely got to outgross *Jaws* or I'm personally impaled up the ass with a sharpened telegraph pole."

Effects men sat slumped against the exterior wall of the bunker (there were few chairs on location) and watched their director rail at the heavens, demanding of God answers that were not forthcoming. Script pages swirled upwards in a spiral, spreading out in a cloud, whipping against the upper trunks of the trees, soaring out over the valley.

"He was worse on *Godfather*," one said.

4

Servants usher Harker into a well-appointed drawing room. A table is set with an informal feast of bread, cheese and meat. Dr Jack Seward, in a white coat with a stethoscope hung around his neck, warmly shakes Harker's hand and leads him to the table. Quincey P. Morris sits to one side, tossing and catching a spade-sized bowie knife.

Lord Godalming, well-dressed, napkin tucked into his starched collar, sits at the table, forking down a double helping of paprika chicken. Harker's eyes meet Godalming's, the nobleman looks away.

Seward: *Harker, help yourself to the fare. It's uncommonly decent for foreign muck.*

Harker: *Thank you, no. I took repast at the inn.*

Seward: *How is the inn? Natives bothering you? Superstitious babushkas, what?*

Harker: *I am well in myself.*

Seward: *Splendid... the vampire, Countess Marya Dolingen of Graz. In 1883, you cut off her head and drove a hawthorn stake through her heart, destroying her utterly.*

Harker: *I'm not disposed just now to discuss such affairs.*

Morris: *Come on, Jonny-Boy. You have a commendation from the church, a papal decoration. The frothing she-bitch is dead at last. Take the credit.*

Harker: *I have no direct knowledge of the individual you mention. And if I did, I reiterate that I would not be disposed to discuss such affairs.*

Seward and Morris exchange a look as Harker stands impassive. They know they have the right man. Godalming, obviously in command, nods.

Seward clears plates of cold meat from a strong-box that stands on the table. Godalming hands the doctor a key, with which he opens the box. He takes out a woodcut and hands it over to Harker.

The picture is of a knife-nosed mediaeval warrior prince.

Seward: *That's Vlad Tepes, called "The Impaler." A good Christian, defender of the faith. Killed a million Turks. Son of the Dragon, they called him. Dracula.*

Harker is impressed.

Mr. Vampire
(USA)

A Night of Horror
(GERMANY; Arthur Robinson)

A Vampire Out of Work
(USA)

A Village Vampire
(USA; Mack Sennett)

1917
The Beloved Vampire
(FRANCE; Méliès)

Maeia
(HUNGARY; Alexander Korda)

The Vamp of the Camp
(USA; Allen Curtis)

Vamping Rueben's Millions
(USA; Dick Smith)

1918
The Vamp
(USA; Jerome Stern)

The Vamp Cure
(USA; Eddie Lyons, Lee Moran)

1919
Vamps and Variety
(USA; Gilbert Pratt)

1920
The Great London Mystery
(UK; Charles Raymond; serial)

The Vampire
(USA)

1921
Drakula
(HUNGARY; Károly Lajthay)

Vamps and Scamps
(USA; Jimmie Davis)

1922
Nosferatu: a Symphony of Horror
(GERMANY; F. W. Murnau; orig.: *Nosferatu: Eine Symphonie des Grauens;* a.k.a. *Nosferatu the Vampire, The Twelfth Hour, Die Zwoelfte Stunde, eine Nacht des Grauens*)

1923
The Blond Vampire
(USA)

Morris: ***Prince Vlad had Orthodox Church decorations out the ass. Coulda made Metropolitan. But he converted, went over to Rome, turned Candle.***

Harker: ***Candle?***

Seward: ***Roman Catholic.***

Harker looks again at the woodcut. In a certain light, it resembles the young Marlon Brando.

Seward walks to a side-table, where an antique dictaphone is set up. He fits a wax cylinder and adjusts the needle-horn.

Seward: ***This is Dracula's voice. It's been authenticated.***

Seward cranks the dictaphone.

Dracula's Voice: ***Cheeldren of the naight, leesten to them. What museek they maike!***

There is a strange distortion in the recording.

Harker: ***What's that noise in the background?***

Seward: ***Wolves, my boy. Dire wolves, to be precise.***

Dracula's Voice: ***To die, to be realllly dead, that must be… gloriousssss!***

Morris: ***Vlad's well beyond Rome now. He's up there, in his impenetrable castle, continuing the crusade on his own. He's got this army of Szgany Gypsies, fanatically loyal fucks. They follow his orders, no matter how atrocious, no matter how appalling. You know the score, Jon. Dead babies, drained cattle, defenestrated peasants, impaled grandmothers. He's god-damned Un-Dead. A fuckin' monster, boy.***

Harker is shocked. He looks again at the woodcut.

Seward: ***The firm would like you to proceed up into the mountains, beyond the Borgo Pass…***

Harker: ***But that's Transylvania. We're not supposed to be in Transylvania.***

Godalming looks to the heavens, but continues eating.

Seward: ***…beyond the Borgo Pass, to Castle Dracula. There, you are to ingratiate yourself by whatever means come to hand into Dracula's coterie. Then you are to disperse the Count's household.***

Harker: ***Disperse?***

Godalming puts down his knife and fork.

5

"What can I say, we made a mistake," Francis said, shrugging nervously, trying to seem confident. He had shaved off his beard, superstitiously hoping that would attract more attention than his announcement. "I think this is the courageous thing to do, shut down and recast, rather than continue with a frankly unsatisfactory situation."

Kate did not usually cover show business, but the specialist press stringers (*Variety*, *Screen International*, *Positif*) were dumbstruck enough to convince her it was not standard procedure to fire one's leading man after two weeks' work, scrap the footage and get someone else. When Keitel

Vamped
(USA)

1925
Vampires of Warsaw
(POLAND; orig.: *Wampiry Warszawy*) Possibly the same as 1914 version.

1926
Vamping Babies
(Use)

1927
London After Midnight
(USA; Tod Browning; production title: *The Hypnotist*)

1928
I Bevitori di Saugue
(ITALY/GERMANY)

Vamping Venus
(USA; Eddie Cline)

The Vampire
(USA)

Vampire a du Mode
(FRANCE)

1931
Dracula
(USA; Tod Browning)

Dracula
(USA; Spanish-language version; George Melford)

Vampyr
(FRANCE/GERMANY; Carl Dreyer)

1933
The Vampire Bat
(USA; Frank Strayer)

1935
Condemned to Live
(USA; Frank Strayer)

Mark of the Vampire
(USA; Tod Browning; production title: *The Vampires of Prague*)

1936
Dracula's Daughter
(USA; Lambert Hillyer)

1939
The Return of Dr. X
(USA; Vincent Sherman)

was sent home, the whole carnival ground to a halt and everyone had to sit around while Francis flew back to the States to find a new star.

Someone asked how far over budget *Dracula* was. Francis smiled and waffled about budgets being provisional.

"No one ever asked how much the Sistine Chapel cost," he said, waving a chubby hand. Kate would have bet that while Michelangelo was on his back with the brushes, Pope Julius II never stopped asking how much it cost and when would it be finished.

During the break in shooting, money was spiralling down a drain. Fred Roos, the co-producer, had explained to her just how expensive it was to keep a whole company standing by. It was almost more costly than having them work.

Next to Francis at the impromptu press conference in the Bucharest Town Hall was Martin Sheen, the new Jonathan Harker. In his mid-thirties, he looked much younger, like the lost boy he played in *Badlands*. The actor mumbled generously about the opportunity he was grateful for. Francis beamed like a shorn Santa Claus on a forced diet and opened a bottle of his own wine to toast his new star.

The man from *Variety* asked who would be playing Dracula, and Francis froze in mid-pour, sloshing red all over Sheen's wrist. Kate knew the title role — actually fairly small, thanks to Bram Stoker and screenwriter John Milius — was still on offer to various possibles — Klaus Kinski, Jack Nicholson, Christopher Lee.

"I can confirm Bobby Duvall will play Van Helsing," Francis said. "And we have Dennis Hopper as Renfield. He's the one who eats flies."

"But who is Dracula?"

Francis swallowed some wine, attempted a cherubic look, and wagged a finger.

"I think I'll let that be a surprise. Now, ladies and gentlemen, if you'll excuse me, I have motion picture history to make."

6

As Kate took her room-key from the desk, the night manager nagged her in Romanian. When she had first checked in, the door of her room fell off as she opened it. The hotel maintained she did not know her own vampire strength and should pay exorbitantly to have the door replaced. Apparently, the materials were available only at great cost and had to be shipped from Moldavia. She assumed it was a scam they worked on foreigners, especially vampires. The door was made of paper stretched over a straw frame, the hinges were cardboard fixed with drawing pins.

She was pretending not to understand any language in which they tried to ask her for money, but eventually they would hit on English and she'd have to make a scene. Francis, light-hearted as a child at the moment, thought it rather funny and had taken to teasing her about the damn door.

1940
The Devil Bat
(USA; Jean Yarbrough; a.k.a. *Killer Bats, Devil Bats*)

1941
Spooks Run Wild
(USA; Phil Rosen)

1943
Dead Men Walk
(USA; Sam Newfield)

Dr. Terror's House of Horrors
(UK; compilation, including sequences from Dreyer's *Vampyr*)

The Return of the Vampire
(USA; Lew Landers)

Son of Dracula
(USA; Robert Siodmak)

Tiger Fangs
(USA; Sam Newfield)

1944
House of Frankenstein
(USA; Erle C. Kenton; production title: *The Devil's Brood*)

1945
The Crime Doctor's Courage
(USA; George Sherman)

House of Dracula
(USA; Erle C. Kenton; production title: *Destiny*)

Isle of the Dead
(USA; Mark Robinson)

The Spider Woman Strikes Back
(USA; Arthur Lubin)

Le Vampire
(FRANCE; Jean Painleve)

The Vampire's Ghost
(USA; Lesley Selander)

1946
Devil Bat's Daughter
(USA; Frank Wisbar)

The Face of Marble
(USA; William Beaudine)

Valley of the Zombies
(USA; Philip Ford)

1948
Abbott and Costello Meet Frankenstein (USA; Charles Barton; production title: *The Brain of Frankenstein;* UK title: *Abbott and Costello Meet the Ghosts*)

1951
Mother Riley Meets the Vampire (UK; John Gilling; a.k.a. *Vampire over London, King Robot;* USA 1964 release: *My Son, the Vampire*)

The Thing from Another World (USA; Christian Nyby and [uncredited] Howard Hawks)

La Traite du Vampire (FRANCE)

Le Vampire (FRANCE; misleading title about fighter planes)

1953
Drakula Instanbul'da (TURKEY; Mehmet Muhtar)

El vampiro negro (ARGENTINA; Roman Vinoly Barreto)

1956
The Devil's Commandment (ITALY; Riccardo Freda, with Mario Bava; orig.: *I vampiri;* a.k.a. *Lust of the Vampire, The Vampire of Notre Dame*)

Kyuketsuki Ga (JAPAN; Nobuo Nakagawa)

Pontianak (MALAYA; B. N. Rao)

Super Giant 2 (JAPAN; a.k.a. *Spaceman contro i vampiri della spazio*)

The Vampire (MEXICO; Fernando Mendez; orig.: *El Vampiro*)

1957
Blood of Dracula (USA; Herbert L. Strock; a.k.a. *Blood Is My Heritage*)

Castle of the Monsters (MEXICO; Julian Soler; orig.: *El castillo de los monstruos*)

Dendhamn Pontianak (MALAYA; B. N. Rao)

Not tired, but glad to be off the streets after nightfall, she climbed the winding stairs to her room, a cramped triangular attic space. Though barely an inch over five feet tall, she could only stand up straight in the dead centre of the room. A crucifix hung ostentatiously over the bed, a looking glass was propped up on the basin. She thought about taking them down but it was best to let insults pass. In many ways, she preferred the camp-site conditions in the mountains. She only needed to sleep every two weeks, and when she was out she was literally dead and didn't care about clean sheets.

They were all in Bucharest for the moment, as Francis supervised script-readings to ease Sheen into the Harker role. His fellow coach-passengers — Fredric Forrest (Westenra), Sam Bottoms (Murray) and Albert Hall (Swales) — had all been on the project for over a year, and had been through all this before in San Francisco as Francis developed John Milius's script through improvisation and happy accident. Kate didn't think she would have liked being a screenwriter. Nothing was ever finished.

She wondered who would end up playing Dracula. Since his brief marriage to Queen Victoria made him officially if embarrassingly a satellite of the British Royal Family, he had rarely been represented in films. However, Lon Chaney had taken the role in the silent *London After Midnight*, which dealt with the court intrigues of the 1880s, and Anton Walbrook played Vlad opposite Anna Neagle in *Victoria the Great* in 1937. Kate, a lifelong theatregoer who had never quite got used to the cinema, remembered Vincent Price and Helen Hayes in *Victoria Regina* in the 1930s.

Aside from a couple of cheap British pictures which didn't count, Bram Stoker's *Dracula* — the singular mix of documentation and wish-fulfilment that inspired a revolution by showing how Dracula could have been defeated in the early days before his rise to power — had never quite been filmed. Orson Welles produced it on radio in the 1930s and announced it as his first picture, casting himself as Harker and the Count, using first-person camera throughout. RKO thought it too expensive and convinced him to make *Citizen Kane* instead. Nearly ten years ago, Francis had lured John Milius into writing the first pass at the script by telling him nobody, not even Orson Welles, had ever been able to lick the book.

Francis was still writing and rewriting, stitching together scenes from Milius's script with new stuff of his own and pages torn straight from the book. Nobody had seen a complete script, and Kate thought one didn't exist.

She wondered how many times Dracula had to die for her to be rid of him. Her whole life had been a dance with Dracula, and he haunted her still. When Francis killed the Count at the end of the movie — if that was the ending he went with — maybe it would be for the last time. You weren't truly dead until you'd died in a motion picture. Or at the box office.

The latest word was that the role was on offer to Marlon Brando. She couldn't see it: Stanley Kowalski and Vito Corleone as Count Dracula. One of the best actors in the world, he'd been one of the worst Napoleons in the movies. Historical characters brought out the ham in him. He was terrible as Fletcher Christian too.

Officially, Kate was still just a technical adviser. Though she had never actually met Dracula during his time in London, she had lived through the period. She had known Stoker, Jonathan Harker, Godalming and the rest. Once, as a warm girl, she had been terrified by Van Helsing's rages. When Stoker wrote his book and smuggled it out of prison, she had helped with its underground circulation, printing copies on the presses of the *Pall Mall Gazette* and ensuring its distribution despite all attempts at suppression. She wrote the introduction for the 1912 edition that was the first official publication.

Actually, she found herself impressed into a multitude of duties. Francis treated a $20,000,000 (and climbing) movie like a college play and expected everyone to pitch in, despite union rules designed to prevent the crew being treated as slave labour. She found the odd afternoon of sewing costumes or night of set-building welcome distraction.

At first, Francis asked her thousands of questions about points of detail; now he was shooting, he was too wrapped up in his own vision to take advice. If she didn't find something to do, she'd sit idle. As an employee of American Zoetrope, she couldn't even write articles about the shoot. For once, she was on the inside, knowing but not telling.

She had wanted to write about Romania for the *New Statesman*, but was under orders not to do anything that might jeopardise the cooperation the production needed from the Ceausescus. So far, she had avoided all the official receptions Nicolae and Elena hosted for the production. The Premier was known to an be extreme vampire-hater, especially since the stirrings of the Transylvania Movement, and occasionally ordered not-so-discreet purges of the undead.

Kate knew she, like the few other vampires with the *Dracula* crew, was subject to regular checks by the *Securitate*. Men in black leather coats loitered in the corner of her eye.

"For God's sake," Francis had told her, "don't *take* anybody local."

Like most Americans, he didn't understand. Though he could *see* she was a tiny woman with red hair and glasses, the mind of an aged aunt in the body of an awkward cousin, Francis could not rid himself of the impression that vampire women were ravening predators with unnatural powers of bewitchment, lusting after the pounding blood of any warm youth who happened along. She was sure he hung his door with garlic and wolfsbane, but half-hoped for a whispered solicitation.

After a few uncomfortable nights in Communist-approved beer-halls, she had learned to stay in her hotel room while in Bucharest. People here had memories as long as her lifetime. They crossed themselves and muttered prayers as she walked by. Children threw stones.

Not of This Earth
(USA; Roger Corman)

The Vampire
(USA; Paul Landres; television title: *Mark of the Vampire*)

The Vampire's Coffin
(MEXICO; Fernando Mendez; orig.: *El ataud del vampiro;* a.k.a. *El retorno del vampiro, El ataud de la muerte*)

1958

Anak Pontianak
(MALAYA; Raymond Estella)

Blood of the Vampire
(UK; Henry Cass)

Dracula
(USA/TV)

First Man Into Space
(UK; Robert Day)

Horror of Dracula
(UK; Terence Fisher; orig.: *Dracula*)

It! The Terror From Beyond Space
(USA; Edward L. Cahn)

The Return of Dracula
(USA; Paul Landres; a.k.a. *Curse of Dracula* [TV], *The Fantastic Disappearing Man* [UK])

Sumpah Pontianak
(MALAYA)

1959

Attack of the Giant Leeches
(USA; Bernard Kowalski; a.k.a. *The Giant Leeches, Demons of the Swamp* [UK])

Curse of the Undead
(USA; Edward Dein)

The Lurking Vampire
(ARGENTINA; orig.: *El vampiro aecheca*)

La nave de los monstruos
(MEXICO; Rogelio A. Gonzales)

Onna Kyuketsuki
(JAPAN; Nobu Nakagawa; French release title: *L'homme vampire*)

Plan 9 from Outer Space
(USA; Edward D. Wood, Jr.)

Uncle Was a Vampire
(ITALY; Stefano Vanzina; orig.: *Tempi duri pea i vampiri;* a.k.a. *Hard Times for Vampires*)

1960
Black Sunday
(ITALY; Mario Bava; orig.: *Le maschera del demonio;* a.k.a. *House of Fright, Mask of the Demon, Revenge of the Vampire*)

The Blood of Nostradamus
(MEXICO; Federico Curiel; orig.: *La sangre de Nostradamus*)

The Brides of Dracula
(UK; Terence Fisher)

Creature of the Walking Dead
(MEXICO; Federico Corte [Fernando Cortes])

The Genii of Darkness
(MEXICO; Frederico Curiel; orig.: *Nostradamus, el Genio de las tinieblas*)

Little Shop of Horrors
(USA; Roger Corman)

Mill of the Stone Women
(ITALY/FRANCE; Georgio Ferroni; orig.: *Il mulino Belle donne di pietra, Le moulin des supplices;* a.k.a. *Drops of Blood, Icon*)

The Monster Demolisher
(MEXICO; Federico Curiel; orig.: *Nostradamus y el destructor de monstruos*)

The Playgirls and the Vampire
(ITALY; Piero Regnoli; orig.: *L'ultima preda del vampiro;* a.k.a. *Curse of the Vampire*)

The Vampire and the Ballerina
(ITALY; Renato Polselli; orig.: *L'amante del vampiro*)

The World of the Vampires
(MEXICO; Alfonso Corona Blake; orig.: *El mundo de los vampiros*)

She stood at her window and looked out at the square. A patch of devastation where the ancient quarter of the capital had been marked the site of the palace Ceausescu was building for himself. A three-storey poster of the Saviour of Romania stood amid the ruins. Dressed like an orthodox priest, he held up Dracula's severed head as if he had personally killed the Count.

Ceausescu harped at length about the dark, terrible days of the past when Dracula and his kind preyed on the warm of Romania. In theory, it kept his loyal subjects from considering the dark, terrible days of the present when he and his wife lorded over the country like especially corrupt Roman Emperors. Impersonating the supplicant undertaker in *The Godfather*, Francis had abased himself to the dictator to secure official co-operation.

She turned on the radio and heard tinny martial music. She turned it off, lay on the narrow, lumpy bed — as a joke, Fred Forrest and Francis had put a coffin in her room one night — and listened to the city at night. Like the forest, Bucharest was alive with noises, and smells.

It was ground under, but there was life here. Even in this grim city, someone was laughing, someone was in love. Somebody was allowed to be a happy fool.

She heard winds in telephone wires, bootsteps on cobbles, a drink being poured in another room, someone snoring, a violinist sawing scales. And someone outside her door. Someone who didn't breathe, who had no heartbeat, but whose clothes creaked as he moved, whose saliva rattled in his throat.

She sat up, confident she was elder enough to be silent, and looked at the door.

"Come in," she said, "it's not locked. But be careful. I can't afford more breakages."

7

His name was Ion Popescu and he looked about thirteen, with big, olive-shiny orphan eyes and thick, black, unruly hair. He wore an adult's clothes, much distressed and frayed, stained with long-dried blood and earth. His teeth were too large for his skull, his cheeks stretched tight over his jaws, drawing his whole face to the point of his tiny chin.

Once in her room, he crouched down in a corner, away from a window. He talked only in a whisper, in a mix of English and German she had to strain to follow. His mouth wouldn't open properly. He was alone in the city, without community. Now he was tired and wanted to leave his homeland. He begged her to hear him out and whispered his story.

He claimed to be forty-five, turned in 1944. He didn't know, or didn't care to talk about, his father- or mother-in-darkness. There were blanks burned in his memory, whole years missing. She had come across that before. For most of his vampire life, he had lived underground, under the Nazis and then the Communists. He was the sole survivor of several resistance movements. His warm comrades had never really trusted him, but his capabilities were useful for a while.

She was reminded of her first days after turning. When she knew nothing, when her condition seemed a disease, a trap. That Ion could be a vampire for over thirty years and never pass beyond the new-born stage was incredible. She truly realised, at last, just how backward this country was.

"Then I hear of the American film, and of the sweet vampire lady who is with the company. Many times, I try to get near you, but you are watched. *Securitate*. You, I think, are my saviour, my true mother-in-the-dark."

Forty-five, she reminded herself.

Ion was exhausted after days trying to get close to the hotel, to "the sweet vampire lady," and hadn't fed in weeks. His body was icy cold. Though she knew her own strength was low, she nipped her wrist and dribbled a little of her precious blood onto his white lips, enough to put a spark in his dull eyes.

There was a deep gash on his arm, which festered as it tried to heal. She bound it with her scarf, wrapping his thin limb tight.

He hugged her and slept like a baby. She arranged his hair away from his eyes and imagined his life. It was like the old days, when vampires were hunted down and destroyed by the few who believed. Before *Dracula*.

The Count had changed nothing for Ion Popescu.

8

Bistritz, a bustling township in the foothills of the Carpathian Alps. Harker, carrying a Gladstone Bag, weaves through crowds towards a waiting coach and six. Peasants try to sell him crucifixes, garlic and other lucky charms. Women cross themselves and mutter prayers.

A wildly-gesticulating photographer tries to stop him slowing his pace to examine a complicated camera. An infernal burst of flash-powder spills purple smoke across the square. People choke on it.

Corpses hang from a four-man gibbet, dogs leaping up to chew on their naked feet. Children squabble over mismatched boots filched from the executed men. Harker looks up at the twisted, mouldy faces.

He reaches the coach and tosses his bag up. Swales, the coachman, secures it with the other luggage and growls at the late passenger. Harker pulls open the door and swings himself into the velvet-lined interior of the carriage.

There are two other passengers. Westenra, heavily moustached and cradling a basket of food. And Murray, a young man who smiles as he looks up from his Bible.

Harker exchanges curt nods of greeting as the coach lurches into motion.

Harker's Voice: ***I quickly formed opinions of my travelling companions. Swales was at the reins. It was my commission but sure as shooting it was his coach. Westenra, the one they called "Cook," was from Whitby. He was***

1961

Atom Age Vampire
(ITALY/FRANCE; Anton Giulio Majano; orig.: *Seddok, L'erede di Santana;* a.k.a. *Seddok*)

The Bad Flower
(SOUTH KOLA; Yongmin Lee; orig.: *Ahhea Khots*)

Blood and Roses
(FRANCE; Roger Vadim; orig.: *Et mourir de plaisir*)

The Brainiac
(MEXICO; Chano Urveta; orig.: *El barón del terror*)

La Cases de los Espantos
(MEXICO; Alfredo B. Crevenna)

Curse of the Crying Woman
(MEXICO; Rafael Baledon; orig.: *La maldición de la llorona*)

Frankenstein, el Vravnpiro y cia
(MEXICO; Benito Alazraki)

Goliath and the Vampires
(ITALY; Giacomo Gentilomo & Sergio Corbucci orig.: *Maciste contro il vampiro*)

The Vampire of the Opera
(ITALY; Renato Polselli; orig.: *Il mostro dell'opera,* a.k.a. Il *vumpiro dell'opera*)

Hercules in the Haunted
(ITALY; Mario Bava; orig.: *Ercole al centro della terra;* a.k.a. *Hercules in the Center of the Earth*)

Invasion of the Vampires
(MEXICO; Miguel Morayta; orig.: *La invasion de los vampiros*)

The Naked Witch
(USA; Andy Milligan)

Samson vs. the Vampire Women
(MEXICO; Alfonso Corona Blake; orig.: *Santo contra las mujeres vampiras*)

Vampiresas 1930
(SPAIN; Jesus Franco; a.k.a. *Vampiresas*)

1962

The Bloody Vampire
(MEXICO; Michael Morata; orig.: *El vampiro sagriente,* a.k.a. *El conde Frankmahausen*)

ratcheted several notches too tight for Wallachia. Probably too tight for Whitby, come to that. Murray, the fresh-faced youth with the Good Book, was a rowing blue from Oxford. To look at him, you'd think the only use he'd have for a sharpened stake would be as a stump in a knock-up match.

Later, after dark but under a full moon, Harker sits up top with Swales. A wind-up phonograph crackles out a tune through a sizable trumpet.

Mick Jagger sings "Ta-Ra-Ra-BOOM-De-Ay."

Westenra and Murray have jumped from the coach and ride the lead horses, whooping it up like a nursery Charge of the Light Brigade.

Harker, a few years past such antics, watches neutrally. Swales is indulgent of his passengers.

The mountain roads are narrow, precipitous. The lead horses, spurred by their riders, gallop faster. Harker looks down and sees a sheer drop of a thousand feet, and is more concerned by the foolhardiness of his companions.

Hooves strike the edge of the road, narrowly missing disaster.

Westenra and Murray chant along with the song, letting go of their mounts' manes and doing hand-gestures to the lyrics. Harker gasps but Swales chuckles. He has the reins and the world is safe.

Harker's Voice: *I think the dark and the pines of Romania spooked them badly, but they whistled merrily on into the night, infernal cake-walkers with Death as a dancing partner.*

9

In the rehearsal hall, usually a people's ceramics collective, she introduced Ion to Francis.

The vampire youth was sharper now. In a pair of her jeans (which fit him perfectly) and a *Godfather II* T-shirt, he looked less the waif, more like a survivor. Her Biba scarf, now his talisman, was tied around his neck.

"I said we could find work for him with the extras. The gypsies."

"I am no gypsy," Ion said, vehemently.

"He speaks English, Romanian, German, Magyar and Romany. He can co-ordinate all of them."

"He's a kid."

"He's older than you are."

Francis thought it over. She didn't mention Ion's problems with the authorities. Francis couldn't harbour an avowed dissident. The relationship between the production and the government was already strained. Francis thought (correctly) he was being bled of funds by corrupt officials, but could afford to lodge no complaint. Without the Romanian army, he didn't have a cavalry, didn't have a horde. Without the location permits that still hadn't come through, he couldn't shoot the story beyond Borgo Pass.

"I can keep the rabble in line, maestro," Ion said, smiling.

Somehow, he had learned how to work his jaws and lips into a smile.

Fantasmagorie
(FRANCE; Patrice Molinard)

La Huellu Macabre
(MEXICO; Alfredo E. Crevenna)

Little Red Riding Hood and the Monsters
(MEXICO; Roberto Rodriguez; orig.: *Capurcitu y Pulgareito vs. los ynonstruos*)

Slaughter of the Vampires
(ITALY; Roberto Mauri; a.k.a. *Curse of the Blood Ghouls*; orig.: *La stvage dei vampiri*)

Vampire Woman
(HONG KONG)

1963
Black Sabbath
(ITALY; Mario Bava)

Castle of Blood
(ITALY; Antonio Margheriti [Anthony Dawson]; TV title: *Castle of Terror;* orig.: *La danza macabra*)

Cave of the Living Dead
(WEST GERMANY/YUGOSLA-VIA; Akos Von Ratony; orig.: *Der Fluch der Gruenen Augen*)

Crypt of Horror
(ITALY/SPAIN; Camillo Mastrocinque)

Insomnia
(FRANCE; Pierre Etaix; orig.: *Insomnie*)

The Kiss of the Vampire
(UK; Don Sharp; TV title: *Kiss of Evil*)

Pontianak kembali
(MALAYA; Ramon Estellia)

El Pueblo Fantasma
(MEXICO; Alfredo E. Crevenna)

Vampires D'Alfama
(FRANCE; Pierre Kast)

Los Vampiros de Oeste
(MEXICO; Juan Ortega)

1964
Batman vs. Dracula
(USA; Andy Warhol)

With her blood in him, he had more control. She noticed him chameleoning a little. His smile, she thought, might be a little like hers.

Francis chuckled. He liked being called "maestro." Ion was good at getting on the right side of people. After all, he had certainly got on the right side of her.

"Okay, but keep out of the way if you see anyone in a suit."

Ion was effusively grateful. Again, he acted the age he looked, hugging Francis, then her, saluting like a toy soldier. Martin Sheen, noticing, raised an eyebrow.

Francis took Ion off to meet his own children — Roman, Gio and Sofia — and Sheen's sons — Emilio and Charlie. It had not sunk in that this wiry kid, obviously keen to learn baseball and chew gum, was in warm terms middle-aged.

Then again, Kate never knew whether to be twenty-five, the age at which she turned, or 116. And how was a 116-year-old supposed to behave anyway?

Since she had let him bleed her, she was having flashes of his past: scurrying through back-streets and sewers, like a rat; the stabbing pains of betrayal; eye-searing flashes of firelight; constant cold and red thirst and filth.

Ion had never had the time to grow up. Or even to be a proper child. He was a waif and a stray. She couldn't help but love him a little. She had chosen not to pass on the Dark Kiss, though she had once, during the Great War, come close and regretted it.

Her bloodline, she thought, was not good for a new-born. There was too much Dracula in it, maybe too much Kate Reed.

To Ion, she was a teacher not a mother. Before she insisted on becoming a journalist, her whole family seemed to feel she was predestined to be a governess. Now, at last, she thought she saw what they meant.

Ion was admiring six-year-old Sofia's dress, eyes bright with what Kate hoped was not hunger. The little girl laughed, plainly taken with her new friend. The boys, heads full of the vampires of the film, were less sure about him. He would have to earn their friendship.

Later, Kate would deal with Part Two of the Ion Popescu Problem. After the film was over, which would not be until the 1980s at the current rate of progress, he wanted to leave the country, hidden in among the rest of the production crew. He was tired of skulking and dodging the political police, and didn't think he could manage it much longer. In the West, he said, he would be free from persecution.

She knew he would be disappointed. The warm didn't really *like* vampires in London or Rome or Dublin any more than they did in Timisoara or Bucharest or Cluj. It was just more difficult legally to have them destroyed.

Devils of Darkness
(UK; Lance Comfort)

Dr. Terror's House of Horrors
(UK; Freddie Francis)

The Last Man on Earth
(ITALY/US; Sidney Salkow and Ubaldo Ragona; orig.: *L'ultimo uomo della terra*)

Pontianak gua musang
(MALAYA; B. N. Rao)

The Secrets of Dracula
(PHILIPPINES; orig.: *Mga manugang ni Drakula*)

Sexy Probitissimo
(ITALY)

Sexy-Super Interdit
(FRANCE)

Le Vampire de Dusseldorf
(FRANCE/ITALY/SPAIN; Robert Hossein; a.k.a. *El asesino de Dusseldorf*)

1965

El Baron Brahola
(MEXICO; Jose Diaz Morales)

Blood Thirst
(PHILIPPINES/USA; Michael du Pont)

Bring Me the Vampire
(MEXICO; Alfredo E. Crevenna; orig.: *Echenme al vampiro*)

Charm de las Calaveras
(MEXICO; Alfredo Salazar)

Incubus
(USA; Leslie Stevens; in Esperanto)

Nightmare Castle
(ITALY; Mario Caiano [Allen Grunewald]; orig.: *Amanti d'oltretomba*)

Planet of the Vampires
(ITALY/SPAIN; Mario Bava; orig.: *Terrore nello spazio;* a.k.a. *Terror en el espacio, Demon Planet, Planet of Blood*)

A Vampire for Two
(SPAIN; Pedro Lazanga Sabater; orig.: *Un vampiro para dos*)

1966
Billy the Kid vs. Dracula
(USA; William Beaudine)

Beast of Morocco
(MOROCCO/UK; Fredric Goale; a.k.a. *The Hand of Night*)

Blood Bath
(USA; Jack Hill & Stephanie Rothman; a.k.a. *Track of the Vampire*)

The Blood Drinkers
(PHILIPPINES; Gerardo de Leon; a.k.a. *Vampire People*)

Blood Fiend
(UK; Samuel Gallu; a.k.a. *Theatre of Death*)

La Bonne Dame
(FRANCE; Pierre Philippe)

Carry on Screaming
(UK; Gerald Thomas)

The Devil's Mistress
(USA; Orville Wanzen)

Dracula, Prince of Darkness
(UK; Terence Fisher)

The Eye of Count Flickenstein
(USA; Tony Conrad)

El Imperio de Dracula
(MEXICO; Federico Curiel)

Island of the Doomed
(SPAIN/WEST GERMANY; Mel Welles [Ernst von Theumer]; orig.: *La isla de la muerte, Das Gebheimnis der Todeninsel;* a.k.a. *Blood Suckers* [UK], *Man Eater of Hydra, Island of Death*)

Munster, Go Home!
(USA; Earl Bellamy)

Queen of Blood
(USA; Curtis Harrington; a.k.a. *Planet of Blood*)

The Wild World of Bat-woman
(USA; Jerry Warren; a.k.a. *She Was a Hippy Vampire*)

1967
Blood Demon
(WEST GERMANY; Harold Reinl; orig.: *Die Schlangenrube and das Pendel;* a.k.a. *Blood Demon, Castle of the Walking Dead, The Torture Chamber of Dr. Sadism*)

10

Back in the mountains, there was the usual chaos. A sudden thunderstorm, whipped up out of nowhere like a djinn, had torn up real and fake trees and scattered them throughout the valley, demolishing the gypsy encampment production designer Dean Tavoularis had been building. About half a million dollars' worth of set was irrevocably lost. The bunker itself had been struck by lightning and split open like a pumpkin. The steady rain poured in and streamed out of the structure, washing away props, documents, equipment and costumes. Crews foraged in the valley for stuff that could be reclaimed and used.

Francis acted as if God were personally out to destroy him.

"Doesn't anybody else notice what a disaster this film is?" he shouted. "I haven't got a script, I haven't got an actor, I'm running out of money, I'm all out of time. This is the goddamned *Unfinished Symphony*, man."

Nobody wanted to talk to the director when he was in this mood. Francis squatted on the bare earth of the mountainside, surrounded by smashed balsawood pine-trees, hugging his knees. He wore a stetson hat, filched from Quincey Morris's wardrobe, and drizzle was running from its brim in a tiny stream. Eleanor, his wife, concentrated on keeping the children out of the way.

"This is the worst fucking film of my career. The worst I'll ever make. The last movie."

The first person to tell Francis to cheer up and that things weren't so bad would get fired and be sent home. At this point, crowded under a leaky lean-to with other surplus persons, Kate was tempted.

"I don't want to be Orson Welles," Francis shouted at the slate-grey skies, rain on his face, "I don't want to be David Lean. I just want to make an *Irwin Allen* movie, with violence, action, sex, destruction in every frame. This isn't Art, this is atrocity."

Just before the crew left Bucharest, as the storm was beginning, Marlon Brando had consented to be Dracula. Francis personally wired him a million-dollar down-payment against two weeks' work. Nobody dared remind Francis that if he wasn't ready to shoot Brando's scenes by the end of the year, he would lose the money and his star.

The six months was up, and barely a quarter of the film was in the can. The production schedule had been extended and reworked so many times that all forecasts of the end of shooting were treated like forecasts of the end of the War. Everyone said it would be over by Christmas, but knew it would stretch until the last trump.

"I could just stop, you know," Francis said, deflated. "I could just shut it down and go back to San Francisco and a hot bath and decent pasta and forget everything. I can still get work shooting commercials, nudie movies, series TV. I could make little films, shot on video with a four-man crew, and show them to my friends. All this D.W. Griffith-David O. Selznick shit just isn't fucking necessary."

He stretched out his arms and water poured from his sleeves. Over a hundred people, huddled in various shelters or wrapped in orange plastic ponchos, looked at their lord and master and didn't know what to say or do.

"What does this cost, people? Does anybody know? Does anybody care? Is it worth all this? A movie? A painted ceiling? A symphony? Is anything worth all this shit?"

The rain stopped as if a tap were turned off. Sun shone through clouds. Kate screwed her eyes tight shut and fumbled under her poncho for the heavy sunglasses-clip she always carried. She might be the kind of vampire who could go about in all but the strongest sunlight, but her eyes could still be burned out by too much light.

She fixed clip-on shades to her glasses and blinked.

People emerged from their shelters, rainwater pouring from hats and ponchos.

"We can shoot around it," a co-associate assistant producer said.

Francis fired him on the spot.

Kate saw Ion creep out of the forests and straighten up. He had a wooden staff, newly-trimmed. He presented it to his maestro.

"To lean on," he said, demonstrating. Then, he fetched it up and held it like a weapon, showing a whittled point. "And to fight with."

Francis accepted the gift, made a few passes in the air, liking the feel of it in his hands. Then he leaned on the staff, easing his weight onto the strong wood.

"It's good," he said.

Ion grinned and saluted.

"All doubt is passing," Francis announced. "Money doesn't matter, time doesn't matter, we don't matter. This film, this *Dracula*, that is what matters. It's taken the smallest of you," he laid his hand on Ion's curls, "to show me that. When we are gone, *Dracula* will remain."

Francis kissed the top of Ion's head.

"Now," he shouted, inspired, "to work, to work."

11

The coach trundles up the mountainside, winding between the tall trees. A blaze of blue light shoots up.

Westenra: *Treasure!*

Harker's Voice: *They said the blue flames marked the sites of long-lost troves of bandit silver and gold. They also said no good ever came of finding it.*

Westenra: *Coachman, stop! Treasure.*

Swales pulls up the reins, and the team halt. The clatter of hooves and reins dies. The night is quiet.

The blue flame still burns.

Westenra jumps out and runs to the edge of the forest, trying to see between the trees, to locate the source of the light.

Blood of Dracula's Castle
(USA; Al Adamson & Jean Hewitt)

Dr. Terror's Galley of Horror
(USA; David L. Hewitt; a.k.a. *The Blood Suckers;* TV title: *Return from the Past*)

Dracula's Wedding Day
(USA; Mike Jackson)

The Fearless Vampire Killers or: Pardon Me, But Your Teeth Are In My Neck
(UK; Roman Polanski; orig.: *Dance of the Vampires*)

Frankenstein's Bloody Terror
(SPAIN; Henry Egan [Enrique L. Eguiluz]; a.k.a. *The Vampire of Dr. Dracula, Hell's Creatures* [UK]; orig.: *La marca del hombre lobo*)

The Killing Game
(FRANCE; Alain Jessua; orig.: *Jeu de massacre*)

Mad Monster Party
(USA/UK; Jules Bass)

The Naked World of Harrison Marks
(UK; Harrison Marks)

A Taste of Blood
(USA; Herschell Gordon Lewis; a.k.a. *The Secret of Dr. Alucard*)

The Vampire Beast Craves Blood
(UK; Vernon Sewell; a.k.a. *The Blood Beast Terror, Deathshead Vampire, Blood Beast from Hell*)

The Vampire's Rape
(FRANCE; Jean Rollin; orig.: *Le viol du vampire, La reine des vampires*)

Vampirisme
(FRANCE; short subject)

Vij
(RUSSIA; a.k.a. *Viy*)

1968

Dracula Meets the Outer Space Chicks
(USA)

Dracula Has Risen From the Grave
(UK; Freddie Francis)

Fangs of the Living Dead
(SPAIN/ITALY; Amando de Ossorio; a.k.a. *Malenka;* orig.: *La nipote del vampiro*)

The Fear Chamber
(MEXICO/USA; John [Juan] Ibanez; orig.: *La camara del terror;* a.k.a. *Torture Chamber, The Torture Zone*)

Goke, Body Snatcher from Hell
(JAPAN; Hajime Sato; orig.: *Kyuketsuki Gokemidoro*)

Night of the Living Dead
(USA; George A. Romero)

Sangre de virgenes
(MEXICO; Emilio Viegra)

Santo y el Tesoro de Dracula
(MEXICO; Rene Cardonna; a.k.a. *El vampiro y el sexo*)

Tower of the Screaming Virgins
(GERMANY/FRANCE/ITALY; Leo Joannon [Francois Legrand]; orig.: *Der turm der verbotenen liebe, La tour de Nesle;* a.k.a. *The Sweetness of Sin*)

Upior (The Vampire)
(POLAND; Stanislaw Lenartowicz; short subject)

Las Vampiras
(MEXICO; Federico Curiel)

A Vampire's Nostalgia
(YUGOSLAVIA)

Yokai Daisenso
(JAPAN; Yoshiyuki Kuroda)

1969
Assignment Terror
(SPAIN/WEST GERMANY/ITALY; Hugo Fregonese and Tulio Demichelli; orig.: *Dracula jagt Frankenstein, Los monstruos del terror;* a.k.a. *Dracula vs. Frankenstein*)

Blood of Dracula's Castle
(USA; Al Adamson)

Dracula
(UK/TV; Patrick Dromgoole)

Dracula (The Dirty Old Man)
(USA; William Edwards)

Drakulita
(PHILIPPINES; Consuelo P. Osorio)

Hiroku Haibyodeu
(JAPAN; Tokuzo Tanaka)

Harker: *I'll go with him.*

Warily, Harker takes a rifle down from the coach, and breeches a bullet.

Westenra runs ahead into the forest, excited. Harker carefully follows up, placing each step carefully.

Westenra: *Treasure, man. Treasure.*

Harker hears a noise, and signals Westenra to hold back. Both men freeze and listen.

The blue light flickers on their faces and fades out. Westenra is disgusted and disappointed.

Something moves in the undergrowth. Red eyes glow.

A dire wolf leaps up at Westenra, claws brushing his face, enormously furred body heavy as a felled tree. Harker fires. A red flash briefly spotlights the beast's twisted snout.

The wolf's teeth clash, just missing Westenra's face. The huge animal, startled if not wounded, turns and disappears into the forest.

Westenra and Harker run away as fast as they can, vaulting over prominent tree-roots, bumping low branches.

Westenra: *Never get out of the coach... never get out of the coach.*

They get back to the road. Swales looks stern, not wanting to know about the trouble they're in.

Harker's Voice: *Words of wisdom. Never get out of the coach, never go into the woods... unless you're prepared to become the compleat animal, to stay forever in the forests. Like him, Dracula.*

12

At the party celebrating the 100th Day of Shooting, the crew brought in a coffin bearing a brass plate that read simply DRACULA. Its lid creaked open and a girl in a bikini leaped out, nestling in Francis's lap. She had plastic fangs, which she spit out to kiss him.

The crew cheered. Even Eleanor laughed.

The fangs wound up in the punch-bowl. Kate fished them out as she got drinks for Marty Sheen and Robert Duvall.

Duvall, lean and intense, asked her about Ireland. She admitted she hadn't been there in decades. Sheen, whom everyone thought was Irish, was Hispanic, born Ramon Estevez. He was drinking heavily and losing weight, travelling deep into his role. Having surrendered entirely to Francis's "vision," Sheen was talking with Harker's accent and developing the character's hollow-eyed look and panicky glance.

The real Jonathan, Kate remembered, was a decent but dull sort, perpetually 'umble around brighter people, deeply suburban. Mina, his fiancée and her friend, kept saying that at least he was real, a worker and not a butterfly like Art or Lucy. A hundred years later, Kate could hardly remember Jonathan's face. From now on, she would always think of Sheen when anyone mentioned Jonathan Harker. The original was eclipsed.

Or erased. Bram Stoker had intended to write about Kate in his book, but left her out. Her few poor braveries during the Terror tended to be ascribed to Mina in most histories. That was probably a blessing.

"What must it have been like for Jonathan," Sheen said. "Not even knowing there were such things as vampires? Imagine, confronted with Dracula himself. His whole world was shredded, torn away. All he had was himself, and it wasn't enough."

"He had family, friends," Kate said.

Sheen's eyes glowed. "Not in Transylvania. *Nobody* has family and friends in Transylvania."

Kate shivered and looked around. Francis was showing off martial arts moves with Ion's staff. Fred Forrest was rolling a cigar-sized joint. Vittorio Storaro, the cinematographer, doled out his special spaghetti, smuggled into the country inside film cans, to appreciative patrons. A Romanian official in an ill-fitting shiny suit, liaison with the state studios, staunchly resisted offers of drinks he either assumed were laced with LSD or didn't want other Romanians to see him sampling. She wondered which of the native hangers-on was the *Securitate* spy, and giggled at the thought that they all might be spies and still not know the others were watching them.

Punch, which she was sipping for politeness's sake, squirted out of her nose as she laughed. Duvall patted her back and she recovered. She was not used to social drinking.

Ion, in a baseball cap given him by one of Francis's kids, was joking with the girl in the bikini, a dancer who played one of the gypsies, his eyes reddening with thirst. Kate decided to leave them be. Ion would control himself with the crew. Besides, the girl might like a nip from the handsome lad.

With a handkerchief, she wiped her face. Her specs had gone crooked with her spluttering and she rearranged them.

"You're not what I expected of a vampire lady," Duvall said.

Kate slipped the plastic fangs into her mouth and snarled like a kitten. Duvall and Sheen laughed.

13

For two weeks, Francis had been shooting the "Brides of Dracula" sequence. The mountainside was as crowded as Oxford Street, extras borrowed from the Romanian army salted with English faces recruited from youth hostels and student exchanges. Storaro was up on a dinosaur-necked camera crane, swooping through the skies, getting shots of rapt faces.

The three girls, two warm and one real vampire, had only showed up tonight, guaranteeing genuine crowd excitement in long-shot or blurry background rather than the flatly faked enthusiasm radiated for their own close-ups.

Kate was supposed to be available for the Brides, but they didn't need advice. It struck her as absurd that she should be asked to tell the actresses

Men of Action Meet Women of Dracula
(PHILIPPINES; Artemio Marquez)

The Nude Vampire
(FRANCE; Jean Rollin; orig.: *La vampire nue*)

Santo and Dracula's Treasure
(MEXICO; Rene Cardona; orig.: *El vampiro y el sexo*)

Santo Contra los Cazadores de Cabezas
(MEXICO; Rene Cardona)

Santo y Blue Demon vs. los monstruos
(MEXICO; Gilberto Martinez Solares; a.k.a. *Santo y Blue Demon contra los monstruos*)

Tales of Blood and Horror
(ITALY; Maurine Dawson)

Taste the Blood of Dracula
(UK; Peter Sasdy)

Valerie and Her Week of Wonders
(CZECHOSLOVAKIA; Jaromil Jires; orig.: *Valerie a tyden divu*)

A Vampire's Dream
(BRAZIL; Ibere Calvacant; orig.: *Um sonho de vampiros*)

La Venganza de las mujeres vampiro
(MEXICO; Federico Curiel; a.k.a. *Santo eve la venganza de las mujeres vampiro*)

1970

Bloodsuckers
(UK; Michael Burrowes [Robert Hartford-Davis]; a.k.a. *Incense for the Damned*)

Count Dracula
(SPAIN/WEST GERMANY/ITALY; Jess Franco; a.k.a. *El conde Dracula, Nachts Wenn Dracula Erwacht, Il conte Dracula, Bram Stoker's Count Dracula*)

Count Yorga, Vampire
(USA; Bob Kelljan)

Countess Dracula
(UK; Peter Sasdy)

Cuadecuc (Vampir)
(SPAIN; Pedro Portabella)

Curse of the Vampires
(PHILIPPINES/USA; Gerardo de Leon; orig.: *Dugo ng vampira;* a.k.a. *Creatures of Evil*)

The Devil's Shin
(SINGAPORE)

Dracula's Vampire Lust
(SWITZERLAND; Mario d'Alcala; orig.: *Dracula's Lusterne Sex-Vampire*)

Garu, the Mad Monk
(USA; Andy Milligan)

Guess What Happened to Count Dracula
(USA; Laurence Merrick)

Hannah, Queen of the Vampires
(TURKEY/USA; Ray Danton, Julio Salvador; orig.: *La tumba de la isla maldita,* a.k.a. *Crypt of the Living Dead, Vampire Woman, Young Hannah, Queen of the Vampires*)

The Horrible Sexy Vampire
(SPAIN; Jim Delavena and Arthur Davidson [Jose Luis Madrid]; orig.: *El vampiro de la autopista*)

Horror of the Blood Monsters
(USA; Al Adamson; a.k.a. *Creatures of the Prehistonic Planet, Creatures of the Red Planet, Flesh Creatures of the Red Planet, The Flesh Creatures, Horror Creatures of the Prehistoric Planet, Space Mission to the Lost Planet, Vampire Men of the Lost Planet*)

House of Dark Shadows
(USA; Dan Curtis)

The House That Dripped Blood
(UK; Peter Duffell)

Jonathan
(WEST GERMANY; Hans W. Geissendorfer; a.k.a. *Jonathan, Vampire Sterbeu Nicht*)

Love—Vampire Style
(WEST GERMANY; Helmut Foernbacher; orig.: *Beiss Mich, Liebling!;* a.k.a. *Bite Me, Darling!*)

Lust for a Vampire
(UK; Jimmy Sangster)

The Murder Mansion
(SPAIN/ITALY; Francisco Lara Polop; orig.: *La mansion de la niebla, Quando Marta urlo della tomba*)

The Night of the Vampire
(JAPAN; Michio Yamamoto; orig.: *Chi o suu niug To;* a.k.a. *The Vampire Doll*)

how to be alluring. The vampire Marlene, cast as the blonde bride, had been in films since the silent days and wandered about nearly naked, exposing herself to the winds. Her warm sisters needed to be swathed in furs between shots.

In a shack-like temporary dressing room, the brides were transformed. Bunty, a sensible Englishwoman, was in charge of their make-up. The living girls, twins from Malta who had appeared in a *Playboy* layout, submitted to all-over pancake that gave their flesh an unhealthy shimmer and opened their mouths like dental patients as false canines were fitted. Their fangs were a hundred times more expensive if hardly more convincing than the joke shop set Kate had kept after the party.

Francis, with Ion in his wake carrying a script, dropped by to cast an eye over the brides. He asked Marlene to open her mouth and examined her dainty pointy teeth.

"We thought we'd leave them as they were," said Bunty.

Francis shook his head.

"They need to be bigger, more obvious."

Bunty took a set of dagger-like eye-teeth from her kit and approached Marlene, who waved them away.

"I'm sorry, dear," the make-up woman apologised.

Marlene laughed musically and hissed, making Francis jump. Her mouth opened wide like a cobra's, and her fangs extended a full two inches.

Francis grinned.

"Perfect."

The vampire lady took a little curtsey.

14

Kate mingled with the crew, keeping out of camera-shot. She was used to the tedious pace of film-making. Everything took forever and there was rarely anything to see. Only Francis, almost thin now, was constantly on the move, popping up everywhere — with Ion, nick-named "Son of Dracula" by the crew, at his heels — to solve or be frustrated by any one of a thousand problems.

The stands erected for the extras, made by local labour in the months before shooting, kept collapsing. It seemed the construction people, whom she assumed also had the door contract at the Bucharest hotel, had substituted inferior wood, presumably pocketing the difference in leis, and the whole set was close to useless. Francis had taken to having his people work at night, after the Romanians contractually obliged to do the job had gone home, to shore up the shoddy work. It was, of course, ruinously expensive and amazingly inefficient.

The permits to film at Borgo Pass had still not come through. An associate producer was spending all her time at the Bucharest equivalent of the Circumlocution Office, trying to get the tri-lingual documentation out of the Ministry of Film. Francis would have to hire an entire local film crew and pay them to stand idle while his Hollywood people did the work. That was the expected harassment.

The official in the shiny suit, who had come to represent for everyone the forces hindering the production, stood on one side, eagerly watching the actresses. He didn't permit himself a smile.

Kate assumed the man dutifully hated the whole idea of *Dracula*. He certainly did all he could to get in the way. He could only speak English when the time came to announce a fresh snag, conveniently forgetting the language if he was standing on the spot where Francis wanted camera track laid and he was being told politely to get out of the way.

"Give me more teeth," Francis shouted through a bull-horn. The actresses responded.

"All of you," the director addressed the extras, "look horny as hell."

Ion repeated the instruction in three languages. In each one, the sentence expanded to a paragraph. Different segments of the crowd were enthused as each announcement clued them in.

Arcs, brighter and whiter than the sun, cast merciless, bleaching patches of light on the crowd, making faces look like skulls. Kate was blinking, her eyes watering. She took off and cleaned her glasses.

Like everybody, she could do with a shower and a rest. And, in her case, a decent feed.

Rumours were circulating of other reasons they were being kept away from Borgo Pass. The twins, flying in a few days ago, had brought along copies of the *Guardian* and *Time Magazine*. They were passed around the whole company, offering precious news from home. She was surprised how little seemed to have happened while she was out of touch.

However, there was a tiny story in the *Guardian* about the Transylvania Movement. Apparently, Baron Meinster, an obscure disciple of Dracula, was being sought by the Romanian authorities for terrorist outrages. The newspaper reported that he had picked up a band of vampire followers and was out in the forests somewhere, fighting bloody engagements with Ceausescu's men. The Baron favoured young get; he would find lost children, and turn them. The average age of his army was fourteen. Kate knew the type: red-eyed, lithe brats with sharp teeth and no compunctions about anything. Rumour had it that Meinster's Kids would descend on villages and murder entire populations, gorging themselves on blood, killing whole families, whole communities, down to the animals.

That explained the nervousness of some of the extras borrowed from the army. They expected to be sent into the woods to fight the devils. Few of them would come near Kate or any other vampire, so any gossip that filtered through was third-hand and had been translated into and out of several languages.

There were quite a few civilian observers around, keeping an eye on everything, waving incomprehensible but official documentation at anyone who queried their presence. Shiny Suit knew all about them and was

Le sadique aux dents rouge
(BELGIUM; Jean-Louis van Belle)

Scars of Dracula
(UK; Roy Ward Baker)

Scream and Scream Again
(UK; Gordon Hessler)

Sex and the Vampire
(FRANCE; Jean Rollin; orig.: *Les frisson des vampires;* a.k.a. *The Vampire's Thrill*)

Valerie and Her Week of Wonders
(CZECHOSLOVAKIA; Jaromil Jires; orig.: *Valerie a Tyden Divu*)

The Vampire Lovers
(UK; Roy Ward Baker)

El vampiro de la autopista
(SPAIN; Jose Luis Madrid)

Vampyros Lesbos die Erbin des Dracula
(WEST GERMANY/SPAIN; Franco Manera [Jesus Franco]; a.k.a. *El signo del vampiro, Las vampiras*)

Web of the Spider
(ITALY/WEST GERMANY/FRANCE; Anthony M. Dawson [Antonio Margheriti]; orig.: *Nella stretta morsa del ragno, Dracula im Schloss des Schreckens, Prisonnier de l'araignée*)

Werewolf vs. the Vampire Woman
(SPAIN/WEST GERMANY; Leon Ylim (Klimovsky); orig.: *La noche de Walpurgis, Nacht der Vampire;* a.k.a. *Blood Moon, Shadow of the Werewolf*)

1971

Blood Pie
(SPAIN; Jose Maria Valles; orig.: *Pastel de sangre*)

Blood Thirst
(PHILIPPINES/USA; Newt Arnold)

The Blue Sextet
(USA)

The Body
(USA/UK; Andy Milligan)

Caged Virgins
(FRANCE; Jean Rollin; orig.: *Requiem pour un vampire;* a.k.a. *Requiem for a Vampire, Vierges et vampires, Virgins and Vampires, Caged Virgins, Crazed Vampires*)

their unofficial boss. Ion kept well away from them. She must ask the lad if he knew anything of Meinster. It was a wonder he had not become one of Meinster's Child Warriors. Maybe he had, and was trying to get away from that. Growing up.

The crowd rioted on cue but the camera-crane jammed, dumping the operator out of his perch. Francis yelled at the grips to protect the equipment, and Ion translated but not swiftly enough to get them into action.

The camera came loose and fell thirty feet, crunching onto rough stone, spilling film and fragments.

Francis looked at the mess, uncomprehending, a child so shocked by the breaking of a favourite toy that he can't even throw a fit. Then, red fury exploded.

Kate wouldn't want to be the one who told Francis that there might be fighting at Borgo Pass.

15

In the coach, late afternoon, Harker goes through the documents he has been given. He examines letters sealed with a red wax "D," old scrolls gone to parchment, annotated maps, a writ of excommunication. There are pictures of Vlad, woodcuts of the Christian Prince in a forest of impaled infidels, portraits of a dead-looking old man with a white moustache, a blurry photograph of a murk-faced youth in an unsuitable straw hat.

Harker's Voice: *Vlad was one of the Chosen, favoured of God. But somewhere in those acres of slaughtered foemen, he found something that changed his mind, that changed his soul. He wrote letters to the Pope, recommending the rededication of the Vatican to the Devil. He had two cardinals, sent by Rome to reason with him, hot-collared — red-hot pokers slid through their back passages into their innards. He died, was buried, and came back…*

Harker looks out of the coach at the violent sunset. Rainbows dance around the tree-tops.

Westenra cringes but Murray is fascinated

Murray: *It's beautiful, the light…*

Up ahead is a clearing. Coaches are gathered. A natural stone amphitheatre has been kitted out with limelights which fizz and flare.

Crowds of Englishmen take seats.

Harker is confused, but the others are excited.

Murray: *A musical evening. Here, so far from Piccadilly…*

The coach slows and stops. Westenra and Murray leap out to join the crowds.

Warily, Harker follows. He sits with Westenra and Murray. They pass a hip-flask between them.

Harker takes a cautious pull, stings his throat.

Into the amphitheatre trundles a magnificent carriage, pulled by a

Chantoc Contra el Tigre y el Vampiro (MEXICO; Gilberto Martinez Solares)

The Curse of the Vampire (SPAIN; Joseph de Lacy [Jose Maria Elorietta]; orig.: *La llamada del vampiro*)

Daughters of Darkness (BELGIUM/FRANCE/WEST GERMANY; Harry Kumel)

Dracula vs. Frankenstein (USA; Al Adamson; a.k.a. *Blood of Frankenstein, Satan's Blood Freaks, Blood of Ghastly Horror*)

Hanno Cambiato Faccia (ITALY; Corrado Farina)

Jupiter (FRANCE; Jean-Pierre Prevost)

Lake of Dracula (JAPAN; Michio Yamamoto; orig.: *Chi o Suu Mee;* a.k.a. *Japula*)

Let's Scare Jessica to Death (USA; John Hancock)

O Macabro Dr. Scivano (BRAZIL; Raul Calhado & Rosalvo Cacador)

The Night Stalker (USA/TV; John Llewellyn Moxey)

The Omega Man (USA; Boris Sagal)

El Retorno de los Vampiros (SPAIN; Jose Maria Zabalza)

The Return of Count Yorga (USA; Bob Kelljan)

The Time of Vampires (YUGOSLAVIA; Nikola Majdak; orig.: *Vrijeme Vampira*)

Tombs of the Blind Dead (SPAIN/PORTUGAL; Armando de Ossorio; orig.: *La noche del terror ciego;* a.k.a. *The Blind Dead*)

Twins of Evil (UK; John Hough)

single, black stallion. The beast is twelve hands high. The carriage is black as the night, with an embossed gold and scarlet crest on the door. A red-eyed dragon entwines around a letter "D."

The driver is a tall man, draped entirely in black, only his red eyes showing.

There is mild applause.

The driver leaps down from his seat, crouches like a big cat and stands taller than ever. His cloak swells with the night breeze.

Loud music comes from a small orchestra.

"Take a Pair of Crimson Eyes," by Gilbert and Sullivan.

The driver opens the carriage door.

A slim white limb, clad only in a transparent veil, snakes around the door. Tiny bells tinkle on a delicate ankle. The toe-nails are scarlet and curl like claws.

The audience whoops appreciation. Murray burbles babyish delight. Harker is wary.

The foot touches the carpet of pine needles and a woman swings out of the carriage, shroud-like dress fluttering around her slender form. She has a cloud of black hair and eyes that glow like hot coals.

She hisses, tasting the night, exposing needle-sharp eye-teeth. Writhing, she presses her snake-supple body to the air, as if sucking in the essences of all the men present.

Murray: *The bloofer lady…*

The other carriage door is kicked open and the first woman's twin leaps out. She is less languid, more sinuous, more animal-like. She claws and rends the ground and climbs up the carriage wheel like a lizard, long red tongue darting. Her hair is wild, a tangle of twigs and leaves.

The audience, on their feet, applaud and whistle vigorously. Some of the men rip away their ties and burst their collar-studs, exposing their throats.

First Woman: *Kisses, sister, kisses for us all…*

The hood of the carriage opens, folding back like an oyster to disclose a third woman, as fair as they are dark, as voluptuous as they are slender. She is sprawled in abandon on a plush mountain of red cushions. She writhes, crawling through pillows, her scent stinging the nostrils of the rapt audience.

The driver stands to one side as the three women dance. Some of the men are shirtless now, clawing at their own necks until the blood trickles.

The women are contorted with expectant pleasure, licking their ruby lips, fangs already moist, shrouds in casual disarray, exposing lovely limbs, swan-white pale skin, velvet-sheathed muscle.

Men crawl at their feet, piling atop each other, reaching out just to touch the ankles of these women, these monstrous, desirable creatures.

Murray is out of his seat, hypnotised, pulled towards the vampires,

Vampire Playgirls
(BELGIUM/ITALY; Jean Brismee; orig.: *La plus longue nuit du diable, La notte piu lunga del diavolo;* a.k.a. *Devil's Nightmare, The Devil Walks at Midnight, Castle of Death, Succubus*)

Vampire Circus
(UK; Robert Young)

The Vampire Happening
(WEST GERMANY; Freddie Francis; orig.: *Gebissen Wird nur Nachts — Happening den Vampire*)

The Velvet Vampire
(USA; Stephanie Rothman; a.k.a. *The Waking Hour*)

1972

Ghost
(USA; Frederic Hobbs)

Angeles y Querubines
(MEXICO; Rafael Corkidi)

Baron Blood
(ITALY; Mario Bava)

Blacula
(USA; William Crain)

The Blood-Spattered Bride
(SPAIN; Vincente Aranda; orig.: *La novia ensangrentada;* a.k.a. *Till Death Do Us Part*)

Capulina contra los Monstruos
(MEXICO; Rene Cardona)

Capulina contra los Vampiros
(MEXICO; Rene Cardona)

Il cavaliere contante nicosia demoniaco ovvero Dracula in Brianza
(ITALY; Lucio Fulci)

The Deathmaster
(USA; Ray Danton)

Dracula's Great Love
(SPAIN; Javier Aguirre; orig.: *El gran amor del conde Draculu;* a.k.a. *Cemetery Girls, Dracula's Virgin Loves, Vampire Playgirls*)

Dracula A.D. 1972
(UK; Alan Gibson)

Grave of the Vampire
(USA; John Hayes)

La Invasion de los Muertos
(MEXICO; Rene Cardona)

Invasion of the Blood Farmers
(USA; Ed Adlum)

The Legend of Blood Castle
(SPAIN/ITALY; Jorge Grau; orig.: *Ceremonies sangriente;* a.k.a. *The Female Butcher, Blood Ceremony*)

Lips of Blood
(FRANCE/SPAIN; Ken Rudder [Alejandro Parti Gelibert & Pierre Chevalier]; orig.: *Le sang des autres, Les chemins de la violence, El segretto de la momia, Perversions sexuelles*)

The Mystery in Dracula's Castle
(USA/TV; Robert Totten)

Night of the Devils
(ITALY/SPAIN; Giorgio Ferroni; orig.: *La notte dei diuvoli, La noche de los diablos*)

Orgy of the Vampires
(SPAIN; Leon Mimovsky; orig.: *La orgia nocturna de los vampiros, La noche de los vampiros,* a.k.a. *The Vampire's Night Orgy*)

Saga of the Draculas
(SPAIN; Leon Ylimovsky; orig.: *La saga de los Dracula;* a.k.a. *The Dracula Saga, Dracula-The Bloodline*)

Santo y Blue Demon Contra Dracula y el Hombre Lobo
(MEXICO; Miguel M. Delgado)

The Screaming Dead
(SPAIN/FRANCE; Jess Franco; orig.: *Dracula contra Frankenstein, Dracula le prisonnier de Frankenstein;* a.k.a. *Dracula vs. Frankenstein, Dracula, Prisoner of Frankenstein*)

Vampire Kung-Fu
(HONG KONG; Li Fai Mon)

Voodoo Heartbeat
(USA; Charles Nizet)

1973
Andy Warhol's Dracula
(ITALY/FRANCE; Paul Morrissey and Antonio Margheriti; orig.: *Dracula cerca sangue di vergine e . . . mori di sete;* a.k.a. *Blood for Dracula*)

eyes mad. Harker tries to hold him back, but is wrenched forward in his wake, dragged like an anchor.

Murray steps over his fallen fellows, but trips and goes down under them.

Harker scrambles to his feet and finds himself among the women. Six hands entwine around his face. Lips brush his cheek, razor-edged teeth drawing scarlet lines on his face and neck.

He tries to resist but is bedazzled.

A million points of light shine in the women's eyes, on their teeth, on their earrings, necklaces, nose-stones, bracelets, veils, navel-jewels, lacquered nails. The lights close around Harker.

Teeth touch his throat.

A strong hand, sparsely bristled, reaches out and hauls one of the women away.

The driver steps in and tosses another vampire bodily into the carriage. She lands face-down and seems to be drowning in cushions, bare legs kicking.

Only the blonde remains, caressing Harker, eight inches of tongue scraping the underside of his chin. Fire burns in her eyes as the driver pulls her away.

Blonde Woman: *You never love, you have never loved...*

The driver slaps her, dislocating her face. She scrambles away from Harker, who lies sprawled on the ground.

The women are back in the carriage, which does a circuit of the amphitheatre and slips into the forests. There is a massed howl of frustration, and the audience falls upon each other.

Harker, slowly recovering, sits up. Swales is there. He hauls Harker out of the mêlée and back to the coach. Harker, unsteady, is pulled into the coach.

Westenra and Murray are dejected, gloomy. Harker is still groggy.

Harker's Voice: *A vampire's idea of a half-holiday is a third share in a juicy peasant baby. It has no other needs, no other desires, no other yearnings. It is mere appetite, unencumbered by morality, philosophy, religion, convention, emotion. There's a dangerous strength in that. A strength we can hardly hope to equal.*

16

Shooting in a studio should have given more control, but Francis was constantly frustrated by Romanians. The inn set, perhaps the simplest element of the film, was still not right, though the carpenters and dressers had had almost a year to get it together. First, they took an office at the studio and turned it into Harker's bedroom. It was too small to fit in a camera as well as an actor and the scenery. Then, they reconstructed the whole thing in the middle of a sound stage, but still bolted together the walls so they couldn't be moved. The only shot Storaro could take was from the ceiling looking down. Now the walls were fly-away enough to allow camera movement, but Francis wasn't happy with the set dressing.

Prominent over the bed, where Francis wanted a crucifix, was an idealised portrait of Ceausescu. Through Ion, Francis tried to explain to Shiny Suit, the studio manager, that his film took place before the President-for-Life came to power and that, therefore, it was highly unlikely that a picture of him would be decorating a wall anywhere.

Shiny Suit seemed unwilling to admit there had ever been a time when Ceausescu didn't rule the country. He kept looking around nervously, as if expecting to be caught in treason and hustled out to summary execution.

"Get me a crucifix," Francis yelled.

Kate sat meekly in a director's chair — a rare luxury — while the argument continued. Marty Sheen, in character as Harker, sat cross-legged on his bed, taking pulls at a hip-flask of potent brandy. She could smell the liquour across the studio. The actor's face was florid and his movements slow. He had been more and more Harker and less and less Marty the last few days, and Francis was driving him hard, directing with an emotional scalpel that peeled his star like an onion.

Francis told Ion to bring the offending item over so he could show Shiny Suit what was wrong. Grinning cheerfully, Ion squeezed past Marty and reached for the picture, dextrously dropping it onto a bed-post which shattered the glass and speared through the middle of the frame, punching a hole in the Premier's face.

Ion shrugged in fake apology.

Francis looked almost happy. Shiny Suit, stricken in the heart, scurried away in defeat, afraid that his part in the vandalism of the sacred image would be noticed.

A crucifix was found from stock and put up on the wall.

"Marty," Francis said, "open yourself up, show us your beating heart, then tear it from your chest, squeeze it in your fist and drop it on the floor."

Kate wondered if he meant it literally.

Marty Sheen tried to focus his eyes, and saluted in slow motion.

"Quiet on set, everybody," Francis shouted.

17

Kate was crying, silently, uncontrollably. Everyone on set, except Francis and perhaps Ion, was also in tears. She felt as if she was watching the torture of a political prisoner, and just wanted it to stop.

There was no script for this scene.

Francis was pushing Marty into a corner, breaking him down, trying to get to Jonathan Harker.

This would come at the beginning of the picture. The idea was to show the real Jonathan, to get the audience involved with him. Without this scene, the hero would seem just an observer, wandering between other people's set-pieces.

Black Vampire
(USA; Bill Gunn; a.k.a. *Ganja and Hess, Blood Couple, Double Possession, Black Evil, Black Out, The Moment of Terror*)

Curse of the Devil
(SPAIN/MEXICO; Carlos Aured; orig.: *El retorno de Walpurgis*)

The Daughter of Dracula
(FRANCE/PORTUGAL; Jesus Franco; orig.: *La fille de Dracula, La bija de Dracula*)

Dead People
(USA; Willard Huyck; a.k.a. *Messiah of Evil, Return of the Living Dead, Revenge of the Screaming Dead*)

The Devil's Plaything
(SWITZERLAND; Joe Sarno)

The Devil's Wedding Night
(ITALY; Paul Solvay [Luigi Batzella]; orig.: *Il plenilunio delle vergine*)

Dracula
(CANADA/TV; Jack Nixon Browne)

Erotikill
(FRANCE/BELGIUM/PORTUGAL; J. P. Johnson [Jesus Franco]; orig.: *La comtesse noire, La comtesse aux seins nus, Les avaleuses* a.k.a. *The Bare Breasted Countess* and *The Loves of Irina*)

Der Fluch der Schwarzen Schwestern
(SWITZERLAND; Joseph W. Sarno)

Geek Maggot Bingo
(USA; Nick Zedd; a.k.a. *The Freak from Suckweasel Mountain*)

The Ghastly Orgies of Count Dracula
(ITALY; Ralph Brown [Renato Polselli]; orig.: *Riti magie nere e segrete orge del trecento;* a.k.a. *Reincarnation of Isabel*)

Horror of the Zombies
(SPAIN; Armando de Ossorio; orig.: *El buque maldito*)

Legacy of Satan
(USA; Gerard Damiano)

Lemora, The Lady Dracula
(USA; Richard Blackburn; a.k.a. *The Legendary Curse of Lemora, Lemora, a Child's Tale of the Supernatural, Lady Dracula*)

Night of the Sorcerers
(SPAIN/PORTUGAL; Armando de Ossorio; orig.: *La noche de los brujos*)

Pepito y Chabelo vs. los Moustruos
(MEXICO; Jose Estrada)

Return of the Evil Dead
(SPAIN; Armando de Ossorio; orig.: *El ataque de los muertos sin ojos;* a.k.a. *Attach of the Blind Dead*)

Satanic Rites of Dracula
(UK; Alan Gibson; a.k.a. *Count Dracula and His Vampire Bride [US], The Rites of Dracula*)

Scream, Blacula, Scream
(USA; Bob Kelljan)

Son of Dracula
(UK; Freddie Francis; reissue title: *Young Dracula*)

Tenderness of Wolves
(WEST GERMANY; Ulli Lommel; orig.: *Die Zartlichkeit der Wolfe*)

Los Vampiros de Coyoacan
(MEXICO; Arturo Martinez)

The Vault of Horror
(UK; Roy Ward Baker; reissue title: *Tales from the Crypt*)

1974
Barry McKenzie Holds His Own
(UK/FRANCE/AUSTRALIA; Bruce Beresford)

The Bat People
(USA; Jerry Jameson)

Blood
(USA; Andy Milligan)

Captain Kronos: Vampire Hunter
(UK; Brian Clemens)

Dracula
(USA/TV; Dan Curtis)

"You, Reed," Francis said, "you're a writer. Scribble me a voice-over. Internal monologue. Stream-of-consciousness. Give me the real Harker."

Through tear-blurred spectacles, she looked at the pad she was scrawling on. Her first attempt had been at the Jonathan she remembered, who would have been embarrassed to have been thought capable of stream-of-consciousness. Francis had torn that into confetti and poured it over Marty's head, making the actor cross his eyes and fall backwards, completely drunk, onto the bed.

Marty was hugging his pillow and bawling for Mina.

All for Hecuba, Kate thought. Mina wasn't even in this movie except as a locket. God knows what Mrs Harker would think when and if she saw *Dracula*.

Francis told the crew to ignore Marty's complaints. He was an actor, and just whining.

Ion translated.

She remembered what Francis had said after the storm, "what does this cost, people?" Was anything worth what this seemed to cost? "I don't just have to make *Dracula*," Francis had told an interviewer, "I have to *be* Dracula."

Kate tried to write the Harker that was emerging between Marty and Francis. She went into the worst places of her own past and realised they still burned in her memory like smouldering coals.

Her pad was spotted with red. There was blood in her tears. That didn't happen often.

The camera was close to Marty's face. Francis was intent, bent close over the bed, teeth bared, hands claws. Marty mumbled, trying to wave the lens away.

"Don't look at the camera, Jonathan," Francis said.

Marty buried his face in the bed and was sick, choking. Kate wanted to protest but couldn't bring herself to. She was worried Martin Sheen would never forgive her for interrupting his Academy Award scene. He was an actor. He'd go on to other roles, casting off poor Jon like an old coat.

He rolled off his vomit and looked up, where the ceiling should have been but wasn't.

The camera ran on. And on.

Marty lay still.

Finally, the camera operator reported "I think he's stopped breathing."

For an eternal second, Francis let the scene run.

In the end, rather than stop filming, the director elbowed the camera aside and threw himself on his star, putting an ear close to Marty's sunken bare chest.

Kate dropped her pad and rushed into the set. A wall swayed and fell with a crash.

"His heart's still beating," Francis said.

She could hear it, thumping irregularly.

Marty spluttered, fluid leaking from his mouth. His face was almost scarlet.

His heart slowed.

"I think he's having a heart attack," she said.

"He's only thirty-five," Francis said. "No, thirty-six. It's his birthday today."

A doctor was called for. Kate thumped Marty's chest, wishing she knew more first aid.

The camera rolled on, forgotten.

"If this gets out," Francis said, "I'm finished. The film is over."

Francis grabbed Marty's hand tight, and prayed.

"Don't die, man."

Martin Sheen's heart wasn't listening. The beat stopped. Seconds passed. Another beat. Nothing.

Ion was at Francis's side. His fang-teeth were fully extended and his eyes were red. It was the closeness of death, triggering his instincts.

Kate, hating herself, felt it too.

The blood of the dead was spoiled, undrinkable. But the blood of the dying was sweet, as if invested with the life that was being spilled.

She felt her own teeth sharp against her lower lip.

Drops of her blood fell from her eyes and mouth, spattering Marty's chin.

She pounded his chest again. Another beat. Nothing.

Ion crawled on the bed, reaching for Marty.

"I can make him live," he whispered, mouth agape, nearing a pulseless neck.

"My God," said Francis, madness in his eyes. "You can bring him back. Even if he dies, he can finish the picture."

"Yesssss," hissed the old child.

Marty's eyes sprang open. He was still conscious in his stalling body. There was a flood of fear and panic. Kate felt his death grasp her own heart.

Ion's teeth touched the actor's throat.

A cold clarity struck her. This undead youth of unknown bloodline must not pass on the Dark Kiss. He was not yet ready to be a father-in-darkness.

She took him by the scruff of his neck and tore him away. He fought her, but she was older, stronger.

With love, she punctured Marty's throat, feeling the death ecstasy convulse through her. She swooned as the blood, laced heavily with brandy, welled into her mouth, but fought to stay in control. The lizard part of her brain would have sucked him dry.

But Katharine Reed was not a monster.

She broke the contact, smearing blood across her chin and his chest

The Legend of the Seven Golden Vampires
(HONG KONG/UK; Roy Ward Baker; a.k.a. *The Seven Brothers Meet Dracula*)

Night of the Walking Dead
(SPAIN; Leon Klimovsky; orig.: *El extraño amor de los vampiros;* a.k.a. *Strange Love of the Vampires*)

Le nosferat ou les eaux glacees du calcul egoïste
(BELGIUM; Maurice Rabinowitz)

Old Dracula
(UK; Clive Donner; orig.: *Vampira*)

Quem Tem Medo de Lobisomem
(BRAZIL; Reginaldo Faria)

Tender Dracula
(FRANCE; Pierre Grunstein; orig.: *La grand trouille, Tendre Dracula*)

Three Immoral Women
(FRANCE; Walerian Borowezyk; orig.: *Contes immoraux;* a.k.a. *Immoral Tales*)

The Thirsty Dead
(PHILIPPINES; Terry Becker; a.k.a. *Blood Cult of Shangri-La*)

Those Cruel Aged Bloody Vampires
(SPAIN; Julio Perez Tabernero; orig.: *Las alegres vampiras de vogel*)

Vampyres: Daughters of Dracula
(UK; Joseph [Jose] Larraz; orig.: *Vampyres;* a.k.a. *Blood Hunger, Satan's Daughters*)

1975

Alucarda
(MEXICO; Juan Lopez Moctezuma; a.k.a. *Sisters of Satan, Innocents from Hell*)

Deafula
(USA; Peter Wechsberg)

The Evil of Dracula
(JAPAN; Michio Yamamoto; orig.: *Chi O Suu Bara*)

El Joveucito Dracula
(SPAIN; Carlos Benpar [Carlos Benito Parra])

Kathavai Thatteeya Mohni Paye
(INDIA; M.A. Rajaramann)

hair. She ripped open her blouse, scattering tiny buttons, and sliced herself with a sharpening thumbnail, drawing an incision across her ribs.

She raised Marty's head and pressed his mouth to the wound.

As the dying man suckled, she looked through fogged glasses at Francis, at Ion, at the camera operator, at twenty studio staff. A doctor was arriving, too late.

She looked at the blank round eye of the camera.

"Turn that bloody thing off," she said.

18

The principles were assembled in an office at the studio. Kate, still drained, had to be there. Marty was in a clinic with a drip-feed, awaiting more transfusions. His entire bloodstream would have to be flushed out several times over. With luck, he wouldn't even turn. He would just have some of her life in him, some of her in him, forever. This had happened before and Kate wasn't exactly happy about it. But she had no other choice. Ion would have killed the actor and brought him back to life as a newborn vampire.

"There have been stories in the trades," Francis said, holding up a copy of *Daily Variety*. It was the only newspaper that regularly got through to the company. "About Marty. We have to sit tight on this, to keep a lid on panic. I can't afford even the rumour that we're in trouble. Don't you understand, we're in the twilight zone here. Anything approaching a shooting schedule or a budget was left behind a long time ago. We can film round Marty until he's ready to do close-ups. His brother is coming over from the States to double him from the back. We can weather this on the ground, but maybe not in the press. The vultures from the trades want us dead. Ever since *Finian's Rainbow*, they've hated me. I'm a smart kid and nobody likes smart kids. From now on, if anybody *dies* they aren't dead until I say so. Nobody is to tell anyone anything until it's gone through me. People, we're in trouble here and we may have to lie our way out of it. I know you think the Ceausescu regime is fascist but it's nothing compared to the Coppola regime. You don't know anything until I confirm it. You don't do anything until I say so. This is a war, people, and we're losing."

19

Marty's family was with him. His wife didn't quite know whether to be grateful to Kate or despise her.

He would live. Really live.

She was getting snatches of his past life, mostly from films he had been in. He would be having the same thing, coping with scrambled impressions of her. That must be a nightmare all of its own.

They let her into the room. It was sunny, filled with flowers.

The actor was sitting up, neatly groomed, eyes bright.

Leonor
(SPAIN/FRANCE/ITALY; Juan Buñuel)

Levres de Sang
(FRANCE; Jean Rollin)

Mary, Mary, Bloody Mary
(MEXICO; Juan Lopez Moctezuma)

Mrs. Amworth
(UK/CANADA/TV; Alvin Rakoff)

Spermula
(FRANCE; Charles Matton)

Terror Beach
(SPAIN; Armando de Ossorio; orig.: *La noche de las gaviotas;* a.k.a. *Night of the Seagulls, Night of the Death Cult*)

1976
Bloodlust
(SWITZERLAND; Marijan Vajda; orig.: *Mosquito der Schaender*)

Dead of Night
(USA; Dan Curtis)

Dracula and Son
(FRANCE; Edouard Molinaro; orig.: *Dracula, père et fils*)

Martin
(USA; George Romero)

El Pobrecito Draculín
(SPAIN; Juan Fortuny)

Rabid
(CANADA; David Cronenberg)

Tiempos Duros Para Dracula
(SPAIN/ARGENTINA; Jorge M. Darnell)

The Vampire
(AUSTRIA; Karin Brandaner)

1977
Blood Relations
(NETHERLANDS/FRANCE; Wim Lindner; orig.: *Bloedverwanten, Les vampires en ont ras le bol*)

Count Dracula
(UK/TV; Philip Saville)

Doctor Dracula
(USA; Al Adamson & Paul Aratow)

"Now I know," he told her. "Now I really know. I can use that in the part. Thank you."

"I'm sorry," she said, not knowing what for.

20

At a way-station, Swales is picking up fresh horses. The old ones, lathered with foamy sweat, are watered and rested.

Westenra barters with a peasant for a basket of apples. Murray smiles and looks up at the tops of the trees. The moon shines down on his face, making him look like a child.

Harker quietly smokes a pipe.

Harker's Voice: *This was where we were to join forces with Van Helsing. This stone-crazy double Dutchman had spent his whole life fighting evil.*

Van Helsing strides out of the mountain mists. He wears a scarlet army tunic and a curly-brimmed top hat, and carries a cavalry sabre. His face is covered with old scars. Crosses of all kinds are pinned to his clothes.

Harker's Voice: *Van Helsing put the fear of God into the Devil. And he terrified me.*

Van Helsing is accompanied by a band of rough-riders. Of all races and in wildly different uniforms, they are his personal army of the righteous. In addition to mounted troops, Van Helsing has command of a couple of man-lifting kites and a supply wagon.

Van Helsing: *You are Harker?*

Harker: *Dr Van Helsing of Amsterdam?*

Van Helsing: *The same. You wish to go to Borgo Pass, Young Jonathan?*

Harker: *That's the plan.*

Van Helsing: *Better you should wish to go to Hades itself, foolish Englishman.*

Van Helsing's Aide: *I say, Prof, did you know Murray was in Harker's crew. The stroke of '84.*

Van Helsing: *Hah! Beat Cambridge by three lengths. Masterful.*

Van Helsing's Aide: *They say the river's at its most level around Borgo Pass. You know these mountain streams, Prof. Tricky for the oarsman.*

Van Helsing: *Why didn't you say that before, damfool? Harker, we go at once, to take Borgo Pass. Such a stretch of river should be held for the Lord. The Un-Dead, they appreciate it not.* Nosferatu *don't scull.*

Van Helsing rallies his men into mounting up. Harker dashes back to the coach and climbs in. Westenra looks appalled as Van Helsing waves his sabre, coming close to fetching off his own Aide's head.

Westenra: *That man's completely mad.*

Harker: *In Wallachia, that just makes him normal. To fight what we have to face, one has to be a little mad.*

Van Helsing's sabre shines with moonfire.

Van Helsing: *To Borgo Pass, my angels… charge!*

Van Helsing leads his troop at a fast gallop. The coach is swept along

Dracula
(USA/TV; Dan Curtis; a.k.a. *Bram Stoker's Dracula*)

Hyocho No Bijo
(JAPAN/TV; Umeji Inoue)

The Incredible Melting Man
(USA; William Sachs)

Lady Dracula
(WEST GERMANY; Franz-Joseph Gottlieb)

McCloud Meets Dracula
(USA/TV; Bruce Kessler)

Le Rouge de Chine
(FRANCE; Jacques Richard)

1978
Dawn of the Dead
(USA; George A. Romero)

A Deusa de Marvnore Escruva do Diablo
(BRAZIL; Rosangela Maldonado)

La Dinastia Dracula
(MEXICO)

Dracula's Dog
(USA; Albert Band; a.k.a. *Zoltan . . . Hound of Dracula*)

Nightmare in Blood
(USA; John Stanley; a.k.a. *Horror Convention*)

Tame re Champo ne Avne Kel
(INDIA; Chandrakant Sangani)

1979
Dracula
(UK/USA; John Badham)

Dracula Blows His Cool
(WEST GERMANY; Carlo Ombra; orig.: *Graf Dracula in Oberbayern*)

Fascination
(FRANCE; Jean Rollin)

Vampire Dracula Comes to Kobe: Evil Makes Women Beautiful
(JAPAN; Hajime Sato; orig.: *Kyuketsuki Dorakyura Kobe ni arawaru: Akuma wa onna wo utsukushiku suru*)

Love at First Bite
(USA; Stan Dragoti)

Mama Drucula
(BELGIUM; Boris Szulzinger)

Nightwing
(USA; Arthur Hiller)

Nocturna, Granddaughter of Dracula
(USA; Harry Tampa [Hurwitz])

Nosferatu the Vampyre
(WEST GERMANY/FRANCE; Werner Herzog; orig.: *Nosferatu: Phantom der Nacht*)

'Salem's Lot
(USA/TV; Tobe Hooper)

Thirst
(AUSTRALIA; Rod Hardy)

Vampire
(USA/TV; E. W. Swackhamer)

The Vampire Hookers
(PHILIPPINES; Cirio H. Santiago; a.k.a. *Sensuous Vampires*)

1980
Les Charlots Contest Dracula
(FRANCE; Jean-Pierre Desagnat)

The Craving
(SPAIN; Jack Molina [Paul Naschy]; orig.: *El retorno del hombre lobo*)

Deadline
(CANADA; Mario Azzopardi; a.k.a. *Anatomy of a Horror*)

Dracula
(JAPAN; Minori Okazaki; animated feature)

Dracula's Last Rites
(USA/FRANCE; Domonic Paris)

The Games of the Countess Dolingen of Gratz
(FRANCE; Catherine Binet; orig.: *Les jeux de la comtesse Dolingen de Gratz*)

The Monster Club
(UK; Roy Ward Baker)

Wolnyoui Han
(KOREA; Kim In Soo)

1981
The Black Room
(USA; Elly Kenner)

in the wake of the uphill cavalry advance. Man-lifting box-kites carry observers into the night air.

Wolves howl in the distance.

Between the kites is slung a phonograph horn.

Music pours forth. The overture to *Swan Lake*.

Van Helsing: *Music. Tchaikovsky. It upsets the devils. Stirs in them memories of things that they have lost. Makes them feel dead. Then we kill them good. Kill them forever.*

As he charges, Van Helsing waves his sword from side to side. Dark, low shapes dash out of the trees and slip among the horses' ankles. Van Helsing slashes downwards, decapitating a wolf. The head bounces against a tree, becoming that of a gypsy boy, and rolls down the mountainside.

Van Helsing's cavalry weave expertly through the pines. They carry flaming torches. The music soars. Fire and smoke whip between the trees.

In the coach, Westenra puts his fingers in his ears. Murray smiles as if on a pleasure ride across Brighton Beach. Harker sorts through crucifixes.

At Borgo Pass, a small gypsy encampment is quiet. Elders gather around the fire. A girl hears the Tchaikovsky whining among the winds and alerts the tribe.

The gypsies bustle. Some begin to transform into wolves.

The man-lifting kites hang against the moon, casting vast bat-shadows on the mountainside.

The pounding of hooves, amplified a thousandfold by the trees, thunders. The ground shakes. The forests tremble.

Van Helsing's cavalry explode out of the woods and fall upon the camp, riding around and through the place, knocking over wagons, dragging through fires. A dozen flaming torches are thrown. Shrieking werewolves, pelts aflame, leap up at the riders.

Silver swords flash, red with blood.

Van Helsing dismounts and strides through the carnage, making head shots with his pistol. Silver balls explode in wolf-skulls.

A young girl approaches Van Helsing's aide, smiling in welcome. She opens her mouth, hissing, and sinks fangs into the man's throat.

Three cavalrymen pull the girl off and stretch her out face-down on the ground, rending her bodice to bare her back. Van Helsing drives a five-foot lance through her ribs from behind, skewering her to the bloodied earth.

Van Helsing: *Vampire bitch!*

The cavalrymen congratulate each other and cringe as a barrel of gunpowder explodes nearby. Van Helsing does not flinch.

Harker's Voice: *Van Helsing was protected by God. Whatever he did, he would survive. He was blessed.*

Van Helsing kneels by his wounded Aide and pours holy water onto the man's ravaged neck. The wound hisses and steams, and the Aide shrieks.

Van Helsing: *Too late, we are too late. I'm sorry, my son.*

With a kukri knife, Van Helsing slices off his aide's head. Blood gushes over his trousers.

The overture concludes and the battle is over.

The gypsy encampment is a ruin. Fires still burn. Everyone is dead or dying, impaled or decapitated or silver-shot. Van Helsing distributes consecrated wafers, dropping crumbs on all the corpses, muttering prayers for saved souls.

Harker sits, exhausted, bloody earth on his boots.

Harker's Voice: *If this was how Van Helsing served God, I was beginning to wonder what the firm had against Dracula.*

The sun pinks the skies over the mountains. Pale light falls on the encampment.

Van Helsing stands tall in the early morning mists.

Several badly-wounded vampires begin to shrivel and scream as the sunlight burns them to man-shaped cinders.

Van Helsing: *I love that smell... spontaneous combustion at daybreak. It smells like... salvation.*

21

Like a small boy whose toys have been taken away, Francis stood on the rock, orange cagoule vivid against the mist-shrouded pines, and watched the cavalry ride away in the wrong direction. Gypsy extras, puzzled at this reversal, milled around their camp set. Storaro found something technical to check and absorbed himself in lenses.

No one wanted to tell Francis what was going on.

They had spent two hours setting up the attack, laying camera track, planting charges, rigging decapitation effects, mixing kensington gore in plastic buckets. Van Helsing's troop of ferocious cavalry were uniformed and readied.

Then Shiny Suit whispered in the ear of the captain who was in command of the army-provided horsemen. The cavalry stopped being actors and became soldiers again, getting into formation and riding out.

Kate had never seen anything like it.

Ion nagged Shiny Suit for an explanation. Reluctantly, the official told the little vampire what was going on.

"There is fighting in the next valley," Ion said. "Baron Meinster has come out of the forests and taken a keep that stands over a strategic pass. Many are dead or dying. Ceausescu is laying siege to the Transylvanians."

"We have an agreement," Francis said, weakly. "These are my men."

"Only as long as they aren't needed for fighting, this man says," reported Ion, standing aside to let the director get a good look at the Romanian

Krvava Pani
(CZECHOSLOVAKIA; Viktor Kubal; animated)

The Munsters' Revenge
(USA; Don Weis)

Saturday the 14th
(USA; Howard R. Cohen)

Vengeful Vampire Girl
(KOREA; Kim In Soo; orig.: *Huphyokwi yanyo*)

1982

Buenas Noches, Señor Monstruo
(MEXICO; Antonio Mercero)

Dracula Rises From His Coffin
(KOREA; Lee Hyoung Pyo)

Ferat Vampire
(CZECHOSLOVAKIA; Juraj Herz; orig.: *Upir z feratu*)

One Dark Night
(USA; Thomas McLoughlin; a.k.a. *Entity Force*)

1983

The Beastmaster
(USA; Don Coscarelli)

La Belle Captive
(FRANCE; Alain Robbe-Grillet)

Dracula Tan Exarchia
(GREECE; Nikos Zervos)

The Hunger
(UK/USA; Tony Scott)

A Polish Vampire in Burbank
(USA; Mark Pirro; a.k.a. *A Polish Vampire*)

Pura Sangre
(COLOMBIA; Luis Ospina)

The Tail
(HONG KONG; Ronny Yu; orig.: *Pao Dan Fei Che*)

1984

Bloodsuckers From Outer Space
(USA; Glen Coburn)

Carne de to Carne
(COLOMBIA; Carlos Mayolo)

Curse of the Wicked Wife
(HONG KONG; Wong King-Fang)

official. Shiny Suit almost smiled, a certain smug attitude suggesting that this would even the score for that dropped picture of the Premier.

"I'm trying to make a fucking movie here. If people don't keep their word, maybe they deserve to be overthrown."

The few bilingual Romanians in the crew cringed at such sacrilege. Kate could think of dozens of stronger reasons for pulling down the Ceausescu regime.

"There might be danger," Ion said. "If the fighting spreads."

"This Meinster, Ion. Can he get us the cavalry? Can we do a deal with him?"

"An arrogant elder, maestro. And doubtless preoccupied with his own projects."

"You're probably right. Fuck it."

"We're losing the light," Storaro announced.

Shiny Suit smiled blithely and, through Ion, ventured that the battle should be over in two to three days. It was fortunate for him that Francis only had prop weapons within reach.

In the gypsy camp, one of the charges went off by itself. A pathetic phut sent out a choking cloud of violently green smoke. Trickles of flame ran across fresh-painted flats.

A grip threw a bucket of water, dousing the fire.

Robert Duvall and Martin Sheen, in costume and make-up, stood about uselessly. The entire camera crew, effects gang, support team were gathered, as if waiting for a cancelled train.

There was a long pause. The cavalry did not come riding triumphantly back, ready for the shot.

"Bastards," Francis shouted, angrily waving his staff like a spear.

22

The next day was no better. News filtered back that Meinster was thrown out of the keep and withdrawing into the forests, but that Ceausescu ordered his retreat be harried. The cavalry were not detailed to return to their film-making duties. Kate wondered how many of them were still alive. The retaking of the keep must have been a bloody, costly battle. A cavalry charge against a fortress position would be almost a suicide mission.

Disconsolately, Francis and Storaro sorted out some pick-up shots that could be managed.

A search was mounted for Shiny Suit, so that a definite time could be established for rescheduling of the attack scene. He had vanished into the mists, presumably to escape the American's wrath.

Kate huddled under a tree and tried to puzzle out a local newspaper. She was brushing up her Romanian, simultaneously coping with the euphemisms and lacunae of a non-free press. According to the paper, Meinster had been crushed weeks ago and was hiding in a ditch somewhere, certain to be beheaded within the hour.

Evils of the Night
(USA; Mardi Rustam; a.k.a. *Space Monsters*)

Haunted Cop Shop
(HONG KONG; Jeff Lau)

I Married a Vampire
(USA; Jay Raskin)

Lust in the Fast Lane
(USA)

Mutant
(USA; John "Bud" Cardos; a.k.a. *Night Shadows*)

1985

Billy the Kid and the Green Baize Vampire
(UK; Alan Clarke)

Blue Lamp in a Winter Light
(HONG KONG; Yao Funa)

Day of the Dead
(USA; George A. Romero)

Dragon Against Vampire
(HONG KONG; Lionel Leung; a.k.a. *Dragon vs. Vampire*)

Fright Night
(USA; Tom Holland)

Hello Dracula
(HONG KONG; Henry Wu-Leung)

I Like Bats
(POLAND; Grzegorz Warchol; orig.: *Lubie nietoperze*)

Lifeforce
(UK/USA; Tobe Hooper)

Loves of the Living Dead
(HONG KONG; Ho Menga; a.k.a. *Heaven Wife, Hell Wife*)

Mixed Up
(HONG KONG; Henry S. Chen)

Mr. Vampire
(HONG KONG; Lau Kun Wai)

Once Bitten
(USA; Howard Storm)

The Tomb
(USA; Fred Olen Ray)

She couldn't help feeling the real story was in the next valley. As a newspaperwoman, she should be there, not waiting around for this stalled juggernaut to get back on track. Meinster's Kids frightened and fascinated her. She should know about them, try to understand. But American Zoetrope had first call on her, and she didn't have the heart to be another defector.

Marty Sheen joined her.

He was mostly recovered and understood what she had done for him, though he was still exploring the implications of their blood link. Just now, he was more anxious about working with Brando — due in next week — than his health.

There was still no scripted ending.

23

The day the cavalry — well, some of them — came back, faces drawn and downcast, uniforms muddied, eyes haunted, Shiny Suit was discovered with his neck broken, flopped half-in a stream. He must have fallen in the dark, tumbling down the precipitous mountainside.

His face and neck were ripped, torn by the sharp thorns of the mountain bushes. He had bled dry into the water, and his staring face was white.

"It is good that Georghiou is dead," Ion pronounced. "He upset the maestro."

Kate hadn't known the bureaucrat's name.

Francis was frustrated at this fresh delay, but graciously let the corpse be removed and the proper authorities be notified before proceeding with the shoot.

A police inspector was escorted around by Ion, poking at a few broken bushes and examining Georghiou's effects. Ion somehow persuaded the man to conclude the business speedily.

The boy was a miracle, everyone agreed.

"Miss Reed," Ion interrupted. She laid down her newspaper.

Dressed as an American boy, with his hair cut by the make-up department, a light-meter hung around his neck, Ion was unrecognisable as the bedraggled orphan who had come to her hotel room in Bucharest.

Kate laid aside her journal and pen.

"John Popp," Ion pronounced, tapping his chest. His J-sound was perfect. "John Popp, the American."

She thought about it.

Ion — no, John — had sloughed off his nationality and all national characteristics like a snake shedding a skin. New-born as an American, pink-skinned and glowing, he would never be challenged.

"Do you want to go to America?"

"Oh yes, Miss Reed. America is a young country, full of life. Fresh blood. There, one can be anything one chooses. It is the only country for a vampire."

Transylvania 6-5000
(USA/YUGOSLAVIA; Rudy De Luca)

Vampire Hunter D
(JAPAN; Tayoo Ashida; animated feature)

Vampires in Havana
(CUBA/SPAIN/WEST GERMANY; Juan Padron; a.k.a. *Vampiros en la Habana!;* animated feature)

Who Is Afraid of Dracula
(ITALY; Neri Parenti; orig.: *Fracchi contra Dracula, Fachia vs. Dracula*)

1986

Anemia
(ITALY; Alberto Abbruzzese)

The Close Encounter of the Vampire
(HONG KONG; Yuen Ping)

The Devil Vendetta
(THAILAND/HONG KONG; L. Chang-Xu)

Haunted Cop Shop 2
(HONG KONG; Jeff Lau)

Kung Fu Vampire Buster
(HONG KONG; Xen Lung Ting; orig.: *New Mr. Vampire*)

Little Shop of Horrors
(UK/USA; Frank Oz)

The Mark of Lilith
(UK; Bruna Fionda, Polly Gladwin, Islling Mack-Nataf; experimental short)

The Midnight Hour
(USA/TV; Jack Bender)

Mr. Vampire II
(HONG KONG; Sung Kan Shing)

Red and Black
(CHINA; Andrew Kam Yuen Wah)

The Seven Vampires
(BRAZIL; Ivan Cardoso; orig.: *As sete vampiros*)

Vamp
(USA; Richard Wenk)

Vampire's Breakfast
(HONG KONG; Wong Chung)

1987
Elusive Song of the Vampire
(TAIWAN; Takako Shira)

Graveyard Disturbance
(ITALY; Lamberto Bava; orig.: *Dentro il cimitrio*)

Graveyard Shift
(CANADA; Gerard Ciccoritti)

Jonathan of the Night
(USA; Buddy Giovinazzo; short subject)

The Lost Boys
(USA; Joel Schumacher)

Love Me, Vampire
(HONG KONG; Irene Wang)

The Monster Squad
(USA; Fred Dekker)

Mr. Vampire III
(HONG KONG; Wong Kee Hung)

My Best Friend Is a Vampire
(USA; Jimmy Huston)

Near Dark
(USA; Kathryn Bigelow)

One Eyebrow Priest
(HONG KONG; Mason Ching; orig.: *New Mr. Vampire 11*)

A Return to 'Salem's Lot
(USA; Larry Cohen)

Toothless Vampires
(HONG KONG; Lee Hun Yu)

The Wicked
(AUSTRALIA; a.k.a. *Outback Vampires*)

Vampire at Midnight
(USA; Gregory McClatchy)

Vampire Knights
(USA; Daniel M. Peterson)

Vampires Live Again
(HONG KONG; Kam Yoo Tu; comedy video feature)

1988
Because the Dawn
(USA; Amy Goldstein)

Kate wasn't sure whether to feel sorry for the vampire youth or for the American continent. One of them was sure to be disappointed.

"John Popp," he repeated, pleased.

Was this how Dracula had been when he first thought of moving to Great Britain, then the liveliest country in the world just as America was now? The Count had practiced his English pronunciation in conversations with Jonathan, and memorised railway time-tables, relishing the exotic names of St Pancras, King's Cross and Euston. Had he rolled his anglicised name — Count DeVille — around his mouth, pleased with himself?

Of course, Dracula saw himself as a conqueror, the rightful ruler of all lands he rode over. Ion-John was more like the Irish and Italian emigrants who poured through Ellis Island at the beginning of the century, certain America was the land of opportunity and that each potato-picker or barber could become a self-made plutocrat.

Envious of his conviction, affection stabbing her heart, wishing she could protect him always, Kate kissed him. He struggled awkwardly, a child hugged by an embarrassing auntie.

24

Mists pool around Borgo Pass. Black crags project from the white sea.

The coach proceeds slowly. Everyone looks around, wary.

Murray: *Remember that last phial of laudanum… I just downed it.*

Westenra: *Good show, man.*

Murray: *It's like the Crystal Palace.*

Harker sits by Swales, looking up at the ancient castle that dominates the view. Broken battlements are jagged against the boiling sky.

Harker's Voice: *Castle Dracula. The trail snaked through the forest, leading me directly to him. The Count. The countryside* was *Dracula. He had become one with the mountains, the trees, the stinking earth.*

The coach halts. Murray pokes his head out of the window, and sighs in amazement.

Swales: *Borgo Pass, Harker. I'll go no further.*

Harker looks at Swales. There is no fear in the coachman's face, but his eyes are slitted.

A sliver of dark bursts like a torpedo from the sea of mist. A sharpened stake impales Swales, bloody point projecting a foot or more from his chest.

Swales sputters hatred and takes a grip on Harker, trying to hug him, to pull him onto the sharp point sticking out of his sternum.

Harker struggles in silence, setting the heel of his hand against Swales's head. He pushes and the dead man's grip relaxes. Swales tumbles from his seat and rolls off the precipice, falling silently into the mists.

Murray: *Good grief, man. That was extreme.*

25

Rising over Borgo Pass was Castle Dracula. Half mossy black stone, half fresh orange timber.

Kate was impressed.

Though the permits had still not come through, Francis had ordered the crew to erect and dress the castle set. This was a long way from Bucharest and without Georghiou, the hand of Ceausescu could not fall.

From some angles, the castle was an ancient fastness, a fit lair for the vampire King. But a few steps off the path and it was a shell, propped up by timbers. Painted board mingled with stone.

If Meinster's Kids were in the forests, they could look up at the mountain and take heart. This sham castle might be their rallying-point. She hummed "Paper Moon," imagining vampires summoned back to these mountains to a castle that was not a castle and a king who was just an actor in greasepaint.

A grip, silhouetted in the gateway, used a gun-like device to wisp thick cobweb on the portcullis. Cages of imported vermin were stacked up, ready to be unloosed. Stakes, rigged up with bicycle seats that would support the impaled extras, stood on the mountainside.

It was a magnificent fake.

Francis, leaning on his stake, stood and admired the edifice thrown up on his orders. Ion-John was at his side, a faithful Renfield for once.

"Orson Welles said it was the best train set a boy could have," Francis said. Ion probably didn't know who Welles was. "But it broke him in the end."

In her cardigan pocket, she found the joke shop fangs from the 100th Day of Shooting Party. Soon, there would be a 200th Day Party.

She snapped the teeth together like castanets, feeling almost giddy up here in the mists where the air was thin and the nights cold.

In her pleasant contralto, far more Irish-inflected than her speaking voice, she crooned "it's a Barnum and Bailey world, just as phoney as it can be, but it wouldn't be make-believe if you believed in me."

26

On foot, Harker arrives at the gates of the castle. Westenra and Murray hang back a little way.

A silent crowd of gypsies parts to let the Englishmen through. Harker notices human and wolf teeth strung in necklaces, red eyes and feral fangs, withered bat-membranes curtaining under arms, furry bare feet hooked into the rock. These are the Szgany, the children of Dracula.

In the courtyard, an armadillo noses among freshly-severed human heads. Harker is smitten by the stench of decay but tries to hide his distaste. Murray and Westenra groan and complain. They both hold out large crucifixes.

A rat-like figure scuttles out of the crowds.

Beverly Hills Vamp
(USA; Fred Olen Ray)

Chillers
(USA; Daniel Boyd)

Dance of the Damned
(USA; Katt Shea Ruben)

Dinner with the Vampire
(ITALY; Lamberto Bava)

Dracula's Widow
(USA; Christopher Coppola)

Fright Night Part II
(USA; Tommy Lee Wallace; French release title: *Vampire? Avez vous-dit vampire?*)

Howl of the Devil
(SPAIN; Paul Naschy; orig.: *El aullido del diablo*)

The Jitters
(CANADA/USA/JAPAN; John M. Fasano)

The Kiss
(USA/CANADA; Pen Densham)

The Lair of the White Worm
(UK; Ken Russell)

Midnight
(USA; Thaddeus Vane)

Mr. Vampire IV
(HONG KONG; Law Lit)

The Mysterious Death of Nina Cherecau
(USA/BELGIUM; Dennis Berry)

Nosferatu in Venice
(ITALY; Augusto Caminito; orig.: *Nosferatu a Venezia;* a.k.a. *Vampire in Venice, Vampires in Venice*)

Not of This Earth
(USA; Jim Wynorski)

Scooby-Doo and the Reluctant Werewolf
(USA; Ray Patterson; animated feature)

Sundown: The Vampire in Retreat
(USA; Anthony Hickox)

Teen Vamp
(USA; Samuel Bradford)

Renfield: ***Are you English? I'm an Englishman. R.M. Renfield, at your service.***

He shakes Harker's hand, then hugs him. His eyes are jittery, mad.

Renfield: ***The Master has been waiting for you. I'm a lunatic, you know. Zoophagous. I eat flies. Spiders. Birds, when I can get them. It's the blood. The blood is the life, as the book says. The Master understands. Dracula. He knows you're coming. He knows everything. He's a poet-warrior in the classical sense. He has the vision. You'll see, you'll learn. He's lived through the centuries. His wisdom is beyond ours, beyond anything we can imagine. How can I make you understand? He's promised me lives. Many lives. Some nights, he'll creep up on you, while you're shaving, and break your mirror. A foul bauble of man's vanity. The blood of Attila flows in his veins. He is the Master.***

Renfield plucks a crawling insect from Westenra's coat and gobbles it down.

Renfield: ***I know what bothers you. The heads. The severed heads. It's his way. It's the only language they understand. He doesn't love these things, but he knows he must do them. He knows the truth. Rats! He knows where the rats come from. Sometimes, he'll say "They fought the dogs and killed the cats and bit the babies in the cradles, and ate the cheeses out of the vats and licked the soup from the cooks' own ladles."***

Harker ignores the prattle and walks across the courtyard. Scraps of mist waft under his boots.

A huge figure fills a doorway. Moonlight shines on his great, bald head. Heavy jowls glisten as a humourless smile discloses yellow eye-teeth the size of thumbs.

Harker halts.

A bass voice rumbles.

Dracula: ***I... am... Dracula.***

27

Francis had first envisioned Dracula as a stick-insect skeleton, dried up, hollow-eyed, brittle. When Brando arrived on set, weighing in at 250 pounds, he had to rethink the character as a blood-bloated leech, full to bursting with stolen life, overflowing his coffin.

For two days, Francis had been trying to get a usable reading of the line "I am Dracula." Kate, initially as thrilled as anyone else to see Brando at work, was bored rigid after numberless mumbled retakes.

The line was written in three-foot tall black letters on a large piece of cardboard held up by two grips. The actor experimented with emphases, accents, pronunciations from "Dorragulya" to "Jacoolier." He read the line looking away from the camera and peering straight at the lens. He tried it with false fangs inside his mouth, sticking out of his mouth, shoved up his nostrils or thrown away altogether.

To Die For
(USA; Deran Sarafian)

The Understudy: Graveyard Shift II
(CANADA; Gerard Ciccoritti)

Vampire Princess Miyu
(JAPAN; Hirano Toshihiro; animated feature)

Vampires
(USA; Len Anthony; a.k.a. *Abadou*)

Vampires on Bikini Beach
(USA)

Vampires Strike Back
(HONG KONG; Kam Yoo Tu; video feature)

Waxwork
(USA; Anthony Hickox)

1989

Daughhter of Darkness
(USA/TV; Stuart Gordon)

Fright House
(USA; Len Anthony; incorporating *Vampires* [1988])

The Lost Platoon
(USA; David A. Prior)

Magic Cop
(HONG KONG; H. Ching; a.k.a. *Mr. Vampire V*)

Mom
(USA; Patrick Rand)

Nick Knight
(USA/TV; Farhad Mann)

Nightlife
(MEXICO/USA/TV; Daniel Taplitz)

Pale Blood
(USA; V. V. Dachin)

Spirit vs. Zombi
(TAIWAN/HONG KONG; Yao Fenpan)

Spooky Family
(HONG KONG; Law Lit)

Transylvania Twist
(USA; Jim Wynorski)

Vampire's Kiss
(USA; Robert Bierman)

Once he came out with a bat tattooed on his bald head in black lipstick. After considering it for a while, Francis ordered the decal wiped off. You couldn't say that the star wasn't bringing ideas to the production.

For two hours now, Brando had been hanging upside-down in the archway, secured by a team of very tired technicians at the end of two guy-ropes. He thought it might be interesting if the Count were discovered like a sleeping bat.

Literally, he read his line upside-down.

Marty Sheen, over whose shoulder the shot was taken, had fallen asleep.

"I am Dracula. I am Dracula. I am Dracula. I am Dracula.

"I am Dracula! I am Dracula?

"Dracula am I. Am I Dracula? Dracula I am. I Dracula am. Am Dracula I?

"I'm Dracula.

"The name's Dracula. Count Dracula.

"Hey, I'm Dracula.

"Me . . . Dracula. You . . . liquid lunch."

He read the line as Stanley Kowalski, as Don Corleone, as Charlie Chan, as Jerry Lewis, as Laurence Olivier, as Robert Newton.

Francis patiently shot take after take.

Dennis Hopper hung around, awed, smoking grass. All the actors wanted to watch.

Brando's face went scarlet. Upside-down, he had problems with the teeth. Relieved, the grips eased up on the ropes and the star dropped towards the ground. They slowed before his head cracked like an egg on the ground. Assistants helped him rearrange himself.

Francis thought about the scene.

"Marlon, it seems to me that we could do worse than go back to the book."

"The book?" Brando asked.

"Remember, when we first discussed the role. We talked about how Stoker describes the Count."

"I don't quite ..."

"You told me you knew the book."

"I never read it."

"You said..."

"I lied."

28

Harker, in chains, is confined in a dungeon. Rats crawl around his feet. Water flows all around.

A shadow passes.

Harker looks up. A gray bat-face hovers above, nostrils elaborately frilled, enormous teeth locked. Dracula seems to fill the room, black cape stretched over his enormous belly and trunk-like limbs.

1990

The Arrival
(USA; David Schmoeller)

Baby Blood
(FRANCE; Alain Robak)

Bandh Darwaza
(INDIA; Tulsi Ramsay and Shyam Ramsay)

Banglow 666
(INDIA)

Crazy Safari
(HONG KONG; Lo Weng-Tung)

Dawn
(UK; Niall Johnson; video feature)

Def by Temptation
(USA; James Bond III)

Encounters of the Spooky Kind II
(HONG KONG; Sammo Hung)

First Vampire in China
(HONG KONG; Yam Chun-Lu)

Howling VI: The Freaks
(USA/UK; Hope Perello)

I Bought a Vampire Motorcycle
(UK; Dirk Campbell)

La Maschera del Demonio
(ITALY; Lamberto Bava)

Red Blooded American Girl
(CANADA; David Blyth)

The Reflecting Skin
(USA/CANADA; Philip Ridley)

Rockula
(USA; Luca Bercovici)

Streets
(USA; Katt Shea Ruben)

Subspecies
(USA/ROMANIA; Ted Nicolaou)

Vampire Buster
(HONG KONG; Norman Law and Stanley Siu)

Vampire Cop
(USA; Donald Farmer)

Vampyre
(USA; Bruce G. Hallenbeck; video feature)

Witchcraft III: The Kiss of Death
(USA; R. L. Tillmann)

1991
Blood-Suckers
(RUSSIA; E. Tatarskiy; a.k.a. *Those Feed on Blood*)

Demon Cop
(USA; Rocco Karega)

Doctor Vampire
(HONG KONG; Q. Xen Lee)

The Malibu Beach Vampires
(USA; Francis Creighton)

Moon Legend
(HONG KONG; Joey Wang)

My Grandpa Is a Vampire
(NEW ZEALAND; David Blyth; a.k.a. *Moonrise, Vampire*)

My Lovely Monster
(GERMANY; Michel Bergmann)

Spooky Family 2
(HONG KONG; Law Lit)

Trilogy of Fear
(USA; Richard L. Fox, Jr.)

Ultimate Vampire
(HONG KONG; Andrew Lau)

Undying Love
(USA; Greg Lambertson)

Valerie
(USA; Jay Lind)

Vampire Trailer Park
(USA; Steve Latshaw)

Vampire's Embrace
(USA; Glen Andreiev)

Vampires Settle on Police Camp
(HONG KONG; Lo Wei Lang)

Waxwork II: Lost in Time
(USA; Anthony Hickox)

1992
Back to the USSR
(FINLAND; Jari Halonen)

Bloodlust
(AUSTRALIA; Richard Wolstencroft and John Hewitt)

Dracula drops something into Harker's lap. It is Westenra's head, eyes white.

Harker screams.

Dracula is gone.

29

An insectile clacking emerged from the Script Crypt, the walled-off space on the set where Francis had hidden himself away with his typewriter.

Millions of dollars poured away daily as the director tried to come up with an ending. In drafts Kate had seen — only a fraction of the attempts Francis had made — Harker killed Dracula, Dracula killed Harker, Dracula and Harker became allies, Dracula and Harker were both killed by Van Helsing (unworkable, because Robert Duvall was making another film on another continent), lightning destroyed the whole castle.

It was generally agreed that Dracula should die.

The Count perished through decapitation, purifying fire, running water, a stake through the heart, a hawthorn bush, a giant crucifix, silver bullets, the hand of God, the claws of the Devil, armed insurrection, suicide, a swarm of infernal bats, bubonic plague, dismemberment by axe, permanent transformation into a dog.

Brando suggested that he play Dracula as a Green Suitcase.

Francis was on medication.

30

"Reed, what does he mean to you?"

She thought Francis meant Ion-John.

"He's just a kid, but he's getting older fast. There's something…"

"Not John. Dracula."

"Oh, him."

"Yes, him. Dracula. Count Dracula. King of the Vampires."

"I never acknowledged that title."

"In the 1880s, you were against him?"

"You could say that."

"But he gave you so much, eternal life?"

"He wasn't my father. Not directly."

"But he brought vampirism out of the darkness."

"He was a monster."

"Just a monster? In the end, just that?"

She thought hard.

"No, there was more. He was more. He was … he is, you know … big. Huge, enormous. Like the elephant described by blind men. He had many aspects. But all were monstrous. He didn't bring us out of the darkness. He was the darkness."

"John says he was a national hero."

"John wasn't born then. Or turned."

"Guide me, Reed."

"I can't write your ending for you."

31

At the worst possible time, the policeman was back. There were questions about Shiny Suit. Irregularities revealed by the autopsy.

For some reason, Kate was questioned.

Through an interpreter, the policemen kept asking her about the dead official, what had their dealings been, whether Georghiou's prejudice against her kind had affected her.

Then he asked her when she had last fed, and upon whom?

"That's private," she said.

She didn't want to admit that she had been snacking on rats for months. She'd had no time to cultivate anyone warm. Her powers of fascination were thinning.

A scrap of cloth was produced and handed to her.

"Do you recognise this?" she was asked.

It was filthy, but she realised that she did.

"Why, it's my scarf. From Biba. I . . . "

It was snatched away from her. The policeman wrote down a note.

She tried to say something about Ion, but thought better of it. The translator told the policeman Kate had almost admitted to something.

She felt distinctly chilled.

She was asked to open her mouth, like a horse up for sale. The policeman peered at her sharp little teeth and tutted.

That was all for now.

32

"How are monsters made?"

Kate was weary of questions. Francis, Marty, the police. Always questions.

Still, she was on the payroll as an advisor.

"I've known too many monsters, Francis. Some were born, some were made all at once, some were eroded, some shaped themselves, some twisted by history."

"What about Dracula?"

"He was the monster of monsters. All of the above."

Francis laughed.

"You're thinking of Brando."

"After your movie, so will everybody else."

He was pleased by the thought.

"I guess they will."

"You're bringing him back. Is that a good idea?"

"It's a bit late to raise that."

"Seriously, Francis. He'll never be gone, never be forgotten. But your

Bloodthirsty
(USA; Robert Guy Barrows)

Bram Stoker's Dracula
(USA; Francis Ford Coppola)

Buffy the Vampire Slayer
(USA; Fran Rubel Kuzui)

Children of the Night
(USA; Tony Randel)

Cronos
(MEXICO; Guillermo del Toro)

Darkness
(USA; Leif Jonker)

Dracula Rising
(USA; Fred Gallo)

Dracula's Hair
(RUSSIA; Vadim Prodan)

Innocent Blood
(USA; John Landis)

In the Midnight Hour
(USA; Joel Bender)

Project Vampire
(USA; Peter Flynn)

The Reluctant Vampire
(USA; Malcolm Marmorstein)

Sherlock Holmes: The Last Vampyre
(UK; Tim Sullivan)

Sleepwalkers
(USA; Mick Garris)

Tale of a Vampire
(UK/JAPAN; Shimako Sato)

To Sleep with a Vampire
(USA; Adam Friedman)

Un vampire au paradis
(FRANCE; Abdelkrim Bahloul)

The Vampire's Rope
(RUSSIA; Nikolai Gibu)

The Vampyr: A Soap Opera
(UK/TV; Nigel Finch)

1993

Bloodlust: Subspecies III
(USA; Ted Nicolaou)

Bloodstone: Subspecies II
(USA; Ted Nicolaou)

Dracula will be powerful. In the next valley, people are fighting over the tatters of the old, faded Dracula. What will your Technicolor, 70 mm, Dolby stereo Dracula *mean*?"

"Meanings are for the critics."

33

Two Szgany gypsies throw Harker into the great hall of the castle. He sprawls on the straw-covered flagstones, emaciated and wild-eyed, close to madness.

Dracula sits on a throne which stretches wooden wings out behind him. Renfield worships at his feet, tongue applied to the Count's black leather boot. Murray, a blissful smile on his face and scabs on his neck, stands to one side, with Dracula's three vampire brides.

Dracula: *I bid you welcome. Come safely, go freely and leave some of the happiness you bring.*

Harker looks up.

Harker: *You... were a Prince.*

Dracula: *I am a Prince still. Of Darkness.*

The brides titter and clap. A look from their Master silences them.

Dracula: *Harker, what do you think we are doing here, at the edge of Christendom? What dark mirror is held up to our unreflecting faces?*

By the throne is an occasional table piled high with books and periodicals. *Bradshaw's Guide to Railway Timetables in England, Scotland and Wales*, George and Weedon Grossmith's *Diary of a Nobody*, Sabine Baring-Gould's *The Book of Were-Wolves*, Oscar Wilde's *Salomé.*

Dracula picks up a volume of the poetry of Robert Browning.

Dracula: *"I must not omit to say that in Transylvania there's a tribe of alien people that ascribe the outlandish ways and dress on which their neighbours lay such stress, to their fathers and mothers having risen out of some subterraneous prison into which they were trepanned long time ago in a mighty band out of Hamelin town in Brunswick land, but how or why, they don't understand."*

Renfield claps.

Renfield: *Rats, Master. Rats.*

Dracula reaches down with both hands and turns the madman's head right around. The brides fall upon the madman's twitching body, nipping at him greedily before he dies and the blood spoils.

Harker looks away.

34

At the airport, she was detained by officials. There was some question about her passport.

Francis was worried about the crates of exposed film. The negative was precious, volatile, irreplaceable. He personally, through John, argued with

Midnight Kiss
(USA; Joel Bender)

Robo Vampire
(HONG KONG; Joe Livingstone)

1994
Interview with the Vampire
(USA; Neil Jordan)

Nadja
(USA; Michael Almereyda)

The Vampire Conspiracy
(USA; Geoffrey De Valois)

1995
The Addiction
(USA; Abel Ferrara)

Dracula: Dead and Loving It
(USA; Mel Brooks)

The Nosferatu Diaries
(USA)

Vampire in Brooklyn
(USA; Wes Craven)

NOTABLE RECENT FILMS

1996
From Dusk till Dawn
(USA; Robert Rodriguez)

1997
Habit
(USA; Larry Fessenden)

Vampire Journals
(ROMANIA/USA; Ted Nicolaou)

1998
From Dusk till Dawn: The Hangman's Daughter
(USA; P. J. Pesce)

From Dusk till Dawn 2: Texas Blood Money
(SOUTH AFRICA/USA; Scott Spiegel)

Vampires
(USA/MEXICO; John Carpenter)

2000
Dracula 2000
(USA; Patrick Lussier)

Shadow of the Vampire
(USA/LUXEMBOURG; E. Elias Merhige)

the customs people and handed over disproportionate bribes. He still carried his staff, which he used to point the way and rap punishment. He looked a bit like Friar Tuck.

The film, the raw material of *Dracula,* was to be treated as if it were valuable as gold and dangerous as plutonium. It was stowed on the aeroplane by soldiers.

A blank-faced woman sat across the desk from Kate.

The stirrings of panic ticked inside her. The scheduled time of departure neared.

The rest of the crew were lined up with their luggage, joking despite tiredness. After over a year, they were glad to be gone for good from this backward country. They talked about what they would do when they got home. Marty Sheen was looking healthier, years younger. Francis was bubbling again, excited to be on to the next stage.

Kate looked from the Romanian woman to the portraits of Nicolae and Elena on the wall behind her. All eyes were cold, hateful. The woman wore a discreet crucifix and a Party badge clipped to her uniform lapel.

A rope barrier was removed and the eager crowd of the *Dracula* company stormed towards the aeroplane, mounting the steps, squeezing into the cabin.

The flight was for London, then New York, then Los Angeles. Half a world away.

Kate wanted to stand up, to join the plane, to add her own jokes and fantasies to the rowdy chatter, to fly away from here. Her luggage, she realised, was in the hold.

A man in a black trenchcoat (*Securitate?*) and two uniformed policemen arrived and exchanged terse phrases with the woman.

Kate gathered they were talking about Shiny Suit. And her. They used old, cruel words: leech, *nosferatu,* parasite. The *Securitate* man looked at her passport.

"It is impossible that you be allowed to leave."

Across the tarmac, the last of the crew — Ion-John among them, baseball cap turned backwards, bulky kit-bag on his shoulder — disappeared into the sleek tube of the aeroplane. The door was pulled shut.

She was forgotten, left behind.

How long would it be before anyone noticed? With different sets of people debarking in three cities, probably forever. It was easy to miss one mousy advisor in the excitement, the anticipation, the triumph of going home with the movie shot. Months of post-production, dialogue looping, editing, rough cutting, previews, publicity and release lay ahead, with box office takings to be crowed over and prizes to be competed for in Cannes and on Oscar night.

Maybe when they came to put her credit on the film, someone would think to ask what had become of the funny little old girl with the thick glasses and the red hair.

THE REAL COPPOLA'S DRACULA

Francis Ford Coppola's $40 million screen extravaganza *Bram Stoker's Dracula* (1992) was the latest in a long string of films that have appropriated Stoker's novel and title character for highly idiosyncratic exercises in filmmaking. Therefore, it was not surprising that the Coppola production emerged as an essential subversion of the 1897 text; what was surprising was the uncritical way that journalists and critics alike swallowed the studio's laughable assertion that *Bram Stoker's Dracula* was the first truly faithful adaptation of the book — a classic love story, no less.

Now, whatever *Dracula* is, it is hardly a love story; it is difficult, in fact, to imagine a more antiromantic narrative. Stoker's Dracula is a cunning Darwinian superman who never seduces — he simply seizes. And while he grows younger as he drinks blood, he never becomes attractive in any physical sense. *Dracula*'s sexuality, or implied sexuality, is both rancid and repellent. The story's nightmarish power derives in large part from the tension between between a highly controlled (i.e., "civilized") Victorian surface narrative clashing with a raging subtext of unsublimated animalism.

Following a string of expensive flops, Coppola was ready to reestablish his credibility with the Hollywood mainstream by producing an on-time, on-budget commercial film for a major studio. James V. Hart's screenplay, originally titled *Dracula: The Untold Story*, was stalled in TV-movie limbo when actress Winona Ryder brought it to Coppola's attention. In one sense, Hart's script was indeed "untold"— by Bram Stoker, at least, since Hart replaced Stoker's ravenous monster with a lovesick revenant

"You are a sympathiser with the Transylvania Movement."

"Good God," she blurted, "why would anybody want to live here?"

That did not go down well.

The engines were whining. The plane taxied towards the runway.

"This is an old country, Miss Katharine Reed," the *Securitate* man sneered. "We know the ways of your kind, and we understand how they should be dealt with."

All the eyes were pitiless.

35

The giant black horse is lead into the courtyard by the gypsies. Swords are drawn in salute to the animal. It whinnies slightly, coat glossy ebony, nostrils scarlet.

Inside the castle, Harker descends a circular stairway carefully, wiping aside cobwebs. He has a wooden stake in his hands.

The gypsies close on the horse.

Harker's Voice: *Even the castle wanted him dead, and that's what he served at the end. The ancient, blood-caked stones of his Transylvanian fastness.*

Harker stands over Dracula's coffin. The Count lies, bloated with blood, face puffy and violet.

Gypsy knives stroke the horse's flanks. Blood erupts from the coat.

Harker raises the stake with both hands over his head.

Dracula's eyes open, red marbles in his fat, flat face. Harker is given pause.

The horse neighs in sudden pain. Axes chop at its neck and legs. The mighty beast is felled.

Harker plunges the stake into the Count's vast chest.

The horse jerks spastically as the gypsies hack at it. Its hooves scrape painfully on the cobbles.

A gout of violently red blood gushes upwards, splashing directly into Harker's face, reddening him from head to waist. The flow continues, exploding everywhere, filling the coffin, the room, driving Harker back.

Dracula's great hands grip the sides of the coffin and he tries to sit. Around him is a cloud of blood droplets, hanging in the air like slo-mo fog.

The horse kicks its last, clearing a circle. The gypsies look with respect at the creature they have slain.

Harker takes a shovel and pounds at the stake, driving it deeper into Dracula's barrel chest, forcing him back into his filthy sarcophagus.

At last, the Count gives up. Whispered words escape from him with his last breath.

Dracula: *The horror... the horror...*

seeking the reincarnation of his fourteenth-century love. The theme was extraordinarily derivative; it had been used by screenwriter Richard Matheson in his 1973 television adaptation of *Dracula*, as an ongoing plot device in the sixties TV soap opera "Dark Shadows," and even in the exploitation flick *Blacula* (1972). And the screen inspiration for all these reincarnation tales was *The Mummy* (1932), with Boris Karloff pursuing a parallel romance across the millennia.

***Bram Stoker's Dracula* was promoted like a steamroller by Columbia Pictures, and had the most extensive merchandising tie-ins of any film before *Jurassic Park*. As a result, there was almost no independent, intelligent reporting on the film's evolution, just an avalanche of sycophantic puff pieces and coffee table books. In the major media, only *Newsweek* called the film's bluff, going so far as to run parallel texts from the novel and the screenplay to reveal the 180-degree switch in sensibility.**

The casting, to say the least, was odd. Gary Oldman as Dracula (succeeding the previously considered — or at least rumored to be considered — Andy Garcia) is more pixie-ish than princely; neither Winona Ryder nor Keanu Reeves as the young lovers is convincingly British; Tom Waits's Renfield (like Oldman) is often vocally incomprehensible; and Anthony Hopkins as a near-crackpot Van Helsing seems manically adrift, almost undirected. The operatically-inclined costumes by Eiko Ishioka are, on the whole, very impressive — but where is the opera? Not every sartorial concept works, however; Lucy Westenra's huge-collared dress was supposedly inspired by the anatomy of a frilled lizard, but creates a far more bizarre effect: the Victorian virgin served up as a white cheese pizza.

36

She supposed there were worse places than a Romanian jail. But not many.

They kept her isolated from the warm prisoners. Rapists and murderers and dissidents were afraid of her. She found herself penned with uncommunicative Transylvanians, haughty elders reduced to grime and resentful new-borns.

She had seen a couple of Meinster's Kids, and their calm, purposeful, blank-eyed viciousness disturbed her. Their definition of enemy was terrifyingly broad, and they believed in killing. No negotiation, no surrender, no accommodation. Just death, on an industrial scale.

The bars were silver. She fed on insects and rats. She was weak.

Every day, she was interrogated.

They were convinced she had murdered Georghiou. His throat had been gnawed and he was completely exsanguinated.

Why her? Why not some Transylvanian terrorist?

Because of the bloodied once-yellow scrap in his dead fist. A length of thin silk, which she had identified as her Biba scarf. The scarf she had thought of as civilisation. The bandage she had used to bind Ion's wound.

She said nothing about that.

Ion-John was on the other side of the world, making his way. She was left behind in his stead, an offering to placate those who would pursue him. She could not pretend even to herself that it was not deliberate. She understood all too well how he had survived so many years underground. He had learned the predator's trick: to be loved, but never to love. For that, she pitied him even as she could cheerfully have torn his head off.

There were ways out of jails. Even jails with silver bars and garlic hung from every window. The Romanian jailers prided themselves on knowing vampires, but they still treated her as if she were feeble-minded and fragile.

Her strength was sapping, and each night without proper feeding made her weaker.

Walls could be broken through. And there were passes out of the country. She would have to fall back on skills she had thought never to exercise again.

But she was a survivor of the night.

As, quietly, she planned her escape from the prison and from the country, she tried to imagine where the "Son of Dracula" was, to conceive of the life he was living in America, to count the used-up husks left in his wake. Was he still at his maestro's side, making himself useful? Or had he passed beyond that, found a new patron or become a maestro himself?

Eventually, he would build his castle in Beverly Hills and enslave a harem. What might he become: a studio head, a cocaine baron, a rock promoter, a media mogul, a star? Truly, Ion-John was what Francis had wanted of Brando, *Dracula* reborn. An old monster, remade for the new

The film operates like a broken, very expensive kaleidoscope, jamming image atop precious image until the whole thing ends up feeling disjointed and insubstantial. Of course, all the film's incongruities and flaws and superficiality were applauded by Coppola partisans as evidence of a brilliant "postmodernist" sensibility. The postmodern defense, not surprisingly, is the last refuge for almost anything these days that has no discernible point of view, borrows egregiously, or simply makes no sense.

Although the film contains some stunning pictorial compositions and set-pieces, they are usually self-conscious references to earlier filmmakers like Cocteau and Dreyer. If there's a real method to Coppola's madness, it remains maddeningly elusive.

***Bram Stoker's Dracula* was a resounding box office success, despite extremely mixed reviews. Here is a sampling of some of the dissenting voices that went unquoted in the film's ad campaign:**

***New York* magazine called *Bram Stoker's Dracula* "an unholy mess, a bombastic kitschfest of whirling, decomposing photography, writhing women, and spurting blood."**

In the *New Yorker*, Terrence Rafferty slammed the studio spin as "an ingenious attempt to duck responsibility for what must be the most aggressively silly version of this famous story ever committed to film." According to Rafferty, the "movie isn't very scary (except, perhaps, as an example of what Hollywood executives will give people tremendous amounts of money to do)." The film, Rafferty wrote, "just keeps coming at us, indefatigably, unstoppably, as if it were pursuing us through

world and the next century, meaning all things, tainting everything he touched.

She would leave him be, this new monster of hers, this creature born of Hollywood fantasy and her own thoughtless charity. With *Dracula* gone or transformed, the world needed a fresh monster. And John Popp would do as well as anyone else. The world had made him and it could cope with him.

Kate extruded a fingernail into a hard, sharp spar, and scraped the wall. The stones were solid, but between them was old mortar, which crumbled easily.

37

Harker, face still red with Dracula's blood, is back in his room at the inn in Bistritz. He stands in front of the mirror.

Harker's Voice: *They were going to make me a saint for this, and I wasn't even in their fucking church any more.*

Harker looks deep into the mirror.

He has no reflection.

Harker's mouth forms the words, but the voice is Dracula's.

The horror... the horror...

eternity, and it leaves us feeling mysteriously drained."

Kenneth Turan, film critic for the *Los Angeles Times*, noted that "given the fact that almost none of these people look comfortable in their period costumes or sound at ease in their earnest accents, the result tends to feel like a $40 million high school play."

Coppola's real attitude toward Stoker may be revealed in his voice-over commentary on the Criterion Collection disc of the film, when he tells us, "Very few people have gotten through the book, if truth be known...it's very hard going...."

For Coppola and company — obviously.

Stoker's Dracula, *or Coppola's? The 1992 movie tie-in edition of the original novel. (New American Library)*

Bela's Plot

CAITLÍN R. KIERNAN
(1997)

HOLLYWOOD IS A VAMPIRE, Magwitch thinks again and sips at his sweet, iced coffee, coffee milk-faded the muddy color of rain water. He stares out at the heat shimmering up from the four-o'clock street, the asphalt that looks wet, that looks like a perfect stream of melted licorice beneath the too-blue sky. Stares out through his dime-store shades and the diner's smudge-tinted plateglass and wishes his nerves would stop jangling, humming livewire from the Ecstasy he dropped with Lark and Crispin the night before.

A very bedraggled hooker crosses Vine, limping slow on a broken heel, squinting at the ripening afternoon. Magwitch watches her, and wonders that she doesn't simply burst into white flame, doesn't dance a fiery jig past the diner window until there's only a little ash and singed wig scraps floating on the liquid blacktop. He closes his eyes and she's still there, and all the wriggling electric lines of static from his head.

When he opens them again, Tam's walking towards him between the tangerine booths, her long-legged, confident stride and layers of lace and ragged fishnet beneath the black vinyl jacket she almost always wears. He smiles and stirs at his coffee, rearranges ice cubes, and Tam draws attention from the fat waitress and the two old men at the counter as she comes. She sits down across from him, and he smells sweat and roses.

"Where's Lark?" she asks, and he hears the impatience rising in her voice, the crackling restlessness she wears more often even than the jacket.

"I told Crispin they could meet us at Holy Cross later on," and he watches her eyes, her gray-green eyes painted in bold Egyptian strokes of mascara and shadow, looks for some hint of reproach that her instructions have not been followed, that anything is different than she planned.

But Tam only nods her head, drums nervous insect sounds against the table with her sharp, black nails and then the fat waitress sets a tumbler of ice water down in front of her and Tam orders nothing. They wait until the woman leaves them alone again, wait until the two old men stop looking and go back to their pie wedges and folded newspaper browsing.

Tam takes a pink packet of saccharin, tears it open and pours the coke-fine powder into her water. It sifts down between the crushed ice, dissolves, and she stirs it hard with her spoon.

"Well?" and she tests the water with the tip of her tongue, frowns and dumps in another packet of Sweet'n Low. "Am I supposed to fucking guess, Maggie?"

ABOUT THE AUTHOR

Caitlín R. Kiernan's first novel, *Silk*, received both the Barnes and Noble Maiden Voyage and International Horror Guild awards for best first novel. Her short fiction has been selected for *The Year's Best Fantasy and Horror* and *The Mammoth Book of Best New Horror* and has been collected in *Candles for Elizabeth*, *Tales of Pain and Wonder*, and *From Weird and Distant Shores*. Trained as a vertebrate paleontologist, Caitlín divides her time between science and her writing. She lives in Birmingham, Alabama.

BELA LUGOSI'S DEAD

Although Bela Lugosi never actually requested to be buried in his Dracula cape (he was frankly terrified of death), his family nonetheless delivered one of his stage costumes to the Hollywood mortuary that prepared his body for the wake and funeral. The image of Lugosi fused in death with the character of Dracula was simultaneously poignant and somewhat bizarre; not surprisingly, it made headlines around the world.

While nothing so outrageous as the incident described in "Bela's Plot" has ever actually happened, eyebrows were definitely raised in the 1980s, when a mail order outfit began selling ghoulish soil samples purportedly taken from Lugosi's grave. The fact of Lugosi's demise and his singular style of interment were popularized by the short lived (1979-1983) but seminal goth band Bauhaus, whose first hit, "Bela Lugosi's Dead" served as its anthem and was featured on the soundtrack of the 1983 film *The Hunger*.

Today, Lugosi's grave is a well-trafficked tourist and unmolested destination at Holy Cross Cemetery in Culver City, California. Dracula's final resting place is adjacent to that of Bing Crosby, and nearby those of Rita Hayworth and Sharon Tate.

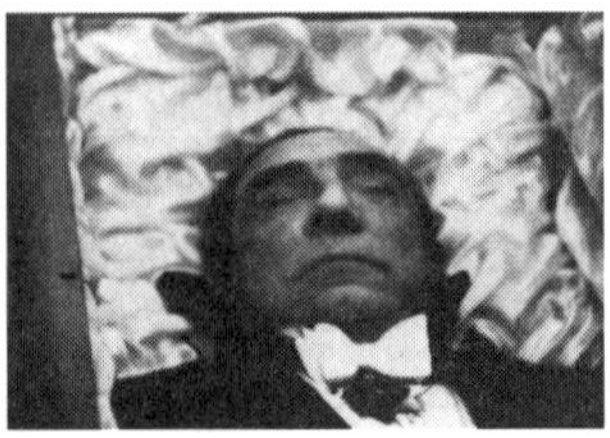

Bela Lugosi: the final curtain call.

"These guys are bad," he says, and he looks away from her mossy eyes, looks back out at the street, into the glare and broil; the prostitute has gone, vaporized, or maybe just limped away towards Hollywood Boulevard.

"These guys are bad motherfuckers," but Magwitch knows there's nothing he can say, and no point in trying, now that she's made up her mind.

You might have lied. You might have tried telling her you never found them, or that they laughed at you and threatened to kick your skinny ass if you ever came poking around again.

"And you're beginning to sound like a real pussy, Maggie." Tam raises the tumbler to her red lips, razored pout, and he watches her marble throat as she swallows. She leaves behind a lipsticky crescent on the clear plastic.

Magwitch sighs, brushes his long, inky bangs back from his face.

You might have told her to go to hell this time.

"Friday night, they'll be at Stigmata," he says, "and they said to bring the money with us if we want to talk."

"Then we're in," she says, and of course it's not a question; the certainty in her voice, the guarded hint of satisfaction at the corners of her painted mouth, make him cringe, down inside where Tam won't see.

And if he feels like a pussy it's only because he's the one that's done more than talk the talk, he's the one that sat very still and kept his eyes on the shit-stained men's room floor while Jimmy DeSade and the Gristle Twins laid down their gospel, the slippery rules that Tam had sent him off to learn.

"Did you drive?" she asks, and he knows that she's seen his rust-cruddy car parked on the street, baking beneath the July sun and gathering parking tickets like bright paper flies. Proves the point by not bothering to wait for his answer. "Christ, Maggie, I hope to fuck you got your AC fixed. I need to make some stops on the way."

Magwitch finishes his coffee, swallows a mouthful of creamslick ice, before he lays two dollars on the table and follows Tam out into the sun.

By the time they reach Slauson Avenue, twilight's gone and all the long shadows have run together into another sticky L.A. night. The gates of Holy Cross Cemetery are closed and locked, keeping the dead and living apart until dawn, but Tam slips the rent-a-cop at the gate twenty bucks and he lets them through on foot. Tam doesn't climb fences or walls, doesn't squeeze herself through chain link or risk glass-studded masonry, and he's seen her talk and bribe and blow her way into every boneyard in the Valley.

Magwitch walks three steps behind her, and their Doc Marten boots clock softly against the paved drive, not quite in step with one another and keeping unsteady time with the fading traffic sounds. A little way more and they leave the drive and start up the sloping lawn, the perfect grass muzzling their footsteps as they pass reflecting pools like mirrors in darkened rooms and the phony grotto and its marble virgin, the stone woman in her stone shawl who kneels forever before the virgin's downcast eyes.

The markers are laid out in tuxedo-neat rows of black and white, one after the next like fallen dominos, and they step past the Tin Man's grave, over the fallen Father of Dixieland Jazz and Mr. Bing Crosby, and there are Lark and Crispin, nervous eyesores, waiting alone with Bela.

They could be twins, Magwitch thinks, fingers this old and well-worn observation. And he remembers how much he loves them, beautiful and interchangeable brothers or sisters in their spider threads and white faces, realizes how afraid he is for them now.

Neither of them says anything, wait to hear in their silk-silent anxiety, and Magwitch steps forward to stand beside Tam. The toes of her boots almost touch the flat headstone and it's much too dark to read the few words cut deep into granite, "Beloved Father," dates of birth and death.

There should be candles, he thinks, because there have always been candles, cinnamon or rose or the warm scent of mystery spices and their faces in the soft amber flicker. Sometimes so many candles at once that he was afraid the cops would see and make the guards run them out. Enough candles that they could sit with Bela and read aloud, taking turns with Mary Shelley or Anne Rice or *The Lair of the White Worm.*

"It's all been arranged," Tam says, her voice big and cavalier in the dark. "Friday night we'll meet them at Stigmata and work out the details."

"Oh," Lark says and this one syllable seems to drift from somewhere far away; Crispin breathes in loudly and stares down at his feet.

"Jesus, you're *all* a bunch of pussies, aren't you?"

Crispin does not raise his head, does not risk her eyes, and when he speaks, it's barely a hoarse whisper.

"Tam, how do you even know that Jimmy DeSade is telling the truth about the cape?"

And Magwitch feels her tense, then, feels the anger gathering itself inside her, sleek anger and insult taut as jungle cat muscles and "She doesn't," he answers, and a heartbeat later, "Neither do I."

"Jesus Christ," Tam hisses and lights one of the pungent brown Indonesian cigarettes she smokes. For an instant, her high cheekbones and pouty lips are nailed in the glow of her lighter and then the night rushes back over them.

"If you guys fuck this up," but her voice trails away into a ghost cloud of smoke and they stand together at the old man's grave, junkie old man cold and in his final casket almost twenty years before Magwitch was born. And there is no thunder and forked lightning, no wolf howling wind, only sirens and the raw squeal of tires on the cooling city streets.

Magwitch sits, smoking, on one rumpled corner of Tam's bed in the apartment he could never dream of affording, that she could never afford either if it was only her part-time at Retail Slut paying the rent. If it weren't for the checks from her mother and father that she pretends she doesn't get once a month, faithful as her period. The sun's coming up outside, the earth rolling around to burn again, and he can see the faintest graying blue

CLOAKS AND CAPES

Where would a vampire be without its black velvet cloak? This most evocative of garments represents concealment, darkness, the secrets and terrors of the night itself. Spread wide, it suggests the wings of a bat, the promise of an exhilarating flight from ordinary human constraints. Capes and cloaks have been associated with theatrical vampires since the 1820s, when the creatures first became stock figures of melodrama and opera. In the twentieth century, the cape most closely associated with the character of Dracula is an opera cloak lined in black satin, especially characterized by a big stand-up collar. The collar was introduced in 1924 by playwright Hamilton Deane to facilitate a stage illusion; in order for Dracula to "vanish" before the eyes of the audience, escaping through a trapdoor or secret panel, it was necessary to fit the cape with a collar large enough to conceal the actor's head when he turned his back to the audience. Such a collar had no real usefulness in the movies, where trick photography could be employed, but the upended collar was so striking visually that it became a permanent fixture of vampire costuming in all media.

Varney the Vampyre shows off his cloak of darkness.

EDITOR'S CHOICE: RECOMMENDED READING

Beyond the books already excerpted in the primary text or cited in sidebar commentary, the editor wishes to recommend a variety of other titles for further reading and research. These selections represent the editor's taste only, but all should point the reader in a variety of provocative directions as he or she continues an exploration of the vampire realm. Happy hunting!

GENERAL VAMPIRE NONFICTION

Auerbach, Nina. *Our Vampires, Ourselves* (Chicago and London: University of Chicago Press, 1995).

Barber, Paul. *Vampires, Burial and Death* (New Haven and London: Yale University Press, 1988).

Bunson, Matthew. *The Vampire Encyclopedia* (New York: Crown, 1993).

Carter, Margaret L. *The Vampire in Literature: A Critical Bibliography* (Ann Arbor, MI: U.M.I. Research Press, 1989).

Copper, Basil. *The Vampire: In Legend, Fact and Art* (London: Hale, 1973).

Dundes, Alan. ed. *The Vampire: A Casebook* (Madison, WI: University of Wisconsin Press, 1998).

Guiley, Rosemary Ellen. *The Complete Vampire Companion* (New York: MacMillan USA, 1994).

Melton, J. Gordon. *The Vampire Book* (Detroit and London: Visible Ink Press, 1999).

rind around the edges of Tam's thick velvet drapes. The only other light comes from her big television, sleek black box and the sound down all the way, Catherine Deneuve and David Bowie at piano and cello.

"I wouldn't even have *thought* of suggesting it," she says, and it takes him a moment to find his place in the conversation, his place between her words. "Not if they weren't acting that way. If they weren't both scared totally shitless."

"It should be their call," he says softly, stubs out his cigarette. He doesn't want the twins along either, wants them clear of this so bad that he's almost willing to hurt them. But almost isn't ever enough, and no matter what Tam might say, he knows that she'll have them there, lovely china pawns by her side, lovelier for the fear just beneath their skin.

"Oh," she says. "Absolutely. I just won't have them going there and fucking everything up, Maggie. This is for real, not cemetery games, not trick or fucking treat."

Magwitch lies back on the cool sheets, satin like slippery midnight skin, and tries not to think about the twins or Jimmy DeSade. Watches Tam at her cluttered antique dressing table, the back of her head, hair blacker than the sheets and her face reflected stark in the Art Deco mirror. Knowing he could hate her if only he were a little bit stronger, if he'd never let himself love her. She selects a small brush and retraces the cupid's bow of her upper lip, her perfect pout, and he looks away, stares up at the ceiling, the dangling forest of dolls hanging there, dolls and parts of dolls, ripe and rotting Barbie fruit strung on wire and twine. The breeze from the air conditioner stirs them, makes them sway, some close enough to bump into others, a leg against an arm, an arm against a plastic torso.

And then Tam comes teasing slow to him, crosses the room in nothing but white, white skin and crimson panties. She crouches on the foot of the bed, just out of reach, a living gargoyle down from Notre Dame, stone made flesh by some unlikely and unforgiving alchemy. Her eyes sparkle in the TV's electric glare, more hungry than John and Miriam Blaylock ever imagined; as she speaks, her left hand drifts absently to the tattoo that entirely covers her heart, the maroon and ebony petals that hide her nipple and areola, thorns that twine themselves tight around her breast and draw inky drops of blood.

"I'm sure Lark and Crispin won't disappoint me," she says and smiles like a wound. "They never do."

And of course she's holding the razor clutched in her right hand, the straight and silver razor with its mother-of-pearl handle, bottomless green and blue iridescence.

"And you won't ever disappoint me, sweet Maggie."

He sits up, no need for her to have to ask, turns his back and grips the iron headboard, palms around cold metal like prison bars. The old springs creak and groan gentle protests as she slips across the bed to him, and Magwitch feels her fingertips, Braille reading the scars on his shoulders, down the length of his spine, one for every night he's spent with her.

"You love me," she says and he closes his eyes and waits for the release of the blade.

Friday night, very nearly Saturday morning now, and Magwitch is waiting with Lark and Crispin at their wobbly corner table on one edge of the matte black circus. Through the weave and writhe of dancers he catches occasional half-glimpses of Tam at the bar, Tam talking excitedly with the two pretty boys in taffeta and shiny midnight latex, the boys who have posed for *Propaganda*, but whose names he can't remember. Lark sips uneasily at her second white Russian and Crispin's eyes never leave the crimson door at the other end of the dance floor. The Sisters of Mercy song ends, bleeds almost imperceptibly into a crashing remix of Fields of the Nephilim's "Preacher Man."

It has been more than a year since the first rumors, the tangled bits of hearsay and contradiction that surfaced in the shell-shock still after the riots. A year since Tam sat across this same wobbly table and repeated every unlikely detail, her sage-green eyes sparking like flints and she whispered to him, "They have the cape, Maggie. They have the fucking *cape*."

He doesn't remember exactly what he said, remembers her leaning close and the glinting scalpel of her voice.

"Hell, Maggie, there was so much going on they didn't even *care* if the cops saw them. They just walked right in and started digging. I mean, Christ, who's gonna give two shits if someone's out digging up dead guys when half the city's burning down?"

He didn't believe it then, still doesn't believe it, but Magwitch knows that playing these head games with the likes of Jimmy DeSade is worse than stupid or crazy. He fishes another Percodan from the inside pocket of his velvet frock coat and washes it down with a warm mouthful of bourbon.

"I don't think he's going to show," Crispin says, and there's just the faintest trace of relief, the slimmest wishful crust around his words. And then the crimson door opens on cue, opens wide and Jimmy DeSade, black leather and silver chains from head to toe, steps into the throb and wail of Stigmata.

And even over the music, the sound of Lark's glass hitting the floor is very, very loud.

Together, they are herded out of the smoky, blacklight pit of the club, up narrow stairs and down the long hall, past the husky old tattoo queen who checks IDs and halfheartedly searches everyone for dope and weapons, out into the night. Jimmy DeSade walks in front and one of the Gristle Twins drives them from behind, and as they leave the sidewalk and slip between abandoned warehouses, an alley like an asphalt paper cut, Magwitch wants to run, tells himself he *would* run if it were only Tam. Would leave her to play out this pretty little horror show on her own; but

Summers, Montague. *The Vampire* (1928; reprint, New York: Dorset Press, 1991). Original title: *The Vampire: His Kith and Kin.*
—*The Vampire in Europe* (1929; reprint, New York: Gramercy Books, 1996).

Twitchell, James B. *The Living Dead: A Study of the Vampire in Romantic Literature* (Durham, NC: Duke University Press, 1981).

VAMPIRE ANTHOLOGIES

Brite, Poppy Z. ed. *Love in Vein* (New York: HarperPrism, 1994).
—-*Love in Vein II* (New York: HarperCollins, 1996).

Dalby, Richard. *Dracula's Brood* (1987; reprint, New York: Dorset Press, 1991).

Datlow, Ellen, ed. *Blood is Not Enough* (New York: Morrow, 1987).
—*A Whisper of Blood* (New York: Morrow, 1991).

Frayling, Christopher. *Vampyres: From Lord Byron to Count Dracula* (London: Faber and Faber, 1992).

Jones, Stephen, ed. *The Mammoth Book of Dracula* (London: Robinson Publishing, 1997).

Keesey, Pam, ed. *Daughters of Darkness: Lesbian Vampire Stories* (Pittsburgh and San Francisco: Cleis Press, 1993).

Ryan, Alan, ed. *The Penguin Book of Vampire Stories* (New York and London: Penguin, 1988). Original title: *Vampires* (1987).

Shepard, Leslie, ed. *The Dracula Book of Great Vampire Stories* (Secaucus, NJ: Citadel, 1977).

Weinberg, Robert, Stefan R. Dziemianowicz, and Martin H. Greenberg, eds. *Weird Vampire Tales: 30 Blood-chilling stories from the weird fiction pulps* (New York: Gramercy, 1992).

DRACULA STUDIES

Belford, Barbara. *Bram Stoker: A Biography of the Author of* Dracula (New York: Alfred A. Knopf, 1996).

McNally, Raymond T., and Radu Florescu, *In Search of Dracula* (Greenwich, CT: New York Graphic Society, 1973).

Madison, Bob, ed. *Dracula: The First Hundred Years* (Baltimore, MD: Midnight Marquee Press, 1997).

Eighteen-Bisang, Robert. *Dracula: An Annotated Bibliography* (White Rock, B. C., Canada: Transylvania Press, 1994).

Leatherdale, Clive. *Dracula: The Novel and the Legend* (1985, revised ed.; Brighton, East Sussex: Desert Island Books, 1993)
—ed. *Bram Stoker's Dracula Unearthed* (Westcliff-on-Sea, Essex: Desert Island Books, 1998).

Miller, Elizabeth, *Reflections on Dracula: Ten Essays* (White Rock, British Columbia: Transylvania Press, 1997)
— ed. *Dracula: The Shade and the Shadow* (Westcliff-on-Sea, Essex: Desert Island Books, 1998).

Skal, David J., *Hollywood Gothic: The Tangled Web of* Dracula *from Novel to Stage to Screen* (New York and London: W.W. Norton & Company, 1990)
—ed. *Dracula: The Ultimate, Illustrated Edition of the World-Famous Vampire Play* (New York: St. Martin's Press, 1993).

Stoker, Bram. *Dracula: A Norton Critical Edition*, Nina Auerbach and David J. Skal, eds. (New York and London: W.W. Norton & Company, 1996).

he looks at Crispin and Lark, who hold whiteknuckled hands, and there's nothing but the purest silver terror in their eyes.

Jimmy DeSade's junkheap car waits for them in the alley, rumbling hulk of a Lincoln, rust bleeding from a thousand dents and scrapes in its puke-green skin, one eye blazing, the other dangling blind. Its front fender hangs crooked loose beneath shattered grille teeth, truculent chrome held on with duct tape and wire. Jimmy DeSade opens a door and shoves them all into the backseat. And the car smells like its own shitty exhaust and stale cigarettes and pot smoke and spilled alcohol, but more than anything else, the sweet, clinging perfume of rot. Magwitch gags, covers his mouth with his hand, and then Tam is screaming.

"Don't mind Fido," Jimmy DeSade says in the Brit accent that no one has ever for a minute believed is real, and then the twisted thing stuffed into the corner with Tam tips over, all bloodcrusty fur and legs bent in the wrong places, and lies stiff across their laps. Its ruined maroon coat is alive, maggots seethe, and parts that belong inside are slipping out between shattered ribs.

"Manny here ran into the poor thing the other day and shit, man, I always wanted a doggy."

Tam screams again and Crispin and Lark vomit in perfect, twinly unison, spray booze and the pork-flavored Ramen noodles they ate before leaving Magwitch's apartment onto the roadkill tangle. Magwitch swallows hard, turns his head to the window as Manny pulls out of the alley into traffic.

"You stupid, sick fuck..." he says, and the Gristle Twins laugh and Jimmy DeSade grins his wide, yellow-toothed hyena-face, winks.

"Oh god," Tam whispers, "Oh god, oh god," and she dumps the mess into the floorboard, leaving their laps glazed with dog and vomit, velvet and satin and silk stained and goreslicked.

When Jimmy DeSade laughs, it sounds like bricks and broken bottles tumbling in a clothes dryer.

"Man, I love you prissy little goth-geek motherfuckers," he says and thumbs through the stack of twenty- and fifty-dollar bills Tam gave him back at Stigmata, three hundred dollars all together, and he sniffs at the money before it's tucked inside his leather jacket.

"You dumb buggers wear that funeral drag and paint your eyes and think you're fucking around with death, think you're the reaper's own harem, don't you?"

"Fuck you," Tam says, whimpers, and "Yeah," Jimmy DeSade says, "Yeah, well, we'll see, babe."

The city rushes past the Lincoln's windows and Magwitch tries to keep track, but nothing out there looks familiar, and he doesn't know if they're driving west, towards the ocean, or east, towards the desert.

"You better have it," Tam says. "I paid you your goddamned money and you better have it."

Jimmy DeSade watches her a moment silently, feral chalk blue eyes, Nazi eyes, and he smiles again slow.

"You got yourself an unhealthy obsession there, little girl," he says, oozes the words across his lips like greasy pearls and the middle Gristle Twin, the one whose name isn't Manny, snickers.

"We made a deal," Tam says, and Magwitch has to stop himself from laughing.

"Don't you worry, Tammy. I got it all right. And *oh*, Tammy, you should'a *seen'im*, the shriveled old hunky fuck," and Jimmy DeSade's eyes sparkle, lightless shine, and he bites at his lower lip with one long canine. "Lying there so bloody peaceful in his penguin suit, the dirt just dribbling down into his face. And that cape, that cape still draped around his bony shoulders right where it was when they closed the casket on him."

Jimmy DeSade sniffles his cokehead's chronic sniffle and wipes his nose, grime-cracked nails and the one on his little finger elegant long and polished candy apple red.

"Nineteen hundred and fifty-six," he purrs, "Terrible long time to go without one sweet breath of night air. That was Elvis's big year, did you know that, Tammy dear? Too damn bad it wasn't poor old Bela's."

The nameless Gristle Twin pops a cassette into the Lincoln's tape deck and a speed metal blur screeches from the expensive, bass-heavy speakers behind their heads. Magwitch tries to ignore the music, tries desperately to concentrate on the buildings and street signs flashing past, as the car leaves Hollywood behind and rolls along through the hungry California night.

In the catacombs beneath downtown, the old tunnels dug a century before for smugglers and Chinese opium dens, the air is not warm and smells like mold and standing water. Magwitch sits on his bare ass on the stone floor; his black jeans are a shapeless wad nearby, but when he reaches for them it wakes up the pain in his ribs and shoulder and he gasps, slumps back against the seeping wall. There is light, row after row of the stark and soulless fluorescent bulbs and so he can see Crispin and Lark, naked and filthy, huddled together on the other side of the chamber.

Tam, wrapped in muddy vinyl and torn pantyhose, stands with her back to them near the door.

There are chains like rusty intestine loops, chains that end in meathook claws, dangling worse things than the road dog in Jimmy DeSade's car.

And Jimmy DeSade sits on the high-backed wooden chair in the center of the room, smoking, watching them. The others have all gone, the Gristle Twins and the woman with teeth filed to sharp piranha points, the man without ears and only a pink, scar-puckered hole where his nose should be.

"Are you finished?" Tam says, "Are you satisfied?" and it frightens Magwitch how little her voice has changed, how ice calm she still sounds.

VAMPIRE CINEMA

Jones, Stephen. *The Illustrated Vampire Movie Guide* (London: Titan Books, 1993).

Melton, J. Gordon. *VideoHound's Vampires on Video* (Detroit, MI: Visible Ink Press, 1997).

Pirie, David. *The Vampire Cinema* (London: Hamlyn, 1977).

Ursini, James, and Alain Silver. *The Vampire Film* (1975; reprint, New York: Limelight Editions, 1993).

Waller, Gregory A. *The Living and the Undead: From Stoker's* Dracula *to Romero's* Dawn of the Dead (Urbana, IL: University of Illinois Press, 1986).

REAL-LIFE VAMPIRES

Noll, Richard. *Vampires, Werewolves, and Demons: Twentieth Century Reports in the Psychiatric Literature* (New York: Brunner/Mazel, Publishers, 1992).

"Will you show me the Dracula cape now?"

Jimmy DeSade chuckles low, shakes his shaggy head, and Magwitch hears a whole laundromat full of dryers, each churning with its load of bricks and shattered soda bottles.

"I'd like to," he says and crushes his cigarette out on the floor, grinds the butt flat with the snakeskin toe of his boot. "I would, Tammy, I really would. But I'm afraid that's no longer something I can do."

Tam turns around slowly, arms crossed and the jacket pulled closed to hide her small breasts. Her face is streaked with blood and makeup smears, and livid bite marks, the welts and clotting punctures, dapple her throat and the backs of her hands. The flesh around her left eye looks pulpy and is swollen almost shut.

One last snarl: Lugosi during his final Dracula *tour* (1951).

"Yeah, well, you see, I let it go last week to a Lugosi freak from Mexico City. This crazy fucker paid me ten Gs for the thing, can you believe that shit?"

For a moment, Tam's old mask holds, indomitable frost and those eyes that betray nothing, show nothing more or less than she wishes. But the moment passes and the mask splinters, falls away, revealing the roil beneath, shifting kaleidoscope of bone and skin, tumbling bright flecks of rage and violation. And Magwitch is almost afraid for Jimmy DeSade.

"Hey, Tammy. It's just biz, right?" And he holds out his hand to her. "These things happen. No bad feelings?"

Her lips part, wet, salmon hint of her tongue between teeth before the cascade of emotion drains away and there's nothing left in its place, and she falls, collapses into herself and gravity's will, crumples like ash to her knees and the uneven cobblestones.

"Sure," Jimmy DeSade says, "I understand," and he stands, pushes the chair aside. "When you guys are done in here, turn off the lights, okay, and close the door behind you?" He points to a switch plate rigged on one wall, a bundle of exposed red and yellow wires.

And they're alone, then, except for the things without eyes, the careless hanging sculptures of muscle and barbed wire, and Magwitch drags himself the seven or eight feet to Tam. The pain in his side strobes violet and he has to stop twice, stop and wait for his head to clear, for the crooning promise of numb and quiet and cool oblivion pressing at his temples to fade.

They do not come here much anymore, visit the old man less and less as summer slips unnoticed into fall. Without Tam, the guards are less agreeable, and they've been caught once already sneaking in, have been warned and threatened with jail. Tonight there is a cough-dry wind and Magwitch has heard on the radio that there are wildfires burning somewhere in the mountains.

Lark finishes the chapter, Mina's account of the sea captain's funeral and the terrified dog, somnambulant Lucy, and Crispin blows out the single votive candle he's held up to the pages while she read. Even such a small flame is a comfort and Magwitch knows they share the same sudden uneasiness when there's only its after-image floating in their eyes, the baby-aspirin orange blot accenting the night.

Lark closes her paperback and lays it on the headstone, sets it neatly between the three white roses and the photocopied snapshot in its cheap, drugstore frame. A dashing, impossibly young Lugosi, 1914 and the shadow of his fedora soft across eyes still clear with youth and ignorance of the future. In the morning, the groundskeeper will clear it all away, will grumble and hastily restore the sterile symmetry of his ghostless, modern cemetery.

"I saw Daniel Mosquera last night," Lark says, and it takes Magwitch a second to place the name with a face. "He said someone saw Tam in San Francisco last week, and she's dancing at the House of Usher now."

And then she pauses and the traffic murmurs through the wordless space she's left; Magwitch imagines he can smell wood smoke on the breeze.

"This is the last time I can come here, Maggie."

"Yeah," he says, "I know," and Crispin nods and puts the stub end of his candle down with everything else.

"But we'll still see each other, right? At DDT, and next month Shadow Project's playing The Roxy and we'll see you then."

"Sure," Magwitch says. He knows better but is careful to sound like he doesn't.

"You sleep tight," Lark whispers and kisses her fingertips, touches them to the grass.

When they've gone, Magwitch lights a cigarette and sits with Bela another hour more.

Dracula's final resting place.

German poster for Taste the Blood of Dracula. *(1969).*

Acknowledgments

IN ADDITION TO the authors who contributed to this collection, I wish to thank the following individuals for their assistance, advice, and encouragement: Mike Ashley, Richard Curtis, Stefan Dziemianowicz, Lokke Heiss, Del and Sue Howison, Stephen Jones, Andrea Plunkett, Bob Postawko, Elias Savada, Christopher Schelling, and Jeanne Youngson.

Thanks to Mary Woronov for use of her stunning artwork, and to the present owners of two of the paintings, Edward Plumb and Buddy Barnett.

Special thanks to Laura Ross, J.P. Leventhal, True Sims and Amy Carothers at Black Dog and Leventhal, Publishers; to my literary agent, Malaga Baldi; to Harriet Eckstein, without whose production and design skills this book might never have happened; and to a most gifted copy editor, Lesley Bruynesteyn.

For generous help with photographs and visuals, I most gratefully thank Ronald V. Borst / Hollywood Movie Posters; Buddy Barnett / Cinema Collectors; Ron and Howard Mandelbaum and Buddy Weiss at Photofest, photographer Donal F. Holway; Erik N. Stogo and Mack Dennard; and Scott MacQueen. Thanks also to Kalia Rork and Coyote Communications for their expert work in scanning hundreds of images.

Research institutions consulted included the Library of Congress, the University of California libraries in Los Angeles and Berkeley, and the Los Angeles Public Library. Newspaper reviews cited in my marginal commentary were drawn primarily from the clippings collections at the Margaret Herrick Library, the Academy of Motion Picture Arts and Sciences, Beverly Hills, California; and the Billy Rose Theatre Collection, New York Public Library for the Performing Arts at Lincoln Center.

ABOUT THE EDITOR

David J. Skal is a respected scholar of all things macabre, and the author of *Hollywood Gothic*, *The Monster Show*, *Screams of Reason*, and, with Elias Savada, *Dark Carnival: The Secret World of Tod Browning*. With Nina Auerbach, he is coeditor of the Norton Critical Edition of Bram Stoker's *Dracula*. A frequent talk show guest, he has appeared on "Charlie Rose," A&E's "Biography," National Public Radio's "All Things Considered" and "Fresh Air," and has lectured at dozens of colleges and universities across the United States. His nonfiction has been nominated for both the Hugo and Nebula Awards. He has written, produced and appeared in a variety of television and DVD documentaries on occult and pop culture subjects. He lives in Los Angeles.

Photo: Steven Speliotis